Conservation of Dragonflies

Sentinels for Freshwater Conservation

Conservation of Dragonflies

Sentinels for Freshwater Conservation

Michael J. Samways

Department of Conservation Ecology and Entomology
Stellenbosch University
South Africa

CABI is proud to present this book as a co-publication with the Royal Entomological Society (RES).

The RES is devoted to the understanding and development of insect science. It was established in 1833 and is dedicated to supporting international collaboration, research, and the publication of insect science discoveries and understanding. Its aim is to show every person how remarkable and valuable insects are.

Michael Samways is a Fellow of the RES and was the recipient of the RES Award for Insect Conservation in 2019.

Find out more about CABI's partnership with the RES here:
https://www.royensoc.co.uk/news/res-handbooks-now-available-via-cabi/

Royal Entomological Society

CABI

CABI is a trading name of CAB International

CABI
Nosworthy Way
Wallingford
Oxfordshire OX10 8DE
UK

Tel: +44 (0)1491 832111
E-mail: info@cabi.org
Website: www.cabi.org

CABI
200 Portland Street
Boston
MA 02114
USA

Tel: +1 (617)682-9015
E-mail: cabi-nao@cabi.org

A catalogue record for this book is available from the British Library, London, UK.

ISBN-13: 9781789248371 (hardback)
9781789248388 (ePDF)
9781789248395 (ePub)

DOI: 10.1079/9781789248395.0000

Commissioning Editor: Ward Cooper
Editorial Assistant: Helen Elliott
Production Editor: Rosie Hayden

Typeset by Straive, Pondicherry, India
Printed and bound in the UK by Bell & Bain, Glasgow

Contents

Preface

Dragonflies are among the most iconic and best-known of all insect groups. They are one of the longest surviving insect groups, with their basic body form largely unchanged since before the appearance of dinosaurs. They belong to the order Odonata, composed of two main suborders, the Anisoptera ('true' dragonflies) and Zygoptera (damselflies), and one small suborder, the Anisozygoptera, with some characteristics of both of the other two.

The adults are conservative compared to most other insect groups in their basic body form. The head bears a pair of large compound eyes and small antennae, while the highly muscular thorax bears four elongated wings and the abdomen is long and slender. In short, the adult is built for efficient and manoeuvring flight.

The young larva emerges from an egg deposited close to or in water. It catches prey in the water, enabling it to develop through several growth stages. Having both an aerial and an aquatic life stage means that dragonflies not only represent freshwater quality and condition but also that of the terrestrial surroundings. They are also one of the most sensitive groups, helping us to understand the effects of short-term environmental change at the local scale as well as long-term climate change at the global scale.

Dragonflies occur throughout the world, except where there is permanently frozen freshwater. Among the 6400 plus known species, there are a wide range of habitat specialities, meaning that some are highly susceptible to human activity while others benefit from it. This range of specialization, combined with the group being comparatively well-known, means that they can be used for assessing freshwaters and their surrounding landscapes at the species level. While specific species can tell us much about how the world is changing, it is the particular sets of species that hold greatest value for freshwater conservation.

Not surprisingly, the charisma of dragonflies along with their great utility means that they are being linked to freshwater assessment to guide environmental improvement around the world. In short, their visibility, accessibility and responsiveness to freshwater condition and change, whether associated with ponds and marshes or small streams and rivers, means that they are being adopted as sentinels for freshwater conservation.

The aims of this book are to embrace dragonflies as conservation subjects in their own right, as well as exploring their utility for freshwater conservation globally. As freshwaters around the world are under severe pressure, and the fate of both dragonflies and humans are entwined, dragonflies are helping us determine freshwater status, for better or for worse. They help guide us toward making the right decisions, and then tell us what we are doing right or wrong. They also indicate how well, or not, a whole lot of other biodiversity, that lives alongside them in the same water bodies and landscapes, is doing.

This book is a synthesis of research findings related to the conservation of dragonflies, while also indicating the best strategies for the preservation and improvement of freshwater bodies. While some clear global perspectives on this topic are emerging, there can be differences relative to the deep biogeographic history, type of human activity, extent of the effects of climate change, and, last but not least, how policy makers and managers are responding to improve matters locally, regionally and globally.

The charisma and tractability of dragonflies means that amateur naturalists and citizen scientists are now playing a huge role in helping make the right decisions for a better future. Therefore, this book is for a wide range of people, including those interested in dragonfly and freshwater conservation in a wide range of capacities and professions in these times where urgent conservation action is required.

Acknowledgements

This book could not have been written without substantial interaction with many freshwater biologists and odonatologists over many years. As regards freshwater conservation, I thank Abby Lynch for driving some important recent initiatives. As for dragonflies, I would especially like to thank Andreas Chovanec, Viola Clausnitzer, Klaas-Douwe Dijkstra, Aleš Dolný, Leandro Juen, Rassim Khelifa, Jens Kipping, Federico Lozano, Andreas Martens, the late Mike May, the late Norman Moore, Javier Muzón, Jürgen Ott, Fabio Roque, Frank Suhling and Jessica Ware for many fascinating discussions. I would also like to thank research students, post-doctoral fellows and staff in my local environment, particularly Charl Deacon, René Gaigher, Gabriella Kietzka, James Pryke, John Simaika and Carlien Vorster for many stimulating exchanges.

Very special thanks to Charl Deacon for many insightful comments on the text and figures, and to the graphic artists Robyn Symons and Megan Louise Potgieter for their amazing illustrative support. Thanks also go to the field photographers and illustrators, each of whom is acknowledged alongside their hard-earned pictures.

Thanks as well to Elton Jacobs and Lorenda Boyd of Stellenbosch University library for tracking down a whole host of research papers.

I also wish to express my great appreciation for the wonderful production support from Rosie Hayden, Lucy Pritchard and Ward Cooper of CABI, and Roger Bullen of CartoZapp.

Lastly, I thank Georgina Ledger for patiently giving me space and an enormous amount time to complete this book.

Some technical notes

- 'Dragonflies' is a general term that refers to all members of the order Odonata, but where it is important to emphasize that both Anisoptera and Zygoptera species are being discussed, then 'odonates' is used.
- Where it is important to distinguish Anisoptera species from Zygoptera species, the expressions 'anisopterans' or 'zygopterans' are used.
- There is ongoing debate about whether 'larva' or 'nymph' should be used to describe the juvenile dragonfly. Either term is acceptable, although throughout 'larva/e' is used here for consistency.
- As a set of dragonfly species is in one taxonomic order, it is technically correct to use 'assemblage/s' for the species set, rather than 'community/ies'.
- The exuvia is technically not a developmental stage, as it is the cast exoskeleton of the last instar larva. However, the exuvia has great relevance for illustrating the transition of an individual from water to air and indicating completed development. Furthermore, in some parts of the world, or in the case of particular species, it is a useful manifestation to assess and monitor, as it is often highly visible, identifiable and quantifiable.
- The scientific species names are given here so that there is clarity on the species being discussed. Vernacular or common names are not given as these can vary from one region to another, and according to the home language of a country. Furthermore, in many cases vernacular names have not yet been assigned. However, vernacular names are important at the local scale in citizen science and for practical conservation action. They are also often necessary when communicating with members of civil society outside the realms of odonatology. The names are usually easily sourced from the local literature and websites.

About the Author

Michael Samways obtained his PhD from London University. He is a Fellow of the Royal Society of South Africa and a Member of the Academy of Science of South Africa, and currently Emeritus Distinguished Professor in the Department of Conservation Ecology & Entomology, Stellenbosch University, South Africa. He focuses on all aspects of insect conservation, both nationally and internationally, with a strong focus on dragonfly and freshwater conservation. Michael is recipient of the John Herschel Medal of the Royal Society of South Africa, the Senior Captain Scott and Gold Medals of the South African Academy of Science and Arts, and the Gold Medal of the Academy of Science of South Africa. He received the lifetime Stellenbosch University Chancellor's Award, and IUCN/Species Survival Commission Chair's Citation of Excellence. He and his team, the Mondi Ecological Networks Programme, were named Winner of the NSTF-South32 Award. Michael also received the Marsh Award from the Royal Entomological Society for outstanding and exemplary lifetime contribution in Insect Conservation.

1 Introduction

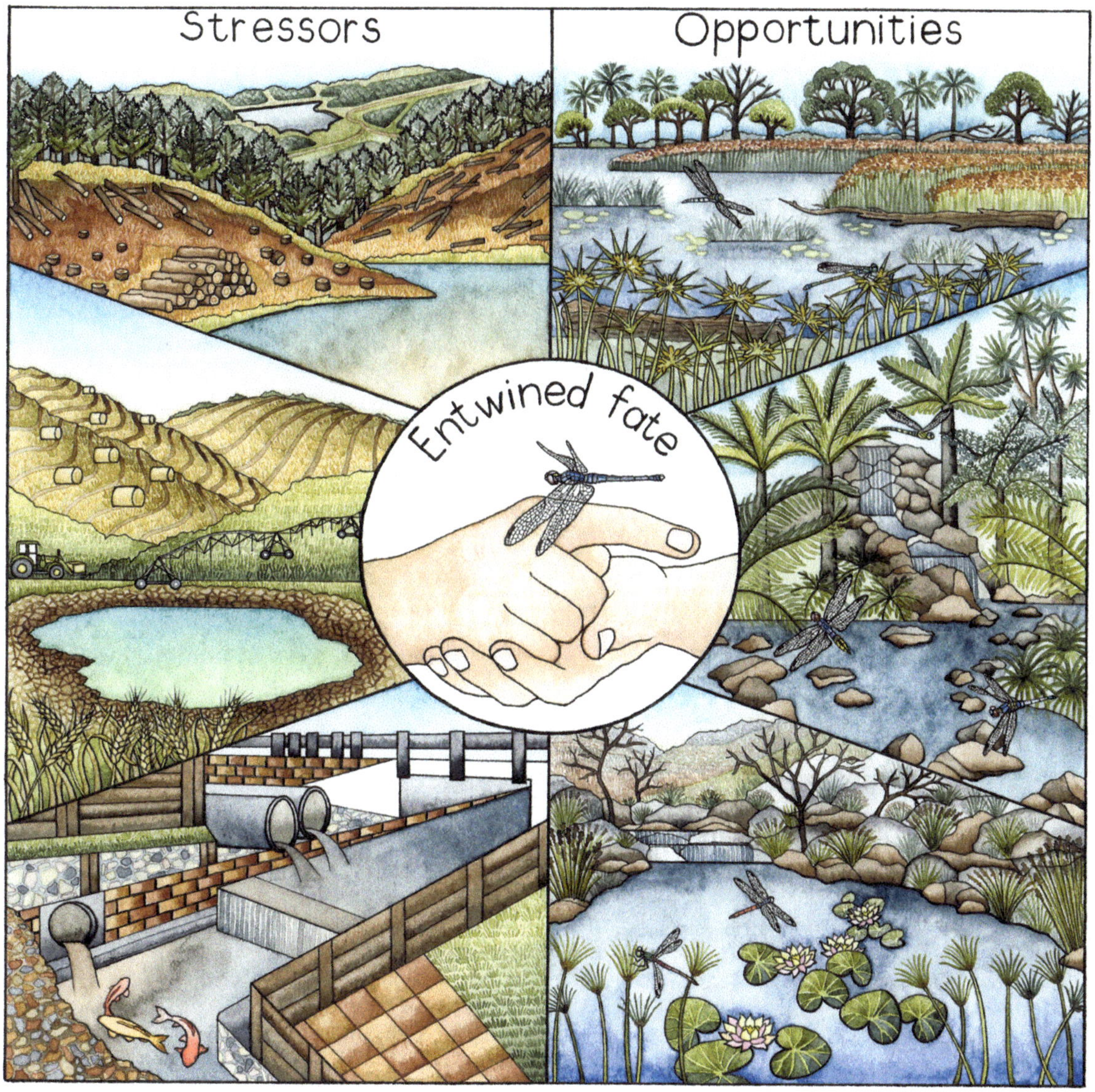

Illustration: Megan Louise Potgieter.

 Conservation of Dragonflies: Sentinels for Freshwater Conservation (M.J. Samways)
DOI: 10.1079/9781789248395.0001intro

Abstract

Dragonflies have deep roots in human cultural history. Their charisma, and as with us, their dependence on freshwater, has fed concerns for their conservation. Their well-being and ours are entwined through the medium of healthy water systems. This has led to the development of dragonfly conservation science.

Dragonflies depend on the land as well as water. This means that the conservation of landscapes associated with freshwater water bodies is central to dragonfly and freshwater conservation. Dragonflies and their habitats have been, and continue to be, affected by various human activities. In essence, their conservation involves reversing environmental damage. Yet some dragonflies benefit from certain human constructions, especially ponds.

1.1 The Dragonfly As a Charismatic Organism

Around the world, we grow up easily recognizing 'a dragonfly'. Apart from the saltiest, driest, highest, coldest and most urbanized environments, dragonflies are familiar subjects in everyday life everywhere. At times they even fly into our homes, making an unmistakable noisy fluttering as they attempt to escape back into their natural realm. Occasionally, some migratory species secure a temporary resting site on ships crossing open water. As both us and dragonflies are attracted to a combination of greenery and water, inevitably we both come into close contact on a regular basis.

Dragonflies imprint on us as children, as they fly so freely and glisten in the sun. Yet they are always furtive enough to be beyond reach, so symbolizing freedom, while also emphasizing our mutual need for water. They also often accompany us in our watery recreational activities, so reaffirm our connection with the natural world.

Our connection with dragonflies has a deep history, in part as dragonflies are so iconic. An icon is a powerful visual representation. Dragonflies in ancient and medieval texts are usually depicted in flight, not resting. The reason for this is that the flying insect is a close-up view of a form representing high manoeuvrability and near-surround vision. The resting dragonfly does not arouse such iconic representation (McCloud, 1998).

In Japan, where dragonflies have been revered for centuries, they represent bravery and strength, as well as loyalty, in view of the dragonfly's propensity to return to the same perch (Kiauta, 1986). In earlier times, the country went by the name of *Akitsu-shima* (island of the dragonfly) (Corbet *et al.*, 1995). In the Zulu language, a dragonfly is called a 'jigamanzi', a water dancer, in recognition of its agility. It is this iconic status that has also led to dragonflies being popular subjects for jewellery (Liu, 2001), and on pottery of some cultures, such as the Mimbreños of south-western North America (Capinera, 1993), as well as on everyday items (Fig. 1.1).

Dragonflies are also iconic as they are conspicuous and often so colourful, contributing to their appeal to humans. It is aesthetics, combined with size, yet without doing harm to humans or livelihoods, or being repugnant, that makes dragonflies the second most popular insects, after butterflies, in civil society today (Barua *et al.*, 2012). Inevitably, they feature strongly in recreation and tourism (Lemelin, 2007).

Icons in conservation are known as flagships. Popular and charismatic species that are icons increase conservation awareness and galvanize conservation action (Jepson and Barua, 2015). Yet among the spectrum of species in the dragonfly order Odonata, there is a range of flagship potential. The large and colourful male Anisoptera ('true dragonflies') (Fig. 1.2) stand in for their less colourful, but just as agile, females. Also, the term 'dragonflies' often represents the less robust, slender, and less conspicuous, yet endearing, damselflies (Zygoptera). However, in terms of conservation, these two suborders display some distinct differences, into which we shall delve deeper in the following chapters.

1.2 A Short History of Dragonfly Conservation

Many traditional nature reserves include dragonfly protection as part of their remit for conserving biodiversity. However, a landmark specifically for dragonfly conservation was in 1993 when the first international symposium dedicated to dragonfly conservation and their habitats took place, appropriately in Japan, with its long-standing reverence for dragonflies (Corbet *et al.*, 1995). Then, in 1997, the International Union for Conservation of Nature/Species Survival Commission (IUCN/SSC) *Dragonflies: Status Survey and Conservation Action Plan* was launched (Moore, 1997). This was the first scientific document to highlight the conservation status of dragonfly species globally. This

Fig. 1.1. Dragonflies are much appreciated by people, even on everyday items such as beer bottles and matchboxes. (Photo: Michael Samways).

Fig. 1.2. Dragonflies are often large, colourful and iconic, and symbolic of loyalty, bravery, strength and agility according to culture. Their fate and ours are entwined through healthy freshwater environments. Pictured here is a young *Diastatops intensa* from north-eastern Argentina. (Photo: Michael Samways).

assessment and action plan emanated from the IUCN/SSC Odonata Specialist Group, known today as the Dragonfly Specialist Group, and composed of about 50 or so expert odonatologists from about 30 countries.

Interest in dragonflies and conservation of their aquatic and terrestrial habitats has increased sharply in recent years. While there were no scientific papers dedicated to dragonfly conservation prior to 1980, these increased to a few between 1980 and 1995, and to tens of papers from 2013 onward (Bried and Samways, 2015). This increase coincided with a surge of interest in dragonflies as a new form of recreation and tourism activity, aligning with greater awareness of them and their plight (Lemelin, 2009).

This increased interest in dragonflies and concerns for their conservation also came about from increasing involvement of citizen scientists, who greatly improved the extent of data gathering over both space and time. This in turn led to dragonflies becoming mainstream, not only as special subjects, but also in freshwater quality assessment and education (Clausnitzer *et al.*, 2017). In turn, citizen science has also led to conservation action. For example, *Leucorrhinia dubia* is highly threatened in the UK, limited to a few acidic bog pools with floating sphagnum moss. Collaboration between certain members of the British Dragonfly Society and Cumbria Wildlife Trust led to the introduction of the species at a site in northern England from other locations, with great success in re-establishment of the species (https://www.cumbriawildlifetrust.org.uk/nature-reserves/foulshaw-moss).

Despite the progress in dragonfly research, there remain some biogeographic gaps, especially in hotter regions. There are also gaps in study focus, such as basic biology (e.g. life cycle, anatomy, physiology, habitat), biogeographic extent of species, parasitism, intra- and inter-specific competition, certain evolutionary and phylogenetic relationships, and dragonfly eggs (Oliveira-Junior *et al.*, 2022).

Among the recent developments are local, national and regional dragonfly species' assessments of their presence, locations, habitats and conservation. In many parts of the world, there has been a recent and great increase in assessments of many individual species. One progressive regional assessment has been the IUCN/SSC project *The Diversity of Life in African Freshwaters: Underwater, Under Threat* (Darwall *et al.*, 2011), which covered an important (in terms of biodiversity) yet relatively poorly explored continent (mainland Africa), while also assessing a range of other freshwater taxa besides dragonflies (Dijkstra *et al.*, 2011; Clausnitzer *et al.*, 2012; Simaika *et al.*, 2013). A more detailed assessment was also made of dragonflies in southern Africa under the same initiative (Suhling *et al.*, 2009), and under the German-sponsored *Biodiversity in Southern Africa* (Suhling *et al.*, 2010).

1.3 Dragonflies in Water and on Land

We tend think of dragonflies as aquatic insects, but importantly from a conservation perspective, they are amphibiotic: they live in water as the sluggish, growing larva (sometimes referred to as a nymph), and in the air as the dispersing, reproductive adult (Fig. 1.3). Both stages are predaceous, mostly on other arthropods or insects available in their respective watery or airy realms. Some may feed on other dragonflies, while a few are highly cannibalistic as larvae.

The larva emerges from an egg, which may have been scattered on the water surface or inserted into living or dead vegetation in or overhanging the water. The larva goes through several growth stages in or on the substratum, or among aquatic plants. There are some species where the larva inhabits moist rather than watery environments. The larva of a *Pseudocordulia* sp., for example, lives in leaf litter in the Queensland rainforest of Australia (Watson, 1982).

Before emergence of the adult, the last instar larva crawls out of the water onto a bank, boulder, or living or dead vegetation. After leaving the water, the larva remains stationary on a firm surface and splits dorsally from which the young adult emerges (the process of emergence), usually at night or in early morning to avoid predators. The larval 'shuck' left behind is the exuvia (plural: exuviae), which

Fig. 1.3. Dragonflies are amphibiotic organisms, meaning that they require good quality habitat both in the water (for the larva) and on land (for the adult). (Illustration: Robyn Symons).

dries into a hard shell while retaining the important larval features valuable in species identification. In most species, the adult emerges within a few metres of the water. Although in the case of some species, such as the Japanese relict dragonfly *Epiophlebia superstes*, the larva can travel many tens of metres for a month or more to possibly complete internal larval development and to locate sufficiently warm conditions for adult emergence (Naraoka, 2016).

The recently emerged adult is soft, and its body and its wings expand from haemolymph pressure. The fully formed adult then begins to harden, and weakly takes flight as the teneral adult. This soft and weakly mobile stage is highly vulnerable to predators. The teneral adult has relatively inconspicuous colouration, usually moves away from the water and remains cryptic among vegetation. Then, a few more days pass before the adult fully hardens and assumes its adult colouration.

Once the adult is reproductively mature, it develops its characteristic high mobility and manoeuvrability. Depending on species and sex, it may stay away from the water to hunt prey, only returning to the vicinity of the water body to mate and oviposit. Males of many species hold territories in association with the water body, usually at its edge. The male and female mate by first coupling and then maintaining a tandem position, and in some species the coupled pair is able to fly together remarkably well. The differences in form of these various life stages, each of which has been honed by local abiotic and biotic environmental selection pressures, is known as developmental polymorphism.

Understanding where these different stages occur and optimally function is important in their conservation. The total spatial area of the locations and conditions for completion of a species' life cycle is its 'habitat' (Dennis, 2010). This is not the same as the physical space where each life stage lives, known as the 'biotope'. Any species' habitat is made up of certain biotopes according to life stage. For species in the Lestidae family, for example, typically the egg has a biotope of vegetation overhanging the water, the larva a biotope of submerged aquatic vegetation, the exuvia occurs in grass/sedge/reed vegetation at the water's edge, the teneral adult often needs bushy or tree vegetation, while the mature adult often requires a variety of tall grasses, herbs and bushes at the water's edge. In turn, the 'habitat' of any one species is the total of all these biotopes.

1.4 Dragonfly Habitats

Dragonflies as a group live in a wide range of freshwater bodies stretching from north of the Arctic treeline (Cannings and Cannings, 1997) to the southern region of South America (Garrison *et al.*, 2006), and are one of the most prominent groups of freshwater organisms (Balian *et al.*, 2010). They are most speciose in association with flowing waters in rainforests, especially in Oriental and Neotropical regions (Kalkman *et al.*, 2008). Dragonflies occur over a wide range of elevations above sea level. Migrations occur in the high Hindu Kush in Central Asia at over 6000 m above sea level (Wojtusiak, 1974). Over a 0–3000 m elevation gradient running directly east to west in South Africa, the number of resident dragonfly species drops from 86 on the coastal plain to only 4 at 2400–3000 m (Samways, 1989).

Dragonflies live in many types of freshwaters (Fig. 1.4), although some species are tolerant of slightly saline conditions (Uboni *et al.*, 2020). Species variously live in microhabitats associated with part of the spectrum: running water (lotic species) through to standing waters (lentic species). Running water first manifests as a small stream responding to gravity and following lines of least resistance, coming into confluence with other streams, with an increasing volume of flowing water. Further confluences lead to an ever-increasing river size and volume associated with drop in elevation. As the river approaches the coastal plain, it slows into a meander before approaching the sea. Some rivers do not flow into the sea at all, as the geology and slightly increased elevation blocks their passage, and so water flows into lakes or swamps. Some swamps are extensive and highly seasonal according to changes in river discharge levels, as in the case of the extensive Okavango Delta in Botswana, which is rich in dragonfly species (Kipping, 2010).

1.4.1 Running water dragonfly habitats

Rivers are dynamic riverine landscapes, where groundwater and flood plains are as much a part of the river as the main channel. Furthermore, the riparian zone along the banks is also part of the river system, supporting vegetation that contributes to the physicochemical character of the river through supply of nutrients, either directly to groundwater or indirectly from leaf fall into the water (Ward, 1998).

Rivers are largely a continuum (Vannote *et al.*, 1980) with a range of different physicochemical

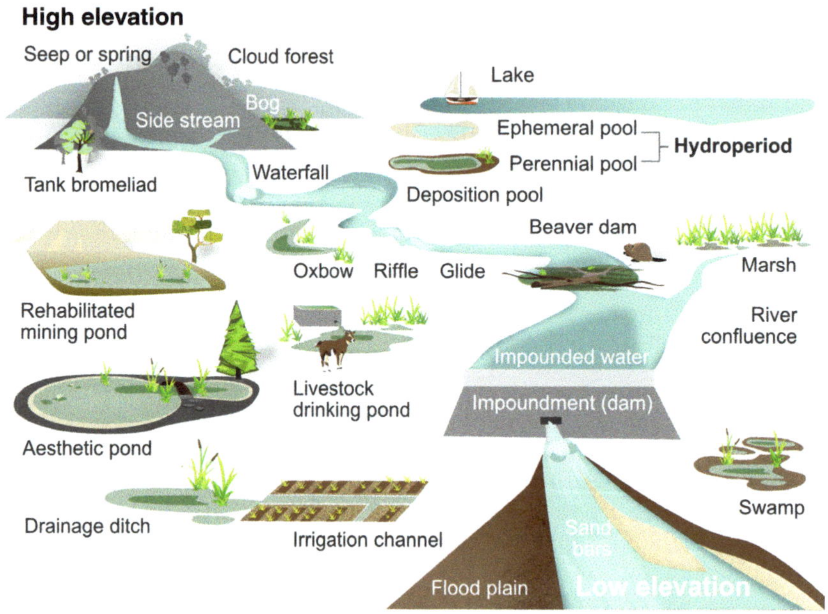

Fig. 1.4. Dragonflies inhabit a wide range of running and standing water habitats across many elevations, including several types of artificial systems. (Illustration: Robyn Symons).

conditions and biotic characteristics along their length. Among the important physical conditions important for dragonfly larvae are water flow velocity and water temperature, determined by climate and elevation, as well as the degree of shade provided by the overhanging vegetation. Aspects of water chemistry that are particularly important are levels of dissolved oxygen and pH, determined mostly by water velocity, geology and organic matter input.

Rivers usually begin as seeps, springs and clear trickles, often rich in settled organic matter. Decreasing elevation and regular confluences, as well as seepage from the banks, leads to the trickle becoming a fast-running stream with a hard substratum. There may be fine inorganic and organic matter in the river-bottom interstices, along the banks and in deposition pools, which are alcoves in the bank or in a riverbed depression. Depending on local geology and topography, these streams and small rivers have a variety of features, such as waterfalls, riffles and glides, as well as deposition pools.

In wide valleys, the river may contain sand bars. The earlier fast-moving rivers, after dropping in elevation, and following several confluences and more inflow, become ever-slower increasingly silt-laden large rivers. During this river journey, there is a change in physicochemical characteristics and vegetation in both the water and along the banks. Where the elevation decline is small, meandering parts of the river may form oxbows, which break away to become pools. These oxbow lakes may or may not open again with flash flooding and/or some geological upheaval. Sometimes, intense flooding may even cause a river to change course.

All these physicochemical conditions and characteristics are strong determinants of the type of vegetation present along the continuum. In turn, these abiotic and biotic characteristics determine which dragonfly species are present. Also, the functional morphology of the larvae is associated with the specific microhabitat in the water. Natural selection pressures are differential upon the larva versus adult, and these combined pressures combined with adaptation among individuals within a population determine which species will be present at any one location. Along the length of a river, other drivers, besides those that are physicochemical and botanical, come into play, such as predation by different species of fish, frogs, some reptiles, such as young crocodiles, and insectivorous birds and mammals, both in the water and on the banks, as well as disease.

1.4.2 Standing water dragonfly habitats

Among the standing waters that dragonflies inhabit are tank bromeliad epiphytes, seeps, and remnant pools in streams and rivers (after drought), through to various types of wetlands from pools, ponds, marshes, bogs (mires) and swamps to lakes. Hydroperiod (the amount of time during the year that a depression is inundated with water) matters. While many pools are permanent, others regularly dry out, and are known as temporary pools (also ephemeral pools, vernal pools). The hydroperiod often determines which species live in a pool, and for how long.

Standing waters are subject to changes in water level, due to high rainfall, either from long in the past that has found its way through soil and rocks over time, and/or from instant run-off in the local area. Some grasslands become flooded during times of very high rainfall, attracting some highly vagile dragonfly species to breed opportunistically. When rivers swell, standing waters often form as flood plains or as flooded forest. Beavers, by building lodges, also create ponds.

1.4.3 Artificial dragonfly habitats

There are several types of artificial bodies occupied by certain dragonflies. These include impoundments (dams), irrigation channels, drainage ditches, livestock water troughs, aesthetic ponds in urban areas, game drinking ponds, gravel pits and rehabilitated mining ponds.

1.5 Human Impact on Freshwater Ecosystems and on Dragonflies

1.5.1 Stressors to freshwater macroinvertebrates

Freshwater systems and their biodiversity are highly threatened by human activity globally (Dudgeon *et al.*, 2006; Vörösmarty *et al.*, 2010; Carpenter *et al.*, 2011; Reid *et al.*, 2019). This is of concern as people need freshwater biodiversity (Lynch *et al.*, 2023). The main threats to freshwater invertebrates in general are, in descending order: pollution, dam and water management, infrastructure development, habitat loss from agriculture, habitat loss from logging, climate change and severe weather events, invasive alien species like fish, and human disturbance such as peat harvesting and a drop in the water table (Darwall *et al.*, 2012) (Fig. 1.5). However, improved water management in recent years has led to a small improvement in freshwater insect abundance, mostly in the northern hemisphere (van Klink *et al.*, 2020). Yet, there is still concern for the species-rich tropics where logging and other stressors are having a major impact on dragonflies.

1.5.2 Threatened dragonflies

As regards the level of threat to dragonflies, an earlier assessment of a random sample of 1500 species, out

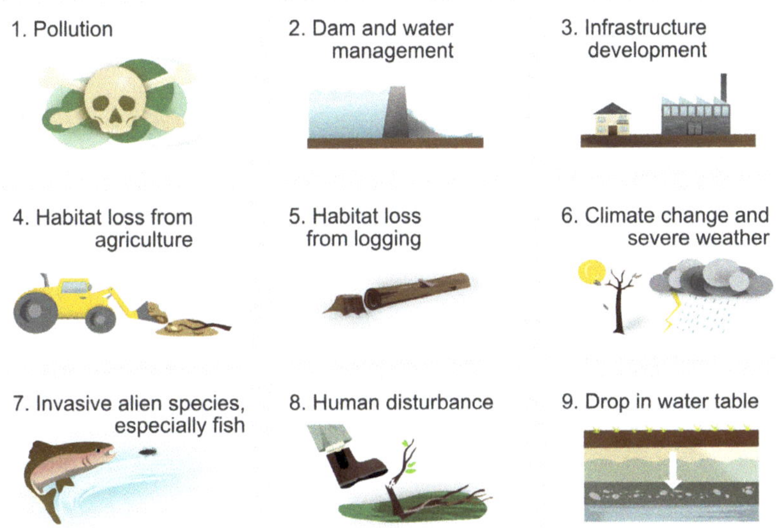

Fig. 1.5. Main threats to freshwater invertebrates in descending order of importance. (Illustration: Robyn Symons).

of the then 5680 described species, found 10% fell into one of the IUCN Red List threat categories (Vulnerable, Endangered, Critically Endangered), with species occupying running waters more threatened than those of standing waters. The actual number of threatened species would likely have been higher had the large rainforest blocks of the Amazon and Congo been better explored, and especially given the rapid rate of deforestation. The actual number of threatened species globally is likely to be about 15% (Clausnitzer *et al.*, 2009).

The IUCN Red List of Threatened Species (www.iucnredlist.org), as of early 2023, has assessed 6223 species, excellent proportionate coverage for an invertebrate order. However, 29.4% are currently Data Deficient (DD) (i.e. not enough data available to assess threat status), while 55.4% are of Least Concern (LC) (i.e. assessed as being currently not threatened). A further 3.8% are Near Threatened and 4.8% are Vulnerable. The species of great concern are those that are Endangered (5.0%) and particularly those that are Critically Endangered (1.5%). One species, *Sympetrum dilatatum*, formerly of St Helena island in the mid-Atlantic, is the only one currently listed as Extinct (Fig. 1.6).

1.5.3 Stressors on dragonfly populations and assemblages

1.5.3.1 *Spatial concerns*

Some species are Regionally Extinct (RE), meaning that they are extinct in a certain region but remain common in others. For example, *Rhyothemis semihyalina* is extinct in Algeria, Israel and on the island of Socotra, due mainly to water drainage and pollution of its habitat, but common in many parts of Africa (Dumont, 1975; www.iucnredlist.org).

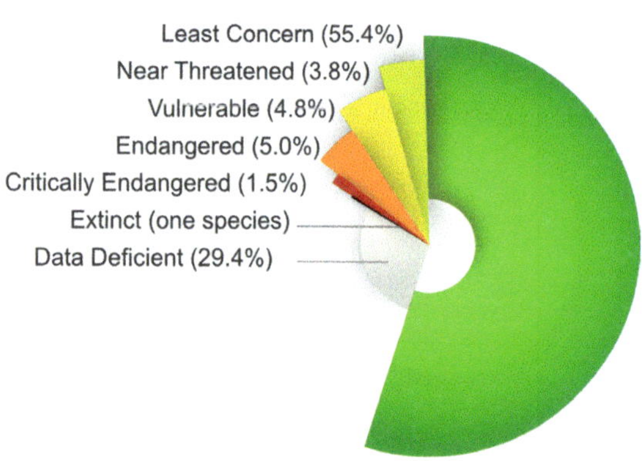

Fig. 1.6. Proportions of IUCN-assessed dragonflies in the various Red List categories. (Illustration: Robyn Symons, based on the IUCN Red List of Threatened Species).

Island dragonflies are particularly threatened, especially on the large and species-rich islands of Madagascar and Sri Lanka, and some archipelagos, such as those of Hawai'i, the Philippines and the granitic Seychelles. Although technically listed as Critically Endangered (possibly Extinct) one likely island dragonfly extinction (as well as *Sympetrum dilatatum* on St Helena) has been the damselfly *Megalagrion jugorum*, last recorded on Lanai island (Hawai'i) in 1917. However, some sky islands have equally high levels of threatened dragonfly species as oceanic islands, as in the Cape Fold Mountains of South Africa (Samways *et al.*, 2011).

1.5.3.2 *The 'biodiversity gap'*

Despite great advances in recent years, there is much we still do not know about dragonflies, both as conservation subjects and as a window to the world of biodiversity conservation in general. This is particularly acute in lesser explored yet species-rich areas of the world (a phenomenon known as the 'biodiversity gap'). Nevertheless, research priorities for dragonflies are currently widening considerably. There is much focus on priorities such as assessment and monitoring at various levels from species through to ecosystems, as well as ecological evaluation and management of both freshwater and terrestrial ecosystems, species, and the extent of ecosystem concordance and complementarity at multiple levels. Dragonflies as a group, sometimes along with other taxa and functional groups, are now playing a major role in this progress. Advances are also being made in landscape planning and how it must encompass climate change at both local and larger spatial scales. This progress is alongside data modelling and simulation of trends from population through to ecosystem levels, as well as rare species conservation (Bried and Samways, 2015).

1.5.3.3 *Stressors on dragonflies in all types of freshwaters*

Some stressors are common to both running and standing waters. Pollution is a major and pervasive issue, which includes nitrification, agricultural and urban pollutant inflow, and pesticides. Both types of freshwaters can also suffer damage to their banks and drops in water levels from the two interrelated factors of over-abstraction of water and extreme drought events. Alien species, such as invasive alien plants, crayfish and fish, are also a major threat in some systems around the world. We discuss all these and more in Chapter 5.

Fragmentation of the landscape, whether from forest loss or wetland destruction, can isolate a previously contiguous dragonfly population. These human activities pose a risk to the future of many species. There are two sides to this issue: environmental, both abiotic and biotic (i.e. loss of potential or actual habitat), and genetic (i.e. loss of genetic resilience).

One aim of dragonfly conservation is to provide maximum opportunities for a species to survive and reproduce, enabling maintenance of an effective population (a sufficiently large population that contributes to the next generation, usually referring to the number of breeding individuals required to maintain a viable population in the long term). The crux of the matter is that large populations have more individuals to offset mortality, while also having more genetic (allelic) diversity. This genetic diversity means more genetic resilience in the face of changing environmental conditions, compared to that in small populations. The upshot is that large populations are preferred to increase the odds that there will be enough individuals and genetic diversity to overcome environmental impacts, whether abiotic, like cold or hot weather extremes, and risk of disease, or biotic, like increased predator pressure. As genetic diversity may vary among subpopulations (or extensive metapopulations), each of which is adapted to its own local conditions, the ideal is to enable local subpopulations to interbreed freely.

While genetic diversity determines the potential fitness of a population in a changing world, when the populations become small, especially when down to just a few breeding individuals, local extinction risk increases. Besides decreasing opportunities for survival through environmental bottlenecks, small populations can suffer from inbreeding depression, where each generation becomes increasingly closely related. This is the result of increased homozygosity (where there is an increasing proportion of identical genes present on both homologous chromosomes). Small populations also accumulate deleterious mutations, and may also experience genetic drift, where there is a change in the frequency of alleles (gene variants) through random effects. Together, this loss of genetic diversity reduces a population's ability to overcome environmental change.

1.5.3.4 Stressors on dragonflies in running waters

Rivers, being linear and flowing systems, are prone to an impact at a single point along the river, such as tree cover loss or a major pollution event, having consequences downstream. However, sometimes dragonfly population recovery can be relatively rapid so long as the stressor is removed. For example, when a stressor like invasive alien trees is eliminated, population recovery can be good but only when a source population is present, for example, upstream. River impoundments, whether large hydroelectric dams or small weirs in either rural or urban environments, can also greatly change the local dragonfly assemblages.

Channelization of streams and rivers can also have adverse consequences, mostly where it increases water velocity, scours the substratum, takes away marginal vegetation, and from a wider perspective, creates a hydrological drought in the surrounding landscape. Among infrastructure impacts, bridges have a variable effect on dragonflies, depending on species. In Germany, individuals of *Calopteryx splendens* are inhibited from flying along the river edge because of bridges (Schutte *et al.* 1997). It is also a widespread phenomenon that bridges can partially impound streams and rivers, locally causing a change from a running water dragonfly assemblage to one characteristic of standing water.

In turn, road building can also affect dragonflies, whether by cutting across feeder streams or through dumping of rubble in the river. Furthermore, roadkill impact on dragonfly populations can be high, with daily casualties of between 10 and 256/km in a Great Lakes coastal wetland, USA (Riffell, 1999), with low-flying species especially susceptible (Soluk *et al.* 2011).

1.5.3.5 Stressors on dragonflies in standing waters

The greatest concern for standing waters is their drainage for agriculture or urban development, whether highland bogs or low-lying wetlands. While artificial ponds can increase the area of occupancy of several dragonfly species, mostly habitat generalists, there can sometimes be downsides. Their conservation value decreases greatly when they experience droughts and concomitant water abstraction, whether in rural or agricultural areas. Furthermore, artificial ponds may function as ecological traps, with certain dragonfly species attracted to, but not able to breed in them through conditions being unsuitable for larval growth, such as high fish densities in aquaculture ponds (Šigutová *et al.*, 2015).

1.5.3.6 Emerging stressors on dragonflies

Artificial lights at night attract both males and females of some species of dragonfly, often resulting in their death. Conversely, dark retreats, such as open windows, attract some species into buildings, especially the female, often later dying at a glass window. However, these behaviours have been exploited to sample dragonfly species, with light sheets at night and Malaise traps during the day detecting the presence of species formerly overlooked by traditional aerial sampling of adults (De Almeida *et al.*, 2013).

Possible emerging threats include the effect of microplastics, which are known to pass from the larva to the adult mosquito (Al-Jaibachi *et al.*, 2018), but the extent to which the plastic particles are passed on to predators like dragonflies or directly absorbed by them requires further research (Guimarães *et al.*, 2021). However, evidence to date suggests that microplastics pollution in freshwater systems is an ever-growing challenge (Sazli *et al.*, 2023). Artificial radiofrequency electromagnetic fields can disorientate various species of wildlife (Balmori, 2015), including bees (Favre, 2011) and ants (Cammaerts and Johansson, 2014), although the affect can vary among insects (Lázaro *et al.*, 2016), and may be an emerging threat for some dragonflies. Artificial noise also has a variable effect on certain insects (Bunkley *et al.*, 2017; Davis *et al.*, 2018), and could possibly affect some dragonflies. In turn, wind turbines have an impact on many flying insects to the extent of 'fouling blades' (Corten and Veldkamp, 2001), with the common colours of white and pale grey having most impact (Long *et al.*, 2011) (Fig. 1.7).

Fig. 1.7. Some major stressors on dragonfly populations and assemblages across the world. (Illustration: Robyn Symons).

1.6 Dragonfly Occupation of Modified Water Systems

Transformed landscapes are not necessarily bad for all dragonflies. Some species thrive in artificial water bodies, especially standing water ones. These standing waters range from barrages across small streams to major dams for urban, agricultural and/or industrial use, and in the case of the large dams, hydroelectric power generation. Small artificial water bodies include ponds supplying water for local irrigation of farmland. Other ponds are aesthetic, providing public amenity in urban environments, or they may grace home gardens. Drinking troughs for domestic livestock and perennial birdbaths also attract a few species.

Rice paddies and yam gardens also create artificial wetlands, attractive to certain species, especially when paddies are abandoned. Some of these abandoned paddies have been redesigned into dedicated dragonfly sanctuaries, especially in Japan, where they are known as Dragonfly Nature Parks (singular: *tomboshizenkouen*). Nakamura Nature Park is one of the best and most famous (Fig. 1.8).

Stormwater ponds constructed to catch water run-off from roads are also home to several dragonfly species. Even vehicle ruts along dirt roads attract some highly mobile and opportunist species which have larvae that grow quickly in warm conditions, such as the pan-global aptly named Globe skimmer *Pantala flavescens*.

These artificial water bodies, while mostly attracting habitat generalists, play an important role in increasing the area of occupancy for many species. This comes about from the dragonflies being provided with extra habitat space across landscapes in a region. Artificial ponds may act as stepping-stone habitats when they are suboptimal, and residential habitats when optimal, both improving metapopulation dynamics, and therefore gene flow.

A collection of ponds, both natural and/or artificial, is known as a 'pondscape' (Boothby, 1997;

Fig. 1.8. Nakamura Nature Park, Japan, is dedicated to increasing awareness and providing extensive information on dragonflies. It is positioned on what was originally a mosaic of rice paddies with a variety of wooded and open habitats. (Photo: Michael Samways).

Hill *et al.*, 2018). An important function of pondscapes is that dragonfly individuals can occupy optimal habitats for breeding, while also enabling population resilience in the face of landscape and global change. This assumes that the ponds are of a good quality and not suffering from adverse drivers like pollution, eutrophication, and/or great water-level fluctuation from severe abstraction combined with impacts from drought.

1.7 Developing Strategies for Dragonfly Conservation

1.7.1 Philosophical underpinning to dragonfly conservation

Conservation of dragonflies is essentially about valuing them. We are motivated to conserve what we value. Value then translates into caring for dragonflies through a process of conservation psychology (Simaika and Samways, 2018), largely reflecting mutual well-being. Dragonflies provide us with a sense of wonder, leading to our care for them. In turn, the caring translates into respect for the water and land, good for both dragonflies and us.

Dragonflies have had a head start over many other insects, as they are so charismatic and familiar to all, with popularity an important driver for species conservation (Mammola *et al.*, 2020). Their greatest rivals are butterflies. But as Clausnitzer *et al.* (2017) have put it, dragonflies are flagships for sustainable use of water resources, while also important in environmental conservation. This means that fundamental to dragonfly conservation is caring for the watery realm as well as the landscape context surrounding the water body.

While the focus here is on dragonflies relative to freshwater conservation, they are not simply delimited items. Like all life, they are open systems that breathe, move, breed and rest. They face challenges from predators, pathogens and competitors, as well as in finding optimal conditions for feeding and growing as larvae, and for feeding, mating and egg-laying as adults. As amphibiotic insects, they must be able to seek out both the best water and terrestrial conditions (i.e. optimal habitats). They must also do so in the face of changing natural and anthropogenic environmental pressures, especially as human needs and those of dragonflies often converge where surface water is present.

The behavioural and ecological dimensions of their existence are expressions of their genetic viability and evolutionary potential. In short, we conserve dragonflies not just ecologically for the present moment, but we also need to ensure that they have a long-term future through opportunities to adapt and evolve. That is, ensuring that their populations have resilience in the face of ongoing environmental change. It also means recovering opportunities for a safe future, mostly through insightful restoration of their habitats. The fundamental challenge is then how to act effectively to maintain and improve the conditions of their habitats to enable long-term survival, especially given that we do not have all the basic information for doing this for all species. This means that we must find strategic ways forward.

1.7.2 Retrospective analysis

While it is tempting to think that times have moved on from extensive data gathering to define the problem, there are two fundamental issues that continue to nag conservation progress. The first is the 'shifting baseline', that is, can we be sure that we are measuring progress against a genuine natural (historic) condition or against an already altered ecosystem state or impoverished condition. The second is the interaction between stressors. There are two factors that come into play here: that of the interplay between various stressors (i.e. the effect of multiple stressors) and whether novel stressors must also be considered, some of which we may not be fully aware yet. In turn, there is the overarching effect of climate change and its spin-off, extreme weather events (Harvey *et al.*, 2023). We discuss these issues in detail in future chapters.

In view of the biodiversity crisis, we need to take urgent action to ensure no further declines in freshwaters and dragonfly populations globally. Today, we need to be more proactive about instigating strategic conservation. We can do this using retrospective analysis (Grose, 2014), a solution-based approach where the attainment of a goal ('the solution') is first defined and visualized. Then only the strategically necessary data are gathered, while also applying known and workable conservation solutions in association with essential biological knowledge of the focal taxon (i.e. dragonflies here). Results are then tested to address the essential question of the extent to which conservation action is working (Fig. 1.9).

Let's now use a hypothetical example. Marshlands have been surveyed for a rare damselfly. Prior knowledge and experience tell us that this species

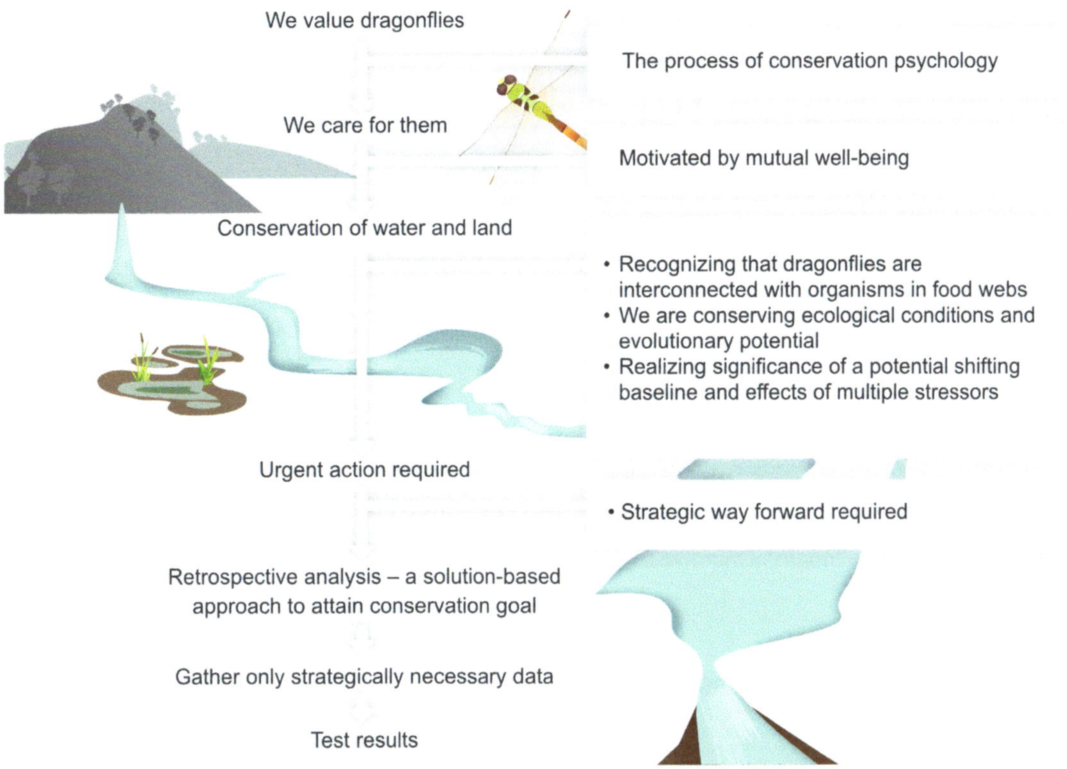

Fig. 1.9. The philosophy of dragonfly and freshwater conservation translated into strategic action for sustaining the insects and their watery and land habitats in the long term. (Illustration: Robyn Symons).

has a certain habitat, say, a shallow marsh, rich in submerged and floating macrophytes, with some marginal trees (among which the teneral adult matures). But we also know that this species is a good umbrella species for many other dragonfly species, and even non-dragonfly species, perhaps a rare frog. We then decide to take expeditious action to locate strong and viable populations of the damselfly. We consider these as source populations and aim to improve the metapopulation dynamics of the species. We then identify marshland between the source populations and aim to restore conditions to the historic state, with the right proportions of open water, macrophytes, nearby trees and healthy, clean water conditions. We also know that the species disperses reasonably well, perhaps even by means of using air currents above the boundary layer, as well as across the land surface. The point is that we know what constitutes an optimal habitat, we recreate it, and the species has the best chance to find these right conditions. When later we survey these restored areas and find that perhaps the damselfly has not yet colonized them, we may then consider translocation of some individuals. Meanwhile, it is a pleasant surprise that several other dragonfly species have already colonized the restored marshland, and even a species or two that we were not expecting. Perhaps, even the frog has found its way there on some moist, sultry night. Then a further survey shows that the damselfly has indeed arrived and established, and so our conservation goal is attained.

To achieve these ends, we must nevertheless take some strategic steps. These steps are intrinsic to a feasibility analysis.

1.7.3 Setting the conservation goals and feasibility analysis

It is so important to be crystal clear on the goal of the freshwater/dragonfly conservation project. What exactly should be achieved? This is not just about prioritizing a particular species and its habitat, or a particular landscape or river system, but also about setting priorities for conservation action (Game *et al*., 2013) while at the same time being aware of practical and financial feasibility (Carwardine *et al*., 2019). Furthermore, there must be preparedness when something goes wrong, new challenges arise, or should time and/or money run out. This means that there must be good yet inclusive and resilient planning. These are some of the details of the considerations needed and the steps required (Samways, 2020) (Fig. 1.10):

1. Cultural underpinning of the stated goal(s).
   - Determine the local belief system, i.e. the perspectives and requirements of the local human community (Palmer and Finlay, 2003).
2. Inclusion of all stakeholders.
   - Discuss the project with all members of the local community right from the beginning of conceptualization of the project and establish how these members might be involved.
   - Engage the local land managers and incorporate their knowledge into planning.
   - Clearly communicate the significance and dimensions of the project.
3. Define the task.
   - Determine the dimensions of the project, its exact goal(s) and intended conservation outcomes.
   - Estimate realistically what is achievable.
   - Decide on priorities for action.
   - Delineate the priorities. For example, is the goal to illustrate a possible way forward for freshwater conservation, or is the intention to actually undertake conservation action? In the last case, we would also need to test whether the action was successful.
4. Resources available.
   - Procure sufficient funding and human resources to complete the task.
   - Reprioritize should full funding not be available.

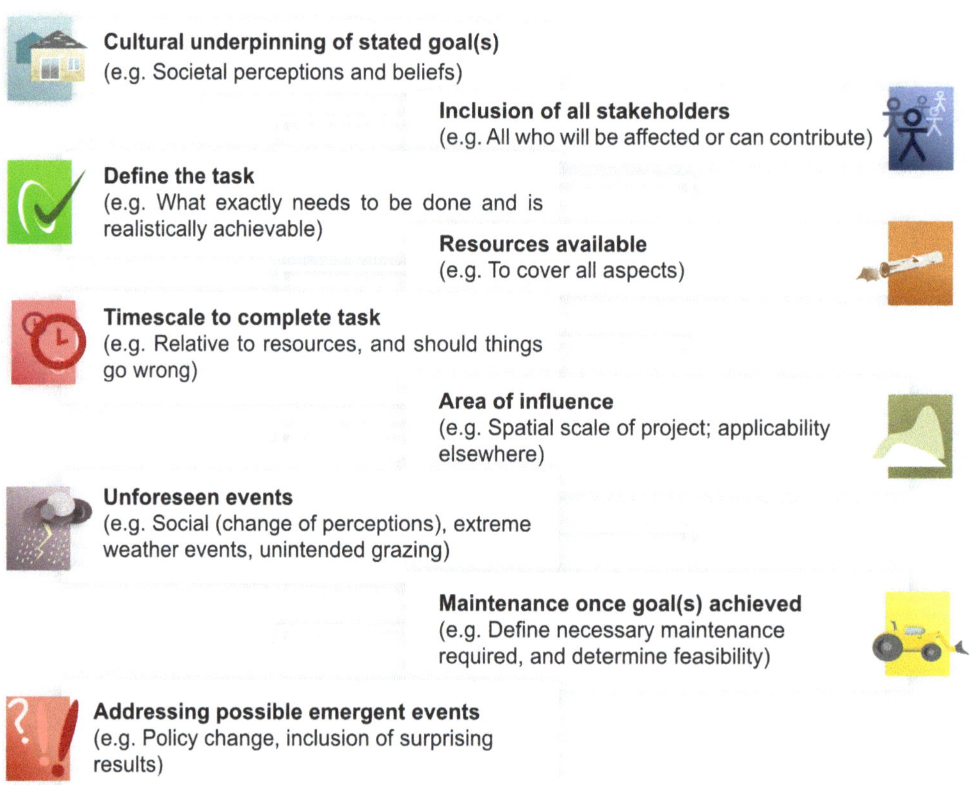

Fig. 1.10. Setting conservation goals and feasibility analysis. (Illustration: Robyn Symons).

5. Timescale to complete the task.
   - Develop a timescale for completion of the task, including contingency plans should time be running short.
6. Area of influence, i.e. spatial scale.
   - Delineate the physical area for the conservation action, bearing in mind the matrix effects, i.e. that of the surrounding area on the focal area.
   - Establish the extent of influence of the results, whether only of local significance, or whether they have wider significance, say, regionally or even globally.
7. Unforeseen events.
   - Prepare for possible unforeseen events. While the timing of events may not be predictable (e.g. weather events, change in political orientation), it is possible to consider what would happen to sites under unusually high rainfall or drought, or fire, while adverse political change can partly be mitigated through citizen science engagement and activity.
8. Maintenance once conservation goal(s) achieved.
   - Determine how the site will be maintained and even improved once the intended goal(s) have been reached.
   - Develop a monitoring scheme to assess the ongoing quality of the site. For dragonflies, it may be the whole assemblage, or perhaps the population level of certain key species, as well as aspects of the vegetation and water quality.
   - Where necessary, plan regular habitat maintenance and enhancement (such as duckboards for viewing, information applications for visitors) at the site over time.
9. Addressing possible emergent events.
   - Be aware that there may be changes in local land use policy and management.
   - Relay your findings via publications, websites and social media so that others at other sites can learn from the experience gained.

These fundamentals are essential if dragonfly conservation is to succeed in terms of the mutual well-being of both dragonflies and us.

References

Al-Jaibachi, R., Cuthbert, R.N. and Callaghan, A. (2018) Up and away: ontogenic transference as a pathway for aerial dispersal of microplastics. *Biology Letters* 14, e20180479. DOI: 10.1098/rsbl.2018.0479.

Balian, E., Harrison, I., Butchart, S., Chambers, P., Cordeiro, J. *et al.* (2010) Diversity of species in freshwater systems. In: Mittermeier, R.A., Farrell, T., Harrison, I.J., Upgren, I.J. and Brooks, T. (eds) *Fresh Water: The Essence of Life*. Conservation International and Cemex Conservation Book Series, Earth in Focus Editions, China, pp. 50–89.

Balmori, A. (2015) Anthropogenic radiofrequency electromagnetic fields as an emerging threat to wildlife orientation. *Science of the Total Environment* 518–519, 58–60. DOI: 10.1016/j.scitotenv.2015.02.077.

Barua, M., Gurdak, D.J., Ahmed, R.A. and Tamuly, J. (2012) Selecting flagships for invertebrate conservation. *Biodiversity and Conservation* 21, 1457–1476. DOI: 10.1007/s10531-012-0257-7.

Boothby, J. (1997). Pond conservation: towards a delineation of pondscape. *Aquatic Conservation* 7, 127–132.

Bried, J.T. and Samways, M.J. (2015) A review of odonatology in freshwater applied ecology and conservation science. *Freshwater Science* 34, 1023–1031. DOI: 10.1086/682174.

Bunkley, J.P., McClure, C.J.W., Kawahara, A.Y., Francis, C.D. and Barber, J.R. (2017) Anthropogenic noise changes arthropod abundances. *Ecology and Evolution* 7, 2977–2985. DOI: 10.1002/ece3.2698.

Cammaerts, M.-C. and Johansson, O. (2014) Ants can be used as bio-indicators to reveal biological effects of electromagnetic waves from some wireless apparatus. *Electromagnetic Biology and Medicine* 33, 282–288. DOI: 10.3109/15368378.2013.817336.

Cannings, S.G. and Cannings, R.A. (1997) Dragonflies (Odonata) of the Yukon. In: Danks, H.V. and Downes, J.A. (eds) *Insects of the Yukon*. Biological Surveys of Canada (Terrestrial Arthropods), Ottawa, Canada, pp. 169–200.

Capinera, J.L. (1993) Insects in art and region: the American Southwest. *American Entomologist* Winter 1993, 221–229.

Carpenter, S.R., Stanley, E.H. and Vander Zanden, M.J. (2011) State of the World's freshwater ecosystems: physical, chemical, and biological. *Annual Review of Environment and Resources* 36, 75–99. DOI: 10.1146/annurev-environ-021810-094524.

Carwardine, J., Martin, T.G., Firn, J., Reyes, R.P., Nicol, S. *et al.* (2019) Priority threat management for biodiversity conservation: a handbook. *Journal of Applied Ecology* 56, 481–490. DOI: 10.1111/1365-2664.13268.

Clausnitzer, V., Kalkman, V.J., Ram, M., Collen, B., Baillie, J.E.M. *et al.* (2009) Odonata enter the biodiversity crisis debate: the first global assessment of an insect group. *Biological Conservation* 142, 1864–1869. DOI: 10.1016/j.biocon.2009.03.028.

Clausnitzer, V., Koch, R., Dijkstra, K.-D.B., Boudot, J.-P., Darwall, W.R.T. *et al*. (2012) Focus on African freshwaters: hotspots of dragonfly diversity and conservation concern. *Frontiers in Ecology and the Environment* 10, 129–134. DOI: 10.1890/110247.

Clausnitzer, V., Simaika, J.P., Samways, M.J. and Daniel, B.A. (2017) Dragonflies as flagships for sustainable use of water resources in environmental education. *Applied Environmental Education and Communication* 16, 196–209. DOI: 10.1080/1533015X.2017.1333050.

Corbet, P.S., Dunkle, S.W. and Ubukata, H. (1995) *Proceedings of the International Symposium on the Conservation of Dragonflies and Their Habitats*. Japanese Society for the Preservation of Birds, Kushiro, Japan.

Corten, G.P. and Veldkamp, H.F. (2001) Insects can halve wind-turbine power. *Nature* 412, 41–42. DOI: 10.1038/35083698.

Darwall, W., Smith, K., Allen, D., Holland, R., Harrison, I. and Brooks, E. (eds) (2011) *The Diversity of Life in African Freshwaters: Underwater, Under Threat*. IUCN, Gland, Switzerland and Cambridge.

Darwall, W., Seddon, M., Clausnitzer, V. and Cumberlidge, N. (2012) Freshwater invertebrate life. In: Collen, B., Böhm, M., Kemp, R. and Baillie, J.E.M. (eds) *Spineless: Status and Trends of the World's Invertebrates*. Zoological Society of London, London, pp. 26–33.

Davis, A.K., Schroeder, H., Yeager, I. and Pearce, J. (2018) Effects of simulated highway noise on heart rates of larval monarch butterflies, *Danaus plexippus*: implications for roadside habitat suitability. *Biology Letters* 14, e20180018. DOI: 10.1098/rsbl.2018.0018.

De Almeida, M.V.O., Pinto, Â.P., Carvalho, A.do L. and Takiya, D.M. (2013) When rare is just a matter of sampling: unexpected dominance of clubtail dragonflies (Odonata, Gomphidae) through different collecting methods at Parque Nacional da Serra do Cipó, Minais Gerais, Brazil. *Revista Brasileira de Entomologia* 57, e4. DOI: 10.1590/S0085-56262013005000042.

Dennis, R.L.H. (2010) *A Resource-based Habitat View for Conservation: Butterflies in the British Landscape*. Wiley-Blackwell, Oxford. DOI: 10.1002/9781444315257.

Dijkstra, K-D.B., Boudot, J.-P., Clausnitzer, V., Kipping, J., Kisakye, J.J. *et al*. (2011) Dragonflies and damselflies of Africa (Odonata): history, diversity, distribution, and conservation. In: Darwall, W.R.T., Smith, K.G., Allen, D.J., Holland, R.A, Harrison, I.J., and Brooks, E.G.E. (eds) *The Diversity of Life in African Freshwaters: Under Water, Under Threat. An analysis of the status and distribution of freshwater species throughout mainland Africa*. IUCN, Cambridge and Gland, Switzerland, pp. 126–177.

Dudgeon, D., Arthington, A.H., Gessner, M.O., Kawabata, Z.-I., Knowler, D.J. *et al*. (2006) Freshwater biodiversity: importance, threats, status and conservation challenges. *Biological Reviews* 81, 163–182. DOI: 10.1017/S1464793105006950.

Dumont, H.J. (1975). Endemic dragonflies of Late Pleistocene age of the Hula Lake area (northern Israel), with notes on the Calopterygidae of the rivers Jordan (Israel, Jordan) and Litani (The Lebanon), and description of *Urothemis edwardsi hulae* subspec. nov. (Libellulidae). *Odonatologica* 4, 1–9.

Favre, D. (2011) Mobile phone-induced honeybee worker piping. *Apidologie* 42, 270–279. DOI: 10.1007/s13592-011-0016-x.

Game, E.T., Kareiva, P. and Possingham, H.P. (2013) Six common mistakes in conservation priority setting. *Conservation Biology* 27, 480–485. DOI: 10.1111/cobi.12051.

Garrison, R.W., von Ellenrieder, N. and Louton, J.A. (2006) *Dragonfly Genera of the New World: An Illustrated and Annotated Key to the Anisoptera*. The John Hopkins University Press, Baltimore, USA. ISBN 0-8018-8446-2.

Grose, M.J. (2014) Thinking backwards can inform concerns about 'incomplete' data. *Trends in Ecology and Evolution* 29, 546–547. DOI: 10.1016/j.tree.2014.07.007.

Guimarães, A.T.B., Rodrigues, A.S.L., Pereira, P.S., Silva, F.G. and Malafia, G. (2021) Toxicity of polystyrene nanoplastics in dragonfly larvae: an insight on how these pollutants can affect benthic macroinvertebrates. *Science of the Total Environment* 752, e141936. DOI: 10.1016/j.scitotenv.2020.141936.

Harvey, J.A., Tougeron, K., Gols, R., Heinen, R., Abarca, M. *et al*. (2023) Scientists' warning on climate change and insects. *Ecological Monographs* 93, e1553. DOI: 10.1002/ecm.1553.

Hill, M.J., Hassall, C., Oertli, B., Fahrig, L., Robson, B.J. *et al*. (2018) New policy directions for global pond conservation. *Conservation Letters* 11, e12447. DOI: 10.1111/conl.12447.

Jepson, P. and Barua, M. (2015) A theory of flagship species action. *Conservation and Society* 13, 95–104.

Kalkman, V.J., Clausnitzer, V., Dijkstra, K.-D.B., Orr, A.G., Paulson, D.R. and van Tol, J. (2008) Global diversity in dragonflies (Odonata) in freshwater. *Hydrobiologia* 595, 351–363. DOI: 10.1007/s10750-007-9029-x.

Kiauta, B. (1986) Dragonfly in Halku. *Odonatologica* 15, 91–96.

Kipping, J. (2010) The dragonflies and damselflies of Botswana: an annotated checklist with notes on distribution, phenology, habitats and Red List status of the species (Insecta: Odonata). *Mauritiana (Altenburg)* S126–204.

Lázaro, A., Chroni, A., Tscheulin, T., Devalez, J., Matsoukas, C. and Petanidou, T. (2016) Electromagnetic radiation of mobile telecommunication antennas affects the abundance and composition of wild pollinators. *Journal of Insect Conservation* 20, 315–324. DOI: 10.1007/s10841-016-9868-8.

Lemelin, R.H. (2007) Finding beauty in the dragon: the role of dragonflies in recreation and tourism. *Journal of Ecotourism* 6, 139–145. DOI: 10.2167/joe161.0.

Lemelin, H. (2009) Goodwill hunting: dragon hunters, dragonflies and leisure. *Current Issues in Tourism* 12, 553–571. DOI: 10/1080/13683500802346169.

Liu, R.K. (2001) The symbolic importance of insects in jewelry. *Transactions of the American Entomological Society* 127, 167–171.

Long, C.V., Flint, J.A. and Lepper, P.A. (2011) Insect attraction to wind turbines: does colour play a role? *European Journal of Wildlife Research* 57, 323–331. DOI: 10.1007/s10344-010-0432-7.

Lynch, A.J., Cooke, S.J., Arthington, A.H., Baigun, C., Bossenbroek, L. *et al*. (2023) People need freshwater biodiversity. *WIRES Water* 10, e1633. DOI: 10.1002/wat2.1633.

Mammola, S., Riccardi, N., Prié, V., Correia, R., Cardoso, P. *et al.* (2020) Towards a taxonomically unbiased European Union biodiversity strategy for 2030. *Proceedings of the Royal Society B* 287, e20202166. DOI: 10.1098/rspb.2020.2166.

McCloud, S. (1998) Understanding comics: the invisible art. *IEEE Transactions on Professional Communication* 42, 66–69.

Moore, N.W. (1997) *Dragonflies: Status Survey and Conservation Action Plan*. IUCN/SSC, Gland, Switzerland.

Naraoka, H. (2016) The factors influencing the length of the terrestrial period in the last instar larvae of *Epiophlebia superstes* (Selys, 1889) (Anisozygoptera: Epiophebiidae). *International Journal of Odonatology* 19, 95–105. DOI: 10.1080/13887890.2015.1129990.

Oliveira-Junior, J.M.B., Rocha, T.S., Vinagre, S.F., Miranda-Filho, J.C., Mendoza-Penagos, C.C. *et al*. (2022) A bibliometric analysis of the global research in Odonata: Trends and gaps. *Diversity* 14, e1074. DOI: 10.3390/d14121074.

Palmer, M. and Finlay, V. (2003) *Faith in Conservation: New Approaches to Religions and the Environment.* The World Bank, Washington, DC.

Reid, A.J., Carlson, A.K., Creed, I.F., Eliason, E.J., Gell, P.A. *et al*. (2019) Emergent threats and persistent conservation challenges for freshwater biodiversity. *Biological Reviews* 94, 849–873. DOI: 10.1111/brv.12480.

Riffell, S.K. (1999) Road mortality of dragonflies (Odonata) in a Great Lakes coastal wetland. *The Great Lakes Entomologist* 32, 63–73.

Samways, M.J. (1989) Taxon turnover in Odonata across a 3000 m altitudinal gradient in southern Africa. *Odonatologica* 18, 263–274.

Samways, M.J. (2020) *Insect Conservation: A Global Synthesis*. CAB International, Wallingford, UK.

Samways, M.J., Pryke, J.S. and Simaika, J.P. (2011) Threats to dragonflies on land islands can be as high as those on oceanic islands. *Biological Conservation* 144, 1145–1151. DOI: 10.1016/j.biocon.2010.12.034.

Sazli, D., Nassouhi, D., Ergönül, M.B. and Atasagun, S. (2023) A comprehensive review on microplastic pollution in aquatic ecosystems and their effects on aquatic biota. *Aquatic Sciences and Engineering* 38, 12–46. DOI: 10.26650/ASE20221186783.

Schutte, G., Reich, M. and Plachter, H. (1997) Mobility of the rheobiont damselfly *Calopteryx splendens* (Harris) in fragmented habitats (Zygoptera: Calypterygidae). *Odonatologica* 26, 317–327.

Šigutová, H., Šigut, M. and Dolný, A. (2015) Intensive fish ponds as ecological traps for dragonflies: an imminent threat to the endangered species *Sympetrum depressiusculum* (Odonata: Libellulidae). *Journal of Insect Conservation* 19, 961–974. DOI: 10.1007/s10841-015-9813-2.

Simaika, J.P. and Samways, M.J. (2018) Insect conservation psychology. *Journal of Insect Conservation* 22, 635–642. DOI: 10.1007/s10841-018-0047-y.

Simaika, J.P., Samways, M.J., Kipping, J., Suhling, F., Dijkstra, K.-D.B., *et al*. (2013) Continental-scale conservation prioritization of African dragonflies. *Biological Conservation* 157, 245–254. DOI: 10.1016/j.biocon.2012.08.039.

Soluk, D.A., Zercher, D.S. and Worthington, A.M. (2011) Influence of roadways on patterns of mortality and flight behavior of adult dragonflies near wetland areas. *Biological Conservation* 144, 1638–1643. DOI: 10.1016/j.biocon.2011.02.015.

Suhling, F., Samways, M.J., Simaika, J.P. and Kipping, J. (2009) The status and distribution of dragonflies (Odonata). In: Darwall, W., Smith, K.G., Tweddle, D. and Skelton, P. (eds) *The Status and Distribution of Freshwater Biodiversity in Southern Africa*. IUCN, Gland, Switzerland and Cambridge, pp. 48–65.

Suhling, F., Samways, M.J., Simaika, J.P., Richter, O., Marais, E. *et al*. (2010) Dragonfly diversity from the Cape to Okavango. In: Schmeidel, U. and Jürgens, N. (eds) *Biodiversity in Southern Africa 2: Patterns and Processes at Regional Scale*. Klaus Hess, Göttingen, Germany & Windhoek, Namibia, pp. 64–69.

Uboni, C., Jugovic, J., Tordoni, E., Pizzul, E., Riservato, E. and Bacaro, G. (2020) Dragonfly (Odonata) diversity patterns in mixohaline coastal wetlands. *Estuaries and Coasts* 43, 375–386. DOI: 10.1007/s12237-019-00687-y.

van Klink, R., Bowler, D.E., Gongalsky, K.B., Swengel, A.B., Gentile, A. and Chase, J.M. (2020) Meta-analysis reveals declines in terrestrial but increases in freshwater insect abundances. *Science* 368 (6489), 417–420. DOI: 10.1126/science.aax9931.

Vannote, R.L., Minshall, G.W., Cummins, K.W., Sedell, J.R. and Cushing, C.E. (1980) The river continuum concept. *Canadian Journal of Fisheries and Aquatic Sciences* 37, 130–137. DOI: 10.1139/f80-017.

Vörösmarty, C.J., McIntyre, P.B., Gessner, M.O., Dudeon, D., Prusevich, A. *et al*. (2010) Global threats to human

water security and river biodiversity. *Nature* 467, 555–561. DOI: 10.1038/nature09440.
Ward, J.V. (1998) Riverine landscapes: biodiversity patterns, disturbance regimes, and aquatic conservation. *Biological Conservation* 83, 269–278.
Watson, J.A.L. (1982) A truly terrestrial dragonfly larva from Australia (Odonata: Corduliidae). *Australian Journal of Entomology* 21, 309–311. DOI: 10.1111/j.1440-6055.1982.tb01826.x.
Wojtusiak, J. (1974) A dragonfly migration in the High Hindu Kush (Afghanistan), with a note on high altitude records of *Aeshna juncea mongolica* Bartenev, and *Pantala flavescens* (Fabricius) (Anisoptera: Aeshnidae and Libellulidae). *Odonatologica* 3, 137–142.

2 Dragonfly Functional Morphology

Illustration: Megan Louise Potgieter.

DOI: 10.1079/9781789248395.0002

Abstract

Among organisms, their form largely follows function. Appreciating dragonfly functional morphology is to see the outcome of countless generations of natural and sexual selection pressure honed by behavioural and ecological pressures.

The adult dragonfly is aerial, while the larva is aquatic. Both must survive according to the medium where each stage feeds, rests, moves, and in the case of the adult, reproduces.

Each life stage faces its own challenges and opportunities, with survival of an individual dependent on overcoming the successive developmental hurdles to survive and then successfully reproduce.

Dragonfly conservation aims to ameliorate the challenges while enabling opportunities for immediate and long-term survival in a new world with novel challenges. However, dragonflies are constrained by their functional morphology, physiology and biology. Some will be winners through having traits suited to survival in this new world, but others will be winners only with concerted conservation effort on our part.

2.1 The First Dragonflies

Ancestors of modern Odonata, the stem superorder Odonatoptera, were the top aerial predators in the Carboniferous and Permian periods, especially about 330–260 million years ago (mya), and existed well before competition from predatory pterodactyls, birds and bats. The largest of them (Meganisoptera, Meganeuridae, 'Griffinflies'), such as *Meganeura monyi* with a wingspan of 70 cm and *Meganeuropsis permiana* with one of 71 cm (Shear and Kukalová-Peck, 1990), were dominant predators in the late Palaeozoic airspace (Fig. 2.1, left).

They probably displayed hawking behaviour like modern aeshnid dragonflies. These large species may also have fed on smaller dragonfly species, as well as on proto-cockroaches, Palaeodictyoptera (Nel *et al.*, 2018), especially given that primary productivity may have been mostly channelled through decomposers and detritivores, rather than via herbivores (Shear and Kukalová-Peck, 1990). Other early flying insects, such as ancestral mayflies (Ephemeroptera) (McCafferty and Edmunds, 1979), may also have been prey, as might insects that were plucked from plant surfaces. They had many ecological opportunities during this wet and warm period with its high levels of oxygen and high air pressure. These conditions physiologically enabled respiration to power large muscles and so provide lift for flight, and at a time when extensive coastal and inland swamps, as well as rivers, were present (Harrison *et al.*, 2010; Ward and Kirchvink, 2015).

The meganeurids had massive, toothed jaws, a robust body and strong legs (Cannell, 2018). A well-preserved specimen of *Meganeurites gracilipes*, with a wingspan of 32 cm, clearly shows well-developed compound eyes. These endowed it with good vision, enabling it to search for prey with great acuity, while also looking out for larger, predatory

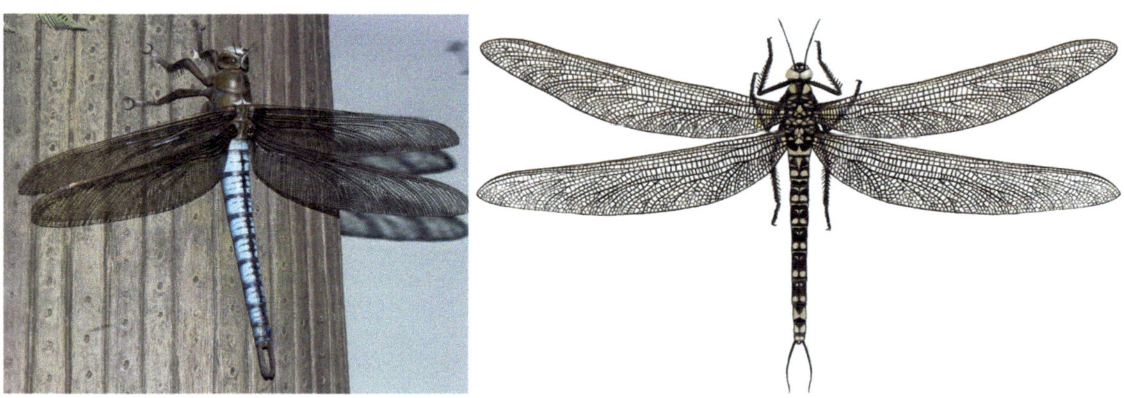

Fig. 2.1. An artistic interpretation of how *Meganeura* may have appeared alive (left), and a reconstruction of *Meganeurites gracilipes* (right). (Illustrations: Yinan Chen (Wikimedia Commons) (left), and Nel *et al.* (2018) (Gaëlle Doitteau, 2017, e-recolnat Project, MNHN) (Wikimedia Commons) (right)).

species, for risk of being eaten. With their large wings, they probably flew in open spaces, rather than in forests (Nel *et al.*, 2018) (Fig. 2.1, right).

Not all Odonatoptera were large. The damselfly-like *Progoneura nobilis* had a wingspan of only 30 mm (Wootton, 1981). There appears to have been considerable diversity among these small proto-zygopterans (Nel *et al.*, 2012). They likely inhabited a wide variety of habitat types, feeding on many types of prey, such as the forerunners of caddisflies, mayflies and flies.

Odonata as an order appeared at least 237 mya, by which time, the two major suborders, Zygoptera and Anisoptera had diverged. The third extant but very small suborder Anisozygoptera (geographically restricted to Asia) branched from the Anisoptera at least 199 mya (Kohli *et al.*, 2016).

Often, the limitations of fossil records make it difficult to assess the impact of major past climatic events on taxonomic diversity alone. One way to address this challenge is to focus on trends in morphological diversity, known as 'disparity'. In the case of trends in odonate wings of the Permian to Jurassic, there is a decreasing trend in disparity, while species richness increased. Both the Permian–Triassic and Triassic–Jurassic transitions are significant as major events associated with strong morphospace restructuring from selective extinction. At each geological transition, recovery occurred through diversification of new forms compensating the loss of others. This means that early representatives continuously evolved new shapes, a pattern contrasting with the classical view that morphospace was fulfilled early and then followed by selective extinctions and specialization within it (Deregnaucourt *et al.*, 2023).

2.2 The Modern Dragonfly Adult As a Functional Organism

2.2.1 Flight

2.2.1.1 Flight performance

The adult dragonfly is a superb flying creature, whether a large species in open environments or a small species manoeuvring in confined spaces (Fig. 2.2). It all depends on the aerial realm necessary for finding prey, suitable habitats and a mate, as well as for scattering or placing of eggs on the water or in association with a substrate, usually vegetation (Bybee *et al.*, 2016; Nakata *et al.*, 2020).

Insect flight probably first appeared in association with the aquatic environment, with the Odonata likely to be a sister group to all other winged insects (Yu *et al.*, 2021). The dragonfly's remarkable flight ability arises from efficient and flexible wing design and articulation, and effective chemico-structural wing components and muscle power (Jongerius and Lentink, 2010; Appel and Gorb, 2011; Donoughe *et al.*, 2011; Bomphrey *et al.*, 2016; Rajabi *et al.*, 2016; Rajabi and Gorb, 2020; Wootton, 2020). The large dragonfly *Austrophlebia costalis* can attain a speed of 34.1 km/hr (Woodall, 1995). Furthermore, having four wings, with accurately timed phasing and positioning of forewings relative to hindwings, is

Fig. 2.2. Large anisopteran dragonflies, like this large and robust 75 mm long *Anax imperator* (left), are adapted to manoeuvring in large open spaces at high speed. In contrast, small damselflies, like this tiny 20 mm long *Agriocnemis ruberrima* (right), are suited only to highly local slow movement in the confines of dense marshland vegetation. (Photos: Michael Samways).

highly efficient in power generation (Usherwood and Lehmann, 2008) and manoeuvrability (Noda *et al.*, 2023).

Flight performance is also associated with body shape (Okamoto *et al.*, 1996). This, in turn, has been evolutionarily crafted by mobility for population persistence, habitat preference, courting, mating, oviposition behaviour, feeding, escape from predators and for advantage in agonistic encounters. Each family has developed its own wing morphology associated with its particular behaviour, habitat and flight mode (Sacchi and Hardersen, 2013). There is ongoing selection pressure to integrate and optimize all these activities for survival in contemporary environments.

In the case of dragonflies chasing flying prey, the flight patterns of both predator and prey are important. When a prey item, such as a fly, performs random turns during flight, the dragonfly accordingly manoeuvres to catch it. Fly behaviour is less erratic when their own population density is high, with dragonflies then having a higher success rate at catching their fly prey (Combes *et al.*, 2012).

Dragonflies are remarkably successful hunters, catching prey up to 95% of the time (Gonzalez-Bellido *et al.*, 2013). This is partly because a dragonfly can adjust its flight path to a point in front of its prey target, so intercepting rather than simply chasing it (Olberg *et al.*, 2007).

Hunting has disadvantages as well as advantages, as time spent hunting also risks the hunter being eaten. This means that each dragonfly species must rapidly assess the value of its prey, both its size and type (Combes *et al.*, 2013).

The activity of catching prey involves a high level of perception, prediction and reaction, normally associated with vertebrates, which astonishingly, the dragonfly achieves. It does this this by not necessarily reacting to a change in prey manoeuvre straight away. Rather, it approaches the prey from below to reduce detection. Then, by moving its head and body independently, it can get an accurate fix on prey position without its own body movement disrupting that fix. It does this by using complex neural integration of its own body movements relative to those of its prey in the air (Mischiati *et al.* 2015) or from a perch (Lin and Leonardo, 2017), improving its hunting efficiency considerably through selection pressure for success (Moore and Biewener, 2015).

Similar procedures are used when conspecifics interact while defending territories. However, there is some overshooting which avoids potentially damaging collisions, while also enabling the competitor time to swerve away (Lohmann *et al.*, 2019) (Fig. 2.3).

Dragonflies, and especially damselflies, also engage in local movements, measured in centimetres, mostly in sheltered environments, while courting, engaging in agonistic behaviour or feeding. The petiolate wings of damselflies are particularly suited to manoeuvring in confined areas, so long as weather conditions are calm. Some species can even maintain a stationary position. Many species can hover while laying eggs in precise positions. Such hovering requires considerable neural integration (Deng and Hu, 2008). The helicopter damselfly group (Pseudostigmatinae) (Toussaint *et al.*, 2019), includes the world's largest damselfly *Megaloprepus caerulatus*, which can hover to pluck small orb-weaver spiders, and even occasionally take wrapped prey, from their webs (Fincke, 1992) (Fig. 2.4). Except for excellent manoeuvrability, they seemingly have no special adaptations to avoid being entangled in a spider web other than the normal waxy covering of dragonfly bodies and wings, in association with a life close to the perils of a watery environment (Gorb, 2019).

Intricate wing manoeuvres take place among rival males displaying intraspecific agonistic behaviour (Hilfert-Rüppell and Rüppell, 2013). Flying backward at such close quarters also has some advantages, and in some anisopteran dragonflies, this can be achieved with considerable force (Bode-Oke *et al.*, 2018). Sophisticated manoeuvring requires an evolutionary trade-off in wing shape. This is especially so in the forewings, where species can be migratory yet are also able to guard their mate at close quarters (Johansson *et al.*, 2009). There can sometimes be differences in wing venation between the male and female, with the male's wings more resistant to wing wear, a result of selection for a more demanding flying lifestyle than that of the female (Hardersen, 2006).

Coupled male and female dragonfly species can fly together as a unit remarkably well, whether in tandem or attached in the 'wheel position' (Corbet, 2004). However, this behaviour has risks, but also advantages when predators are present. In the case of *Coenagrion puella*, more solitary females are killed by the water bug *Notonecta glauca* than tandem females (Zeiss *et al.*, 1999). In the case of tandem pairs of *Ischnura elegans*, both sexes flap their wings but at different frequencies. Tandem males increase their flapping frequency by 19% but decrease flapping amplitude by 17% in comparison

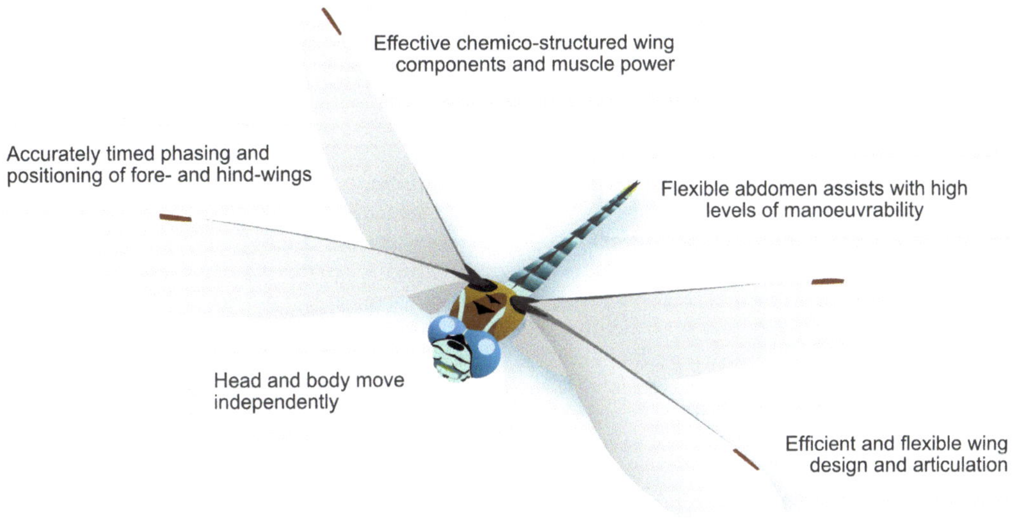

Fig. 2.3. The dragonfly is a highly efficient flying insect at catching prey while avoiding being eaten by its predators or challenged by rivals. (Illustration: Robyn Symons).

Fig. 2.4. The helicopter damselfly *Megaloprepus caeluratus* hovering in front of a leaf it is inspecting. (Photos: Nagi Aboulenein).

with solitary males. Females reduce effort, while males increase aerodynamic output, meaning that they carry not only their own weight but also some of that of the female (Davidovich and Ribak, 2016).

2.2.1.2 Thermal physiology and behaviour

The huge and complex set of muscles in the thorax that power both wings and legs (Büsse *et al.*, 2013) require chemical energy for optimal functioning at a time relative to the behavioural activity being undertaken. The flight muscles also need to be at an optimal temperature. Their temperature is also dependent on that of other parts of the body, especially the abdomen and head (May, 1998), with the haemolymph (blood) especially important in transferring heat (Heinrich and Casey, 1978). Thermoregulation is achieved by active behaviours and physiological responses to ambient temperature, involving changes in metabolic heat production and/or heat exchange with the immediate environment (May, 1991).

Three characteristics affect the ability of anisopteran dragonflies to thermoregulate. These include being predominantly diurnal (but they may also be crepuscular or nocturnal in warm climates), being strong fliers and being large insects. There is a broad functional dichotomy among anisopterans, between the 'fliers' that spend much of the daytime in the air and the 'perchers', which perch most of the daytime while making short forays to procure food and defend territories. Each category has a set of characteristic thermoregulatory strategies and energy demands. However, within each category there is also a continuum of capacity to heat the body ectothermically (their body temperature is determined by that of the immediate environment, notwithstanding some behavioural responses) by irradiation from the sun, or endothermically metabolizing for heat production, according to body size (Corbet and May, 2008). Dragonflies from the two categories also typically differ in shape, with the fliers having a long and proportionately thin abdomen, and the perchers having a stockier build.

Many species may seek shade when too hot. Perchers, however, will also adjust their perching position while also changing the angle of their abdomen relative to the sun. To cool down, some percher dragonflies raise their abdomen upward to various degrees, sometimes to a vertical position – the 'obelisk position' (Fig. 2.5). Others may exploit optimal thermal times during the day to forage and mate, which include resting on warm rocks in the evening to extend foraging time in the cooler temperatures (McGeoch and Samways, 1991).

Among anisopterans on the Brazilian Plateau in Minas Gerais there are differences in activity patterns to temperature and cloud cover related to body size. There is a minimum size for species able to regulate using sun exposure rather than temperature, associated with body colour, preferred substrate and perch posture (De Marco and Resende, 2002). In the lower elevation of the Atlantic Forest region of Brazil, the largest anisopteran species spend much time patrolling and in territorial defence, while the smaller species largely perch. Larger species with good thermoregulatory abilities can spend more time performing reproductive activities. These behaviours suggest that the simple binary classification of fliers versus perchers

Fig. 2.5. The libellulid 'percher' dragonfly *Urothemis edwardsii* in a partial obelisk position (left), and *Trithemis stictica* in a vertical position (right). The legs, as well as the thorax and abdomen, enable these positions to be maintained. (Photos: Michael Samways).

should be refined, as behavioural patterns among the species in the assemblage are more related to body size and behavioural variation between species. A more realistic view is a continuum associated with perchers, as large perchers, like fliers, spend much of their time in flight, including patrolling. Translating these behavioural activities into interspecific interactions can lead to a change in assemblage structure at any one location (De Marco *et al.*, 2005).

Many forest damselflies typically seek sunflecks in which to bask and receive increased insolation. In turn, many desert species are well adapted to remaining active at temperatures that would be lethal to temperate species (Polcyn, 1994). In an extremely cold climate, species are physiologically adapted by having a low minimum flight temperature, and while all species generate physiological heat to some extent, it is a particularly effective adaptation in large species (Sformo and Doak, 2006).

Species with a wide geographic range, such as *Ischnura elegans*, have different physiological and behavioural tolerances to environmental temperatures at different locations (Hilfert-Rüppell, 1998). Overall, selection of optimal microhabitats is a complex mix of behaviour and morphology, including body colour, preferred substrate and perching posture (De Marco and Resende, 2002).

While all dragonflies are ectothermic, some are endothermic when the need arises. They generate heat by wing whirring, involving vibrating their wings without taking off, to heat up their bodies. Typically, this behaviour takes place in the early morning, but also at dusk, to optimize foraging (May, 1976a).

Flight has a convective cooling effect on the body. Cooling is achieved by gliding. With muscles not being at full power and not generating heat when gliding, it is also an effective energy-saving behaviour, complemented by the wings being pleated, so reducing drag (Vargas *et al.*, 2008).

Gliding in migrant species is possible by having large wings with an expanded hindwing discoidal area, and longer and denser thoracic pilosity (Suárez-Tovar and Sarmiento, 2016). Long-distance travellers and continuous fliers like *Tramea virginia* are able to spend five hours in continuous flight (Ishida, 1958), likely only possible with intermittent gliding. Wing shape is so finely tuned that there can be differences among species in the same location and within a single genus (*Libellula*) according to their habitat preferences and dispersal characteristics (McCauley, 2013).

The wings of many dragonfly males are strongly coloured. In the Neotropical savanna, the wings of *Zenithoptera lanei* are bright metallic blue above and blackish below. The wings are covered with wax nanocrystals that reflect ultraviolet light and infrared radiation. The wings are also permeated by an intricate network of tracheae. The dorsal wing surface lowers body temperature while the ventral surface raises it, seemingly to enable the species to warm up or cool down according to wing position while perching (Guillermo-Ferrero and Gorb, 2021) (Fig. 2.6).

2.2.1.3 Wing colouring and sexual selection

Wing colouration plays a particularly strong role in sexual behaviour and selection. In the wings of

Fig. 2.6. Upper side (top) and lower side (bottom) of the wings of *Zenithoptera lanei*. (Photos: Stanislav N. Gorb and Rhainer Guillermo-Ferreira).

calopterygid damselflies, colouring is associated with certain hindwing shapes, and being broader where the colour patches and bars are positioned, enhances sexual signalling (Outomuro *et al.*, 2013a) (Fig. 2.7). Similarly, in the dragonfly genus *Trithemis*, wide hindwing bases are associated with more colouration, as well as phenotypic correlation of wing shape between intraspecific males and females. Although there is no correlation between water body type and wing shape, male forewings and female hindwings are broader in open compared to forested habitats. These findings suggest that natural and sexual selection are acting partially independently, despite correlation of wing shape between males and females (Outomuro *et al.*, 2013b).

Among species in the Libelluloidea, wing colouring as a trait has diversified faster than the morphological traits of body size and wing size. This suggests that evolutionary responses to disparate reproductive demands may be fundamental to the persistence and coexistence of closely related species that we see today (Hersch and Moore, 2023).

2.2.1.4 Risks of having an active lifestyle and membranous wings

As a dragonfly's wings are large and the membranes delicate, it means that they are prone to being snagged on spiny plant surfaces, especially those with minute hooks and barbs (trichomes), and particularly given the momentum of dragonfly flight. Young and old male libellulids are occasionally entangled on these trapping surfaces. The function of the hooks and barbs for plant seeds is

Fig. 2.7. Males of different calopterygid damselflies have a wide range of wing colours and extent. Some, like *Phaon iridipennis* have clear but glistening wings (top left). Others, such as *Hetaerina rosea* (top right) and *Hetaerina (Mnesarete) pudica* (bottom left), have different extents of red colouring on the wings. In turn, *Calopteryx virgo* (bottom right) has almost completely dark blue coloured wings. (Photos: Michael Samways (top left and right), Frederico Lencioni (bottom left), Jens Kipping (bottom right)).

principally to gain dispersal by adhering to fur or feathers (epizoochory), with the entanglement of dragonflies likely incidental. While entrapment does not benefit the dragonfly, its carcass could provide some plant nutrients, assuming ants do not remove it first (Samways, 1991; Martens and Suhling, 2003). Dragonflies also occasionally become entangled in spider webs, partly attributed to their proportionately large wings (Fig. 2.8).

2.2.2 Vision

2.2.2.1 Value of high visual acuity

An important feature closely associated with high levels of flight manoeuvrability, is good vision, along with effective neural integration (Gonzalez-Bellido *et al.*, 2013; Evans *et al.* 2019). Dragonflies have exceptionally large eyes among insects in terms of the number of ommatidia (up to nearly 29,000 in some species of aeshnids). This attribute bestows dragonflies with remarkable visual acuity (Pritchard, 1966; Sherk, 1978). The visual field is almost spherical, except immediately behind the head, where the body obstructs vision (Fig. 2.9).

Dragonflies can also detect the plane of light polarization, an important visual cue for finding suitable freshwater sites for mating and oviposition (Kriska *et al.*, 2009). In addition, they have colour vision, including ultraviolet wavelengths (Bybee *et al.* 2012). The dorsal zone of the compound eye is well developed for visually locating the position of small prey at a distance against the sky (Labhart and Nilsson, 1995). The behavioural and ecological versatility of dragonflies is also due to great visual acuity and sensitivity associated with high genetic diversity, with 15–33 opsin genes in dragonflies compared to only 3 in mammals and birds (Futahashi *et al.*, 2015).

Flight manoeuvrability is strongly associated with excellent vision in dragonflies when engaging in motion camouflage, where the pursuer is hidden from the pursued individual's sight, even though the pursuer is in motion. This can occur, for example, among rival males. Motion camouflage means that the pursued dragonfly is deceived into perceiving the approaching dragonfly as if it is stationary at infinity. Among rival males of the Australian *Hemianax papuensis*, the pursuer takes a flight path where it appears to loom ever larger until they encounter each other, with the pursuer at an advantage (Mizutani *et al.*, 2003).

2.2.2.2 Vision, prey capture and mate selection

Eye morphology and function also vary according to the habitat in which the species lives, whether a damselfly searching for prey at close quarters, or a large aeshnid seizing prey out of the air, even in dim light (Pritchard, 1966; Sherk, 1978). *Aeshna grandis* avoids capturing large prey that possess warning colouration

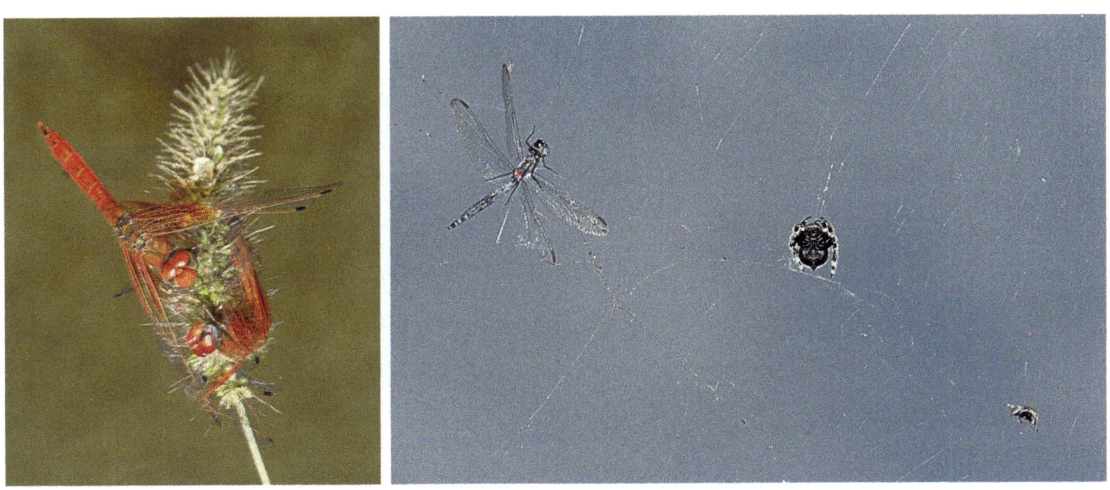

Fig. 2.8. Two individuals of *Trithemis kirbyi* entangled in a spiky and hooked seedhead (left). An individual of *Nesciothemis farinosa* entangled in a spider web (right). (Photos: Michael Samways).

(e.g. wasps) under natural conditions (Kauppinen and Mappes, 2003). However, another study found through experiment that a variety of dragonfly species, mainly perchers, are not deterred by potential prey warning colouration, as smaller prey with wasp-like colouring and patterns are captured as often as dull-coloured prey (Rashed *et al.* 2005). It appears that these different results stem from differences in the dragonfly species involved, prey size and perhaps also from wasp-like prey not being so profitable under field conditions. Dragonflies are also highly selective in the size of prey taken (Duong *et al.*, 2017).

Visual recognition is the primary cue in dragonfly mate recognition (Winfrey and Fincke, 2017), although when close, potential mates also use olfaction (Frati *et al.*, 2015). Similarly, adults can detect the odour of potential prey by olfaction using antennal sensilla (Piersanti *et al.*, 2014) (Fig. 2.9).

2.2.3 Leg function

Dragonfly legs are not for walking but for catching prey, grasping surfaces when resting (including maintaining the obelisk position; Fig. 2.5), wiping clean their compound eyes, repelling competitive males during agonistic encounters and courtship signalling in some species. Species in the Aeshnidae, Coduliidae and Lestidae generally use their mandibles to catch prey (Büsse *et al.*, 2021; Josten *et al.*, 2022), but most other species catch their flying prey on the wing using their outstretched front legs. Captured prey is repositioned by action of the legs moving the prey toward the mandibles, frequently in flight. However, helicopter damselflies in the Pseudostigmatidae are accomplished at hovering and can pluck spiders from their webs using their forelegs (Fincke, 1992).

Effectiveness of prey capture is enhanced through the front legs possessing tibial spines. These spines form a basket-like catching surface yet with little wind resistance. These spines are long in some species (Fig. 2.10).

In some dragonflies, such as *Leucorrhinia rubicunda*, both males and females, and tandem pairs, use their legs to repel advancing competitive males (Rüppell, 1989). Large body size among *Celithemis*

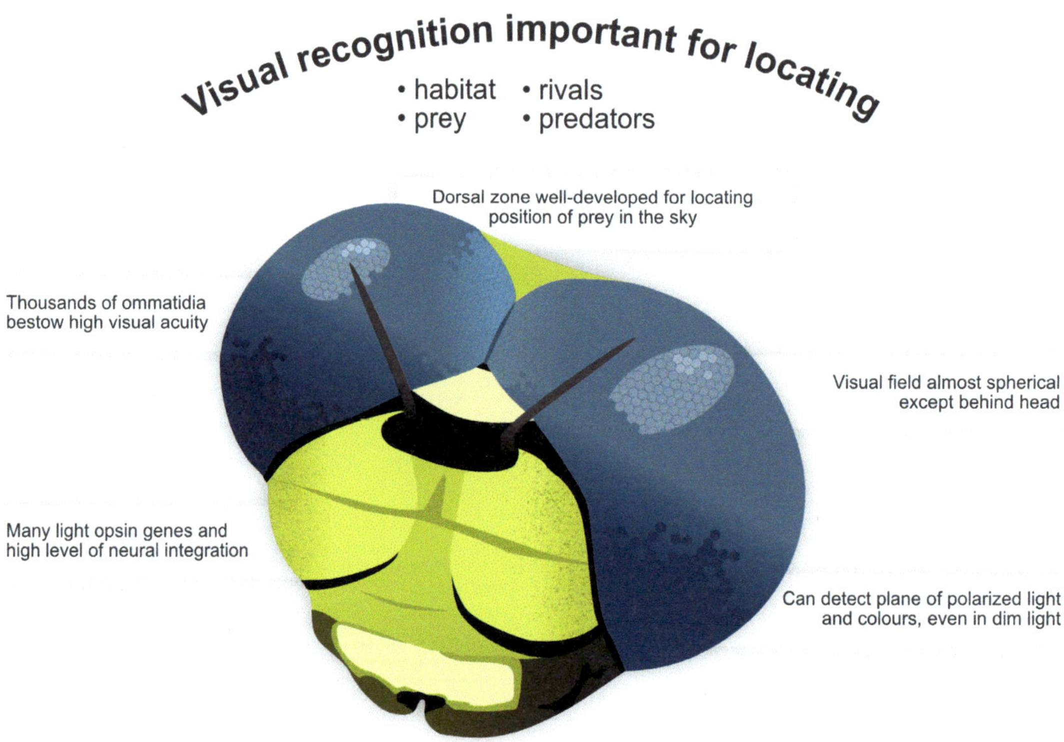

Fig 2.9. The marvel of the dragonfly eye. (Illustration: Robyn Symons).

Fig. 2.10. Legs are used for perching, yet they are braced and ready for rapid take-off. They also have other functions, such as having long tibial spines (inset) that are positioned to make a basket for catching prey in flight. In this species, *Platycypha caligata*, the tibial areas of the male are red on the outside and white on the inside and used in courtship displays. (Photo: Charl Deacon, and inset: Michael Samways).

species confers a competitive advantage for doing this (Worthen and Morrow, 2016).

The legs of certain male damselflies, like *Azuragrion vansomereni* (Martens and Grabow, 1994), *A. nigridorsum* (Samways, 1994) and *Spesbona angusta* (Deacon and Samways, 2017), are used for floating on the water surface, of both standing and shallow slow-running waters. In the *Azuragrion* species, floating appears to be territory-guarding behaviour, especially when perches are in short supply. In the case of *A. nigridorsum*, it flies upwind and then drifts downstream by sailing in the breeze on the water surface, continually repeating the process. *Spesbona angusta* males float on the water surface while guarding ovipositing females. They do so more often when many competing males occupy pond margins (Fig. 2.11).

Visual identification of oviposition sites is the initial and important step in egg deposition, followed by tactile assessment to make the final selection of a suitable site. Males of *Perithemis mooma*, undertake visual selection of high-quality sites suitable for oviposition by the female, and do so by touching the soft ground through leg dabbing. The female, once in the area, then also tests the quality of the substrate with her legs before ovipositing (Wildermuth, 1992).

2.2.4 Reproductive morphology

Dragonfly reproductive success, and hence evolution, involves many complex and varied behaviours according to various ecological settings. While the overall dragonfly body plan is constraining, it also presents opportunities. Remarkably, flight ability coupled with good vision and a long abdomen, has led to a wide range of behaviours, all based on some basic themes.

Many male dragonflies are territorial, guarding potential oviposition sites of the females. When a female arrives, the male will often guard the female from rival males, while also sometimes displaying strongly agonistic behaviour toward its conspecific rivals, such as diving down on competitors, or raising

Fig. 2.11. *Azuragion nigridorsum* floating on the water surface (top), and a male *Spesbona angusta* floating while in tandem with an ovipositing female (bottom). (Photos: Charl Deacon).

its wings or legs when in tandem. Males recognize their mates on their movement patterns, as well as body and wing colour and patterning. Males approach females, often so vigorously that the female is toppled. Mating takes place after either some time devoted to courtship behaviour or with minimal courtship activity, depending on species. Male individuals of *Diastatops obscura* frequently interact, sometimes physically, to gain or retain territories. Resident males usually win the contest, and those with the highest living biomass have a distinct advantage in territorial disputes (Lopes Junior and Peixoto, 2013).

The males of anisopterans fly up to a potential mate, and in one swift movement, grab the back of the female's head with their terminal abdominal appendages. In the case of male damselflies, they clasp the female's prothorax. The fine morphologies associated with these mating tactics are highly species-specific, and often it is the definitive morphology used to distinguish one species from another, especially when the overall body appearance is similar. In short, the copulatory apparatuses of males and females are finely sculptured and matched through natural and sexual selection. However, there is a trade-off, as both sexes are adapted in a way that enables each sex to maintain optimal reproductive success. While in many damselflies the females avoid male harassment by behavioural adaptations (Cordero-Rivera and Rivas-Torres, 2019), the female of *Enallagma cyathigerum*, for example, has a spine on her reproductive apparatus that agitates the male during mating so that she can control copulation duration to optimize reproductive success without harming the male (Rivas-Torres *et al.*, 2023).

Coupling of many anisopteran species is entirely in the air, although they often perch together. In this stationary position, the female is often highly camouflaged against the background substrate. However, most damselflies couple while perching on a substrate, usually vegetation. Time of mating varies among species, ranging from a few seconds to many hours.

Prior to coupling, often in the early morning, the male curls his abdomen down underneath his body and transfers sperm from the ninth abdominal segment to pouch-like secondary genitalia below the second and third abdominal segments, either in flight or while perching. The male is then prepared for a mating encounter with the female's genital apparatus on the underside of her eighth and ninth segments. After grasping the female, and assuming she is ready to mate, the female curves her abdomen around and under, so locking with the male's accessory genitalia. This is the familiar wheel-position, mentioned above. Copulation then takes place with rapid transfer of sperm from the sperm-storage sac of the male to the reception apparatus of the female. During copulation, the penis performs two tasks. First, it scrapes out any sperm from a previous encounter with a rival male. Only then, is it involved in sperm transfer (Waage, 1979a).

By way of guarding the female from the advances of other males, both remain coupled for various amounts of time, from a few seconds to more than an hour (Fig. 2.12). In many species, the female disengages with the male's genitalia but continues to remain attached to the male while he clasps the female anterior with his terminal abdominal appendages. This flying in tandem occurs with well-coordinated manoeuvres with

Fig. 2.12. A mating pair of the anisopteran *Orthetrum julia capicola*, with the blue male above and the brown female below, camouflaged against the background substrate (left). A mating pair of the damselfly *Lestes pallidus*, with the male guarding the female while she lays eggs in a plant stem (right). (Photos: Michael Samways).

the male assured that he is protecting the female that he has impregnated.

Tandem flight ensures that the male protects the female from advances from other solitary males. In many damselflies, the male continues to remain attached to the female while she lays her eggs. It is a plastic behaviour and involves adaptive learning, with the pair optimizing on what each does relative to prevailing microhabitat conditions (Helebrandová *et al.*, 2019).

During the tandem activity, the female may extend her abdomen below the water surface where egg laying takes place. At other times, the female, by holding on to submerged plants, will descend several tens of centimetres below the surface, often with the male hovering above the water, at the spot where the female descended. Among some species, the pair, still attached to each other, descend together to well below the water surface. While doing so, they continue to breathe by enveloping their bodies in an air sac, the plastron, made possible through water-repellent setae on the body. This often occurs at sites where water levels are prone to drop greatly (Fig. 2.13, left). It is an activity that insures against the eggs being exposed to air and drying out during droughts. Some female dragonflies deposit their eggs well below the water surface while the head and thorax remain above the water ready for a speedy escape if need be (Fig. 2.13, right).

Some dragonflies lay their eggs inside a substrate, usually living or dead plant tissue (endophytic oviposition), while others lay eggs on the outside of a substrate, usually plant tissue (epiphytic oviposition). Still others, drop their eggs on the water surface, or the eggs are washed from the tip of the abdomen during flight, without the eggs being placed directly into or onto a solid substrate (exophytic oviposition) (Corbet, 2004).

Both the ovipositor and abdomen tip are endowed with fine musculature and morphologies. They are also rich in sensilla for carefully placing eggs in the appropriate microhabitat for egg hatching, and subsequent larval development (e.g. Matushkina, 2011; Matushkina and Lambret, 2011).

2.2.5 Body colour

Many adult dragonflies gradually change colour with age. The pterostigmas may change from muted colours in the young adult to distinct colours on maturity. Some damselflies, such as *Chlorolestes* species, have greyish pterostigmas in young adults, becoming distinctly bicoloured black and white with age. Others, like *Proischnura rotundipennis*, develop pterostigmas with a blue margin, which fades on death.

As the young, teneral adult matures, the body begins to attain its fully mature colour as the

Fig. 2.13. A pair of *Pseudagrion salisburyense* submerged beneath the water surface with the female laying eggs in plant tissue (left). A female *Pinheyschna subpupillata* laying eggs in the substrate well below the water surface (right). Both are strategies to compensate for possible drops in water level. (Photos: Michael Samways).

exoskeleton hardens. This may include development of a waxy pruinescence on the body surface, and strikingly so on the abdomen of some species, such as many *Orthetrum* species. This pruinescence is a strong reflector of ultraviolet light (Futahashi, 2020).

Once developed, the colours are usually fixed. However, later the colours normally become duller and darker in old individuals. Similarly, the wings change from being clear when young to often becoming light amber and even brownish with the onset of old age, usually associated with increased wing wear. Females of some species, especially in the Libellulidae, tend toward male colouration in old age.

Colour patterns of the body can play a role in thermoregulation (May, 1976b), including physiological colour change (through pigment granules moving within chromatocytes in the cuticle) in response to temperature change, for example, in the spots of male *Aeshna caerulea* (Sternberg, 1987). *Mnais costalis* males that have longer exposure to the sun have a warmer thorax, and can court more females, while also indicating warmer and more optimal oviposition sites (Tsubaki *et al.*, 2010).

Both sexes of the highly threatened *Spesbona angusta* rapidly change colour two ways: blue↔brown, even as a cloud passes over, presumably to thermoregulate, as the colour change is associated with temperature change (Deacon and Samways, 2016) (Fig. 2.14). In *Indolestes peregrinus*, bluing takes place in both sexes in minutes as the ambient temperature increases, but when the temperature decreases, browning then occurs but over several hours (Hasebe *et al.*, 2023).

There is also a large-scale, regional-scale component to dragonfly colour, with darker species favoured in cooler climates and lighter species favoured in warmer climates (Zeuss *et al.*, 2014). There can also be differences in colour with season. *Hetaerina titia* has lighter wing pigmentation earlier in the year, not only leading to more mating attempts but also to more agonistic encounters with a sympatric species such as *H. occisa*, which it closely resembles. However, conflict later in the year decreases as *H. titia* darkens and becomes less like *H. occisa* (Drury *et al.*, 2015).

In male *H. americana* territorial disputes, individuals that are middle-aged have the highest muscle and fat values. In turn, winners of territorial conflicts have larger red spot areas on the wings and higher fat reserves, therefore making the spot area an energy-reflecting trait communicated during territorial engagements (Contreras-Garduño *et al.*, 2008). Also in this species, reduced black pigmentation signals high gregarine gut parasite levels (Córdoba-Aguilar *et al.*, 2007). Similarly, *Hetaerina* (*Mnesarete*) *pudica*, males with more opaque wings and larger red spots are more likely to win territorial contests. Rivals possessing and best able to assess wing pigmentation are the ones that hold the high-value territories (Guillermo-Ferreira *et al.*, 2015).

There appears to be value in having increased wing pigmentation in certain libellulid dragonflies. *Pachydiplax longipennis* males with more extensive wing melanization are in better condition and tend

Fig. 2.14. *Spesbona angusta* undergoes rapid and reversible colour change in both sexes relative to temperature. (Photos: Charl Deacon).

to win more disputes, indicating wing melanization is a condition-related signal of fighting ability (Moore and Martin, 2016). However, use of wing pigmentation does not appear to relay such information in the polythorid damselfly *Euthore fasciata* (Palacino Rodríguez *et al*., 2016), suggesting differences among taxa.

Sometimes ultraviolet light, which creates fluorescence signals in *Hetaerina* (*Mnesarete*) *pudica*, is positively synergistic with pigmentation in eliciting agonistic behaviour in males and more pronounced sexual responses by females (Guillermo-Ferreira *et al*., 2014). However, in the case of young males, they display both female mimicry and crypsis which helps them avoid both predation and potentially damaging agonistic interactions (Cezário *et al*., 2022a).

The gigantic *Megaloprepus caerulatus* flashes ultraviolet-blue bands to signal to potential mates across forest light gaps. The white wing patches also serve in mate recognition, possibly indicating male condition and territorial status (Schultz and Fincke, 2009), with ultraviolet reflectance playing a powerful role in the competitive outcome of male contests (Xu and Fincke, 2015). There can also be additional roles attributed to wing iridescence. The blue iridescent wings of the libellulid *Zenithoptera lanei* also confer an advantage in terms of being camouflaged against the bright background of the water surface when searching for dipteran prey or avoiding detection by bird predators (Cezário *et al*., 2022b). In this species, males that are large, with high levels of abdominal and thoracic body fat, are the individuals that are most active territorially and have the highest mating success (Valdivia *et al*., 2020).

The damselfly *Chalcopteryx scintillans* inhabits streams in the dense forest of the Amazon. The male has colourful wings, the dorsal surface of which flashes yellow-orange at certain angles, while the ventral surface reflects ultraviolet light. The male defends fallen log oviposition sites in forest gaps. Courtship involves wing clapping, with the dynamic iridescent signals used for communicating with conspecifics while at the same time obscuring the signal to both its prey and bird predators, especially under the dark conditions of the rainforest (Cezário *et al*., 2021).

Interestingly, *Hetaerina* species display lekking behaviour, where males form groups to advertise and to attract females. Males choose well-sunlit places for doing this, and the males with the largest pigmented areas on the wings and with a larger body size succeeding in the greatest number of mating events. However, oviposition then takes place outside the lekking centres, perhaps to avoid male harassment (Córdoba-Aguilar *et al.*, 2009).

Body colour, as opposed to wing colour, can also pertain to mate recognition and territorial behaviour. In the coenagrionid *Tigriagrion aurantinigrum*, an increase in orange pigmentation on the head behind the eyes signals male condition and predicts success over rival males in competitive encounters (Vilela *et al.*, 2017).

As dragonflies have colour vision, it is not surprising that colour plays a major role in species and sexual signalling, species recognition, sexual selection and predation risk, especially from birds (Svensson and Waller, 2013; Kuchta and Svensson, 2014), to greatest effect when the light is brightest (Schultz *et al.* 2008). However, many species have cryptic colours, especially in the female, for camouflage on substrates, such as on tree trunks or rocks, and among vegetation.

In many species, the male body is distinctly bright in colouring. This is the case for *Platycypha caligata*, where the male's flattened tarsi are also coloured bright white on the inside, but bright red on the outside. Signalling is enhanced by expanded tibia when displayed by the male in a territorial courtship dance while flying in front of the dun-coloured female, alternating the visible angle of the tibia (Robertson, 1982) (Fig. 2.10).

Wing colour can vary in extent across the geographical range of a species. For example, populations of *Chlorolestes tessellatus* in different geographical areas may have clear wings or black-banded wings, with weak banding in some other geographical areas (Samways, 2006) (Fig. 2.15). From a biogeographical study on two species of *Calopteryx* damselflies, sexual selection is the prime driver of wing pigmentation, rather than any change due to latitude, elevation or thermoregulation (Hassall, 2014), notwithstanding that at the local level there is also character displacement, based mainly on wing colouration (Waage, 1979b).

Female-based polychromatism (females but not the males have alternative colour forms) is common among many damselflies. Fifty-four percent of genera of Zygoptera species in Europe have at least one other colour morph (Cordero and Andrés, 1996). The basis for polychromatism is genetic (Takahashi *et al.*, 2021), and appears to also have involved some hybridization and introgression (Sánchez-Guillén *et al.*, 2005), but selection pressures vary biogeographically (Sánchez-Guillén *et al.*, 2011) and according to local habitat types (Fincke *et al.*, 2005).

The genus *Ischnura* is estimated to have arisen about 23.4–13.8 mya, with the ancestral state of

Fig. 2.15. While *Chlorolestes tessellatus* is endemic to South Africa, there are different colour morphs in different regions of the country. In the south-west, the male wings are colourless (left), while in the north-east they are moderately banded (centre), and in the south-east they are strongly banded (right). (Photos: Michael Samways).

the females likely to have been of just one colour and distinctly different from the males. Polychromatism appears to have arisen several times in the past, resulting in male lookalikes (Fig. 2.16). Furthermore, polychromatism is correlated with larger population sizes, with monochromatic females more usually associated with smaller geographic range sizes and island endemism, and possible higher risk of extinction (Blow *et al.* 2021).

There are trade-offs associated with polychromatism. Alternative colour morphs have advantages and disadvantages from a selective point of view. Male-mimicking females are easily detected by predators. More so than are brown morphs camouflaged among grassy vegetation. Male-mimicking females can join a group of males, where they blend in with the crowd, reducing risk from predation. Nevertheless, they are easily detected by the males, through visual and olfactory cues, whereafter the males soon attempt to mate with the male-mimicking female. However, this safety in the crowd is less effective when male density is low, and so the camouflaged morph becomes the favoured mate, even though hidden among vegetation. In short, the selective trade-off is 'being lost in the crowd or hidden in the grass' (Schultz and Fincke, 2013) (Fig. 2.17).

2.3 The Egg As a Functional Organism

While the adult dragonfly is highly mobile and can move into optimal habitats, this is not the case for the stationary egg, which depends on the adult's good choice of habitat and the egg's inherent qualities to survive. In turn, the fate of the larva first depends on where the adult female lays her eggs. While selection of an optimal microhabitat is visual, the female can also detect optimal water chemistry for its larvae (Kietzka *et al.*, 2017).

Egg laying carries high risk for females, as they are often stationary, or nearly so, and therefore relatively easy targets for certain birds, frogs and fish. Egg scattering on the water surface is one way to reduce risk, as the female is on the move and close to the water surface. This localized air space is freer from many bird predators, which would otherwise risk snagging their wings on the water surface.

There are other challenges. The female has little time to select the right oviposition site, with risks escalating the more time she spends in careful site selection. There is also the risk of harassment by males. Whereas the female has a somewhat limited choice among potential male mates, there is strong sexual selection on males to maximize the potential number of fertilizations per male, resulting in various types of mating systems (Conrad and Pritchard, 1992).

Mating among odonates is highly diverse, with most species being promiscuous, and both sexes mating several times during their lifetime. However, females of *Ischnura hastata* in the Galápagos Islands mate only once in their short lifespan, creating intense sexual conflict for mating rate, and females resisting attempts by other males to mate (Cordero-Rivera *et al.*, 2023).

Fig. 2.16. Female-based polychromatism in *Ischnura senegalensis*. The female below in this mating pair is similar in its bluish colour to the male above (left), while other females are rust-coloured (right). (Photos: Michael Samways).

Fig. 2.17. The male of *Pseudagrion furcigerum* is blue and rests conspicuously on flat boulders in streams, as do the blue females (top). However, the rust-coloured females inhabit the vegetated stream margins (bottom). (Photos: Michael Samways).

Tandem flight ensures high fidelity with a mate. In species like *Pantala flavescens* and *Sympetrum fonscolombii*, eggs are scattered over several ponds, so ensuring that at least some eggs are placed in the right habitat. In contrast, those species that mate guard, such as the *Orthetrum chrysostigma* and *Trithemis kirbyi*, select specific microhabitats in certain ponds but have more challenges from rival males. For females engaged in tandem pairing, egg size varies among clutches and the subsequent larvae also vary in size. In contrast, females of guarding pairs produce the largest eggs and have subsequent larvae of larger size in the earlier clutches, reducing disturbance risk from rival males, and hence improved progeny survival (Schenk *et al.*, 2004).

The egg is another developmental morph besides the larva and adult. As for the other life stages, loss of eggs to an adverse impact could reduce the size of the effective population. In central Brazil, forest loss changes the local habitat and therefore oviposition sites. Exophytic species (laying eggs on the water surface) and epiphytic species (laying eggs on exposed surfaces of rocks, leaves, trunks, etc., protruding from the water surface) are much more affected than the endophytic species which lay their eggs within plant tissue (Rodrigues *et al.*, 2019).

An additional consideration with respect to egg survival is that they are susceptible to parasitism by minute hymenopterans, with parasitism rates sometimes reaching 90% or more (Corbet, 2004). Another mortality factor is infertility and unhatchability, which in the laboratory for *Pyrrhosoma nymphula* reaches 25% (Bennett and Mill, 1995).

Dragonfly eggs are typically oval to elongated with rounded ends. They are generally pointed at the end with the micropyles, areas through which sperm cells enter the egg (Corbet, 2004). Eggs typically have three main layers, a vitelline layer around the embryo, then an endochorion and on the outside an exochorion. Eggs laid inside vegetation (endophytic eggs) typically have a homogenous vitelline layer, a multilayered endochorion, and a pliable exochorion with many small pores and a jelly-like outer coat. Eggs laid on surfaces have a very thick vitelline layer, but a thin endochorion, and a highly water-absorbent jelly-like layer that expands greatly as it swells with water (Sahlén, 1995).

There is a trade-off between egg size and clutch size, at least among some European libelluilids. When more eggs are present in a clutch, the eggs are smaller. Larger clutch size is also associated with larger adult female size, but egg size is not (Sahlén and Suhling, 2002).

The embryo uses fat reserves inside the egg for development. However, the jelly-like outer layer absorbs water, stimulating embryonic development. Eggs lacking this layer, may possibly absorb rainwater to encourage embryonic development (Corbet, 2004).

Embryonic development may take just a few days, but the process of hatching is sometimes delayed by many days. Many temperate species have delayed embryonic development as a means of overwintering. However, when laid in the spring, there may be summer embryonic development. The continuously high temperatures of the tropics provide a longer opportunity for development, with many tropical species having two or more generations per year.

Hatching depends on the right stimulus being present. This stimulus is environmental, differing among species, and may be a change in the light/dark cycle, immersion in water or wetting following rainfall, or reduced oxygen levels (Corbet, 2004). These responses play an important role in determining geographic distributions of species. *Pantala flavescens* has a critical low temperature for egg development of 14.3°C, and a total effective temperature for the egg stage of 80 degree-days, which prevents this highly migratory yet predominantly tropical species from establishing permanently in most of Japan (Ichikawa *et al.* 2017).

While the egg is a sessile stage, it can be dislodged due to severe storms and flooding. These extreme weather events detach leaves, flatten marginal vegetation and wash away logs, silt and sand, taking the eggs downstream. Presumably, this has happened many times in the past, and natural selection has been a filter, but today it may be synergistic with adverse human impacts.

2.4 The Larva As a Functional Organism

2.4.1 Functional comparison of adult and larva

From a functional perspective the immature dragonfly, the larva, is a very different organism from the adult, albeit that both are predatory in feeding habits. Furthermore, the selection pressures are markedly different in the larva's watery realm compared to the adult's aerial realm. This means that different larval taxa may be similar in body form when living in the same microhabitat. Larvae are also adapted to the various predators that are part of the same community, especially fish and other, mostly larger, dragonflies. Large larvae of certain species, especially among the aeshnids, also feed on certain vertebrates such as small fish and tadpoles.

Unlike the male adults, larvae are dull in colour and match their immediate background environment. They are cryptic to avoid visually active predators, while also enabling them to stalk prey themselves. This crypsis is often enhanced by their body accumulating small items of aquatic debris and detritus. Unlike the adult, larval legs are used for short distance walking to search for more optimal microhabitats, procure a more cryptic location and to creep up on prey. While they have well-developed eyes for detecting prey in particular, they also have longer antennae than the adult, mostly used for sensing the immediate physicochemical environment.

Importantly, while the adult might be associated with a particular habitat, such as a fast-moving stream, the larva is often associated with quieter microhabitats. These less turbulent microcosms may be the interstices between and among cobbles on the riverbed, or much less turbulent local areas like deposition pools or among the submerged parts of marginal vegetation. Whatever the microhabitat, the larvae must be adapted to deal with extreme weather events, whether flood or drought. They must also be adapted to changing fortunes in terms of the interactions they have with other organisms, both their prey and their own predators.

2.4.2 Larval development and habitat selection

Larvae may take several days to several years to develop through by ecdysis ('moulting', 'emerging') into an adult. The very few species that have partial or completely terrestrial larvae are likely to have developed this lifestyle secondarily.

On hatching from the egg, the first instar is a prolarva, lasting from a few seconds to a few minutes according to species. The first moult may be delayed for some hours if it has to move across land to find water, a highly vulnerable stage in the dragonfly's life cycle (Corbet, 2004). On reaching the water, it floats on the water surface before moulting and then submerging to begin its series of growth stages through to adulthood.

Respiration among larvae takes place by maximizing some aspects of the body's surface area, mostly by external or internal appendages or internal baskets with a large surface area. In damselflies, the main respiratory appendages are at the tip of the abdomen, and in most families, they are present as conspicuous caudal lamellae. In the case of anisopterans, as well as anisozygopterans (see Fig. 3.3), the main respiratory apparatus is in the form of rectal gills inside the abdomen (Fig. 2.18).

Like the adults, all larvae are predatory. However, as in the case of all predators, they will inevitably and inadvertently, take in other, microscopic organisms and some undigested plant material with their prey, possibly providing some essential nutrients.

All larval odonates seize their prey with a hinged, extensible and extremely fast-moving labium (labial mask). At the distal end of the labium are moveable hooks not only for capturing prey but also for handling it. These are two separate movements, one a straight-line movement and the other requiring great dexterity to manoeuvre the prey. These two actions are made possible by a remarkably flexible and switchable joint where the labium is attached to the head (Koehnsen *et al.*, 2023).

Fig. 2.18. Damselflies respire through caudal lamellae at the end of the abdomen (top), while anisopterans do not have these protrusions, and respire by passing water over their rectal gills inside the abdomen (bottom). (Photos: Michael Samways).

Most captures by larvae are by sighting moving prey, with tactile and chemical cues involved. However, sometimes dead or stationary prey are eaten. Prey may be ambushed from a stationary position or by slowly approaching it. Then, through very rapid extension of the labium through stored elastic energy, the prey item is seized by the hooks and drawn in toward the mandibles (Büsse *et al.*, 2021). Larvae also possess light polarization sensitivity, important for detecting prey. This capability enhances visual contrast by selectively filtering polarized light scattered by the underwater environment (Sharkey *et al.* 2015).

Prey type largely depends on what is available in the immediate vicinity. However, when prey is scarce, there may be cannibalism, common among *Tramea* spp. and *Pantala flavescens* in small warm, often temporary, pools. In extreme cases, sometimes one small pool ends up with just one fully-grown *P. flavescens* larva. In turn, there can also be intraspecific competition, with feeding activity and growth rate of larvae both reduced when population density is high (Van Buskirk, 1993).

Prey type also depends on the age and size of the larva. Small larvae will feed on large protists and small crustaceans. The late stages eat insect larvae of various sorts, as well as some adult insects. Soft parts of snails and tadpoles, and occasionally small or young fish, may also be consumed. Overall, dragonfly larvae are largely opportunistic feeders, feeding on any potential prey available at any one time. Some species can go without prey for many weeks, especially when water temperatures are low.

Dragonfly larvae among species in both Zygoptera and Anisoptera have a conservative body plan, making them instantly recognizable in any part of the world. However, they can vary greatly in details in their functional morphology, having undergone natural selection to survive in their various specific microhabitats. Furthermore, some display genetic polymorphism, as well as environmentally induced polyphenism, as shown by *Coenagrion armatum* (Johansson and Nilsson-Örtman, 2013).

2.4.3 Zygopteran larva functional morphology

Morphological variation among zygopteran larvae is mostly associated with their various microhabitats. These highly localized micro-environments can be very different from the general appearance of an entire ecosystem. For example, the larvae of many 'stream' species actually live in the deposition pools of the streams. In turn, river species may shelter, hide and feed among woody debris on the riverbed, in the shelter of plant roots, or in the interstices among pebbles or boulders.

In adapting to these various microhabitats, damselfly larvae species, although having a similar general body plan, differ largely in details such as the shape and dimensions of the caudal lamellae (which are unusually large, robust and spiny in the Neotropical genera *Cora* and *Polythore* in the Polythoridae (Novelo-Gutiérrez and Bota-Sierra 2023)), slenderness versus stoutness of the abdomen, length of legs, degree of setation and shape of the labium (Fig. 2.19).

2.4.4 Anisopteran larva functional morphology

While some lineages among anisopterans, such as aeshnids and libellulids, have relatively standard

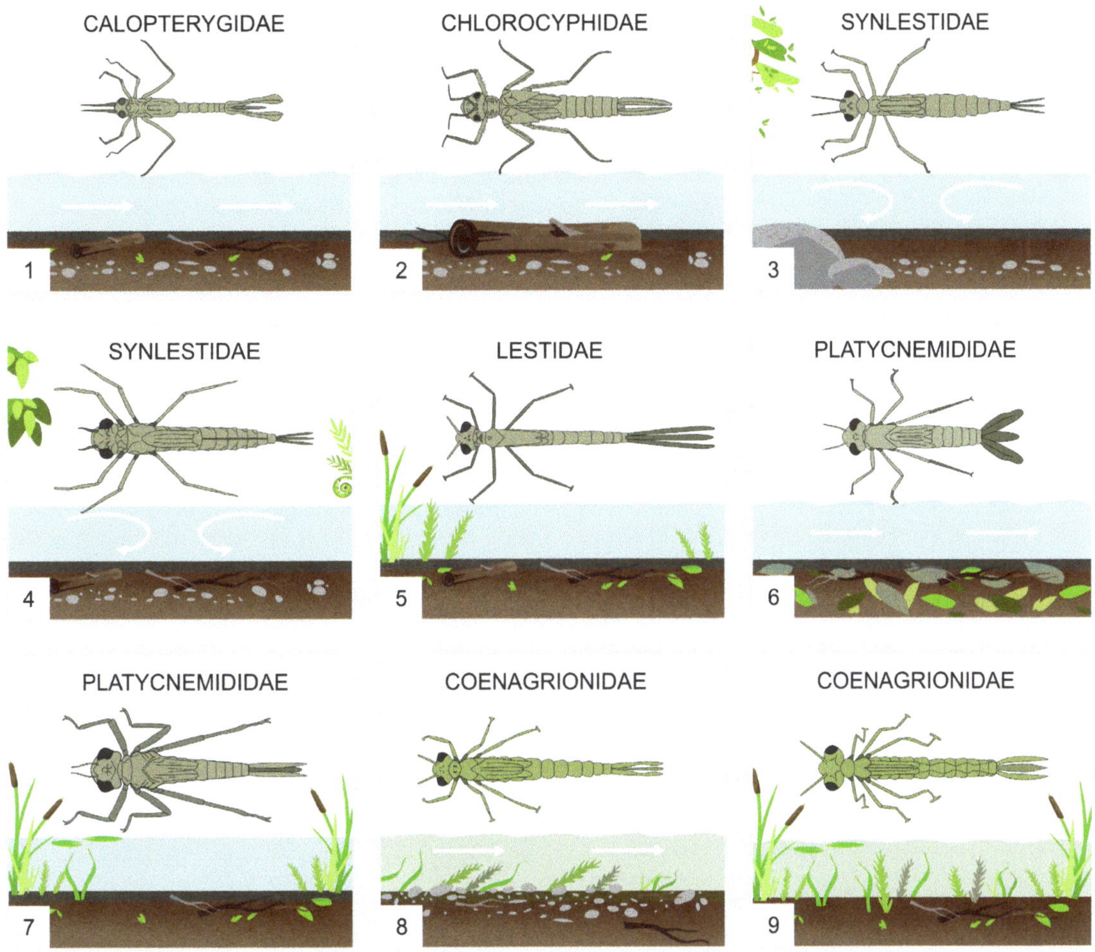

Fig. 2.19. Some African zygopteran larvae in various families and the specific microhabitats of certain species: 1) *Phaon iridipennis* – meandering shallow rivers with leaf and twig litter; 2) *Platycypha caligata* – fairly fast flowing rivers with much woody debris; 3) *Chlorolestes fasciatus* – streams with boulders and deposition pools; 4) *Ecchlorolestes nylephtha* – forested streams with deposition pools rich in leaf and twig litter; 5) *Lestes plagiatus* – ponds rich in marginal emergent vegetation, aquatic plants and leaf litter; 6) *Allocnemis leucosticta* – shallow streams with a deep bed of leaf and twig litter; 7) *Mesocnemis singularis* – rapid rivers with quiet side pools rich in marginal and aquatic plants; 8) *Pseudagrion spernatum* – fast streams with cobbles and quieter microcosms of aquatic plants; and 9) *Africallagma glaucum* – pools with much marginal and aquatic vegetation. (Illustration: Robyn Symons).

gross morphology, the gomphids display a wide variety of proportions among body parts, a manifestation of adaptation to their specific microhabitats (Fig. 2.20).

Anisopteran functional morphology can be broadly divided into the following types (Corbet, 2004). 1) The 'claspers', that maintain a firm hold on the substrate with strong legs and an elongate body, and may be active among aquatic plants near the water surface, or are sedentary among stones or plants on the bottom. 2) The 'sprawlers', which have a squat abdomen and long legs, live mostly among aquatic plants and may be either active or sedentary hunters. 3) The 'hiders', with a flattened abdomen, that hide among leaf litter or boulders. 4) The 'shallow burrowers', which live just beneath

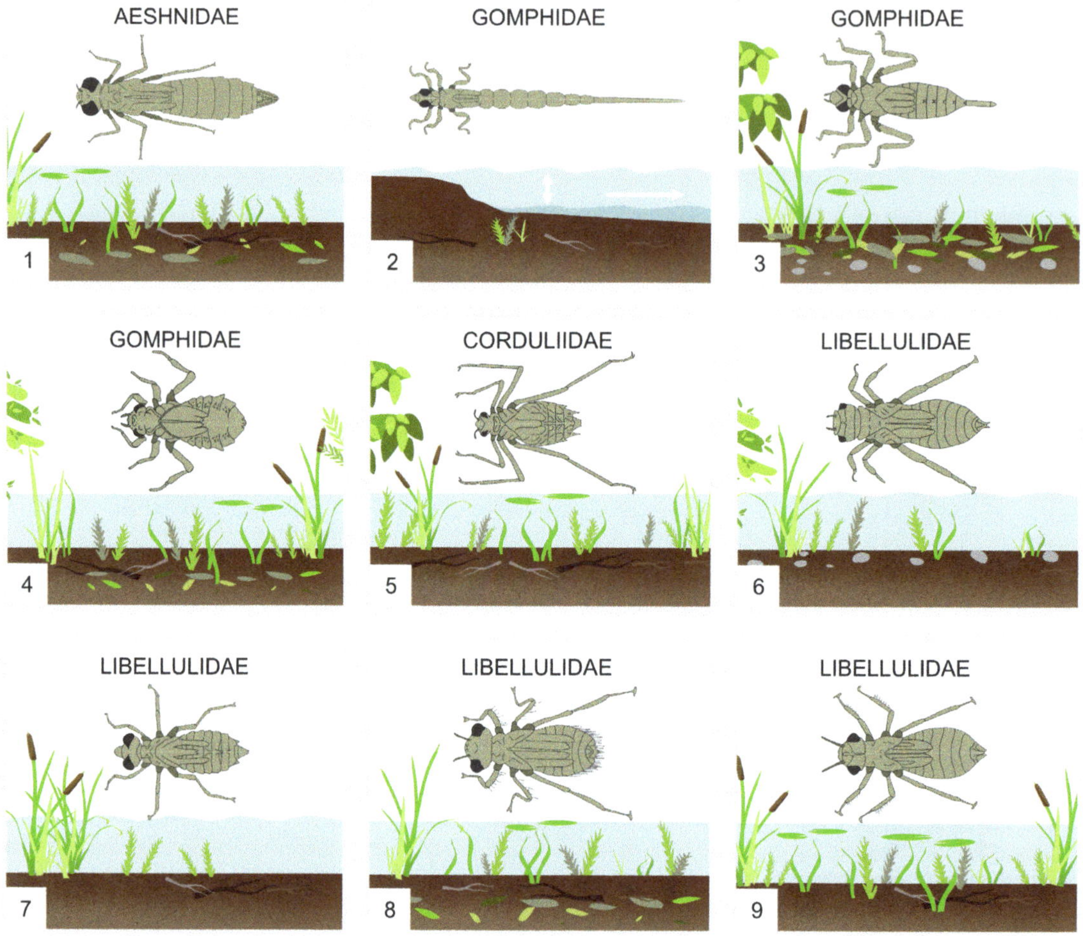

Fig. 2.20. Some African anisopteran larvae in various families and the specific microhabitats of certain species: 1) *Anax imperator* – ponds and lakes with well-vegetated margins and much aquatic vegetation; 2) *Neurogomphus* sp. – large meandering rivers prone to water-level drops and with muddy margins; 3) *Lestinogomphus angustus* – rapid rivers with quiet marginal pools; 4) *Ictinogomphus ferox* – lakes and marshes with much vegetation; 5) *Phyllomacromia picta* – well-vegetated still reaches of large rivers and lakes; 6) *Orthetrum caffrum* – rapid rivers with still side pools; 7) *Trithemis werneri* – rapid rivers with many reeds; 8) *Crocothemis erythraea* – still reaches of rivers and ponds with much marginal and aquatic vegetation; and 9) *Sympetrum fonscolombii* – small ponds with much marginal and aquatic vegetation. (Illustration: Robyn Symons).

the substratum when resting, but hunt on the substratum surface and are often associated with a substrate that has a characteristic particle size. They have either an elongate or squat abdomen, with those that are elongate being either highly setose or smooth. 5) The 'deep burrowers', which make their own burrows, and can be many centimetres long. They either forage on the substratum surface, often at night, or sieve the substrate for prey. Some have a highly elongate last abdominal segment, used as a breathing tube.

2.4.5 Dragonfly larval development

In lowland tropical areas where rainfall is seasonal, larval development usually takes 1.5–4 months during the rainy season, with several generations per year. This contrasts with high elevation temperate regions where larval development can take up to 10 years, indicating the role of temperature and water availability in determining dragonfly development (Corbet, 2004). These responses have implications for dragonfly survival under climate change.

After several moults, and several and a variable number of instars (according to species and environmental conditions, including food availability and abiotic conditions, such as water temperature and photoperiod), the adult emerges from the last instar larva. Emergence from water into the aerial realm usually takes place at a time between sunset and sunrise, although daytime emergence takes place in some species. If air temperatures are too low, the larvae of some species return to the water, to repeat the activity the following day, or shortly after.

Fig. 2.21. The freshly emerged, soft and largely immobile adult (top: *Pseudagrion citricola*) is vulnerable to predators and pathogens. Once it has emerged from the larval exoskeleton, the young must then expand and harden its wings (bottom left: *Phyllomacromia contumax*) and usually moves away from the water where it continues to develop its adult colouring. The shuck (exuvia) left behind (bottom right: *Anax imperator*) has great value in certain circumstances for determining the residency and abundance of some species. (Photos: Michael Samways).

Fig. 2.22. *Orthetrum trinacria* eating a *Trithemis arteriosa* (top), *O. icteromelas* eating a *Rhyothemis semihyalina* (bottom left), and *Acisoma variegatum* eating an *Ischnura senegalensis* (bottom right). (Photos: Marienne De Villiers (top), Michael Samways (bottom left), Charl Deacon (bottom right)).

After emergence from the water, the larva takes a firm hold onto the substrate, when the dorsal region of the cuticle splits and the young adult emerges (Fig. 2.21). This takes place usually within 1 m of where the larva emerged from the water, although it can be up to many metres either horizontally or vertically in some species and individuals.

Ecdysis to adulthood is a time of great risk relative to the amount of time that it takes to complete. Mortality rates range between 3–28%, although localized higher levels have been recorded. Besides mortality due to a cold snap, wind or rain, predatory spiders, birds and reptiles also take their toll. Fungal pathogens and gregarine protist, trematode, cestode and acarine parasites also infect or attach to the emerging adult (Corbet, 2004).

The teneral adult, being frail and weak of flight, is highly susceptible to extreme weather events. It is also at risk from a suite of predators, including birds, frogs, lizards and spiders, as well as certain larger and opportunistic dragonflies. For example, the damselfly *Ceriagrion glabrum* may consume teneral or immature individuals of other damselflies. *Orthetrum icteromelas* and *O. trinacria* will readily eat young and mature adults of other dragonflies, including relatively large species, such as *Rhyothemis semihyalina* and *Trithemis arteriosa*, while some small dragonflies like *Acisoma variegatum* will eat damselflies (Fig. 2.22).

References

Appel, E. and Gorb, S.N. (2011) Resilin-bearing wing vein joints in the dragonfly *Epiophlebia superstes*. *Bioinspiration and Biomimetics* 6, e046006. DOI: 10.1088/1748-3182/6/4/046006.

Bennett, S. and Mill, P.J. (1995) Lifetime egg production and egg mortality in the damselfly *Pyrrosoma nymphula* (Sulzer) (Zygoptera: Coenagrionidae). *Hydrobiologia* 310, 71–78.

Blow, R., Willink, B. and Svensson, E.I. (2021) A molecular phylogeny of forktail damselflies (genus *Ischnura*) reveals a dynamic macroevolutionary history of female colour polymorphisms. *Molecular Phylogenetics and Evolution* 160, e107134. DOI: 101016/j.ympev.2021.107134.

Bode-Oke, A.T., Zeyghami, S. and Dong, H. (2018) Flying in reverse: kinematics and aerodynamics of a dragonfly in backward free flight. *Journal of the Royal Society Interface* 15, e20180102. DOI: 10.1098/rsif.2018.0102.

Bomphrey, R.J., Nakata, T., Henningsson, P. and Lin, H.-T. (2016) Flight of dragonflies and damselflies. *Philosophical Transactions of the Royal Society B* 371, e20150389. DOI: 10.1008/rstb.2015.0389.

Büsse, S., Genet, C. and Höornschemeyer, T. (2013) Homologization of the flight musculature of Zygoptera (Insecta: Odonata) and Neoptera (Insecta). *PLoS ONE* 8, e55787. DOI: 10.1371/journal.pone.0055787.

Büsse, S., Tröger, H.-L. and Gorb, S.N. (2021) The toolkit of a hunter: functional morphology of larval mouthparts in a dragonfly. *Journal of Zoology* 315, 247–260. DOI: 10.1111/jzo.12923.

Bybee, S.M., Kaihileipihamekeola, J., Gering, E.J., Whiting, M.F. and Crandall, K.A. (2012) All the better to see you with: a review of odonate color vision with transcriptomic insight into the odonate eye. *Organisms Diversity and Evolution* 12, 241–250. DOI: 10.1007/s13127-012-0090-6.

Bybee, S., Córdoba-Aguilar, A., Duryea, M.C., Futahashi, R., Hansson, B. *et al.* (2016) Odonata (dragonflies and damselflies) as a bridge between ecology and evolutionary genomics. *Frontiers in Zoology* 13, e46. DOI: 10.1186/s12983-016-0176-7.

Cannell, A.E.R. (2018) The engineering of the giant dragonflies of the Permian: revised body mass, power, air supply, thermoregulation and the role of air density. *Journal of Experimental Biology* 221, 1–7. DOI:10.1242/jeb.185405.

Cezário, R.R., Lopez, V.M., Gorb, S. and Guillermo-Ferreira, R. (2021) Dynamic iridescent signals of male copperwing damselflies coupled with wing-clapping displays: the perspective of different receivers. *Biological Journal of the Linnean Society* 134, 229–239. DOI: 10.1093/biolinean/blab068.

Cezário, R.R., Lopez, V.M., Gorb, S.N. and Guillermo-Ferreira, R. (2022a) Camouflage by counter-brightness: the blue wings of Morpho dragonflies *Zenithoptera lanei* (Anisoptera: Libellulidae) match the water background. *Journal of Zoology* 317, 92–100. DOI: 10.1111/jzo.12955.

Cezário, R.R., Therézio, E.M., Mauletta, A., Gorb, S.N. and Guillermo-Ferreira, R. (2022b) Ontogenetic colour change of a sexual ornament in males of a damselfly: female mimicry, crypsis or both? *The Science of Nature* 109, e2. DOI: 10.1007/s00114-021-01775-5.

Combes, S.A., Rundle, D.E., Iwasaki, J.M. and Crall, J.D. (2012) Linking biomechanics and ecology through predator-prey interactions: flight performance of dragonflies and their prey. *Journal of Experimental Biology* 215, 903–913. DOI: 10.1242/jeb.059394.

Combes, S.A., Salcedo, M.K., Pandit, M.M. and Iwasaki, J.M. (2013) Capture success and efficiency of dragonflies pursuing different types of prey. *Integrative and Comparative Biology* 53, 787–798. DOI: 10.1093/icb/ict072.

Conrad, K.F. and Pritchard, G. (1992) An ecological classification of odonate mating systems: relative influence of natural, inter- and intra-sexual selection on males. *Biological Journal of the Linnean Society* 45, 255–269.

Contreras-Garduño, J., Buzatto, B.A., Serrano-Meneses, M.A., Nájera-Cordero, K. and Córdoba-Aguilar, A. (2008) The size of the red wing spot of the American rubyspot as a heightened condition-dependent ornament. *Behavioral Ecology* 19, 724–732. DOI: 10.1093/beheco/arn026.

Corbet, P.S. (2004) *Dragonflies: Behavior and Ecology of Odonata*, 2nd ed. Cornell University Press, Ithaca, NY, USA.

Corbet, P.S. and May, M.L. (2008) Fliers and perchers among Odonata: dichotomy or a multidimensional continuum? A provisional appraisal. *International Journal of Odonatology* 11, 151–171. DOI: 10.1080/13887890.2008.9748320.

Cordero, A. and Andrés, J.A. (1996) Colour polymorphism in odonates: females that mimic males? *Journal of the British Dragonfly Society* 12, 50–60.

Cordero-Rivera, A. and Rivas-Torres, A. (2019) Sexual conflict in water striders, dragonflies and diving beetles. In: K. Del-Claro, K. and Guillermo, R. (eds) *Aquatic Insects*. Springer, Switzerland, pp. 295–320. DOI: doi.org/10.1007/978-3-030-16327-3_11.

Cordero-Rivera, A., Rivas-Torres, A., Encalada, A.C. and Lorenzo-Carballa, M.O. (2023) Sexual conflict and the evolution of monandry: the case of the damselfly

Ischnura hastata (Odonata: Coenagrionidae) in the Galápagos Islands. *Ecological Entomology* 48, 336–346. DOI: 10.1111/een.13225.

Córdoba-Aguilar, A., Lesher-Treviño, A.C. and Anderson, C.N. (2007) Sexual selection in *Hetaerina titia* males: a possible key species to understand the evolution of pigmentation in calopterygid damselflies (Odonata: Zygoptera). *Behaviour* 144, 931–952. DOI: 10.1163/156853907781492672.

Córdoba-Aguilar, A., Raihini, G., Serrano-Meneses, M.A. and Contreras-Garduño, J. (2009) The lek mating system of *Hetaerina* damselflies (Insecta: Calopterygidae). *Behaviour* 146, 189–207. DOI: 10.1163/156853909X410739.

Davidovich, H. and Ribak, G. (2016) Flying with eight wings: inter-sex differences in wingbeat kinematics and aerodynamics during copulatory flight of damselflies (*Ischnura elegans*). *The Science of Nature* 103, e65. DOI: 10.1007/s00114-016-1390-z.

De Marco Jr, P. and Resende, D.C. (2002) Activity patterns and thermoregulation in a tropical dragonfly assemblage. *Odonatologica* 31, 129–138.

De Marco Jr, P., Latini, A.O. and Resende, D.C. (2005) Thermoregulatory constraints on behavior: patterns in a Neotropical dragonfly assemblage. *Neotropical Entomology* 34, 155–162.

Deacon, C. and Samways, M.J. (2016) Conservation of a phenomenon: rapid, reversible colour change in both sexes of one of the world's most threatened damselflies. *Journal of Insect Conservation* 20, 497–504. DOI: 10.1007/s10841-016-9882-x.

Deacon, C. and Samways, M.J. (2017) Conservation planning for the extraordinary and Endangered Spesbona damselfly. *Journal of Insect Conservation* 21, 121–128. DOI: 10.1007/s10841-017-9960-8.

Deng, X. and Hu, Z. (2008) Wing-wing interactions in dragonfly flight. *The Neuromorphic Engineer* 3pp. DOI: 10.2417/1200811.1269.

Deregnaucourt, I., Bardin, J., Villier, L., Julliard, R. and Béthoux, O. (2023) Disparification and extinction trade-offs shaped the evolution of Permian to Jurassic Odonata. *iScience* 26, e107420. DOI: 10.1016/j.isci.2023.107420.

Donoughe, S., Crall, J.D., Merz, R.A. and Combes, S.A. (2011) Resilin in dragonfly and damselfly wings and its implications for wing flexibility. *Journal of Morphology* 272, 1409–1421. DOI: 10.1002/jmor.10992.

Drury, J.P., Anderson, C.N. and Grether, G.F. (2015) Seasonal polyphenism in wing coloration affects species recognition in rubyspot damselflies (*Hetaerina* spp.). *Journal of Evolutionary Biology* 28, 1439–1452. DOI: 10.1111/jeb.12665.

Duong, T.M., Gomez, A.B. and Sherratt, T.N. (2017) Response of adult dragonflies to artificial prey of different size and coloration. *PLoS ONE* 12, e179483. DOI: 10.1371/journal.pone.179483.

Evans, B.J.E., O'Carroll, D.C., Fabian, J.M. and Wiederman, S.D. (2019) Differential adaptation to visual motion allows robust encoding of optic flow in the dragonfly. *The Journal of Neuroscience* 39, 8051–8063. DOI: 10.1523/JNEUROSCI.0143-19.2019.

Fincke, O.M. (1992) Behavioural ecology of the giant damselflies of Barro Colorado Island (Odonata: Zygoptera: Pseudostigmatidae). In: Quintero, D. and Aiello, A. (eds) *Insects of Panama and Mesoamerica*. Oxford University Press, Oxford, pp. 102–113.

Fincke, O.M., Jödicke, R., Paulson, D.R. and Schultz, T.D. (2005) The evolution and frequency of female color morphs in Holarctic Odonata: why are male-like females typically the majority? *International Journal of Odonatology* 8, 183–212. DOI: 10.1080/13887890.2005.9748252.

Frati, F., Piersanti, S., Conti, E., Rebora, M. and Salerno, G. (2015) Scent of a dragonfly: sex recognition in a polymorphic coenagrionid. *PLoS ONE* 10, e0136697. DOI: 10.1371.0136697.

Futahashi, R. (2020) Diversity of UV reflection patterns in Odonata. *Frontiers in Ecology and Evolution* 8, e201. DOI: 10.3389/fevo.2020.00201.

Futahashi, R., Kawahara-Miki, R., Kinoshito, M., Yoshitake, K., Yajima, S. *et al.* (2015) Extraordinary diversity of opsin genes in dragonflies. *PNAS* E-1247-E1256. DOI: 10.1073/pnas.1424670112.

Gonzalez-Bellido, P.T., Peng, H., Yang, J., Georgopoulos, A.P. and Olberg, R.M. (2013) Eight pairs of descending visual neurons in the dragonfly give wing motor centers accurate population vector of prey direction. *PNAS* 110, 696–701. DOI: 10.1073/pnas.1210489109.

Gorb, S. (2019) Wing surface in the damselfly *Mecistogaster ornata* (Zygoptera, Pseudostigmatidae): interactions between nanoscale wax and sticky spider webs. *International Journal of Odonatology* 22, 51–57. DOI: 10.1080/13887890.2019.1566097.

Guillermo-Ferreira, R. and Gorb, S.N. (2021) Heat-distribution in the body and wings of the morpho dragonfly *Zenithoptera lanei* (Anisoptera: Libellulidae) and a possible mechanism of thermoregulation. *Biological Journal of the Linnean Society* 133, 179–186. DOI: 10.1093/biolinnean/blaa216.

Guillermo-Ferreira, R., Therézio, E.M., Gehlen, M.H., Bispo, P.C. and Marletta, A. (2014) The role of wing pigmentation, UV and fluorescence as signals in a Neotropical damselfly. *Journal of Insect Behavior* 27, 67–80. DOI: 10.1007/s10905-013-9406-4.

Guillermo-Ferreira, R., Gorb, S.N., Appel, E., Kovalev, A. and Bispo, P.C. (2015) Variable assessment of wing colouration in aerial contests of the red-winged damselfly *Mnesarete pudica* (Zygoptera, Calopterygidae). *The Science of Nature* 102, e13. DOI: 10.1007/s00114-015-1261-z.

Hardersen, S. (2006) Sexual dimorphism in wing cell patterns in *Xanthocnemis zealandica* McLachlan (Zygoptera: Coenagrionidae). *Odonatologica* 35, 143–149.

Harrison, J.F., Kaiser, A. and VandenBrooks, J.M. (2010) Atmospheric oxygen level and the evolution of insect body size. *Proceedings of the Royal Society B* 277, 1937–1946. DOI: 10.1098/rspb.2010.0001.

Hasebe, Y., Nagano, Y. and Yokoi, T. (2023) Rapid bluing and slow browning: reversible body color change according to ambient temperature in damselfly *Indolestes peregrinus* (Ris, 1916). *Entomological Science* 26, e12537. DOI: 10.1111/ens.12537.

Hassall, C. (2014) Continental variation in wing pigmentation in *Calopteryx* damselflies related to presence of heterospecifics. *PeerJ* 2, e438. DOI: 10.7717/peerj.438.

Heinrich, B. and Casey, T.M. (1978) Heat transfer in dragonflies: 'fliers' and 'perchers'. *Journal of Experimental Biology* 74, 17–36.

Helebrandová, J.B., Pyszka, P. and Dolný, A. (2019) Behavioural phenotypic plasticity of submerged oviposition in damselflies (Insecta: Odonata). *Insects* 10, e124. DOI: 10.3390/insects10050124.

Hersch, K. and Moore, M.P. (2023) Ornamentation diversified faster than eco-morphology across Nearctic dragonflies. *Biological Journal of the Linnean Society* 139, 70–78. DOI: 10.1093/biolinnean/blad008.

Hilfert-Rüppell, D. (1998) Temperature dependence of flight activity of Odonata by ponds. *Odonatologica* 27, 45–59.

Hilfert-Rüppell, D. and Rüppell, G. (2013) Do coloured-winged damselflies and dragonflies have flight kinematics different from those with clear wings? *International Journal of Odonatology* 16, 119–134. DOI: 10.1080/13887890.2013.763332.

Ichikawa, Y., Yokoi, T. and Watanabe, M. (2017) Thermal factors affecting egg development in the wandering glider dragonfly, *Pantala flavescens* (Odonata: Libellulidae). *Applied Entomology and Zoology* 52, 89–95. DOI: 10.1007/s13355-016-0457-9.

Ishida, S. (1958) Observations on the habits of *Tramea virginia* Degeer in the vicinity of Yokkaichi, Mie Prefecture, central Japan. *Tombo* 1, 3–6.

Johansson, F. and Nilsson-Örtman, V. (2013) Predation and the relative importance of larval colour polymorphisms and colour polyphenism in a damselfly. *Evolutionary Ecology* 27, 579–591. DOI: 10.1007/s10682-01-9617-8.

Johansson, F., Söderquist, M. and Bokma, F. (2009) Insect wing shape evolution: independent effects of migratory and mate guarding flight on dragonfly wings. *Biological Journal of the Linnean Society* 97, 362–372. DOI: 10.1111/j.1095-8312.2009.01211.x.

Jongerius, S.R. and Lentink, D. (2010) Structural analysis of a dragonfly wing. *Experimental Mechanics* 50, 1323–1334. DOI: 10.1007/s11340-010-9411-x.

Josten, B., Gorb, S.N. and Büsse, S. (2022) The mouthparts of the adult dragonfly *Anax imperator* (Insecta: Odonata), functional morphology and feeding kinematics. *Journal of Morphology* 283, 1163–1181. DOI: 10.1002/jmorph.21497.

Kauppinen, J. and Mappes, J. (2003) Why are wasps so intimidating: field experiments on hunting dragonflies (Odonata: *Aeshna grandis*). *Animal Behaviour* 66, 505–511. DOI: 10.1006/anbe.2003.2225.

Kietzka, G.J., Pryke, J.S. and Samways, M.J. (2017) Aerial dragonflies are highly sensitive to in-water conditions across an ancient landscape. *Diversity and Distributions* 23, 14–26. DOI: 10.1111/ddi.12493.

Koehnsen, A., Gorb, S.N. and Büsse, S. (2023) A switchable joint in the head of dragonfly larvae (Insecta: Odonata) as key to the multifunctionality of the prehensile labial mask. *Journal of Experimental Zoology A: Ecological and Integrated Physiology* 339, 644–654. DOI: 10.1002/jez.2706.

Kohli, M.K., Ware, J.L. and Bechly, G. (2016) How to date a dragonfly: Fossil calibrations for odonates. *Palaeontologia Electronica* 19.1.1FC, 1–14. DOI: 10.26879/576.

Kriska, G., Bernáth, B., Frarkas, R. and Horváth, G. (2009) Degrees of polarization of reflected light eliciting polarotaxis in dragonflies (Odonata), mayflies (Ephemeroptera) and tabanid flies (Tabanidae). *Journal of Insect Physiology* 55, 1167–1177. DOI: 10.1016/j.jinsphys.2009.08.013.

Kuchta, S.R. and Svensson, E.I. (2014) Predator-mediated natural selection on the wings of the damselfly *Calopteryx splendens*: Differences in selection among trait types. *The American Naturalist* 184, 91–109. DOI: 10.1086/676043.

Labhart, T. and Nilsson, D.-E. (1995) The dorsal eye of the dragonfly Sympetrum: specialization for prey detection against the blue sky. *Journal of Comparative Physiology A* 176, 437–453.

Lin, H.-T. and Leonardo, A. (2017) Heuristic rules underlying dragonfly prey selection and interception. *Current Biology* 27, 1124–1137. DOI: 10.1016/j.cub.2017.03.010.

Lohmann, A.C., Corcoran, A.J and Hedrick, T.L. (2019) Dragonflies use underdamped pursuit to chase conspecifics. *Journal of Experimental Biology* 222, e190884. DOI: 10.1242/jeb.190884.

Lopes Junior, R.S. and Peixoto, P.E.C. (2013) Males of the dragonfly *Diastatops obscura* fight according to predictions from game theory models. *Animal Behaviour* 85, 663–669. DOI: 10.1016/j.animbehav.2012.12.033.

Martens, A. and Grabow, K. (1994) Males of *Enallagma vansomereni* Pinhey settling on water surface (Zygoptera: Coenagrionidae). *Odonatologica* 23, 169–174.

Martens, A. and Suhling, F. (2003) The barbed inflorescences of the grass *Setaria verticillata* (L.) Palisot de Beauvois (Poaceae) as a lethal trap for dragonflies (Odonata). *Cimbebasia* 18, 243–246.

Matushkina, N.A. (2011) Morphology of exophytic ovipositors in dragonflies (Odonata: Gomphidae, Coduliidae, Libellulidae), with particular reference to ovipositor muscles and sensilla. *International Journal of Odonatology* 14, 233–248. DOI: 10.108013887890.2011.613736.

Matushkina, N.A. and Lambret, P.H. (2011) Ovipositor morphology and egg laying behaviour in the dragonfly

Lestes macrostigma (Zygoptera: Lestidae). *International Journal of Odonatology* 14, 69–82. DOI: 10.1080/13887890.2011.568190.

May, M.L. (1976a) Thermoregulation and adaptation to temperature in dragonflies (Odonata: Anisoptera). *Ecological Monographs* 46, 1–32. DOI: 10.2307/1942392.

May, M.L. (1976b) Physiological color change in New World damselflies (Zygoptera). *Odonatologica* 5, 165–171.

May, M.L. (1991) Thermal adaptations of dragonflies, revisited. *Advances in Odonatology* 5, 71–88.

May, M.L. (1998) Body temperature regulation in a late-season dragonfly, *Sympetrum vicinum* (Odonata: Libellulidae). *International Journal of Odonatology* 1, 1–13. DOI: 10.1080/13887890.1998.9748090.

McCafferty, W.P. and Edmunds, G.F. (1979) The higher classification of the Ephemeroptera and its evolutionary basis. *Annals of the Entomological Society of America* 72, 5–12.

McCauley, S.J. (2013) Relationship between morphology, dispersal and habitat distribution in three species of *Libellula* (Odonata: Anisoptera). *Aquatic Insects* 34, 195–204. DOI: 10.1080/01650424.2013.800557.

McGeoch, M.A. and Samways, M.J. (1991) Dragonflies and the thermal landscape: implications for their conservation (Anisoptera). *Odonatologica*, 20, 303–320.

Mischiati, M., Lin, H.-T., Herold, P., Imler, E., Olberg, R. and Leonardo, A. (2015) Internal models direct dragonfly interception steering. *Nature* 517, 333–338. DOI: 10.1038/nature14045.

Mizutani, A., Chahl, J.S. and Srinivasan, M.V. (2003) Motion camouflage in dragonflies. *Nature* 423, 604. DOI: 10.1038/423604a.

Moore, M.P. and Martin, R.A. (2016) Intrasexual selection favours an immune-correlated colour ornament in a dragonfly. *Journal of Evolutionary Biology* 29, 2256–2265. DOI: 10.1111/jeb.12953.

Moore, T.Y. and Biewener, A.A. (2015) Outrun or outmaneuver: predator-prey interactions as a model system for integrating biomechanical studies in a broader ecological and evolutionary context. *Integrative and Comparative Biology* 55, 1188–1197. DOI: 10.1093/icb/icv074.

Nakata, T., Henningsson, P., Lin, H.-T. and Bomphrey, R.J. (2020) Recent progress on the flight of dragonflies and damselflies. *International Journal of Odonatology* 23, 41–49. DOI: 10.1080/13887890.2019.1688502.

Nel, A., Bechly, G., Prokop, J., Béthoux, O. and Fleck, G. (2012) Systematics and evolution of Paleozoic and Mesozoic damselfly-like Odonatoptera of the 'protozygopteran' grade. *Journal of Paleontology* 86, 81–104.

Nel, A., Prokop, J., Pecharová, M., Engel, M.S. and Garrouste, R. (2018) Palaeozoic giant dragonflies were hawker predators. *Scientific Reports* 8, e12141. DOI: 10.1038/s41598-018-30629-w.

Noda, R., Liu, X., Hefler, C., Shyy, W. and Qiu, H.H. (2023) The interplay of kinematics and aerodynamics in multiple flight modes of a dragonfly. *Journal of Fluid Mechanics* 967, eA31. DOI:10.1017/jfm.2023.471.

Novelo-Gutiérrez, R. and Bota-Sierra, C.A. (2023) The larvae of *Cora inca* Selys, 1873 and *Polythore gigantea* (Selys, 1853) from Colombia (Odonata: Polythoridae), with a larval diagnoses of some genera in the family. *Zootaxa* 5254 (4), 517–533. DOI: 10.11646/zootaxa.5254.4.4.

Okamoto, M., Yasuda, K. and Azuma, A. (1996) Aerodynamic characteristics of the wings and body of a dragonfly. *Journal of Experimental Biology* 199, 281–294.

Olberg, R.M., Seaman, R.C., Coats, M.I. and Henry, A.F. (2007) Eye movements and target fixation during dragonfly prey-interception flights. *Journal of Comparative Physiology A* 193, 685–693. DOI: 1007/s00359-007-0223-0.

Outomuro, D., Adams, D.C. and Johansson, F. (2013a) The evolution of wing shape in ornamented-winged damselflies (Calopterygidae, Odonata). *Evolutionary Biology* 40, 300–309. DOI: 10.1007/s11692-012-9214-3.

Outomuro, D., Dijkstra, K.-D.B. and Johansson, F. (2013b) Habitat variation and wing coloration affect wing shape evolution in dragonflies. *Journal of Evolutionary Biology* 26, 1866–1874. DOI: 10.1111/jeb.12203.

Palacino-Rodríguez, F., Munguía-Steyer, R., Mesa-Blanco, L.V. and Córdoba-Aguilar, A. (2016) Survival is predicted by territorial status but not wing pigmentation in males of a polythorid damselfly, *Euthore fasciata* (Odonata: Zygoptera: Polythoridae). *International Journal of Odonatology* 19, 183–190. DOI: 10.1080/13887890.2016.1209137.

Piersanti, S., Frati, F., Conti, E., Gaino, E., Rebora, M. and Salerno, G. (2014) First evidence of the use of olfaction in Odonata behavior. *Journal of Insect Physiology* 62, 26–31. DOI: 10.1016/j.jinsphys.2014.01.006.

Polcyn, D.M. (1994) Thermoregulation during summer activity in Mojave Desert dragonflies (Odonata: Anisoptera). *Functional Ecology* 8, 441–449.

Pritchard, G. (1966) On the morphology of the compound eyes of dragonflies (Odonata: Anisoptera), with special reference to their role in prey capture. *Proceedings of the Royal Entomological Society of London, A* 41, 1–8.

Rajabi, H. and Gorb, S.N. (2020) How do dragonfly wings work? A brief guide to functional roles of wing structural components. *International Journal of Odonatology* 23, 23–30. DOI: 10.1080/13887890.2029.1677515.

Rajabi, H., Ghoroubi, N., Darvizeh, A., Appel, E. and Gorb, S.N. (2016) Effects of multiple vein microjoints on the mechanical behaviour of dragonfly wings: numerical modelling. *Royal Society Open Science* 3, e150610. DOI: 10.1098/rsos.150610.

Rashed, A., Beatty, C.D., Forbes, M.R. and Sherratt, T.N. (2005) Prey selection by dragonflies in relation to prey size and wasp-like colours and patterns. *Animal Behaviour* 70, 1195–1202. DOI: 10.1016/janbehav.2005.03.009.

Rivas-Torres, A., Di Pietro, V. and Cordero-Rivera, A. (2023) Sex wars: a female genital spine forces male

damselflies to shorten copulation duration. *Evolution* 77, 1659–1666. DOI: 10.1093/evolut/qpad073.
Robertson, H.M. (1982) Mating behaviour and its relationship to territoriality in *Platycypha caligata* (Selys) (Odonata: Chlorocyphidae). *Behaviour* 74, 11–26.
Rodrigues, M.E., De Oliveira Roque, F., Guillermo-Ferreira, R., Saito, V.S. and Samways, M.J. (2019) Egg-laying traits reflect shifts in dragonfly assemblages in response to different amount of tropical forest cover. *Insect Conservation and Diversity* 12, 231–240. DOI: 10.1111/icad.12319.
Rüppell, G. (1989) Fore legs of dragonflies used to repel males. *Odonatologica* 18, 391–396.
Sacchi, R. and Hardersen, S. (2013) Wing length allometry in Odonata: differences between families in relation to migratory behavior. *Zoomorphology* 132, 23–32. DOI: 10.1007/s00435-012-0172-1.
Sahlén, G. (1995) Transmission electron microscopy of the eggshell of five damselflies (Zygoptera: Coenagrionidae, Megapodagrionidae, Calopterygidae). *Odonatologica* 24, 311–318.
Sahlén, G. and Suhling, F. (2002) Relationship between egg size and clutch size among European species of Sympetrinae (Odonata: Libellulidae). *International Journal of Odonatology* 5, 181–191. DOI: 10.1080/13887890.2002.9748185.
Samways, M.J. (1991) Dragonfly death by entanglement on hooked and barbed plant surfaces (Anisoptera: Libellulidae). *Odonatologica* 20, 79–83.
Samways, M.J. (1994) 'Sailing' on the water surface by adult male *Enallagma nigridorum* Selys (Zygoptera: Coenagrionidae). *Odonatologica* 23, 175–178.
Samways, M.J. (2006) Open and banded wings: hypotheses on damselfly wing position (Zygoptera: Odonata). *Odonatologica* 35, 67–73.
Sánchez-Guillén, R.A., Van Gossum, H. and Cordero-Rivera, A. (2005) Hybridization and the inheritance of intrasexual polymorphism in two ischnurid damselflies (Odonata: Coenagrionidae). *Biological Journal of the Linnean Society* 85, 471–481. DOI: 10.1111/j.1095-8312.2005.00506.x.
Sánchez-Guillén, R.A., Hansson, B., Wellenreuther, M., Svensson, E.I. and Cordero-Rivera, A. (2011) The influence of stochastic and selective forces in the population divergence of female colour polymorphism in damselflies of the genus *Ischnura*. *Heredity* 107, 513–522. DOI: 10.1038/hdy.2011.36.
Schenk, K., Suhling, F. and Martens, A. (2004) Egg distribution, mate guarding intensity and offspring characteristics in dragonflies (Odonata). *Animal Behaviour* 68, 599–606. DOI: 10.1016/j.anbehav.2003.12.010.
Schultz, T.D. and Fincke, O.M. (2009) Structural colours create a flashing cue for sexual recognition and male quality in a Neotropical damselfly. *Functional Ecology* 23, 724–732. DOI: 10.1111/j.1365-2435.2009.01584.x.
Schultz, T.D. and Fincke, O.M. (2013) Lost in the crowd or hidden in the grass: signal apparency of female polymorphic damselflies in alternative habitats. *Animal Behaviour* 86, 923–931. DOI: 10.1016/j.anbehav.2013.08.008.
Schultz, T.D., Anderson, C.N. and Symes, L.B. (2008) The conspicuousness of colour cues in male pond damselflies depends on ambient light and visual systems. *Animal Behaviour* 76, 1357–1364. DOI: 10.1016/j.anbehav. 2008.04.024.
Sformo, T. and Doak, P. (2006) Thermal ecology of Interior Alaska (Odonata: Anisoptera). *Functional Ecology* 20, 114–123. DOI: 10.11111/j.1365-2435.2006.01064.x.
Sharkey, C.R., Partridge, J.C. and Roberts, N.W. (2015) Polarization sensitivity as a visual contrast enhancer in the Emperor dragonfly larva, *Anax imperator*. *Journal of Experimental Biology* 218, 3399–3405. DOI: 10.1242/jeb.122507.
Shear, W.A. and Kukalová-Peck, J. (1990) The ecology of Paleozoic terrestrial arthropods: the fossil evidence. *Canadian Journal of Zoology* 68, 1807–1834. DOI: 10.1139/z90-262.
Sherk, T.E. (1978) Development of the compound eyes of dragonflies (Odonata): adult compound eyes. *Journal of Experimental Zoology* 203, 61–80.
Sternberg, K. (1987) On reversible, temperature-dependent colour change in males of *Aeshna caerulea* (Ström, 1783) (Anisoptera: Aeshnidae). *Odonatologica* 16, 57–66.
Suárez-Tovar, C.M. and Sarmiento, C.E. (2016) Beyond the wing platform: morphological differentiation between migratory and non-migratory dragonfly species. *Journal of Evolutionary Biology* 29, 690–703. DOI: 10.1111/jeb.12830.
Svensson, E.I. and Waller, J.T. (2013) Ecology and sexual selection: evolution of wing pigmentation in calopterygid damselflies in relation to latitude, sexual dimorphism, and speciation. *The American Naturalist* 182, E174–E195. DOI: 10.5061/drad.rd0f6.
Takahashi, M., Okude, G., Futahashi, R., Takahashi, Y. and Kawata M. (2021) The effect of the *doublesex* gene in body colour masculinization of the damselfly *Ischnura senegalensis*. *Biology Letters* 17, e20200761. DOI: 10.1098/rsbl.2020.0761.
Toussaint, E.F.A., Bybee, S.M., Erickson, R.J. and Condamine, F.L. (2019) Forest giants on different evolutionary branches: ecomorphological convergence in helicopter damselflies. *Evolution* 73, 1045–1054. DOI: 10.1111/evo.13695.
Tsubaki, Y., Samejima, Y. and Siva-Jothy, M.T. (2010) Damselfly females prefer hot males: higher courtship success in males in sunspots. *Behavioral Ecology and Sociobiology* 64, 1547–1554. DOI: 10.1007/s00265-010-0968-2.
Usherwood, J.R. and Lehmann, F.-O. (2008) Phasing of dragonfly wings can improve aerodynamic efficiency by removing swirl. *Journal of the Royal Society Interface* 5, 1303–1307. DOI: 10.1098/rsif.2008.0124.

Valdivia, F.G.A., Alves-Silva, E. and Del-Claro, K. (2020) Differences in size and energy content affect the territorial status and mating success of a neotropical dragonfly. *Austral Ecology* 45, 748–758. DOI: 10.1111/aec.12891.

Van Buskirk, J. (1993) Consequences of larval crowding in the dragonfly *Aeshna juncea*. *Ecology* 74, 1950–1958.

Vargas, A., Mittal, R. and Dong, H. (2008) A computational study of the aerodynamic performance of a dragonfly wing section in gliding flight. *Bioinspiration and Biomimetics* 3, e026004. DOI: 10.1088/1748-3182/3/2/026004.

Vilela, D.S., Tosta, T.A.A., Rodrigues, R.R., Del-Claro, K. and Guillermo-Ferreira, R. (2017) Colours of war: visual signals may influence the outcome of territorial contests in the tiger damselfly, *Tigriagrion aurantinigrum*. *Biological Journal of the Linnean Society* 121, 786–795. DOI: 10.1093/biolinnean/blx024.

Waage, J.K. (1979a) Dual function of the damselfly penis: sperm removal and transfer. *Science* 203 (4383), 916–918. DOI: 10.1126/science.203.4383.916.

Waage, J.K. (1979b) Reproductive character displacement in *Calopteryx* (Odonata: Calopterygidae). *Evolution* 33, 104–116.

Ward, P. and Kirschvink, J. (2015) *A New History of Life*. Bloomsbury, New York.

Wildermuth, H. (1992) Visual and tactile stimuli in choice of oviposition substrates by the dragonfly *Perithemis mooma* Kirby (Anisoptera: Libellulidae). *Odonatologica* 21, 309–321.

Winfrey, C. and Fincke, O.M. (2017) Role of visual and non-visual cues in damselfly mate recognition. *International Journal of Odonatology* 20, 43–52. DOI: 10.1080/13887890.2017.1297259.

Woodall, P.F. (1995) Notes on the habitat, flying speed and behaviour of *Austrophlebia costalis* (Tillyard) (Odonata: Aeshnidae) in Brisbane Forest Park, Queensland. *Australian Entomologist* 22, 33–36.

Wootton, R.J. (1981) Palaeozoic insects. *Annual Review of Entomology.* 26, 319–344.

Wootton, R.J. (2020) Dragonfly flight: morphology, performance and behaviour. *International Journal of Odonatology* 23, 31–39. DOI: 10.1080/13887890.2019.1687991.

Worthen, W.B. and Morrow, P.H. (2016) Perch selection by three co-occurring species of *Celithemis* (Odonata: Libellulidae): testing for competitive hierarchy among similar species. *Psyche* 2016, e9028105. DOI: 10.1155/2016/9028105.

Xu, M. and Fincke, O.M. (2015) Ultraviolet wing signal affects territorial contest outcome in a sexually dimorphic damselfly. *Animal Behaviour* 101, 67–74. DOI: 10.1016/j.anbehav. 2014.12.018.

Yu, D.-N., Yu, P.P., Zhange, L.P., Storey, K.B., Gao, X.Y. and Zhange, J.-Y. (2021) Increasing 28 mitogenomes of Ephemeroptera, Odonata and Plecoptera support Chiastomyaria hypothesis with three different outgroup combinations. *PeerJ* 9, e11402. DOI:7717/peerj.11402.

Zeiss, C., Martens, A. and Rolff, J (1999) Male mate guarding increases females' predation risk? A case study on tandem oviposition in the damselfly *Coenagrion puella* (Insecta: Odonata). *Canadian Journal of Zoology* 77, 1013–1016.

Zeuss, D., Brandl, R., Brändle, M., Rahbek, C. and Brunzel, S. (2014) Global warming favours light-coloured insects in Europe. *Nature Communications* 5, e3874. DOI: 10.1038/ncomms4874.

3 Dragonfly Diversity and Distribution

Illustration: Megan Louise Potgieter.

DOI: 10.1079/9781789248395.0003

Abstract

There are two major groups of dragonflies, the zygopteran damselflies and the anisopteran dragonflies, and one minor group, the anisozygopterans. Over 6400 species are described.

Dragonflies likely arose in the tropics, and later colonized colder environments. Today, they remain most speciose in the wet tropics. Those of north temperate regions contracted and expanded their geographic ranges with the advance and retreat of glacial ice sheets. With isolation of other species on mountains, on islands and in permanent special habitats, new species evolved, often with narrow geographic ranges and specialist characteristics.

Being highly mobile, dragonflies disperse across landscapes in search of optimal habitats, especially pond species in times of drought. With recent rapid climate change, many species have expanded their geographic ranges poleward. Some specialist species of high mountains and those in bogs are experiencing biogeographic range shrinkage. In general, habitat-specialist species face the greatest risk of local extinction.

3.1 Current Diversity and Phylogeny

3.1.1 Extant Odonata species and their early evolution

The three extant dragonfly (Odonata) suborders, Anisoptera, Anisozygoptera and Zygoptera are each monophyletic (Bybee *et al.* 2021; Sandall *et al.*, 2022). Together, they comprise 6406 described species (Paulson *et al.*, 2024), with this list likely to increase as new field discoveries are made and other species are erected following genetic analysis.

There are roughly equal numbers of damselfly (Zygoptera) and dragonfly (Anisoptera) species. Zygoptera phylogeny has been the subject of much debate (Dijkstra *et al.*, 2014), currently with about 34 extant families. The phylogeny of the Anisoptera has more resolution, consisting of 11 extant families (Bybee *et al.*, 2021).

Introgression (the transfer of genetic material from one species into the gene pool of another by the repeated backcrossing of an interspecific hybrid with one of its parent species) appears to have been an important evolutionary force at various taxonomic levels in the Odonata. The Anisozygoptera retain phenotypic characteristics of the other two suborders, having undergone high levels of introgression likely coming from zygopteran genomes. There is also a strong case for introgression also having occurred in the early ancestry of the odonate superfamilies (Suvarov *et al.*, 2022).

Molecular evidence also suggests that the Zygoptera and the Anisoptera + Anisozygoptera diverged in the mid-Triassic, about 250 mya. By the late Triassic, about 200 mya, the Anisoptera and Anisozygoptera had diverged. The major odonate lineages originated in the Cretaceous, 140–66 mya (Suvarov *et al.*, 2022).

3.1.2 The Zygoptera

All Zygoptera have widely separated eyes on either side of a widened head, a long, slender body and similarly stalked (petiolate) fore- and hindwings. While most rest with their wings closed along their abdomens, some, such as most Lestidae and the Synlestidae, rest with their wings open. The geographically widespread superfamily Lestoidea comprises the families Hemiphlebiidae, Perilestidae, Synlestidae, Perilestidae and Lestidae (Bybee *et al.* 2021), although morphological and genetic evidence suggests that Perilestidae should be merged into the Synlestidae (Simaika *et al.*, 2020) (Fig. 3.1).

The Lestoidea are a sister group to the remaining 93% of damselfly species (Suhling *et al.*, 2015). The Platystictoidea, with its one family Platystictidae, has many species across all temperate and warm

Fig. 3.1. *Chlorolestes umbratus*: a representative of the Synlestidae. (Photo: Michael Samways).

continents except Africa, and is a sister group to most of the other Zygoptera superfamilies, except Lestoidea (Bybee *et al.*, 2021). In turn, 58% of all species belong to the superfamily Coenagrionoidea. It has three families: Isostictidae (with few species and confined to Australasia), Platycnemididae (with about 400 species in the Old World only) and the Coenagrionidae (with about 1200 species widely spread across the world) in a wide range of ecosystems in many regions.

The current superfamily 'Calopterygoidea' still requires some taxonomic resolution, and for now, consists of four groups (Bybee *et al.*, 2021). It is a significant superfamily, as it makes up about 27% of damselfly species, many of which are well known for their wing colouration used in mating and male agonistic behaviours (Fig. 2.7). In terms of species numbers, complex behaviours and often strong responses to habitat transformation, the Hetaerininae is an important subfamily. Genetic studies indicate that the genera *Mnesarete* and *Ormenophlebia* are nested within *Hetaerina*, with forest habitat the ancestral state for Hetaerininae, and transitions to generalist species of *Hetaerina* mostly present in Mesoamerica (Standring *et al.*, 2022).

The remaining Zygoptera families are composed of relatively few species, whose phylogeny still require resolution (Suhling *et al.*, 2015; Bybee *et al.*, 2021).

3.1.3 The Anisoptera

The Anisoptera are distinctly more robust in body form than the Zygoptera, with highly muscular thoraces, large eyes associated with their high mobility, and with stiff yet flexible wings for great manoeuvrability. With these credentials, it is not surprising that they are such icons of free air space. The hindwings are wide at the base, and much more so than in the forewings. The eyes either meet in the middle of the top of the head or are widely spaced, but always a prominent feature of the head.

The Anisoptera are monophyletic and comprise five superfamilies (Bybee *et al.* 2008). The Aeshnoidea is composed of the Aeshnidae, with its many large species and widespread distribution across the world, and the Austropetaliidae, with a few very large species in southern Australia and the southern tip of South America.

The superfamily 'Petaluroidea', with its one family Petaluridae and just 10 species, has a wide but disjunctive distribution from Chile, Australia and New Zealand to Japan and North America. Sister to Petaluridae is the large family Gomphidae (Bybee *et al.*, 2021; Kohli *et al.*, 2021b), numbering about 1000 species, widely spread across the world. They all have separated eyes, and often have distinctly flanged or widened abdomen tips.

The superfamily Cordulegastroidea covers the families Cordulegastridae, mainly in the northern hemisphere, and the Chlorogomphidae in the Oriental biogeographic region, each with nearly 50 species. The third family Neopetaliidae has only one species, *Neopetalia punctata*, confined to the southern tip of South America (Fig. 3.2).

In terms of species numbers, the Libelluloidea, comprising 'Corduliidae' (with complex phylogenetic relationships; Bybee *et al.*, 2021), Libellulidae, Macromiidae and 'Synthemistidae' (which requires revision), together make up the most speciose anisopteran superfamily, with over 1300 species, and distributed across much of the world, although 'Synthemistidae' is confined to the Australasian biogeographic region. Suhling *et al.* (2015) provide a comprehensive table of the number of described odonate genera and species per family per biogeographic region.

3.1.4 The Anisozygoptera

The third suborder, Anisozygoptera, has just one relict family, Epiophlebiidae, with one genus, *Epiophlebia*, of three extant species, confined to east Asia (Büsse *et al.* 2012; Carle 2012; Büsse and Ware, 2022). Although intermediate between damselflies (Zygoptera) and dragonflies (Anisoptera), there has been considerable debate over whether the group is more closely related to Zygoptera (Wang *et al.*, 2014) or Anisoptera (Kim *et al.*, 2014), or even included in the Anisoptera (Davis *et al.*, 2011). However, latest evidence suggests that *Epiophlebia* is sister to Anisoptera (Bybee *et al.*, 2021) (Fig. 3.3).

3.1.5 Ongoing new dragonfly species discoveries

There are still some geographical areas of the world where important new dragonfly records are highlighting threats, as for example in Acre state, Brazil, an important transition zone between the Amazonian and Andean regions (Miranda Filho *et al.*, 2022). In turn, many new species are being recognized by more thorough and extensive geographical exploration, as well as detailed examination of fine aspects of morphology (Dijkstra *et al.*, 2015).

Fig. 3.2. *Neopetalia punctata*: the only species in the Neopetaliidae. (Photo: Michael Samways).

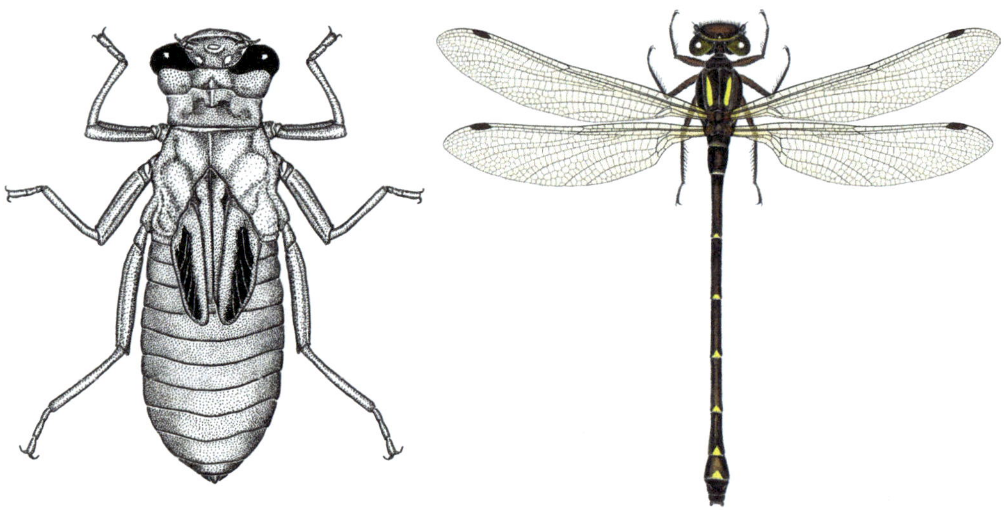

Fig. 3.3. Larva (left) and adult (right) of the anisozygopteran *Epiophlebia laidlawi* (larva and adult not to scale). (Illustrations: Albert G. Orr).

3.1.6 Genetic elucidation of new dragonfly species

Among insects in general, many new species are being discovered through splitting of morphologically determined species following DNA analysis. This was highlighted early on when the familiar Neotropical skipper butterfly *Astraptes fulgerator* turned out to be a complex of ten cryptic species following detailed DNA work (Hebert *et al.*, 2004). The reason for this large number of cryptic species is that the human interpretation of the morphological world is largely visual, whereas among Lepidoptera, for instance, pheromones play a major role in species recognition.

Dragonflies have a complex mating system and there can be rapid evolution of reproductive barriers through strong sexual selection (Cordero Rivero *et al.*, 2004), and fine changes in genital structures and wing colouration (Wellenreuther and Sánchez-Guillén, 2016) can favour sexual selection rather than ecological niche-based natural selection, at least in some species.

Newly identified dragonfly species through DNA analysis have high value for determining phylogeny relative to biogeography in species-rich areas such as the Amazon (Carvalho *et al.*, 2022). Using a database containing reliable DNA barcodes greatly enhances the discovery of unknown species, with diagnostic characters for geographic clusters of specimens being valuable for ecological assessment, resolving speciation debates, identifying larvae and assisting conservation management. However, these activities are more effective when DNA barcoding uses a combination of genetic markers (ND1 and CO1), rather than either one alone (Bergmann *et al.*, 2013).

Among Odonata species of the Upper Plata Basin of Brazil there is a 79–94% match between genetically versus morphologically diagnosed species (Koroiva *et al.* 2017). An in-depth analysis of Italian odonate species showed that, even for such a supposedly well-known fauna, there was only 85% correspondence between molecular and morphological species, with Zygoptera showing only a 71% match and Anisoptera at a much higher level of 93%. Among these species, 32.3–38.7% of Zygoptera and 8.8–15.8% of Anisoptera were designated as 'warnings' (i.e. showing a mismatch between molecular and morphological species). These differences between the two taxonomic groups could in part be due to the weaker dispersal abilities of Zygoptera, leading to more isolation and diversification among them, compared to Anisoptera. One conclusion is that Italy is an important biogeographic and evolutionary centre for odonate diversification in the overall Mediterranean area (Galimberti *et al.*, 2021). The Italian endemic *Coenagrion mercuriale castellanii* has now been raised to the level of species, *C. castellani* (La Porta *et al.*, 2023).

As both sexual and natural selection is associated with great responsiveness to ecological change, speciation can occur without any major changes in morphology (Fig. 3.4). However, these cryptic species can be uncovered with an integrated approach, where character-based DNA barcoding is combined with ecological, biogeographic and fine-scale morphological information (Damm *et al.*, 2010). In turn, there needs to be a single standard threshold model for equating cryptic species with putative species. At least in the case of European moths, this appears to be the General Mixed Yule Coalescent. It is especially valuable in poorly inventoried biogeographic regions where historical events and factors have promoted genetic diversity and singularity (Gaytán *et al.*, 2020).

3.2 Dragonfly Macroecology

3.2.1 Biogeographic range dynamics and dragonfly conservation

Conservation of a species is essentially about maintaining, and even better, improving the effective population size, while preventing any attrition of its biogeographic range. However, populations and geographic ranges are not only dynamic entities, but also dynamic processes. A species feeds, breeds and rests across a constantly changing landscape. These activities are affected by both short-term weather events and long-term climate variation and topography. Weather events, such as fire and floods, as well as changed responses among other organisms with which dragonflies share their local environment, are another layer of influence.

It is not really 'the species', but the collective of individuals making up the populations and metapopulations, that is facing the brunt of these environmental stressors. Sometimes though, these stressors also provide opportunities, as in the case of warm weather fronts for migrants or rain pools for opportunistic species.

Individuals making up a population experience natural selection (and sexual selection) from *current* pressures. However, there is also the past to consider. Populations of individuals possess a set of traits honed and filtered by past events that have exerted an earlier set of selection pressures. Among the individuals composing all the populations of a species, there is an adaptive range of survivability, with certain genomes surviving at one time and place, and others at other times and places. In short, there is a continual push and shove by the environment, countered by adaptation among all the individuals to survive and produce offspring. In other words,

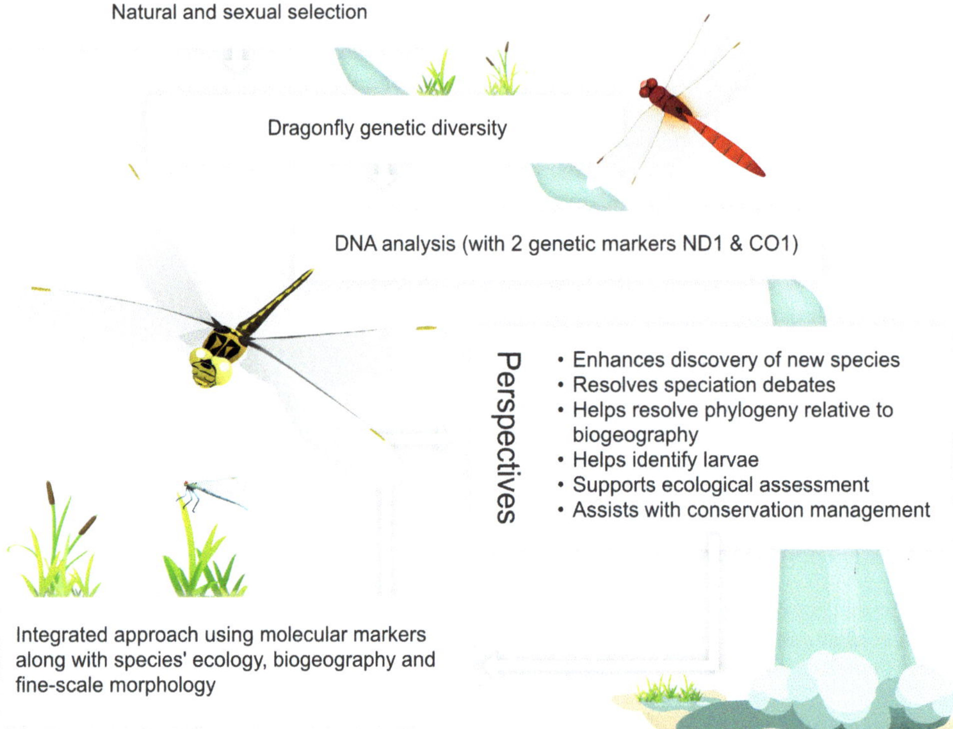

Fig. 3.4. Natural and sexual selection are both important in determining dragonfly genetic diversity. When this diversity is adequately analysed with two genetic markers, the results have value from various perspectives. In turn, when molecular markers are used along with a species' ecological preferences, its biogeography, and its fine-scale morphology, it leads to an integrated approach which feeds back into the value of the various perspectives. (Illustration: Robyn Symons).

for a dragonfly species to maintain an effective population, it is essential that it contributes maximally with quality individuals to the next generation. Additionally, each individual also has inherent phenotypic plasticity which itself is honed by past events and is tested by current ones (Chapter 4).

Spatial and temporal scales are an additional and important consideration when focusing on an effective population. An individual is living within a particular microhabitat, within a biotope, and within an ecosystem in a region. In turn, a species' *habitat* is the area necessary for it to carry out all its life functions to maintain an effective population. As genetic diversity is so important for equipping a species to survive changing environmental conditions, both abiotic and biotic, there must be good functional connectivity across the landscape enabling gene flow to take place for maintaining high genetic diversity, notwithstanding beneficial new mutations.

However, there is a limit to the extent that a species can survive changing ambient conditions. To recap, surviving individuals making up an effective population do so through natural selection, sexual selection and/or phenotypic plasticity. Their survival depends largely on whether they stay or move: stay and tolerate prevailing conditions, or move to find better local conditions. If they stay, then the stressors must not be severe for too long. If they move, then they must be able to reach a suitable new habitat. For this mobility to enhance population survival, the movements need to be in the right direction, and the habitat needs to be reachable. Both are relative to their powers of survivability while on the move, and their dispersal capability.

In the case of dragonflies today, the problem is that despite all the hurdles they have had to endure and survive in the past, it is the new challenges dealt to them by humans that is the crux of why we need conservation. Easily said, but not so easily done, as there are a multitude of new challenges many of which act together, so increasing the pressure on populations. Not only that, but the conservation action must also, ideally, be effective over the whole biogeographic range. However, this is not an easy task for simply logistical reasons. As there are various and varying human pressures of one sort or another across their geographic range, conservation action cannot be carried out everywhere all at once. This is why we must carefully strategize what we do in conservation for maximum effectiveness and long-term sustainability. In short, we must have strategic management involving resilience foresight based in part on what we know about dragonfly biogeographic ranges, coupled with a knowledge of their traits across their ranges (Fig. 3.5).

3.2.2 Past climates and current dragonfly biogeographic ranges

Climate has always been, and still is, a major driver of dragonfly biogeographic ranges. Bechly *et al.* (2021) found that for North American damselflies, there is strong evidence of former terrestrial interchange of species between North America, Asia and Europe in the warm Eocene, about 50 mya. These fascinating findings suggest that many odonates at the time were sufficiently mobile to be able to move across wide areas, albeit in incremental steps by movement of adults per generation over time. However, in the case of five species (*Aeshna juncea*, *A. subarctica*, *Sympetrum danae* (which appears now to be two species), *Libellula quadrimaculata* and *Somatochlora sahlbergi*) from across the Holarctic in more modern times, genetic evidence suggests that North American and European populations are genetically distinct, having been separated for more than 400,000 years. Fluctuating sea levels over this time and the effect of mountain ranges appear to have influenced dispersal routes and the positions of refugia, with Beringia a transit zone rather than a refugium. In turn, *Somatochlora sahlbergi* (Fig. 3.6) shows little genetic variation across the area, while appearing to have diverged from *S. alpestris* about 1.87 mya (Kohli *et al.* 2021a).

Global dragonfly species richness follows climate patterns, especially evapotranspiration, which is the sum of evaporation from both the land surface and transpiration from plants (Keil *et al.*, 2008; Bush *et al.*, 2013). This is notwithstanding the local importance of various types of freshwater bodies in which the different species breed, and appropriate structural vegetation among which adults hunt for food. This is particularly striking in the case of dragonfly species richness across Africa, where it is greatest in the wet tropical rainforest belt (Clausnitzer *et al.*, 2012), with the tropical forests of the Oriental and Neotropical regions being even more speciose (Kalkman *et al.*, 2008).

3.2.3 Early species diversification drivers and opportunities

Larval habitat has played a major role in early diversification of dragonflies, especially for species with large biogeographic ranges. Genetic studies suggest that the original primary habitat for dragonflies was running water in the tropics. Aeshnidae, Corduliidae and Libellulidae species then colonized a wide variety of standing water types, leading to a high rate of speciation, especially among aeshnids and libellulids. Meanwhile, the standing water groups remained relatively conservative among higher level taxa. Species with large biogeographic ranges are more likely to diversify as barriers to movement take effect, so isolating populations, and leading to speciation events. Large range size also means potentially being in contact with the typically wide range of standing water habitat types. Extra niches would then be available for occupation, leading to further diversification (Letsch *et al.*, 2016).

Hypothetical rules relating to species' geographic ranges are useful as models to elucidate past and present drivers of dragonfly distributions. Rapoport's rule (RR) proposes that the size of the geographic range of a species of the same lineage is smaller toward the equator. A test of this rule in Brazil shows that the shape of a landmass and the climatic history as manifested by certain biomes today have a strong effect. In this large country, RR applies to the Amazon biome but not the drier Amazon–Cerrado Transition Forest and Cerrado biomes. In turn, the Latitudinal Diversity Gradient (LDG) rule, which hypothesizes that there is an increase in species richness toward the equator, does not hold for all odonates together, but it does for damselflies alone. The reason for these deviations from the RR and LDG appears to be due to two barrier factors. The first is related to the geometry

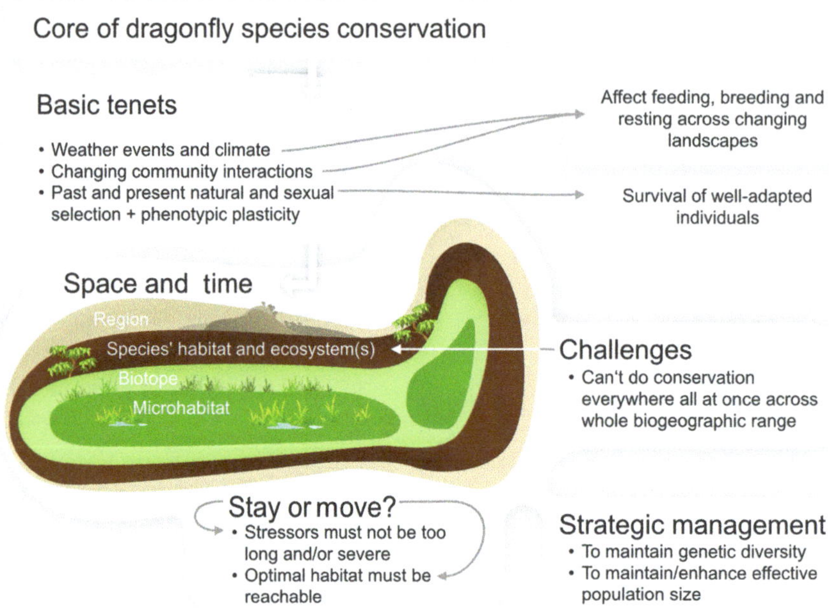

Fig. 3.5. At the core of dragonfly species conservation is resilience foresight to strategically manage populations so as to conserve genetic diversity and maintain, or even enhance, effective population levels. (Illustration: Robyn Symons).

Fig. 3.6. A young male *Somatochlora sahlbergi*. (Photo: Magnus Billqvist).

of the country, which is wider in the north than in the south, limiting geographic expansion in the south. Second, these geographic characteristics seem to interact strongly with the species traits, especially those related to thermoregulation and dispersal, with the damselflies benefitting from a long history of geoclimatic stability in the Amazon resulting in greater niche packing among members of this group (Miguel *et al.*, 2022) (Fig. 3.7).

3.2.4 Geographic range grain size and population dynamics

Field guides, for practical convenience, usually portray a species' biogeographic range as contiguous, which rarely accurately reflects all occupied versus non-occupied locations across the illustrated range. Furthermore, often there is mapping bias according to administrative borders, especially around the transition zone from tropical to sub-tropical regions (Hughes *et al.*, 2021). In reality, within the biogeographic range, there are lacunae where resident populations of any particular species are absent or vice versa. Absence in a particular area may be for one or various reasons, such as local climatic conditions being unsuitable for permanent establishment, as in the case of a cold peak of a mountain, or an unusually dry area. Long-term presence, on the other hand, indicates suitable habitat for residence and breeding. In short, the size of the grain we use to portray a species' range matters and depends on our macroecological and/or conservation question. The helicopter damselfly *Mecistogaster modesta* (Fig. 3.8) grows up in tank bromeliads, abundant throughout much of the Neotropics. Yet the species cannot occupy all the available habitats, as the Andean mountains, as well as cold and dry areas, prevent its dispersal (Amundrud *et al.*, 2018).

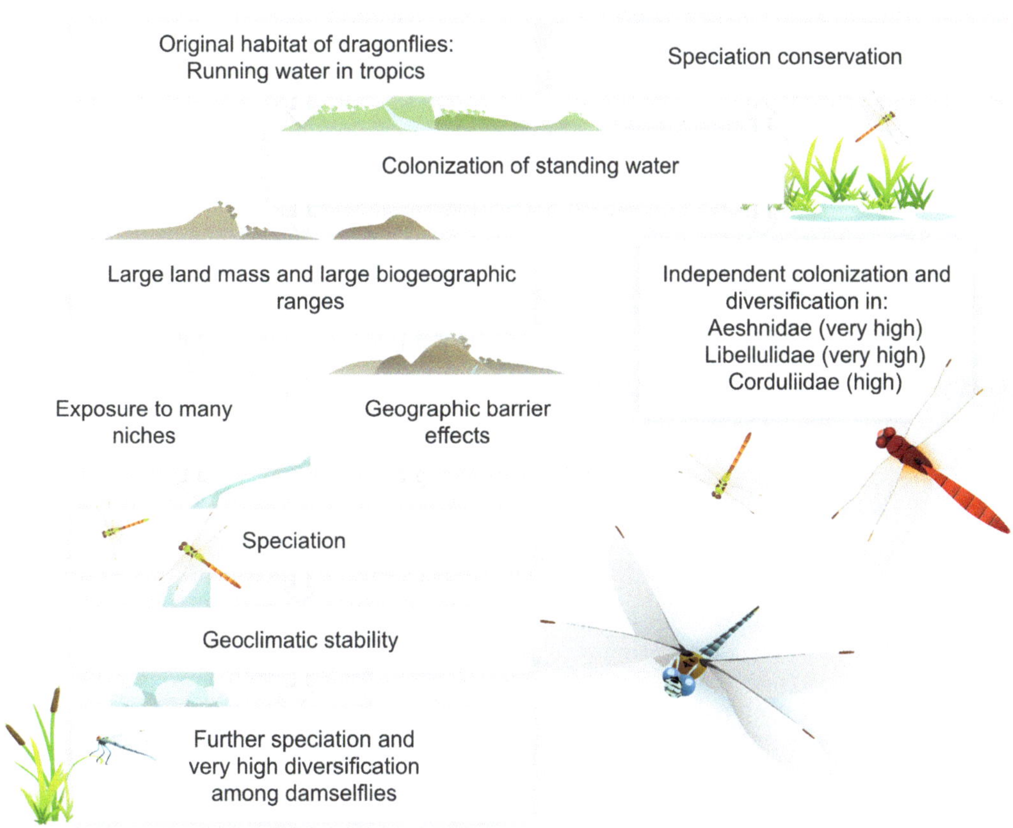

Fig. 3.7. Hypothesized early diversification drivers and opportunities for dragonflies. (Illustration: Robyn Symons).

Fig. 3.8. An inhabitant of tank bromeliads, *Mecistogaster modesta* (left) and a cluster of epiphytic bromeliad plants (right). (Photos: Bernard Dupont (Wikimedia Commons) (left), Leandro Juen (right)).

While the collective population of individuals making up each dragonfly species is adapted to a particular array of climatic conditions, it is at the local level where habitat occupancy is determined by the traits prevailing among local individuals (Chapter 4). In turn, the set of traits varies from area to area, a result of historical drivers. For example, some individuals may be more cold- or warm-tolerant than others according to latitude or elevation. There is also the question of the length of time that individuals of a particular species can tolerate certain adverse conditions. For example, some individuals may tolerate a cold spell, and others not. Yet the local population may not survive an excessively long extremely cold period. Another species may tolerate a drought, with even the temporary drying out of its habitat. But tolerance has a limit. If rains do not come, and the dry period is extended, the remaining individuals, whether as egg, larva or adult, will die off. There are also contingent issues. Extended drought may also change the plant community, and so change the habitat. The result can be a change, and even replacement, of the composition of the dragonfly assemblage. There can also be a change back again when extended rains bring back the earlier plant community, although this may take some years and assumes that individuals are close enough to re-establish.

North-eastern South Africa and Mozambique experience strong El Niño–Southern Oscillation (ENSO) events. This area has extreme weather associated with the oscillation, continually and regularly shifting local dragonfly geographic ranges (Samways, 2010). The various species in the area have traits enabling them to inhabit a wide range of elevations as well as certain special habitats alternately during wet and dry ENSO phases (Samways and Niba, 2010). Two species illustrate these extremes. The large and highly mobile *Orthetrum robustum* retreats to occupy only the few permanent, richly vegetated coastal lakes during droughts, returning to ephemeral pools and lakes, and even artificial ponds, during the wet phase. Sometimes, however, recolonization by a particular species does not occur, and geographic ranges become partly reduced (Fig. 3.9).

Aciagrion dondoense, which had never been previously recorded in South Africa, suddenly appeared there in the year 2000, following extensive and massive storms and flooding in Mozambique and north-eastern South Africa. At some sites, it even became the dominant species in some coastal shallow marshes in 2001. When the ENSO dry phase returned, it started to disappear, and was completely gone by 2006. Being a weak flyer that normally lives a sheltered life among rushes and sedges, this is

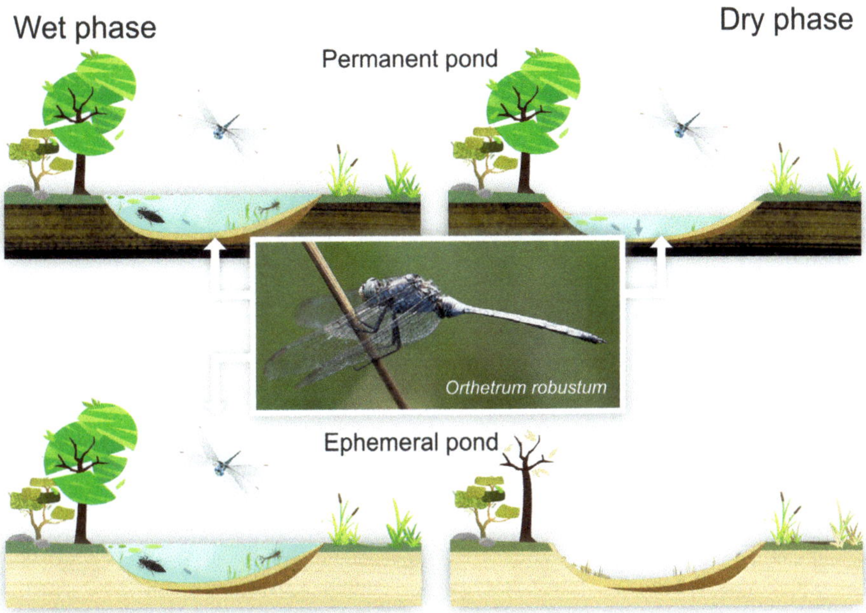

Fig. 3.9. El Niño–Southern Oscillations events result in wet and dry climatic phases. Permanent ponds retain water, albeit at a slightly lower level, when the dry phase follows the wet one. In contrast, shallow ponds may be ephemeral and lose their water during the dry phase. Some species, such as *Orthetrum robustum*, occupy both permanent and ephemeral ponds during the wet phase but retreat to occupy only the permanent ponds during the dry phase. (Photo: Michael Samways; Illustration: Robyn Symons).

an extraordinary phenomenon, with its ephemeral appearance in South Africa due to cyclonic winds, yet later it was extirpated by drought (Samways and Deacon, 2022) (Fig. 3.10).

3.2.5 Influence of glacial retreat

With the retreat of the northern hemisphere glaciers at the end of the Last Glacial Maximum (LGM) about 22,000 years ago, but with an intervening cool period of the Younger Dryas about 12,900–11,700 years ago, general global temperatures have increased. Glacial recession provided opportunities for the movement north of dragonfly species, though it is not a simple case of all dragonfly species' distributions shifting northward *en masse*. During the Pleistocene, there were species range expansion phases as conditions warmed, alternating with retraction phases as conditions cooled. Expansions from glacial refugia depended on three basic factors: climate, topography and species traits, especially thermal tolerance.

In Europe, the expansions were along the wide river valleys in geotectonic depressions, redirected by mountain chains and narrow river valleys (Sternberg, 1998). Furthermore, phylogenetic conservatism, manifested by attrition of phylogenetic diversity farther north, has occurred. This takes place in closely related species having similar thermal adaptations, where only certain lineages cope with the harsher and more variable conditions in the north. This likely has taken place in combination with differential dispersal and colonization capabilities, and together with phylogenetic conservatism, has left a distinct footprint on current distributions.

While *Aeshna cyanea* originated from a glacial North African refugium, dispersing to the Iberian Peninsula and then throughout the rest of Europe (Simonsen *et al.*, 2020), other species moved in the opposite direction. Large populations of European *Cordulegaster bidentata* and *C. boltonii* survived in diverse refugia during the ice ages, twice colonizing North Africa from the Iberian Peninsula, with one colonization by a common ancestor of *C. princeps* and *C. boltonii*, and more recently by *C. boltonii* (Froufe *et al.*, 2014).

In turn, species phylogenetic diversity and narrow-range endemism is greatest in formerly unglaciated

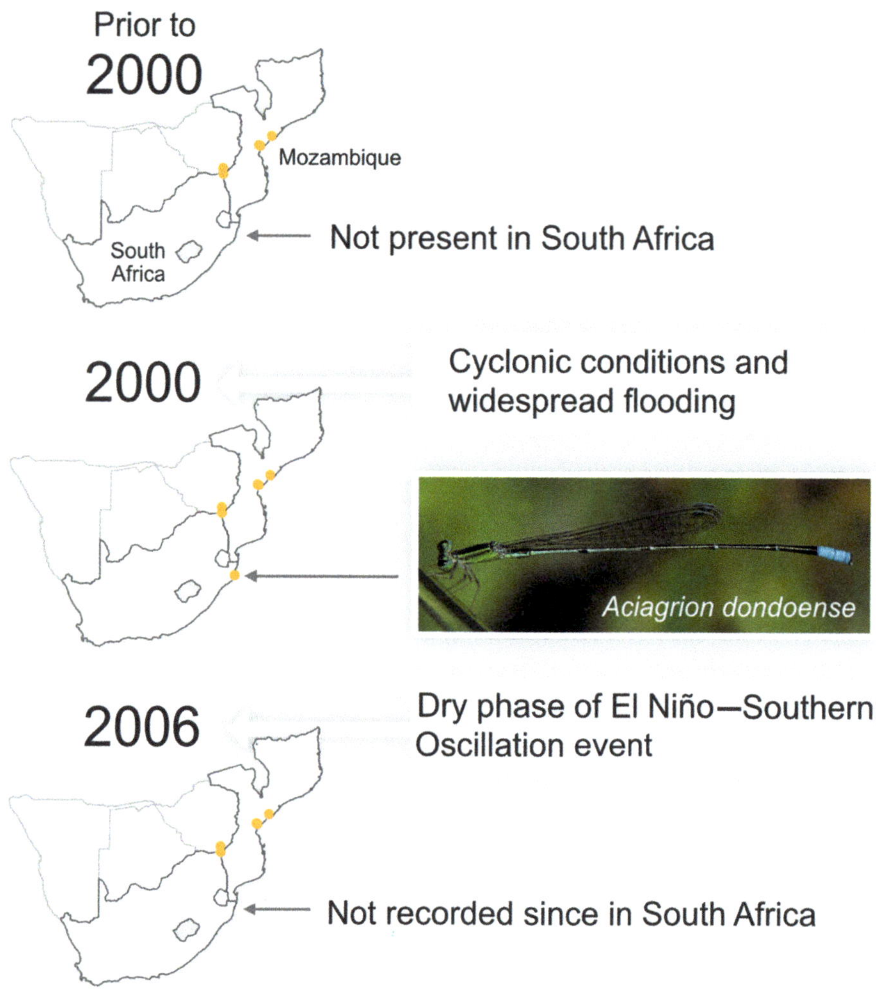

Fig. 3.10. *Aciagrion dondoense* had never been recorded in South Africa prior to the year 2000. That year, extreme cyclonic conditions and widespread flooding led to its appearance in the country, where it became the dominant species at some marshes in 2001. By the year 2006, and the dry phase of an El Niño–Southern Oscillation event had taken effect, the species disappeared completely, and has not been recorded in the country since. (Photo: Michael Samways; Illustration: Robyn Symons).

areas south of the LGM 0°C isotherm compared to the areas that were glaciated (Pinkert *et al.*, 2018). The northern distributions can be highly disjunctive in some species, as is seen today among the eastern Siberian dragonfly fauna (Kosterin, 2005). Nevertheless, dragonflies in previously glaciated areas tend to be relatively homogenous, being made up of closely related species. They are also predominantly standing water species, which have larger latitudinal spans, and more northerly range boundaries and distribution centres. This is the case in North America as well as Europe (Hof *et al.*, 2006).

There has also been a converse situation. Some cold-adapted species suffered range retractions in the post-LGM warming world. These species remained behind in some cold mountainous areas, where some of them still remain today. The anisozygopterans *Epiophlebia* spp. in Asia likely had a wide distribution during the last ice age when suitable cold-water stream habitats were more widely distributed.

However, during the subsequent warming phase, the ranges of the species in the genus contracted, resulting in their disjunctive distributions that we see today (Büsse *et al.*, 2012) (Fig. 3.11).

3.2.6 Glacial refugia

The effects of the ice ages were not restricted just to the north temperate zone, but also cooled many other areas of the world. The influence of cold temperatures on dragonflies in South-east Asia has resulted in their evolution and survival in many regional refugia (Heiser and Schmitt, 2013), also indirectly supporting the concept of a tropical forest origin for dragonflies (Paulson, 2006). The South-east Asian dragonfly faunal hotspots appear to be due to a combination of the effect of mountain ranges, moderate ice age climatic oscillations and location in the wet tropics. This has led to the dragonfly fauna being largely stationary in South-east Asia, especially for damselflies, which are more location- and island-specific than anisopterans (Heiser and Schmitt, 2013).

Two hotpots for dragonfly diversity in Mexico, Huasteca Potosina and Los Tuxtlas, are also tropical forested areas. Besides being species-rich and sharing many species, these hotspots are complementary through each supporting several species that are geographically range-restricted (González Soriano *et al.*, 2011).

In the western Palaearctic, the highest dragonfly species richness is in central Europe, a result of

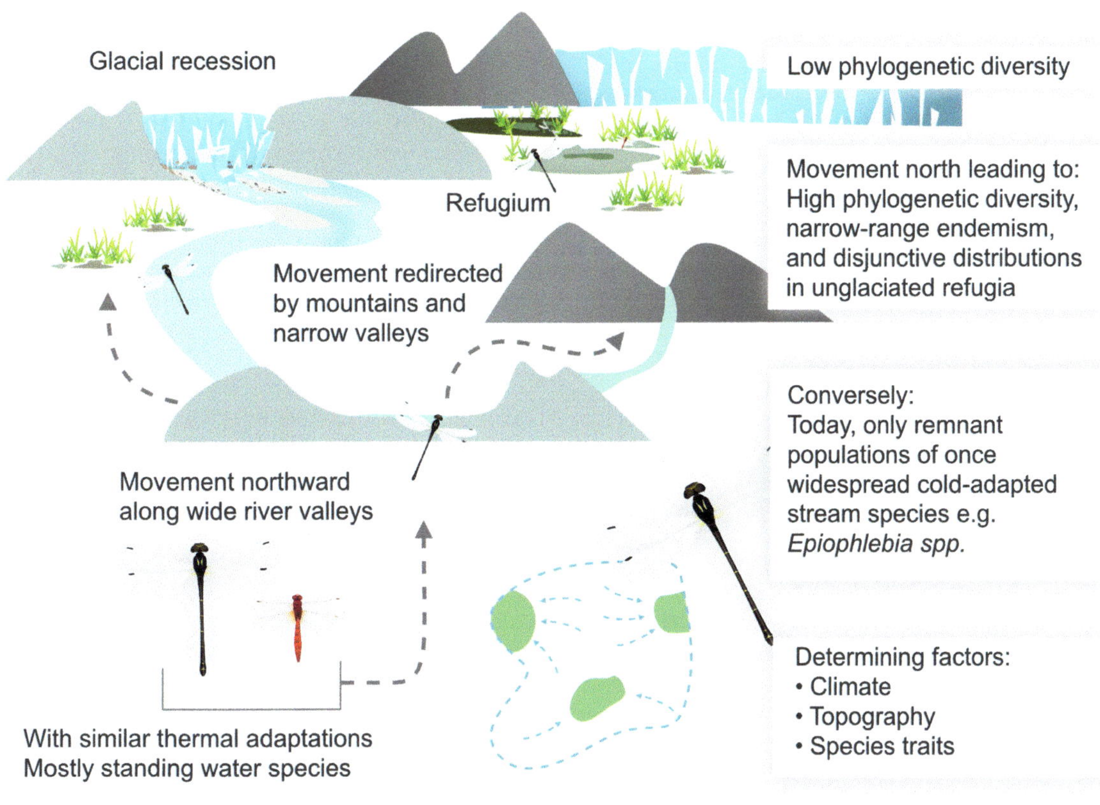

Fig. 3.11. Glacial advances and recessions have had a major direct effect on dragonflies in the north temperate zone. With glacial recession, the movement path northward by certain species with similar thermal adaptations, and largely associated with standing waters, has been largely determined by topographic features. Among the species arriving in the north is low phylogenetic diversity. However, phylogenetic diversity and endemism, with associated disjunctive distributions remain high in refugia that were not covered by ice. In turn, some cold-adapted stream species have declined in geographic range since the last glacial maximum. (Illustration: Robyn Symons).

intermixing of dragonflies of different origins and with varying distributional histories. Farther north is species poorer, and represented only by postglacial colonizers, the western Palaearctic endemic *Cordulegaster boltonii* being the exception. However, in Fennoscandia and the British–Irish Isles, sea transgressions have led to some distinct faunas. Furthermore, the Fennoscandian dragonfly fauna is richer in the south of the area, while the northern species also have larger geographic ranges (Korkeamäki *et al.*, 2018).

The Mediterranean dragonfly fauna is not as species-rich as central Europe, and poorer in the west than the east. However, there are clear differences between species richness and endemism. The western Mediterranean is biogeographically distinct, and has the highest number of local endemics, in both Iberia and the northern African Maghreb. The Strait of Gibraltar has been a major barrier to zygopterans, but not for anisopterans with their stronger flight capability (Heiser and Schmitt, 2010). The eastern Mediterranean is poorer in endemics. Nevertheless, the Balkan Peninsula is species-rich, and likely to have been an important centre of postglacial expansion. Today, in terms of the dragonfly fauna, it is one of the few places where the Alpine biogeographical region meets the Mediterranean region (De Knijf *et al.*, 2013), made possible in part by a wide variety of topographies and ecosystems (Kulijer *et al.*, 2013).

Across the Palaearctic region (excluding China and the Himalayan region), odonates representing all habitat types show a clear pattern of decreasing species diversity longitudinally, with species richness decreasing in the eastern half of Europe and remaining low throughout much of Russia, then increasing again toward Russia's Far East and Korea. However, there are differences when species of standing waters are compared with those of running waters. Standing water species dominate the colder and more arid areas. Geographic areas with high numbers of threatened species are largely restricted to the Mediterranean, south-western Asia and Japan, with distinct hotspots in the Levant and the southern half of Japan. Diversity of species, genera and families is higher in southern Japan than in areas at a similar latitude in the western Mediterranean. This is likely due to the more humid climate of Japan, supporting a higher diversity of freshwater habitats, as well as the greater impact of glacial events in the western Palaearctic in combination with the dry Sahara restricting tropical African species from dispersing northward (Kalkman *et al.*, 2022).

The extensive grasslands of Canada support a postglacial cold-adapted set of dragonfly species, largely with wide biogeographic distributions (Cannings, 2014). However, north-eastern USA has a current species-rich dragonfly fauna at the crossroads of ancient and recent odonate evolution, made up of separate boreal, temperate and tropical elements. The area has been both a conduit and a generator of odonate diversity, with the ancient Appalachian Mountains being an important centre for freshwater diversity (Corser *et al.*, 2014). However, there are different historical biogeographic patterns between Zygoptera and Anisoptera. The zygopteran species are mainly of Neotropical origin from about 60 mya, with post-Eocene tropical conservatism. However, there was also a surge of Pliocene–Pleistocene (20–5 mya) speciation in some groups, such as *Lestes* and *Enallagma*, as evidenced by different morphologies among subspecies, hybrid zones, genetic bottlenecks, and range retraction and expansion, all emanating from repeated glacial cycles over the last 100,000 years (Donnelly 2003, 2004; Turgeon *et al.*, 2005). However, among them are also some subtle differences in habitat preference, e.g. *Lestes disjunctus* and *L. forcipatus* in Canada (Cannings and Simaika, 2005).

The Anisoptera in the North American region is composed of crown endemic taxa, as well as ancient taxa that show temperate conservatism, having occupied refuges in the Appalachian Mountains dating back more than 180 mya (Carle and Cook, 1984; Carle, 1995; Corser *et al.*, 2014). Since the LGM, there have been northward range expansions of North American odonates, especially associated with protective forested environments, leading to colonization of northern latitudes (Cannings and Cannings, 1994; Corser, 2010), and the disassembly and reassembly of glacially generated genotypes (McPeek and Gavrilets, 2006). However, not all species have established permanently. The North American *Rhionaeschna mutata* rapidly expanded its range north-westward in 2009 but did not establish permanently, possibly due to fish predation (Schilling *et al.*, 2019).

Across the entire Nearctic region (Canada, USA, Mexico) there is greater odonate species richness east of the Rocky Mountains, owing to high aquatic habitat diversity at micro- and macroscales, promoting niche partitioning and specialization. The south-eastern USA has the highest number of localized endemic species, likely due to glacial refuges providing opportunities for the evolution of a rich and unique biota (Abbott *et al.*, 2022) (Fig. 3.12).

In turn, certain species have isolated populations in cold areas, where the species have colonized unusually warm spot locations. In north-east Asia, warm geothermal springs have enabled four species (*Mnais costalis*, *Anotogaster sieboldii*, *Orthetrum melania* and *O. albistylum*) to colonize localities beyond their general range margins, but whether these populations are relicts from warmer times, or are the result of recent colonization is currently unknown (Aksenova *et al.*, 2020). In Canada, *Octogomphus specularis* specializes in breeding specifically in stream outlets of lakes, which are several degrees warmer than the inflow streams. This enables the species to occupy habitats at the far northern part of its geographic range (Chalmers *et al.*, 2023).

3.2.7 Types of islands and dragonfly occupation

Islands are areas of elevated land rising above the matrix. In the case of oceanic islands, the matrix is the sea, whereas for sky islands (mountains), the matrix is largely a different terrestrial ecosystem. Islands play a major role in conservation, as they are often areas where relict or isolated dragonfly populations coincide with human activity.

Islands are also significant at the small spatial scale. Landscapes often have a particular ecosystem surrounded by a different, more extensive, ecosystem. These landscape islands may, for example, be a pond or marsh in a grassland matrix, a waterfall with its deposition zone along a tropical forest river, a spring in the desert or a clump of trees that provide shelter from harsh surrounding environmental conditions.

Islands at the large spatial scale are particularly significant from an evolutionary point of view, and besides being refugia, are often generators of novel species through isolation and evolution over time.

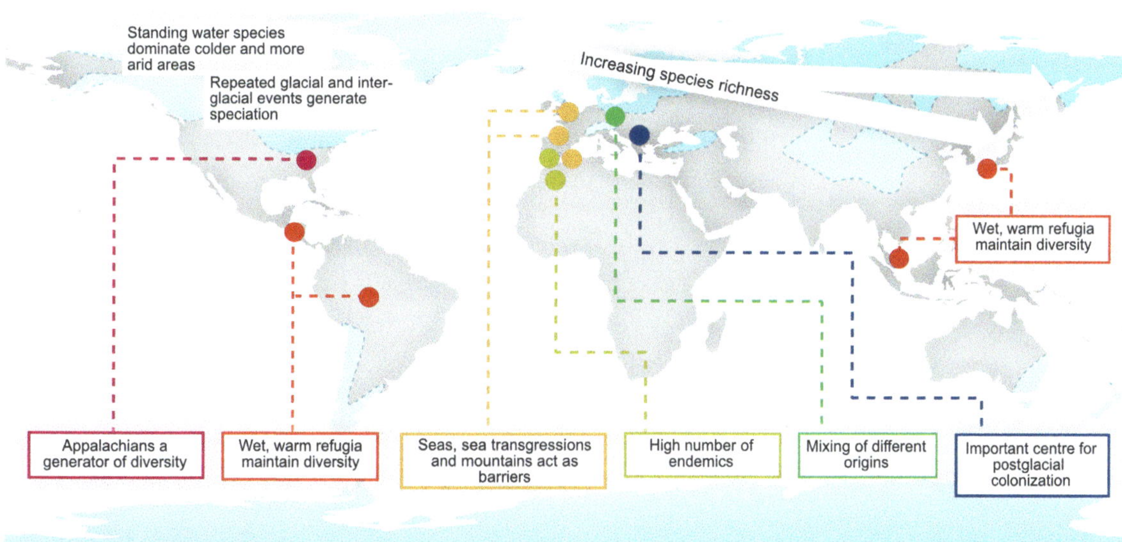

Fig. 3.12. Periodic advances and retreats of glacial sheets have left a profound legacy on biota across the world. While at their maximum, as shown here for the last ice age, large glacial ice sheets caused global cooling. Nevertheless, relatively wet, warm conditions at lower latitudes became major refugia during maximal ice coverage. The return of warm conditions triggered range expansion among many species, although movement was greatly influenced by barriers such as seas, sea transgressions and mountain ranges. Overall, dragonfly response and speciation were the result of the interacting effect of alternate cooling and warming events combined with geographic barriers. Speciation co-occurred with movement, with some areas being important centres for postglacial colonization of increasingly ice-free areas. However, not all species expanded their geographic ranges, leading to localized endemism. Standing water species dominated the cooler and more arid areas. (Illustration: Robyn Symons).

In contrast, island patches at the small spatial scale are important ecologically and behaviourally for local species, such as *Zygonyx natalensis* breeding in waterfalls. These patches may be significant habitats, such as springs for geographically widespread species, like *Trithemis kirbyi*, in the Namibian desert (Suhling *et al.*, 2009). They may also be a necessary component of certain species' habitat. Clumps of trees on the edge of marshland in savanna Africa, for example, are retreats for tenerals, immature adults and mature adults of species like *Lestes uncifer.* Tree clumps also provide cool and shady retreats for normally open-biotope species, such as *Trithemis annulata*. On exceptionally hot days, individuals of *Pseudagrion coeleste umsingaziense* may occur in their dozens in the shade of one large, isolated tree next to their typical grassy pond habitat (Fig. 3.13).

Fig. 3.13. At the large spatial scale there are essentially two island types: oceanic islands surrounded by sea, and prominent sky islands surrounded by flat land. At the small spatial scale, there are various island types. For example, standing water bodies can act as refuges for certain species. These may be temporary or permanent, with some species, such as many *Lestes* species, specializing in locating in these areas: for example, *Lestes uncifer* (left) requiring pools surrounded by bushes in a hot and dry matrix. Waterfalls are islands along a river, with some strong fliers, such as *Zyngonyx natalensis* (centre left), breeding in them. Springs in the desert can attract and provide a home to certain wandering dragonflies, such as *Trithemis kirbyi* (centre right). Tree clumps can be an important local island, especially in open landscapes, where adults can find shade when air temperatures are very high, as in the case of *Trithemis annulata* and *Pseudagrion coeleste umsingaziense* (right). (Photos: Michael Samways; Illustration: Robyn Symons).

3.2.8 Oceanic islands as refugia and species generators

Oceanic islands as ecological entities are defined in terms of their timelines. An oceanic island today may have been of a very different size in the past when sea levels were either higher or lower. Smaller islands can disappear as they drown with rising sea levels. Others can explode, as did volcanic Anak Krakatoa, Indonesia in 2018.

About 20,000 years ago, the central granitic Seychelles were one large island when sea level was 130 m lower than today, whereas now it is an archipelago of several small islands. Being a remnant of Gondwanaland, yet mountainous and covered by tropical forest, these central islands support some island endemic dragonfly species now confined to what is in effect the mountain peaks of the LGM (Samways, 2003) (Fig. 3.14).

Other oceanic islands are the result of volcanic eruptions in the sea. As volcanoes rise in elevation over a long time, they become crucibles of evolutionary novelty following the gradual arrival of

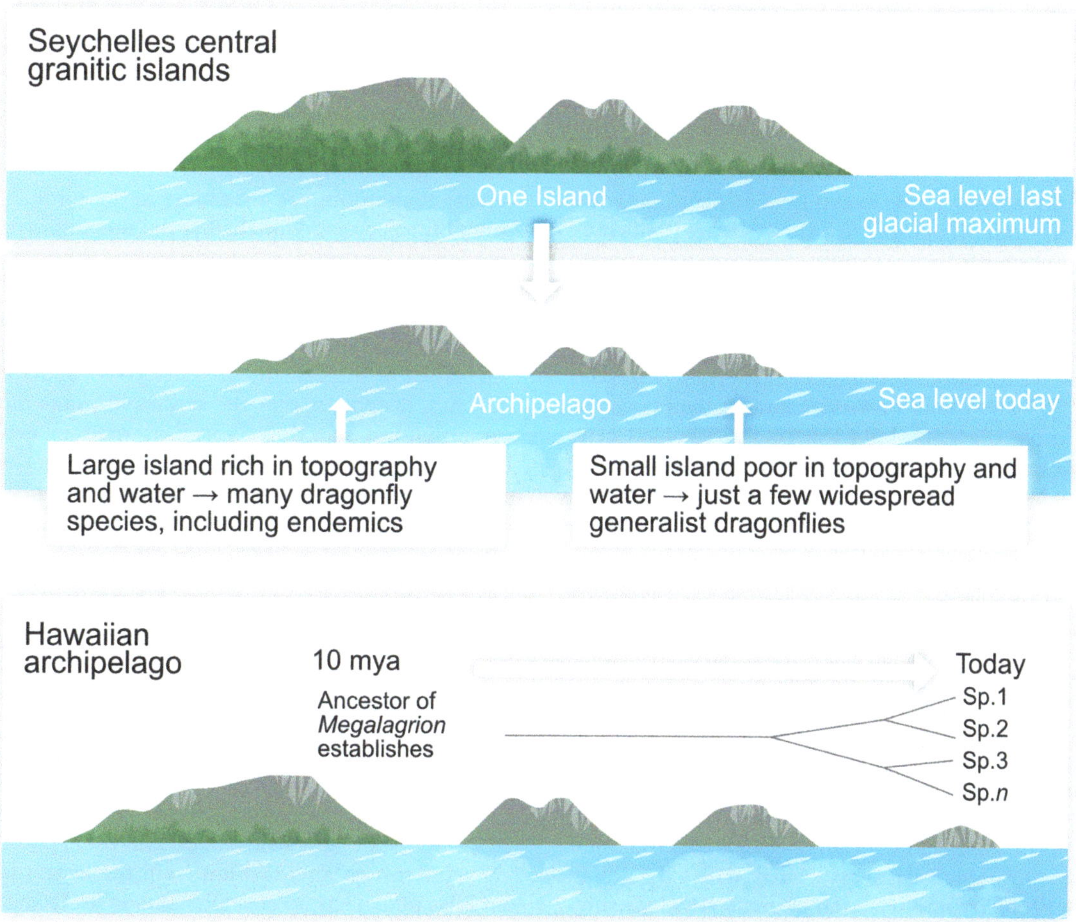

Fig. 3.14. Islands are dynamic entities that change in appearance and extent over time. The central granitic islands of the Seychelles were one island during the last glacial maximum, while today and after a rise in sea level since then, they are now a group of small islands. They differ in their dragonfly fauna mainly due to a combination of island size and topography. The Hawaiian Islands are volcanic, with a long geological history. A founder population of an ancestor of the current *Megalagrion* established itself about 10 mya, and then evolved over time into several species, each with considerable habitat specialization. (Illustration: Robyn Symons).

species from a mainland source. Not all species that arrive are able to establish, as there may not be suitable receiving habitats. Then there is also the founder effect to consider. When a species establishes on an island, it usually does so in only very low numbers and with a reduced level of genetic diversity. This sets the trajectory for evolutionary novelty.

Dragonflies that do establish on new islands have less opportunity to move far and find optimal conditions compared to the mainland. This inevitably means that selection pressure and adaptation to small and localized conditions is strong, while only being equipped with a small genetic pool. From this genetically humble start, there is then fine-scale local adaptation over time as individuals undergo natural selection to thrive under the various island-typical local conditions. This natural selection process is then accompanied by sexual selection, further contributing to species divergence.

Hawai'i is a prime example of ancient damselfly diversification among *Megalagrion* species (Polhemus and Asquith,1996; Daigle, 2000). The founder population, which established about 10 mya, over time, evolved into several new species, which also involved some hybridization. The changes in the Pleistocene glacial stages led alternately to land bridges and island formation among the Hawaiian volcanic island peaks. Land bridges enabled gene flow, while higher sea levels and associated oceanic island formation reduced gene flow, and led to associated founder effects and genetic bottlenecks, as well as reduced habitat availability. Trade winds funnelled between the volcanoes may have aided dispersal to some islands (Jones and Jordan 2015).

This evolutionary process on the Hawaiian archipelago led to some extreme habitat specialization among the endemic genus *Megalagrion*, with its 24 species. The larval habitats of the various species include living in rain-fed leaf axils, among leaf litter under a particular fern, and in certain types of seeps, as well as in fast streams (Jordan *et al.*, 2003, 2005). The fine-scale microhabitat specialization is also associated with changes in eye geometry and selection of great visual sensitivity (Scales and Butler, 2016). In turn, extent of light exposure (degree of canopy cover) has also played a role, with red eyes being more predominant in open-canopy local environments. Interestingly, sexual colour dimorphism among *Megalagrion* species is not just the result of sexual selection, but also the effects of separate and strong ecological microhabitat selection by the two sexes (Cooper *et al.*, 2016) (Fig. 3.15).

Not all oceanic islands are the stage for dragonfly genetic divergence. For example, *Euphaea formosa* shows very little genetic diversity across Taiwan despite an abundance of habitat and considerable topographic variation, possibly due to genetic bottlenecks in the past (Kanke *et al.*, 2021).

Sometimes, we can glimpse genetic change taking place on remote islands. Although *Pantala flavescens* is the ultimate migrant with panmictic populations (Troast *et al.*, 2016; Low *et al.*, 2017), and the most compact mitogenome of all anisopterans (David *et al.*, 2021), it is the only dragonfly species resident on the remote Pacific volcanic island of Rapa Nui (Easter Island) (164 km^2), 3512 km from the South American mainland. The individuals there have small, more asymmetrical and darker wings, and they differ behaviourally by flying close to the ground. These morphological and behavioural adaptations are a response to the confines of island life. The individuals that did not conform to life within the island's boundaries and flew out to sea, likely perished (Samways and Osborn, 1998). Males also have longer abdomens, which may facilitate adoption of the tandem position and increase the success and frequency of copulation. Overall, it appears that the species has undergone some natural selection as well as random evolutionary processes, including founder effects and genetic drift, leading to morphological and behavioural changes, modifying its migratory disposition to that of an island resident (Alvial *et al.*, 2019).

The volcanic Cape Verde Islands (total area: 4033 km^2) are relatively arid and positioned in the mid-Atlantic, about 600–850 km from the African mainland. Despite several million years of history, there are no endemic dragonfly species there. Nevertheless, the islands have attracted 14 dragonfly species, mostly anisopterans, all of which are far ranging species typical of arid parts of the African mainland (Aistleitner *et al.*, 2008). Similarly, Puerto Rico (9104 km^2), which is 3500 km from mainland North America and 380 km from the large island of Hispaniola, has no endemics among its 48 species on this wet island. However, the island was severely degraded by agricultural activities in earlier decades, although today it has now regained much of its forest. It is possible that there were island endemics which were lost in the earlier agriculturally intensive period, although we shall never really know (Ramírez *et al.*, 2020).

Fig. 3.15. The ancient origin of the volcanic islands of Hawai'i has led to diversification within both species and the genus *Megalagrion*. Some, like *M. nigrohamatum nigrohamatum* (top left) and *M. nigrohamatum nigrolineatum* (top right) are listed as Least Concern on the IUCN Red List, while others are highly threatened, like the Endangered *M. pacificum* (bottom left) and the Critically Endangered *M. nesiotes* (bottom right). (Photos: Karl Magnacca (top left), Bill Haines (top right, bottom left), Dan Polhemus/public domain of United States Fish and Wildlife Service (bottom right)).

Cyprus (9251 km^2) is also an island with no endemic dragonfly species, but 37 species overall. This compares with similar-sized Crete (8450 km^2, 160 km from mainland), with two endemic odonates. It seems that Cyprus, being only 70 km from the mainland, may be subject to regular genetic exchange with colonizing individuals, lessening the chances of incipient evolution (Sparrow *et al.*, 2021, 2022). This point is emphasized by *Orthetrum ransonnetii* which recently established on the island of Fuerteventura (1660 km^2), in the Canary Islands, about 100 km from the African mainland (Nowak and Weihrauch, 2022).

The Canary Islands, along with the Azores, Madeira, the Savage Islands and the Cape Verde Islands, make up an ancient volcanic island group, Macaronesia. It is relatively poor in odonate species, with about 23 species (Weihrauch *et al.*, 2016). However, it is an interesting island group, as it is a mixture of one endemic species (*Sympetrum nigrifemur*) with the other species having arrived from three continents (Africa, Europe and North America) (Weihrauch, 2011).

The ancient volcanic Galapagos Islands (7880 km^2) in the Pacific Ocean and about 1000 km from the South American mainland have ten odonate species, with two endemics (*Rhionaeschna galapagoensis* and *Tramea darwini*) (Lorenzo-Carballa *et al.*, 2021). This archipelago is dry today but may in the past have endured droughts.

Ischnura hastata, normally considered a North American species, is abundant in the mid-Atlantic Azores archipelago (total area: 2346 km^2), about 3900 km away. The population there is composed of only females. As no genetic recombination is apparent, its arrival in the archipelago is likely to have been one event. It reproduces by thelytokous parthenogenesis (females produce only female offspring), which prevents formation of lethal recessive mutations, and has seemingly led to the species overcoming the negative effects that would have arisen from lack of sexual reproduction (Lorenzo-Carballa and Cordero-Rivera, 2009) (Fig. 3.16).

The small archipelago of Malta (316 km^2) is 288 km from the African mainland and 191 km from Sicily. It has recorded about 16 odonate species.

Fig. 3.16. Species from the Canary Islands: the endemic *Sympetrum nigrifemur* (left), and a female of *Ischnura hastata* (right). (Photos: Adolfo Cordero-Rivera).

Three have declined in recent years (*Orthetrum cancellatum*, *O. brunneum* and *Sympetrum striolatum*) while three pan-African species (*Orthetrum trinacria*, *Trithemis annulata* and *T. arteriosa*) have now established in the archipelago (Ebejer *et al.*, 2008; Balzan, 2012). Malta may be a stepping-stone area for range extension northward.

The dragonfly fauna along the Croatian coast displays some interesting trends among the islands within a few kilometres of the mainland. While the largest island (Krk: 406 km^2) combined with a wide variety of habitat types supports the most species (41 in total), one of the small islands (Veliki Brijun: 5.7 km^2) still supports 14 species, stemming from it having diverse water microhabitats and well-developed aquatic vegetation. In turn, another large island (Cres: also 406 km^2) only supports 16 species, and a moderate-sized island (Lošinj: 74 km^2) supports only two species, a manifestation of low levels of suitable habitat. The overall effect is widely differing species richness per square kilometre among the four islands: Veliki Brijun (2.44), Krk (0.1), Cres (0.04) and Lošinj (0.03). Furthermore, the presence of suitable habitat can be attractive to certain rare species such as *Anax parthenope* on Veliki Brijun island (Štih and Koren, 2022).

Santa Catarina island (424 km^2) is close enough to the southern Brazilian mainland to be connected by constructed bridges. Even so, it displays the typical insular effect of a reduced number of dragonfly genera and species compared to the mainland. In streams on the island, substrate (organic and inorganic) and mesohabitat (riffle and pool) types are important in driving the species composition of the larvae while indicating the role of climate and habitat structure in influencing the local subtropical stream odonate assemblages (Pires *et al.*, 2020).

Islands of coral origin are largely home to only widespread species. This is because these cays and atolls are of geologically recent origin, are flat, usually small and have little, often only temporary, freshwater. In the Indian Ocean, they are less species-rich than other small islands of granitic origin and of comparative size (Samways *et al.*, 2011). The dynamic nature of widespread, vagile species is seen in the coralline Chagos Archipelago in the central Indian Ocean. While there is a core of ten species on the largest island (Diego Garcia), there are some new records of *Rhyothemis variegata* and *Anax ephippiger* arriving at the archipelago, as well as range expansions of five species (*Anax guttatus*, *Diplacodes trivialis*, *Macrodiplax cora*, *Tramea limbata* and *Zyxomma petiolatum*) within the archipelago of 55 islands. *Pantala flavescens* is a regular visitor (Carr, 2022). Species in the genus *Hemicordulia*, originating from East Asia, have speciated across some of the wetter islands in the Indian Ocean, while also appearing on the African mainland. The agent of movement appears to be prevailing westerly winds (Dijkstra, 2007).

The Gulf of Guinea islands of Príncipe (128 km^2), São Tomé (836 km^2) and Annobón (17 km^2) lie about 300 km off the central African mainland and arose from volcanic activity 31–4.8 mya. To date, 22 named species have been recorded from the islands, with one endemic, *Trithemis nigra*, on Príncipe. While further collecting may reveal more species, including the possibility of two more suspected endemics, the dragonfly fauna is remarkably poor (Dijkstra *et al.*, 2022). The reasons for this odonate faunal paucity are unclear, but past extreme climatic events, especially droughts, may be one reason.

In summary, there are clearly identifiable features associated with the dragonfly fauna of relatively small oceanic islands. Remoteness from the mainland, along with age, topography and local climate, are highly influential factors that determine the number and type of dragonfly species present. For example, the ancient yet remote mountainous Hawaiian Islands are rich in endemic species within the genus *Megalagrion*, whereas the extremely isolated Rapa Nui has only one widespread species (*Pantala flavescens*), which is showing incipient evolution. The low-lying coralline Chagos Archipelago (which sits on the submerged peaks of volcanoes) has only a few widespread species. Islands near the mainland, such as those in the Mediterranean Sea, and which have a similar geology to the mainland, essentially have only a subset of the species that occur on the mainland, with the occasional island endemic species, as on Crete. While size of the island or island group plays a role, it is the presence of consistently available freshwater that is highly significant in the long term (Fig. 3.17).

In the western Indian Ocean, dragonfly diversity increases with oceanic island size (Samways *et al.*, 2011). This classic species-area effect stems from a greater number of habitat types associated with

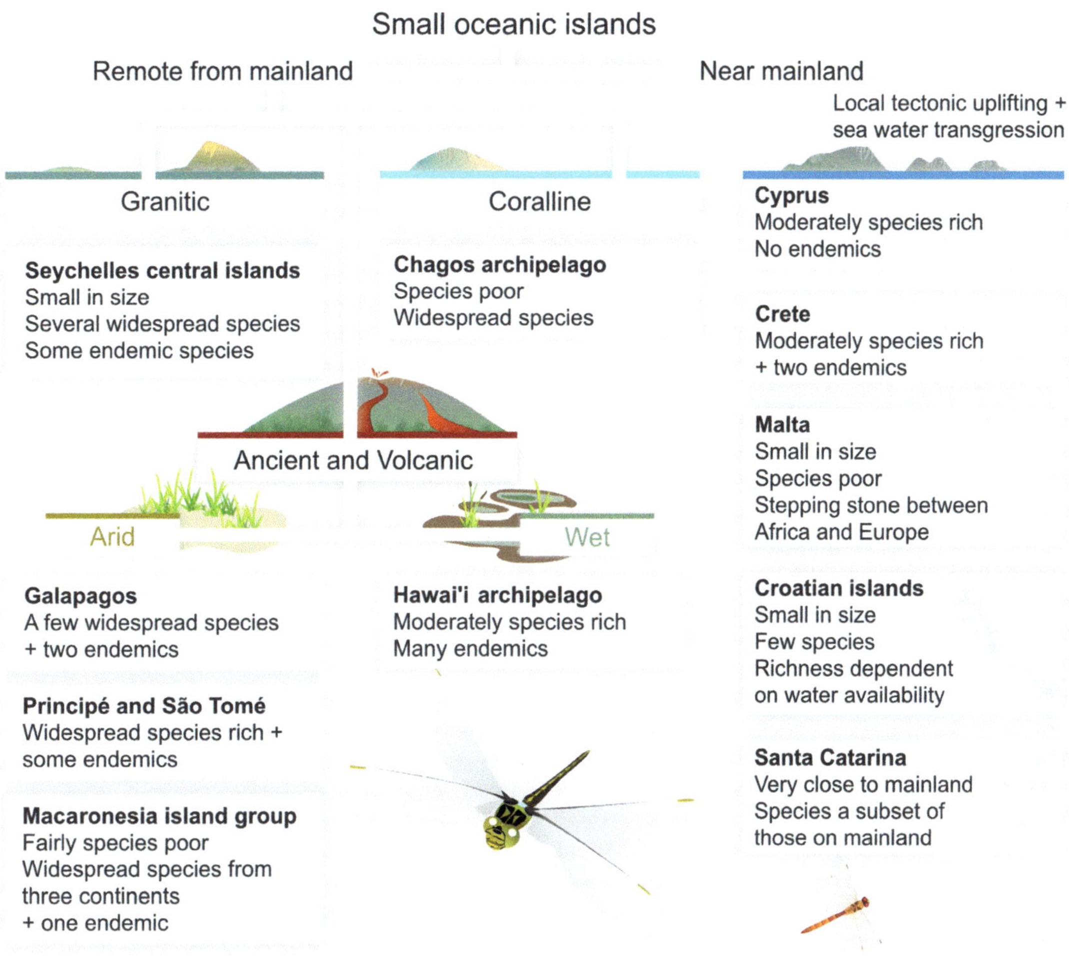

Fig. 3.17. A summary of the main geographic and environmental variables shaping dragonfly assemblages on relatively small oceanic islands. (Illustration: Robyn Symons).

more predictable water provision and varied topography on the largest islands. The smallest islands of the Seychelles, whether granitic or coralline, are low-lying and have scarce watery habitats. The most species-rich Seychelles islands (Mahé, Praslin, Silhouette) are granitic but with significantly high elevations enabling feeder streams to flow to some degree all year round. This contrasts with the large Aldabra atoll and the Chagos Archipelago which are low-lying and water poor, supporting a relatively poor dragonfly assemblage.

Across the sea from the Seychelles is the African mainland, and the sky islands of the Cape Floristic Region. These sky islands are overall more species-rich than the oceanic islands, due in large part to perennial springs and feeder streams, especially on the wetter slopes. The exception is the Cape Peninsula, and when seen on its own, it is poorer in species than the other, more connected sky island, the Cape Fold Mountains. This is a classical peninsula effect, where a peninsula is having effects on the fauna somewhere between a mainland and an island.

A larger area compared to a smaller one provides the platform for more genetic rescue effects. This occurs especially through tourist individuals providing new alleles, driven in part by more landscape heterogeneity at various spatial scales in both horizontal and vertical dimensions. On small islands, these environments are less prevalent, while also being susceptible to extreme weather events. Furthermore, small islands, with their limited resources, are also low in plant productivity compared to mainland areas of similar size. This is due in part to being restricted by the sea, which takes away some interdependencies that maintain the diversity, productivity and competitiveness present on the mainland (Leigh Jr. *et al.*, 2009).

It seems that larger oceanic islands with tropical or subtropical climates, like Sri Lanka and Madagascar, with their ancient history and complex topography, have slightly higher species richness than predicted, and perform more like 'small mainlands', rather than large versions of smaller islands, a manifestation of opportunities provided by not only the species-area effect but also by a very wide range of niches available (Samways *et al.*, 2011) (Fig. 3.18).

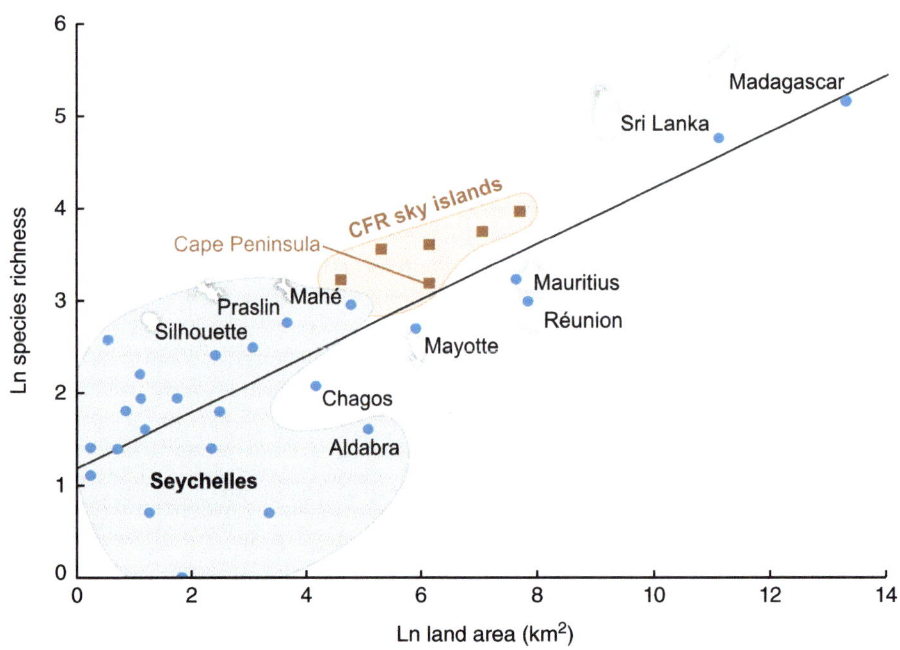

Fig. 3.18. Species-area effects on the dragonfly faunas on oceanic islands in the western Indian Ocean and sky islands on the African mainland. CFR = Cape Floristic Region. (Illustration: Robyn Symons).

3.2.9 Dragonfly survival on sky islands

Where warm sky islands are above the reach of the Quaternary sea-level transgressions, they provide enough time and varying local climates to generate considerable local diversity. Well-vegetated tropical mountain streams are strong determinants of macroinvertebrate beta diversity, often associated with high oxygen levels (Callisto *et al.*, 2021).

During the Pleistocene, sea levels in southern Africa varied overall between 120 m lower and 11 m higher than they are today. This has resulted in some coastal mountains being alternately connected and disconnected. The Cape Floristic Region's ancient Cape Fold Mountains, that uplifted and folded about 250 mya, are rich in localized endemic dragonflies. Among them is the genus *Syncordulia*. The four extant species in the genus diverged between about 60–45 mya (Ware *et al.*, 2009). The interesting point is that these species have survived not only the Eocene maximum temperatures but also the regular temperature oscillations of the Pleistocene, including the glacial maxima in relatively recent geological time. One species, *Syncordulia gracilis*, is the most ancient species of all, with a pedigree dating back 59.57 mya. At some time in the distant past, it dispersed to other mountain ranges, suggesting that this species, as well as the others in the genus, have survived for so long by making use of suitable habitats that span a wide range of elevations, and so overcoming sea transgressions and regressions (Samways and Deacon, 2022).

These *Syncordulia* species, as well as many other localized endemic species, notably *Chlorolestes* and *Ecchlorolestes* species, are all inhabitants of mountain streams – stream habitats with deposition pools present across a wide range of elevations (Simaika *et al.*, 2020) (Fig. 3.19).

The 1000 km^2 Kogelberg Biosphere Reserve (KBR) in the heart of the Cape Floristic Region supports about 1880 plant species of which 77 species are endemic to the reserve. The KBR also supports 38 dragonfly species, of which 26% are national endemics. However, what is extraordinary is that 53% of the total recorded individuals are members of regional and even local endemic species, compared to 47% being geographically widespread species. This emphasizes how montane refugia can support high diversity of local species with geographically small but effective populations, whether plants or dragonflies, a result of continuous provision of suitable habitats over a very long time. However, congruent isolation of the plant communities and dragonfly assemblages is more indirect rather than directly causal, albeit that the plants are oviposition sites for many of the locally endemic species (Grant and Samways, 2007) (Fig. 3.20).

3.2.10 Dragonfly assemblage turnover in response to barriers and drivers

High levels of endemism can also occur in some areas with only modest topography. In the Amazon basin, the various dragonfly assemblages are maintained by both environmental and biogeographic factors. Drivers include local regions, annual mean temperature, annual rainfall, and habitat integrity and intactness, while large rivers act as natural dispersal barriers (Alves-Martins *et al.*, 2019). The poorer-dispersing Amazonian damselflies show higher levels of localization across the interfluves (the areas between the major rivers) than the more mobile anisopterans. The Guiana interfluve in the north of the area has the highest level of endemism, supporting 66 recorded endemic odonate species, due to the legacy of past environments as well as current ones (Juen and De Marco Jr, 2012).

There are other legacies associated with relatively flat areas that have stimulated species turnover. In the warm flat pampas of Brazil, the topographically homogenous environment has different dragonfly assemblages, seemingly driven by underlying geology (Renner *et al.*, 2019).

It is interesting to take a metacommunity perspective. A metacommunity is a set of local communities connected by species dispersal. This approach uncovers geographic patterns and determines the extent to which environmental gradients shape the occurrence and abundance of the component species. As mentioned above, temperature, rainfall, and habitat integrity and intactness drive species turnover at the large spatial scale across the Amazon region. This relatively gradual turnover is known as a Clementsian pattern of species distribution, where whole assemblages show turnover along particular environmental gradients (Alves-Martins *et al.*, 2019). In British Columbia, Canada, anisopterans follow a similar pattern, but following a temperature and latitude gradient, while being relatively insensitive to types of vegetation cover, unlike the situation in the tropical Amazon (Cerini *et al.*, 2021).

In South Africa, biological communities have been shaped by both lack of glaciations over the

Fig. 3.19. Extremely localized endemism and habitat specialization occurs among the ancient *Syncordulia*, *Chlorolestes* and *Ecchlorolestes* species in the Cape Fold Mountains at the southern tip of Africa. *Syncordulia gracilis* (top left), *S. legator* (top right), *S. serendipator* (centre left), *S. venator* (centre right), *Chlorolestes conspicuus* (bottom left) and *Ecchlorolestes nylephtha* (bottom right). (Photos: Michael Samways).

Fig. 3.20. The Kogelberg Biosphere Reserve (left) in the Cape Floristic Region has extremely high levels of local endemism among many taxa, including dragonflies, like *Ecchlorolestes peringueyi* (right). (Photos: Michael Samways).

last 250 million years and by sea currents, with the warm Agulhas current running southward down the east coast and the cold Benguela current running northward up the west coast. Dragonfly species-diversity patterns in the region are complex, with strong latitude and longitude gradients, patterns of endemism and assemblage composition. Inevitably, climate patterns are important, especially average seasonal rainfall, average ambient temperature and average seasonal solar radiation. Underlying geology also plays a role, especially soil type and drain rate. Nevertheless, dragonfly assemblage patterns are gradual rather than sharp and strongly driven by a topography which shapes local climatic patterns and has led to patterns of species richness and endemism being weakly concordant. In turn, aridity reduces both richness and endemism, with vegetation cover and water permanence being strong determinants of local distributional patterns (Deacon *et al.*, 2020). Interestingly, in the same region, there is a direct connection between climatic regime and overall wetland species diversity. High rainfall and water volume across landscapes makes for more groundwater, interflow and/or surface water. In turn, this leads to larger numbers of wetland habitats containing steeper environmental gradients and more microhabitats for wetland plants, leading to higher aquatic species diversity (Sieben *et al.*, 2021) (Fig. 3.21).

Wet tropics mountain ranges, even those of more recent origin than the current Mediterranean-type climate of the Cape Floristic Region, are particularly rich in localized endemic dragonfly species, both zygopterans and anisopterans. This endemic richness has come about because these well-forested regions provide a huge range of microhabitats, with catchments often with their own characteristics and being relatively isolated from others according to species dispersal capacity. This pattern is also apparent in the well-watered mountain ranges of tropical South America (Paulson, 2006; Vivas-Santeliz and De Marmels, 2017; Bota-Sierra *et al.*, 2018), Angola (Kipping *et al.*, 2023), the Afrotropics (Dijkstra and Clausnitzer, 2006), South Asia (Bedjanič, 2004; Bose and Kakkassery, 2019), and the South-east Asian mainland and large islands (Hämäläinen, 2004a, b; Orr, 2004, 2006; Oppel, 2005; Jumawan *et al.*, 2012; Ramos *et al.*, 2020).

Topography is an overlay on the latitudinal gradient. Leaving aside topography, at least among trees, there is higher beta diversity in tropical compared to temperate forests. This comes about from beta diversity contributing to the latitudinal species-richness gradient as a component of gamma-diversity. Underlying this contribution of beta to gamma diversity, is that in the tropics, both niche specialization and niche marginality (a measure of niche spacing along an environmental gradient: in the tropics, niche spaces are narrowly defined along an environmental gradient) increase toward the equator. The outcome is that in the tropics there is tighter species packing and larger niche space (i.e. in a particular area with an environmental gradient, there are more occupied niches per area than in the temperate zone) (Cao *et al.*, 2021). This is likely to be the case with dragonflies, but this has yet to be rigorously tested at the global scale.

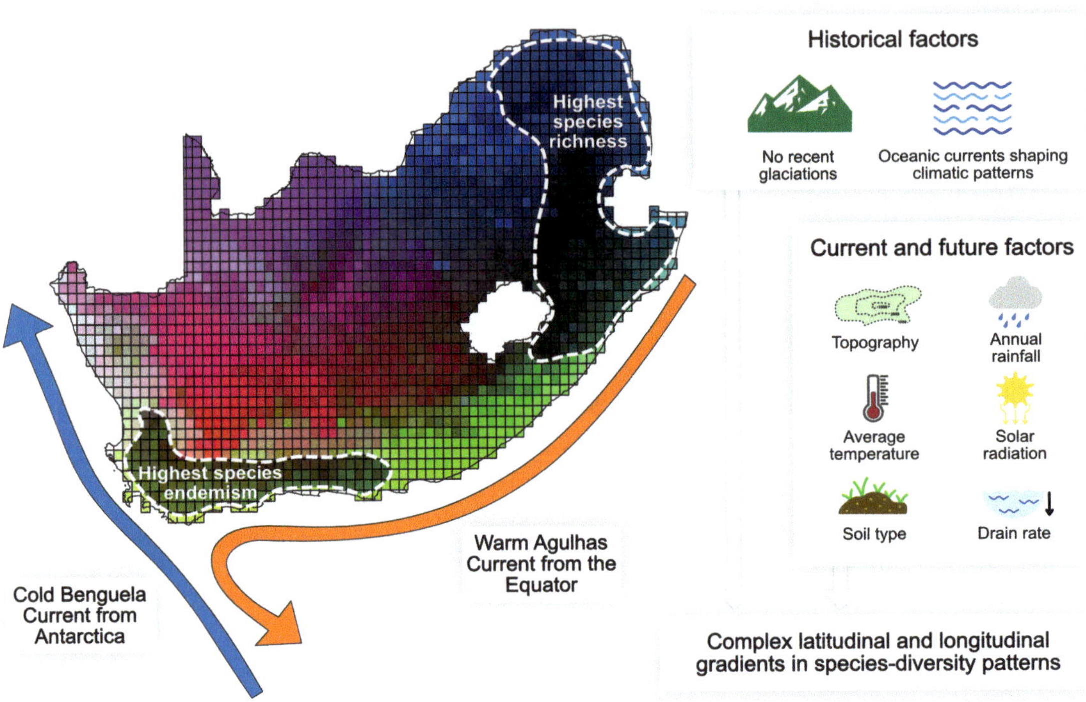

Fig. 3.21. Dragonfly diversity in South Africa is shaped by various historic and modern factors. The overall outcome is that the area with very high endemism is in the Cape Floristic Region in the south-west, whereas highest species richness is in the north-east. The white areas are neighbouring countries. (Illustration: Charl Deacon).

3.3 Recent Climate-driven Geographic Range Shifts

3.3.1 Significance of defining the spatial scale of dispersal

Conrad *et al.* (1999) showed that among seven species of odonates in the UK, most individuals do not disperse beyond 800 m from their natal pond. These results though, cannot be generalized for all species across the world. In arid Namibia, some individuals of three species of dragonfly (excluding the well-known migrants) and even one damselfly, can disperse up to several hundred kilometres (Suhling *et al.*, 2017). As pointed out by Termaat *et al.* (2019), at least some individuals in a population are likely to disperse over large distances even though most do not. Furthermore, unusual dispersal events can be triggered by adverse abiotic or biotic conditions, such as drought or overcrowding.

The effects of anthropogenic climate change, along with its increasingly severe impacts on ecosystems, are now well established. The important point regarding impacts on dragonfly diversity is the extent to which the effects will be distributed regionally, and the levels of change in rainfall and other events such as extent and severity of heatwaves, droughts, floods and wildfires (Stainforth and Calel, 2020; Harvey *et al.*, 2023).

Dragonflies are 'climate canaries' for the integrity of freshwater ecosystems. The basis for this thinking is that shifts in the functional composition of assemblages can inform us about the magnitude and direction of climate and ecosystem change against appropriate reference conditions. This approach can then help develop appropriate conservation strategies for freshwater biota in general, leading to some important further considerations. First, climate is not a sole driver of local dragonfly distributions, albeit a major one. Other extrinsic factors play a role, not least the relative quality of the home habitat versus the quality of the receiving habitat at the times that individuals are on the move. Second, movement of individuals will likely be in waves: there will be some pioneers, and if conditions continue to remain favourable, then others will independently follow in due course. Third, these movements are likely to take place in steps. Once a new suitable habitat is located, individuals will likely settle there and breed, with perhaps some individuals undergoing incidental long-distance dispersal. However, should the habitat become suboptimal, by a drought event for example, they may fail to produce a next generation. Fourth, local movement at least is likely to be in all directions, although with wind direction being important in biasing dispersal direction (Bush and Hoskins, 2017). For instance, strong April winds, coinciding with adult emergence, have a strong effect on the direction of dispersal among Iranian damselflies (Barzoki *et al.*, 2021).

Overall, however, the process of range shifting due to climate change and extreme weather events can be idiosyncratic. Details of shifted geographic ranges are also relative to environmental conditions and the availability of receiving sites for establishment according to trait variation among individuals coupled with the phenotypic plasticity inherent among them (Chapter 4).

How we view these range shifts depends then on the spatial scale of the observations. At the large spatial scale and over long time-periods, it will appear, for example, as if there is a distinct march poleward. At the smaller spatial scale, however, there may be some range reversals during cold spells and local increases in the area of occupancy as regional habitat improvement progresses with restoration. In short, interpreting macroecological manifestations must also look toward small spatial and temporal scales, and must incorporate population dynamics, as individuals ebb and flow across the landscape.

3.3.2 The concept of neonatives

Essl *et al.* (2019) proposed the term 'neonative' to describe species that have expanded their residential geographical ranges beyond trivial or seasonal movements, but not those that have now reoccupied an area from which they were earlier extirpated through human agency. They propose that the term be applied to a latitudinal shift of at least 100 km, and an elevation shift of several hundreds of metres, which is reasonable enough for dragonflies. The most objective temporal baseline for describing a neonative species is the early Neolithic, although of course, we cannot be sure where each species was positioned then (Fig. 3.22).

3.3.3 Geographic range shifts relative to time periods

At the species level, British non-migratory dragonflies shifted their geographic ranges between two time-periods: 1960–1970 and 1985–1995. All but two of the 37 species increased their range size. All but three shifted northward at their northern range margin by an average of 74 km, while southern species shifted their northern range margin by 88 km overall. The greatest shift was by *Sympetrum striolatum*, which increased its northern margin by 346 km. While improved water quality across Britain between the two time-periods may have played a role in assisting increases in range size, the northern shift of margins is mainly driven by climate change. Among the four northern species, two (*Somatochlora arctica* and *Coenagrion hastulatum*), while having expanded their overall range size, also retracted northward at their southern margin. Two other northern species (*Aeshna caerulea* and *Leucorrhinia dubia*) were the only two species out of the total to show a decline in range size yet also a shift northward. In general, the northward shift of the northern boundary has been faster than the northward shift of the southern range margin (Hickling *et al.*, 2005).

Also in the UK, between 1980 and 2012, dragonflies showed some distinct site occupancy trends, with 22% declining while 53% increased, due mostly to improved water quality combined with climate change. Two of the southern species (*Ischnura pumilio* and *Gomphus vulgatissimus*) showed no increase in range, apparently due to lack of suitable habitat. Overall, those species with the widest habitat breadth benefitted most, while the habitat specialists were disadvantaged. There was also the added factor of geographic position. Southern species benefitted more than oceanic and widespread species, suggesting that warmer temperatures are benefitting the southern species disproportionately (Powney *et al.*, 2015).

In the Palaearctic, about 40 odonate species have been recorded north of the Arctic Circle. This number recently increased, likely the result of recent accelerated climate change, with *Lestes sponsa*, *Coenagrion pulchellum*, *Pyrrhosoma nymphula*, *Erythromma najas*, *Somatochlora flavomaculata* and *Aeshna cyanea* reaching the region in the last few years. Furthermore, some species, such as *Enallagma cyathigerum*, are now more common in the Arctic Circle than they were some years ago. However, there

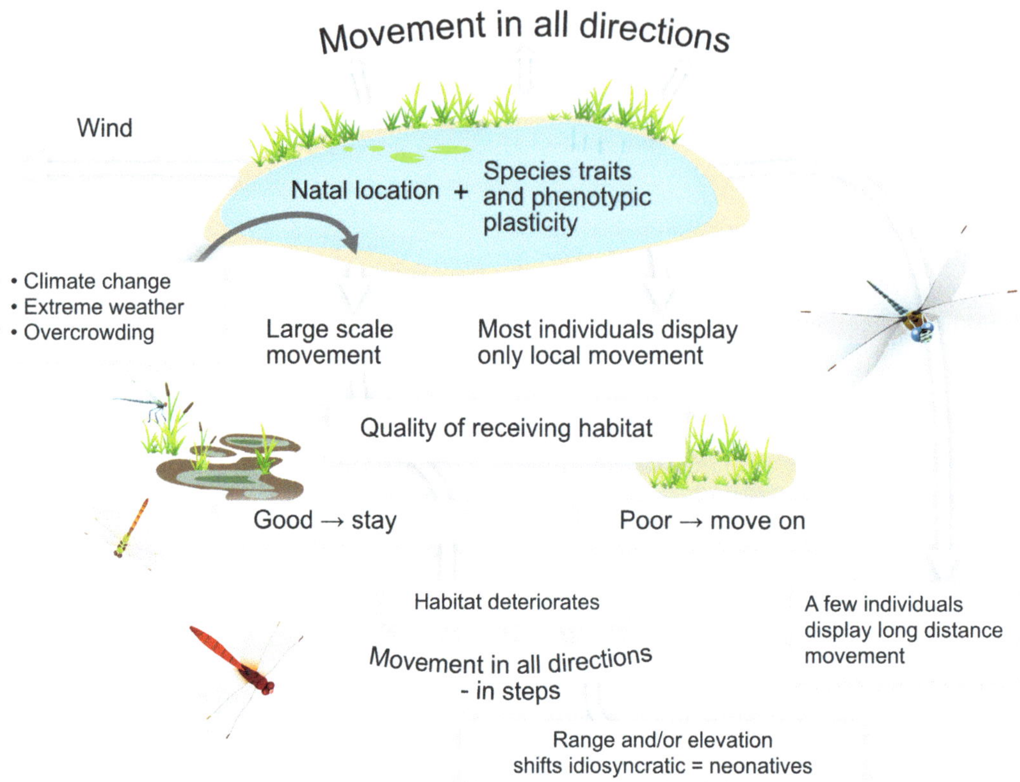

Fig. 3.22. Under relatively stable environmental conditions, most adult dragonfly individuals do not move far, although a few may fly away much farther. They disperse in all directions but may move more with wind direction. But when environmental conditions change, or there is overcrowding, many if not all individuals move on. They only settle where habitat conditions are good. However, if conditions later deteriorate, they (usually a later generation) move away, again in all directions. The overall result is that there may be range and/or elevation shifts which in reality are idiosyncratic. Once established in a new area they are known as neonatives. (Illustration: Robyn Symons).

is concern that cold-adapted species, especially *Somatochlora sahlbergi* in Fennoscandia, may be displaced from their habitats by more competitive species, such as *S. metallica*, as the local climate warms (Brockhaus, 2022).

Northward margin shifts have occurred among many European species (Boudot and Kalkman 2015). In another broad-scale study, covering 10 regions largely in Western Europe, and extending from the Mediterranean to all of Sweden, Termaat *et al.* (2019) showed that among the 87 species that they studied, 55 increased their range from 1990 to 2015, while 32 were stable and none declined. They found clear effects among warm-adapted species, and more so than among cold-adapted species in six of the regions (Bavaria, Britain, Flanders, France, Netherlands and Wallonia). In Sweden, both cold and warm species were stable between 1990 and 2001, but both increased their ranges to the same extent after 2002, with most shifting their northern range limits. Across Europe, warm species in general increased in proportion to cold species, changing assemblage composition in many areas. As range changes overall lagged behind temperature increases, they accumulated what has been described as a 'climatic debt' (Devictor *et al.*, 2012), which varies among species and regions.

Across the Netherlands, dragonflies declined severely in geographic range size during the period 1951–1990, compared to the earlier period 1850–1950, due to physical and chemical deterioration of water systems. However, between 2000–2018 they increased greatly, especially the running water species as streams and rivers were restored. The exceptions were the moorland pool species that never returned to the 1850–1950 levels, owing mostly to drainage of their habitats. After 1991, and especially after 2000, both climate change and habitat restoration contributed to the recovery of many species (van Strien and van Grunsven, 2023).

The northern position of Sweden is interesting in that dragonfly range changes have been strong. It is where species at the edge of their northern range are more selective in habitat and occur at lower densities than those well inside their range (Sahlén and Ekestubbe, 2001). In northern Uppland, between 1996–2006, among the lake-inhabiting dragonfly species, one third of the species disappeared, while among the survivors, a third changed their ecological and habitat preferences, while some formerly unrecorded species appeared. These changes were not due to forestry practices but due to climate change. During that short time of 10 years, rainfall patterns changed, with late summer and autumn being drier and longer, extending the dragonfly growth season. Formerly rare species became more frequent and occurred in lakes with a wider quality range, while some formerly widespread species became more selective in choice of water habitat, seemingly also due to shifts in niche space (Flenner and Sahlén, 2008).

In Germany between 1980–2016, analysis of an extensive and detailed database of dragonfly occupancy and range changes gives a complex picture of positive and negative effects, as well as linear and non-linear changes in dragonfly species distribution. The effects of climate change entwine with those of habitat change and environmental management. Five patterns of dragonfly responses, habitat type and traits relative to climate change were identified. First, species showing a distinct increase in range, tailed off in the most recent years. Second, those that were initially stable but had increased since 2000, included some of the most warm-adapted species. Third, bog-inhabiting species declined from 2000 onward. Fourth, cold-adapted bog species declined in the 1980s and 1990s. And fifth, bog-inhabiting species with short wings continued to decline over the whole period. Of the 77 species studied, 37 showed a significant increase in range, while 18 showed a significant decrease. Of the remaining species, 20 showed some range changes. The two remaining species were formerly rare warm-adapted species (*Crocothemis erythraea* and *Erythromma viridulum*) that had increased their ranges substantially. *Anax imperator*, being a large and powerful flier and a relatively habitat-generalist species, also benefitted greatly from climate change. Overall, the threatened, red-listed species declined. In general, the cold-adapted species of standing water were affected more negatively than those of running water due to physiological stress under warming conditions coupled with loss of bog or moorland habitats. Among these has been *Sympetrum danae* and *Coenagrion hastulatum*. Interestingly, the proportions of dragonflies versus damselflies, either increasing or decreasing, were not significantly different throughout the investigated time-period (Bowler *et al.*, 2021).

In a study covering three countries (Mauritania, Tunisia, Sweden) habitat generalists benefitted over specialists under recent climate change, with a northward expansion of the generalists. Species of cool temporary ponds, especially those surrounded by dense woody vegetation, were most prone to local extinction, being replaced by pioneer species (Cerini *et al.*, 2020).

3.3.4 Interaction between climate and landscape change

When Swedish forested and agricultural landscapes were compared between 2002 and 2011–2012, and where mean seasonal air temperature increased by 0.5°C over the 10-year period, dragonfly species richness in agriculture increased by 17%, but decreased by 13% in forestry. Changes in occupancy in the two landscapes were similar at around 70%, as were average local extinction rates at around 40%. However, changes in species composition in the forested areas were 29% greater than in the agricultural areas, seemingly due to the regional species pool in the surrounding areas stabilizing extinction rates, especially in forestry (Koch *et al.*, 2014).

The lower Alpine region of Italy is seeing similar compound effects. About a third of odonate species have disappeared or have declined strongly in recent years, with most being standing or running temporary water specialists or cold-adapted species. The 12% of species that have increased, or are new to the area, are specialists of permanent standing waters or warm-adapted species. These changes overall have led to homogenization of the odonate fauna, and it seems to have been driven by a combination of land-use change, drainage, land-use intensification, reclamation or abandonment, environmental pollution and climate change (Assandri, 2021).

In Morocco, some damselflies have been lost due to multiple drivers, including excessive water abstraction. However, several odonates, especially dragonflies, appear to have expanded their range (Taybi *et al.*, 2019). Overall, habitat generalists rather than specialists have expanded their ranges the most. A reason for this is that climate warming combined with a suitable receiving habitat have been the most important drivers favouring range expansion (Platts *et al.*, 2019).

Olsen *et al.* (2022) found that among European dragonflies, running water species have more specialized and narrower niches, with a more fixed niche position than standing water species. Running water species also have smaller geographic ranges, and between 1988–2010, some displayed more range contraction than those in standing or temporary waters. This makes running water species especially at risk from habitat loss and climate change. Furthermore, temporary water species are much more habitat generalist with broader niches, and do not have such fixed niche positions, unlike permanent water species. In other words, temporary water species track climate change better than permanent water species. Overall, this suggests that ecological specialization may cause contemporary range shifts to remain behind climate change and resources (i.e. there is a climatic debt), as well as indicating that climate change and human transformation of the landscape together cause biotic homogenization, where specialists are replaced by generalists (Fig 3.23).

3.4 Predicted Range Shifts with Further Climate Change

3.4.1 Considerations of large spatial scales

Past climatic changes and their filters provide us with some insight into how we might deal with ongoing climate change (Nogués-Bravo *et al.* 2018). This means that projections on future dragonfly distributions largely reflect what has already occurred in recent decades. While recent and ongoing global warming is among the foremost of drivers affecting local and geographic distributions of mobile insects like dragonflies, there are several important concurrent effects. These include rapidly increasing sea levels, leading to the salting of coastal wetlands. There is also increased likelihood of stronger ENSO events (Cai *et al.* 2018), and more heat and wildfire extremes (Fasullo *et al.* 2018), as well as extended droughts that affect the quality of various ecosystems from rainforests (Tng *et al.* 2018) to ponds (Deacon *et al.*, 2019). There has also been a substantial increase in extreme precipitation events and flooding in recent years (Alifu *et al.*, 2022).

Furthermore, climate change is often synergistic with other adverse effects, such as landscape fragmentation and habitat loss, as well as the effect of invasive alien organisms (Domisch *et al.*, 2011). All these drivers together modify microhabitats and affect functional connectivity among their most suitable habitats. This means that we must consider large-scale contingency plans and possible management options across a wide range of spatial scales from local landscapes through to regions. In doing this, we must also bear in mind that source populations may become more remote and even non-existent, especially in the lowland tropics where species may already be at their thermal extreme (Colwell *et al.*, 2008).

3.4.2 Complexity of climate change effects

Even when considering just climate warming, there are challenges for insects. While some species deal with ongoing climate warming by behavioural adaptations (e.g. seeking shade more often) and movement, long-term survival may depend on a combination of genetic adaptation and phenotypic plasticity. However, plastic responses are limited, and may not on their own be enough to ensure survival. Rapid evolutionary adaptation may then be the only option (González-Tokman *et al.*, 2020). Meanwhile, understanding the dispersal capabilities of specific species, relative to the trajectory and intensity of climate change, is critical for the conservation of threatened species into the future (Jaeschke *et al.*, 2013).

Among the important effects of climate change that are predicted for the end of this century are alterations to water flow. Changes in flow may decrease the abundance of aquatic biota in European catchments by as much as 42% with assemblage changes in 98% of the river reaches (Kakouei *et al.*, 2018). In contrast, and in the case of high-elevation European ponds, there is likely to be a significant *increase* of 83–150% in species richness of aquatic biota, associated with a warmed climate, land use and changes in water quality. At low elevations

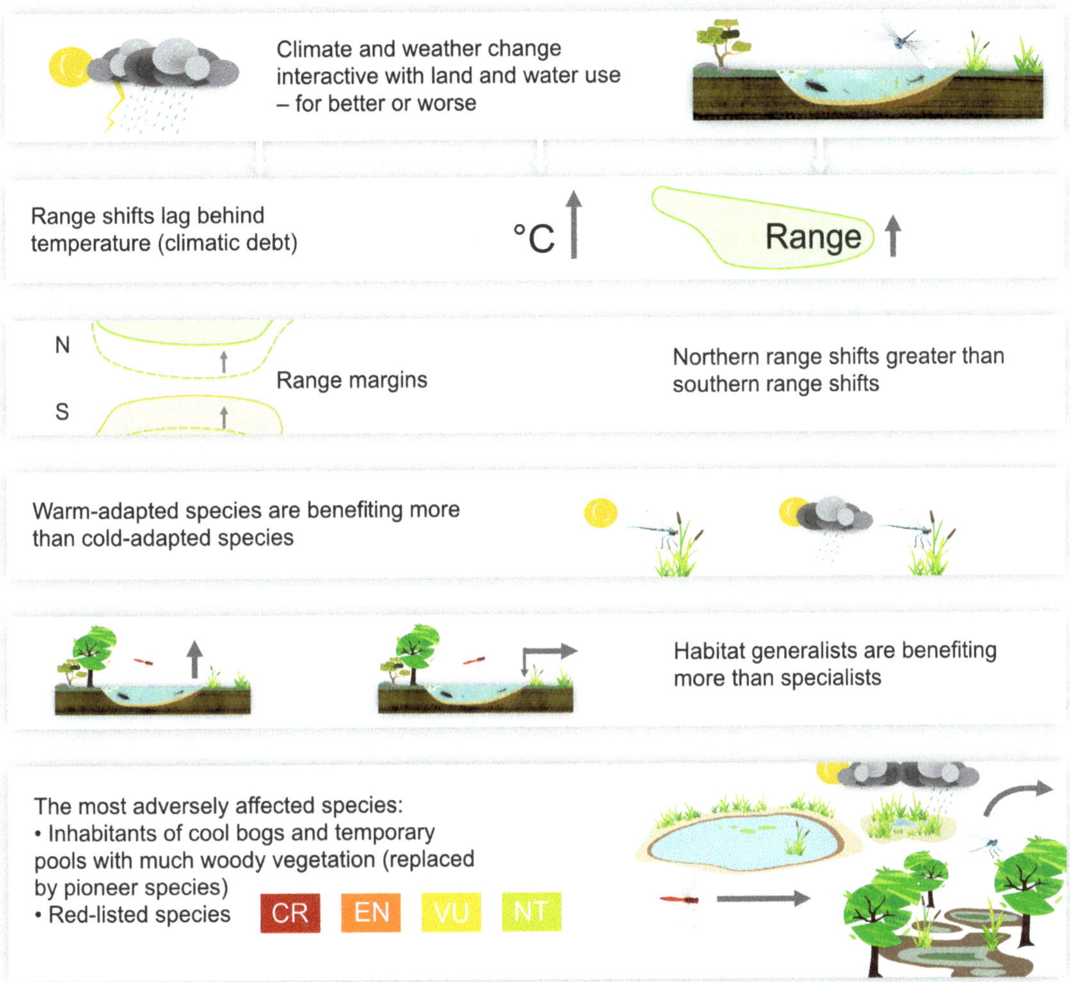

Fig. 3.23. The main outcomes of climate change and attendant weather events on dragonfly distributions in the northern hemisphere in recent years. The meteorological drivers are interactive with land- and river-use change and intensity, which in the case of rivers, has now improved in some areas. (Illustration: Robyn Symons).

changes in water quality, such as increases in nutrient loads, may limit the increases in species richness, along with differential effects on the various taxa, including some local extinctions, leading to ecosystem change (Rosset *et al.*, 2010).

Based on climate models alone (i.e. excluding possible adverse synergistic effects), most or all (97–100%) benthic macroinvertebrates are likely to survive until 2080. Range changes are more likely to contract among species in streams with low annual mean temperatures but expand among species in currently warmer streams. Additionally, there is likely to be a reorganization of species composition and community structure along the river continuum, with headwater species decreasing in population size and a possible associated drop in species diversity (Domisch *et al.*, 2011). Overall, endemic species are likely to lose more climatically suitable areas compared to the more geographically widespread species (Domisch *et al.*, 2013).

Across western and central Asia, there are likely to be strong declines in odonate diversity, especially in western Turkey, the Levant and Azerbaijan, but not such strong declines in parts of Iran and southern

Central Asia. Regions with fewer than 15 species are also likely to see a decline. In contrast, there is likely to be an increase in diversity in eastern Turkey and at high elevations in Central Asia, with possibly some increase across the Arabian Peninsula. Furthermore, odonate species composition is likely to experience strong shifts, with Palaearctic species declining and Oriental and Afrotropical species increasing. But interestingly, there is no evidence that overall there will be any strong trend differences between running and standing water species, although there may be differences from one area to another (Cadena *et al.*, 2023).

3.4.3 Range shifts and elevation

Among various pond taxa across Switzerland, dragonflies that are warm thermal specialists and thermal generalists are likely to be the winners from future climate change, with as many as 63% of species being winners. This is already starting to happen, with *Sympetrum meridionale*, *S. fonscolombii*, *Crocothemis erythraea*, *Coenagrion scitulum* and *Aeshna affinis* having already colonized the country in recent years. The losers will be the cold and cool thermal specialists under a future warming climate (Rosset and Oertli, 2011).

Climate warming in recent years has seen the appearance of several European dragonflies at higher elevations in Germany than recorded previously (Ott, 2010). However, in Switzerland, while the number of alpine water bodies is increasing, there is concern for some high-elevation, cold-adapted species, particularly *Aeshna caerulea*, *A. juncea*, *A. subarctica*, *Coenagrion hastulatum*, *Leucorrhinia dubia*, *Somatochlora alpestris* and *S. arctica* (Oertli, 2010).

Remnant populations of the boreo-alpine specialist *Somatochlora alpestris* in the Carpathian Mountains, Romania, have elevation as a major environmental variable. With a rise of 1.5°C, there is likely to be an upward shift in elevation occupancy of 200 m and a decrease in area of occupancy of 40%. With a rise of 3°C, the predicted upward shift will be 600 m, along with a decrease in area of occupancy of 90%. However, the effects of climate warming may be more severe as the bog habitat of the species may become suboptimal during the warming process (De Knijf *et al.*, 2011).

Future climates suggest there will be a change in the geographic range of suitable areas for *Cordulegaster bidentata* and *C. heros*, both of which inhabit certain mountainous areas in Europe. The suitable area for both species is likely to be strongly influenced by bioclimatic variables and show an upward shift toward higher elevations, along with loss of suitable areas for *C. bidentata* but a large gain for *C. heros* (Fekete *et al.*, 2023).

In the case of the anisozygopteran *Epiophlebia laidlawi*, predictions are that it will shift up in elevation between 294–374 m by 2050, and by 342–599 m by 2080. This would mean that in the future, and from a thermal perspective, it would be even more relictual than it is today, and only be able to survive in the highest mountains of east Nepal (Shah *et al.*, 2012).

Interestingly, *Sympetrum meridionale* has increased its elevation range as well as its geographical range in the Indian region. Previous records of the species were all below 2,500 m above sea level but it is now at 3,500 m above sea level (Kumar *et al.*, 2021).

As there is much elevation variation across South Africa, almost certainly some species will occupy higher elevations in the future, as they appear to have shifted in elevation in the deep past (Deacon *et al.*, 2020) and are already primed to disperse from being adapted to ENSO events (Samways and Deacon, 2022). Nevertheless, further climate change in South Africa is likely to see complex and non-linear effects on dragonfly local distributions via shifts in bioregions, some of which will expand while others will contract. Of concern, is that the current protected area network, which today is particularly effective in conserving many bioregion-specific endemic species, may not be able to do so in the future (Basel *et al.*, 2021).

Positive response to ENSO events may possibly be an advantage for dragonflies under climate change in the western Peruvian Andes. However, with overall increasing temperatures, there is an indication that some species now occur at higher elevations than previously recorded. Of particular concern is the increased ultraviolet light in the area (Hoffmann, 2010).

3.4.4 Rate of range shifting

Li and Park (2020) found that environmental and habitat preferences explain geographic range shifts among European odonates, with mobility traits being less significant. They suggest that overall, the leading edge of these species in the future is more likely to shift poleward at a slower rate (1.13–2 km/year) than the retraction of their southern, trailing margins (2.38–10.34 km/year). Furthermore, existing recent range shifts among UK odonates appear to be strongly associated to thermal adaptation. This is seen among the warm-adapted species which have already shifted northward to a greater extent than the cooler-adapted species (Hickling et al., 2005). However, species at the southern margin may hold their range limits in part due to some populations remaining in increasingly isolated remnants of suitable habitat, especially at high elevations, as in the case of the Carpathian *Somatochlora alpestris* mentioned above.

Not all species will respond to the overall effect of climate change equally. This suggests that not all species will follow the thermal climate window to the same degree. Responses will be in accordance with changes in environmental variables such as water temperature and velocity (Li and Park, 2020), as well as relative to their physiological, behavioural and ecological traits. However, environmental conditions and habitat preferences may override dispersal ability, especially as dispersal is constrained by landscape fragmentation (Estrada *et al.*, 2015), and availability of suitable receiving habitats may be reduced. This could lead to a climatic debt, with the isotherm edge likely to shift by about 8 km/year, compared to the dragonfly assemblage shift of 4.5 km/year. The response of the European dragonfly assemblage to climate change is likely not to be linear, with overall distribution ranges likely to increase until the 2050s, then decrease until the 2080s (Li and Park, 2020).

Some dragonfly species show different preferences in microhabitat across their geographic range. Such intrinsic plasticity can also help certain species colonize new geographic areas. In the case of *Crocothemis erythraea*, when it first arrived and established in Germany, landscape transformation in the form of gravel pits and clay pits was of benefit to the species, enabling it to get a toehold at certain localities where, for example, rehabilitation of urban ponds was taking place (Ott, 1988). Since these early appearances, the species has now established in various water types, increasing its area of occupancy, both in Germany (Ott, 2010) and in Ukraine (Khrokalo, 2010) (Fig. 3.24). It has also been recorded as having crossed the English Channel into Britain, with the potential for establishment (Parr, 2020). Another far-ranging wanderer from more southerly regions, *Sympetrum fonscolombii*, also appears to have established in Germany (Ott, 2023).

3.4.5 Barriers versus opportunities in terms of habitat occupancy

In the case of Australian dragonflies, by 2085, the overall extent of habitat may decline by 56–69%, affecting 153–187 species, with a third of the species assessed (270 species, 85% of the Australian fauna) as being vulnerable or highly vulnerable. Loss of suitable habitats leading to reduced landscape connectivity will likely be a major problem. Species in south-east Australia are at particularly high risk where they are not able to disperse to refuges in Tasmania in view of the Bass Strait. In turn, many endemic species of the Pilbara, Western Australia, will depend on continued availability of permanent fresh water to avoid extinction (Bush *et al.*, 2014a).

Species richness of Australian dragonflies is concentrated in the wet tropics, central-north Australia, and south-east Queensland, as well as some other local areas. Models suggest that there will be major shifts in composition, although little if any loss of evolutionary diversity, across most of the east coast in response to climate change. The presence of remaining suitable habitat may play a significant role in these shifts (Bush *et al.*, 2014b).

In South Africa, by 2050, at least two dragonfly species of the 14 studied species will be lost from the area as they have narrow elevation ranges. However, most of this species set (12 of 14) thrive in artificial waters that act as stepping-stone areas for dispersal or are suitable breeding habitat. This suggests that functional connectivity for survival will not be restricting. But two of the studied stream species in the area are habitat specialists and are especially vulnerable due to potential loss of habitat. In turn, South Africa, being continental, is likely to receive species from farther north in Africa, suggesting that there will not be a decrease in species richness, but species replacement is a possibility (Simaika and Samways, 2015).

3.4.6 Optimal growth challenges in a warming world

Experimental evidence suggests that the abundant and widespread North American *Enallagma civile* will be susceptible to elevated water temperature as the climate warms. While it may benefit from a small increase in temperature shortening the period of its life cycle, large increases (reaching 38–41°C)

Fig. 3.24. *Crocothemis erythraea* (top left), a climate change winner that used gravel pits to establish at new northerly localities before then establishing in various types of water bodies. The gravel pit where the species was first discovered as breeding in south-western Germany (top right). It also began to occupy rehabilitated clay pits (bottom left) and urban mitigation ponds (bottom right). (Photos: Michael Samways (top left); Jürgen Ott (top right, bottom left and right)).

may lead to smaller larvae and adults at emergence and decreased survivorship overall. The concern is that if this is the case for such a habitat and thermal generalist, climate warming will have severe effects on small-range thermal specialists (Starr and McIntyre, 2020).

Ongoing climatic warming, besides simulating poleward shifts in odonate species ranges, may also shift environmental temperatures toward optima for larval growth, especially among habitat generalists. Habitat specialists may be at a disadvantage compared to generalists. Between 16–28°C, the growth rate of habitat specialist *Leucorrhinia dubia* increases only threefold, compared to the tenfold increase in generalist *Crocothemis erythraea*. The point being that there will likely be changes in community structure in aquatic systems as the climate continues to warm, with shifts in competitive interactions, as well as in predation and parasitism (Suhling *et al.*, 2015).

3.5 Dragonflies as Invasive Organisms

While insects in general feature strongly as invasive alien organisms, there are very few aquatic insects that have expanded their range from human agency. In general, there is not the economic incentive to transport aquatic insects like dragonflies. Furthermore, dragonflies are not intimately associated with plants to the same extent as many invasive terrestrial insects, nor do they have the same high reproductive rate overall. Also, dragonflies, being amphibiotic, would have to establish in two new environments, aquatic and terrestrial, at the same time. Those that are adapted to running waters are well adapted to certain local and changing conditions, and species would require very similar conditions if they were to establish at a new location (Fenoglio *et al.*, 2016).

Nevertheless, there are some invasive alien dragonflies, especially in Hawai'i, where new arrivals are easily detectable from the indigenous fauna (Polhemus and Asquith, 1996). Several species (*Enallagma civile*, *Ischnura posita, I. ramburii*, *Crocothemis servilia*, *Orthemis ferruginea*, *Tramea abdominalis* and *T. lacerata*) have newly arrived in the Hawaiian Islands, and today are present on one to several of the major islands. There is no evidence to date that they pose a threat to the endemic *Megalagrion* species (Englund, 1999). *Crocothemis servilia* was the last alien species to be recorded (in 1994) in Hawai'i until the present time. The species is also present as an alien species in Florida (Paulson, 1978) and Cuba (Flint, 1996).

The main route of invasion in Hawai'i has been the importation of aquatic plants for aquaculture, possibly coupled with release of unwanted aquarium contents into the wild. The warm Hawaiian climate has likely also been instrumental for the new arrivals which are largely from warm countries. Arrival by ship in earlier decades, and by plane in later times, may have also contributed to certain alien dragonfly arrivals. The recent non-arrival of more alien odonates is due in no small part to much stricter quarantine measures today than in the past. This, coupled with strong public awareness campaigns and the pet trade freely taking in unwanted water biota to prevent them from being discarded into local freshwater systems, and less necessity/desire to import further water plants, has also prevented more alien odonate arrivals (Dan Polhemus, personal communication). Strict quarantine as well as containment of alien dragonflies remains a priority, as the Hawaiian streams remain susceptible to further invasion (Moody *et al*., 2017).

Arrival as eggs or larvae in association with water plants destined for the aquarium or aquaculture industry has also been the main route of human-assisted invasion by dragonflies in Europe. However, of the 40 or so alien odonate species recorded in Europe, none have invaded local natural ecosystems (Laister *et al*., 2014; Adriaens and De Knijf, 2015).

Two alien damselfly species have been recorded in Japan: *Pseudagrion australasiae* (Karube *et al*., 2011), and *Ceriagrion auranticum ryukyuanum*, which has been controlled to remain at low densities (Karube *et al*. 2014).

While most of the adult odonate aliens have been intercepted as young and emerging stages in indoor environments, there is always the risk of inadvertent escape of any emerging adults (Móra and Sebteoui, 2020). Evidence to date suggests that climate similarity, especially temperature, of the receiving area presents a risk, as has been the case in Hawai'i. Of concern, with respect to climate warming, is that alien dragonflies of subtropical or tropical provenance that arrive in warmed environments have a greater chance of survival should they escape into the wild. This means that with a warming climate, currently cooler countries will have to be increasingly vigilant about intercepting new arrivals.

References

Abbott, J.C., Bota-Sierra, C.A., Guralnick, R., Kalkman, V., González-Soriano, E. *et al*. (2022) Diversity of Nearctic dragonflies and damselflies (Odonata). *Diversity* 14, 575. DOI: 10.3390/d14070575.

Adriaens, T. and De Knijf, G. (2015) A first report of introduced non-native damselfly species (Zygoptera: Coenagrionidae) for Belgium. *Belgium Journal of Zoology* 145, 76–80.

Aistleitner, E., Barkemeyer, W., Lehmann, G. and Martens, A. (2008) A checklist of the Odonata of the Cape Verde Islands. *Mitteillungen des Internationalen Entomologischen Vereins* 33, 45–57.

Aksenova, O.V., Potapov, G.S., Bespalaya, Y.V., Kolosova, Y.S., Vikhrev, I.V. *et al*. (2020) Dragonflies from hot springs in Russia with a country-level checklist of species known to occur in geothermal environments. *Ecologica Montenegrina* 34, 49–63. DOI:10.37828/em.2020.34.6.

Alifu, H., Hirabayashi, Y., Imada, Y. and Shiogama, H. (2022) Enhancement of river flooding due to global warming. *Scientific Reports* 12, e20687. DOI: 10.1038/s41598-022-25182-6.

Alves-Martins, F., Brasil, L.S., Juen, L., De Marco Jr, P., Stropp, J. and Hortal, J. (2019) Metacommunity patterns of Amazonian Odonata: the role of environmental gradients and major rivers. *PeerJ* 7, e6472. DOI: 10.7717/peerj.6472.

Alvial, I.E., Vargas, H.A., Marinov, M., Esquivel, C., Araya, J. *et al*. (2019) Isolation on a remote island: genetic and morphological differentiation of a cosmopolitan odonate. *Heredity* 122, 893–905. DOI: 10.1038/s41437-018-0165-z.

Amundrud, S.L., Videla, M. and Srivasta, D.S. (2018) Dispersal barriers and climate determine the geographic distribution of the helicopter damselfly *Mecistogaster modesta*. *Freshwater Biology* 63, 214–223. DOI: 10.1111/fwb.13054.

Assandri, G. (2021) Anthropogenic-driven transformations of dragonfly (Insecta: Odonata) communities of low elevation mountain wetlands during the last century. *Insect Conservation and Diversity* 14, 26–39. DOI:10.1111/icad.12439.

Balzan, M.V. (2012) Associations of dragonflies (Odonata) to habitat variables within the Maltese Islands: a spatiotemporal approach. *Journal of Insect Science* 12, e87. DOI: 10.1673/031.012.8701.
Barzoki, Z.E., Ebrahimi, M., Kiany, M. and Sadeghi, S. (2021) Ecological drivers of Odonata beta diversity in arid and semi-arid regions of the Central Plateau of Iran. *Insect Conservation and Diversity* 14, 40–51. DOI: 10.1111/icad.12464.
Basel, A.M., Simaika, J.P., Samways, M.J., Midgley, G.F., MacFadyen, S. and Hui, C. (2021) Assemblage re-organization of South African dragonflies due to climate change. *Diversity and Distributions* 27, 2542–2558. DOI: 10.1111/ddi.13422.
Bechly, G., Garrouste, R., Aase, A., Karr, J.A., Grande, L. and Nel, A. (2021) The damselfly palaeofauna from the Eocene of Wyoming and Colorado, USA (Insecta, Odonata, Zygoptera). *Papers in Palaeontology* 7, 1373–1402. DOI: 10.1002/spp2.1346.
Bedjanič, M. (2004) Odonata fauna of Sri Lanka: research and threat status. *International Journal of Odonatology* 7, 279–294.
Bergmann, T., Rach, J., Damm, S., DeSalle, R., Scheirwater, B. and Hadrys, H. (2013) The potential of distance-based thresholds and character-based DNA barcoding for defining problematic taxonomic entities by CO1 and ND1. *Molecular Ecology Resources* 13, 1069–1081. DOI: 10.1111/1755-0998.12125.
Bose C.N. and Kakkassery, F.K. (2019) Study on the diversity and abundance of odonates in three different geographical divisions in Kerala. Research and Postgraduate Department of Zoology, St Thomas' College, Thrissur, India, pp. 74–78.
Bota-Sierra, C.A., Corso, A., Janni, O., Sandoval-H., J. and Viganò, M. (2018) Seventeen new dragonfly records from Colombia and the confirmation of the synonymy of *Philogenia monotis* and *P. tinalandia* (Insecta: Odonata). *International Journal of Odonatology* 21, 115–127. DOI: 10.1080/13887890.2018.1462262.
Boudot, J.-P. and Kalkman, V.J. (eds) (2015) *Atlas of the European Dragonflies and Damselflies*. KNNV Publishing, Zeist, Netherlands.
Bowler, D.E., Eichenberg, D., Conze, K.J., Suhling, F., Baumann, K. *et al.* (2021) Winners and losers over 35 years of dragonfly and damselfly distributional change in Germany. *Diversity and Distributions* 27, 1353–1366. DOI: 10.1111/ddi.13274.
Brockhaus, T. (2022) The Palaearctic dragonfly (Insecta: Odonata) fauna of the lands north of the Arctic Circle: a critical synopsis. *Eurasian Entomological Journal* 21, 142–152. DOI: 10.15298/euroasentj.21.3.02.
Bush, A. and Hoskins, A.J. (2017) Does dispersal capacity matter for freshwater biodiversity under climate change? *Freshwater Biology* 62, 382–396. DOI: 10.1111/fwb.12874.
Bush, A., Theischinger, G., Nipperess, D., Turak, E. and Hughes, L. (2013) Dragonflies: climate change canaries for river management. *Diversity and Distributions* 19, 86–97. DOI: 10.1111/ddi.12007.
Bush, A.A., Nipperess, D.A., Duursma, D.E., Theischinger, G., Turak, E. and Hughes, L. (2014a) Continental-scale assessment of risk to the Australian Odonata from climate change. *PLoS ONE* 9, e88958. DOI: 10.1371/journal.pone.0088958.
Bush, A., Nipperess, D.A., Theischinger, G., Turak, E. and Hughes, L. (2014b) Testing for taxonomic bias in the future diversity of Australian Odonata. *Diversity and Distributions* 20, 1016–1028. DOI: 10.1111/ddi.12196.
Büsse, S. and Ware, J.L. (2022) Taxonomic note on the species status of *Epiophlebia diana* (Insecta, Odonata, Epiophlebiidae), including remarks on biogeography and possible species distribution. *Zookeys* 1127, 79–90. DOI: 10.3897/zookeys.1127.83240.
Büsse, S., Grumbkow, P., Hummel, S., Shah, D.N., Tachamo Shah, R.D. *et al.* (2012) Phylogenetic analysis elucidates the influence of the ice ages on the disjunct distribution of relict dragonflies in Asia. *PLoS ONE* 7, e38132. DOI: 10.1371/journal.pone.0038132.
Bybee, S.M., Ogden, T.H., Branham, M.A. and Whiting, M.F. (2008) Molecules, morphology and fossils: a comprehensive approach to odonate phylogeny and evolution of the odonate wing. *Cladistics* 24, 477–514. DOI: 10.1111/j.1096-0031.2007.00191.x.
Bybee, S.M., Kalkman, V.J., Erickson, R.J., Frandsen, P.B., Breinholt, J.W. *et al.* (2021) Phylogeny and classification of Odonata using targeted genomics. *Phylogenetics and Evolution* 160, e107115. DOI: 10.1016/j.ympev.2021.107115.
Cadena, J.T., Boudot, J.-P., Kalkman, V.J. and Marshall, L. (2023) Impacts of climate change on dragonflies and damselflies in West and Central Asia. *Diversity and Distributions* 29, 912–925. DOI: 10.1111/ddi.13704.
Cai, W., Wang, G., Dewitte, B., Wu, L., Santoso, A. *et al.* (2018) Increased variability of eastern Pacific El Niño under greenhouse warming. *Nature* 564, 201–206. DOI: 10.1038/s41586-018-0776-9.
Callisto, M., Linares, M.S., Kiffer Jr, W.P., Hughes, R.M., Moretti, M.S. *et al.* (2021) Beta diversity of aquatic macroinvertebrate assemblages associated with leaf patches In neotropical montane streams. *Ecology and Evolution* 11, 2551–2560. DOI: 10.1002/ece3.7215.
Cannings, R.A. (2014) The dragonflies and damselflies (Odonata) of Canadian grasslands. In: Cárcamo, H.A. and Giberson, D.J. (eds) *Arthropods of Canadian Grasslands*. Biological Survey of Canada, Canada, pp. 231–269. DOI: 10.3752/9780968932162.ch8.
Cannings, S.G. and Cannings, R.A. (1994) The Odonata of the northern cordilleran peatlands of North America. *Memoirs of the Entomological Society of Canada* 169, 89–110.
Cannings, R.A. and Simaika, J.P. (2005) *Lestes disjunctus* and *forcipatus* (Odonata: Lestidae): an evaluation of status and distribution in British Columbia. *Journal of the Entomological Society of British Columbia* 102, 57–63.

Cao, K., Condit, R., Mi, X., Chen, L., Ren, H. *et al.* (2021) Species packing and the latitudinal gradient in beta-diversity. *Proceedings of the Royal Society B* 288, e20203045. DOI: 10.1098/rspb.2020.3045.

Carle, F.L. (1995) Evolution, taxonomy, and biogeography of ancient Gondwanian libelluloids, with comments on anisopteroid evolution and phylogenetic systematics. *Odonatologica* 24, 383–424.

Carle, F.L. (2012) A new *Epiophlebia* (Odonata: Epiphlebioidea) from China with a review of epiophlebian taxonomy, life history, and biogeography. *Arthropod Systematics and Phylogeny* 70, 75–83.

Carle, F.L. and Cook, C. (1984) A new *Neogomphus* from South America, with extended comments on the phylogeny and biogeography of the Octogomphini trib. nov. (Anisoptera: Gomphidae). *Odonatologica* 13, 55–70.

Carr, P. (2022) Odonata of the Chagos Archipelago, central Indian Ocean: an update. *Notulae odonatologicae* 9, 229–235. DOI: 10.5281/zenodo.4268581.

Carvalho, F.G., Duarte, L., Seger, G.D.S., Nakamura, G., Cordero-Rivera, A. and Juen, L. (2022) Detecting Darwinian shortfalls in the Amazon Odonata. *Neotropical Entomology* 51, 404–412. DOI: 10.1007/s13744-022-00961-y.

Cerini, F., Stellati, L., Luiselli, L. and Vignoli, L. (2020) Long-term shifts in the communities of Odonata: effect of chance or climate change? *North-Western Journal of Zoology* 16, e191101.

Cerini, F., Bombi, P., Cannings, R. and Vignoli, L. (2021) Odonata metacommunity structure in northern ecosystems is driven by temperature and latitude. *Insect Conservation and Diversity* 14, 675–685. DOI: 10.1111/icad.12507.

Chalmers, P., Grace Melchers, G., Tikka, K. and Richardson, J.S. (2023) Distribution of the grappletail dragonfly (*Octogomphus specularis*), a lake outlet stream specialist at its northern range limit. *Fundamental and Applied Limnology* 197, 1–8. DOI: 10.1127/fal/2023/1502.

Clausnitzer, V., Dijkstra, K.-D.B., Koch, R., Boudot, J.-P., Darwall, W.R.T. *et al.* (2012) Focus on African freshwaters: hotspots of dragonfly diversity and conservation concern. *Frontiers in Ecology and Environment* 10, 129–134. DOI: 10.1890/110247.

Colwell, R.K., Brehm, G., Cardelús, C.L., Gilman, A.C. and Longino, J.T. (2008) Global warming, elevational range shifts and lowland biotic attrition in the wet tropics. *Science* 322 (5899), 258–261.

Conrad, K.F., Willson, K.H., Harvey, I.F., Thomas, C.J. and Sharratt, T.N. (1999) Dispersal characteristics of seven odonate species in an agricultural landscape. *Ecography* 22, 524–531.

Cooper, I.A., Brown, J.M. and Getty, T. (2016) A role for ecology in the evolution of colour variation and sexual dimorphism in Hawaiian damselflies. *Journal of Evolutionary Biology* 29, 418–427. DOI: 10.1111/jeb.12796.

Cordero-Rivera, A., Andrés, J.A., Córdoba-Aguilar, A. and Utzeri, C. (2004) Postmating sexual selection: allopatric evolution of sperm competition mechanisms and genital morphology in calopterygid damselflies (Insecta: Odonata). *Evolution* 58, 249–359. DOI: 10.1111/j.0014-3820.2004.tb01650.x.

Corser, J.D. (2010) Status and ecology of a rare Gomphid dragonfly at the northern extent of its range. *Northeastern Naturalist* 17, 341–345.

Corser, J.D., White, E.L. and Schlesinger, M.D. (2014) Odonata origins, biogeography, and diversification in an Eastern North American hotspot: multiple pathways to high temperate forest insect diversity. *Insect Conservation and Diversity* 7, 393–404. DOI: 10.1111/icad.12065.

Daigle, J.J. (2000) The distribution of the Odonata of Hawaii. *Bulletin of American Odonatology* 6, 15.

Damm, S., Schierwater, B. and Hadrys, H. (2010) An integrative approach to species discovery in odonates: from character-based DNA barcoding to ecology. *Molecular Ecology* 19, 3881–3893. DOI: 10.1111/j.1365-294X.2010.04720.x.

David, F.J., Herzog, R., Bielke, A., Bergjürgen, N., Osigus, H.J. and Hadrys, H. (2021) The first complete genome of the migratory dragonfly *Pantala flavescens* Fabricius, 1798 (Libellulidae: Odonata). *Mitochondrial DNA Part B* 6, 808–810. DOI: 10.1080/23802359.2021.1882914.

Davis, R.B., Nicholson, D.B., Saunders, E.L.R. and Mayhew, P.J. (2011) Fossil gaps inferred from phylogenies after the apparent diversification in dragonflies and their relatives. *BMC Evolutionary Biology* 11, e252. DOI: 10.1186/1471-2148-11-252.

De Knijf, G., Flenker, U., Vanappelghem, C., Manci, C.O., Kalkman, V.J. and Demolder, H. (2011) The status of two boreo-alpine species, *Somatochlora alpestris* and *S. arctica*, in Romania and their vulnerability to the impact of climate change (Odonata: Corduliidae). *International Journal of Odonatology* 14, 111–126. DOI: 10.1080/13887890.2011.578565.

De Knijf, G., Vanappelghem, C. and Demolder, H. (2013) Odonata from Montenegro, with notes on taxonomy, regional diversity and conservation. *Odonatologica* 42, 1–29.

Deacon, C., Samways, M.J. and Pryke, J.S. (2019) Aquatic insect decline in abundance and occupy low-quality artificial habitats to survive hydrological droughts. *Freshwater Biology* 64, 1643–1654. DOI: 10.1111/fwb.13360.

Deacon, C., Samways, M.J. and Pryke, J.S. (2020) Determining drivers of dragonfly diversity patterns and the implications for conservation in South Africa. *Biological Conservation* 245, e108548. DOI: 10.1016/j.biocon.2020.108548.

Devictor, V., van Swaay, C., Brereton, T., Brotons, L., Chamberlain, D. *et al.* (2012) Differences in the climatic debts of birds and butterflies at a continental scale. *Nature Climate Change* 2, 121–124. DOI: 10.1038/nclimate1347.

Dijkstra, K.-D.B. (2007) Gone with the wind: westward dispersal across the Indian Ocean and island speciation in *Hemicordulia* dragonflies (Odonata: Corduliidae). *Zootaxa* 1438: 27–48.
Dijkstra, K.-D. and Clausnitzer, V. (2006) Thoughts from Africa: how can forest influence species composition, diversity and speciation in tropical Odonata? In: Cordero Rivera, A. (ed.) *Forests and Dragonflies*. Pensoft, Sofia, Bulgaria, pp. 127–151.
Dijkstra, K.-D.B., Kalkman, V.J., Dow, R.A., Stokvis, F.R. and van Tol, J. (2014) Refining the damselfly families: a comprehensive molecular phylogeny of Zygoptera (Odonata). *Systematic Entomology* 39, 68–98. DOI: 10.1111/syen.12035.
Dijkstra, K.-D.B., Kipping, J. and Mézière, N. (2015) Sixty new dragonfly and damselfly species from Africa (Odonata). *Odonatologica* 44, 447–678. DOI: 10.5281/zenodo.35388.
Dijkstra, K.-D.B., Tate, R.B. and Papazian, M. (2022). Dragonflies and damselflies (Odonata) of Príncipe, São Tomé, and Annobón. In: Ceríaco, L.M.P., de Lima, R.F., Melo, M. and Bell, R.C. (eds) *Biodiversity of the Gulf of Guinea Oceanic Islands*. Springer, Cham, Switzerland. DOI: 10.1007/978-3-031-06153-0_14.
Domisch, S., Jähnig, S.C. and Haase, P. (2011) Climate-change winners and losers: stream macroinvertebrates of a submontane region in Central Europe. *Freshwater Biology* 56, 2009–2020. DOI: 10.1111/j.1365-2427.2011.02631.x.
Domisch, S., Araújo, M.B., Bonada, N., Pauls, S.U., Jähnig, S.C. and Haase, P. (2013) Modelling distribution in European stream macroinvertebrates under future climates. *Global Change Biology* 19, 752–762. DOI: 10.1111/gcb.12107.
Donnelly, T.W. (2003) *Lestes disjunctus*, *forcipatus* and *australis*: a confusing complex of North American damselflies. *Argia* 15, 10–13.
Donnelly, T.W. (2004) Distribution of North American Odonata: Part III: Calopterygidae, Lestidae, Coenagrionidae, Protoneuridae, Platystictidae. *Bulletin of American Odonatology* 8, 33–99.
Ebejer, M.J., Degabriele, G. and Sciberras, A. (2008) An annotated checklist of Odonata of the Maltese Islands, with evidence for a recent influx of species. *Libellula* 27, 133–145.
Englund, R.A. (1999) The impact of introduced poeciliid fish and Odonata on the endemic *Megalagrion* (Odonata) damselflies of Oahu Island, Hawaii. *Journal of Insect Conservation* 3, 225–243.
Essl, F., Dullinger, S., Genovesi, P., Hulme, P.E., Jeschke, J.M. *et al.* (2019) A conceptual framework for range-expanding species that track human-induced environmental change. *BioScience* 69, 908–920. DOI: 10.1093/biosci/biz101.
Estrada, A., Meireles, C., Morales-Castilla, I., Poschlod, P., Vieites, D. *et al.* (2015) Species' intrinsic traits inform the range limitations and vulnerability under environmental change. *Global Ecology and Biogeography* 24, 849–858. DOI: 10.1111/geb.12306.
Fasullo, J.T., Otto-Bliesner, B.L. and Stevenson, S. (2018) ENSO's changing influence on temperature, precipitation, and wildfire in a warming climate. *Geophysical Research Letters* 45, 9216–9225. DOI: 10.1029/2018GL079022.
Fekete, J., De Knijf, G., Dinis, M., Padisák, J., Boda, P. *et al.* (2023) Winners and losers: *Cordulegaster* species under the pressure of climate change. *Insects* 14, e348. DOI: 10.3390/insects14040348.
Fenoglio, S., Bonada, N., Guareschi, S., López-Rodríguez, M.J., Millán, A. and Tierno de Figueroa, J.M. (2016) Freshwater ecosystems and aquatic insects: a paradox in biological invasions. *Biology Letters* 12, e20151075. DOI: 10.1098/rsbl.2015.1075.
Flenner, I. and Sahlén, G. (2008) Dragonfly community re-organisation in boreal forest lakes: rapid species turnover driven by climate change? *Insect Conservation and Diversity* 1, 169–179. DOI: 10.1111/j.1752-4598.2008.00020.x.
Flint, O.S. Jr (1996) The Odonata of Cuba, with a report on a recent collection and checklist of the Cuban species. *Cocuyo* 5, 17–20.
Froufe, E., Ferreira, S., Boudot, J.-P., Alves, P.C. and Harris, D.J. (2014) Molecular phylogeny of the Western Palaearctic *Cordulegaster* taxa (Odonata: Anisoptera; Cordulegastridae). *Biological Journal of the Linnean Society* 111, 49–57. DOI: 10.1111/bij.12190.
Galimberti, A., Assandri, G., Maggioni, D., Ramazzotti, F., Baroni, D. *et al.* (2021) Italian odonates in the Pandora's box: a comprehensive DNA barcoding inventory shows taxonomic warnings at the Holarctic scale. *Molecular Ecology Resources* 21, 183–200. DOI: 10.1111/1755-0998.13235.
Gaytán, Á., Bergsten, J., Canelo, T., Pérez-Izquierdo, C., Santoro, M. and Bonal, R. (2020). DNA barcoding and geographical scale effect: the problem of undersampling genetic diversity hotpots. *Ecology and Evolution* 10, 10754–10772. DOI: 10.1002/ece3.6733.
González Soriano, E., Noguira, F. and Oñate Ocaña, L. (2011) A biodiversity hotspot for odonates in Mexico: the Huasteca Potosina, San Luis Potosí. *Odonatologica* 40, 179–190.
González-Tokman, D., Córdoba-Aguilar, A., Dáttilo, W., Lira-Noriega, A., Sánchez-Guillén, R.A. and Villalobus, F. (2020) Insect responses to heat: physiological mechanisms, evolution and ecological implications in a warming world. *Biological Reviews* 95, 802–821. DOI: 10.1111/brv.12588.
Grant, P.B.C. and Samways, M.J. (2007) Montane refugia for endemic and red listed dragonflies in the Cape Floristic Region biodiversity hotspot. *Biodiversity and Conservation* 16, 787–806. DOI: 10.1007/s10531-005-6201-3.

Hämäläinen, M. (2004a) Critical species of Odonata in the Philippines. *International Journal of Odonatology* 7, 295–304.

Hämäläinen, M. (2004b) Critical species of Odonata in Thailand and Indochina. *International Journal of Odonatology* 7, 305–310.

Harvey, J.A., Tougeron, K., Gols, R., Heinen, R., Abarca, M. *et al.* (2023) Scientists' warning on climate change and insects. *Ecological Monographs* 93, e1553. DOI: 10.1002/ecm.1553.

Hebert, P.D.N., Penton, E.H., Burns, J.M., Janzen, D.H. and Hallwachs, W. (2004) Ten species in one: DNA barcoding reveals cryptic species in the neotropical skipper butterfly *Astraptes fulgerator*. *Proceedings of the National Academy of Sciences of the USA* 101, 14812–14817. DOI: 10.1073/pnas.0406166101.

Heiser, M. and Schmitt, T. (2010) Do different dispersal capacities influence the biogeography of western Palaearctic dragonflies (Odonata)? *Biological Journal of the Linnean Society* 99, 177–195.

Heiser, M. and Schmitt, T. (2013) Tracking the boundary between the Palaearctic and the Oriental region: new insights from dragonflies and damselflies (Odonata). *Journal of Biogeography* 40, 2047–2058. DOI: 10.1111/jbi.12133.

Hickling, R., Roy, D.B., Hill, J.K. and Thomas, C.D. (2005). A northward shift of range margins in British Odonata. *Global Change Biology* 11, 502–506. DOI: 10.1111/j.1365-2486.2005.00904.x.

Hof, C., Brändle, M. and Brandl, R. (2006) Lentic odonates have larger and more northern ranges than lotic species. *Journal of Biogeography* 33, 63–70. DOI: 10.1111/j.1365-2699.2005.01358.x.

Hoffmann, J. (2010) Do climate changes influence dispersal and population dynamics of dragonflies in the western Peruvian Andes? *BioRisk* 5, 47–72. DOI: 10.3897/biorisk.5.842.

Hughes, A.C., Orr, M.C., Yang, Q. and Qiao, H. (2021) Effectively and accurately mapping global biodiversity patterns for different regions and taxa. *Global Ecology and Biogeography* 30, 1375–1388. DOI: 10.1111/geb.13304.

Jaeschke, A., Bittner, T., Reineking, B. and Beierkuhnlein, C. (2013) Can they keep up with climate change? Integrating specific dispersal abilities of protected Odonata in species distribution modelling. *Insect Conservation and Diversity* 6, 93–103. DOI: 10.1111/j.1752-4598.2012.00194.x.

Jones, B.R. and Jordan, S. (2015) Genetic consequences of Pleistocene sea-level change in *Megalagrion* damselflies. *Journal of Heredity* 106, 618–627. DOI: 10.1093/jhered/esv036.

Jordan, S., Simon, C. and Polhemus, D. (2003) Molecular systematics and adaptive radiation of Hawaii's endemic damselfly genus *Megalagrion* (Odonata: Coenagrionidae). *Systematic Biology* 52, 89–109. DOI: 10.1080/10635150390132803.

Jordan, S., Simon, C., Foote, D. and Englund, R.A. (2005) Phylogeographic patterns of Hawaiian *Megalagrion* damselflies (Odonata: Coenagrionidae) correlate with Pleistocene island boundaries. *Molecular Ecology* 14, 3457–3470. DOI: 10.1111/j.1365-295X.2005.02669.x.

Juen, L. and De Marco Jr, P. (2012) Dragonfly endemism in the Brazilian Amazon: competing hypotheses for biogeographical patterns. *Biodiversity and Conservation* 21, 3507–3521. DOI: 10.1007/s10531-012-0377-0.

Jumawan, K.M., Medina, M.N.D. and Villanueva, R.J.T. (2012) Annotated list of Odonata from Mainit Hot Spring Protected Landscape, Compostela Valley, Mindanao Island, Philippines. *Philippines Journal of Systematic Biology* 6, 14–27.

Kakouei, K., Kiesel, J., Domisch, S., Irving, K.S., Jähnig, S.C. and Kail, J. (2018) Projected effects of climate-change-induced flow alterations on stream macroinvertebrate abundances. *Ecology and Evolution* 8, 3393–3409. DOI: 10.1002/ece3.3907.

Kalkman, V.J., Clausnitzer, V., Dijkstra, K.-D.B., Orr, A.G., Paulson, D.R. and van Tol, J. (2008) Global diversity in dragonflies (Odonata) in freshwater. *Hydrobiologia* 595, 351–363. DOI: 10.1007/s10750-007-9029-x.

Kalkman, V.J., Boudot, J.-P., Futahashi, R., Abbott, J.C., Bota-Sierra, C.A. *et al.* (2022) Diversity of Palaearctic dragonflies and damselflies (Odonata). *Diversity* 14, e966. DOI: 10.3390/d14110966.

Kanke, E., Suzuki, K., Sekiné, K., Suzuki, T., Hatta, K., Yang, M.-M. and Tojo, K. (2021) Unexpected population genetic structure in two closely related euphaeid damselflies from the Yaeyama and Taiwan Islands (Odonata: Euphaeidae). *Biological Journal of the Linnean Society* 134, 214–228. DOI: 10.1093/biolinnean/blab061.

Karube, H., Futahashi, R., Odajima, T. and Odajima, K. (2011) An occurrence of a southeastern Asiatic species, *Pseudagrion australasiae* Selys, in Japan: a possible case of accidental introduction. *Tombo* 53, 111–114.

Karube, H., Sano, S., Nagasaki, K., Nagasaki, M. and Futahashi, R. (2014) Distributional expansion of alien species *Ceriagrion auranticum ryukyuanum* in southeastern Kanto region and trial of extermination. *Aeschna* 50, 139–143.

Keil, P., Simova, I., Hawkins, B.A. (2008) Water energy and geographical species richness pattern of European and North African dragonflies (Odonata). *Insect Conservation and Diversity* 1, 142–150. DOI: 10.1111/j.1752-4598.2008.00019.x.

Khrokalo, L.A. (2010) Expansion of *Crocothemis erythraea* in Ukraine. *BioRisk* 5, 211–223. DOI: 10.3897/biorisk.5.851.

Kim, M.J., Jung, K.S., Park, N.S., Wan, X., Kim, K.-G. *et al.* (2014) Molecular phylogeny of the higher taxa of

Odonata (Insecta) inferred from CO1, 16S rRNA, 28S rRNA, and EF1-α sequences. *Entomological Research* 44, 65–79. DOI: 10.1111/1748-5967.12051.

Kipping, J., Clausnitzer, V. and Dijkstra, K.-D.B. (2023) The highlands and escarpments of Angola as an endemism hotspot for African dragonflies and damselflies (Insects: Odonata). *Namibian Journal of Environment* 8 (Monograph on Endemism in the Highlands and Escarpments of Angola and Namibia), 173–186.

Koch, K., Wagner, C. and Sahlén, G. (2014) Farmland versus forest: comparing changes in Odonata species composition in western and eastern Sweden. *Insect Conservation and Diversity* 7, 22–31. DOI: 10.1111/icad.12034.

Kohli, M., Djernæs, M., Sanchez Herrera, M., Sahlen, G., Pilgrim, E. *et al.* (2021a) Comparative phylogeography uncovers evolutionary past of Holocene dragonflies. *PeerJ* 9, e11338. DOI: 10.7717/peerj.11338.

Kohli, M., Letsch, H., Greve, C., Béthoux, O., Deregnaucourt, I. *et al.* (2021b) Evolutionary history and divergence times of Odonata (dragonflies and damselflies) revealed through transcriptomics. *iScience* 24, e103324. DOI: 10.1016/j.isci.2021.103324.

Korkeamäki, E., Elo, M., Sahlén, G., Salmela, J. and Suhonen, J. (2018) Regional variations in occupancy frequency distribution patterns between odonate assemblages in Fennoscandia. *Ecosphere* 9, e02192. DOI: 10.1002/ecs2.2192.

Koroiva, R., Pepinelli, M., Rodrigues, M.E., Roque, F. de O., Lorenz-Lemke, A.P. and Kvist, S. (2017) DNA barcoding of odonates from the Upper Plata Basin: Database creation and genetic diversity distribution. *PLoS ONE* 12, e0182283. DOI: 10.1371/journal.pone.0182283.

Kosterin, O.E. (2005) Western range limits and isolates of eastern odonate species in Siberia and their putative origins. *Odonatologica* 34, 219–242.

Kulijer, D., De Knijf, G. and Franković, M. (2013) Review of the Odonata of Bosnia and Herzegovina. *Odonatologica* 42, 109–123.

Kumar, C.S., George, B.S. and Anooj, S.S. (2021) Range extension and high altitudinal record of *Sympetrum meridionale* (Selys) (Odonata: Libellulidae). *Indian Journal of Entomology*, e20353. DOI: 10.55446/IJE.2021.301.

La Porta, G., Landi, F., Leandri, F. and Assandri, G. (2023) The new checklist of the Italian fauna: Odonata. *Biogeographia – The Journal of Integrative Biogeography* 38, uc1009. DOI: 10.21426/B638158781.

Laister, G., Lehmann, G. and Martens, A. (2014) Exotic Odonata in Europe. *Odonatologica* 43, 125–135.

Leigh Jr., E.G., Vermeij, G.J. and Wikelski, M. (2009) What do human economies, large islands and forest fragments reveal about the factors limiting ecosystem evolution? *Journal of Evolutionary Biology* 22, 1–12. DOI: 10.1111/j.1420-9101.2008.01624.x.

Letsch, H., Gottsberger, B. and Ware, J.L. (2016) Not going with the flow: a comprehensive time-calibrated phylogeny of dragonflies (Anisoptera: Odonata: Insecta) provides evidence for the role of lentic habitats on diversification. *Molecular Ecology* 25, 1340–1353. DOI: 10.1111/mec.13562.

Li, F. and Park, Y.-S. (2020) Habitat availability and environmental preference drive species range shifts in concordance with climate change. *Diversity and Distributions* 26, 1343–1356. DOI: 10.1111/ddi.13126.

Lorenzo-Carballa, M.O. and Cordero-Rivera, A. (2009) Thelytokous parthenogenesis in the damselfly *Ischnura hastata* (Odonata, Coenagrionidae): genetic mechanisms and lack of bacterial infection. *Heredity* 103, 377–384. DOI: 10.1038/hdy.2009.65.

Lorenzo-Carballa, M.O., Garrison, R.W., Encalada, A.C. and Cordero-Rivera, A. (2021) Darwin returns to the Galapagos: genetic and morphological analyses confirm the presence of *Tramea darwini* at the archipelago (Odonata, Libellulidae). *Insects* 12, e21. DOI: 10.3390/insects12010021.

Low, V., Rashid, Y., Yusoff, A., Siow, W., Prakash, B. *et al.* (2017) Pleistocene and high gene flow in the Globe skimmer dragonfly *Pantala flavescens* Fabricius (Odonata: Libellulidae) in Peninsula Malaysia. *Zoologischer Anzeiger* 266, 23–27. DOI: 10.1016/j.jcz.2016.10.002.

McPeek, M.A. and Gavrilets, S. (2006) The evolution of female mating preferences: differentiation from species with promiscuous males can promote speciation. *Evolution* 60, 1967–1980.

Miguel, T.B., Calvão, L.B., Alves-Martins, F., Batista, J.D., Rodrigues, M.E. *et al.* (2022) Odonates in warm regions of South America largely do not follow Rapoport's rule. *Biodiversity and Conservation* 31, 565–584. DOI: 10.1007/s10531-021-02350-0.

Miranda Filho, J.C., Mendoza-Penagos, C.C., Calvão, L.B., Miguel, T.B., Bastos, R.C. *et al.* (2022) Checklist of damselflies and dragonflies (Odonata) from Acre state, and the first record of *Drepanoneura loutoni* von Ellenrieder & Garrison, 2008 for Brazil. *Biota Neotropica* 22, e20211320. DOI: 10.1590/1676-0611-BN-2021-1320.

Moody, K.N., Gagne, R.B., Heim-Ballew, H., Alda, F., Hain, E.F. *et al.* (2017) Invasion hotspots and ecological saturation status of streams across the Hawaiian archipelago. *Cymbium: International Journal of Ichthyology* 41, 127–156.

Móra, A. and Sebteoui, K. (2020) *Trithemis arteriosa* (Burmeister, 1839) (Odonata: Libellulidae) in Hungary: can aquarium trade speed up the area expansion of Mediterranean species? *North-Western Journal of Zoology* 16, e207204.

Nogués-Bravo, D., Rodríguez-Sánchez, F., Orsini, L., de Boer, E., Jansson, R. *et al.* (2018) Cracking the code of biodiversity responses to past climate change. *Trends in Ecology and Evolution* 33, 765–776. DOI: 10.1016/j.tree.2018.07.005.

Nowak, M. and Weihrauch, F. (2022) *Orthetrum ransonnetii* has gained a foothold in the Canary Islands (Odonata: Libellulidae). *Notulae Odonatologicae* 9, 291–295. DOI: 10.5281/zenodo.4746218.

Oertli, B. (2010) The local species richness of dragonflies in mountain waterbodies: an indicator of climate warming? *BioRisk* 5, 243–251. DOI: 10.3897/biorisk.5.853.

Olsen, K., Svenning, J.-C. and Balslev, H. (2022) Niche breadth predicts geographical range size and northern range shift in European dragonfly species (Odonata). *Diversity* 14, e719. DOI: 10.3390/d14090719.

Oppel, S. (2005) Habitat associations of an Odonata community in a lower montane rainforest in Papua New Guinea. *International Journal of Odonatology* 8, 243–257. DOI: 10.1080/13887890.2005.9748256.

Orr, A.G. (2004) Critical species of Odonata in Malaysia, Indonesia, Singapore and Brunei. *International Journal of Odonatology* 7, 371–384.

Orr, A.G. (2006) Odonata in Bornean tropical rain forest formations: diversity, endemicity and implications for conservation management. In: Cordero Rivera, A. (ed.) *Forests and Dragonflies*. Pensoft, Sofia, Bulgaria, pp. 51–78.

Ott, J. (1988) Beiträge zur biologie und zum status von *Crocothemis erythraea* (Brullé, 1832). *Libellula* 7, 1–25.

Ott, J. (2010) Dragonflies and climate change: recent trends in Germany and Europe. *BioRisk* 5, 253–286. DOI: 10.3897/biorisk.5.857.

Ott, J. (2023) Hat *Sympetrum fonscolombii* (Selys, 1840) am kranichwoog bei hütschenhausen (Rheinland-Pfalz, Landstuhler Moorniederung) als larve überwintert (Odonata: Libellulidae)? *Libellula* 42 (1/2) 41, 69–78.

Parr, A.D. (2020) Migrant and dispersive dragonflies in Britain during 2019. *Journal of the British Dragonfly Society* 36 (2), 67–83.

Paulson, D.R. (1978) An Asiatic dragonfly, *Crocothemis servilia* (Drury), established in Florida (Anisoptera: Libellulidae). *Notulae odonatologicae* 1, 9–10.

Paulson, D.R. (2006) The importance of forests to Neotropical dragonflies. In: Rivera, A.C. (ed.) *Forests and Dragonflies.* Pensoft, Sofia, Bulgaria, pp. 79–102.

Paulson, D.R., Schorr, M. and Deliry, C. (2024) Available at:https://www.pugetsound.edu/puget-sound-museum-natural-history/biodiversity-resources/insects/dragonflies/world-odonata-list (accessed 6 March 2024).

Pinkert, S., Dijkstra, K.-D.B., Zeuss, D., Reudenbach, C., Brandl, R. and Hof, C. (2018) Evolutionary processes, dispersal limitation and climatic history shape current diversity patterns of European dragonflies. *Ecography* 41, 795–804. DOI: 10.1111/ecog.03137.

Pires, M.M., Siegloch, A.E., Hernández, M.I.M. and Petrucio, M.M. (2020) Environmental drivers and composition of assemblages of immature odonates (Insecta) in a subtropical island in southern Brazil. *Acta Limnologica Brasiliensia* 32, e2. DOI: 10.1590/S2179-975X8017.

Platts, P.J., Mason, S.C., Palmer, G., Hill, J.K., Oliver, T.H. *et al.* (2019) Habitat availability explains variation in climate-driven range shifts across multiple taxonomic groups. *Scientific Reports* 9, e15039. DOI: 10.1038/s41598-019-51582-2.

Polhemus, D. and Asquith, A. (1996) *Hawaiian Damselflies: A Field Identification Guide*. Bishop Museum Press, Honolulu, USA.

Powney, G.D., Cham, S.S.A., Smallshire, D. and Isaac, N.J.B. (2015) Trait correlates of distribution trends in Odonata of Britain and Ireland. *PeerJ* 3, e1410. DOI: 10.7717/peerj.1410.

Ramírez, A., Maldonado-Benítez, N., Mariani-Ríos, A. and Figueroa-Santiago, J. (2020) Dragonflies and damselflies (Odonata) from Puerto Rico: a checklist with notes on distribution and habitat. *PeerJ* 8, e9711. DOI: 10.7717/peerj.9711.

Ramos, K.A.M., Nuñeza, O.M. and Villanueva, R.J.T. (2020) Species diversity of Odonata in Mimbilisan Protected Landscape, Misamis Oriental, Philippines. *Asian Journal of Conservation Biology* 9, 280–289. ISSN: 2278-7666.

Renner, S., Périco, E., Dalzochio, M.S. and Sahlén, G. (2019) Ecoregions within the Brazilian Pampa biome reflected in Odonata species assemblages. *Austral Ecology* 44, 461–472. DOI: 10.1111/aec.12680.

Rosset, V. and Oertli, B. (2011) Freshwater biodiversity under climate warming pressure: identifying the winners and losers in temperate water bodies. *Biological Conservation* 144, 2311–2319. DOI: 10.1016/j.biocon.2011.06.009.

Rosset, V., Lehmann, A. and Oertli, B. (2010) Warmer and richer? Predicting the impact of climate warming on species richness in small temperate waterbodies. *Global Change Biology* 16, 2376–2387. DOI: 10.1111/j.1365-2486.2010.02206.x.

Sahlén, G. and Ekestubbe, K. (2001) Identification of dragonflies (Odonata) as indicators of general species richness in boreal forest lakes. *Biodiversity and Conservation* 10, 673–690.

Samways, M.J. (2003) Conservation of an endemic dragonfly fauna in the Seychelles archipelago. *Odonatologica* 32, 177–182.

Samways, M.J. (2010) Impacts of extreme weather and climate change on South African dragonflies. *BioRisk* 5, 73–84. DOI: 10.3897/biorisk.5.843.

Samways, M.J. and Deacon, C. (2022) Extinction reprieve for the ancient and imperiled dragonflies at the southern tip of Africa. In: DellaSala, D.A. and Goldstein, M.I. (eds) *Imperiled: The Encyclopedia of Conservation*, vol. 1. Elsevier, pp. 471–484. DOI: 10.1016/B978-0-12-821139-7.00047-7.

Samways, M.J. and Niba, A.S. (2010) Climate and elevational range of a South African dragonfly assemblage. *BioRisk* 5, 85–107. DOI: 10.3897/biorisk.5.844.

Samways, M.J. and Osborn, R. (1998) Divergence in a transoceanic circumtropical dragonfly on a remote island. *Journal of Biogeography* 25, 935–946.

Samways, M.J., Pryke, J.S. and Simaika, J.P. (2011) Threats to dragonflies on land islands can be as high as those on oceanic islands. *Biological Conservation* 144, 1145–1151. DOI: 10.1016/j.biocon.2010.12.034.

Sandall, E.L., Pinkert, S., and Jetz, W. (2022). Country-level checklists and occurrences for the world's Odonata (dragonflies and damselflies). *Journal of Biogeography* 49, 1586–1598. DOI: 10.1111/jbi.14457.

Scales, J.A. and Butler, M.A. (2016) The relationship between microhabitat use, allometry and functional variation in the eyes of *Megalagrion* damselflies. *Functional Ecology* 30, 30, 356–368. DOI: 10.1111/1365-2435.12479.

Schilling, E.G., Lawrenz, R. and Kundel, H. (2019) As assessment of the geographic distribution and status of the rare dragonfly, *Rhionaeschna mutata*, at the northeastern edge of its range. *Northeastern Naturalist* 26, 523–536. DOI: 10.1656/045.026.0305.

Shah, R.D.T., Shah, D.N. and Domisch, S. (2012) Range shifts of a relict Himalayan dragonfly in the Hindu Kush Himalayan region under climate change. *International Journal of Odonatology* 15, 209–222. DOI: 10.1080/13887890.2012.697399.

Sieben, E.J.J., Glen, R.P., van Deventer, H. and Dayaram, A. (2021) The contribution of wetland flora to regional floristic diversity across a wide range of climatic conditions in southern Africa. *Biodiversity and Conservation* 30, 575–596. DOI: 10.1007/s10531-020-02104-4.

Simaika, J.P. and Samways, M.J. (2015) Predicted range shifts of dragonflies over a wide elevation gradient in the southern hemisphere. *Freshwater Science* 34, 1133–1143. DOI: 10.1086/682686.

Simaika, J.P., Ware, J.L., Garrison, R. and Samways, M.J. (2020) Phylogeny of the Synlestidae (Odonata: Zygoptera), with special emphasis on *Chlorolestes* Selys and *Ecchlorolestes* Barnard. *Scientific Reports* 10, e15088. DOI: 10.1038/s41598.

Simonsen, T.J., Olsen, K. and Djernæs, M. (2020) The African-Iberian connection in Odonata: mtDNA and ncDNA based on phylogeography of *Aeshna cynanea* (Müller, 1764) (Odonata: Aeshnidae) in Western Palaearctic. *Arthropod Systematics and Phylogeny* 78, 309–320. DOI: 10.26049/ASP78-2-2020-06.

Sparrow, D.J., De Knijf, G. and Sparrow, R.L. (2021) Diversity, status and phenology of the dragonflies and damselflies of Cyprus (Insecta: Odonata). *Diversity* 13, e352. DOI: 10.3390/d13110532.

Sparrow, D.J., Makris, C., Sparrow, R., Michaelides, M., Konis, D. and De Knijf, G. (2022) First records of *Aeshna isoceles* and the rediscovery of *Lestes barbarus* on Cyprus (Odonata: Lestidae, Aeshnidae). *Notulae odonatologicae* 9, 185–195. DOI: 10.5281/zenodo.3823257.

Stainforth, D.A. and Calel, R. (2020) New priorities for climate science and climate economics in the 2020s. *Nature Communications* 11, e3864. DOI: 10.1038/s41467-020-16624-8.

Standring, S., Sánchez-Herrera, M., Guillermo-Ferreira, R., Ware, J.L., Vega-Sánchez, Y.M. *et al.* (2022) Evolution and biogeographic history of rubyspot damselflies (Heterininae: Calopterygidae: Odonata). *Diversity* 14, e757. DOI: 10.3390/d14090757.

Starr, S.M. and McIntyre, N.E. (2020) Effects of water temperature under predicted climate change on the development and survival of *Enallagma civile* (Odonata: Coenagrionidae). *Environmental Entomology* 49, 230–237. DOI: 10.1093/ee/nvz138.

Sternberg, K. (1998) The postglacial colonization of central Europe by dragonflies, with special reference to southwestern Germany (Insecta, Odonata). *Journal of Biogeography* 25, 319–337.

Štih, A. and Koren, T. (2022) Dragonfly fauna (Insecta: Odonata) of the Brijuni National Park, Croatia. *Natura Croatica* 31, 19–30. DOI: 10.20302/NC.2022.31.2.

Suhling, F., Martens, A. and Marais, E. (2009) How to enter a desert: patterns of Odonata colonisation of arid Namibia. *International Journal of Odonatology*, 12, 287–308. DOI: 10.1080/13887890.2009.9748346.

Suhling, F., Sahlén, G., Gorb, S., Kalkman, V.J., Dijkstra, K.-D.B. and van Tol, J. (2015) Order Odonata. In: Thorp, J.H. and Rogers, D.C. (eds) *Thorp and Covich's Freshwater Invertebrates*, 4th ed. Elsevier, pp. 893–932. DOI: 10.1016/B978-0-12-385026-3.00035-8.

Suhling, F., Martens, A. and Suhling, I. (2017) Long-distance dispersal in Odonata: Examples from arid Namibia. *Austral Ecology* 42, 544–552. DOI: 10.1111/aec.12472.

Suvarov, A., Scornavacca, C., Fujimoto, M.S., Bodily, P., Clement, M. *et al.* (2022) Deep ancestral introgression shapes evolutionary history of dragonflies and damselflies. *Systematic Biology* 71, 526–546. DOI: 10.1093/sysbio/syab063.

Taybi, A.F., Mabrouki, Y., Berrahou, A., Sbaa, M. and Brochard, C. (2019) New data on the dragonfly fauna (Odonata) of the Moulouya River Basin and Oriental Region, Morocco. *Arxius de Miscellània Zoològica* 17, 85–108. DOI: 10.32800/amz.2019.17.0085.

Termaat, T., van Strien, A.J., van Grusven, R.H.A., De Knijf, G., Bjelke, U. *et al.* (2019) Distribution trends of European dragonflies under climate change. *Diversity and Distributions* 25, 936–950. DOI: 10.1111/ddi.12913.

Tng, D.Y.P., Apgaua, D.M.G., Ishida, Y.F., Mencuccini, M., Lloyd, J. *et al.* (2018) Rainforest trees respond to drought by modifying their hydraulic architecture. *Ecology and Evolution* 8, 12479–12491. DOI: 10.1002/ece3.4601.

Troast, D., Suhling, F., Jinguji, H., Sahlén, G. and Ware, J. (2016) A global population genetic study of *Pantala flavescens*. *PLoS ONE* 11, e0148949. DOI: 10.1371/journal.pone.0148949.

Turgeon, J., Stoks, R., Thum, R.A., Brown, J.M. and McPeek, M.A. (2005) Simultaneous Quaternary radiations of three damselfly clades across the Holarctic. *The American Naturalist* 165, E78–E107.

van Strien, A.J. and van Grunsven, R.H.A. (2023) In the past 100 years dragonflies declined and recovered by habitat restoration and climate change. *Biological Conservation* 277, e109865. DOI: 10.1016/j.biocon.2022.109865.

Vivas-Santeliz, J. and De Marmels, J. (2017) Current knowledge of Odonata in Venezuela: diversity and distribution of endemic taxa. *Odonatologica* 46, 35–54. DOI: 10.5281/zenodo.572355.

Wang, J.-F., Chen, M.-Y., Chaw, S.-M., Morii, Y., Yoshimura, M. *et al.* (2014) Complete mitochondrial genome of an enigmatic dragonfly, *Epiophlebia superstes* (Odonata, Epiphlebiidae). *Mitochondrial DNA* 1940–1744. DOI: 10.3109/19401736.2013.845756.

Ware, J.L., Simaika, J.P. and Samways, M.J. (2009) Biogeography and divergence time estimation of the relic Cape dragonfly genus *Syncordulia*: global significance and implications for conservation. *Zootaxa*, 2216, 22–36.

Weihrauch, F. (2011) A review of the distribution of Odonata in the Macaronesian Islands, with particular reference to the *Ischnura* puzzle. *Journal of the British Dragonfly Society* 27, 28–46.

Weihrauch, F., Vieira, V., Cordero-Rivera, A., and de Santos Loureiro, N. (2016) Update on the zoogeography of Odonata in the Macaronesian Islands. *Boletín Rola* 8, 9–22.

Wellenreuther, M. and Sánchez-Guillén, R.A. (2016) Nonadaptive radiation damselflies. *Evolutionary Applications* 9, 103–118. DOI: 10.1111/eva.12269.

4 A Trait Perspective on Dragonfly Conservation

Illustration: Megan Louise Potgieter.

DOI: 10.1079/9781789248395.0004

Abstract

Conserving dragonflies means understanding their traits relative to the changing environment. All traits operating together determine the fate of individuals, populations and species in a changing environment and according to their combined genetics and trait plasticity.

Each developmental stage, from egg through larva to adult, must survive into the next stage, and produce a viable next generation. Any later stage must live with legacies of events that shaped an earlier stage, the grist of natural and adult sexual selection.

Selection is survival of the combined effects of abiotic conditions and intra- and inter-specific interactions. Non-dragonfly species can be beneficial, whether certain plant types, presence of enough prey, internal gut symbionts or beaver dam building. Other species are harmful, be they competitors or predators and parasites.

Successful dragonfly conservation aims to provide the right watery environment and adequate quality land space, while reducing mortality from all adverse factors, both abiotic and biotic.

4.1 The Dragonfly As an Amphibiotic Organism

4.1.1 Concept of habitat relative to traits

Dragonflies, like all organisms, are open systems where energy and matter flows into and out of them. They have an outer and inner environment. Outside the body there are both abiotic and biotic environmental variables to which they are subject, and to which they respond. Inside the body, there is a milieu of biochemistry and physiology, as well as symbionts which benefit, and parasites which harm, the individual (Fig. 4.1).

Dragonfly traits, while broadly characteristic of each species, are constrained by their phylogenetic past, honed by events, experiences and natural selection in both deep and recent time. In turn, among individuals within any one species there is a range of genetic diversity, including various degrees of genetic plasticity, relating to their traits and their responses to changing environmental conditions.

Extant relative abundance of a species reflects in part the lasting influence of more recent refugia on demographic dynamics. In contrast, phylogenetic composition reflects the influence of more ancient habitat change. Species are lost over time due to extinction events, either through direct impact and/or genetic deterioration, especially among rare specialists. These extinction events result in reduced assemblage composition, particularly when environmental bottlenecks are long and intense (Barthelemy *et al.*, 2021).

A dragonfly's habitat is the area necessary for it to carry out all its life functions. Life functions include development, feeding, mating and breeding, overall involving both the aquatic and terrestrial realms. In turn, individuals seek optimal habitat for any one of these life functions at any one time. The larva may seek the cover of aquatic plants or, in the case of other species, the substrate, where it can find choice prey in adequate abundance, and where it will be most safe from predators. The emerging larva then has to find the right vegetation or a hard surface on which to crawl out of the water, usually under cover of darkness to reduce risk of being eaten. The teneral, soft but well-camouflaged, seeks the shelter of vegetation nearby. The young adult shows little interest in the water body, and may disperse far from water, as far as 20 km in the case of *Trithemis hecate* in the dry and open landscape of Botswana.

On maturing, the adult, according to species and sex, may or may not return to water, according to its needs at the time. A further component of habitat are the locations where both sexes feed and breed, which may be close to, or away from, the water. Females must return to the vicinity of the water to lay eggs (Fig. 4.2).

As the various individuals in any one population vary slightly in their genotype, the optimal habitat also varies slightly from one individual to the next, according to their genetically determined requirements, preferences and responses – the dual processes of natural and sexual selection. Different populations of some species at far apart locations may reside in slightly different habitats. An additional perspective is that the collective of individuals making up a population provides that population with some variation in genotypes that equip the population with genetic and phenotypic resilience in times of environmental change, whether natural or driven by human activity. This means that 'the habitat' of a species across its geographic range is the sum of the actual habitats occupied by the various populations of a species present in each defined area. Its ability to occupy such a range of particular habitats is its realized adaptive capacity, which for some species becomes greater under human-induced landscape change, but for others it is reduced.

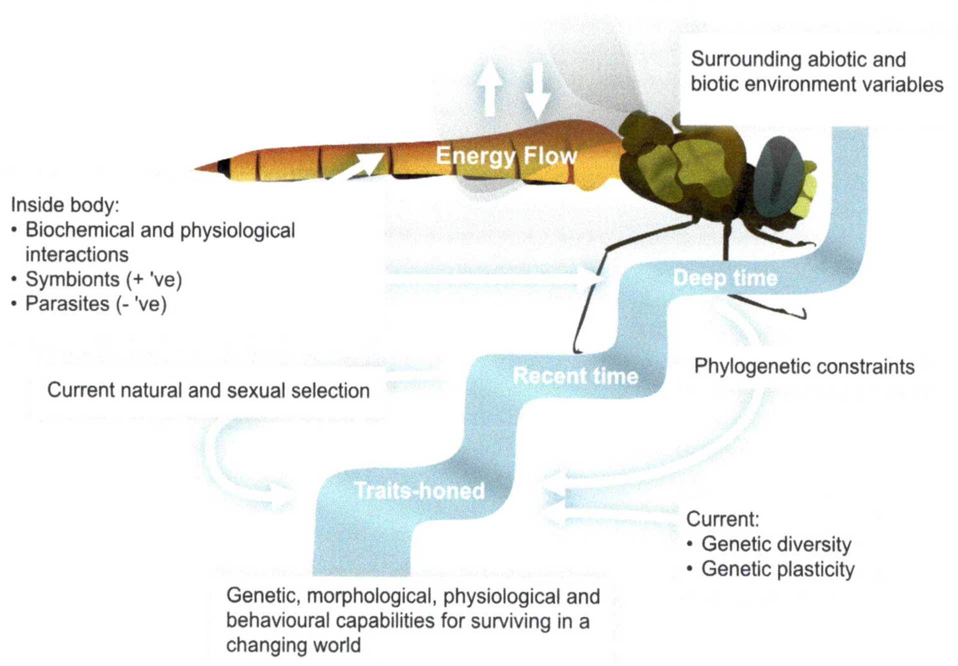

Fig. 4.1. The dragonfly is an open system, through which energy flows in and out of its body via various biochemical and physiological processes. While a host of surrounding abiotic and biotic variables impinge on its body, it reacts to them according to its traits to survive. However, other organisms are also involved, with symbionts positively (+ 've) affecting its survival and parasites having a negative (– 've) effect. As time passes, the dragonfly evolves but retains the legacy of past events in the form of phylogenetic constraints. Survival in current times depends on how it is equipped with genetic diversity and plasticity, honed by natural and sexual selection, against a background of the phylogenetic constraints. This genetic base manifests as morphological, physiological and behavioural capabilities – its traits – for surviving in an ever-changing world. (Illustration: Robyn Symons).

The importance of specific habitat requirements is emphasized when among some closely related and morphologically similar species, there are nevertheless distinct differences in habitat preference. Such disparate habitat preferences, even within the same landscape, alert us to the possibility of two or more cryptic biological species with similar morphology being present, as in the case of some African *Trithemis* species (Damm *et al.*, 2010a).

When we say 'the habitat of a species', we are really referring to all its requirements and preferences necessary to complete its life cycle and maintain effective populations. This may include parts of the local or wider landscape occupied only temporarily. A dragonfly may forage over an agricultural field where prey is abundant at the time, even when this field does not constitute part of its natural habitat. But being such a mobile organism, a dragonfly may, for a brief while at least, be able to fly over this artificial landscape to its benefit before moving on to a preferred natural area to undertake other activities. In effect, it is percolating across the landscape seeking optimal habitat for its current needs.

Visualization of the dynamic activities of a dragonfly enables us in our conservation thinking to always consider the functional connectivity between the habitat patches of a species. This connectivity is determined in part by the environment, but also by the traits of the individuals when on the move. However, connectivity is more than just about the phenotype and the landscape. From a conservation perspective, importantly the phenotype/landscape interaction must support and promote genetic diversity

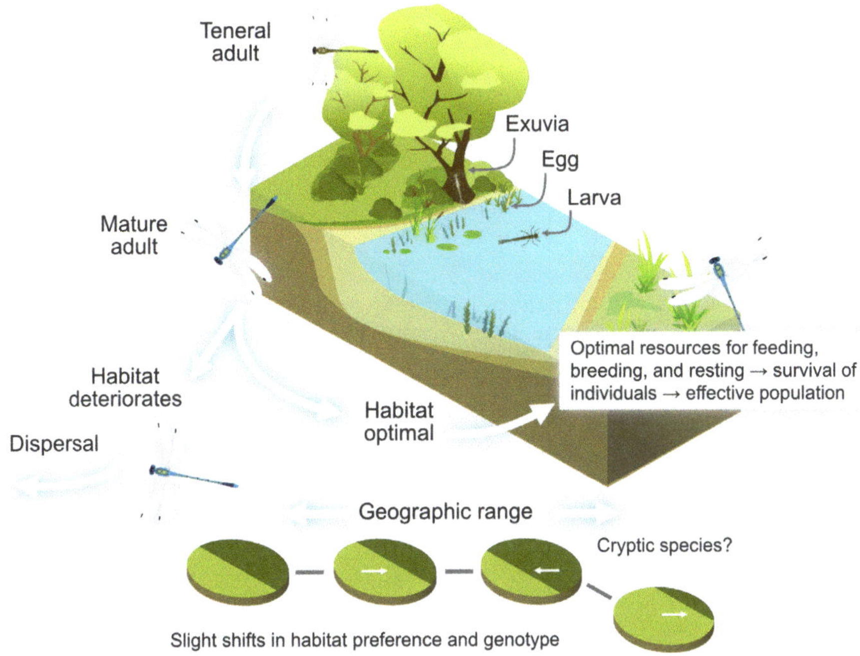

Fig. 4.2. The habitat is the total dynamic space where a resident dragonfly can harness resources and successfully complete its life cycle. At the population level, this means maintaining an effective population. While this is the case for an optimal habitat, environmental conditions may change, and the habitat deteriorates, leading to dispersal. Across a species' geographic range there are sometimes small shifts in habitat preference and its genotype, represented here by proportions of light green versus dark green in the circles. Sometimes though, while there is little apparent difference in morphology, there may be a distinct difference in habitat preference, which may indicate that a cryptic species is present (bottom far right). (Illustration: Robyn Symons).

and hence local survivability of a species. This is especially so in rapidly changing landscapes.

When a habitat becomes suboptimal, through drought or severe agricultural disturbance, for example, individuals need to be given the opportunity to move in search of better conditions. We can even help them on their way by deliberately providing appropriate stepping-stone habitats, such as artificial ponds. These temporary habitats act as conduits or stop-over points to help dispersing dragonflies on their way to their new residential areas.

In dragonfly conservation, it is sometimes useful to distinguish between 'habitat' and 'biotope'. A biotope is the small, defined physical area where a dragonfly individual is present, with the term used to describe the structure of the vegetation and physical features, expressed, for example, as 'a well-vegetated pond with a dense margin of sedges', or 'the biotope where the damselfly was observed was an isolated clump of trees'. Usually, biotopes are singular physical parts of the total habitat.

4.1.2 Types of traits

4.1.2.1 Traits relative to species history

As traits are the characteristics of an organism, they represent the survivability of individuals of a species that enable it to maintain an effective population. Traits are determined genetically, although made operational by physiology, behaviour and ecology relative to prevailing environmental conditions, though honed by phylogenetic constraints. Traits are also phenotypically plastic, emanating from different allele expressions among individuals in a population. This phenotypic plasticity is especially important when environmental conditions are extreme. For instance, among a set of dragonfly conspecific larvae that experience a short-term elevation in water temperature, some will survive the heatwave while others will not. Those that survive may do so because there is phenotypic plasticity in the population (i.e. the phenotypic variation in response to different environments). What this means is that there is sufficient variation in their collective genomes (genetic plasticity), and their physiology, behaviour and morphology enable them to survive certain environmental changes.

Traits are also honed by past environmental conditions and particular events to which the organism had responded and survived through natural and sexual selection retaining favourable mutations. Those that did not survive may have suffered from retroposition, deleterious mutations and genetic drift in small populations. These environment–genetic interactions are ongoing, varying among populations at different geographic locations, but can also vary within one small area, driven by natural and sexual selection and random genetic change.

However, the extent to which extant traits equip a species to survive into the future, given current climate and environmental change, is largely still to be determined. In other words, some traits will be favourable and others not, depending on future environmental conditions, abiotic and biotic, at a given location. Yet inferences can be gleaned by investigating how the effect of past events, especially those relating to range retraction versus range expansion, have shaped current evolutionary and ecological trait diversity. This means that past events should be assessed relative to extinction risk in the future (Rocha-Ortega *et al.*, 2023).

Furthermore, focusing on organisms and their traits enables predictions on the spatio-temporal distribution of groups of organisms in terms of how metabolic energy is distributed over areas, time and resources, so structuring ecological communities and determining ecosystem dynamics (Musters *et al.*, 2023).

4.1.2.2 Operational trait suites

It is the total of all the traits, operating synchronously, that determine the survival and reproductive success of a local population. However, certain traits may be critical at one time and place. For example, individuals with high sensitivity to impending catastrophic environmental change and have good flight ability may be the ones that soon leave a drying pond to seek a new one with perennial water. In turn, suitable individual traits enable an individual and species to overcome certain transient environmental bottlenecks in any one area at one time. Certain individuals of a species may survive devastation of their natural habitat by a severe storm or fire, while other individuals may not.

The proportion of individuals surviving a particular extreme environmental event, will vary among species, and according to the type, severity and longevity of the event. There may even be additional environmental pressures at the same time. These impacts may be novel, such as a severe freeze, pesticide application or local climate warming, with the synergism between events perhaps being too much for long-term survival of a local population. Certain individuals of the population at one location and time may not have the right traits enabling them to respond well to these novel stressors and may perish. Other individuals, however, may leave the unsuitable habitat. This dispersal activity could possibly involve the use of temporary stepping-stone habitats, to seek new, more optimal, habitat.

4.1.2.3 Biological and ecological traits

Traits may be categorized as biological or ecological. Biological traits are inherent characteristics associated with the functioning of a species, related to life history and mobility. They may be physiological, behavioural or morphological. For example, 'the dragonfly species has biological traits that enable it to develop quickly as puddle water temperatures increase and its pool begins to dry out, and being a powerful flyer, readily disperses to colonize distant potential habitats (e.g. *Pantala flavescens*)'. In turn, ecological traits refer to biotope specialization and habitat preference. For example, 'the damselfly species has ecological traits that enable it to survive best in small, perennial ponds, rich in submerged macrophytes, lily pads, and algal mats (e.g. *Pseudagrion massaicum*)'.

Biological and ecological traits are useful categories when describing the innate survivability of a species under certain prescribed conditions, such as climate-induced warming of montane bog ponds. However, operationally the two types of traits entwine. Together all the traits characterize the innate ability of a population or a species to survive and leave viable offspring given certain environmental conditions in space and time.

Another view related to traits is the probability of occurrence of a specific species at a locality relative to the presence of other species (species co-occurrence) given a set of certain environmental/climatic variables. The particular suites of functional traits of two species may reduce their co-occurrence arising from high niche overlap that drives negative interactions. Alternatively, co-occurrence is promoted when there is niche separation following trait-mediated environmental filtering. Discovering which of the two is most important is being tackled using dragonflies. The upshot is that ecological species traits can predict species co-occurrence. This is achieved using a Hierarchical Modelling of Species Communities framework to quantify the pairwise species co-occurrences after controlling for environmental and climatic factors, followed by machine learning to generate models which are capable of accurately predicting the observed co-occurrence patterns from species functional traits (Cerini *et al.*, 2023).

4.1.2.4 Modelling the importance of phenotypic plasticity

Although phenotypic plasticity is the ability of a single genotype to produce different phenotypes according to prevailing environmental conditions, there is the issue of continuously fluctuating environmental conditions. Yet organisms, having evolved strategies to survive these fluctuations, are potentially genetically equipped for future evolutionary dynamics. Lalejini *et al.* (2021) investigated adaptive phenotypic plasticity using digital organisms in the form of self-replicating computer programs. They found that populations with adaptive phenotypic plasticity undergo less evolutionary change than non-plastic populations, which rely on new mutations to provide genetic variation to continuously adapt to environmental fluctuation. The benefit of phenotypic plasticity is that it stabilizes populations against environmental fluctuation. This in turn, enables plastic populations to retain novel adaptive traits to a greater degree in comparison with the non-plastic organisms. In sum, evolution of adaptive phenotypic diversity shifts evolutionary dynamics toward that of a population in a static environment, with greater opportunities for continuance of a viable population (Fig. 4.3).

4.1.2.5 Aspects of traits in dragonfly biology, ecology and conservation

There are phylogenetic constraints on dragonfly traits. For example, across the UK, zygopteran wing length decreases as temperature increases and as the season advances. In contrast, anisopteran wing length in general does change with increasing temperature. Zygopterans, which generally have a smaller body size than anisopterans, show greater sensitivity to temperature than anisopterans. Furthermore, these phylogenetic constraints are stronger than traits relating to generation times. Also, zygopteran individuals flying late in the season tend to be smaller than those of the same species flying early in the season, and adults tend to be smaller in years with warm seasonal temperatures during larval development through acceleration of the developmental rate of the larvae at the cost of smaller adult body size. This is not the case for anisopterans (Wonglersak *et al.*, 2020).

Bergmann's rule refers to endotherms (animals that physiologically generate heat to warm their bodies) having larger body size in colder environments, whether toward the poles or with increasing elevation. A counterbalance is the resource availability hypothesis, which states that a species' size is determined by its energetic requirements, so that regions with greater resource availability, as in the tropics, should support larger species than regions with lower productivity. In a spatially comparative analysis, Mähn *et al.* (2023) investigate these drivers in the largely ectothermic odonates at the global scale. Using temperature (mean annual temperature and elevation) and the enhanced vegetation index (as a proxy for productivity), they found strong but contrasting effects for temperature between Anisoptera and Zygoptera and consistent positive effects for productivity that explained 35–57% of body size variation. They conclude that temperature, productivity and thermal conservatism relative to size-based thermoregulation variously determine the geographic distribution of odonates. As there is a strong phylogenetic signal, and as some species are thermally constrained while others will likely disperse, the implication for conservation, in an ongoing warming world, is that there will be species re-sorting at the local level.

4.1.2.6 Significance of suites of traits in dragonflies

Suites of traits also encompass a certain degree of genetic plasticity associated with each of the various traits. However, the genotypes underlying traits can be complex. For example, the genetic architecture underlying thermal responsiveness, both plastic and selective, involves a substantial number (about 2000) of genes and orthologous groups (genes in different species that stem from vertical descent from a single gene in the last common ancestor). Responses to thermal environments include differential expression and genotypic differences where each gene has a small effect. This supports the omnigenic concept of complex traits, where heritability is driven by peripheral genes that transregulate core genes (Herrmann and Yampolsky, 2021).

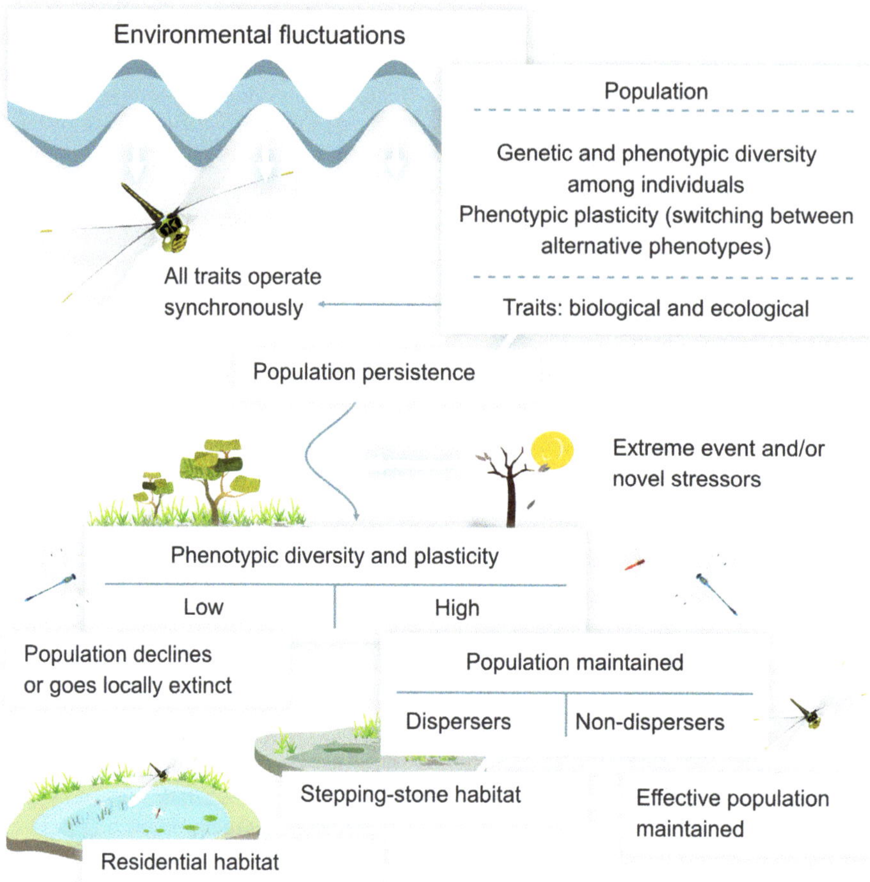

Fig. 4.3. Population persistence in the face of environmental fluctuations depends on how the individuals in that population are equipped in terms of their genetic and phenotypic diversity, as well as phenotypic plasticity. Together these features constitute the traits present in the population. While all traits, both biological and ecological, operate synchronously to enable populations to persist, they are continually tested by extreme environmental events and/or novel stressors. Where phenotypic diversity and plasticity are low, a population is at risk of decline and even local extinction. In contrast, when the diversity of these characteristics is high, the population is more likely to be maintained. Nevertheless, the environmental pressures might still lead to some individuals dispersing via stepping-stone habitats to establish in a new residential habitat. Other individuals may be sufficiently adapted to deal with the novel factors and so maintain an effective population. (Illustration: Robyn Symons).

The geographic range of *Erythromma viridulum* has expanded in northern Europe. During the process of rapid expansion there is high gene flow. Despite this high gene flow potentially having a homogenizing effect, there is extensive phenotypic variation among the invading individuals with a strong, positive correlation between high latitude and body size which is not explained by variation in length of larval development (voltinism). The adaptive significance of the size change is unclear, but is nevertheless underpinned by phenotypic plasticity, permitting a response to environmental correlates of latitude during range expansion (Hassall *et al.* 2014).

4.2 The Local Environment

4.2.1 Dimensions of the local environment

4.2.1.1 The developmental filter

During development, the dragonfly passes through egg (and prolarva), larva (with associated risky moults), emergence, teneral-immature adult and mature adult stages. Each stage is in effect a series of hurdles, all of which must be overcome in sequence to survive to the next generation.

Traditionally, this sequence is portrayed as life tables, showing the proportion of individuals surviving versus dying from one stage to the next, with the proportion surviving illustrated as survivorship curves. One of the main aims of conservation is to ensure that survivorship is as high as possible across all life stages. Once mature, an adult must feed, locate a mate, undergo the mating process and then the female must lay her eggs (Fig. 4.4).

4.2.1.2 Dynamics of habitat occupation

The geographic range of a species (Chapter 3) is the overall extent across a specified area where the species lives and breeds successfully. Within that range are pockets of suitable habitat, manifested as the areas of occupancy (AOO). Geographic ranges are not static entities, but dynamic manifestations as individuals ebb and flow across landscapes to occupy suitable habitats or vacate unsuitable ones. Often it is those individuals with pioneering traits that first reach potentially good new habitat. Establishment then depends on other traits that enable residency. Regular migrants have taken this process of movement and residency to a new level where the traits for this behavioural ecology are finely honed. By definition, their geographic ranges are large. But importantly all the successful migrant species are common, suggesting that regular migration is a successful life strategy.

The geographic range may also be seen as an overlapping series of local environments, each supporting individuals of a species with a set of genotypes and traits enabling them to live in any one local environment at a particular time. If that local environment is optimal for them, then it pays to stay. If suboptimal, there is the option to move to a better environment. However, continuing to stay in a suboptimal environment carries risks. If environmental conditions do not improve, the population is at risk of dying out.

The options are different for the various life stages. The egg, unless associated with a mobile substrate, whether flowing water or a drifting log, cannot move. The larva can move to a small extent to seek better localized micro-environmental conditions, where its prey is abundant enough to sustain it, and where its predators are given as few opportunities as possible. The larva, when it is about to undergo emergence to an adult, has few options where to rest as an exuvia, while facing risks like predation and being cramped by obstacles such as adjacent vegetation. The soft and vulnerable teneral individuals, having modest flight ability, must select shelter carefully. While the mature adult has more mobility options, it must still feed, mate and rest, and importantly, eggs must be placed where the successive life stages have the greatest chance of survival.

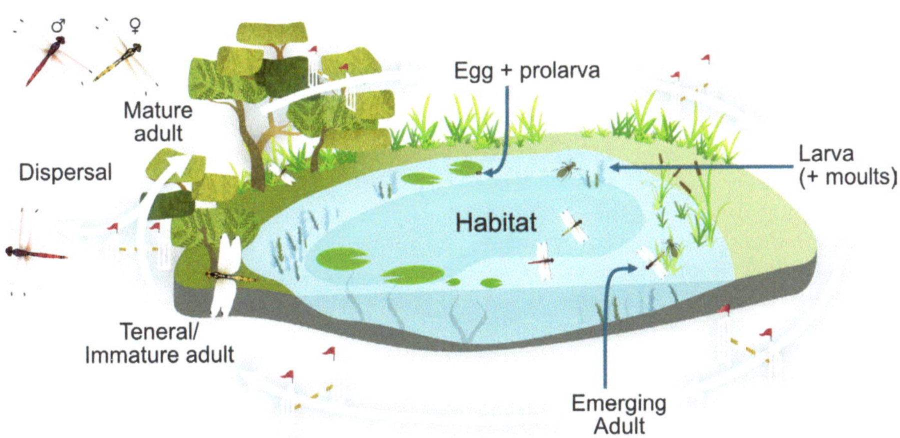

Fig. 4.4. Conservation of dragonflies implies a full understanding that a population is made up of individuals that must possess traits that enable them to clear the various successive developmental hurdles from egg to mature adult. The more individuals that survive, the greater the chances of an effective population. All environmental living conditions for ongoing residency at a location constitute the habitat. When environmental conditions within the habitat deteriorate, or when there is overcrowding, one option is to disperse in search of a more optimal habitat. (Illustration: Robyn Symons).

4.2.2 Habitat conditions relative to traits

4.2.2.1 *Total habitat size*

While a water body is largely a discrete entity to the human eye, its influence in terms of water-dependent food webs associated with it extend far beyond the bank into the neighbouring terrestrial realm (Muehlbauer *et al.*, 2014). A dragonfly's total habitat, for example, is several orders of magnitude greater than the natal habitat where the larva develops. In the case of the threatened *Sympetrum depressiusculum*, the adult requires dense vegetation rich in food resources and resting sites up to 1 km from the emergence site. This emphasizes that dragonfly conservation must include high-quality terrestrial vegetation in addition to the protection of the water zone (Dolný *et al.*, 2014).

While most dragonflies usually spend most of their time in the larval stage, the right conditions for the adult stage are just as important as for the larva (Harabiš and Dolný, 2010). This reminds us again that all the hurdles relating to each developmental stage must be cleared on an ongoing basis if a dragonfly species is to survive in a locality in the long term.

4.2.2.2 *Developmental plasticity*

Traits associated with the interactive effect of environmental conditions and the genotype of an earlier stage can carry over into the adult stage, a process known as developmental plasticity. This process leads to a permanent state, unlike behaviours which are reversible. In the case of environmental temperature, and when a dragonfly population is exposed to altered environmental conditions, the degree of phenotypic plasticity might help or hinder both immediate survival and future adaptation. However, as climate change involves more than just a rise in temperature, it is important to consider the combined effects of temperature along with other environmental drivers on trait expression (thermal plasticity) as well as trait evolution (thermal adaptation). When temperatures rise as time passes, there may at first be tolerance and adaptation by the larva to the increased temperature. However, traits and phenotypic plasticity for tolerating reduced oxygen will also come into play as the water level drops. Perhaps then certain individuals have a trait for early emergence and dispersal to locate a new, more optimal habitat.

The genomics of thermal plasticity involves identifying the genes whose expression depends on temperature (and might underlie temperature-induced change in developmental outcome), and in terms of the genes with allelic variants, contribute to inter-genotype variation in plasticity which causes evolution of thermal plasticity (Rodrigues and Beldade, 2020) (Fig. 4.5).

4.2.2.3 *Dragonfly developmental plasticity*

In the case of *Ischnura elegans*, larvae that grow faster and have a different body shape affect the biological traits of the adult (Abbott and Svensson, 2008), and for *Anax imperator*, it is body size (Minot *et al.*, 2019). From an evolutionary perspective, among 340 species in North America, there are species whose larvae inhabit relatively unstable habitats such as pools where there can be low oxygen levels and temperature extremes. Direction and rate of adult size evolution, which are fixed at maturity, are restricted in the case of larvae of species that develop in these harsh environments, manifested as smaller adult size and slower rates of adult size evolution. In contrast, ornament expression (darkening of wing cells), which continues to develop at maturity, and rates of ornament evolution, are not affected. This indicates that selection on larvae can affect the evolution of adult traits prior to maturity (Moore, 2021).

These carry-over effects are significant, as colour expression has an important role in sexual behaviour and geographic range. In the case of the widespread North American *Pachydiplax longipennis*, temperature both promotes and constrains the evolution of sexual colouration in males. Wing colouration heats males under cool conditions. These thermal effects translate into greater territory acquisition on thermally variable days. Conversely, sexually selected wing colouration is greatly reduced in the hottest part of the range. In areas that have very hot days, this dragonfly is forced to abandon territories to seek shade, affecting its local survivability and may possibly genetically isolate populations, likely exacerbated by a warming climate (Moore *et al.*, 2019) (Fig. 4.6).

In this era of rapid climate change, evidence of exposure to heatwaves provides important insight for the future. A heatwave experienced by the larva of *Ischnura elegans* carries over to the locomotor ability of the adult. The heatwave shortens the larval stage, reduces adult size-related traits and modifies wing shape, although emergence success, relative fat content and muscle mass are not affected. The reduced flight capability could eventually diminish

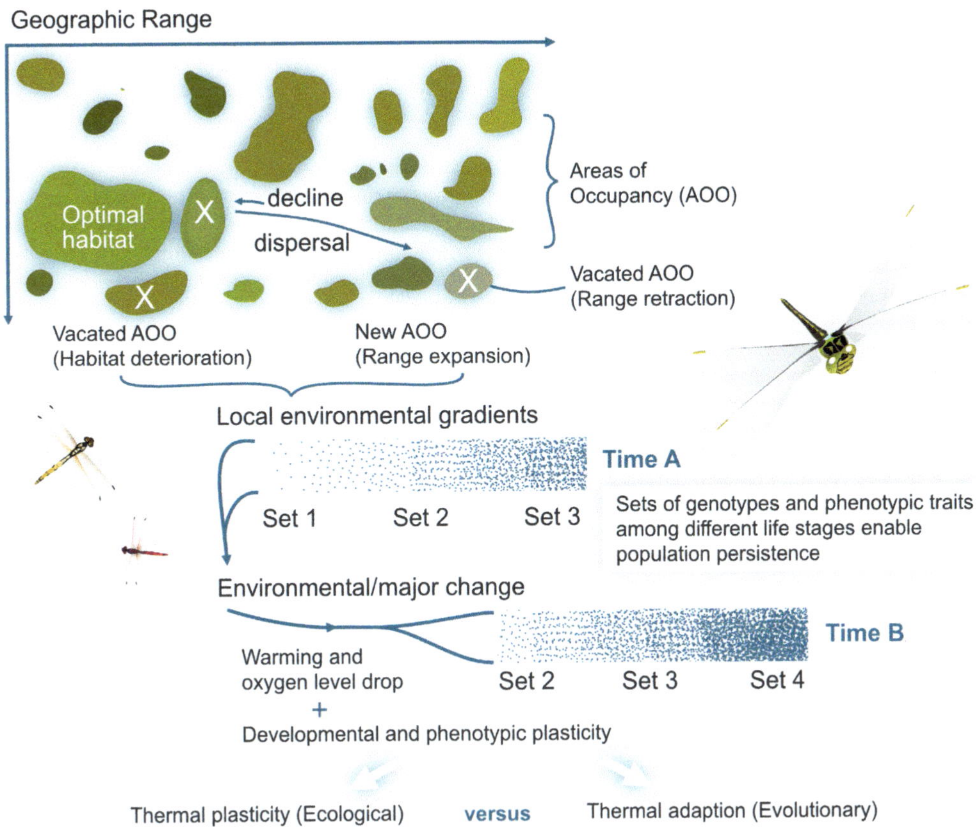

Fig. 4.5. Geographic ranges, along with habitat quality and extent, are dynamic, changing over time. Ranges are not simply blocks, but rather a mosaic of areas of occupancy (AOO) of good habitat. While long-standing optimal habitat retains resident populations, precipitously declining habitats lose populations (X), often through dispersal to new and better habitats. Any physical area has local environmental gradients where different sets of genotypes and phenotypic traits exist among different life stages. When this is the case, populations may persist along the gradients despite fluctuations in environmental conditions. However, a major (or novel) environmental change, such as warming of its habitat and the associated drop in oxygen levels, may not allow all species sets (assemblages) to survive intact, with set 1 here lost. Meanwhile, species sets 2 and 3 remain, as the habitat is still suitable to sustain them. But another set (species set 4 here) may be able to survive the changed conditions. Survival is enabled by developmental and phenotypic plasticity. In the case of warming, it is due to thermal plasticity. In contrast, there may instead be, or in addition, thermal adaptation, which is an evolutionary response. (Illustration: Robyn Symons).

its ability to disperse, and therefore reduce its potential to shift its geographic range poleward with further climate warming. This means that to survive in the long term, it may have to remain largely localized and adapt accordingly (Arambourou *et al.* 2017).

Larvae of *Coenagrion puella* exposed to ultraviolet (UV) radiation increase levels of melanin in their cuticle, causing them to metamorphose later and have smaller size than normal. This UV exposure impairs the adult's immune response, meaning that while the larva invests in melanin production, the cost of this carries over into the adult's health (Debecker *et al.*, 2015). However, small size, through sexual selection, does not necessarily mean a disadvantage. In the territorial *Chalcolestes viridis*, the three morphometric traits of small size combined with high relative thorax mass and low wing asymmetry confer improved mating success (De Block and Stoks, 2007).

Alternative survival strategies can confer advantages in extreme environments, such as drought. Adults of *Coenagrion mercuriale* in North Africa have two cohorts, one that emerges in spring and the other in late summer/autumn. Larvae of the

Fig. 4.6. *Pachydiplax longipennis*, a species that is ecologically and evolutionarily sensitive to ambient temperatures. (Photo: Michael Moore).

spring cohort develop slowly, producing adults with a large body size, in keeping with the temperature-size rule and duration of development. However, there is greater variation in size and age at maturity, and there is higher survival probability, as well as increased lifespan and mating success among these spring slow developers. Why then is a late season fast pace of life cohort present? It appears to be associated with having a flexible life history, with rapid development and adaptation to drought in the wetlands: 'grow up fast and move on quickly to colonize new habitat' (Khelifa *et al.* 2019).

The temperature-size rule suggests that adult sizes will be smaller in warm, low-latitude environments than in cool, high latitudes. In the case of the threatened *Nannophya koreana* in Korea, a local flagship species for wetland conservation, larval body size is smaller in a cold-water-temperature geographical area than in a warm-water-temperature area, contradicting the temperature-size rule. The reason for this is that the water temperature rather than the air temperature differs between the two areas, determining the relationship between temperature and body size, with the species compensating for low water temperature by maintaining its univoltine life cycle (Lee *et al.*, 2022).

Pachydiplax longipennis larvae that mount a strong immune response to physical injury have reduced survivorship in the presence of larval aeshnid predators. These immune-challenged larvae emerge later and are small with less sexually significant wing colouration. In short, when having to mount a strong immune response, there is a great effect both within and across life stages (Moore and Martin, 2018).

In *Leucorrhinia intacta* there is significant intraspecific variation in its response to warming in both egg development time and hatchling size, which are correlated with date of oviposition. The significance of these responses is that high levels of intraspecific variation may help populations survive the effects of climate warming (Frances *et al.*, 2017).

In the case of *Lestes sponsa* in Europe, there are phenotypic differences in male wing shape among populations, although population differentiation does not follow a simple linear relationship with latitude. Wing shape differences across populations are composed of genetic and plastic responses to the larval development environment. It appears that phenotypic plasticity is evolving both in magnitude and direction across populations, with the plastic response being much stronger in northern compared to southern populations, while also differing in its direction. While some comparisons between genetic and plastic responses indicate an alignment, many comparisons indicate that plasticity is not related to genetic differences, as direction of plasticity can go in opposite ways. In sum, support for an alignment between plasticity and genetic differentiation is mixed, suggesting that damselfly wings show evolutionary responses across environmental gradients which are difficult to predict based solely on plasticity. Overall, genetic differences in plasticity across populations, in both direction and magnitude, are important contributors to clinal variation (Johannsson *et al.*, 2023).

4.2.2.4 Traits relative to hydroperiod

Short hydroperiods (limited times when freshwater is present) of aquatic systems place heavy demands on their inhabitants. In rivers that drop in flow velocity and discharge during prolonged times of little water replenishment, there is a decrease in surface area and residual depth. In extreme cases, the residual water retention areas change from a running water to a standing water system. The remnant pools may even completely dry out. During these hydromorphological and hydrodynamic changes, there is a change in physicochemical conditions, especially a drop in dissolved oxygen and an increase in water temperature. During the drying out phase, there is a change in community structure, often associated with an increase in predation as water volume decreases (Bonada *et al.*, 2020).

As regards the drying out of small streams, two endemic Seychelles zygopterans (*Allolestes maclachlani* and *Leptocnemis cynops*) have traits well-honed to survive drought. They inhabit forested streams that stop flowing, leaving only muddy pools at best during the dry season. Teneral individuals emerge when the streams are dry. In the case of *A. maclachlani*, adult males then engage in vigorous territorial disputes, even while the stream has no freshwater, to stake their claims in readiness for the coming rains (Samways, 2003).

For *Calopteryx virgo* (see Fig. 2.7), reduced oxygen levels limit the insect's ability to tolerate high temperatures. However, while there is a thermal plasticity component, there is also a longer-term response. Low oxygen levels trigger an acclamatory response in the form of protective heat shock proteins or through metabolic reorganization that changes the capacity to supply oxygen to tissues or by altering the oxygen demands of the tissues (Verberk and Calosi, 2012).

Permanent and temporary (ephemeral) standing water bodies differ in their physical and biotic characteristics (Wellborn *et al.*, 1996). At the assemblage level, pond permanence is a major determinant of the suite of dragonfly species likely to be present. Water permanence also affects the physicochemical conditions and the composition and structure of plants both in the water and at the margin. This leads to sets of species that characterize a particular pond type at a certain place and time (Schindler *et al.*, 2003).

When the traits of the occupying species are aligned with pond characterization, the assemblage can be considered as a 'representative spectrum of Odonata' (Schmidt, 1985). However, this 'characterization' is far from being a fixed entity. Various drivers are at play, including fluctuating water levels, time of year, dragonfly abundance levels, etc. In short, while it is a useful first-stop concept from a conservation perspective, it does not take cognisance of all the subtleties of metapopulation and metacommunity dynamics associated with fluctuating water levels.

This means that in a changing environment, it is essential to consider a whole range of pond types that make up a viable pondscape. This is emphasized among odonate larvae that occupy Michigan ponds and lakes. Adaptive trade-offs limit the distributions of species across the habitat gradient. Dispersal is also important as it shapes species distributions where species occupy habitats with a wide range of hydroperiods and predator pressure. In turn, successful occupation of new sites depends on the flight ability and the length of the adult flight season. This means that many new sites of varying stability can be colonized rather than occupying just a few stable sites (McCauley *et al.*, 2008).

Dragonfly assemblages at ponds in Austria are determined not only by floating macrophytes, but also by hydroperiod (Schindler *et al.*, 2003). The longer the duration of freshwater, the higher the number of species present, with the appearance of species typical of permanent water, but losing these when the water level drops (Chovanec *et al.*, 2003). Often though, when water levels drop, the remaining species are a subset of those when water levels are high (Vander Vorste *et al.*, 2021).

Temporary waters are limiting for certain dragonfly species, as freshwater may not be available for long enough to complete larva development. In turn, permanent water bodies are limiting in that predation and competition is usually high. This means that traits adapted to each end of the hydroperiod spectrum cannot be maximized at the same time, and so there must be some adaptive trade-off among individuals of different species that occupy both temporary and permanent waters.

Temporary water specialists such as *Pantala flavescens* and *Sympetrum fonscolombii*, which can complete their life cycles in as little as two months, have higher prey capture rates and growth rates than the permanent residents *Crocothemis erythraea* and *Trithemis kirbyi* in the same arid region of southern Africa (Johansson and Suhling, 2004). Importantly, temporary waters, such as rain pools, can be rich in potential opportunistic food items, especially dipteran larvae (Fischer *et al.*, 2000).

In Australia, two abundant dragonfly (*Orthetrum caledonicum* and *Hemicordulia tau*) and one damselfly (*Xanthagrion erythroneurum*) species emerge at smaller body sizes in summer than spring from both temporary and perennial wetlands. This suggests that waning photoperiod and warmer summer temperatures, rather than wetland drying, probably leads to size reduction. The fitness cost of emerging at a smaller adult size may be outweighed by the increased likelihood of successfully emerging in drying water bodies in summer (Mackintosh and Robson, 2023).

The increased warmth associated with temporary ponds attracts some relatively rare species in colder climates. Stormwater and temporary ponds in Germany attract many species, particularly those with traits favouring sparsely vegetated, shallow and sunny ponds. This has conservation significance, as the assemblages include threatened but pioneer species such as *Ischnura pumilio* and *Libellula depressa* (Holtmann *et al.*, 2019).

In Europe, *Lestes macrostigma* is a threatened species of brackish pools. Oviposition plant availability does not drive species larval abundance. Instead, seasonally late flooding of ponds in autumn/winter leads to higher larval abundance the following spring. However, aeshnid larvae mediate the response. Larval abundance of *L. macrostigma* also increases with decreasing pond surface area. In turn, adults are larger when they emerge from ponds with high water levels, likely because deeper waters have lower water temperatures. Overall, these responses help explain the strong inter-annual variation in the population abundance of the species. The conservation directive is to create a network of small-sized and deep temporary ponds (Lambret *et al.*, 2023).

The ability of male *Leucorrhinia intacta* to establish territories at ponds depends on both environmental conditions and traits associated with larval development in ponds with different hydroperiods. Individuals at ponds that dried out the previous year show differences in morphology and parasitic mite loads compared with those at standing ponds. Males colonizing previously dried out ponds are consistently larger and in better condition through low levels of parasitism. This means that large males with fewer parasites readily disperse. Furthermore, males from previously dried ponds are more similar in body size and mite loads than are individuals at permanent ponds where males differ from each other according to local pond condition. However, as permanent ponds lead to more variation among males, there are some individuals with good dispersal capabilities, which also contribute to the dispersal set. Overall, what is taking place here is a bet-hedging strategy among the individuals occupying dried out ponds. There is spreading of reproductive investment across habitats with various levels of risk while at the same time reaping potential fitness rewards. In turn, this strategy is reinforced by the behaviour of top dispersers from the permanent ponds (McCauley, 2005).

Ischnura elegans displays different behaviours in response to different types of ponds drying out. While pond characteristics do not influence survival in this habitat generalist, the landscape context greatly affects movement both within and among ponds. Individuals from open-field ponds move over greater distances than those from urban ponds, with movements from one pond to another in response to pond drying only taking place among individuals associated with open-field ponds (Le Gall *et al.*, 2017).

Lestes eurinus inhabits permanent ponds, and compared with other *Lestes* species of temporary ponds, has developed the trait of stronger compensatory growth (i.e. regains full body mass) when stressed by time constraints. It is the degree of water permanence that allows the species to do this, whereas the temporary pond species have been selected to develop before the pond dries out (De Block *et al.*, 2008a). However, there are other traits related to compensatory mechanisms that also play into this narrative, such as response to predation and food availability, which together show that there is a shift in life history plasticity as species adapt to living in permanent versus temporary pools (De Block *et al.*, 2008b).

These trait characteristics are more than just of academic interest. Both habitat selection and life history plasticity are significant when combined with the environmental factor of changes in pond permanence associated with climate change, especially in the case of rare and threatened species (Zettlemoyer and Peterson, 2021).

Identifying important traits and steering conservation action in a rapidly changing world are going to be critical for ensuring the survival of certain species. For example, *Lestes macrostigma* prefers to oviposit in the plant *Bolboschoenus maritimus* which grows in the deeper parts of temporary ponds. This behaviour begets higher hatching success and faster embryonic development, suggesting that managing populations of the host plant and water levels will go a long way to procuring a future for this rare species (Lambret *et al.*, 2018).

Tetrathemis polleni, in the Kenyan dry season, lays its eggs on dry stems up to 2 m above the prevailing water level of turbid pools. The eggs only hatch when the water level rises after the onset of the rainy season. At this time the larvae can then benefit from the appearance of mosquito larvae prey, which also appear at the same time (McCrae and Corbet, 1982).

Traits related to habitat selection and predation pressure along a pond permanence gradient can lead to evolutionary change over a long time.

In North America, while *Enallagma* species likely first evolved in permanent lakes where fish are the top predators, *Lestes* began in temporary ponds where large dragonflies are the top predators. From these ancestral habitats, species in each of these two genera have spilled over into the habitat already occupied by the other, leading to convergence in how they deal with fish and dragonfly predation. In *Lestes*, this has also involved shifting from a life history with diapausing eggs to one with directly developing eggs, although *Lestes* species invading fish lakes do not develop a swimming speed and morphology that matches those of *Enallagma*, owing to phylogenetic constraints. While *Lestes* species in permanent lakes would seem to be at an advantage by being larger and having high growth rates, they have lost the ability to recognize dragonfly predators as a threat. Furthermore, *Enallagma* species have adapted to, and have diversified greatly, in the new predatory environment by evolving behavioural, morphological and biochemical features that enable them to swim fast, while losing their ability to recognize fish as predators. In short, these ecological and adaptive shifts are not necessarily evolutionary opposites, with *Enallagma* diversifying much more than *Lestes* species that have been constrained by their phylogenetic inertia (Stoks and McPeek, 2006).

In arid Namibia, larval anisopterans in the genera *Crocothemis*, *Orthetrum* and *Trithemis* also have behavioural traits, such as activity and foraging, that are influenced by phylogenetic inertia rather than by habitat variables. In contrast, larval growth rates are shaped by habitat, which is crucial in this region where pools often dry out (Suhling *et al.* 2005) (Fig. 4.7).

4.2.2.5 *Latitude and traits*

Environmental variation interacts with species life histories in terms of the latitudinal position of dragonfly species. It is not simply a case of high-latitudinal generalization versus low-latitude specialization. Traits like voltinism and diapause, for example, can buffer the environmental variation experienced by individuals. This makes thermal specialization an advantage at high latitudes under certain conditions, because without these traits, high latitude environments would be inhospitable to thermal specialists (Nilsson-Örtman *et al.*, 2012).

Among *Enallagma cyathigerum* adults from north to south in Europe, there is latitude-specific thermal adaptation and voltinism to survive winter. In the northern latitudes, the species is semivoltine (takes more than one season to complete its life cycle) and survives the winter by growing slowly before winter to build up energy reserves and to invest in improved immune function (Verheyen *et al.*, 2018).

Environmental variables and various traits, such as response to temperature and photoperiod, act together in determining age and size among the sexes of *Lestes viridis* at emergence. This means that life history plasticity is adaptive, and emphasizes that various traits must be considered together, and not in isolation, to produce environmentally dependent sexual size dimorphism (De Block and Stoks, 2003). Also in this species, there is an immunological cost in the adult stage resulting from larval short-term food stress. The associated compensatory larval growth strongly suggests that physiological costs indicate hidden carry-over effects bridging metamorphosis. The message here is that the larval and adult stages in odonates should be considered together, as trade-offs may span metamorphosis, although age and mass at metamorphosis do not completely translate into adult fitness (De Block and Stoks, 2008a, b) (Fig. 4.8).

4.3 Dispersal

4.3.1 Migration

4.3.1.1 *Flight and the aerial environment*

The atmospheric environment enhances or restricts flight direction and distance of the dragonfly flight path. Flight influences its ability to efficiently find food, locate a mate and find resting and oviposition sites, as well as extend its population range, and enhance genetic diversity. Flight initiation depends on air temperature, wind speed, wind direction, air pressure and air density. Air pressure is the pressure within the Earth's atmosphere. Air density is the mass per unit volume of the atmosphere, and as with air pressure, decreases with increasing altitude. Air density also changes with atmospheric pressure, temperature and humidity. With decreasing air density, drag on the body is reduced, although lift per wingbeat decreases. This means that wingbeat frequency must be increased to compensate for decreasing air density, while also allowing an increase in air speed with increasing flight altitude. However, the efficiency of gliding helps overcome the need for the high energy consumption required by powered flight.

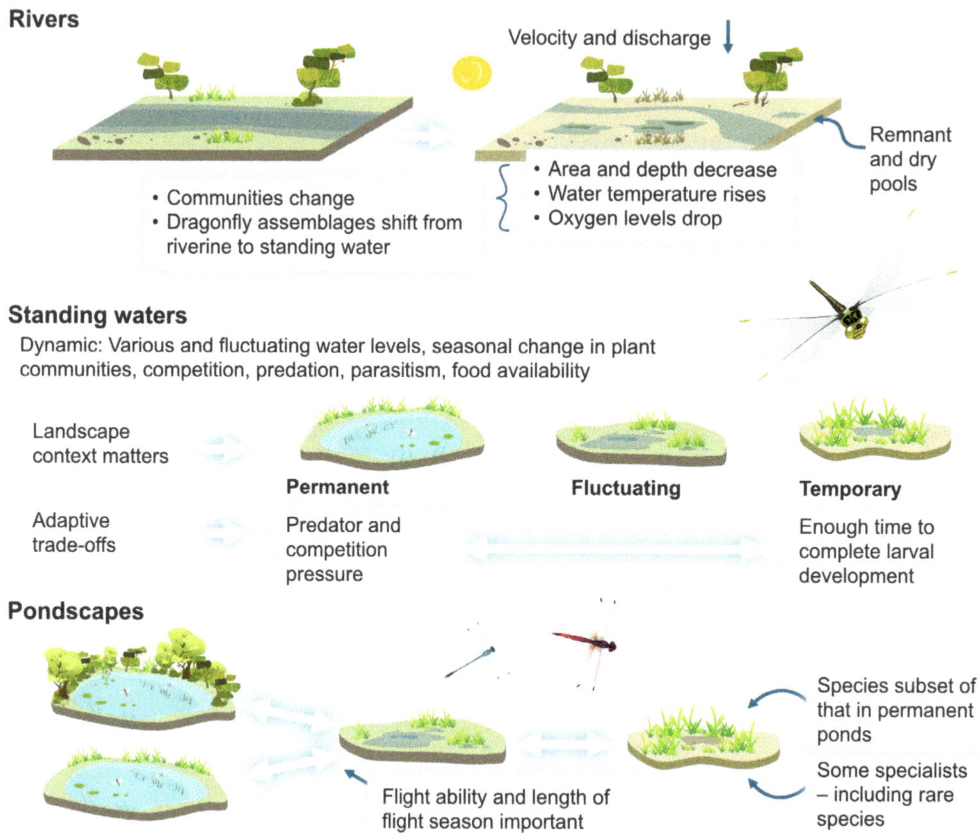

Fig. 4.7. Changes in hydroperiod have profound effects on both running and standing water systems and their biological communities. When a river dries down, there is a change in physicochemical conditions, which besides changing the local community also shifts the dragonfly assemblage from one of running water toward that of standing water. Standing waters are also dynamic, with changes in many abiotic and biotic variables. Ponds may be relatively permanent, fluctuate in water levels, or are temporary, each of which has its characteristic but changing dragonfly assemblage. The assemblages differ according to landscape context, such as open versus closed canopy conditions. There are adaptive trade-offs among the species living in ponds with different hydroperiods. Species inhabiting permanent ponds have to contend with high predation and competition pressure, while for those of temporary ponds there must be enough time for them to complete larval development. Often the assemblage in greatly fluctuating or temporary ponds is just a subset of those of permanent ponds. However, sometimes rare specialists of these dynamic conditions are present. Each pond in a group of ponds (known as a pondscape) may support a particular set of dragonfly species according to type of pond, flight ability and length of flight season. (Illustration: Robyn Symons).

Flight distance and direction depends on prevailing wind speed and direction, with dragonflies using tailwinds to add to their flight speed. They can change direction by altering their flight angle relative to the wind. But while airborne, they are also possibly subject to predation by birds.

Long-distance flight is deflected, or may be terminated, by mid-latitude fronts, gust fronts and heavy rainfall. Flight distance across the landscape may be considered as a continuum of spatial scales, the microscale (≤2 km), the mesoscale (2 to several hundred km) and the macroscale (≥1000 km), and dependent on short-term weather events and long-term climate change (Westbrook and Eyster, 2017) (Fig. 4.9).

4.3.1.2 Major migrations in certain species

Migration in dragonflies is a response and an adaptation to seasonally unfavourable climates, such as

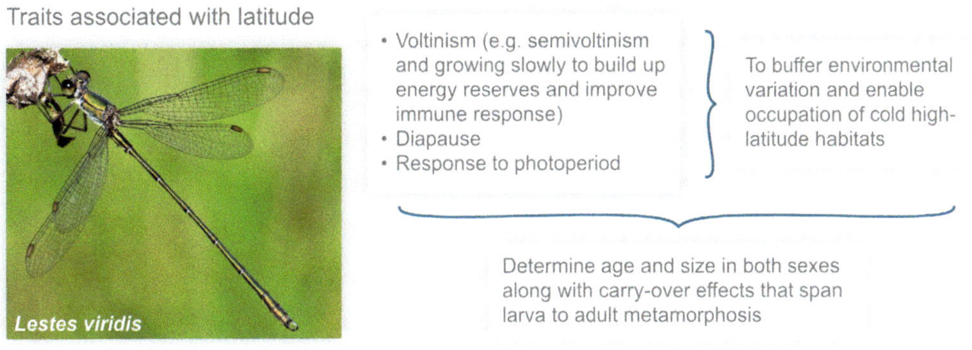

Fig. 4.8. Dragonflies have traits associated with latitude. This enables them to survive variable environmental conditions and permits them to occupy cold, high-latitude habitats. Among these traits are those associated with voltinism, diapause and response to photoperiod. In *Lestes viridis* these traits determine age and size in both sexes, as well as being associated with carry-over effects during development and metamorphosis from larva to adult. (Photo: Jens Kipping; Illustration: Robyn Symons).

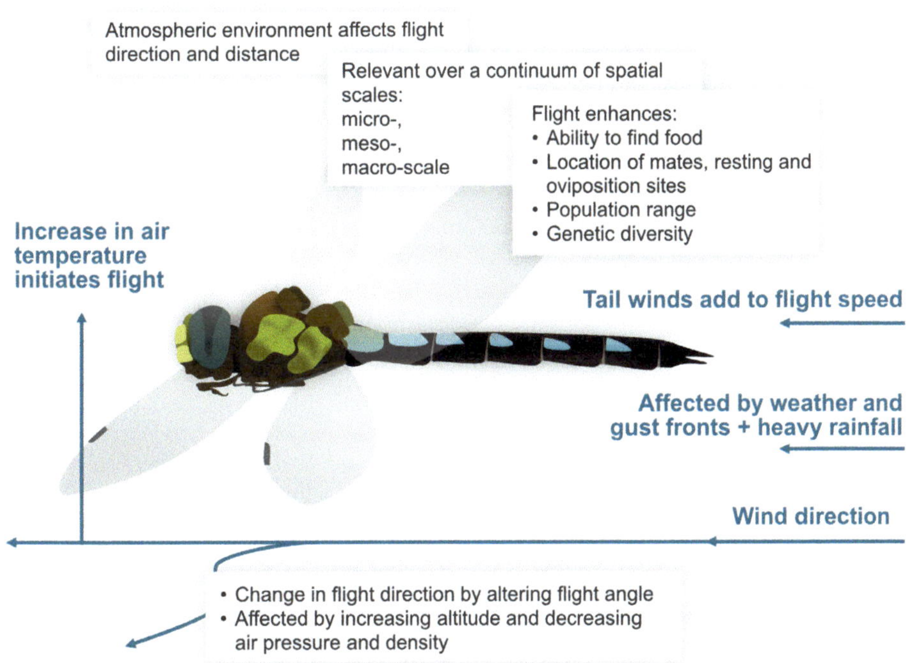

Fig. 4.9. The atmospheric environment affects dragonfly flight direction and distance travelled. Flight is relevant in the life of a dragonfly over various spatial scales, from the microscale to the macroscale, and enhances various life functions as well as population range and genetic diversity. Flight is often initiated by increased air temperatures during cool conditions, and once airborne, tailwinds increase flight speed. The flight can be affected by weather and gust fronts, as well as by heavy rainfall. Large and powerful dragonflies high up can change direction by altering their flight angle. However, with increasing altitude and concurrent decreasing air pressure and density, their powered flight agility is less, which is why long-distance travellers are able to glide to save on energy consumption. (Illustration: Robyn Symons).

extreme weather events, particularly the drying out of habitat, leading to mass emergence followed by travel over long distances. Migration routes are flexible in that dragonfly migrants may change direction when encountering an unfavourable environment such as the sea or a very large lake. In temperate climates, migration may be viewed as an alternative life strategy to diapause among the young stages in very cold times (May, 2013).

Pantala flavescens has remarkable flight ability, producing a total aerodynamic force equal to about 4.3 times its body weight (Su *et al.*, 2020). Each year, millions of dragonflies arrive in the Maldives, an archipelago in the mid-Indian Ocean, and most of the individuals are *P. flavescens*. They leave from India on the way to East Africa, passing over the Maldives, where they arrive from October onward. They follow the rain-laden intertropical convergence zone, apparently flying over 500–1000 km at an altitude of >1000 m, while making use of tailwinds. There is a smaller, second wave of these visitors in May with the south-west monsoon, suggesting a return journey from East Africa on the way to India, where they arrive in June–July, having travelled over 3500 km from East Africa (Anderson, 2009).

Some individuals of *P. flavescens* engage in high-altitude flight, moving over the Himalayas, suggesting that the round trip can be as much as 14,000–18,000 km (Hobson *et al.*, 2012a). There is also a migration route from East Africa to Central Asia, where they arrive in spring as mature breeding individuals. In autumn, a summer generation then departs southward, likely over two generations (Borisov *et al.*, 2020).

Migrating individuals of *P. flavescens* tend to have broader wing bases and more slender wings than non-migrant individuals, although their total wing area is similar (Johansson *et al.*, 2022). The global migratory population shows considerable geographic coverage, with some individuals even reaching as far north as the margin of the Baltic Sea (Shapoval *et al.*, 2022) and New Brunswick, Canada (Troast *et al.*, 2016). It has also been recorded in eastern Russia (Onishko *et al.*, 2023) and as far south as Victoria, Australia (Endersby, 2014).

The species is the only odonate species resident on Rapa Nui (Easter Island) (Samways and Osborn, 1998). It likely arrived on the remote Ngulu island (315 km from the nearest island) in Micronesia in association with strong winds (Buden, 2010). It has also been recorded as a visitor to the remote subantarctic Amsterdam Island (37°50'S) in the southern Indian Ocean (Devaud and Lebouvier, 2019). As there is a high rate of gene flow among populations in South and North America and Asia, the global population can be considered as globally panmictic (Troast *et al.*, 2016; Ware *et al.*, 2022), with Quaternary glacial events having shaped the species' genetic diversity (Low *et al.*, 2017).

Pantala flavescens is not just at the mercy of wind direction as it, as well as *P. hymenaea*, can compensate for wind drift, likely using power flight (Srygley, 2003). *P. flavescens* has also been recorded in northern China, displaying power flight, and not just passively going with the wind direction. They fly for 9–10 hours per night, covering 150–400 km in that time at altitudes of up to 1000 m, but in the highest concentrations at 200–300 m and at 500 m altitude (Feng *et al.*, 2006).

Anderson (2009) noted migratory birds as potential predators arriving at the same times as the dragonflies in the Maldives, which may explain why they fly at night, as bird predation on insects is predominantly visual. However, it may be that the migratory birds are simply taking advantage of the same environmental conditions as the dragonflies.

There is an important conservation perspective to *P. flavescens* migration with regards to the Indian Ocean. The multigenerational migratory circuit of the species across the Indian Ocean is only achievable through having specific physiological, behavioural and morphological traits for sustained flight endurance coupled with enhanced wind selection ability. As migration over the Indian Ocean is dependent on favourable directional winds which occur only at certain times of the year, the migration pattern of *P. flavescens* is highly susceptible to climate change (Hedlund *et al.*, 2021) (Fig. 4.10).

The southward migration of *Anax junius* and some other species along the Chicago waterfront is composed of an estimated 1.2 million individuals, with swarms 10 m deep, and within 110 m of the ground (Russell *et al.*, 1998). Detailed study of *A. junius* in the USA, using miniaturized radio transmitters, showed the species to have a distinct migration pattern (Fig. 4.11). Swarms migrate about every 3 days in a general southward direction, and after stopping to rest, continue to advance about 60 km every 6 days or so. They resume flight only after a couple of cool nights, when wind speeds are less than 25 km/hr (Wikelski *et al.*, 2006). Individuals can travel up to 122 km in a day, with increased temperature assisting the pace of migration, and rainfall not slowing them down (Knight *et al.*, 2019). Some individuals of

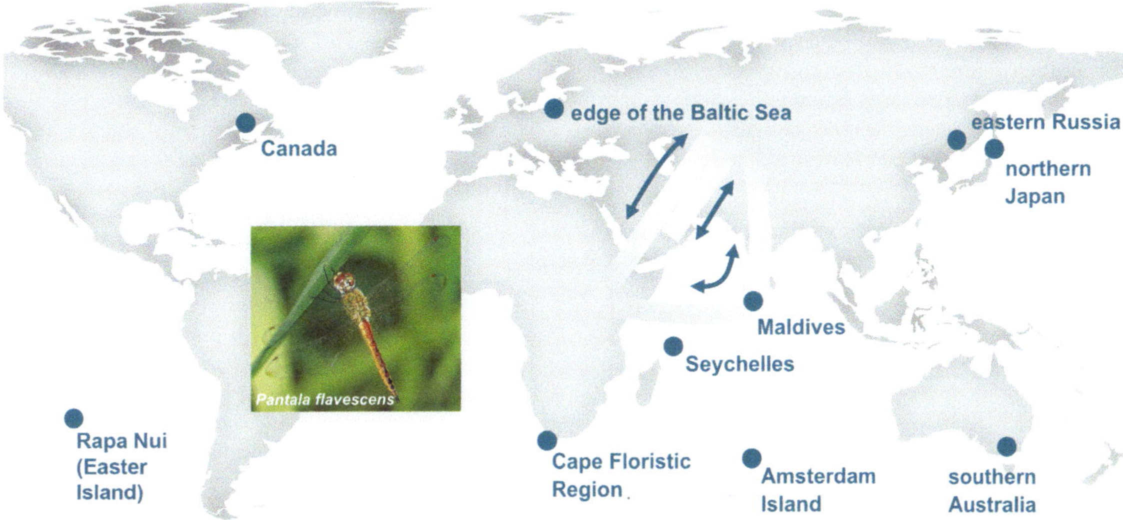

Fig. 4.10. *Pantala flavescens* has remarkable flight ability, and along with other traits such as rapid larval development in mostly small ephemeral rain pools, it has colonized the tropics and subtropics across the world. It is able to travel large distances and breed by making use of prevailing winds and seasonal rains. Some migration routes that have been proposed include Central Asia via the Indian subcontinent and to and from East Africa, Central Asia to and from the Arabian Peninsula, and Central Asia to and from East Africa. These are multigenerational movements. The species is also a wanderer, with regular visits to the Seychelles, and occasional individuals being recorded as far north as Canada and the edge of the Baltic Sea, and as far south as Amsterdam Island and the Cape Floristic Region. It has also been recorded from eastern Russia, northern Japan and southern Australia. On the remote Rapa Nui (Easter Island), it is the only species that occurs there and is resident. Its global population exhibits widespread gene flow. (Photo: Michael Samways; Illustration: Robyn Symons).

Fig. 4.11. Attaching (left) a 300 mg radio transmitter to the thorax of *Anax junius* and releasing the tagged insect (right) to investigate its migration route along the east coast of the USA. (Photos: Christian Ziegler).

A. junius that were likely blown off course were recorded in the UK in September and October 1998 (Corbet, 2000).

Not all individuals of *A. junius* are migrants. Populations in Texas are a mix of local residents and distant migrants from northern USA (Hobson *et al.*, 2012b). Genetic analysis of the species points to its developmental pathways being plastic, allowing it to occupy a wide range of environmental conditions and enabling it to have a wide distribution (Freeland *et al.*, 2003).

In Kazakhstan, three species (*Anax ephippiger*, *Pantala flavescens*, *Sympetrum fonscolombii*) arrive in cooler latitudes from the southern warmer parts of their geographic range to breed in spring. Their offspring then return south in autumn. *Anax parthenope* has a different strategy. Some individuals remain in parts of the range, while other individuals are regular migrants within it (Borisov *et al.*, 2022).

4.3.2 Local movements and metapopulation dynamics

One of the essential advantages of dispersal is to maintain genetic diversity and avoid genetic bottlenecks. The rare European *Leucorrhinia caudalis* is largely sedentary but makes flight excursions of up to 5 km. It displays metapopulation dynamics with strong source populations enhancing genetic diversity in nearby small and isolated populations. However, where small populations are more distant (up to 30 km from the source population), genetic diversity is low, with some indication of genetic bottlenecks (Keller *et al.*, 2010). From a conservation perspective, this illustrates the importance of maintaining historical functional connectivity through strings of optimal habitat, or at least intermediate stepping-stone habitat, to ensure genetically advantageous metapopulation dynamics.

In the case of a rare *Coenagrion mercuriale*, populations need to be within about 2 km of each other for adequate genetic exchange. There must also be no barriers preventing free movement, such as scrub margins, agricultural land or high ground. Quality habitat connectivity then ensures adequate metapopulation dynamics and avoids genetic impoverishment (Purse *et al.*, 2003; Watts *et al.*, 2006). Some individuals move as far as 4.5 km, which may also improve genetic connectivity by bequeathing genetic vigour to the local population (Keller and Holderegger, 2013). However, individuals of another subspecies (or perhaps the newly designated *Coenagrion castellani*; La Porta *et al.*, (2023)) making up a large and dense population in high-quality habitat in central Italy are highly sedentary (La Porta and Goretti, 2020), suggesting that individuals are more likely to disperse when habitat quality deteriorates.

Threatened *Calopteryx exul* of north-eastern Algeria, while capable of dispersing over 1 km, mostly only moves less than 100 m within certain patches where it forages and mates. However, each suitable remaining patch exceeds its flight distance capability, exacerbated by drought. Conservation strategies should aim to improve environmental conditions and design networks of high-quality habitat patches with minimal human disturbance. However, the challenge is the increase in agricultural pressure through water abstraction for irrigation, construction of dams and deviation of watercourses, as well as pollution by fertilizers and pesticides (Mahdjoub *et al.*, 2023b).

It is also important from a population viability perspective not just to focus on demographic dispersal distances, but to also include genetic estimates of dispersal which are less affected by habitat heterogeneity, stochastic events and/or selective processes (Watts *et al.*, 2007). Restricted dispersal in *C. mercuriale* in the UK leads to significant genetic differentiation throughout most of its local range, except between areas of continuous habitat, or isolation by distance even though core populations are separated by less than 10 km. Isolation of these populations by urban areas has a strong effect on the spatial genetic structure. Average pairwise relatedness between individuals is positive at short distances, reflecting fine-scale genetic clustering and isolation by distance both within and between habitat patches. Individuals in isolated patches in fragmented landscapes have higher average kinship than those from a large continuous population, probably because of poorer dispersal and localized breeding in land fragments. The overall spatial pattern of genetic variation matches closely with that expected from direct observations of movement (Watts *et al.*, 2004).

Similar situations arise when a species shifts its range. *Erythromma viridulum* in the UK has expanded its range northward in recent years, but this expansion has led to the species having less genetic diversity in marginal populations (Watts *et al.*, 2010) (Fig. 4.12).

4.3.2.1 Phylogenetic constraints and ecological responses

Co-occurring species do not all have the same dispersal characteristics, based partly on their body size and niche requirements. Among Amazonian streams that are both natural and disturbed, there is a difference in assemblage composition, especially in zygopterans that generally have distinct microhabitat preferences and narrow niches. Zygopterans are also demanding of environmental conditions, often manifested by intense intra- and inter-specific competition. In contrast to the zygopterans, traits of anisopterans in the same area have wider microhabitat and niche preferences. When the local environment

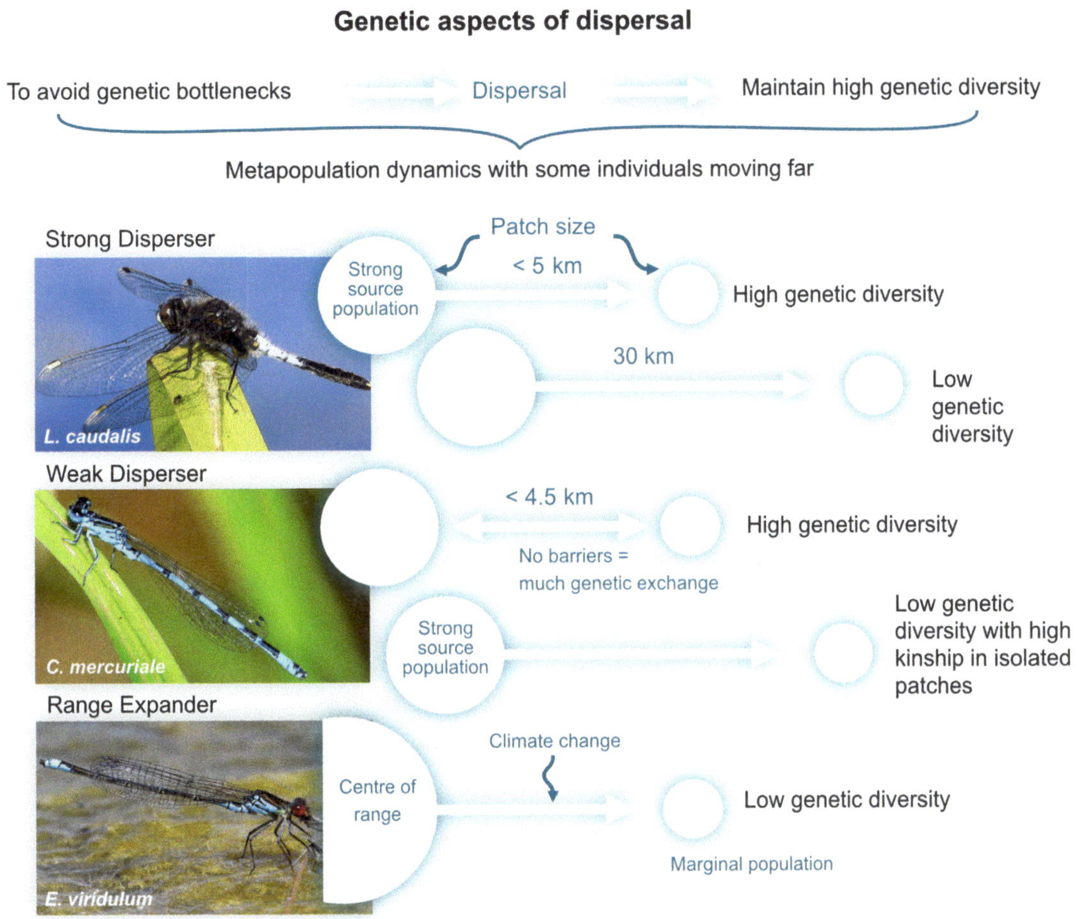

Fig. 4.12. The genetic aspects of dispersal are fundamental to dragonfly conservation. Dispersal is essential to avoid genetic bottlenecks. In turn, movement is a basic facet of metapopulation genetics, and recognizes that there are core populations in high-quality habitats where genetic diversity is high. While most individuals stay at or close to their natal locations, some individuals move far. But how far depends on traits associated with their dispersal ability. In the case of *Leucorrhinia caudalis* (top), while it is largely sedentary, some individuals move far away. Individuals in a population with high genetic diversity in a large patch may disperse up to 5 km away and occupy a small patch, where they confer high genetic diversity. However, small patches that are isolated by 30 km are too far away to be genetically reinvigorated, and so have low genetic diversity. *Coenagrion mercuriale* (centre) is a weak disperser with high genetic exchange only occurring within 4.5 km, and usually less, while also assuming there are no barriers preventing movement. Populations in small patches that are more distant tend to be genetically isolated and characterized by low genetic diversity and high levels of kinship within the patches. *Erythromma viridulum* (bottom) is a species that has capitalized on climate change by shifting its geographic range but has done so at a genetic price. The marginal populations tend to be low in genetic diversity. (Photos: Jens Kipping; Illustration: Robyn Symons).

deteriorates and/or competition becomes too intense, and being larger than zygopterans, they readily move to secure optimal habitat elsewhere and where they can amplify their niches (Oliveira-Junior *et al.* 2021). Phylogenetic constraints also limit expression of sets of morphological traits and can determine how groups of dragonfly species respond to landscape change and its history in Amazonian streams (Bastos *et al.*, 2021).

Among three co-occurring North American *Libellula* species, two (*L. luctuosa* and *L. pulchella*) have wings with a relatively low aspect ratio and wider basal proportions, as well as larger thoraces. They also have higher rates and distances of dispersal, and a broad range of larval habitats. In contrast, *L. incesta* occupies only permanent lakes and has limited dispersal capability (McCauley, 2013). The two more robust and mobile species also have a much wider geographical distribution than the more gracile species. However, when more species of North American libellulids are considered, some different perspectives emerge. While migratory species and those with large basal hindwing areas are the most widely distributed, body size and larval habitat are not correlated with large range size, or latitudinal and longitudinal ranges. After considering phylogenetic constraints, it is valuable from a conservation and restoration point of view to consider different aspects of geographic range, which includes range area, as well as extents of latitude and longitude, in addition to other ecological and behavioural nuances (Outomuro and Johansson, 2019) (Fig. 4.13).

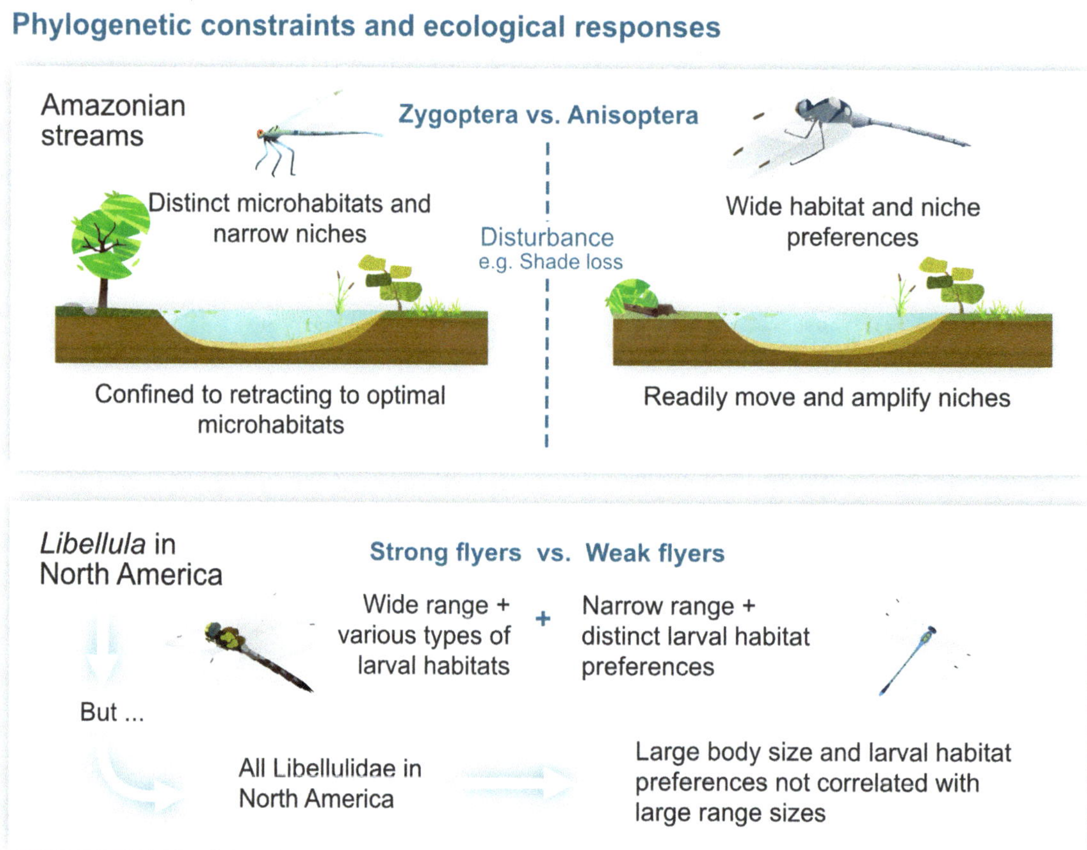

Fig. 4.13. Phylogenetic constraints play a major role in determining how different taxa respond to human disturbance on ecosystems. In Amazonian streams, zygopteran and anisopteran species overall respond differently to disturbance like tree removal along the riparian zone. Zygopterans retract into remaining forested zones, while anisopterans readily move and amplify their niches in disturbed areas. Within genera, such as *Libellula* in North America, two robust species with good powers of flight have a wide geographic range and, as larvae, occupy a wide range of habitat types. Another species that is more gracile and a relatively weak flyer has a narrow geographical range and a distinct habitat preference for permanent lakes. However, when the Libellulidae as a whole are considered, large body size and larval habitat preferences are not correlated with large range size. This indicates there are many species with deviations among their traits associated with body size and habitat preference. (Illustration: Robyn Symons).

4.3.2.2 *Movement across complex landscapes*

Functional connectivity for long-term survival of dragonfly species is essentially about maintaining gene flow across the landscape and waterscape. The flow in turn depends on a combination of flight ability and navigation across the terrain that is fractal in essence. Understanding gene flow pathways in the local extinction/re-colonization process is a fundamental aspect of dragonfly conservation, especially in this era of human-induced environmental change.

Traits pertaining to movement relate to both the local scale and that of the larger landscape. In the case of *Calopteryx aequabilis* and *C. maculata*, probability of crossing a patch boundary (patch boundary permeability) and rate of movement in a habitat patch (patch viscosity) are important determinants of damselfly movement across heterogeneous landscapes, arising from behaviours at multiple spatial scales (Jonsen and Taylor, 2000). *C. aequabilis* moves equally well through both forest and over open ground (pasture), while *C. maculata* more readily flies over open ground than in forest. However, there is also an historical component. Experimentally transplanted individuals that were native to open areas moved more readily through forested areas than their forest-inhabiting counterparts. This suggests that earlier selection pressure for greater movement capability can override the influence of the immediate environment. Overall, there was little difference in flight behaviour for the two sexes of both species, although there was some suggestion that *C. aequabilis* males move with greater ability through forest than females, while in open ground, females rather than males moved more easily (Pither and Taylor, 1998).

Navigating heterogeneous landscapes can differ among populations according to whether they live in forested versus open areas. In fragmented landscapes, *C. maculata* flies across pastures to reach foraging sites in forest patches, with morphological plasticity conferring a selective advantage in those individuals that can cross open ground to reach optimal forested habitat (Taylor and Merriam, 1995). There is a set of simple rules governing this movement. In open landscapes, female *C. maculata* gravitate to forest associated with streams. In contrast, in the forested landscape, movement is along the stream to locate optimal oviposition sites (Taylor, 2006).

Genetic flow among riverine species is constrained by the branched nature of river networks, while the spatial genetic structure of dragonfly populations is largely determined by the characteristics of the catchment as well as land-use activities. Although larvae may drift downstream, adults disperse more widely. *Calopteryx splendens* disperses mainly along river courses but also overland, with isolation by distance important in both cases. Landscape transformation can trigger individuals to move overland as well as along the river course. The implication of this is that functional connectivity in terms of movement in times of change must be maintained. This is achieved in practical terms by ensuring that there are stepping-stone habitats, such as wetlands in the case of *C. splendens*, to maintain effective metapopulation dynamics for long-term survival and resilience of the population (Chaput-Bardy *et al.*, 2008).

In the case of *Leucorrhinia hudsonica* and its movement in timber-harvested versus forested landscapes, there is an interactive effect between distance and the type of matrix around peatland ponds. This means that functional patch isolation is relative to matrix structure. Specifically, in open harvested matrices rather than in forested ones, individuals are more likely to move over distances of more than 700 m. However, at distances of less than 700 m, proportionately fewer individuals move over open ground, preferring forest in which to move. Peatland size also plays a role, with the likelihood of moving becoming less as the size of the home peatland increases, where there are likely more resources, especially territories, than are available in small ponds. The species readily crosses boundaries from pond to forest where it can forage and roost relatively free of predators. However, an individual may have to leave the pond as conditions deteriorate, with a concurrent shortage of territories or oviposition sites, or perhaps simply because its habitat has dried up. If it then encounters forest, it continues to wander through the forest until it finds another peatland. In contrast, should the matrix be open, an individual will likely fly more directly to acquire territory and other resources while also minimizing risk of predation (Chin and Taylor, 2009).

Important for dragonfly conservation is that evidence to date suggests that female odonates tend to disperse more than males (Angelibert and Giani, 2003) and also over longer distances (Conrad *et al.*, 2002) (Fig. 4.14).

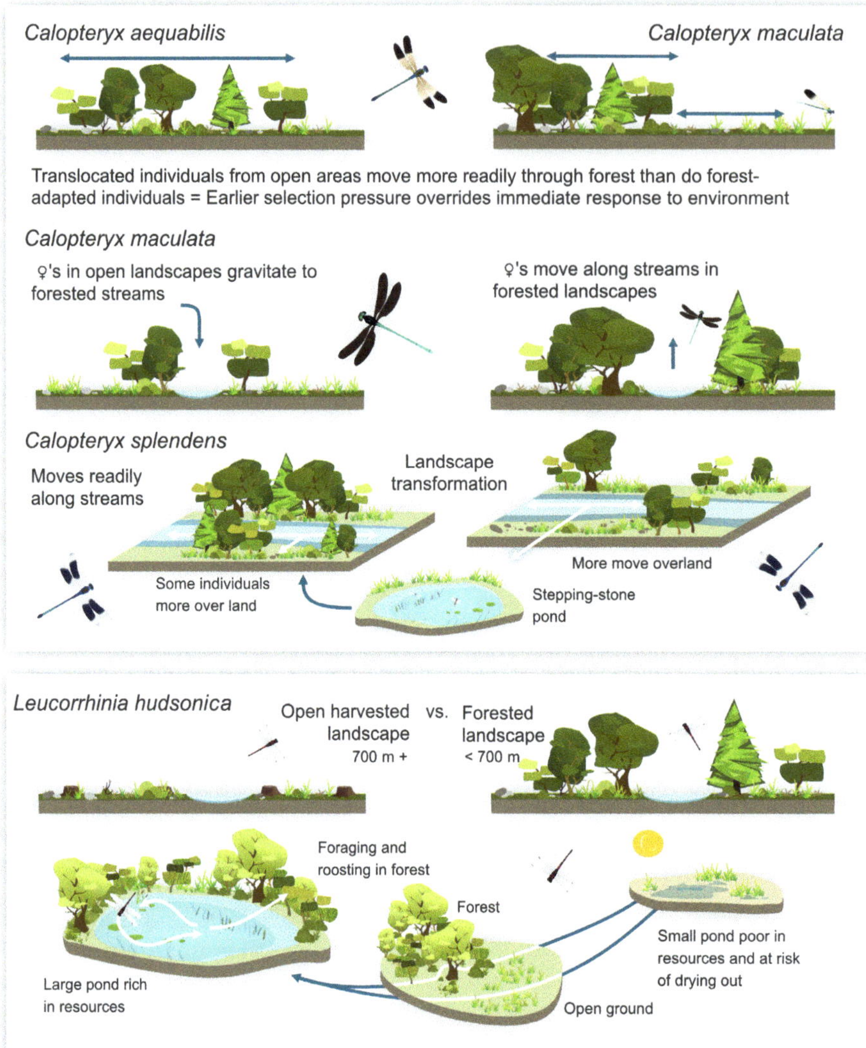

Fig. 4.14. There is an interaction between patch type, spatial grain size and the mobility traits of particular dragonfly species. While *Calopteryx aequabilis* moves readily over both forested and open ground, *C. maculata* moves more over open than forested areas. Translocated individuals from open areas, when introduced into forest, move more readily through the forest than the forest-adapted individuals. This suggests that earlier selection pressure overrides immediate response to the prevailing environmental conditions. Females of *C. maculata* in open, forest-harvested landscapes gravitate toward forested streams, while in forested landscapes they tend to move along the streams to seek optimum oviposition sites. In turn, *C. splendens* naturally moves along forested streams, although some individuals disperse over more open land. However, agricultural transformation of the landscape leads to more individuals moving overland. In doing so, they find suboptimal stepping-stone pond habitats from which they then disperse to more optimal habitats along forested streams. *Leucorrhinia hudsonica* that occupies ponds in open tree-harvested landscapes readily flies to over 700 m. However, in forested landscapes, it tends to linger more and fly less than 700 m. When ponds are large and of good quality, individuals largely stay at the pond, while also using nearby forest to forage and roost. However, at smaller ponds and where resources are poorer and there is risk of the pond drying out, they readily disperse to seek better habitats. When they do so through forest, they wander more than they would when they encounter open ground, where they fly more directly to better avoid predators. (Illustration: Robyn Symons).

4.3.2.3 *Nocturnal roosting*

Roosting site selection can be important for certain species, and becomes significant for rare and threatened species, such as *Calopteryx exul* in North Africa, which displays competition for its specialized roosting sites (Mahdjoub *et al.*, 2023a).

Nocturnal group roosting behaviour occurs among a few species of odonates in various parts of the world. Mostly these nocturnal groups are of a single species, but occasional mixed species groups have been seen, such as *Palpopleura lucia* alongside *Acisoma inflatum* in Nigeria (Hassan, 1976). It seems that the roosting site is selected before nightfall, and the roosts may be all males or a mix of the sexes. The selective advantage of this behaviour is not clear, although in the case of females avoiding male groups, it may be to evade predators attracted to the bonanza of males (Paulson *et al.*, 2020; Paulson, 2021) (Fig. 4.15).

4.3.2.4 *Optimal habitat for adult activities*

The optimal local environment for the adult can override that of the larva. In Canada, *Amphiagrion abbreviatum* and *Argia vivida* can co-occur at the same geothermally heated springs, but their relative abundance depends on the extent of tree cover at the springs. *A. abbreviatum* avoids heavily treed springs where there are few perching sites, and where it has great competition for these sites from *A. vivida* (Pritchard and Kortello, 1997). However, *A. vivida* does not live in open sites, preferring heterogeneous canopy closure, where there are trees in which it can roost at night, then warm up, forage and display breeding behaviour during the day in canopy openings. In the case of *A. vivida*, from a conservation perspective, this means that there should be a landscape mosaic of forest and small cleared areas in association with its breeding ponds to ensure its local persistence (Kortello and Ham, 2010).

Fig. 4.15. A male and two females of *Orthemis levis* roosting together at night in Costa Rica. (Photo: Frank Fleming).

Similarly, *Sympetrum infuscatum* in Japan requires not only forest, but also open areas to feed and breed (Watanabe *et al.*, 2004). Landscape mosaics of lowland forest and heathland are also important for supporting a high diversity of odonates in Borneo (Hendriks *et al.*, 2023).

The very rare and microhabitat/habitat-specialist damselfly *Spesbona angusta* is extremely temperature sensitive at various spatial scales from microhabitat to landscape. Not only does it move locally to continually find the optimal microclimates, but also seasonally across the landscape, as determined by thermal imaging (Deacon and Samways, 2017). Flight toward optimal temperatures is enhanced by both the males and females rapidly changing colour (brown↔blue), even as a cloud passes over (Deacon and Samways, 2016).

4.3.2.5 *Traits of desert species*

Deserts present challenges for dragonflies, as risk of their habitat drying out is high. Additionally, another suitable habitat must be reachable to sustain the local population. In the arid region of Central Asia, dragonfly life cycles are synchronous with seasonal climatic changes where they optimize on water and temperature patterns. Reproduction is restricted to the rainy season, emerging as adults either early in the season, and/or late in the year after passing the hottest times as immature stages. However, elevation moderates these emergence patterns. Below 1500 m above sea level, adults appear from early May, but only in June at 2800 m above sea level. In turn, the building of artificial courses and bodies, such as canals, irrigation and overflow ditches, and rice terraces, have benefitted many species, as these artificial waters are richer in food resources, warmer and slower flowing than the natural habitat of snow and glacier meltwaters. Two species, *Platycnemis dealbata* and *Calopteryx intermedia*, have to date only been recorded from these artificial watercourses. Overall, the species in this region possess traits associated with great flexibility in mobility, phenology and habitat preference, not only to live in this challenging environment but also predisposing them to take advantage of human-modified landscapes (Simpkin *et al.*, 2000; Borisov, 2006).

The Namibian desert in south-west Africa has a modest topography compared to Central Asia. Nevertheless, it is surprisingly rich in dragonfly species (about 75 species), with 56 having adapted to living in the desert. The total number of species consists of six main categories (Suhling *et al.*, 2009). The first category includes widespread species that live permanently in the desert, breeding there. Namibian populations of these species have evolved traits that enable them to live under desert conditions, such as remaining on the wing as adults during drought. The second category also live permanently in the desert but are not desert-biased in that they occur in various types of ecosystems, often not even arid, and far from the desert. The third category includes those species that enter the desert seasonally on rain-laden wind currents or with mountain-mediated winds, or along river basins. They may or may not breed temporarily. The fourth category includes species that immigrate to the desert from the tropical regions of the north or north-east, or from the temperate regions to the south. Sometimes they breed locally. The fifth category consists of species that have isolated breeding populations widely separated from potential source populations, likely remnant populations from former climatically wet periods. The final category includes species associated with the two large perennial rivers bordering the desert. Individuals of some of the species can fly several hundred kilometres and occasionally establish new populations, an example of jump dispersal leading to range expansion (Suhling *et al.*, 2017) (Fig. 4.16).

4.4 Habitat Selection, Development and Local Species Sorting

4.4.1 Habitat selection

4.4.1.1 Assembly at newly created habitats

Creation of new ponds can test whether assembly rules prevail. That is: Does the arrival of one dragonfly species in any way preclude the establishment of any subsequent species? At new ponds in South Africa, it is negligible in comparison with other drivers of assemblage patterns. The combined effect of certain abiotic and biotic environmental variables such as water quality, vegetation type, microhabitat heterogeneity, sunlight/shade and marginal/submerged vegetation gradients, act together as the primary drivers of both larva and adult presence. But pond size is not significant. Species richness is greatest in shallow, well-vegetated ponds, with clear, well-oxygenated water. It is species traits and responses that are the primary determinants of species presence and assemblage structure at any one time. However, with pond temporal succession and environmental fluxes assemblage structure shifts dynamically over time (Osborn and Samways, 1996).

It is not unusual that young larvae are eaten by older ones of the same or a different species. These priority effects are seen during experimental studies between *Sympetrum fonscolombii* and *Trithemis kirbyi* where early arrival at new sites and rapid larval growth are advantageous. As *S. fonscolombii* has a faster growth rate than *T. kirbyi*, the latter only wins when it starts growth earlier than its competitor. How then does *T. kirbyi* survive the presence of the other species in the field? It does so by always being present at ponds, while *S. fonscolombii* only arrives with the onset of the rainy period in the Namib desert. *T. kirbyi* is also more active and arrives at newly created experimental ponds first, then occurring at higher larval densities. However, *S. fonscolombii* in the wild is at an advantage by being a migrant on rain fronts associated with the intertropical convergent zone, moving far and quickly to colonize new and natural water-filled ponds (Padeffke and Suhling, 2003) (Fig. 4.17).

Similarly, in simplified mesocosms (brick drinking troughs for cattle) in Italy there is competition and predation among the species, especially the various larval stages of the sprawler guild, such as *Crocothemis erythraea* and *Sympetrum* spp. Intraguild predation among larvae at different development stages results in an effect on the relative densities of the larvae, making competition one of the processes that shape the assemblage structure in these small structures (Cerini *et al.*, 2019).

Using standardized experimental artificial ponds, McCauley (2006) showed that in the USA at the multispecies level, ability to disperse to and colonize novel habitats may depend not so much on the

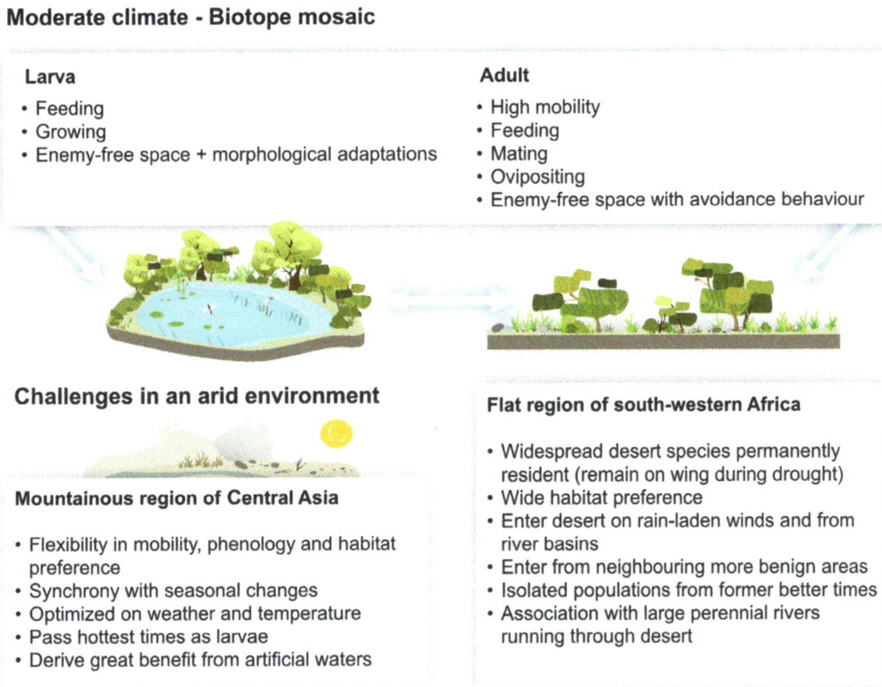

Fig. 4.16. In moderate climates, completion of the dragonfly life cycle depends on a mosaic of biotopes, from the very small microhabitat of tiny species, through to a patch and landscape mosaic for large mobile species. While the larva has traits associated with aquatic feeding and growing, the adult has traits associated with high mobility, aerial feeding, roosting and reproduction. Both larva and adult behave to have as much enemy-free space as possible, although the larva can have morphological adaptations such as spines, and the adult has specific predator avoidance tactics and cryptic roosting behaviour. In arid environments, there are great challenges to survival with many dragonflies nevertheless having adapted to these harsh conditions. Their responses differ according to regional topography. In mountainous Central Asia important traits are associated with flexibility in mobility, phenology and habitat preference. These include being in synchrony with seasonal climatic changes, optimization on water and temperatures, and avoidance of the hottest season by remaining a larva. Together, these traits enable successful colonization of artificial water bodies. In relatively flat south-western Africa, and where there are a surprisingly high number of species, assemblages are diverse and highly changeable based on species with six different modes of living. (Illustration: Robyn Symons).

Fig. 4.17. Female *Trithemis kirbyi* (left) perching on grass next to a pool in an arid area ready to breed early in the season, and a male *Sympetrum fonscolombii* (right), with worn wings after being on the wing for many days. (Photos: Michael Samways).

flight distance capability of species as on their behavioural limitations. Dispersal and recruitment limitations affect the assemblage composition in the various ponds, in part because adult arrival does not translate into the subsequent presence of larvae. Isolation also plays a role, with distant ponds more dissimilar in their odonate assemblages despite a near identical environment. Overall, this emphasizes that functional connectivity of habitats cannot be presumed based simply on flight dispersal capability.

There is also the issue of the circle effect: although a dragonfly may travel linearly 100 m to arrive at a single point (the radius), it must explore 628 m (the circumference) to encounter all potential arrival points 100 m from the departure point. This even assumes it flies in a straight line and then a perfect circle, not a realistic proposition. The point here is that flight is more a 'random walk' than a linear activity, especially as an individual is also potentially distracted by mobile food items, mates, rivals and predators in addition to its reactions to landscape structure.

When clean effluent was consistently pumped into a dry USA riverbed, seven dragonfly species colonized the new water body the same day, and subsequently established. Within three months, the number of species in the novel water body equalled that in the same river 10 km away (Bogan *et al.*, 2020). However, any subsequent drying out of a riverbed can change the community composition of the aquatic species, setting the ecosystem on a new trajectory, emphasizing its dynamic nature (Bogan and Lytle, 2011) (Fig. 4.18).

Nevertheless, based on historical records of adult European anisopterans, there are assembly rules shaping assemblage composition as a result of morphological differences among species and interspecific competition that are associated with the territorial reproductive biology of dragonflies, especially at locations and seasonal times under warm conditions and high niche packing. However, this is not the case for zygopterans (Novella-Fernandez *et al.*, 2023).

4.4.1.2 Importance of canopy cover

Riparian vegetation canopy cover (vegetation structure) plays a major role in determining adult dragonfly presence, and for most species, much more so than vegetation species composition. This can take place down to just a few metres, although at this very fine scale, there is also some influence from the adjacent dominant vegetation type (Smith *et al.*, 2007). However, oviposition sites can be very specific for some species, making plant composition significant in these cases.

The assemblage structure of benthic macroinvertebrates in small Amazonian streams is also determined by local environmental variables such as riparian cover, channel morphology, and variation in size and composition of the substrate. Organismal traits related to species interactions then shape the functional composition of the aquatic insect assemblages, including odonates (Nicacio *et al.*, 2020).

Dragonfly response to canopy cover (i.e. degree of shading of the understorey) is relative to species traits and climate, as well as the relative percentage (levels) of shade measured at various spatial scales. For central European odonates, local vegetation variables influence zygopterans, while anisopterans are more sensitive to larger-scale landscape variables. Zygopteran diversity decreases with increasing bush and tree cover, while anisopteran diversity decreases with the increasing degree of land-use intensification but increases with the length of watercourses. The amount of farmland affects all odonates at the large (5 km) spatial scale, and anisopterans at the small (1.25 km) and medium (2.5 km) scales. In general, landscape homogenization with few buffer strips, bushes and forest patches, as well as possibly fertilizer input, all lead to less space for maturation, feeding and resting (Nagy *et al.*, 2019).

In the UK, canopy cover over a wide range of open versus closed canopy leads to different responses among the species (Hofmann and Mason, 2005). In the case of the sunny climate of South Africa, dense shade, for all but a few species, is generally inhibitory, with species largely adapted to survive in well-lit environments (Steytler and Samways, 1995). Even among the few deep forest species in the country, like the damselfly *Ecchlorolestes nylephtha* (Fig. 3.19), the male actively seeks sunflecks, where it perches and waves the bright tip of its abdomen up and down catching the sunlight, when otherwise it is highly cryptic. Some very small anisopteran males, like *Notiothemis jonesi* will even keep shifting their perching position to remain in even a small sunny spot, yet the female remains highly cryptic among poorly lit dense vegetation. While both sexes of *Phaon iridipennis* regularly occur in densely wooded areas they also occupy adjacent open reedy habitats at the water's edge (Fig. 4.19).

Damselfly males in sunflecks are also warmer, and more attractive to females (Tsubaki *et al.* 2010), bearing in mind that there is a trade-off between

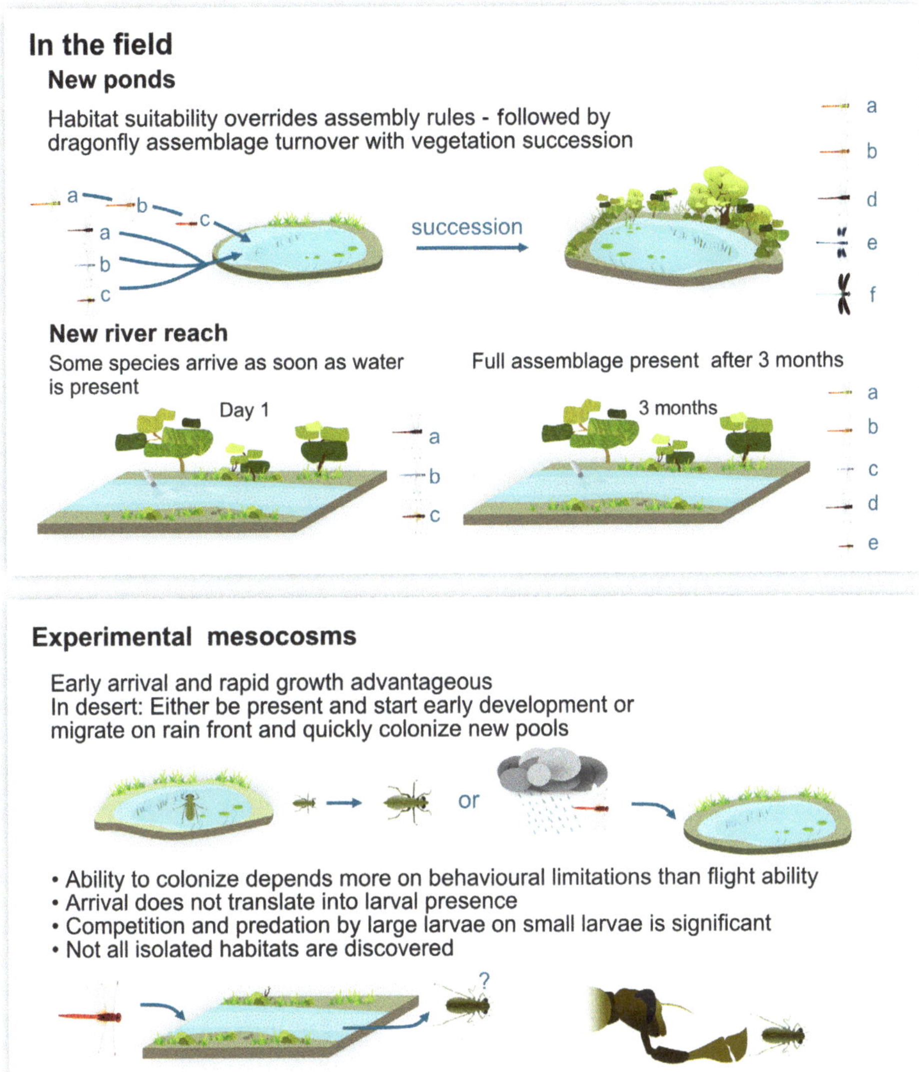

Fig. 4.18. Field studies and experimental mesocosms provide complementary evidence on what happens when dragonflies occupy newly available water systems. Colonization of new field ponds is determined by habitat suitability rather than assembly rules of precedence. A new river reach draws some species the first day that water is present, with the full assemblage appearing after just three months. What mesocosms show is that at the scale of behavioural interactions, colonization strategy followed by rapid larval growth is important, particularly in the desert with intermittent rains. In wetter climates, colonization ability depends more on behavioural limitations than flight ability, with arrival not necessarily translating into larval presence. Furthermore, some isolated habitats are simply not discovered by dispersing adults. Within the mesocosm, there is competition and predation of smaller larvae by larger ones, possibly magnified in the confines of a small mesocosm. (Illustration: Robyn Symons).

gaining thermal advantage in sunflecks versus camouflage in the darker areas of the forest (Shelly, 1982). These findings overall suggest that besides the traits of the species concerned and their phylogenetic constraints, we must also consider what we mean by 'forest cover', its geographical location, extensiveness of canopy and its grain at various spatial scales.

Fig. 4.19. While there are some forest specialists that live in low light levels, many during the day occupy sunflecks, openings in the forest canopy. Among them is the male of the tiny dragonfly *Notiothemis jonesi*, while in contrast the female remains cryptic in the dense and dark lower canopy of the vegetation. (Photos: Michael Samways; Illustration: Robyn Symons).

Experiments in South Africa using shade cloth show that shade alone can influence dragonfly presence independent of any other environmental variables. Shade is largely avoided by anisopteran adults, and the denser the shade, the greater the level of avoidance (Remsburg *et al.*, 2008). In the north of the country in a savanna ecosystem, measured river biotope variables accounted for 82% species membership of any one site, with 44% accounted for by level of canopy cover alone. The other two major variables were microhabitat permanency and river flow rate (Clark and Samways, 1996).

In eastern Canada, experimental mesocosms show that more natural canopy cover decreases the species richness of adult visitors, with canopy cover being an important environmental filter on dragonfly local distribution (French and McCauley, 2018). In drier areas of the USA, shaded habitats are not as species-rich as the more open habitats, although forest specialists gravitate to denser canopies (Ball-Damerow *et al.*, 2014a). In Portugal, three anisopterans, *Cordulegaster boltonii*, *Orthetrum coerulescens* and *O. nitidinerve*, are distinctly shade-loving (Fulan *et al.*, 2008).

A high proportion of odonate species in wooded environments appears to be more typical of tropical than temperate environments where accumulating heat units is less acute. In Puerto Rico, for example, both species richness and abundance are higher in shrubland and forests than in any other biotope (Díaz *et al.*, 2019).

In the tropical Cerrado region of Brazil, there are assembly rules among the odonates based on thermal requirements. Forest cover acts as a filter determining which species successfully colonize any one site. Narrow streams have reduced light yet are also a more stable thermal local environment. These conditions favour the small species, especially the zygopterans, which predominate. In contrast, the anisopterans are more dependent on direct sunlight exposure to thermoregulate, and prefer the wider channels (De Marco *et al.*, 2015).

In a high temperature environment in Mexico, loss of vegetation exposes odonates to even higher (by about 1°C) ambient temperatures. Physiological response by zygopterans versus by anisopterans to canopy cover differs, indicating a phylogenetic signal. The anisopterans have a higher body temperature than the zygopterans in both closed and open canopies. While thoracic and ambient temperatures are positively related, thoracic temperature is higher than ambient temperature. When the zygopteran *Argia pulla* is compared to the anisopteran *Orthemis ferruginea*, *A. pulla* but not *O. ferruginea*, has a higher critical thermal maximum in open sites. Overall, it appears that changes in local environment and ambient temperature filter odonate species, especially among the more vulnerable zygopterans (Castillo-Pérez *et al.*, 2022).

Deep history also plays a role against which current shade preferences are measured. Speciation among the 40 or so species of *Trithemis* in Africa has arisen through changing climates, natural fragmentation and allopatry. The red-coloured species prefer open habitats such as standing water and even temporary pools. They are also highly mobile and have large geographic ranges yet show limited radiation. In contrast, the dark species inhabit forests and are less mobile, while showing much more radiation. They inhabit lowland streams, highland streams or swamps. However, over the last few million years of *Trithemis* evolution, there have been many shifts from open to forested habitats, and in the other direction from forested to open habitats. This is interesting as it shows that traits are flexible and can change over time, driven by changes in environmental conditions (Damm *et al.*, 2010b) (Fig. 4.20).

Similarly, red colour is associated with open habitats among the Hawaiian *Megalagrion* damselflies (Fig. 3.15), with canopy cover being a strong predictor of body colour in both sexes of the species in this group (Cooper *et al.*, 2016).

Evolution of visual systems is a response to visual requirements demanded by the environment, the size of an insect's eyes and the trade-off between resolution and sensitivity. Different microhabitats place different demands on visual systems, strongly influencing eye morphology and geometry in Hawaiian *Megalagrion* damselflies which live in a range of different microhabitats. This has led to three damselfly groups on the islands: large stream species with good powers of flight; medium-sized species of seeps, phytotelmata and terrestrial habitats; and forest pool species with low mobility. Among these species, eye size correlates with body size, and visual resolution plays a greater role than light sensitivity. In poorly lit, forested water-pool microhabitats, mates, prey and potential predators

Deep history has seen shifts in traits and body colour among *Trithemis* species

Red inhabitants of pools, highly mobile and limited phylogenetic radiation

Blue inhabitants of streams and swamps, less mobile and more phylogenetic radiation

Fig. 4.20. The interaction between forest cover and dragonfly traits has a deep history of millions of years during which time changing climate and natural forest fragmentation has led to shifting traits and body colour, such as red versus blue in African *Trithemis* species (e.g. *T. arteriosa* (left) and *T. dorsalis* (right)). (Photos: Michael Samways).

must be distinguished from the background. In these microhabitats, visual resolution in terms of colour, motion and neural processing all come into play to distinguish objects of interest from the background. This is the result of changes to eye geometry that improve feed on a wide variety of insect and spider prey eye performance despite their small size. In contrast, in open well-lit microhabitats, object detection is easier (Scales and Butler, 2016).

Among some species, it is not necessarily a high canopy that is significant, but a low one such as grassland. For example, in Japan the threatened anisopteran *Sympetrum pedemontanum elatum*, which lives in floodplain riverside pools during its larval stage, is a relatively weak disperser as an adult. There are only small genetic differences over a wide area, with some genetically unique populations associated with paleobasins which segregated during ancient tectonic events and now comprise important genetic conservation units. There is very weak gene flow among the populations, suggesting that there is widespread population isolation. Genetic diversity is positively correlated with the amount of grassland within a 1 km buffer, meaning that the preservation of grasslands surrounding its aquatic habitat enhances the immigration and colonization of the adults (Higashikawa *et al.*, 2023).

The process of some species passing across a barrier, whether natural or human, while others do not, is known as species filtering. However, as it is the traits of the species that are responsible for the filtering it can also be considered as trait filtering (Fig. 4.21).

4.4.1.3 Significance of forest edges

Anisopterans released at the forest edge move more frequently over open ground, while individuals released in forest have shorter flight times, but this is species-specific. While open ground facilitates dispersal, forests are not dispersal barriers. What this means is that aiming to improve functional connectivity must incorporate the flight behaviour characteristics (as opposed to flight performance) of a species into landscape planning (French and McCauley, 2019). Whether a species has a short or a long pre-reproductive phase appears to make little difference in how the subsequent adults respond to landscape structures across the landscape (Kadoya *et al.*, 2008).

In Canada, anisopterans have a complex relationship with forest edges, which vary in value for the different species. Forests, as a structural barrier in an open landscape, may impede the movement and dispersal of some species but not others, depending on their requirements at the time, such as foraging, resting or mating. However, ponds that are immediately adjacent to forests, may not be attractive to some species. Spatial scale also matters. At large spatial scales (many hundreds of metres) forests may inhibit dispersal. However, forests at intermediate scales (less than 300 m) do not inhibit dispersal and can offer complementary supporting habitat (Leite *et al.*, 2021).

4.4.1.4 Water barriers

Barriers are sometimes also the water itself rather than the land structures. In the Amazon, channel width (as well as depth) is a significant variable in determining beta diversity patterns among damselflies, largely due to their limited dispersal ability, although this limitation can be overcome, given enough time in a stable environment. However, other factors may also come into play, such as interspecific competition and possibly even predation (Juen and De Marco, 2011).

While *Anax imperator* readily disperses among ponds in continental Europe, the English Channel is a barrier to interchange of individuals between Britain and the mainland, likely though, with occasional individuals crossing the water barrier (Minot and Husté, 2022).

4.4.1.5 Significance of macrophytes for dragonfly assemblages

Density of emergent macrophytes can determine adult species richness (Fig. 4.22). A 34-species study of perennial rivers in eastern Europe showed that both local and landscape variables affect species differentially. One variable in particular, emergent vegetation cover, has a significant effect on odonate species richness and diversity, especially that of zygopterans. However, zygopteran diversity decreases with too much emergent vegetation, while length of watercourse is important for anisopterans (Müller *et al.*, 2003).

Among ponds in South Africa, there is an increase in overall odonate species with an increase in sedge (*Cyperus* spp.) cover in spring and bulrush (*Typha capensis*) cover in summer. However, too much cover adversely affects odonate diversity. There is an increase in species richness in up to 70% sedge cover, and up to 40% bulrush cover,

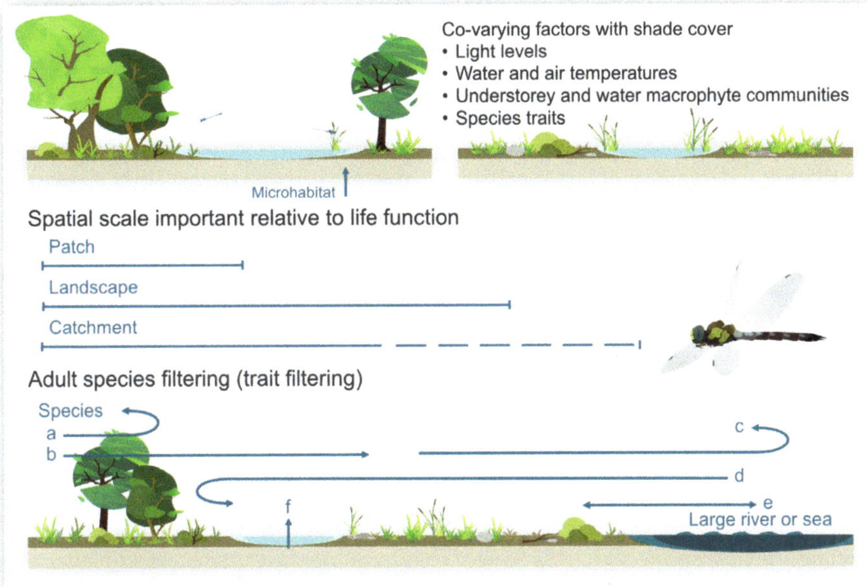

Fig. 4.21. There are several co-varying factors associated with shade cover. These include the abiotic environmental variables of light and temperature, and the biotic variables of understorey and macrophyte community cover. Dragonfly species traits also co-vary with the environmental variables, often with differences according to sex. However, the environmental drivers together and their interactions with traits can vary according to spatial scale, from localized patch to catchment. One of the reasons for these scale effects is that the various landscape elements in effect filter the species as individuals fly or not across different landscapes, and where there are changes in environmental conditions, such as with elevation. As the responses are a manifestation of traits, this filtering may be described as trait filtering. Certain individuals of species 'a' steer away from large trees, while those of 'b' pass through a clump of trees. In turn, species 'c' is deterred from crossing a large river or a sea. Species 'd' readily wanders across a wide range of open conditions, while 'e' readily moves across large water bodies, a characteristic of many migrants. Species 'f' flies along the river course, which is often the case when there are trees along the bank, and it is searching for more open patches. (Illustration: Robyn Symons).

Fig. 4.22. Aquatic macrophytes are a major driver of dragonfly assemblages. Their effect is not just about the macrophyte species present but also their density. When their density is very high, they can totally clog (left) or change (right; notice the still water in the foreground and the wind-churned water in the background) the local environment, for example, the margin of this large river in northern Argentina, where *Pontederia crassipes* is a dominant species. (Photos: Michael Samways).

above which there is a decline in species richness (Briggs *et al.*, 2019).

At the very large spatial scale, such as the south Ural lakes in Russia, the highest dragonfly species richness is found where the perimeter of reed beds is larger and the extent of the wetlands is greater, which together provide a wide range of larval habitats (Mashkova *et al.*, 2018).

Habitat selection by adults across canopy gradients plays an important role in structuring larval assemblages. This selection process is, in turn, an environmental filter determining distributions across the landscape (French and McCauley, 2018). The macrophyte *Pontederia azurea* in Brazil has great morphological plasticity. This species, along with other macrophytes, affects the libellulid dragonflies living in association with it. Adults use *P. azurea* for oviposition, perching and holding territories. However, in the case of the larvae, the physicochemical water variables, especially temperature, pH, dissolved oxygen, oxygen reduction potential and turbidity affect larval abundance. The physical complexity of the microhabitat greatly affects both larval species richness and abundance via predator avoidance traits (Brito *et al.*, 2020).

4.4.1.6 Oviposition

Microhabitat selection is finely tuned in terms of traits for oviposition. The process of egg laying has three phases: selection of a landing or water site, choice of egg positioning microsite, and egg deposition. Selection of a landing or water site depends on various factors, including water surface reflectance, shape and background of plant structure, size and location of the substrate (in some cases), sturdiness of the substrate to maintain purchase (in some cases), and interference factors such as the close presence of conspecific individuals or predators, or those associated with adverse weather conditions.

Once within a suitable biotope, where the level of sunlight versus shade is significant, visual cues then play an important role in females undertaking primary selection of oviposition sites. This is especially important for selecting sites where predatory frogs are least abundant and male harassment is reduced. This activity is then followed by selection of the precise oviposition site using sensilla in the region of the ovipositor (Michiels and Dhondt, 1990). However, among some species that have had a long history with certain environmental conditions, such as the streams in the Cape Fold Mountains, females can also select sites based on water characteristics, especially water temperature, pH, dissolved oxygen and conductivity (Kietzka *et al.*, 2017).

The eggs are usually placed where offspring have the best chance of survival and where predation risk is minimized. The act of laying should be as short as possible to avoid the female being eaten by a predator, and to maximize on other activities such as feeding and searching for other oviposition sites. This would suggest that there is strong selection pressure to recognize, choose and engage in oviposition as efficiently as possible (Martens, 2001).

Female choice of optimal oviposition sites is a strong trait among odonates (Fincke, 1997). Two European damselflies, *Sympecma fusca* and *Coenagrion caerulescens*, lay in the tandem mating position, mostly in moist floating dead plant material. In its selection of landing sites, *S. fusca* does not distinguish between living and dead material, while *C. caerulescens* prefers dead plant material. *S. fusca* is an early flight season species, laying its eggs when floating material is mostly dead, or living substrate though this is difficult to find. It has no need to distinguish between living versus dead substrates. In contrast, *C. carulescens*, a summer flyer, oviposits when green and dead plant material are mixed side by side on the water surface, and so it is at an advantage when it can instantly recognize its preferred dead material, wasting little time in searching for the right substrate (Martens, 2001).

Females select oviposition sites that are most suitable for the subsequent larvae. *Aeshna juncea* is attracted to light reflections on a dark background, particularly when associated with the vertical structure of *Carex* sp. stems (Wildermuth, 1993). Some species select a certain plant species, such as the sedge *Eleocharis* sp. in the case of the Neotropical damselfly *Oxyagrion microstigma*. Males defend the area around the plant, increasing both male density and that of ovipositing females, which is also where the larva lives (Guillermo-Ferreira and Del-Claro, 2011).

Females of the giant Neotropical helicopter damselfly *Megaloprepus coerulatus* mate with larger males defending suitable tree hole larval habitat, especially scarce large tree holes in forest gaps. What this shows is that both mate acquisition and procurement of superior larval habitat have arisen via evolutionary selection of male territory choice. However, because even large tree holes are not adequate to accommodate a female's entire egg production, as well as the potential presence of conspecific or heterospecific larvae, selection favours the females that partition their eggs in several tree holes (Fincke, 1992).

Large size in some damselflies is related to overall greater total egg production. In *Coenagrion puella*, longevity can account for 70% of reproductive success, with clutch rate production accounting for another 20%, and total clutch size another 10% (Banks and Thompson, 1987). What these results show is that eating and growing well as a larva leads to a large size as a female, living a long time and spreading the risk for eggs, with all traits together promoting a high survival rate among individuals of the population (Fig. 4.23).

4.4.2 Local species sorting

4.4.2.1 Maturation and sex differences

Advantages to breeding away from the natal site include lessening the chances of inbreeding. The local spatial distribution of the rare Algerian clubtail *Gomphus lucasii* varies according to sex and age. The mean distance from emergence to maturation site is almost 600 m. Maturation takes place in grassland or dense wheat fields, with females preferring grassland. Overall, mature females are farther from water than are males, which remain largely close to water, but occasionally up to 450 m, with one individual 3 km from any breeding site. The mean dispersal distance between reproductive sites is 181 m. While both sexes prefer fast flowing water, females oviposit in small watercourses. These findings emphasize the importance of the much wider landscape beyond the water, for maturation, foraging and mating. Importantly, however, long-term survival of this species is favoured by the dragonfly's ability to disperse along watercourses as well as over terrestrial areas, providing it with an alternative option in the face of landscape transformation. This underscores that understanding the traits of a dragonfly species for determining its habitat characteristics and size is crucial for its conservation (Zebsa *et al.*, 2015).

Different behaviours by males and females remind us that traits pertain to gender as well as species. The rare and threatened species *Sympetrum depressiusculum* has a scattered distribution in Europe. Its sensitivity to the terrestrial part of its habitat is of conservation concern. Its dispersal characteristics differ between the sexes and throughout its three-month flight period. The male is not territorially

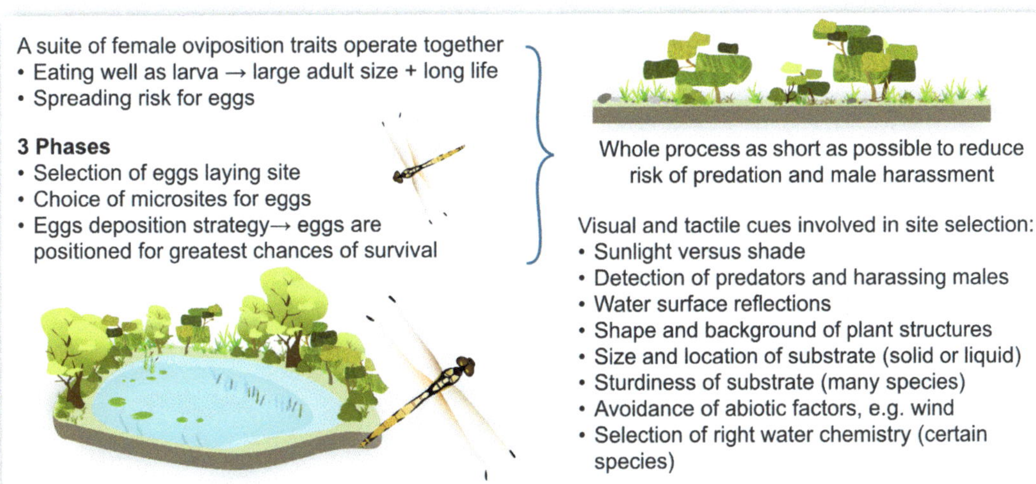

Fig. 4.23. A successful oviposition strategy depends on a suite of traits operating together, starting with the female larva. She needs to eat well to grow and have a large adult size and live a long life. This then enables her to spread the risk for eventual egg survival. There are three phases to oviposition: 1) selection of the landing or water site, 2) choice of egg microsite (in or on living or dead plant material, or on the water surface), and 3) having an egg deposition strategy where the eggs are placed for best chances of survival. The whole process is kept as short as possible to reduce risk to her from predators and conspecific male harassment. Both visual and tactile cues are used to assess various aspects of the local environment. (Illustration: Robyn Symons).

displaced as in many odonate species, and many males may gather at breeding sites. However, both males and females mature and forage regularly over their long flight periods, on average 481 m for males and 446 m for females away from their natal sites. Importantly, they prefer only remnant natural patches and abandoned fields, avoiding agricultural fields, including hay meadows. These behaviours again emphasize that knowledge of a species' traits is critical for its conservation (Dolný *et al.*, 2014) (Fig. 4.24).

For many dragonflies, vegetation structure is important as a perch from which to forage, defend territories and seek mates, as well as on which to thermoregulate. Individuals of *Sympetrum* species males prefer wide diameter, stable and widely spaced perches from which they forage and maintain territories. Height of perch is not that important, so long as it is sturdy. Females prefer thinner, denser perch sites, possibly to avoid male harassment and to be able to forage more freely (Hykel *et al.*, 2020).

Somatochlora hineana is a threatened species in the USA, and as with all threatened species, detailed knowledge of the species' use of its habitat is essential for conservation action to be effective. The adults of this species have a highly biased sex ratio toward males at the breeding habitat. The males mostly use the wetland areas, whereas the females mostly occupy dry meadows and marginal breeding habitat to forage, only returning to water to breed. Despite the greater abundance of fly prey at the wetland, females largely stay away from the wetland breeding habitat to avoid male harassment, so that they can forage free from disturbance (Foster and Soluk, 2006).

Among damselfly populations, there is often a male bias. In *Lestes sponsa* this appears mainly to be due to lower survival probability in females, which experience higher predation levels (Stoks, 2001) (Fig. 4.25).

4.4.2.2 Territory holding and dispersal

In the damselfly *Nannophya longipennis*, males hold territories and guard the egg-laying females. They are evenly spaced across the marshy habitat, whereas the females show an aggregated distribution. At the time of most abundance in mid-summer, territorial disputes cause subordinate males to disperse to neighbouring areas (Fujita *et al.*, 1978).

Conflicts over the monopolization of resources, whether sexual partners, shelters or territories, involve individuals expending energy, suffering injuries or increasing their exposure to predation. Dragonflies can adopt behaviours that optimize the balance between the benefits of resource acquisition and the costs of disputes. Large-sized residents generally win the disputes for unmated females but tend to lose when guarding a mated female irrespective of body size difference between competitors. In this context, both fighting ability and high levels of activity determine the winner of disputes. Therefore, there is some advantage for individuals that can predict the outcome of a dispute (i.e. that can eavesdrop on agonistic interactions between conspecifics to obtain additional information about the competitive ability of potential rivals). This information allows individuals to estimate the resource-holding potential of rivals. In *Erythrodiplax fusca*, eavesdropping on rivals is a low-cost strategy to obtain

Fig. 4.24. The *Sympetrum depressiusculum* male (left) and female (right) forage at different average distances away from emergence sites. Furthermore, they require either remnant natural sites or abandoned fields, avoiding agricultural fields and hay meadows. (Photos: Aleš Dolný and Dan Barta).

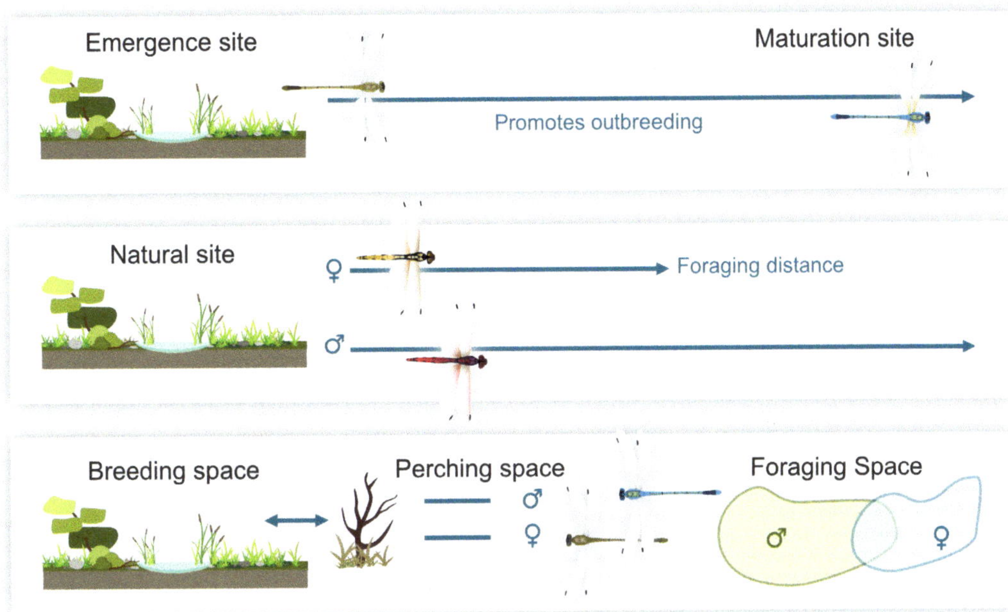

Fig. 4.25. Among adult dragonflies there are often differences in where they are positioned relative to level of maturity and according to sex. Pre-reproductive individuals (tenerals and immatures), despite the seemingly high risk of predation, often disperse very far from their emergence site. One advantage is that it appears to promote outbreeding, and therefore increases genetic diversity across the landscape. Mature males often forage farther than females from their natal sites, where the males can at least optimally forage and pass on genes over greater distances. While the breeding space is necessary for development, males and females may perch and forage in different spaces. One advantage for the female is that she experiences less harassment from conspecific males. Despite the female usually being more cryptically coloured than the male, females overall have lesser levels of survivability than males. (Illustration: Robyn Symons).

information on territory owners and can prevent males from entering fights they would certainly lose, and so they only enter disputes with weaker territory owners (Pinto *et al.*, 2022).

In *Pachydiplax longipennis*, when large males are present and holding territories, small males are more likely to disperse. However, if a receiving habitat is not available for these subordinate males, some genetic variation may be lost. Even if suitable habitat is available, functional connectivity between populations may be less than that estimated from absolute dispersal rates (McCauley, 2010).

Males of the thaumatoneurid damselfly *Paraphlebia zoe* have wings with two colour forms: males that have black-and-white tipped wings that hold territories, and those that have clear wings and are satellites. These satellite individuals can act as sneakers and/or be ready to occupy a male's territory should it become weak or die. Genetically, these are stable and fixed traits. Sometimes, two individuals that are coloured, one dominant and one subordinate, share a territory. There is a correlation between body size and survival in both sexes, along with a positive effect of territory quality on longevity and mating success. However, holding a territory is energetically expensive, and only males of the highest quality can defend territories for long. Dominance decreases with time, and younger, fresher males may take over and become the dominant holders (Rivas-Torres *et al.*, 2019).

Large body size in odonates is often associated with increased lifetime reproductive success. Among territorial species, larger males live longer and have greater mating rates, contributing to the improved success. However, large males of non-territorial species, although not having greater mating rates, do have longer reproductive lifespans, and therefore overall greater mating success. In the case of females, evidence suggests that they too benefit from being large (Sokolovska *et al.*, 2000).

Among a series of anisopterans and zygopterans, territory size increases with body mass, especially among species with spotted wings, a manifestation of high fitness levels. Larger body mass also improves territorial defence capability through increased flying time, rather than spending time perching (Aromaa *et al.*, 2019).

The advantage of large size in healthy individuals begs the question why do males not become very large over evolutionary time? There must be a trade-off. One possible reason is that smaller non-territorial males are more agile and able to court and subdue females (Serrano-Menses *et al.*, 2007). There are likely other environmental pressures, such as resource availability and predation pressures, that also constrain size. Among the resources necessary for holding territories is the intake of sufficient food to build up fat reserves for energy alongside a good immune response, making holding of a territory highly diet-dependent (Córdoba-Aguilar and Mungaía-Steyer, 2015).

There are several species of the endemic genus *Nesobasis* on Fiji (Ferguson *et al.*, 2023). They vary in degree of territoriality and relative rarity. Species with abundant males are highly territorial, and ready to mate with fresh females in optimal habitat. In the species with rare males, females move widely to oviposit. Although it is unknown why males of some species are rare, it is essential from a conservation perspective to understand the reason for these sex ratios in these geographically restricted species (Van Gossum *et al.*, 2007).

Developmental plasticity affects population divergence via gene flow. While male *Calopteryx splendens* species recognition of the female is fixed at emergence, females are more plastic and learn to recognize heterospecific *C. virgo*. This leads to pronounced population divergence in females. What this means for conservation is that with ongoing environmental change and landscape fragmentation, there is a greater chance of formerly allopatric species coming into close association with the focal species. However, developmental plasticity of mate preferences in either or both sexes might enable coexistence and prevent local extinction (Svensson *et al.*, 2014).

Increased phenotypic diversity via female-limited polymorphism increases per individual productivity. This enhances population performance, which is of benefit to the population's stability and persistence. This phenotypic diversity comes about from lower harassment of females by males. When a population has only one type of female morph, males do not detect all the females they encounter, as many are outside the male's search image for the one type of female. Furthermore, different female morphs are not necessarily part of the search image of potential predators. Overall, this means that the polymorphic condition mitigates costly agonistic interactions, mitigates predation pressure and reduces genetic load. The outcome potentially lowers extinction risk through higher effective population sizes (Takahashi *et al.*, 2014). Furthermore, polymorphism can be highly dynamic, and in the case of *Ischnura elegans*, there are stable morph frequencies at the core of the species' geographic range but fluctuating morph dynamics at the range limits (Gosden *et al.*, 2011) (Fig. 4.26).

4.4.3 Traits and genetic plasticity associated with larval development

4.4.3.1 Developmental plasticity

While each dragonfly developmental stage is a survival hurdle to clear when confronting environmental challenges, morphological, physiological and behavioural pathways can generate carry-over effects across the various stages throughout development. However, there are compensatory mechanisms in a later stage to overcome stress on an earlier stage.

Stressful environmental conditions at the egg stage may have fitness implications for the larva, including extended embryonic development time and reduced hatchling size. There can even be carry-over effects from egg right across to adult, like survival to reproductive maturity. In turn, larval environmental conditions can affect adult size, energy storage (as fat reserves), flight performance and various behavioural activities.

However, the adult can overcome some of these setbacks, such as increasing body mass and flight muscles through intense feeding, so long as bad weather does not set them back. Harsh conditions experienced by the adult can reduce lifetime egg production leading later to lower egg and larval densities. The overall message here is that all life stages, and type and degree of compensatory mechanisms, must be considered together to ensure a strong effective population (Stoks and Córdoba-Aguilar, 2012).

An interesting question in evolutionary biology is to what extent, and under what conditions, does

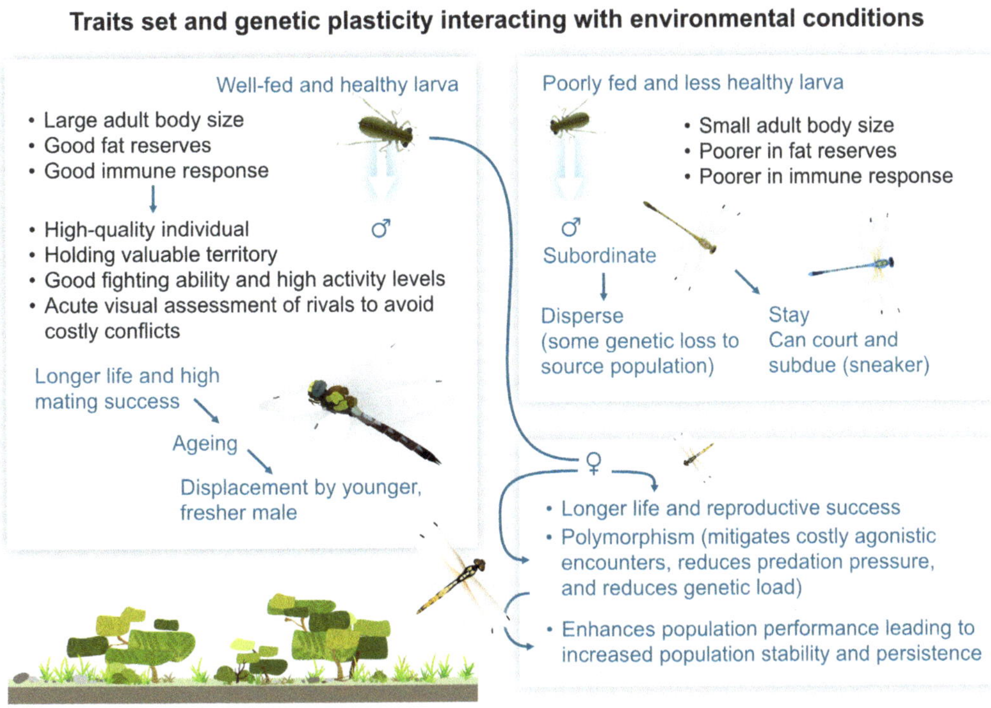

Fig. 4.26. Sets of traits and genetic plasticity are interactive with prevailing environmental conditions. A well-fed and healthy larva leads to a large-sized adult with good fat reserves (representing energy availability) and a good immune response. This is not the case for a poorly fed and less healthy larva. A high-quality male tends to hold a valuable territory, has good fighting ability and high activity levels. Acute visual assessment of rivals avoids costly conflicts. These qualities then lead to a longer life and high levels of mating success. But with age, this male will likely be displaced by a younger, fresher male. A less vigorous, subordinate male might disperse, taking with it some of the genetic diversity of the source population. In some species, subordinate individuals stay in the source population where they can surreptitiously court and subdue a female. These individuals are known as sneakers. Well-fed and large females tend to have longer lifespans and greater reproductive success than lesser quality females. In some species, there is polymorphism among the females. This trait mitigates costly agonistic encounters, reduces predation pressure and reduces genetic load, which enhances population performance leading to increased population stability and persistence in the longer term. (Illustration: Robyn Symons).

natural selection of one life stage affect the evolution of another life stage? In libellulid dragonflies, larval and adult body shapes evolve largely independently. There is however, a functionally important interrelationship among selective pressures on adult wing size associated with increased larval wing sheaths (which house the developing wings) and adult wing size. This suggests that evolution of this aspect of larval morphology at least is sensitive to selection on related adult structures (Moore and Martin, 2021).

However, in cases where evolution is linked to a particular biochemical pathway, there may be a trade-off in response to different selective pressures. In other words, adaptation in one stage appears to impede a species' ability to adapt in other stages. This is the case in species where adults with more wing melanization (important in sexual selection through holding of optimal territories) have weaker melanin immune defences in the larval stage. As aquatic parasites and pathogens are a major threat to dragonfly larvae, there must be a strong immune response to these pathogenic organisms, but this might come about by alternative immune pathways, while arguably allowing the more important process of highly active breeding to contribute most to the following generation (Moore and Martin, 2021) (Fig. 4.27).

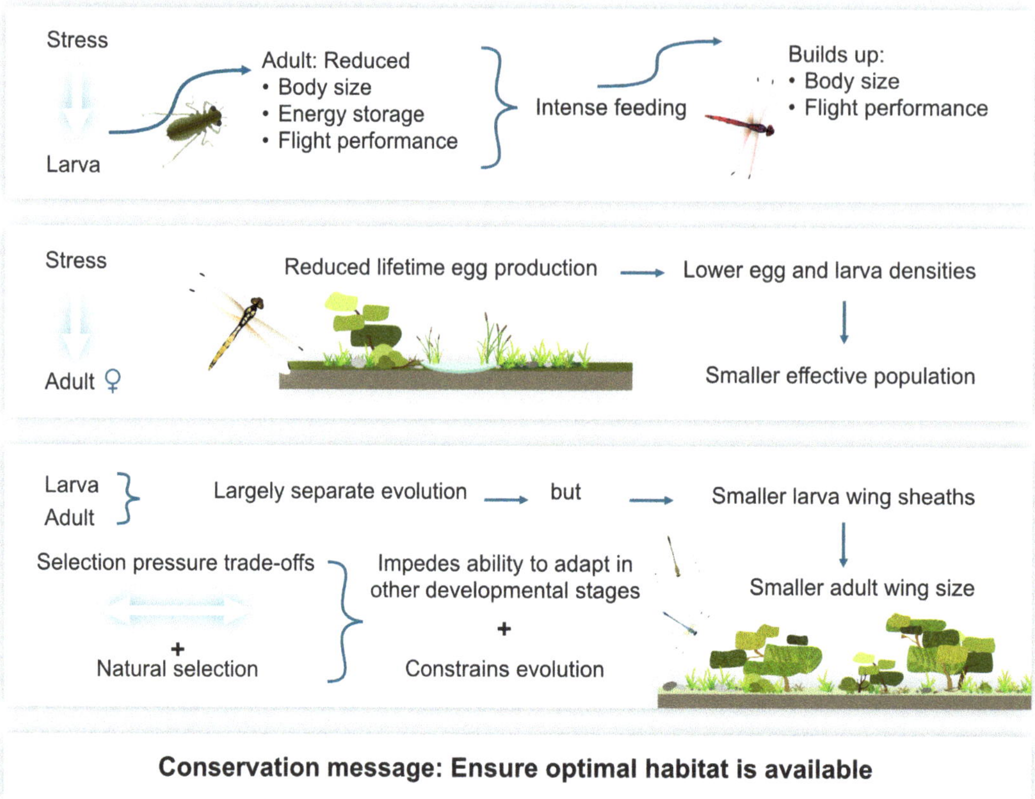

Fig. 4.27. There are compensatory mechanisms among later developmental stages to stress experienced by younger stages. While stress on the larva can reduce adult size, energy storage and flight performance, the adult can overcome larval setbacks by intense feeding that increases body mass and flight muscles. However, an adult female experiencing harsh conditions may have reduced lifetime egg production, leading to lower egg and larva densities, causing a drop in effective population size. While the body shapes of larva and adult evolve largely independently, larval wing sheath development can carry over into adult wing size. There is a trade-off in response to different selection pressures, with adaptation in one stage impeding a species' ability to adapt in other stages. This means that collectively, natural selection in one stage imposes trait-dependent constraints on evolution in other stages. These various responses indicate the conservation value of growing up in optimal habitat while also being able to face the stressors imposed by human activity. (Illustration: Robyn Symons).

4.4.3.2 Habitat selection among bottom-living larvae

Flow velocity generates sediment type, with high velocities leaving behind the coarsest sediments and largest interstices between the solid geological units. The undersurface of pebbles, for example, is a microshelter zone where small sediment particles, organic matter, and hence benthic organisms, including certain dragonfly larvae, accumulate. In contrast, where water velocity is low, fine-grained soft sediments develop expansively, attracting dragonfly larvae that bury themselves, often with an ambush feeding strategy. While certain substrata attract specific species, the subsequent sorting of individuals can be according to both the size of the larva and competition among species (Worthen and Horacek, 2015). This is seen among bottom-dwelling riverine gomphid larvae. The larvae of *Gomphus simillimus* and *Cordulegaster boltonii* affect the microdistribution of the larva of *Onychogomphus uncatus*, forcing it to use suboptimal habitat. Surprisingly, the congener *O. forcipatus* does not affect it (Suhling, 1996).

Specialized traits are also seen among dragonfly larvae in how they respond to the substrate. While many gomphids bury themselves in the bottom substrate, they do not excavate burrows as do most of the petalurids. True burrowing provides a relatively stable environment for the long-lived petalurid larvae. The burrows typically extend below the water table in soft, peaty or organic-rich soils in mires, seepages or along stream margins where their burrowing habitat provides protection from fire, drought, floods and above-ground predators, while also maintaining refuges to employ ambush predation (Baird, 2019) (Fig. 4.28).

As rivers are dynamic in response to quantity of rainfall, no one substratum is always the same over time. Floods may dislodge larvae, which nevertheless are not necessarily flushed far downstream. They possess response traits enabling them to find nearby calm retreats until the flood subsides (Samways, 1989a).

4.4.3.3 Specialist traits associated with different life stages

Traits that characterize an amphibiotic organism, like a dragonfly, result in specialist responses to abiotic or biotic variables associated with a particular life stage, and even the age of the stage (Calvão *et al.*, 2020). *Pantala flavescens* is a globally widespread species often seen as an adult flying and foraging across many types of landscapes, both natural and transformed. The females readily lay eggs in small temporary grassy pools that appear with rain fronts. According to age, the larvae feed on various prey that appear in new pools immediately after rain. Prey types range from crustaceans to tadpoles, and it is also highly cannibalistic. While a key trait of the adult stage is its remarkable flight ability and extensive dispersal, often far away from water, the larva is relatively specialist in its choice of habitat, making the two life stages very different in their lifestyle yet highly complementary (Fig. 4.29).

4.4.3.4 Water chemistry and larval traits

Among the least mobile species that inhabit Canadian lakes, the local geographical distribution of the adults is largely driven by larvae that inhabit lakes with different levels of acidification. Some species are even restricted to lakes with a low pH (below 5.8) (Pollard and Berrill, 1992). However, other factors also come into play in these lakes, including predation by fish and larger odonates, as well as possible changes in their food base (Carbone *et al.*, 1998).

The wetlands along the northern Adriatic coastline are a mix of freshwater and saline conditions leading to a waterscape mosaic rich in dragonfly species, many in high abundance. Dragonfly assemblages

Fig. 4.28. The larva of the enormous *Petalura gigantea* (left) and its burrow opening (right) in *Sphagnum* and restioid-heath vegetation, adjacent to a shallow seepage-fed streamlet in a montane mire in the Blue Mountains, New South Wales, Australia, illustrating the dependence of the species on a high water table and saturated peaty substrate for ovipositing and larval burrow establishment (arrow). (Photos: Ian Baird).

Fig. 4.29. The adult *Pantala flavescens* (top) is highly generalist in terms of landscapes across which it feeds, often far away from water. In contrast, the larva is a relative specialist of small temporary pools that appear with rain fronts (bottom). Shown here is a captured female immediately releasing copious eggs. (Photos: Michael Samways).

show high species turnover relative to level of salinity. Some species, such as *Aeshna mixta*, *Sympetrum fusca*, *S. meridionale*, *Ischnura elegans* and *Erythromma viridulum*, can breed across a wide range of high levels of salinity, and are described as polyhaline. Other species, like *Aeshna affinis*, *Anax ephippiger*, *Brachythron pratense*, *Orthetrum cancellatum* and *Sympetrum fonscolombii*, are oligohaline, and are characteristic of mid-range salinities. In turn, strictly freshwater species include *Aeshna cyanea*, *Libellula depressa*, *Orthetrum albistylum*, *Sympetrum striolatum*, *S. vulgatum*, *Lestes parvidens* and *Coenagrion scitulum*. The range of traits and genetic plasticity among these various dragonflies enables many species to co-occur in these wetlands, making them areas of high conservation value (Uboni *et al.*, 2020).

Water bodies, particularly in warm and dry areas where erosion and evaporation are important, geochemical processes and water chemistry associated with salinity levels can play a major role in determining the presence and abundance of dragonfly species. Among various water bodies in Pakistan, both running and still, species and genera are associated with certain physicochemical conditions of the water. When calcium + magnesium, sodium, carbonates, sodium absorption ratio (SAR), residual sodium carbonate (RSC) and electrical conductivity are measured against the presence of dragonfly larvae, there are some strong correlations. In terms of species responses, *Crocothemis* tolerates high levels of sodium, bicarbonates, electrical conductivity, SAR and RSC, while *Ischnura*, *Pseudagrion* and *Ceriagrion* are tolerant of a wide range of salinities, with *Sympetrum* and *Gomphidia* less so (Zia *et al.*, 2018). Interestingly, *I. senegalensis* in Africa, the same species as in Pakistan, is tolerant of organic pollution as it has traits evolutionarily honed to inhabit pools and water holes rich in hippo and buffalo dung (Samways, 2008).

Water chemistry is interactive with degree of water permanence (hydroperiod), which also determines the vegetation type present. Among Austrian dragonfly assemblages, and besides those species that are habitat generalists, there are habitat-sensitive species which fall into two broad groups: those species of temporary ponds with rush and reed vegetation versus those of permanent ponds with floating macrophytes (Schindler *et al.* 2003).

Rockpool odonate inhabitants often experience much variation in pond permanence, with some species accelerating their larval development to survive the pool drying out. An experimental study among larvae from Colombian granite rockpools that are subject to elevated concentrations of water salts and reductions in water level focused on cues used to increase developmental rate. Libellulidae species, all along the permanence gradient, accelerated their growth in response to an increase in dissolved salts, while *Anax amazili* did so in response to lower water levels. However, the damselfly *Telebasis simulata* is restricted to deeper long-lived pools and does not respond to these cues. Overall, this leads to different species having certain life-history strategies and niche differentiation. In sum, developmental plasticity in response to hydroperiod-related cues enables the various species to avoid mortality and exploit short-lived habitats (Aristizábel-Botero *et al.*, 2023).

Ecological traits of dragonfly larvae in the tufa-depositing habitats of a Croatian Dinaric karst system are related to strong preferences for distinct microhabitats. While water temperature, ammonium and oxygen concentrations greatly influence assemblage composition, highly favoured microhabitats are tufa barriers and inorganic substrates of cobbles and sand associated with low water velocity. Dragonflies largely avoid microhabitats of mosses associated with strong currents (Vilenica, 2017) (Fig. 4.30).

4.4.3.5 Macrophyte species and larval traits

The importance of certain macrophytes relative to dragonfly larval traits lies in the specific structure of those plants for the larvae living among them. In short, plant traits mediate predator–prey relations in the water for the young stages, as the specific structures of terrestrial plants do for the adult stages – teneral, immature and mature.

In north-western USA, three *Enallagma* species (*recurvatum*, *laterale* and *pictum*) are variously and strongly associated with certain species of emergent and floating macrophytes. One species, *E. recurvatum*, is of conservation concern, and has metapopulation dynamics, making it essential that other ponds with the right vegetation (*Juncus militaris* in this case) are within 2 km to ensure its persistence (Gibbons *et al.*, 2002). In the same region, Butler and deMaynadier (2008) emphasize that to maintain the diversity and composition of the lakeside damselfly assemblage, including the threatened habitat specialists (*E. laterale* and *E. pictum* at their sites), it is essential to maintain the richness and abundance of emergent and floating macrophytes, as well as benthic substrate granularity and lake productivity. This means that to conserve the whole damselfly assemblage, including the threatened habitat specialists, human disturbance of the riparian zone must be kept to an absolute minimum.

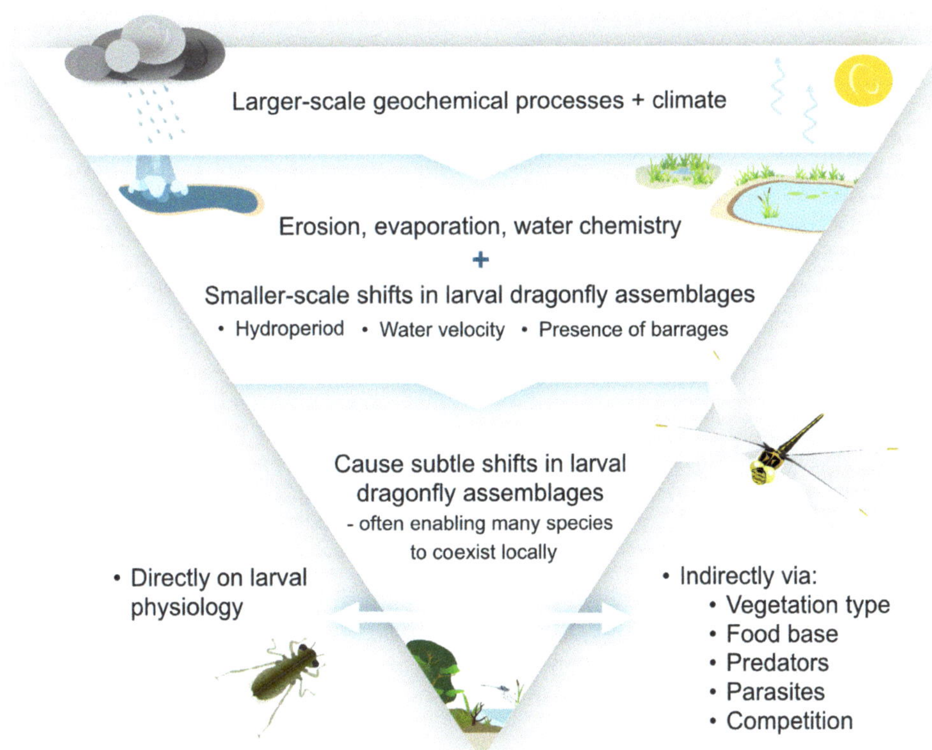

Fig. 4.30. There are top-down processes that shape local larval dragonfly assemblages. Larger scale geochemical processes relative to climate influence factors like erosion, evaporation and water chemistry. When combined with smaller scale factors like hydroperiod, water velocity and presence of barrages, they cause subtle local shifts in the composition of larval assemblages. This often enables many species to coexist in any one local area. How this coexistence comes about is both directly via each species' physiology and/or indirectly via other factors like vegetation type, food base, predator pressure levels (e.g. fish and larger dragonflies), parasite loads and competition from other larvae. (Illustration: Robyn Symons).

Similar findings come from the wet and warm climate of West Bengal, India (Sharma *et al.*, 2016), indicating that the presence of a specific set of macrophytes is important for odonate species assemblages more widely.

The submerged larva has to contend with its predators yet must also forage for prey. Then there is the age and hence size of the larva to consider, as well as its preferred prey at any one stage in its development. Young larvae of *Rhionaeschna multicolor*, for example, prefer certain cladoceran species over other potential prey. But the prey find refuge among the fronds of the submergent macrophyte *Egeria densa*, reducing the larva's prey catch rate when the plant is present. However, with the larva increasing in size, it consumes a greater prey biomass, mostly through consuming larger prey items that are not so able to seek refuge among the plant's framework (Rivera-De la Parra *et al.*, 2016).

Details of the local plant environment and the structure of certain plant species can make a difference on the development rate of odonates. *Enallagma civile* develops better and has greater overall larval biomass when living among *Potamogeton nodosus* than among *Heteranthea dubia* (Schad *et al.*, 2016). In turn, the threatened *Aeshna viridis* is commonly associated with the macrophyte *Stratiotes aloides*, which provides the larva with a refuge from the predatory European perch *Perca fluviatilis* (Rantala *et al.*, 2004). However, it can also face intense competition from its congener *A. grandis*. This has led to fine-scale spatial partitioning of the two species of larva associated with *S. aloides* rosettes. Nevertheless, the intensity of the competition is of concern for the threatened *A. viridis* (Suutari *et al.*, 2004).

4.4.3.6 Cannibalism among larvae

Cannibalism is not surprising in a largely generalist feeder like the dragonfly larva, especially as many individuals are likely to be developing together following selection of optimal oviposition sites by their mothers. While cannibalism among larvae would seem to be disadvantageous to a species, evidence to date seems to suggest that while certain individuals suffer an untimely fate, both the population and the species can benefit. It may well be more widespread under wild conditions than realized. Nevertheless, those species like the libellulids that oviposit on the water surface, would seem to lessen the risk of cannibalism by the developing larvae as they are spread far and wide, often across many water bodies or river reaches. Indeed, risks of any adverse effects in general would appear to be mitigated by this behaviour.

Young larvae of the corduliid *Epitheca cynosura* interact at high densities after hatching from the egg masses. While cannibalism is rare among the same instars, it is common when there is a one-instar difference, and certain when there is a two-instar difference. Together, these responses regulate population size by synchronous emergence to reduce mortality (Hopper *et al.*, 1996).

Cannibalism in *Lestes viridis* is more frequent under time stress (progress of the season) and food shortage. The behaviour is a survival mechanism with benefits that include increased progression of development, faster growth rates and larger body mass at emergence. These benefits are greatest the earlier they occur in the larva's development. But there are also interactive factors, with cannibalism reducing direct competition for food. Cannibalism therefore has a rescue effect that enables cannibalistic larvae to survive harsh environmental conditions, while also being a compensatory mechanism that keeps life history variables near optimal, especially in times of stress. However, there would seem to be some disadvantages, such as possible disease transmission, but this seems not to be the case in this species (De Block and Stoks, 2004).

Lestes sponsa larvae are more active and display greater cannibalism in higher northern latitudes than among individuals from the south. This appears to be the result of more time constraints related to photoperiod in the north compared to the south (Sniegula *et al.*, 2017). Also in this species, size-mediated benefits shape life history traits such as survival and growth, as well as development rates (Sniegula *et al.*, 2019). In turn, there are trait-dependent effects, with phenology, time constraints and latitude provenance having strong effects on life history traits. However, only the time constraint component affects traits related to physiology (energy storage, protein content and investment in immune function) (Raczyński *et al.*, 2021).

Temperature fluctuations associated with climate change are increasingly likely around mean temperatures. When the aspect of temperature fluctuation alone is simulated at three realistic but different levels of amplitude, they do not affect mortality, growth or cannibalism among *Enallagma cyathigerum* larvae. The suggestion here is that an increase in amplitude of temperature fluctuation per se will not necessarily affect growth and cannibalism, compared to increases predicted relative to mean temperatures. This is because exposure to the temperature variations remain within thermal optima for the species (Holzmann *et al.*, 2022).

Cannibalism affects individual fitness, making it important to detect the presence of other cannibals, such as by using conspecific chemical cues. But it is important to distinguish potential prey from predator so as to eat as opposed to being eaten. Among larvae of *Ischnura elegans*, cues from older and larger conspecific larvae cause younger larvae to decrease their activity and shift their respiratory-related behaviour in accordance with an anti-predator defence mechanism. This behaviour balances catching prey with staying alive (Sysiak *et al.*, 2023) (Fig. 4.31).

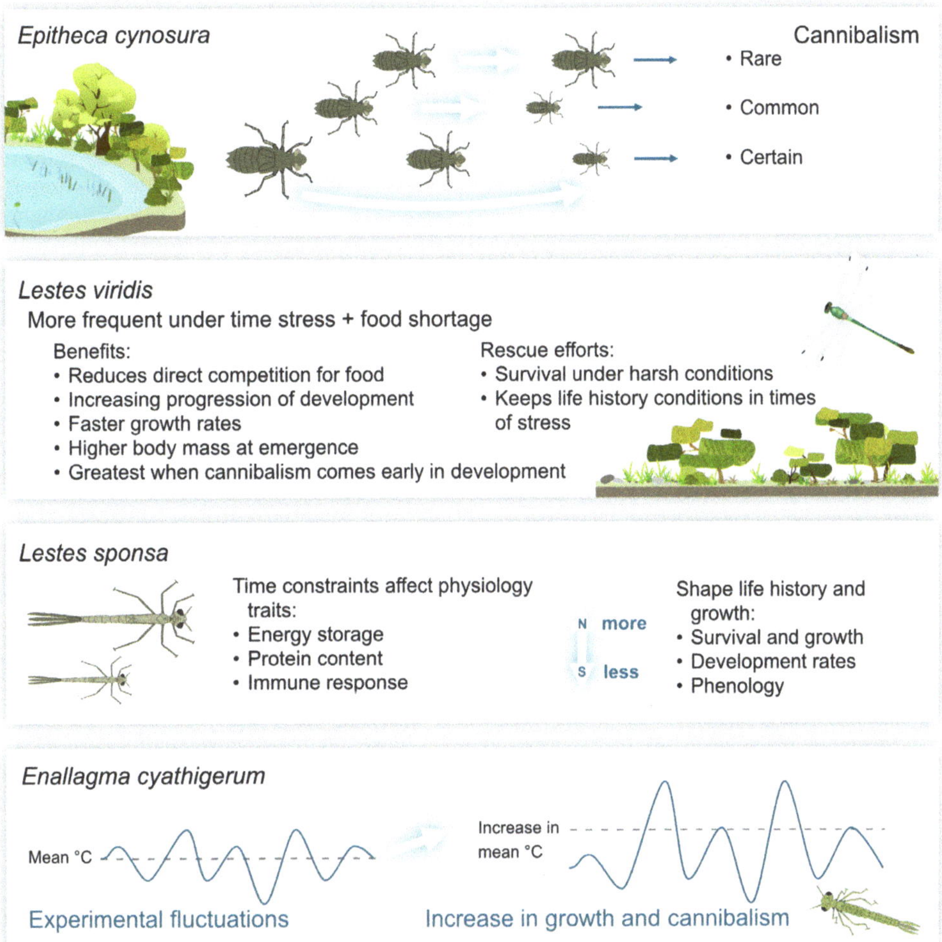

Fig. 4.31. Cannibalism is a significant component of dragonfly larval behaviour, particularly when there is a size difference among the larvae, as in the case of *Epitheca cynosura*. In *Lestes viridis*, cannibalism is more frequent when there is time stress as the season progresses and when there is a food shortage. Rather counter-intuitively, there are benefits to cannibalism, at least to the population as a whole and the species. The behaviour leads to a rescue effect, where there is survival of the population as a whole under harsh conditions, while also keeping life history variables optimal in times of stress. In *Lestes sponsa*, cannibalism is more prevalent at northern latitudes than southern ones as a result of time constraints on traits associated with physiology. In this species, cannibalism also shapes various life history traits. In *Enallagma cyathigerum*, different levels of environmental fluctuation do not affect mortality, growth or cannibalism. But when the mean temperature changes, both growth and levels of cannibalism may increase. (Illustration: Robyn Symons).

4.4.4 Emergence

Emergence success plays an important long-term role in maintaining population levels, especially as this activity exposes the larva and emerging adult to substantial risk. There are many abiotic and biotic variables that must be just right for not only successful emergence but also dispersal by the teneral adult to a safe place.

Among Canadian anisopterans, significant water factors for successful emergence are water temperature, dissolved oxygen levels and long-term aquatic conditions (baseflow). These aquatic conditions are closely associated with appropriate riparian conditions for overall successful emergence. Extent and type of canopy cover attract larvae to an emergence location, and provide structure and shelter for emergence, influencing dragonfly assemblage composition and abundance. Specific terrestrial features include tree position, bark roughness and tree diversity. The important point here is that all these conditions, both aquatic and freshwater, must be present for maintaining the local dragonfly assemblage (O'Malley *et al.*, 2020).

Similar results come from lakeside ecosystems in the same geographic region. Both larval and adult dragonflies are influenced by local abiotic, as well as biotic conditions, such as structure and type of aquatic and riparian vegetation. This emphasizes that from a conservation perspective, focus must be on the interface between both aquatic and terrestrial environments (Van Remsburg and Turner, 2009).

In the case of *Cordulegaster heros*, a threatened species, the larva leaves the water and climbs vertically straight up tree trunks in high-quality natural forest. This appears to be a genetically programmed trait, which must be recognized in terms of natural forest management to ensure the species' survival into the future (Boda *et al.*, 2015).

Emergence times depend on the water temperature that the larva experiences. There are inevitably differences in emergence timing according to latitude of the water body where the larva developed. However, temporal variations at specific locations also play into the genetic plasticity of a species, whose wing shape and morphology can change according to the environmental conditions to which the larvae are exposed (Stewart and Vodopich, 2018).

At large regional spatial scales, photoperiod and associated change in temperature are also factors determining presence and abundance of dragonfly species (De Block and Stoks, 2003; Johansson *et al.*, 2010). These response traits vary among and within species, affecting characteristics such as larval density (Banks and Thompson, 1987), emergence patterns (Farkas *et al.*, 2012), sex ratios (Farkas *et al.* 2013), wing spot allometry (Hardersen, 2010) and voltinism (Corbet *et al.*, 2006; Braune *et al.*, 2008), all of which must be seen against risk factors that the larva experiences at time of emergence, such as high winds (Jakob and Suhling, 1999; Khelifa *et al.*, 2013) and deformity and predation (Zebsa *et al.*, 2014).

At their northern geographic range margins in Britain, the two rare species *Coenagrion mercuriale* and *Ceriagrion tenellum*, are highly responsive to the environmental trigger of duration of sunlight. They emerge more synchronously than they do at warmer more southern locations, emerging in a short time frame in response to sunlight duration the previous day (Purse and Thompson, 2003).

Telemetry using harmonic radar tags and receivers provides much more insight into the behavioural traits of freshly emerged adult dragonflies than does visual observation. This technology established that individuals of *Libellula fulva* fly about 35 m on the first day after emergence, and perch only on trees and bushes 1.8–31 m above the ground, avoiding open microhabitats. In the first 3–4 days, they occupy a relatively small area of 480–2500 m^2. This contrasts with *Aeshna mixta* individuals, which emerge and within a few hours fly over 200 m, beyond which they are no longer detectable by radar (Hardersen, 2007). In contrast, both sexes of the North African endemic *Calopteryx exul* emerge and mature only within 5 m of the water's edge (Khelifa, 2017).

Teneral individuals of male *Calopteryx maculata* leave stream emergence sites to forage and mature in forest light gaps at an average distance of 140 m away. These locations provide interference-free space from other males. They are also areas where prey density is high, enabling them to build up the energy reserves necessary for later attaining and holding streamside territories (Kirkton and Schultz, 2001) (Fig. 4.32).

4.5 Traits Associated with Geographic Range Shifts

4.5.1 Significance of traits associated with their geographic range

There are many traits that could be considered significant for insects responding to changes in environmental

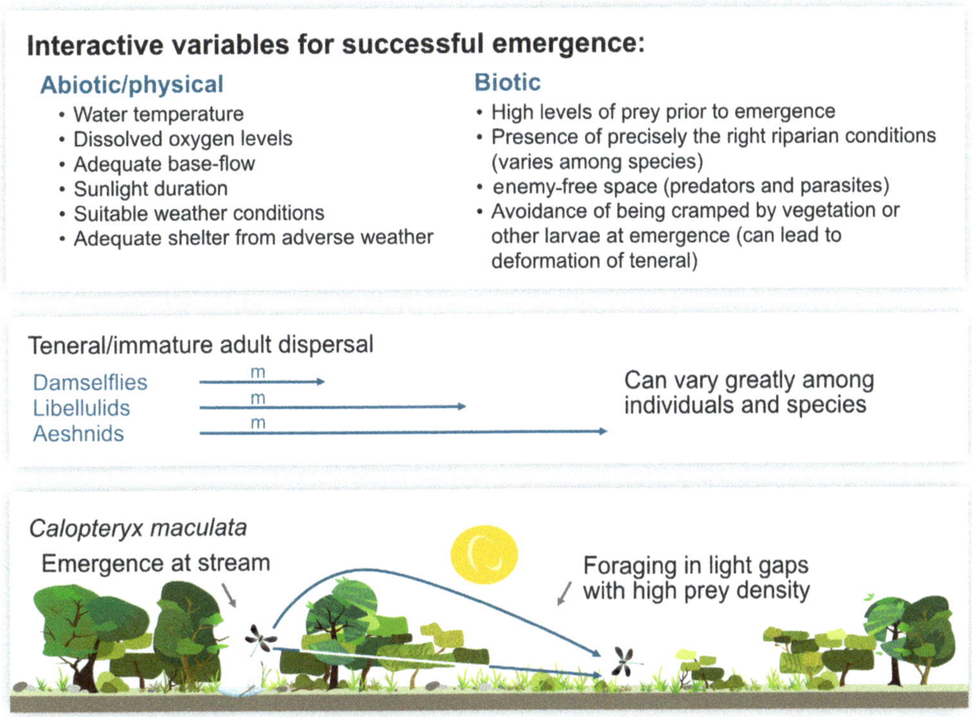

Fig. 4.32. Teneral adult emergence from the larval stage is a time of high risk. It is also demanding in that both aquatic and terrestrial conditions must be just right for the process to be successfully completed. There are various interactive factors that are significant at this time. These are both abiotic/physical and biotic. While aeshnid and libellulid tenerals generally disperse farther than those of damselflies, there are often some individuals of any species that disperse farther than the majority. Once the teneral has hardened its body and is an immature adult, it often disperses even farther away from water, including among many damselflies. Male individuals of *Calopteryx maculata*, for example, leave the stream where they emerged and make their way to forest light gaps to forage in locations with high prey density and where they are free from interference by other males. (Illustration: Robyn Symons).

conditions and ecosystem processes. These traits fall into broad groups related to morphology, feeding, life history, physiology and behaviour (Estrada *et al.*, 2016; Moretti *et al.*, 2017). However, it is essential to be aware of the extent to which evolution has led to current constraints while also determining putative future changes in species geographic ranges. While history may impose phylogenetic constraints on range shift responses, compensatory mechanisms of phenotypic plasticity and adaptive evolution can modulate the range shift responses (Diamond, 2018).

Importantly, movement response to climate change is strongly determined by dispersal ability, but not necessarily with mobility capability. But movement is not the only option. Individuals in any one local population may tolerate or adapt to changing conditions. If none of these two options is possible, colonization of new locations is then the only way to survive, assuming suitable habitats can be reached. However, these may or may not be present, and if they are, then they must be of sufficient quality and size to enable long-term genetic viability without any small-population effects, such as genetic impoverishment and drift. But as any one species interacts with many others in its community, it must also adapt to new situations in terms of intra- and interspecific competitors, predators and parasites, while there must also be enough suitable prey at the right time to sustain its population.

There would seem to be sound reasons for using species' traits to predict shifts in geographic range. Traits such as greater dispersal ability, reproductive potential and ecological generalization intuitively appear to lend themselves to range expansion in the face of climate change. However, a large-scale meta-analysis suggests that these perspectives do not necessarily hold up (MacLean and Beissinger, 2017).

The effects of climate change will also affect aquatic communities which differ across a region. There is an interaction between climate warming, water flow and the traits of species making up North American stream benthic communities. This interaction will cause the various streams and their communities to be differentially affected by climate change (LeRoy Poff *et al.*, 2010).

Let us now look at an example of both larval and adult traits combined that have determined the current distribution of dragonflies in a region. Dragonflies in sub-Saharan Africa have a long history of enduring climatic fluctuations but without glaciers having had a major direct impact. These dragonflies have ecological traits related to habitat type preference and habitat specialization, and biological traits embracing features like mobility and life history. These traits vary greatly across the taxonomic spectrum present in the region. Species occupancy of a wide range of habitat types is weakly correlated with wide geographic ranges, although those species with longer and later breeding seasons have wider ranges. However, overall habitat preference (running versus standing water) and mobility traits are of low importance. In sum, ecological and biological traits are interactive in determining the current geographic ranges of these dragonflies. Failing to recognize the interaction between ecological and biological traits can lead to inaccurate predictions of range variation among these species, as well as a misunderstanding of the mechanisms underlying range size variation. However, while there is a relationship between habitat specialization and range size, it is constrained by phylogenetic history and idiosyncratic responses by certain species to prevailing environmental conditions. For effective conservation of as many dragonfly species as possible in the region in the face of landscape transformation and ongoing climate change, we need to focus on the suites of interactive and relevant traits possessed by each of the various species to make informed decisions (Deacon *et al.*, 2021) (Fig. 4.33).

Many dragonfly species have already shown strong macroecological responses to climate change, especially in the northern hemisphere. But as the different species can have complementary multimodal responses to the same environmental variables, responses to a changing climate are also likely to

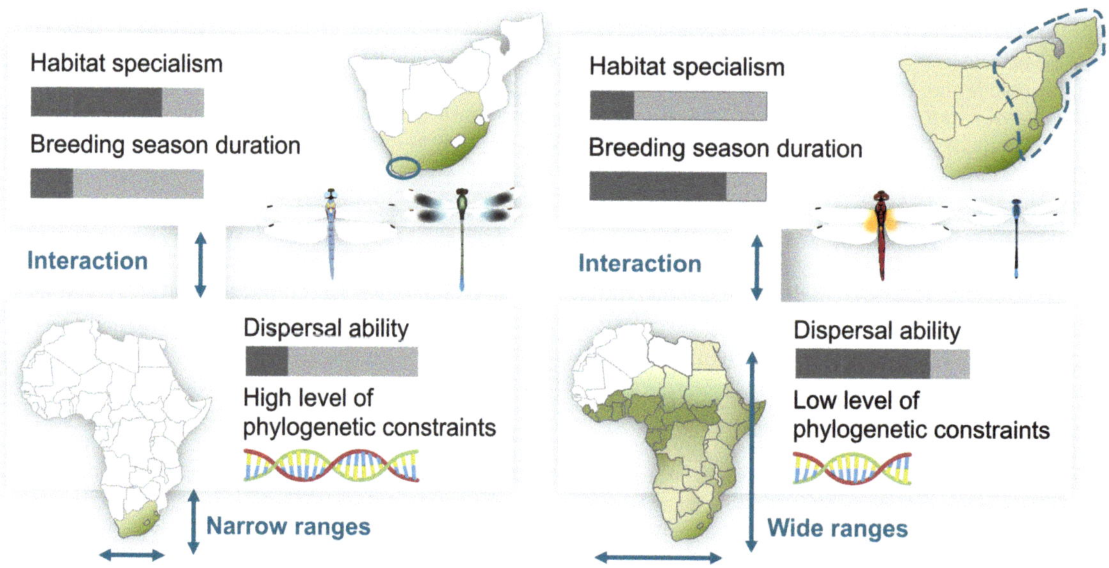

Fig. 4.33. In sub-Saharan Africa, there are sets of species that as both larvae and adults are narrow range endemics. In terms of traits, they are typified by degree of habitat specialism and a short and specific breeding season, which is interactive with a low tendency to disperse, all based on strong phylogenetic constraints. These species contrast with those that are habitat generalists, have a long breeding season, have good powers of dispersal and possess moderate phylogenetic constraints. (Illustration: Charl Deacon).

vary among taxa. This means that to make sense of the full extent of climate change, it is necessary to obtain data on many species (Hassall, 2015).

4.5.2 Phenology and voltinism

Many dragonfly species show a strong phenological response to climate warming. Between 1960–2004, British dragonflies advanced the onset of their flight period by just over three days per 1°C rise in temperature (Hassall *et al.*, 2007). By 2005, Dutch odonates had advanced their flight dates due to the complex effects of changing temperatures. While flight dates did not respond to changes in autumn/winter temperatures, they did respond to increased spring temperatures which advanced the flight period. Complexity appears to be associated with the detrimental effects of high temperatures on egg and larval survival and development, which is possibly why increased summer temperatures, as opposed to spring temperatures, delayed rather than advanced the phenology. This suggests that dragonfly responses to climate warming may vary greatly at different geographical locations around the world (Dingemanse and Kalkman, 2008).

In Norway, recent climate warming has led to the expansion of the odonate flight season in both the Zygoptera and Anisoptera, as well as in species in over half of the families. Yet few species have advanced their dates of emergence, but rather are now on the wing for several more days later in the season. Furthermore, there is little evidence that changes in voltinism lead to a broader flight season, which rather is interactive with use of human-modified habitats, such as a decrease in shading associated with timber extraction. Other possible mechanisms that link warming with the broadening of the flight season include protracted emergence and cohort splitting (Patten and Benson, 2023).

In Japan, the first appearance of adult dragonflies shifted to *later* in the spring, in contrast to elsewhere where there is a shift towards earlier emergence. The response here is likely the result of increased air temperatures in summer and autumn providing longer bivoltine periods and faster growth rates, allowing the second generation, which previously hatched in summer, to emerge in the autumn, causing the size of the population that emerged in spring to decrease (Doi, 2008).

In the case of *Ischnura elegans*, exposed to warming during development, increased investment in immune response is not associated with effects on age or mass at emergence or growth rate. Rather, life history traits are mainly changed by temperature and voltinism, resulting in decreased development time and increased growth rate in univoltine females, but decreased growth rate in univoltine males. This indicates that life history traits, which differ between the sexes, are more strongly affected by warming and voltinism than the effects of emergence phenology. This suggests that in a warming world, shifts in phenology will likely shape physiological and life history traits (Raczyński *et al.*, 2022) (Fig. 4.34).

Dragonfly voltinism in Europe decreases with latitude. In other words, there are overall fewer generations per year in the north compared to the south. In northern Europe, the dragonfly fauna is dominated by larger species. However, not all species use all available resources and time, and have one generation per year, while remaining small. The larvae of dragonflies, being carnivorous on a high-nitrogen diet, often develop to maturity over several years, and during that time, are highly dependent on ambient water temperature, where heat conduction is high and oxygen consumption is dependent on temperature. This extended generation time, compared with butterflies, for example, is a great advantage for dragonflies, especially in colder climates (Zeuss *et al.*, 2017).

The European *Gomphus vulgatissimus* has been extensively researched in its response to further climate change. Models suggest that the species will expand its range northward, although the southern margin is likely not to shift that much (Söndgerath *et al.*, 2012). There is the likelihood of shorter generation times in the northern parts of the range. In the southern areas, there is the possibility of an earlier appearance of adults during the year, but not in the north due to restraints of photoperiod. This may be accompanied by an increase in the length of the flight season, with the possibility of two flight periods per year. Furthermore, assuming sufficient prey is present, there may also be more overlap of larval size classes (Braune *et al.*, 2008). Simulating the effects of different winter temperatures on the emergence of *Gomphus vulgatissimus* indicate an acceleration of emergence by 6–7 days per 1°C temperature rise. However, adequate prey must be present, and parasite activity low, for populations to survive through the vulnerable emergence stage (Richter *et al.*, 2008) (Fig. 4.35).

Seasonal response by *Orthetrum cancellatum* is genetically determined. Rising temperatures in

Fig. 4.34. The *Ischnura elegans* male (left) and female (right) possess slightly different traits in response to warming conditions. (Photos: Jens Kipping).

Fig. 4.35. *Gomphus vulgatissimus*: a species likely to show multiple responses to climate change, including changes in range limits, development characteristics, adult phenology and length of flight period. (Photo: Jens Kipping).

Europe would not allow for more than one generation per year in the northern population. However, the southern population will likely be more flexible. A change in voltinism in the central and northern regions would require northward dispersal of the Mediterranean genotype or an evolutionary change in photoperiodic response in the higher latitude populations (Flenner *et al.*, 2010).

To better predict the interaction between climate change and predation risk on prey and ecosystems, it is necessary to understand both latitudinal patterns in growth rate and growth plasticity to temperature and predation risk, relative to underlying behaviour (food intake) and digestive physiology (growth efficiency). Using northern and southern populations of *Ischnura elegans* and manipulating temperature and predation risk, higher growth rates associated with latitude, temperature and predation risk are mediated by a higher growth efficiency. This suggests that energy allocation trade-offs and associated physiological costs are important in shaping large-scale geographic variation in growth rates while also shaping growth-rate plasticity. Furthermore, efficiency of energy transfer in aquatic food chains, where damselfly larvae are important intermediate predators, will be higher in southern populations at higher temperatures and under greater predation risk. The overall outcome is likely a lengthening of food chains through type of prey affecting predation risk (Stoks *et al.*, 2012).

4.5.3 Body size

Body size comes with phylogenetic constraints. Over the last 65 million years, shifts in dragonfly body size have been overall neutral, resulting from some taxonomic groups increasing in richness while others have decreased. It seems that local environmental conditions have been the major driver, as well as conflicting evolutionary selection on the adult versus the larva. A large adult requires a long-lived larval stage, which has all the risks attached to being a non-breeding stage, of low mobility and vulnerability to various mortality factors, especially predation. But adults of modern dragonflies, at least compared to the ancient times of *Meganeura*, are today also experiencing more modest oxygen levels as well as being subject to bird predation (Waller and Svensson, 2017). Furthermore, it seems that the situation for dragonflies changed greatly in the early Jurassic through to the early Cretaceous, with the widespread appearance of dragonflies with camouflaging disruptive wing colouration, along with diversification of potentially insectivorous pterosaurs and then small, feathered dinosaurs and birds (Jouault *et al.* 2022).

Phenology, voltinism and body size can vary together. For example, individuals of some damselfly species can vary in morphology with latitude. Individuals of *Enallagma cyathigerum* in Europe are largest at both 42°N and 70°N, but smaller between these two latitudes, at about 55°N. This may be because the warmer climate of southern Europe, where the growing season is longer, leads to larger individuals. In the intermediate latitudes, the growing season extends beyond one year, and there is a change from one generation per year to just one generation per two years. In the far north, a generation can take three years, and over this time, individuals can grow to a large size (Johansson, 2003).

Large-sized damselflies are currently more threatened and suffering greater risk of extinction than small-sized species. The threats appear unrelated to sexual size dimorphism or mating system, but rather to anthropogenic pressures, ranging from landscape modification and habitat loss, through to water quality deterioration and climate change. The human impacts may also be reducing prey numbers, affecting the bulkier damselflies the most, especially the larvae that must have sufficient prey to complete development (Suárez-Tovar *et al.*, 2019).

Changing environmental variables drive population dynamics and community and assemblage composition over time. A fundamental question then is: Do different traits come into play as environmental conditions change? Over a 20-year period, Swedish dragonfly assemblages at first (between 1997–2008) showed little change in trait composition, but then changed abruptly between 2008–2015. During the early period, assemblage trait composition was dominated by small species, short adult lifespans, and emergence in spring. This changed in the later period, with assemblages dominated by large species, long adult lifespans, and extended emergence times. However, other traits such as the overwintering stage, larval development time, microhabitat use and larval activity did not change over the whole period. In the later period, species with a medium-fast life cycle were replaced by slow- and fast-growing ones. This suggests that life cycle changes have led to a change in species' relative abundance due to environmental temperature increase, followed by elevated pH and possibly, to a small extent, changes in land use (Dalzochio *et al.*, 2020).

4.5.4 Dispersal ability

Fluctuations in Quaternary environments, and advance and retreat of glaciers, have driven large population fluctuations and dispersal, leading to substantial species radiation among North American damselflies (Turgeon *et al.*, 2005). Among the *Enallagma* species in the region, wing size is an important trait. Species with larger wings, whether absolute or relative to body size, tend to be ones that are more widely distributed. Likely this is one of several important traits, including adult wing shape, flight period and behaviour, and habitat specificity, as well as trade-offs with other traits, such as reproductive output, that have played a role in dispersal, establishment and proliferation at new sites (Rundle *et al.*, 2007).

Similar findings come from *Coenagrion* species across the northern hemisphere. Irrespective of species' phylogenetic ages, their geographic ranges are larger in the northern compared to southern latitudes. The most significant trait is higher flight ability (based on wing aspect ratio) and dispersal capability (Swaegers *et al.*, 2014).

In the UK, the *Coenagrion puella* has recently expanded its range northward beyond its former range in England and is now present in Scotland. In contrast, *Pyrrhosoma nymphula* has been established in Scotland for over a century. The two species, while showing spatially correlated variation in body size across locations, differ in their dispersal-related morphology. Range-stationary *P. nymphula* shows no significant trends between latitude and body mass, wing length and three dispersal-related traits. In contrast, recent range-shifting *C. puella* exhibits changes only in the traits of body mass and two dispersal traits, indicating investment by this species in improved dispersal capability at its northern range margin. There is also a trend toward longer development time through some individuals growing over two seasons in the far northern latitudes (Hassall *et al.*, 2008).

Improved dispersal capability is also the main trait associated with the rate of poleward expansion of *Calopteryx splendens* and *Pyrrhosoma nymphula* in Britain (Hassall *et al.*, 2009). Other correlative traits associated with range change include oviposition method. For example, exophytic British dragonflies have shifted farther north than endophytic species (Angert *et al.*, 2011). This is best described as an establishment-enabling trait at a new locality rather than a dispersal trait per se.

4.5.5 Body colour

Dragonflies across Europe are overall darker in the north compared to the south (Zeuss *et al.*, 2014). In both Europe and North America, another study confirmed this latitudinal colour tendency in anisopteran species to be darker coloured in cold regions and lighter in warm regions, but consistently more so in Europe (Pinkert *et al.*, 2016). Traits related to colour and size across Europe also relate to habitat type. Colour lightness of assemblages increases while body size decreases with increasing temperature. The mechanism underlying these variations is of equal evolutionary significance for both still and running water species. However, general higher dispersal ability of standing water species enables them to better cope with historical climatic changes, and by extension, ongoing climate change (Acquah-Lamptey *et al.*, 2020).

Darker colour in cold regions puts individuals at an advantage, as they heat up quicker than light ones at the same level of solar radiation. There are also small-scale environmental effects. For example, *Mnais costalis* males at higher temperatures display higher flight agility. Importantly, it is the combination of body size and temperature that enables dominant individuals to hold territories in sunny spots (Samejima and Tsubaki, 2010).

4.5.6 Habitat generalization versus specialization

Improved water quality across Britain and Ireland in recent decades, as well as climate warming, has had a significant effect in permitting range shifts to occur. Additionally, other baseline factors, such as geographic region and local distribution of dragonfly species, have also played a role. One of the key traits required for a dragonfly species is habitat specialization, with some evidence that habitat generalists are replacing specialists (Powney *et al.*, 2015).

Experimental studies suggest that a 2°C rise in temperature will favour habitat generalist *Crocothemis erythraea* over specialist *Leucorrhinia dubia* through asymmetric competition (Suhling and Suhling, 2013). This finding supports the concept of biotic homogenization where habitat generalists overwhelm specialists as environmental conditions change. A comparative survey spanning the early twentieth century to the early twenty-first century in western USA indicates that dragonfly habitat specialists, as well as species with a diapause stage, occur less often in the warmer and urbanized areas, but more often in wetter and less disturbed areas. However, overall there has been a replacement of habitat specialists by generalists (Ball-Damerow et al., 2014a, b).

In north-eastern USA, 18% of the odonate species are threatened, based on a combination of range extent, area of occupancy, habitat specificity, relative change in range size and vulnerability of occupied habitats. On this basis, there is a disproportionate number of species threatened that occupy peatlands, low-gradient seeps and streams, high-gradient headwaters and larger rivers (White *et al.*, 2015).

4.5.7 Dragonflies of running versus standing waters

Standing waters are overall less persistent than flowing waters. This leads to species with strong dispersal traits that inhabit standing waters dispersing more readily than those of streams and rivers. Traits for ready dispersal also enable still water species to track climate change better than riverine species, so long as receiving habitats are available (Hof *et al.*, 2012). The northern margins of southern European standing water species moved on average 115 km northward per decade between 1988 and 2006. Neither the southern riverine species, nor species of either aquatic system of the northern group, have shown any consistent range shift. Southern range margins also showed no consistent change, at least prior to 2006. The shifts that did occur appeared to be through a diffusion process across the landscape in response to thermal opportunities relative to various innate traits of the focal species. Opportunistic pond species are more likely than river-adapted species to drift northward as environmental conditions warm. However, barriers like mountain ranges, seas and cities may act as obstacles to the diffusion process (Grewe *et al.*, 2013).

Small Finnish streams with a mosaic of both running and still water support various damselflies and dragonflies. Local population extinction risk is highest among the still water species, particularly the habitat specialists. This group of species displays metapopulation dynamics of extinction and recolonization as conditions deteriorate or improve over time (Korkeamäki and Suhonen, 2002).

In south-western France, three rare and threatened species (*Macromia splendens*, *Oxygastra curtisii*, *Gomphus graslinii*) occur in both rivers and artificial lakes. There is asynchrony of emergence among the three species, as well as morphological variability between the lake and river habitats. Individuals from the lake emerge earlier and are smaller than those from rivers, as lakes are warmer than rivers. Concerns associated with a warming world are that the asynchrony of emergence among neighbouring populations may adversely affect metapopulation functioning and the fitness and fate of early-emerging individuals (Denis *et al.*, 2018).

4.5.8 Extinction risk

Long adult lifespan determined by photoperiod is related to large geographic range size in the damselflies of New York State (Corser *et al.*, 2015). In turn, except in the case of the three most common species, poor quality habitat in Finland prior to 2003 has led to increased local extinction (Suhonen *et al.*, 2010), while the species with the widest geographic ranges are those less likely to go locally extinct (Suhonen *et al.*, 2014).

However, against the background that habitat loss is an undeniable threat for many local odonate populations, there are nuances involved here related to traits. Across a wide range of both dragonfly and damselfly species in central Finland, closely related species have similar hindwing length, length of larval period and habitat of larvae. Even with phylogenetic correction, large geographic range size indirectly reduces risk of local extinction. However, among these Finnish species, habitat specialists, surprisingly, are not facing higher extinction risk than habitat generalists, assuming of course that any special habitats are not in short supply. After controlling for evolutionary relatedness, species with long wings are least at risk of local extinction, while long larval development time presents a greater risk. Overall, species with relatively small geographic range sizes are an accurate estimate of increased extinction risk (Suhonen *et al.*, 2022).

Among 11 species of North American libellulid dragonflies, extinction and colonization dynamics are important in relating dispersal capability to geographic range size. Local extinction events are lower in species with larger ranges. This suggests that metapopulation dynamics interact with dispersal behaviour, determining overall range size. As this interaction is especially significant at range margins, it sets range limits and ultimately range size (McCauley *et al.*, 2014).

Among 603 North American odonates, extinction risk is related to breeding in running rather than standing waters, and to narrow thermal limits. While larger damselflies are more at risk than smaller ones, this is not the case for anisopterans. Southern species are more climate tolerant than northern species and have smaller geographic ranges. All these traits together suggest that the higher elevation species and those living on the plains are most at risk (Rocha-Ortega *et al.*, 2020) (Fig. 4.36).

4.5.9 Responses to elevation

Difference in elevation is either a barrier or provides opportunities for dragonflies according to their traits when set against variable environmental conditions. In turn, steepness of slopes determines the size and flow rate of the streams, while local geological features produce waterfalls and deposition pools. Streams then flow into larger rivers, with decreasing velocity when a coastal plain is present.

Recent range expansion of many species over higher latitudes and elevations implies that the expanding species are able to cope with new biotic interactions at the leading edge. Using the climate adaptable *Sympetrum striolatum* larva as a prey item, Zebsa *et al.* (2022) investigated whether its growth, behavioural (food intake) and morphological (eighth and ninth abdominal lateral spines) responses differed when in the presence of dragonfly predators that dominate either low-elevation (*Aeshna cyanea*) or high-elevation (*A. juncea*) habitats at two temperatures

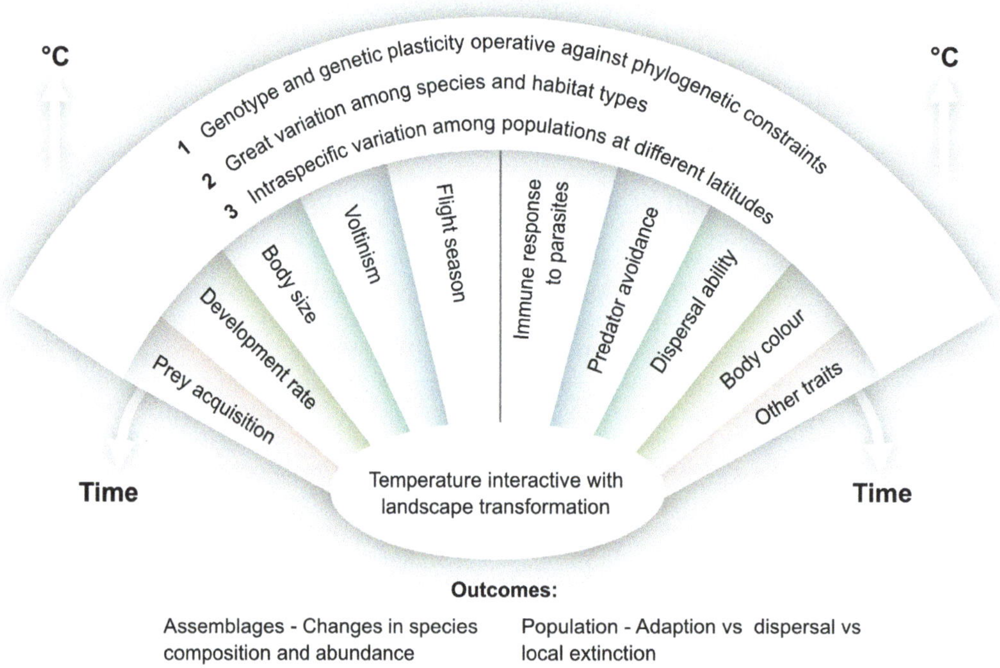

Fig. 4.36. A symbolic 'trait fan' in this age of rapid climate warming. The framework represents three interrelated basic tenets 1–3. The lamina of the fan represents various interactive major traits. There are other traits likely also involved (e.g. response of gut symbionts, changes in prey base, interspecific competition at different locations) about which we currently have little knowledge. While traits for survival were honed by the natural environment along with its historic temperature changes, today there is the overlay of the effects of human-induced landscape transformation. Depending on the type and intensity of the changes there are both losers and winners. As different species will be responding differentially to changing conditions, the inevitable outcome is a shift in species composition and relative abundance within assemblages at various locations. Outcomes from the suite of environmental changes at the level of a population fall broadly into three alternative categories: thermal plasticity and perhaps longer-term adaptation by certain individuals to local conditions, dispersal by individuals to more optimal habitat or local extinction of the population. (Illustration: Robyn Symons).

(20°C and 24°C). The non-consumptive response of growth (rate and efficiency), behaviour (food intake) and morphology (abdominal spines) of *S. striolatum* prey to different predators including low- (familiar) and high-elevation (non-familiar) predatory dragonflies showed that the prey dragonfly larvae responded similarly in growth and food intake to both predators. Yet they responded in morphology to the familiar dragonfly predator. Overall, it appears that the elevation-expanding *S. striolatum* may have succeeded in colonizing new areas at higher elevations because it responds to dominant high-elevation predators in a similar way to the more familiar low-elevation predators.

Lestes virens, in the mountains of Algeria, displays cohort splitting through temporal isolation mediated by elevation (5–1012 m above sea level). There are two cohorts, and although they emerge at the same time in spring, the second cohort aestivates in summer in the local forest and reproduces in autumn. The timing of emergence is delayed with increasing elevation, 2.2 days for the summer cohort, and 0.7 days for the autumn one, per 100 m increase in elevation. The differences in phenology with elevation leads to greater potential for overlap of the summer and autumn cohorts, especially when combined with local movement across elevations, and with climate warming (Amari *et al.*, 2019).

Elevation can also play a role in influencing wing dimensions. Two Spanish populations of *Libellula quadrimaculata,* 95 km apart but at 782 m and 1909 m above sea level, show differences in both wing length and area. Those of the lower elevation plateau are suited for greater mobility, while those in the mountains are more like a refuge or island population (Casanueva *et al.*, 2017).

In gallery forest and savanna in Colombia, anisopterans are much more sensitive to changes in elevation than are zygopterans. It seems that the small streams and rivers provide a wide range of flow types, from riffle to deposition pool, and hence many narrowly defined microhabitats, enabling damselflies to be less influenced by elevation than dragonflies (Palacino-Rodríguez *et al.*, 2020a).

In Kazakhstan, seasonal elevation migrations occur. On emergence, and to escape the hot summer conditions at low elevations, seven species (*Sympecma fusca*, *S. gobica*, *S. paedisca*, *Aeshna mixta*, *Sympetrum arenicolor*, *S. meridionale*, *S. striolatum pallidum*) fly to the mountains where they aestivate until cool autumnal conditions return. They then return to the lower elevations to breed, although avoid hot days by seeking wooded areas (Borisov *et al.*, 2022).

Dragonfly larvae of artificial ponds in subtropical South Africa are remarkably elevation tolerant over a 1250 m gradient (100–1350 m above sea level) at the same latitude. They are mostly generalist species drawn from the regional species pool, with very little species turnover with elevation, due in large part to similar physicochemical conditions, and aquatic and emergent vegetation in the ponds. This elevation resilience appears to be a response to historical and ongoing El Niño–Southern Oscillation (ENSO) wet and dry cycles, having selected for traits related to opportunistic seeking of, and ready establishment in, suitable ponds (Niba and Samways, 2006a). Species making up this overall assemblage contain both temperate (extending to high elevations) and subtropical (extending to sea level) species. As a group, they are predisposed to tolerate seasonal change through having traits that have been honed by many millennia of past climatic fluctuation (Niba and Samways, 2010).

On the east coast of South Africa, where there have been no major glacial events for over 200 million years, elevations range from sea level to 3000 m above sea level at the same latitude (29°S). Across this wide elevation gradient, which supports 117 species inhabiting a wide variety of still and running water bodies, the coastal plain (0–200 m) supports 86 species. Above those elevations, there is constant species richness of 43 species between 200–1400 m, dropping to only 4 species between 2400–3000 m. The various families show some distinct elevation preferences. Libellulidae make up 44% of all species, dropping from 50% at 0–200 m to 25% at 2400–3000 m. Coenagrionidae accounts for 24% of all species and remains in almost the same proportions over the whole 3000 m range. Aeshnidae makes up 9% of the total and stays at that level up to 1600 m, with one of the 4 species reaching 3000 m. All these three families have traits enabling them to occupy various habitats across a wide elevation range. However, Gomphidae proportional species richness remains level at 8% from 0–1400 m but the family drops out by 1600 m. In turn, Lestidae are 5% of the total from 0–1400 m, but zero above 1600 m. These two families can be considered low- to medium-elevation families. In contrast, Synlestidae species make up only 1% of the total at sea level, but 25% of the total above 2–400 m, and are a major component of high montane streams with deposition pools. Chlorocyphidae,

Calopterygidae, Protoneuridae and Corduliidae are consistently rare low-elevation families, while the Platycnemidae is a rare mid-elevation family (Samways, 1989b) (Fig. 4.37).

4.6 Dragonfly Responses to Seasonal and Weather Conditions

4.6.1 Significance of changes in environmental conditions

Ambient conditions at the local scale can make a great difference as to whether a species survives at larger spatial scales. There is also a time dimension, as weather conditions vary, and sometimes extremely so. The failure of rain to arrive for a long period leads to drought, especially affecting smaller standing water bodies. Both physical and chemical changes occur as time passes, including drops in water level, water temperature, dissolved oxygen levels, turbidity, conductivity and pH, all impacting the various life stages of dragonflies directly and indirectly through aquatic, marginal and sometimes even riparian plant quality.

Dragonfly species have various options for survival in one place at one time: stay and literally weather the storm, move locally when the extreme event is spatially explicit, or move far away and establish elsewhere. The suite of traits that individuals possess is a crucial determinant affecting the chances that populations have of surviving to contribute to the next generation. Having genetic plasticity will also be a contributory factor.

Human activity has exacerbated extreme weather events related to climate change. There are now more extreme floods and many extended droughts. There are also some major cold and hot events. In terms of dragonfly conservation, these events are more significant in the northern than southern hemisphere, as that is where there is much more land and freshwater present, more extreme continental climates, and where many dragonflies live, notwithstanding their huge diversity straddling the equator.

In turn, other human activities, including agricultural intensification, infrastructure development, environmental contamination and introduction of alien species, must also be considered, especially when they also act together with climate change. These stressors are discussed in Chapter 5. We now focus on traits relative to seasonal change and varying weather relative to the traits of dragonfly species and their conservation.

4.6.2 Change of season

While adult dragonfly species richness is overall higher in the temperate northern and southern hemisphere summers, it is not necessarily the case in warmer areas, where seasonal rains and dry periods play a major role. In Uttarakhand state in northern India, although certain species are present throughout the year, adult species richness is higher in the rainy season (July–October) than in the drier November–February and March–June periods (Dayakrishna, 2015; Farooq and Arya, 2018), whereas species richness, as well as overall abundance is highest in Maharashtra state immediately after the rainy season (Saha and Marathe, 2017) in the case of zygopterans, but not anisopterans. In West Bengal, dragonfly abundance is greatest during the monsoon and post-monsoon wet seasons. But during the dry season, from March to May, many species are absent. The highest number of species are present when relative humidity is 80–90% and air temperatures are 22–28°C (Mallick and Mondal, 2024).

In the more continental montane climate of Mansehra District, Pakistan, adult species richness is far greater in June–September, and much less in March–April, followed by October–November, largely a more typical north temperate temporal pattern (Mehmood *et al.*, 2020). Similar findings come from the montane, strongly continental Swat District in Pakistan (Attaullah *et al.* 2021).

In the southern Brazilian state of Espírito Santo, the bromeliad-inhabiting damselfly *Leptagrion perlongatum* peaks in abundance in the rainy season (February) but is scarce in the austral dry winter period (June–October), a time when bromeliad plants are prone to dry out (De Marco and Furieri, 2000). Yet in the case of the continental climate of Minas Gerais, greatest species richness is in the dry austral autumn and winter (March–September), with lower richness in the rainy season (October–February), although some rare and undescribed species of conservation significance are present only at the dry time of year (De Souza *et al.* 2017).

In central Brazilian deciduous forest, rainy and dry periods affect odonate assemblage composition. However, they do not affect species richness and abundance, as permanent lagoons associated with a major river are present in a state park and which provide reliable refuge (Pereira de Gouvêa *et al.*, 2022).

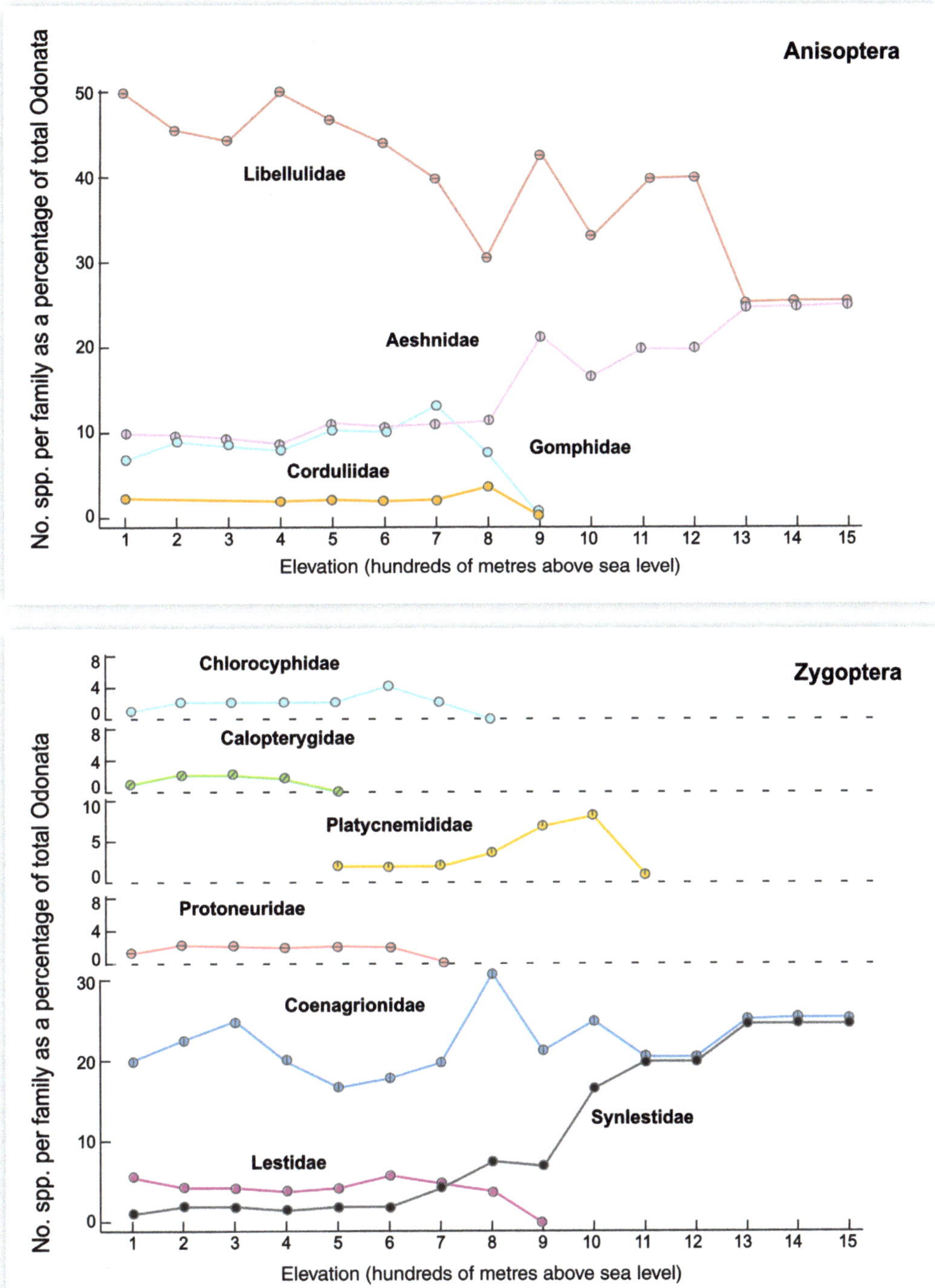

Fig. 4.37. There is a strong phylogenetic signal associated with the elevation range of the various odonate families across a 3000 m gradient at one latitude in southern Africa. (Illustration: Robyn Symons).

When larvae are compared between the dry (July–December) and wet (January–June) seasons in the Cerrado–Caatinga transition zone in northeast Brazil, for all species combined, both species richness and abundance are slightly higher in the wet season, especially among the Anisoptera which are much more species-rich and abundant than Zygoptera. However, when the two suborders are considered separately, zygopteran larvae richness and abundance have much lower variation than among the dominant anisopteran assemblages that display traits enabling them to survive much temporal and spatial variation in this climatically unpredictable area (Moura *et al.*, 2020). In the warm transition zone between the Cerrado and the Amazon forest, seasonal fluctuation in temperature affects abundance of both suborders. However, zygopteran species richness is affected mostly by rainfall, while for anisopterans it is mostly by temperature. Timing of odonate emergence is synchronized with the months preceding the rainy season, when density of larvae is highest. This is a strategy to avoid the effects of climatic variation in water levels while capitalizing on high levels of terrestrial prey. Anisopteran larvae abundance is lowest in the rainy season when the substrate is washed away, removing larvae microhabitats. To avoid being washed away, temperature accelerates or delays the emergence of adults (Resende *et al.*, 2021).

One of the reasons that there are such differences in traits between the South American zygopterans and anisopterans is due to ecophysiological differences (De Marco *et al.*, 2015; Mendes *et al.*, 2015). Anisoptera have greater capacity than Zygoptera for adaptation relative to the timing of adult emergence (Resende *et al.* 2021). This enables Anisoptera to adapt and respond to unfavourable environmental changes. Furthermore, zygopteran adults are more intimately associated with the type of riparian vegetation than are anisopterans. As forest anisopterans largely occupy open areas, heavy rainfall has a direct effect on everyday activities (Juen *et al.*, 2014; Palacino-Rodríguez *et al.*, 2020b), although in the Peruvian Amazon certain species continue to fly on sunless days with light rainfall, suggesting they are adapted to rainy conditions in this warm wet tropical climate (Medina-Espinoza, 2022).

Weather can determine the number of days that a dragonfly species is able to remain airborne (Angelibert and Giani, 2003). Traits relating to flight ability in rain can also vary among populations. *Pantala flavescens* on Rapa Nui (Easter Island) does not fly in the rain and only in fully sunlit conditions, close to the ground. In contrast, on the African mainland, it will fly in drizzle and usually at a height of several metres above the ground (Samways and Osborn, 1998).

Good timing of anisopteran emergence benefits from mass emergence of prey such as termites, which in South America appear prior to the start of the rainy season (da Silva *et al.*, 2011). Density of larvae is associated with high temperatures, as are adults, suggesting that larvae reflect the ecophysiological needs of the adults. This is seen in various biological and ecological traits. Zygopterans, in contrast to anisopterans, have a high surface area/body ratio and are sensitive to high solar radiation, especially under dry conditions. These zygopteran traits have led to microhabitat specialization, endophytic oviposition, and delicate larval and adult morphology. These traits make them particularly sensitive to great seasonal change, shaping assemblage composition according to the differential traits among the various damselfly species (Resende *et al.*, 2021).

At one species-rich location across the same elevation transect at 790 m above sea level, the dragonfly assemblage was dynamic over a 32-year period. While there is a set of 'core resident species' that are always present during the austral summer (Niba and Samways, 2006b), several additional species appear at this elevation from both higher and lower elevations according to seasonal conditions. This leads to an ebb and flow of species across the elevation gradient (Kietzka *et al.*, 2021).

Artificial ponds in the Cape Floristic Region in South Africa are used to retain water during the hot dry summers. They are usually deeper than natural ponds. In times of drought, they may be the only open water remaining after the natural ponds have dried out. These artificial ponds attract dragonflies, although significantly can attract local regional endemic species that are not normally associated with artificial water bodies, such as *Pseudagrion draconis*, *P. furcigerum*, *Zosteraeschna minuscula*, *Pinheyschna subpupillata* and *Syncordulia venator*. This is not to suggest that they breed in these ponds, but rather they have a temporary place of residence for maintaining themselves until adequate rain falls again before returning to their natural habitats. This trait of retreating to suboptimal habitats must have been honed by many past drought events where these species would have to seek temporary watery refuges, such as deposition and remnant pools in dry riverbeds, but now artificial ponds attract them (Deacon *et al.*, 2019).

In the same area, among various artificial ponds with varying degrees of water-level fluctuation, and therefore different environmental conditions, larvae of different widespread dragonfly species have traits related to the specific physicochemical conditions of the water. For example, *Crocothemis erythraea* and *Ischnura senegalensis* are associated with high conductivity and most other species with low conductivity. In terms of pH, *C. erythraea* is associated with alkaline conditions, *Ceratogomphus pictus*, *Trithemis arteriosa*, *T. dorsalis* and *T. stictica* close to neutral pH, and *Africallagma glaucum*, *I. senegalensis*, *Anax epphippiger*, *A. imperator* and *A. speratus* across a wide pH gradient (Jooste *et al.*, 2020) (Fig. 4.38).

4.6.3 Adult overwintering

Winter is a critical period for both aquatic and terrestrial insects, and as a way of avoiding competition by adjusting the timing of life activities. Winter survival of adult dragonflies in regions with moderate climates appears to be strongly associated with geographically widespread habitat generalists. In the northern savanna summer rainfall area of South Africa (22°S) eight species commonly take advantage of sunny microhabitats despite low water levels in both rivers and pools. Of these, seven libellulids and one gomphid are a core species set poised to exploit the first rains (Van Huyssteen and Samways, 2009). In the southern winter rainfall region of South Africa (38°S), where there are also frequent storms and high winds, a largely different set of eight species overwinter, sunning themselves and foraging

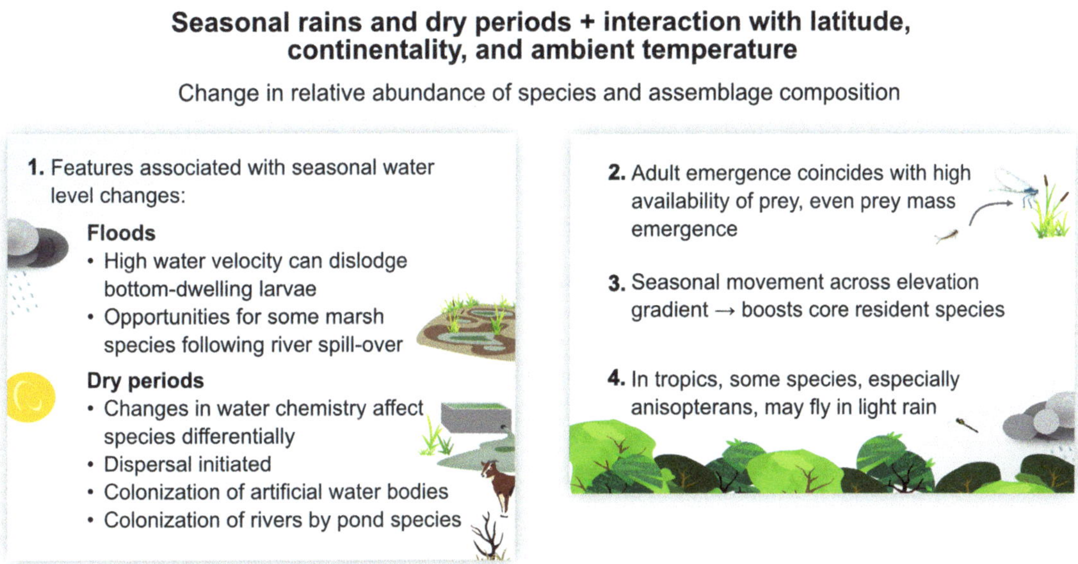

Fig. 4.38. Seasonal changes involve cycles of rains and dry periods, which are interactive with latitude, degree of continentality and ambient temperature. All these abiotic variables have a great effect on adult relative abundance and on assemblage composition with the passing of the seasons. Flooded rivers with high velocity can dislodge some larvae on the riverbed, yet provide opportunities for species that inhabit the flooded, marshy spill-over areas. Dry periods change water chemistry, with a differential effect on the various species. These dry times also result in dispersal, with some species seeking refuge in artificial water bodies, and some standing water species occupying river habitats. There are also changes in assemblage species composition. Adult emergence often coincides with high prey density, and even with mass emergence of prey species. While there may be a subset of all local species that are always present during summer, their numbers may be boosted according to season as dispersers arrive from other elevations. Although in temperate regions total time in flight may be determined by number of sunshine hours, a benefit of living in the tropics is that anisopterans, in particular, can continue to remain airborne in light rain. (Illustration: Robyn Symons).

during breaks in cool rainy conditions, yet retreating to dense vegetation during adverse conditions. Four species are libellulids and four zygopterans. Two overwintering species, *Trithemis arteriosa* and *T. furva*, are common to both regions. Interestingly though, the four damselflies are all national endemics – two synlestids, one protoneurid and one coenagrionid. Three of them (*Chlorolestes conspicuus*, *C. umbratus* and *Elattoneura frenulata*) are endemic to the southern Cape region and appear to be well-adapted to the overall wet winter local climatic conditions (Van Schalkwyk *et al.*, 2014). However, it is not unusual to find occasional individuals of other species also overwintering as adults, especially the national endemic synlestid *Chlorolestes tessellatus* and the widespread *Anax imperator*.

In both regions (savanna and southern Cape), the geographically widespread and common *Trithemis arteriosa* in particular is abundant in spring and has a head start by breeding early as seasonal temperatures begin to rise. The same is likely for the endemic zygopterans that overwinter as adults in the challenging wet and windy weather conditions in the mountains of the southern Cape. It appears that there has been strong selection pressure for traits that enable overwintering, seemingly also associated with oviposition early in the spring ahead of the hot and dry summers.

In Tunisia, 14 adult odonate species (two ischnurids, three aeshnids, one gomphid and eight libellulids) overwinter (Jödicke, 2003). All except *Sympetrum sinaiticum*, are multivoltine, and five, all anisopterans (*Anax imperator*, *Orthetrum chrysostigma*, *Crocothemis erythraea*, *Trithemis annulata* and *T. kirbyi*), are the same species as also recorded as overwintering at the other end of their huge African ranges in South Africa. This suggests that the ability to overwinter is a geographically widespread feature irrespective of season inversion.

While most far northern dragonflies overwinter in the immature stages, three lestid damselfly species in the genus *Sympecma* can pass the northern winter as adults. This appears to be a legacy of evolution among these species in the semi-arid environment of Central Asia where they became adapted to surviving water scarcity. This overwintering strategy remains strong enough to remain in place as the species have expanded their geographic ranges to wetter areas, despite very high mortality of adults while overwintering (Manger and Dingemanse, 2009). *Sympecma* adults are cryptic in coloration and behaviour. They also have a short larval development period which is out of phase with the peak of abundance of other, potentially competitive zygopterans. In short, the adult, by being cryptic and having a highly cold-tolerant physiology, coupled with the larva having time-based niche separation from potential competitors, has developed a successful life strategy arising from its evolutionary legacy (Harabiš *et al.*, 2012).

Sympecma fusca has some other significant traits which are of conservation significance. During the autumnal pre-reproductive period, the adult frequents sunlit ecotones between forest and grassland or meadow. This behaviour enables it to survive the harsh winter by gaining a thermal advantage, notwithstanding high mortality among the adult population at this time of year. Then, during the spring reproductive period, it occupies habitats close to breeding sites, only using the ecotones when weather conditions become inclement. However, ecotones are only a small part of the landscape, making them a critical resource for long-term survival, and must be conserved (Harabiš, 2016) (Fig. 4.39).

4.7 Dragonfly Traits Relative to Other Organisms

4.7.1 Predators

4.7.1.1 Adult dragonfly prey selection

DNA analysis has shown that both sexes of adult *Leucorrhinia intacta* show remarkably similar prey preferences and consume the same array of prey type (Morrill *et al.*, 2021). Similarly, the selection of prey types of both male and female adult *Lestes sponsa*, *Enallagma cyathigerum* and *Sympetrum danae* greatly overlaps, suggesting opportunistic feeding behaviour, although in these species, mostly on chironomid flies (Kaunisto *et al.*, 2017b). However, at least within a group of local Zygoptera, less than 2% of the chironomid population is consumed, along with species from several other taxa, indicating that these damselfly species make little impact on their currently available prey populations (Kaunisto *et al.*, 2020).

4.7.1.2 Carnivorous plants

Submerged carnivorous plants, such as *Utricularia* spp., capture young odonate larvae using their

Fig. 4.39. An unusual member of the Lestidae: *Sympecma fusca* male (left) and female (right), a species that overwinters as an adult in the Palaearctic. (Photos: Jens Kipping).

trapdoor bladders. The plant not only provides concealment but also presents a predation risk. Odonate prey recorded to date include *Coenagrion puella*, *Ischnura elegans* (Martens and Grabow, 2011) and an *Aeshna* sp. (Herzog and Martens, 2013). Experiments with *I. elegans*, *Crocothemis erythraea* and *Libellula quadrimaculata* show mortality among these species ranging from 19–45%. As *Utricularia* spp. can be very abundant, they may pose a threat to the local dragonfly assemblage (Manasov and Suhling, 2020).

4.7.1.3 Vertebrate predators

While dragonfly larvae are prey for some other predatory insects, especially dytiscid diving beetles, as well as larger dragonflies, they are also fed on by certain vertebrates. Several bird, frog and reptile species feed on dragonfly larvae, as well as the adults. Young crocodiles will readily feed on large dragonfly larvae.

Adult dragonflies have strategies to escape frogs. Usually, the flying individual swerves away to avoid the leaping frog, although it sometimes flies backward. Perching individuals will fall to one side, even sometimes landing in the water. Size matters, with large anisopterans escaping more often than small zygopterans. *Anax imperator* females sometimes escape after capture by using strong wingbeats, even underwater (Rüppell *et al.*, 2020).

Tandem pairs of *Pyrrhosoma nymphula* reduce predation risk from frogs by aggregating. Besides the already present groups indicating suitable oviposition microsites, they are also a magnet to new pairs arriving, as they signal a temporary refuge from frogs (Rehfeldt, 1990). The Cane toad *Bufo marinus* is well-known as a high-risk biological control agent of insect pests. Within its native range of Venezuela, 7% of its diet by biomass consists of odonates, but interestingly, the larvae rather than the adults (Evans and Lampo, 1996).

Important bird predators of adult *Calopteryx splendens* are wagtails (*Motacilla*), which mediate wing traits in the damselfly. Predation has only weak effects on features relating to flight performance (size and shape of the wings), suggesting that an adaptive peak has been reached. However, predation has a strong effect on conspicuous secondary sexual traits (wing patch characteristics), indicating a trade-off between natural and sexual selection (Kuchta and Svensson, 2014). Changes in behavioural traits in *C. splendens* can also vary relative to local environmental conditions. This leads to different but erratic behavioural profiles across latitudes, especially associated with adult activity. While courtship effort is highest at mid-latitudes, boldness (level of activity among territorial males, measured in time units) is greatest in the north, but does not differ between mid and southern latitudes (Golab *et al.*, 2022) (Fig. 4.40).

4.7.1.4 Fish predation

The most important vertebrate predators are fish, which can have a major impact on population levels of dragonfly larvae, notwithstanding that some

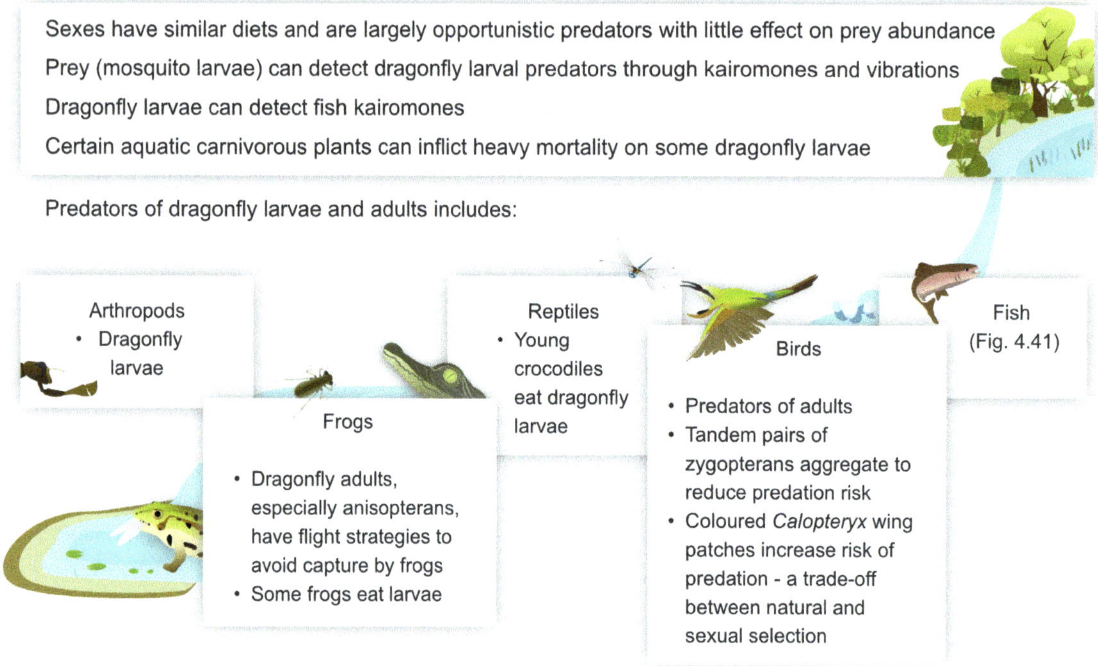

Fig. 4.40. There is a sensitive and highly adapted interplay between dragonflies and other organisms, whether they are prey or predators. (Illustration: Robyn Symons).

microhabitats are typically free of fish, such as tank bromeliads and potholes scoured out of the rock substratum over long periods of time along rapid rivers (Samways *et al.*, 2020).

Overall, the effect of fish can change the local assemblage composition, which in turn varies with time of year (Morin, 1984). While the impact of fish can be substantial for some species of larvae, it can also benefit some others (e.g. *Perithemis tenera*). In contrast, *Leucorrhinia dubia* is particularly susceptible to fish predation, affecting its local distribution in Sweden (Henrikson, 1988). Though species richness in these lakes remains the same with or without fish, there can be noticeable differences in abundance. While *Aeshna juncea* and *Leucorrhinia dubia* are less abundant in lakes with fish, *Erythromma najas* and *Cordulia aenea* are more abundant, with no difference in abundance of *Coenagrion hastulatum*, *Libellula quadrimaculata* and *Leucorrhinia rubicunda* (Johansson and Brodin, 2003). In terms of activity traits, there is higher abundance of species with active foraging in fishless lakes but higher abundance of passive foragers in lakes with fish (Johansson *et al.*, 2006).

North American libellulid larvae show a gradient of mobility and response to type of top predator, as well as occupying different microhabitats, according to species. At one end of the spectrum, highly mobile larvae are associated with temporary ponds with invertebrates or small fish as the top predators. At the other end, less-mobile larvae are associated with large fish in permanent ponds which enable a long larval stage. In turn, length of larval period is associated with high levels of contact with vegetation, although this is not as important as hydroperiod or type of top predator (McCauley, 2008).

Many species of odonate larvae respond to the presence of predatory fish, with some even able to discriminate between fish species. Yet there is no significant relationship between behavioural change and predation risk. This suggests that anti-predator responses and predation risk are finely tuned according to the specific predator and odonate prey species combination (Wohlfahrt *et al.*, 2006). Fish behaviour is also important. The rainbow trout (*Oncorhynchus mykiss*) is highly predatory on various invertebrates. Yet in South Africa where it is alien, it takes very few *Chlorolestes* spp. larvae living on the bottom of deposition pools in montane streams. It appears that the fish is largely focused on obtaining insect food that has fallen on the water surface (Shelton *et al.*, 2015). Nevertheless, trout are known from isotope analysis to feed on some other odonate larvae (Maimela *et al.*, 2022).

Interestingly, among *Leucorrhinia* species, predation affects patterns of multiple trait co-variation when experimentally moved from lake mesocosms with fish as top predators to fishless ones with dragonflies as top predators. The move is accompanied by partial loss and reduction in length of larval spines, reduced burst swimming and associated energy fuelling enzymic activity, which leads to them changing the dynamics of food webs (Mikolajewski *et al.*, 2010).

Fish, as well as some other predators, can exert high selection pressure on larvae, modifying form, affecting colour and patterning, and leading to behavioural changes. Larvae armed with large spines are a feature of dragonfly species subject to high fish predation (Johansson and Mikolajewski, 2008). Among coenagrionid damselflies, some species can only occupy fishless lakes, others in lakes with fish and a third group lives in both types of lake. In fishless lakes, larvae are subject to predation by large aeshnid dragonfly larvae, toward which they show anti-predation behaviour such as rapidly swimming away, which in other lakes would make them vulnerable to fish predators. These responses are also reflected in their morphology (McPeek, 1995). In the lakes with fish, they remain largely immobile to avoid being eaten by the fish, but they are still highly vulnerable to the aeshnid larvae (McPeek, 1990).

Complex networks of traits underlie predator avoidance behaviours in response to larvae facing different predation pressures, as from aeshnid larvae versus fish. In the case of libellulids, there is higher integration of larval abdominal traits leading to high burst-swimming speed in their response to fish, while this behaviour is relaxed among the larvae facing aeshnid predators (Mikolajewski *et al.* 2015).

It appears as if some predators develop a search image for larvae emerging from the water. This is illustrated by choice of trout fishing flies, which mimic certain emerging larvae, and the most effective lures being the ones that best match the species mass emerging at the time, such as *Lestes plagiatus*, *Ischnura senegalensis* and *Trithemis dorsalis* in artificial trout lakes in South Africa at elevations around 1000 m above sea level.

The predatory European perch *Perca fluviatilis* has a strong effect on developing *Lestes sponsa* larvae. While activity levels of the young larvae drop in the presence of the fish, older larvae respond strongly to high densities of their zooplankton prey. The outcome is that with fish present, development time is shorter and size at emergence larger, the result of thinning of the larval population in combination with abundant food (Brodin and Johansson, 2002).

Several fish species occupy Swedish lakes, often dominated by European perch *P. fluviatilis* or common roach *Rutilus rutilus*. Dragonfly diversity in these lakes is reduced by these fish. Four species, *Leucorrhinia caudalis*, *L. dubia*, *Somatochlora flavomaculata* and *Sympetrum striolatum*, are absent in lakes dominated by perch, and one species, *Sympetrum flaveolum*, where roach dominated. In contrast, although *Erythromma najas*, *Ischnura elegans* and *Cordulia aenea* were present in most of the lakes, none of the 24 species was recorded in all the 24 lakes studied. Overall, these results indicate that these food webs are complex and highly variable, driven by fish species and density relative to the traits of the dragonfly larvae (Wittwer *et al.*, 2010).

There is sensitivity on the part of prey to predation by dragonflies, which varies among prey species. Prey, such as mosquito larvae, can detect the presence of kairomones (Roberts, 2012) as well as vibrations (Roberts, 2017) emanating from the dragonfly larva, changing their feeding and movement behaviour accordingly. In turn, dragonfly larvae can detect kairomones from fish (Hopper, 2001) (Fig. 4.41).

4.7.1.5 Metacommunity effects

Larval dragonfly assemblages can structure benthic communities. However, the process is complex and dependent on factors like prey refuges and non-selective predation with patch switching, leading to a greater effect on prey density than on community diversity (Thorp and Cothran, 1984).

Predation can be intense across some food webs, having major consequences for community dynamics, which can reverberate across ecosystems (Cummins and Klug, 1979). Dragonfly larvae are part of the aquatic food web, and adults the terrestrial one. Fish, such as centrarchid sunfishes in the USA, through reducing the larval abundance of *Anax junius* and *Erythemis simplicicollis*, can lead

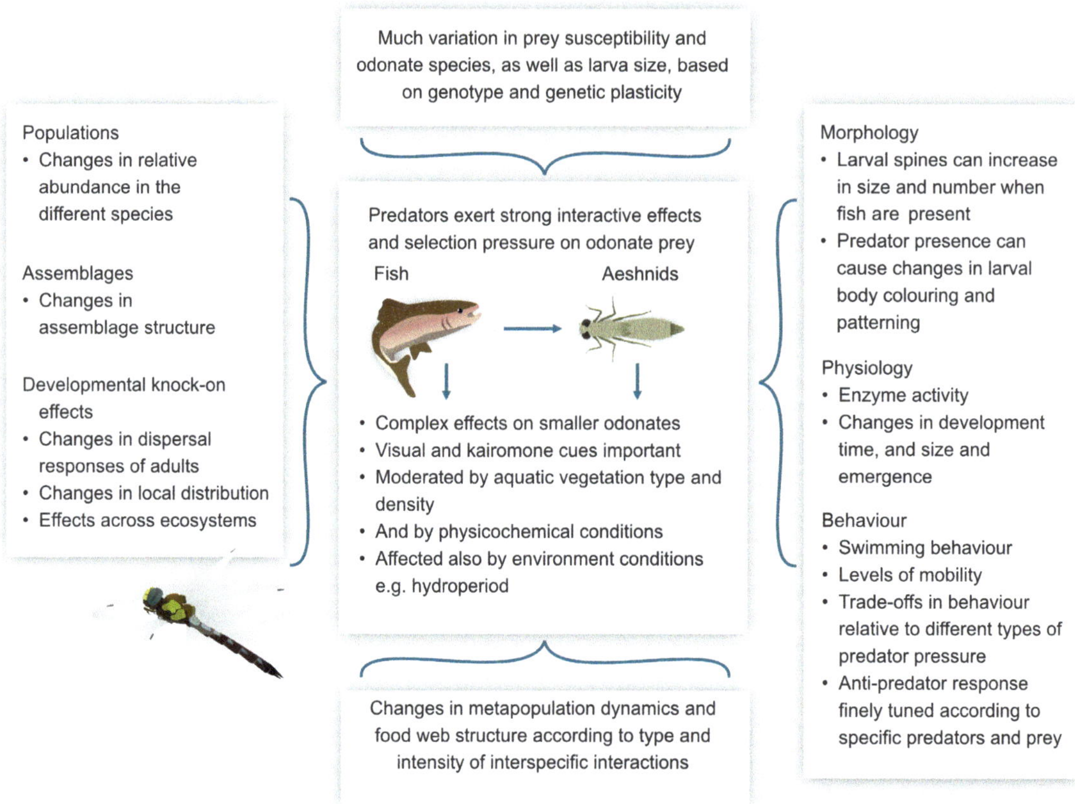

Fig. 4.41. Fish predators exert great pressure on odonate populations, often interactive with predation by large odonates such as aeshnids. The effect on the smaller odonates can be complex, with both visual detection and kairomones playing a role in both predator and prey. The interactive behaviours are moderated by type and density of aquatic vegetation as well as the physicochemical conditions of the water. Changing environmental conditions, like hydroperiod, can also play a role in affecting the biotic interactions. Overall, there is much variation in prey susceptibility among the different odonate species, based on genotype and genetic plasticity. Predation on odonates affects morphology, physiology and behaviour, often changing metapopulation dynamics and food web structure according to the type and intensity of the interactions. At the population level, there are often differential shifts in relative abundance of the odonate prey, which change assemblage structure. These interactive effects are not confined to the larvae, having a developmental knock-on effect on the adults as well. (Illustration: Robyn Symons).

to fewer dragonfly adults on the banks of ponds where the pondside plant *Hypericum fasciculatum* lives. However, when there are no fish, and consequently an abundance of larvae leading to more dragonfly adults, there are fewer pollinators of the plant present. The pollinators avoid plants when many dragonflies are present as they may be eaten. The outcome is that plants growing at fishless ponds experience less pollination events (Knight *et al.*, 2005).

In Estonia, *Orthetrum cancellatum* and some other dragonflies, can exert high mortality on various grassland butterfly species, especially when dragonflies are abundant. However, the impact varies among habitat patches and with time during the season. When considering sites for butterfly conservation, it is best to avoid the sites with high levels of dragonfly activity (Sang and Teder, 2011).

In a metacommunity of fish and aquatic insects, including dragonflies, inhabiting drainage ditches of rice paddies in Japan, agricultural water management and flood events associated with paddies lead to seasonal flow changes. While fish are mainly influenced by local environmental factors such as water depth and distance to downstream rivers, they are restricted in their movements to times of higher water levels to occupy preferred habitats. In contrast, dragonflies, as well as aquatic beetles and bugs, are restricted only by overland distance. Overall, this suggests that regional and landscape variables can influence metacommunity patterns more than the characteristics of the local environment, and interactively influence metacommunity structures in temporally connected habitats (Iwamoto *et al.*, 2022), with the implication that dragonfly flight enables them to find enemy-free space.

4.7.2 Parasitism and disease

4.7.2.1 *Parasitoids*

For a dragonfly population to survive, it must clear the hurdle of parasitism at each developmental stage, starting with the egg. Although the eggs of some aquatic insect groups, such as mayflies, have not yet been recorded as being parasitized, dragonfly eggs can be attacked but not to the extent of some fully terrestrial orders (Fatouros *et al.*, 2020).

Dragonfly egg concealment is one way to escape attack, particularly small eggs deep in plant tissue that are relatively inaccessible to small hymenopteran parasitoids. Investment in an egg protection strategy has benefits. For example, *Chalcolestes viridis* eggs laid in stiff tree bark suffer less parasitism from the hymenopterans *Prestwichia* and *Aprostocetus* (1% of eggs parasitized) than *Lestes sponsa* eggs laid in soft herbaceous tissue (4.7% parasitized). The risk to the *C. viridis* female is that its long oviposition time dealing with a stiff substrate increases exposure time for predator attack. Other benefits for *C. viridis* eggs in bark may be better protection from lethal ice crystals in winter, and by being away from the water they are in a relatively enemy-free space.

In turn, *Sympecma fusca* eggs were recorded as not parasitized at all. Its eggs are laid in dead floating plant tissue, difficult for non-aquatic enemies to reach, as well as being a semiochemical-free environment compared to living plant tissue that would otherwise attract host-seeking parasitoids. *S. fusca* also has other traits that may contribute toward its eggs avoiding parasitism. Overwintering and rapid larval development enables it to be ahead in the season relative to the generalist parasitoids that have many other hosts to attack when they are active later in the season (Harabiš *et al.*, 2019).

As regards dragonfly eggs underwater, they are not as safe as they would seem, as some adult hymenopterans are adapted to move and search underwater, using their wings as paddles. For example, *Anagrus obscurus* (Mymaridae) parasitized 2% of *Calopteryx haemorrhoidalis* and 6% of *Platycnemis pennipes* eggs on the submerged aquatic plant *Potamogeton* sp. (Santolamazza *et al.*, 2011). In turn, *Aeshna tuberculifera* eggs inserted into *Iris versicolor*, suffered 18.4% of eggs parasitized by three parasitoid species (Cebulski and O'Brien, 2013).

Polyphagous insect egg parasitoids can be very abundant in *Lestes sponsa* habitat, and in decreasing order of abundance, they belong to Scelionidae, Myrmaridae, Eulophidae and Trichogrammatidae (Harabiš *et al.*, 2013). Considering these studies alongside those covered by Corbet (2004), where parasitism rates of over 90% have been recorded, parasitoids can be a significant mortality factor for dragonfly eggs in certain locations.

4.7.2.2 *Parasitism by mites and other smaller organisms*

Disease is caused by an infectious biological agent that disrupts the normal physiology of an organism, affecting its fitness in terms of production of viable offspring, or even killing its host. Despite limited knowledge characterizing disease processes in natural communities, it is evident that disease is a powerful driver of population and community dynamics of aquatic insects (Holt *et al.*, 2021).

The two main categories of infectious agents for dragonfly larvae and adults are external parasites such as water mites and internal parasitic microorganisms like viruses, bacteria and protists, especially gregarines which are a heterogeneous protozoan group in the Phylum Apicomplexa. Gregarines parasitize both larval and adult odonates (Hecker *et al.*, 2002), shortening adult lifespan (Åbro, 1971) or lowering adult fat content, with negative effects on host reproduction (Siva-Jothy and Plaistow, 1999). Dragonflies also act as hosts for internally parasitic microsporidia (spore-forming fungal unicellular parasites) (Larsson, 1990; Sokolova *et al.*, 2006; Nakamura *et al.*, 2021). Other internal parasites include helminth flatworms.

Ectoparasitic water mites of the genus *Arrenurus* can reduce female fitness in *Coenagrion puella* (Thompson *et al.*, 2011). The same genus can infect larvae, and the proportion of odonate individuals per species infected can range from 1% (*Perithemis mooma*) to 55% (*Ischnura capreolus*) (Bizarro *et al.*, 2020).

In the Kogelberg Biosphere Reserve in the Cape Floristic Region in South Africa there are extremely high levels of endemic odonate species, as well some common and geographically widespread species. Of the 38 species there, adults of two common species carry *Arrenurus* and *Leptus* mites, with 3.5% of *Ceriagrion glabrum* and 38% of *Ischnura senegalensis* individuals infested. Oddly, not one individual of any of the rare and threatened endemic species carried any mite loads (Grant and Samways, 2007), in keeping with the geographic range hypothesis (Price *et al.*, 1988). As these endemic species have a long evolutionary history at the species level, they have also had a long time to develop resistance in this geographical area. However, populations of the two widespread species, which normally occur as metapopulations, have been isolated in these mountains perhaps just long enough to lose some of their genetic diversity (Fig. 4.42).

This perspective can be seen against the genetics of *Calopteryx splendens*, where individuals with less genetic diversity are much more prone to carrying high gregarine loads (Kaunisto *et al.*, 2013). In turn, the more geographically widespread *C. virgo* is more likely to be parasitized by gregarine parasites than the less widespread *C. splendens*, especially when populations of the two species are separate. However, in mixed populations, *C. virgo* is still more often parasitized than *C. splendens*. The reasons for *C. virgo* being so susceptible to infection may possibly be that it has high levels of interspecific aggression, low genetic diversity and lower levels of immunity than in *C. splendens* (Ilvonen *et al.*, 2011).

A meta-analysis by da Silva *et al.* (2021) showed odonate ectoparasites, such as water mites, decline at lower latitudes, possibly due to elevated water temperatures toward the equator eliciting a stronger immune response. However, there are no differences in gregarine loads with latitude.

In Canada, *Sympetrum internum* has some resistance to *Arrenurus planus* water mites, unlike other dragonfly species in the area (Forbes *et al.* 1999). While water mites on adult *Coenagrion hastulatum* do not reduce the number of eggs produced, gregarine parasites do. This may be due to gregarines taking space which otherwise could have been devoted to eggs, or possibly because immune energy reserves are directed to an immune response instead of eggs (Kaunisto *et al.*, 2017a). This last possibility is plausible as higher gregarine loads cause the host to produce less energetically expensive wing pigmentation in *Calopteryx splendens* (Siva-Jothy, 2000) and *C. haemorrhoidalis* (Córdoba-Aguilar *et al.*, 2003), with the amount of parasite load also playing a role in determining population levels (Rolff and Siva-Jothy, 2004). However, some caution is required in the interpretation of results, as a species' relative ability to resist parasites (as measured by total phenol oxidase or its activity, a commonly used approach) does not explain interspecific variation among species (Mlynarek *et al.*, 2015).

Most male *Libellula pulchella* territory holders have both superior flight ability and lower gregarine loads. Although the female is focused on oviposition site selection, she also mates with the agile and

Fig. 4.42. A cluster of *Arrenurus* mites infesting a male *Ceriagrion glabrum*. (Photo: Michael Samways).

parasite-free males, which assumes some degree of acquisition of some genetically based parasite resistance (Marden and Cobb, 2004).

Gregarine and water mite loads among damselflies are higher in males than females, with loads increasing with adult age. This comes about as males invest resources in thoracic fat production to sustain intense territory defence and mating activities, improved by greater food intake. Females on the other hand invest in mating-related resource allocation such as increased abdominal weight (Córdoba-Aguilar and Munguía-Steyer, 2013). Interestingly, male-mimicking morphs in polymorphic damselflies have higher parasite loads than other female morphs. This is possibly because these male-like morphs pay higher fecundity costs through production of more body mass and protein, compromising the immune system (Sánchez-Guillén *et al.*, 2013).

The number of *Ischnura elegans* larvae infected with microsporidia increased when exposed to higher rearing temperatures and predator cues from the invasive alien crayfish (*Pacifastacus leniusculus*). Infected larvae also displayed distorted wing development, leading to death before emergence. It appears that increasing temperatures and stress caused by the presence of the predator are both factors that have negative fitness consequences following microsporidian infection (Antoł *et al.*, 2022).

Open pasture landscapes in Canada cause *Calopteryx maculata* to fly over greater areas than they do in continuous forest. More extensive flying in open landscapes is beneficial as the species accumulates a lower gregarine parasite load than in the more closed forest environment, seemingly due to a lower encounter rate between the damselfly host and its parasite (Taylor and Merriam, 1996).

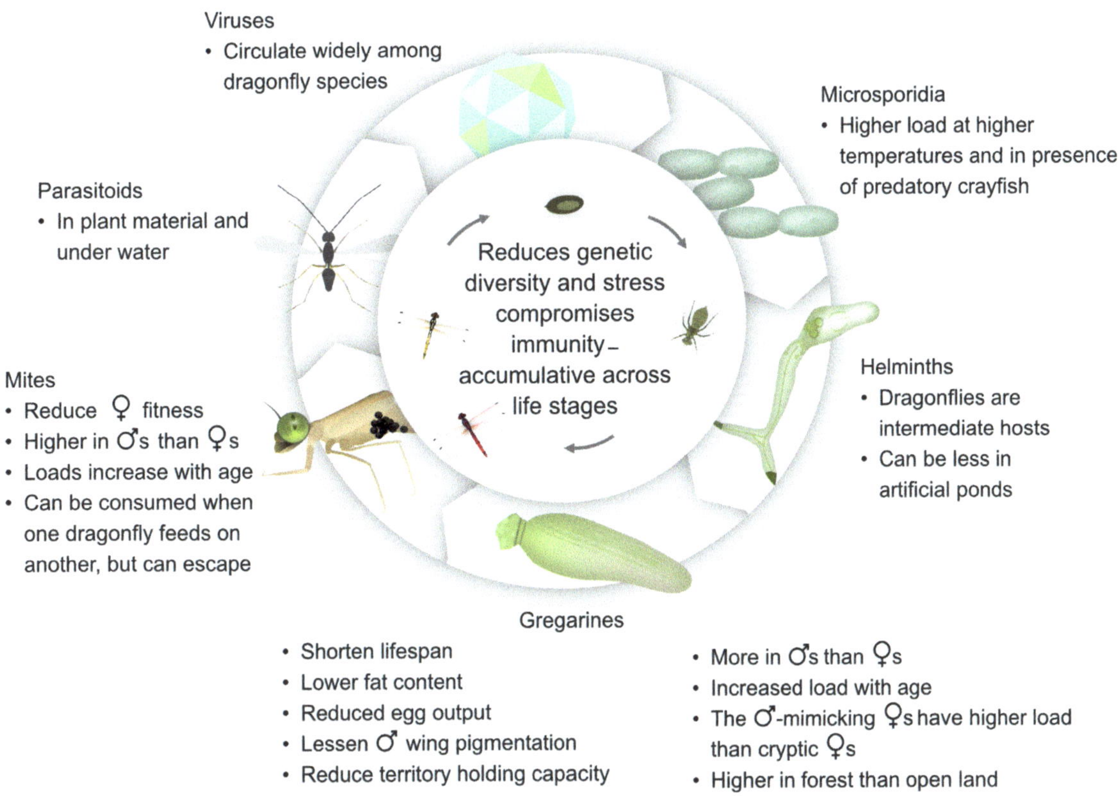

Fig. 4.43. There is a wide range of parasitoids and pathogens adversely affecting dragonfly survival. The effects of these agents are accumulative from one life stage to another (i.e. earlier survivors may then be prone to attack by other agents in later stages). Shown here are some of the known agents and their effects on dragonfly individuals. (Illustration: Robyn Symons).

Dragonflies such as *Libellula* are not put off from eating *Lestes* damselflies that have conspicuous red larval water mite loads. If the mites are only partially engorged, they are eaten along with the damselfly, but in the case of fully engorged mites, half of them detach themselves. The chances of a partially engorged mite surviving to later reproduce is probably zero, but fully engorged individuals have a chance, either when the detached mite falls into the water, or in the case of temporary ponds, it can survive desiccation until water returns (Nagel *et al.*, 2011).

4.7.2.3 *Parasitism by helminths*

Among the Platyhelminthes, trematodes (Trematoda), tapeworms (Cestoda: Cyclophyllidea) and hairworms (Aschelminthes: Nematomorpha) are associated with dragonflies. In the case of trematodes, dragonfly larvae are usually the secondary or intermediate host, with amphibians, birds or mammals being the primary host. Many species of dragonflies, and at certain locations sometimes all species present, can be hosts to trematodes. Some dragonfly larvae can carry very high trematode cercariae loads (summarized by Corbet, 2004; Li *et al.*, 2023).

Agricultural transformation of the landscape can alter the relationship between trematode parasites and co-occurring hosts, such as frog tadpoles and dragonfly larvae. Tadpoles in agricultural ponds are more likely to be parasitized but have lower infection intensities, especially far away from natural forest. There is the opposite tendency in dragonfly larvae, and as they readily colonize the open agricultural ponds, they are relatively parasite-free, an encouraging message for dragonfly conservation (Koprivnikar *et al.*, 2017).

4.7.2.4 *Viruses*

Viruses circulate widely among dragonfly species (Rosario *et al.*, 2012). About 20 different putative viruses were recorded from the larvae of *Leucorrhinia dubia* and *Coenagrion puella* (Johnston *et al.*, 2015) and 31 from larvae of seven odonate species, and they possibly circulate in the ecosystem via faecal matter (Dayaram *et al.*, 2015). The extent to which viral infections have an adverse effect on dragonfly populations is currently not known (Fig. 4.43).

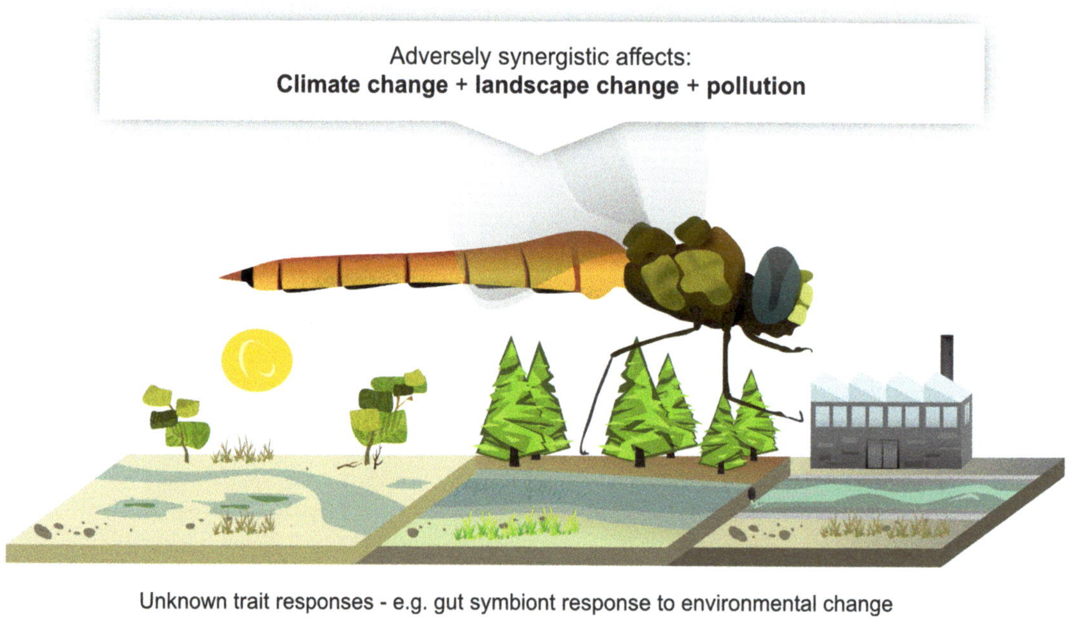

Fig. 4.44. Sensitivity of gut symbionts emphasizes that there are many traits about which we know very little. Indeed, there may be many significant traits that are important for dragonfly survival that we are yet to discover. Some of these traits may be adversely synergistic with, for example, climate change, landscape transformation and pollution by toxic compounds. (Illustration: Robyn Symons).

4.7.3 Symbionts and commensals

4.7.3.1 *Gut symbionts*

The gut microbiota of wild organisms can have a significant effect on the organism's ecology and evolution. Among seven libellulid species in India, geographic location plays the dominant role followed by host species identity in determining the gut bacterial community. The prey profile in the gut of each dragonfly species is distinct, with its own bacterial community, reflecting strong prey choice. Furthermore, the abundance and distribution of 72% of the bacterial taxa are consistent with neutral community assembly. In short, the gut bacterial microbiome of these dragonflies appears to be largely determined by specialized host diet via the processes of microbial dispersal, nutrient availability and colonization (Deb *et al.*, 2019). Additionally, among different species of dragonflies in the same local area host specificity and diet influences the composition of the intestinal microbiome, but the degree of influence may depend on the specific host and environmental conditions (Lim *et al.*, 2023) (Fig. 4.44).

4.7.3.2 *Commensal organisms*

The generally slow-moving behaviour of odonate larvae means that a whole range of exobiota can attach themselves to the submerged body. These include, for example, chironomid larvae and hydra (Maynou and Martín, 2021). Perhaps they obtain food items such as organic particles that become attached to the larva's setae-covered body surface, or perhaps there is some protection from predators or sudden floods. Another possibility is that there is some benefit for the commensal related to dispersal to new local areas where organic material is more abundant. However, it may also simply be an inadvertent settling on the body of the larva. Whatever the case, there is no evidence to date that any harm is done to the odonate larva, with perhaps some advantage conferred as these adventitious forms of life may improve the camouflage of the larva, making it appear more like an inert object. If there is any benefit, it is likely to favour the odonate, with the commensals at risk of dehydration once the larva leaves the water to emerge as an adult.

References

Abbott, J.K. and Svensson, E.I. (2008) Ontogeny of sexual dimorphism and phenotypic integration in heritable morphs. *Evolutionary Ecology* 22, 103–121. DOI: 1007/s10682-007-9161-0.

Åbro, A. (1971) Gregarines: their effects on damselflies (Odonata: Zygoptera). *Insect Systematics and Evolution* 2, 294–300. DOI: 10.1163/187631271X00301.

Acquah-Lamptey, D., Brändle, M., Brandl, R. and Pinkert, S. (2020) Temperature-driven color lightness and body size variation scale to local assemblages of European Odonata but are modified by propensity for dispersal. *Ecology and Evolution* 10, 8936-8948. DOI: 10.1002/ece3.6596.

Amari, H., Zebsa, R., Lazli, A., Bensouilah, S., Mellal, M.K. *et al.* (2019) Differential elevational cline in the phenology and demography of two temporally isolated populations of a damselfly: not two but one taxon? *Ecological Entomology* 44, 93–100. DOI: 10.1111/een.12680.

Anderson, R.C. (2009) Do dragonflies migrate across the western Indian Ocean? *Journal of Tropical Ecology* 25, 347–358. DOI: 10.1017/S0266467409006087.

Angelibert, S. and Giani, N. (2003) Dispersal characteristics of three odonate species in a patchy habitat. *Ecography* 26, 13–20.

Angert, A.L., Crozier, L.G., Rissler, L.J., Gilman, S.E., Tewksbury, J.J. and Chunco, A.J. (2011) Do species' traits predict recent shifts at expanding range edges? *Ecology Letters* 14, 677–689. DOI: 10.1111/j.1461-0248.2011.01620.x.

Antoł, A., Labecka, A.M., Larsson, J.I.R. and Sniegula, S. (2022) First record of microsporidia infection in the damselfly *Ischnura elegans* larvae: temperature and predator cue effects on the host's life history. *Diversity* 14, e428. DOI: 10.3390/d14060428.

Arambourou, H., Sanmartín-Villar, I. and Stocks, R. (2017) Wing shape-mediated carry-over effects of a heat wave during the larval stage on post-metamorphic locomotor activity. *Oecologia* 184, 279–291. DOI: 10.1007/s00442-017-3846-z.

Aristizábel-Botero, A., Snoeks, J., Realpe, E. and Vanschoenwinkel, B. (2023) Conductivity and water level modulate developmental plasticity and explain distribution patterns in a diverse neotropical Odonata assemblage. *Freshwater Biology* 68, 1558–1571. DOI: 10.1111/fwb.14151.

Aromaa, S., Iloven, J.J. and Suhonen, J. (2019) Body mass and territorial defence strategy affect the territory size of odonate species. *Proceedings of the Royal Society B* 286, e20192398. DOI: 10.1098/rspb.2019.2398.

Attaullah, M., Ullah, I., Ali, M., Maula, F., Ilahi, I. *et al.* (2021) Diversity of Anisoptera & Zygoptera (Odonata: Insecta) of Swat, Pakistan. *Brazilian Journal of Biology* 83, e251958. DOI: 10.1590/1519-6984.251958.

Baird, I.R.C. (2019) Establishment of larval pits by *Tachopteryx thoreyi* (Odonata: Petaluridae): habitat modification by a non-burrowing petalurid. *International Journal of Odonatology* 22, 135–146. DOI: 10.1080/1388790.2019.1636889.

Ball-Damerow, J.E., M'Gonigle, L.K. and Resh, V.H. (2014a) Local and regional factors influencing assemblages of dragonflies and damselflies (Odonata) in California and Nevada. *Journal of Insect Conservation* 18, 1027–1036. DOI: 10.1007/s10841-014-9709-6.

Ball-Damerow, J.E., M'Gonigle, L.K. and Resh, V.H. (2014b) Changes in occurrence, richness, and biological traits of dragonflies and damselflies (Odonata) in California and Nevada over the past century. *Biodiversity and Conservation* 23, 2107–2126. DOI: 10.1007/s10531-014-0707-5.

Banks, M.J. and Thompson, D.J. (1987) Lifetime reproductive success of females of the damselfly *Coenagrion puella*. *Journal of Animal Ecology* 56, 815–832. DOI: 10.2307/4950.

Barthelemy, E., Fortunel, C., Jaunatre, M. and Munoz, F. (2021) Imprints of past habitat reduction on extant taxonomic, functional, and phylogenetic composition. *Frontiers in Ecology and Evolution* 9, e634413. DOI: 10.3389/fevo.2021.634413.

Bastos, R.C., Brasil, L.S., Oliveira-Junior, J.M.B., Carvalho, F.G., Lennox, G.D. *et al.* (2021) Morphological and phylogenetic factors structure the distribution of damselfly and dragonfly species (Odonata) along an environmental gradient in Amazonian streams. *Ecological Indicators* 122, e107257. DOI: 10.1016/j.ecolind.2020.107257.

Bizarro, G.L., Périco, E., Dalzochio, M., da Silva, G.L., Ferla, N.J. and Johann, L. (2020) Aquatic larva[e] of the genus *Arrenurus* (Trombidiformes: Parasitengonina: Arrenuridae) associated with Odonata species from Pampa biome, Brazil. *Biota Neotropica* 21, e20201157.

Boda, R., Bereczki, C., Ortmann-Ajkai, A., Mauchart, P., Pernecker, B. and Csabai, Z. (2015) Emergence behaviour of the red listed Balkan Goldenring (*Cordulegaster heros* Theischinger, 1979) in Hungarian upstreams: vegetation structure affects the last steps of the larvae. *Journal of Insect Conservation* 19, 547–557. DOI: 10.1007/s10841-015-9776-3.

Bogan, M.T. and Lytle, D.A. (2011) Severe drought drives novel community trajectories in desert stream pools. *Freshwater Biology* 56, 2070–2081. DOI: 10.1111/j.1365-2427.2011.02638.x

Bogan, M.T., Eppehimer, D., Hamdhani, H. and Hollien, K. (2020) If you build it, they will come: rapid colonization by dragonflies in a new effluent-dependent river reach. *PeerJ* 8, e9856. DOI: 10.7717/peerj.9856.

Bonada, N., Cañedo-Argüelles, M., Gallart, F., von Schiller, D., Fortuño, P. *et al.* (2020) Conservation and management of isolated pools in temporary rivers. *Water* 12, e2870. DOI: 10.3390/w12102870.

Borisov, S.N. (2006) Adaptations of dragonflies (Odonata) under desert conditions. *Entomological Review* 86, 534–554. DOI: 10.1134/S0013873806050058.

Borisov, S.N., Iakovlev, I.K., Borisov, A.S., Ganin, M.Y. and Tiuno, A.V. (2020) Season migration of *Pantala flavescens* (Odonata: Libellulidae) in Middle Asia and understanding of the migration model in the Afro-Asian region using stable isotopes of hydrogen. *Insects* 11, e890. DOI: 10.3390/insects11120890.

Borisov, S.N., Kazenas, V.L. and Borisov, A.S. (2022) Dragonflies (Odonata) of Syrdarya Karaau and the Arys River valley (southern Kazakhstan) with notes on seasonal latitudinal and altitudinal migrations. *Euroasian Entomological Journal* 21, 78–90.

Braune, E., Richter, O., Söndgerath, D. and Suhling, F. (2008) Voltinism flexibility of a riverine dragonfly along thermal gradients. *Global Change Biology* 14, 470–482. DOI: 10.1111/j.1365-2486.2007.01525.x.

Briggs, A., Pryke, J.S., Samways, M.J. and Conlong, D.E. (2019) Macrophytes promote insect conservation in artificial ponds. *Aquatic Conservation: Marine and Freshwater Ecosystems* 29, 1190–1201. DOI: 10.1002/aqc.3157.

Brito, J.G., Roque, F.O., Martins, R.T., Nessiman, J.L., Oliveira, V.C. *et al.* (2020) Small forest losses degrade stream macroinvertebrate assemblages in the eastern Brazilian Amazon. *Biological Conservation* 241, e108263. DOI: 10.1016/j.biocon.2019.108263.

Brodin, T. and Johansson, F. (2002) Effects of predator-induced thinning and activity changes on life history in a damselfly. *Oecologia* 132, 316–322. DOI: 10.1007/s00442-002-0938-0.

Buden, D.W. (2010) *Pantala flavescens* (Insecta: Odonata) rides west winds into Ngulu Atoll, Micronesia: Evidence of seasonality and wind-assisted dispersal. *Pacific Science* 64, 141–143.

Butler, R.G. and deMaynadier, P.G. (2008) The significance of littoral and shoreline habitat integrity to the conservation lacustrine damselflies (Odonata). *Journal of Insect Conservation* 12, 23–36. DOI: 10.1007/s10841-006-9059-0.

Calvão, L.B., Paiva, C.K.S., Brito, J.S., Fares, A.L., Maia, C. *et al.* (2020) Influence of biotic and abiotic factors on adult Odonata (Insecta) in Amazon streams. *Animal Biology* 71, 1–18. DOI: 10.1163/15707563-bja10047.

Carbone, J., Keller, W. and Griffiths, R.W. (1998) Effects of changes in acidity on aquatic insects in rocky littoral habitats of lakes near Sudbury, Ontario. *Restoration Ecology* 6, 376–389.

Casanueva, P., Requena, J.-F.S., Hernández, M.A., Ortega, S., Nunes, L.F. and Campos, F. (2017) Altitudinal variation of wing length and wing area in *Libellula quadrimaculata* (Odonata: Libellulidae). *Odonatologica* 46, 227–240. DOI: 10.5281zenodo.1040305.

Castillo-Pérez, E.U., Suárez-Tovar, C.M., González-Tokman, D., Schondube, J.E. and Córdoba-Aguilar, A. (2022) Insect thermal limits in warm and perturbed habitats: dragonflies and damselflies as study cases. *Journal of Thermal Biology* 103, e103164. DOI: 10.1016/j.jtherbio.2021.103164.

Cebulski, B.C. and O'Brien, M.F. (2013) Observations on egg parasitism of *Aeshna tuberculifera* (Odonata: Aeshnidae) by Eulophidae, Trichogrammatide and Myrmaridae (Hymenoptera) in Alger County, Michigan. *The Great Lakes Entomologist* 46, 145–153.

Cerini, F., Bologna, M.A. and Vignoli, L. (2019) Dragonfly community assembly in artificial habitats: glimpses from field and manipulative experiments. *PLoS ONE* 14, e0214127. DOI: 10.1371/journal.pone.0214127.

Cerini, F., Vignoli, L., Blust, M. and Strona, G. (2023) Functional traits predict species co-occurrence patterns in a North American Odonata metacommunity. *Ecosphere* 14, e4732. DOI: 10.1002/ecs2.4732.

Chaput-Bardy, A., Lemaire, C., Picard, D. and Secondi, J. (2008) In-stream and overland dispersal across a river network influences gene flow in a freshwater insect, *Calopteryx splendens*. *Molecular Ecology* 17, 3496–3505. DOI: 10.1111/j.1365-294X.2008.03856.x.

Chin, K.S. and Taylor, P.D. (2009) Interactive effects of distance and matrix on the movements of a peatland dragonfly. *Ecography* 32, 714–722. DOI: 10.1111/j.1600-0587.2009.05744.x.

Chovanec, A., Fesl, C. and Kollar, H.P. (2003) Notes on the dragonfly community of a temporary pond near Vienna, Austria (Odonata). *Opuscula Zoologica Fluminensia* 215, 1–9.

Clark, T.E. and Samways, M.J. (1996) Dragonflies (Odonata) as indicators of biotope quality in the Kruger National Park, South Africa. *Journal of Applied Ecology* 33, 1001–1012.

Conrad, K.F., Willson, K.H., Whitfield, K., Harvey, I.F., Thomas, C.J. and Sherratt, T.N. (2002) Characteristics of dispersing *Ischnura elegans* and *Coenagrion puella* (Odonata): age, sex, size, morph and ecoparasitism. *Ecography* 25, 439–445.

Cooper, I.A., Brown, J.M. and Getty, T. (2016) A role for ecology in the evolution of colour variation and sexual dimorphism in Hawaiian damselflies. *Journal of Evolutionary Biology* 29, 418–427. DOI: 10.1111/jeb.12796.

Corbet, P.S. (2000) The first recorded arrival of *Anax junius*, Drury (Anisoptera: Aeshnidae) in Europe: a scientist's perspective. *International Journal of Odonatology* 3, 153–162. DOI: 10.1080/13887890.2000.9748146.

Corbet, P.S. (2004) *Dragonflies: Behavior and Ecology of Odonata*, 2nd ed. Cornell University Press, Ithaca, NY, USA.

Corbet, P.S., Suhling, F. and Soendgerath, D. (2006) Voltinism of Odonata: a review. *International Journal of Odonatology* 9, 1–44.

Córdoba-Aguilar, A. and Munguía-Steyer, R. (2013) The sicker sex: understanding male biases in parasitic infection, resource allocation and fitness. *PLoS ONE* 8, e76246. DOI: 10.1371/journal.pone.0076246.

Córdoba-Aguilar, A. and Munguía-Steyer, R. (2015) To be or not to be? Mating success and survival trade-offs when switching between alternative reproductive strategies. *Journal of Evolutionary Biology* 28, 2119–2124. DOI: 10.1111/jeb.12721.

Córdoba-Aguilar, A., Salamanca-Ocaña, J.C. and Lopezaraiza, M. (2003) Female reproductive decisions and parasite burden in a calopterygid damselfly (Insecta: Odonata). *Animal Behaviour* 66, 81–87. DOI 10.1006/anbe.2003.2198.

Corser, J.D., White, E.L. and Schlesinger, M.D. (2015) Adult activity and temperature preference drive region-wide damselfly (Zygoptera) distributions under a warming climate. *Biology Letters* 11, e20150001. DOI: 10.1098/rsbl.2015.0001.

Cummins, K.W. and Klug, M.J. (1979) Feeding ecology of stream invertebrates. *Annual Review of Ecology and Systematics* 10, 147–172.

da Silva, G.G., Poulin, R. and Guillermo-Ferreira, R. (2021) Do latitudinal and bioclimatic gradients drive parasitism on Odonata? *International Journal for Parasitology* 51, 463–470. DOI; 10.1016/j.ijpara.2020.11.008.

da Silva, N.A.P., Frizzas, M.R. and de Oliveira, C.M. (2011) Seasonality in insect abundance in the "Cerrado" of Goiás State, Brazil. *Revista Brasiliera de Entomologia* 55, 79–87. DOI: 10.1590/S0085-56262011000100013.

Dalzochio, M.S., Périco, E., Dametto. N. and Sahlén, G. (2020) Rapid functional traits turnover in boreal dragonfly communities (Odonata). *Scientific Reports* 10, e15411. DOI: 10.1038/s41598-020-71685-5.

Damm, S., Dijkstra, K.-D.B. and Hadrys, H. (2010b) Red drifters and dark residents: the phylogeny and ecology of a Plio-Pleistocene dragonfly radiation reflects Africa's changing environment (Odonata, Libellulidae, *Trithemis*). *Molecular Phylogenetics and Evolution* 54, 870–882. DOI: 10.1016/j.ympev.2009.12.006.

Damm, S., Schierwater, B. and Hadrys, H. (2010a) An integrative approach to species discovery in odonates: from character-based DNA barcoding to ecology. *Molecular Ecology* 19, 3881–3893. DOI: 10.1111/j.1365-294X.2010.04720.x.

Dayakrishna, M.K.A. (2015) Study on the abundance and diversity of dragonflies and damselflies (Insecta: Odonata) of Corbett Tiger Reserve, Uttarakhand, India. *Journal of Entomology and Zoology Studies* 3, 467–472.

Dayaram, A., Potter, K.A., Pailes, R., Marinov, M., Rosenstein, D.D. and Varsani, A. (2015) Identification of diverse circular single-stranded DNA viruses in adult dragonflies and damselflies (Insecta: Odonata) of Arizona and Oklahoma, USA. *Infection, Genetics and Evolution* 30, 278–287. DOI: 10.1016/j.meegid.2014.12.037.

De Block, M. and Stoks, R. (2003) Adaptive sex-specific life history plasticity to temperature and photoperiod in a damselfly. *Journal of Evolutionary Biology* 16, 986–995. DOI: 10.1046/j.1420-9101.2003.00581.x.

De Block, M. and Stoks, R. (2004) Life history responses depend on timing of cannibalism in a damselfly. *Freshwater Biology* 49, 775–786. DOI: 10.1111/j.1365-2427.2004.01224.x.

De Block, M. and Stoks, R. (2007) Flight-related body morphology shapes mating success in a damselfly. *Animal Behaviour* 74, 1093–1098. DOI: 10.1016/j.anbehav.2007.01.023.

De Block, M. and Stoks, R. (2008a) Short-term larval food stress and associated compensatory growth reduce adult immune function in a damselfly. *Ecological Entomology* 33, 796–801. DOI: 10.1111/j.1365-2311.2008.01024.x.

De Block, M. and Stoks, R. (2008b) Compensatory growth and oxidative stress in a damselfly. *Proceedings of the Royal Society B* 275, 781–785. DOI: 10.1098/rspb.2007.1515.

De Marco Júnior, P. and Furieri, K.S. (2000) Ecology of *Leptagrion perlongum* Calvert, 1909: a bromeliad-dweller odonate species. *Boletim do Museu de Biologia Mello Leitão: Nova Série* 11/12, 135–148.

De Marco Júnior, P., Batista, J.D. and Cabette, H.S.R. (2015) Community assembly of adult odonates in tropical streams: an ecophysiological hypothesis. *PLoS ONE* 10, e0123023. DOI: 10.1371/journal.pone.0123023.

De Souza, M.M., Pires, E.P., Brunismann, Â.G., Milani, L.R. and Pinto, Â.P. (2017) Dragonflies and damselflies (Odonata) from the wetland of the Rio Pandeiros, northern region of Minas Gerais State, Brazil, with a description of the male of *Archaeogomphus vanbrinki* Machado (Anisoptera: Gomphidae). *International Journal of Odonatology* 20, 13–26. DOI: 10.1080/13887890.2017.1281848.

Deacon, C. and Samways, M.J. (2016) Conservation of a phenomenon: rapid, reversible colour change in both sexes of one of the world's most threatened damselflies. *Journal of Insect Conservation* 20, 497–504. DOI: 10.1007/s10841-016-9882-x.

Deacon, C. and Samways, M.J. (2017) Conservation planning for the extraordinary and Endangered Spesbona damselfly. *Journal of Insect Conservation* 21, 121–128. DOI: 10.1007/s10841-017-9960-8.

Deacon, C., Samways, M.J. and Pryke, J.S. (2019) Aquatic insect decline in abundance and occupy low-quality artificial habitats to survive hydrological droughts. *Freshwater Biology* 64, 1643–1654. DOI: 10.1111/fwb.13360.

Deacon, C., Samways, M.J. and Pryke, J.S. (2021) Relative importance of ecological versus biological traits in driving range sizes of African dragonflies. *Journal of Biogeography* 48, 1309–1321. DOI: 10.1111/jbi.14077.

Deb, R., Nair, A. and Agashe, D. (2019) Host dietary specialization and neutral assembly shape gut bacterial communities of wild dragonflies. *PeerJ* 7, e8058. DOI: 10.7717/peerj.8058.

Debecker, S., Sommaruga, R., Maes, T. and Stoks, R. (2015) Larval UV exposure impairs adult immune function through a trade-off with larval investment in cuticular melanin. *Functional Ecology* 29, 1292–1299. DOI: 10.1111/1365-2435.12435.

Denis, A.S., Payet, O., Danflous, S., Gouix, N., Santoul, F. *et al.* (2018) Intraspecific variability of the phenology and morphology of three protected dragonflies between natural and artificial habitats. *Journal of Insect Conservation* 22, 419–431. DOI: 10.1007/s10841-018-0070-z.

Devaud, M. and Lebouvier, M. (2019) First record of *Pantala flavescens* (Anisoptera: Libellulidae) from the remote Amsterdam Island, southern Indian Ocean. *Polar Biology* 42, 1041–1046. DOI: 10.1007/s00300-019-02479-3.

Diamond, S.E. (2018) Contemporary climate-driven range shifts: putting evolution back on the table. *Functional Ecology* 32, 1652–1665. DOI: 10.1111/1365-2435.13095.

Díaz, A., Rodríguez-Flores, D.J., Rivera-Vega, A.G., Cruz-Marcano, E., Velázquez-Oliveras, I.M. and Villanueva-Cubero, L.A. (2019) Ecological factors and the distribution of adult Odonata in Puerto Rico. *Life: The Excitement of Biology* 7, 115–147. DOI: 10.9784/LEB7(3)Diaz.01.

Dingemanse, N.J. and Kalkman, V.J. (2008) Changing temperature regimes have advanced the phenology of Odonata in the Netherlands. *Ecological Entomology* 33, 394–402. DOI: 10.1111/j.1365-2311.2007.00982.x.

Doi, H. (2008) Delayed phenological timing of dragonflies emerging in Japan over five decades. *Biology Letters* 4, 388–391. DOI: 10.1098/rsbl.2008.0234.

Dolný, A., Harabiš, F. and Mižičova, H. (2014) Home range, movement, and distribution patterns of the threatened dragonfly *Sympetrum depressiusculum* (Odonata: Libellulidae): a thousand times greater territory to protect? *PLoS ONE* 9, e100408. DOI: 10.1371/journal.pone.0100408.

Endersby, I. (2014) Additional records for Victorian dragonflies (Insecta: Odonata). *Victorian Entomologist* 44(3), 58–64.

Estrada, A., Morales-Castilla, I, Caplat, P. and Early, R. (2016) Usefulness of species traits in predicting range shifts. *Trends in Ecology and Evolution* 31, 190–203. DOI: 10.1016/j.tree.2015.12.014.

Evans, M. and Lampo, M. (1996) Diet of *Bufo marinus*. *Journal of Herpetology* 30, 73–76.

Farkas, A., Jakab, T., Tóth, A., Kalmár, A.F. and Dévai, G. (2012) Emergence patterns of riverine dragonflies (Odonata: Gomphidae) in Hungary: variations between habitats and years. *Aquatic Insects* 34, 77–89. DOI: 10.1080/01650424.2012.643030.

Farkas, A., Jakab, T., Müller, O., Móra, A., Lejter, I. and Dévai, G. (2013) Sex ratio in Gomphidae (Odonata) at emergence: is there a relationship with water temperature? *International Journal of Odonatology* 16, 279–287. DOI: 10.1080/13887890.2013.825937.

Farooq, F. and Arya, M.K. (2018) An observation on the taxonomic composition, relative abundance and seasonal variation of odonate (dragonflies and damselflies) fauna of Chaubatia Garden Ranikhet, Almara, Uttarakhand, India. *Journal of Environmental and Bio-Sciences* 32, 253–257.

Fatauros, N.E., Cusumaro, A., Bin, F., Polaszek, A. and van Lenteren, J.C. (2020) How to escape from egg parasitoids: a review of potential factors explaining parasitoid absence across the Insecta. *Proceedings of the Royal Society B* 287, e20200344. DOI: 10.1098/rspb.2020.0344.

Feng, H.-Q., Wu, K.-M., Ni, Y.-X., Cheng, D.-F. and Guo, Y.-Y. (2006) Nocturnal migration of dragonflies over the Bohai Sea in northern China. *Ecological Entomology* 31, 511–520. DOI: 10.1111/j.1365-2311.2006.00813-x.

Ferguson, D.G., Marinov, M., Saxton, N.A., Rashni, B. and Bybee, S.M. (2023) Phylogeny and classification of *Nesobasis* Selys, 1891 and *Vanuatubasis* Ober & Staniczek, 2009 (Odonata: Coenagrionidae). *Insect Systematics & Evolution* 54, 1555–1572. DOI: 10.1163/1876312X-bja10049.

Fincke, O.M. (1992) Consequences of larval ecology for territoriality and reproductive success of a neotropical damselfly. *Ecology* 73, 449–462. DOI: 10.2307/1940752.

Fincke, O.M. (1997) Conflict resolution in the Odonata: implications for understanding female mating patterns and female mate choice. *Biological Journal of the Linnean Society* 60, 201–220. DOI: 10.1111/j.1095-8312.1997.tb01492.x.

Fischer, S., Marinone, M.C., Fontanarrosa, M.S., Mariela Nieves, M. and Schweigmann, N. (2000) Urban rain pools: seasonal dynamics and entomofauna in a park of Buenos Aires. *Hydrobiologia* 441, 45–53.

Flenner, I., Richter, O. and Suhling, F. (2010) Rising temperature and development in dragonfly populations at different latitudes. *Freshwater Biology* 55, 397–410. DOI: 10.1111/j.1365-2427.2009.02289.x.

Forbes, M.R., Muma, K.E. and Smith, B.P. (1999) Parasitism of *Sympetrum* dragonflies by *Arrenurus planus* mites: maintenance of resistance particular to one species. *International Journal for Parasitology* 29, 991–999. DOI: 101016/S0020-7519(99)00061-2.

Foster, S.E. and Soluk, D.A. (2006) Protecting more than a wetland: the importance of biased sex ratios and habitat segregation for conservation of Hine's emerald dragonfly, *Somatochlora hineana* Williamson. *Biological Conservation* 127, 158–166. DOI: 10.1016/j.biocon.2005.08.006.

Frances, D.N., Moon, J.Y. and McCauley, S.J. (2017) Effects of environmental warming during early life history on libellulid odonates. *Canadian Journal of Zoology* 95, 373–382. DOI:10.1139/cjz-2016-0233.

Freeland, J.R., May, M., Lodge, R. and Conrad, K.F. (2003) Genetic diversity and widespread haplotypes in a migratory dragonfly, the common green darner *Anax junius*. *Ecological Entomology* 28, 413–421. DOI: 10.1046/j.1365-2311.2003.00521.x.

French, S.K. and McCauley, S.J. (2018) Canopy cover affects habitat selection by adult dragonflies. *Hydrobiologia* 818, 129–143. DOI: 10.1007/s10750-018-3600-5.

French, S.K. and McCauley, S.J. (2019) The movement responses of three libellulid dragonfly species to open and closed landscape cover. *Insect Conservation and Diversity* 12, 437–447. DOI: 10.1111/icad.12355.

Fujita, K., Hirano, K., Kawanishi, M., Ohsaki, N., Ohtaishi, M. *et al.* (1978) Ecological studies on a dragonfly, *Nannophya pygmaea* Ramber (Odonata: Libellulidae) I. Seasonal changes of adult population and its distribution in a habitat. *Population Ecology* 19, 209–221. DOI: 10.1007/BF02518828.

Fulan, J.A., Raimundo, R. and Figueiredo, D. (2008) Habitat characteristics and dragonfly (Odonata) diversity and abundance in the Guadiana River, eastern of the Alentejo, Portugal. *Boletín de la Asociacíon Española de Entomologia* 32, 327–340.

Gibbons, L.H., Reed, J.M. and Chew, F.S. (2002) Habitat requirements and local persistence of three damselfly species (Odonata: Coenagrionidae). *Journal of Insect Conservation* 6, 47–55. DOI: 10.1023/A:1015754015603.

Golab, M.J., Sniegula, S. and Brodin, T. (2022) Cross-latitude behavioural axis in an adult damselfly *Calopteryx splendens* (Harris, 1780). *Insects* 13, e342. DOI: 10.3390/insects13040342.

Gosden T.P., Stoks, R. and Svensson, E.I. (2011) Range limits, large-scale biogeographic variation, and localized evolutionary dynamics in a polymorphic damselfly. *Biological Journal of the Linnean Society* 102, 775–785.

Grant, P.B.C. and Samways, M.J. (2007) Ectoparasitic mites infest common and widespread but not rare and red-listed dragonfly species. *Odonatologica* 36, 255–262.

Grewe, Y., Hof, C., Dehling, D.M., Brandl, R., and Brändle, M. (2013) Recent range shifts of European dragonflies provide support for an inverse relationship between habitat predictability and dispersal. *Global Ecology and Biogeography* 22, 403–409. DOI: 10.1111/geb.12004.

Guillermo-Ferreira, R. and Del-Claro, K. (2011) Oviposition site selection in *Oxyagrion microstigma* Selys, 1876 (Odonata: Coenagrionidae) is related to aquatic vegetation structure. *International Journal of Odonatology* 14, 275–279. DOI: 10.1080/13887890.2011.621109.

Harabiš, F. (2016) The value of terrestrial ecotones as refuges for winter damselflies (Odonata: Lestidae). *Journal of Insect Conservation* 20, 971–977. DOI: 10.1007/s10841-016-9929-z.

Harabiš, F. and Dolný, A. (2010) Ecological factors determining the density-distribution of Central European dragonflies (Odonata). *European Journal of Entomology* 107, 571–577.

Harabiš, F., Dolný, A. and Šipoš, J. (2012) Enigmatic adult overwintering in damselflies: coexistence as weaker intraguild competitors due to niche separation in time. *Population Ecology* 54, 549–556. DOI: 10.1007/s10144-012-0331-8.

Harabiš, F., Šigut, M. and Dolný, A. (2013) Density-distribution patterns of egg parasitoids in freshwater habitats. *Časopis Slezského zemského muzea Opava (A)* 62, 65–72. DOI: 10.2478/cszma-2013-0006.

Harabiš, F., Rusková, T. and Dolný, A. (2019) Different oviposition strategies of closely related damselfly species as an effective defense against parasitoids. *Insects* 10, e26. DOI: 10.3390/insects10010026.

Hardersen, S. (2007) Telemetry of Anisoptera after emergence: first results (Odonata). *International Journal of Odonatology* 10, 189–202. DOI: 10.1080/13887890.2007.9748299.
Hardersen, S. (2010) Seasonal variation of wing spot allometry in *Calopteryx splendens* (Odonata: Calopterygidae). *Ethology, Ecology and Evolution* 22, 365–373. DOI: 10.1080/03949370.2010.510042.
Hassall, C. (2015) Odonata as candidate barometers of global climate change. *Freshwater Science* 34, 1040–1049. DOI: 10.1086/682210.
Hassall, C., Thompson, D.J., French, G.C. and Harvey, I.F. (2007) Historical changes in the phenology of British Odonata are related to climate. *Global Change Biology* 13, 933–941. DOI: 10.1111/j.1365-2486.2007.01318.x.
Hassall, C., Thompson, D.J. and Harvey, I.F. (2008) Latitudinal variation in two sympatric damselfly species with contrasting range dynamics (Odonata: Coenagrionidae). *European Journal of Entomology* 105, 939–944.
Hassall, C., Thompson, D.J. and Harvey, I.F. (2009) Variation in morphology between core and marginal populations of three British damselflies. *Aquatic Insects* 31, 187–197. DOI: 10.1080/01650420902776708.
Hassall, C., Keat, S., Thompson, D.J and Watts, P.C. (2014) Bergmann's rule is maintained during a rapid range expansion in a damselfly. *Global Change Biology* 20, 475–482. DOI: 10.1111/gcb.12340.
Hassan, A.T. (1976) Studies on the roosting behaviour of *Palpopleura lucia* (Drury) and *Acisoma panorpoides inflatum* Selys (Anisoptera: Libellulidae). *Odonatologica* 5, 323–329.
Hecker, K.R., Forbes, M.R. and Léonard, N.J. (2002) Parasitism of damselflies (*Enallagma boreale*) by gregarines: sex biases and relations to adult survivorship. *Canadian Journal of Zoology* 80, 162–168.
Hedlund, J.S.U., Lv, H., Lehmann, P., Hu, G., Anderson, R.C. and Chapman, J.W. (2021) Unraveling the world's longest non-stop migration: the Indian Ocean crossing of the Globe Skimmer dragonfly. *Frontiers in Ecology and Evolution* 9, e698128. DOI: 10.3389/fevo.2021.698128.
Hendriks, J.A., Mariaty, Maimunah, S. Anirudh, N.B., Holly, B.A., Erkens, R.H.J. and Harrison, M.E. (2023) Odonata (Insecta) communities in a lowland mixed mosaic forest in Central Kalimantan, Indonesia. *Ecologies* 4, 55–73. DOI: 10.3390/ecologies4010006.
Henrikson, B.-I. (1988) The absence of antipredator behaviour on the larvae of *Leucorrhinia dubia* (Odonata) and the consequences for their distribution. *Oikos* 51, 179–183. DOI: 10.2307/3565640.
Herrmann, M. and Yampolsky, L.Y. (2021) False and true positives in arthropod thermal adaptation candidate gene lists. *Genetica* 149, 143–153. DOI: 10.1007/s10709-021-00122-w.
Herzog, J. and Martens, A. (2013) Larvae von *Aeshna* spec. (Odonata: Aeshnidae) als südlichen wasserschlauchs *Utricularia australis*. *Libellula* 32, 181–185.
Higashikawa, W., Yoshimura, M., Nagano, A.J. and Maeto, K. (2023) Conservation genomics of an endangered floodplain dragonfly, *Sympetrum pedemontanum elatum* (Selys), in Japan. *Conservation Genetics* (in press). DOI: 10.1007/s10592-023-01595-2.
Hobson, K.A., Anderson, R.C., Soto, D.X. and Wassenaar, L.I. (2012a) Isotopic evidence that dragonflies (*Pantala flavescens*) migrating through the Maldives come from the northern Indian subcontinent. *PLoS ONE* 7, e52594. DOI: 10.1371/journal.pone.0052594.
Hobson, K.A., Soto, D.X., Paulson, D.R., Wassenaar, L.I. and Matthews, J.H. (2012b) A dragonfly (δ^2H) isoscape for North America: a new tool for determining natal origins of migratory aquatic emergent insects. *Methods in Ecology and Evolution* 3, 766–772. DOI: 10.1111/j.2041-210X.2012.00202.x.
Hof, C., Brändle, M., Dehling, D.M., Munguía, M., Brandl, R. *et al.* (2012) Habitat stability affects dispersal and the ability to track climate change. *Biology Letters* 8, 639–643. DOI: 10.1098/rsbl.2012.0023.
Hofmann, T.A. and Mason, C.F (2005) Habitat characteristics and the distribution of Odonata in a lowland river catchment in eastern England. *Hydrobiologia* 539, 137–147. DOI: 10.1007/s10750-004-3916-1.
Holt, G., Dwyer, G.K., Bourke, C., Matthews, T.G., Macqueen, A. and Lester, R.E. (2021) Characteristics and consequences of a disease outbreak in aquatic insects. *Freshwater Biology* 66, 1276–1281. DOI: 10.1111/fwb.13715.
Holtmann, L., Brüggeshemke, J., Juchem, M. and Fartmann, T. (2019) Odonatae assemblages of urban stormwater ponds: the conservation value depends on pond type. *Journal of Insect Conservation* 23, 123–132. DOI: 10.1007/s10841-018-00121-x.
Holzmann, K.L., Charrier, C. and Johansson, F. (2022) Weak effects on growth and cannibalism under fluctuating temperatures in damselfly larvae. *Scientific Reports* 12, e12910. DOI: 10.1038/s41598-022-17192-1.
Hopper, K.R. (2001) Flexible antipredator behavior in a dragonfly species that coexists with different predator types. *Oikos* 93, 470–476.
Hopper, K.R., Crowley, P.H. and Kielman, D. (1996) Density dependence, hatching synchrony, and within-cohort cannibalism in young dragonfly larvae. *Ecology* 77, 191–200. DOI: 10.2307/2265668.
Hykel, M., Růžicková, J. and Dolny, A. (2020) Perch selection in *Sympetrum* species (Odonata: Libellulidae): importance of vegetation structure and composition. *Ecological Entomology* 45, 90–96. DOI: 10.1111/een.12778.
Ilvonen, S., Ilvonen, J.J., Kaunisto, K.M., Krams, I. and Suhonen, J. (2011) Can infection by eugregaraine parasites mediate species coexistence in *Calopteryx* damselflies? *Ecological Entomology* 36, 582–587. DOI: 10.1111/j.1365-2311.2011.01307.x.
Iwamoto, H., Tahara, D. and Yoshida, T. (2022) Contrasting metacommunity patterns of fish and aquatic insects in

drainage ditches of paddy fields. *Ecological Research* 37, 635–646. DOI: 10.1111/1440-1703.12334.

Jakob, C. and Suhling, F. (1999) Risky times? Mortality during emergence in two species of dragonflies (Odonata: Gomphidae, Libellulidae). *Aquatic Insects* 21, 1–10. DOI: 10.1076/aqin.21.1.1.4537.

Jödicke, R. (2003) Mid-winter occurrence of dragonflies in southern Tunisia (Insecta: Odonata). *Darmstädter Beiträge zur Naturgeschichte* 12, 119–128.

Johansson, F. (2003) Latitudinal shifts in body size of *Enallagma cyathigerum* (Odonata). *Journal of Biogeography* 30, 29–34.

Johansson, F. and Brodin, T. (2003) Effects of fish predators and abiotic factors on dragonfly community structure. *Journal of Freshwater Ecology* 18, 415–423. DOI: 10.1080/02705060.2003.9663977.

Johansson, F. and Mikolajewski, D.J. (2008) Evolution of morphological defenses. In: Córdoba-Aguilar (ed.) *Dragonflies & Damselflies: Model Organisms for Ecology and Evolution* (pp. 127–137). Oxford University Press, Oxford.

Johansson, F. and Suhling, F. (2004) Behaviour and growth of dragonfly larvae along a permanent to temporary water habitat gradient. *Ecological Entomology* 29, 196–202.

Johansson, F., Englund, G., Brodin, T. and Gardfjell, H. (2006) Species abundance models and patterns in dragonfly communities: effects of fish predators. *Oikos* 114, 27–36. DOI: 10.1111/j.2006.0030-1299.14495.x.

Johansson, F., Śniegula, S. and Brodin, T. (2010) Emergence patterns and latitudinal adaptations on development time of Odonata in north Sweden and Poland. *Odonatologica* 39, 97–106.

Johansson, F., Hedström, T.K., Divakaran, P.K. and Kakkassery, F.K. (2022) Wing shape differences along a migration route of the long-distance migrant Globe Skimmer *Pantala flavescens*. *Journal of Tropical Ecology* 38, 17–24. DOI: 10.1017/S0266467421000444.

Johansson, F., Berger, D., Outomuro, D., Sniegula, S., Tunon, M. *et al.* (2023) Mixed support for an alignment between phenotypic plasticity and genetic differentiation in damselfly wing shape. *Journal of Evolutionary Biology* 36, 368–380. DOI: 10.1111/jeb.14145.

Johnston, P.R., Mikolajewski, D.J. and Rolff, J. (2015) Identification of viruses associated with larvae of the dragonfly *Leucorrhinia dubia* and damselfly *Coenagrion puella* from RNA sequencing data. *International Journal of Odonatology* 18, 81–88. DOI: 10.1080/13887890.2015.1018345.

Jonsen, I. and Taylor, P.D. (2000) *Calopteryx* damselfly dispersions arising from multiscale responses to landscape structure. *Conservation Ecology* 42, e4. Available at: http:/www.consecol.org/vol4/iss2/art4/ (accessed 21 April 2024).

Jooste, M.L., Samways, M.J. and Deacon, C. (2020) Fluctuating pond water levels and aquatic insect persistence in a drought-prone Mediterranean-type climate. *Hydrobiologia* 847, 1315–1326. DOI: 10.1007/s10750-020-04186-1.

Jouault, C., Tischlinger, H., Henrotay, M. and Nel, A. (2022) Wing coloration patterns in the Early Jurassic dragonflies as potential indicator of increasing predation pressure from insectivorous reptiles. *Palaeoentomology* 5, 305–318. DOI: 10.11646/palaeoentomology.5.4.3.

Juen, L. and De Marco Jr, P. (2011) Odonate biodiversity in terra-firme forest streamlets in Central Amazonia: on the relative effects of neutral and niche drivers at small geographical extents. *Insect Conservation and Diversity* 4, 265–274. DOI: 10.1111/j.1752-4598.2010.00130.x.

Juen, L., de Oliveira-Júnior, J.M.B., Shimano, Y., Mendes, T.P. and Cabette, H.S.R. (2014) Composition and richness of Odonata (Insecta) in streams with different levels of conservation in a Cerrado-Amazonian forest ecotone. *Acta Amazonica* 44, 223–233. DOI: 10.1590/S0044-59672014000200008.

Kadoya, T., Suda, S-i., Tsubaki, Y. and Washitani, I. (2008) The sensitivity of dragonflies to landscape structure differs between life-history groups. *Landscape Ecology* 23, 149–158. DOI: 10.1007/s10980-007-9151-1.

Kaunisto, K.M., Viitaniemi, H.M., Leder, E.H. and Suhonen, J. (2013) Association between host's genetic diversity and parasite burden in damselflies. *Journal of Evolutionary Biology* 26, 1784–1789. DOI: 10.1111/jeb.12177.

Kaunisto, K.M., Kaunisto, P., Ilvonen, J.J. and Suhonen, J. (2017a) Parasitism immune response, and egg production of the spearhead bluet (*Ceriagrion hastulatum*) damselfly. *Canadian Journal of Zoology* 95, 367–372. DOI: 10.1139/cjz-2016-0146.

Kaunisto, K.M., Roslin, T., Sääksjärvi, I.E. and Vesterin, E.J. (2017b) Pellets of proof: first glimpse of the dietary composition of adult odonates as revealed by metabarcoding of feces. *Ecology and Evolution* 7, 8588–8598. DOI: 10.1002/ece3.3404.

Kaunisto, K.M., Roslin, T., Forbes, M.R., Morrill, A., Sääksjärvi, I.E. *et al.* (2020) Threats from the air: damselfly predation on diverse prey taxa. *Journal of Animal Ecology* 89, 1365–1374. DOI: 10.1111/1365-2656.13184.

Keller, D. and Holderegger, R. (2013) Damselflies use different movement strategies for short- and long-distance dispersal. *Insect Conservation and Diversity* 6, 590–597. DOI: 10.1111/icad.12016.

Keller, D., Brodbeck, S., Flöss, I., Vonwil, G. and Holderegger, R. (2010) Ecological and genetic measurements of dispersal in a threatened dragonfly. *Biological Conservation* 143, 2658–2663. DOI: 10.1016/j.biocon.2010.07.008.

Khelifa, R. (2017) Partial bivoltinism and emergence patterns in the North African endemic damselfly *Calopteryx exul*: conservation implications. *African Journal of Ecology* 55, 145–151. DOI: 10.1111/aje.12332.

Khelifa, R., Zebsa, R., Amari, H. and Mellal, M.K. (2013) Does wind affect emergence site selection in Odonata? *African Entomology* 21, 383–387. DOI: 10.4001/003.021.0213.

Khelifa, R., Zebsa, R., Amari, H., Mellal, M.K. and Mahdjoub, H. (2019) Field estimates of fitness costs of the pace-of-life in an endangered damselfly. *Journal of Evolutionary Biology* 32, 943–954. DOI: 10.1111/jeb.13493.

Kietzka, G.J., Pryke, J.S. and Samways, M.J. (2017) Aerial dragonflies are highly sensitive to in-water conditions across an ancient landscape. *Diversity and Distributions* 23, 14–26. DOI: 10.1111/ddi.12493.

Kietzka, G.J., Pryke, J.S., Gaigher, R. and Samways, M.J. (2021) 32 years of essential management to retain value of an urban dragonfly awareness pond. *Urban Ecosystems* 24, 1295–1304. DOI: 10.1007/s11252-021-01115-5.

Kirkton, S.D. and Schultz, T.D. (2001) Age-specific behavior and habitat selection of adult male damselflies, *Calopteryx maculata* (Odonata: Calopterygidae). *Journal of Insect Behavior* 14, 545–556. DOI: 101023/A:1011180207852.

Knight, S.M., Pitman, G.M., Flockhart, D.T.T. and Norris, D.R. (2019) Radio-tracking reveals how wind and temperature influence pace of daytime insect migration. *Biology Letters* 15, e20190327. DOI: 10.1098/rsbl.2019.0327.

Knight, T.M., McCoy, M.W., Chase, J.M., McCoy, K.A. and Holt, R.D. (2005) Trophic cascades across ecosystems. *Nature* 437, 880–883. DOI: 10.1038/nature03962.

Koprivnikar, J., Urichuk, T.M.Y. and Szuroczki, D. (2017) Influences of habitat and arthropod density on parasitism in two co-occurring host taxa. *Canadian Journal of Zoology* 95, 589–597. DOI: 10.1039/cjz-2016-0210.

Korkeamäki, E. and Suhonen, J. (2002) Distribution and habitat specialization of species affect local extinction in dragonfly Odonata populations. *Ecography* 25, 459–465.

Kortello, A.D. and Ham, S.J. (2010) Movement and habitat selection by *Argia vivida* (Hagen) (Odonata, Coenagrionidae) in fuel-modified forest. *Journal of Insect Conservation* 14, 133–140. DOI: 10.1007/s10841-009-9233-2.

Kuchta, S.R. and Svensson, E.I. (2014) Predator-mediated natural selection on the wings of the damselfly *Calopteryx splendens*: differences in selection among trait types. *The American Naturalist* 184, 91–109. DOI: 10.1086/676043.

La Porta, G. and Goretti, E. (2020) Movement and demography of Southern damselfly (*Coenagrion mercuriale*, Odonata) in a Mediterranean lotic system. *Ethology Ecology and Evolution* 32, 107–121. DOI: 10.1080/03949370.2019.1680446.

La Porta, G., Landi, F., Leandri, F. and Assandri, G. (2023) The new checklist of the Italian fauna: Odonata. *Biogeographia – The Journal of Integrative Biogeography* 38, uc1009. DOI: 10.21426/B638158781.

Lalejini, A., Ferguson, A.J., Grant, N.A. and Ofria, C. (2021) Adaptive phenotypic plasticity stabilizes evolution in a fluctuating environment. *Frontiers in Ecology and Evolution* 9, e715381. DOI: 10.3389/fevo.2021.715381.

Lambret, P., Rutter, I., Grillas, P. and Stoks, R. (2018) Oviposition plant choice maximizes offspring fitness in an aquatic predatory insect. *Hydrobiologia* 823, 1–12. DOI: 10.1007/s10750-018-3663-3.

Lambret, P., Jeanmougin, M. and Stoks, R. (2023) Factors driving larval abundance and adult size of the threatened *Lestes macrostigma* (Odonata): keys for water management and habitat restoration. *Journal of Insect Conservation* 27, 389–402. DOI: 10.1007/s10841-023-00461-3.

Larsson, J.I.R. (1990) On the cytology and taxonomic position of *Nudispora biformis* N. G., N. Sp. (Microspora, Thelohaniidae), a microsporidian parasite of the dragonfly *Coenagrion hastulatum* in Sweden. *The Journal of Protozoology*, 37, 310–318. DOI: 10.1111/j.1550-7408.1990.tb01151.

Le Gall, M., Chaput-Bardy, A. and Husté, A. (2017) Context-dependent local movements of the blue-tailed damselfly, *Ischnura elegans*: effects of pond characteristics and the landscape matrix. *Journal of Insect Conservation* 21, 243–256. DOI: 10.1007/s10841-017-9971-5.

Lee, C.Y., Kim, M.K. and Kim, D.-G. (2022) Ecological responses of *Nannophya koreana* (Odonata: Libellulidae) to temperature: following converse Bergmann's Rule. *Biology* 11, e830. DOI: 10.3390/biology11060830.

Leite, L.M.S.N., French, S.K., Searcy, C.A. and McCauley, S.J. (2021) Forest edges and their effects on the arrival of dragonflies at north-temperate experimental ponds. *International Journal of Odonatology* 24, 38–50. DOI: 10.23797/2159-6719_24_2.

LeRoy Poff, N., Pyne, M.I., Bledsoe, B.P., Cuhaciyan, C.C. and Carlisle, D.M. (2010) Developing linkages between species traits and multiscaled environmental variation to explore vulnerability of stream benthic communities to climate change. *Journal of the North American Benthological Society* 29, 1441–1458. DOI: 10.1899/10-030.1.

Li, B., Lan, Z., Guo, X.-R., Zhang, A.-H., Wei, W. *et al.* (2023) Survey of the *Prosthogonimus* spp. metacercariae infection in the second intermediate host dragonfly in Heilongjiang Province, China. *Parasitology Research* 122, 2859–2870. DOI: 10.1007/s00436-023-07975-4.

Lim, S.-h., Park, J.-K., Park. W.-B., Won, D.-H. Kim, M.S. *et al.* (2023) Gut microbiome of three species of Odonata. *Entomological Research* 53, 167–172. DOI: 10.1111/1748-5967.12641.

Low, V., Rashid, Y., Yusoff, A., Siow, W., Prakash, B. *et al.* (2017) Pleistocene and high gene flow in the Globe skimmer dragonfly *Pantala flavescens* Fabricius (Odonata: Libellulidae) in Peninsula Malaysia.

Zoologischer Anzeiger 266, 23–27. DOI: 10.1016/j.jcz.2016.10.002.
Mackintosh, L.A. and Robson, B.J. (2023) Seasonal drying influences odonate adults and nymphs in wetlands in an urban Mediterranean-climate landscape. *Austral Ecology* 48, 1383–1404. DOI: 10.1111/aec.13384.
Maclean, S.A. and Beissinger, S.R. (2017) Species' traits as predictors of range shifts under contemporary climate change: a review and meta-analysis. *Global Change Biology* 23, 4094–4105. DOI: 10.1111/gcb.13736.
Mahdjoub, H., Zebsa, R., Amari, H., Bensouilah, S., Zouaimia A. *et al.* (2023a) Potential fitness consequences of roosting spatiotemporal selection in an endangered endemic damselfly: conservation implications. *Frontiers in Ecology and Evolution* 11, e1267705. DOI: 10.3389/fevo.2023.1267705.
Mahdjoub, H., Zebsa, R., Kahalerras, A., Amari, H., Bensouilah, S. *et al.* (2023b) Condition-dependent survival and movement behavior in an endangered endemic damselfly. *Scientific Reports* 13, e21819. DOI: 10.1038/s41598-023-48162-w.
Mähn, L.A., Hof, C., Brandl, R. and Pinkert, S. (2023) Beyond latitude: temperature, productivity, and thermal niche conservatism drive body size variation in Odonata. *Global Ecology and Biogeography* 32, 656–667. DOI: 10.1111/geb.13661.
Maimela, L.T., Chimimba, C.T. and Zengeya, T.A. (2022), Niche expansion by indigenous fish species following the introduction of rainbow trout (*Oncorhynchus mykiss*) in a subtropical river system, upper Blyde River, South Africa. *Frontiers in Environmental Science* 10, e949353. DOI: 10.3389/fenvs.2022.949353.
Mallick, M.A.I. and Mondal, A. (2024) A preliminary assessment: seasonal variations of damselfly and dragonfly abundance in Serampore, Jolkol, West Bengal, India. *Biodiversity* 25, 11–16. DOI: 10.1080/14888386.2023.2271446.
Manasov, S. and Suhling, F. (2020) Hiding among traps? Mortality of early instar odonate larvae in the presence of bladderwort plants. *International Journal of Odonatology* 23, 103–111. DOI: 10. 1080/13887890.2020.1718784.
Manger, R. and Dingemanse, N.J. (2009) Adult survival of *Sympecma paedisca* (Brauer) during hibernation (Zygoptera: Lestidae). *Odonatologica* 38, 55–59.
Marden, J.H. and Cobb, J.R. (2004) Territorial and mating success of dragonflies that vary in muscle power output and presence of gregarine gut parasites. *Animal Behaviour* 68, 857–865. DOI: 10.1016/j.anbehav.2003.09.019.
Martens, A. (2001) Initial preference of oviposition sites: discrimination between living and dead plant material in *Sympecma fusca* and *Ceriagrion caerulescens* (Odonata: Lestidae, Coenagrionidae). *European Journal of Entomology* 98, 121–123.
Martens, A. and Grabow, K. (2011). Early stadium dragonfly larvae (Odonata: Coenagrionidae) as prey of an aquatic plant. *International Journal of Odonatology* 14, 101–104. DOI: 10.1080/13887890.2011.568191.
Martín, R. and Maynou, X. (2021) *Paratanytarsus* sp. (Diptera: Chironomidae) as commensal on larval *Calopteryx* (Odonata: Calopterygidae). *Notulae odonatologicae* 9, 367–371.
Mashkova, I.R., Krupnova, T.G., Kostryukova, A.M. and Vlasov, N.E. (2018) Distribution of dragonflies (Odonata: Insecta) in South Ural lakes. *Biodiversitas* 19, 202–207. DOI: 10.13057/biodiv/d190127.
May, M.L. (2013) A critical overview of progress in studies on migration of dragonflies (Odonata: Anisoptera), with emphasis on North America. *Journal of Insect Conservation* 17, 1–15. DOI: 10.1007/s10841-012-9540-x.
McCauley, S. (2005) Differential dispersal propensities between individuals in male *Leucorrhinia intacta* (Odonata: Libellulidae). *International Journal of Odonatology* 8, 223–232. DOI: 10.1080/13887890.2005.9748254.
McCauley, S.J. (2006) The effects of dispersal and recruitment limit on community structure of odonates in artificial ponds. *Ecography* 29, 585–595.
McCauley, S.J. (2008) Slow, fast and in between: habitat distribution and behaviour of larvae in nine species of libellulid dragonfly. *Freshwater Biology* 53, 253–263. DOI: 10.1111/j.1365-2427.2007.01889.x.
McCauley, S.J. (2010) Body size and social dominance influence breeding dispersal in male *Pachydiplax longipennis* (Odonata). *Ecological Entomology* 35, 377–385. DOI: 10.1111/j.1365-2311.2010.01191.x.
McCauley, S.J. (2013) Relationship between morphology, dispersal and habitat distribution in three species of *Libellula* (Odonata: Anisoptera). *Aquatic Insects* 34, 195–204. DOI: 10.1080/01650424.2013.800557.
McCauley, S.J., Davis, C.J., Relyea, R.A., Yurewicz, K.L., Skelly, D.K. and Werner, E.E. (2008) Metacommunity patterns in larval odonates. *Oecologia* 158, 329–342. DOI: 10.1007/s00442-008-1141-8.
McCauley, S.J., Davis, C.J., Werner, E.E. and Robeson, M.S. (2014) Dispersal, niche breadth and population extinction: colonization ratios predict range size in North American dragonflies. *Journal of Animal Ecology* 83, 858–865. DOI: 10.1111/1365-2656.12181.
McCrae, A.W.R. and Corbet, P.S. (1982) Oviposition behaviour of *Tetrathemis polleni* (Selys): a possible adaptation to life in turbid pools (Anisoptera: Libellulidae). *Odonatologica* 11, 23–31.
McPeek, M.A. (1990) Behavioral differences between *Enallagma* species (Odonata) influencing differential vulnerabilities to predators. *Ecology* 71,1714–1726. DOI: 10.2307/1937580.
McPeek, M.A. (1995) Morphological evolution mediated by behavior in the damselflies of two communities. *Evolution* 49, 749–769. DOI: 10.1111/j.1558-5646.1995.tb02311.x.

Medina-Espinoza, E.F. (2022) Abundance of Odonata in different microhabitats at an oxbow lake in the Peruvian Amazon. *Acta Amazonica* 52, 236–240. DOI: 10.1590/1809-4392202200161.

Mehmood, S.A., Zia, A., Ahmed, S., Panhwar, W.A., Khan, W. *et al*. (2020) Seasonal abundance and distribution of dragonflies in upper Siran Valley of District Mansehra Pakistan. *Brazilian Journal of Biology* 81, 785–791. DOI: 10.1590/1519-6984.231538.

Mendes, T.P., Cabette, H.S.R. and Juen, L. (2015) Setting boundaries: environmental and spatial effects on Odonata larvae distribution (Insecta). *Anais da Academia Brasileira da Ciências* 87, 239–249. DOI: 10.1590/0001-3765201520130477.

Michiels, N.K. and Dhondt, A.A. (1990) Costs and benefits associated with oviposition site selection in the dragonfly *Sympetrum danae* (Odonata: Libellulidae). *Animal Behaviour* 40, 668–678. DOI: 10.1016/S0003-3472(05)80696-7.

Mikolajewski, D.J., De Block, M., Rolff, J., Johansson, F., Beckerman, A.P. and Stoks, R. (2010) Predator-driven trait diversification in a dragonfly genus: co-variation in behavioral and morphological antipredator defense. *Evolution* 64, 3327–3335. DOI: 10.1111/j.1558-5646.2010.01078.x.

Mikolajewski, D.J., Rüsen, L., Mauersberger, R., Johansson, F. and Rolff, J. (2015) Relaxed predation results in reduced phenotypic integration in a suite of dragonflies. *Journal of Evolutionary Biology* 28, 1354–1363. DOI: 10.1111/jeb.12658.

Minot, M. and Husté, A. (2022) Genetic diversity and structure of *Anax imperator* Leach, 1815 populations (Odonata: Aeshnidae) in ponds at regional and European scales. *Diversity* 14, e68. DOI: 10.3390/d14020068.

Minot, M., Le Gall, M. and Husté, A. (2019) Biometry of the large dragonfly *Anax imperator* (Odonata: Aeshnidae): a study of traits from larval development. *European Journal of Entomology* 116, 269–280. DOI: 10.14411/eje.2019.031.

Mlynarek, J.J., Iserbyt, A., Nagel, L. and Forbes, M.R. (2015) Differential water mite parasitism, phenoloxidase activity, and resistance to mites are unrelated across pairs of related damselfly species. *PLoS ONE* 10, e0115539. DOI: 10.1371/journal.pone.0115539.

Moore, M.P. (2021) Larval habitats impose trait-dependent limits on the direction and rate of adult evolution in dragonflies. *Biology Letters* 17, e20210023. DOI: 10.1098/rsbl.2021.0023.

Moore, M.P. and Martin, R.A. (2018) Trade-offs between larval survival and adult ornament development depend on predator regime in a territorial dragonfly. *Oecologia* 188, 97–106. DOI: 10.1007/s00442-018-4171-x.

Moore, M.P. and Martin, R.A. (2021) Natural selection on adults has trait-dependent consequences for juvenile evolution in dragonflies. *The American Naturalist* 197, 677–689. DOI: 10.1086/714048.

Moore, M.P., Lis, C., Ghergel, I. and Martin, R.A. (2019) Temperature shapes costs, benefits and geographical diversification of sexual coloration in a dragonfly. *Ecology Letters* 22, 437–446. DOI: 10.1111/ele.13200.

Moretti, M., Dias, A.T., de Bello, F., Altermatt, F., Chown, S.L. *et al*. (2017) Handbook of protocols for standardized measurement of terrestrial invertebrate functional traits. *Functional Ecology* 31, 558–567. DOI: 10.1111/1365-2435.

Morin, P.J. (1984) Odonate guild composition: experiments with colonization history and fish predation. *Ecology* 65, 1866–1873.

Morrill, A., Kaunisto, K.M., Mlynarek, J.J., Sippola, E., Vesterinen, E.J. and Forbes, M.R. (2021) Metabarcoding prey DNA from fecal samples of adult dragonflies shows no predicted sex differences, and substantial inter-individual variation, in diets. *PeerJ* 9, e12634. DOI: 10.7717/peerj.12634.

Moura, L.P., Couceiro, S.R.M., Juen, L. and Veras, D.S. (2020) Congruence of the composition of Odonata between dry and rainy seasons in the Maranhense Cerrado. *International Journal of Odonatology* 23, 305–314. DOI: 10.1080/13887890.2020.1779826.

Muehlbauer, J.D., Collins, S.F., Doyle, M.W. and Tockner, K. (2014) How wide is a stream? Spatial extent of the potential "stream signature" in terrestrial food webs using meta-analysis. *Ecology* 95, 44–55. DOI: 10.1890/12-1628.1.

Müller, Z., Jakab, T., Tóth, A., Dévai, G., Szállassy, N. *et al*. (2003) Effect of sports fisherman activities on dragonfly assemblages on a Hungarian river floodplain. *Biodiversity and Conservation* 12, 167–179.

Musters, C.J.M., DeAngelis, D.L., Harvey, J.A., Mooij, W.M., van Bodegom, P.M. and de Snoo, G.R. (2023) Enhancing the predictability of ecology in a changing world: a call for an organism-based approach. *Frontiers in Applied Mathematics and Statistics* 9, e1046185. DOI: 10.3389/fams.2023.1046185.

Nagel, L., Zanuttig, M. and Forbes, M.R. (2011) Escape of parasitic water mites from dragonfly predators attacking their damselfly hosts. *Canadian Journal of Zoology* 89, 213–218. DOI: 10.1139/Z10-112.

Nagy, H.B., László, Z., Szabó, F., Dévai, G. and Tóthmérész, B. (2019) Landscape-scale terrestrial factors are also vital in shaping Odonata assemblages of watercourses. *Scientific Reports* 9, e18196. DOI: 10.1038/s41598-019-54628-7.

Nakamura, H., Kurimoto, N., Imura, Y. and Hatakeyama, Y. (2021) The first isolation of microsporidia from dragonflies in Japan. *Japanese Journal of Applied Entomology and Zoology* 65, 29–34.

Niba, A.S. and Samways, M.J. (2006a) Remarkable elevational tolerance in an African Odonata larval assemblage. *Odonatologica* 35, 265–280.

Niba, A.S. and Samways, M.J. (2006b) Development of the concept of 'core resident species' for quality assurance of an insect reserve. *Biodiversity and Conservation* 15, 4181–4196. DOI: 10.1007/s10531-005-3554-6.

Niba, A.S. and Samways, M.J. (2010) Climate and elevational range of a South African dragonfly assemblage. *BioRisk* 5, 85–107. DOI: 10.3897/biorisk.5.855.

Nicacio, G., Cunha, E.J., Hamada, N. and Juen, L. (2020) How habitat filtering can affect taxonomic and functional composition of aquatic insect communities in small Amazonian streams. *Neotropical Entomology* 49, 652–661. DOI: 10.1007/s13744-020-00780-z.

Nilsson-Örtman, V., Stoks, R., De Block, M. and Johansson, F. (2012) Generalists and specialists along a latitudinal transect: patterns of thermal adaptation in six species of damselflies. *Ecology* 93, 1340–1352. DOI: 10.1890/11-1910.1.

Novella-Fernandez, R., Chalmandrier, L., Brandl, R., Pinkert, S., Zeuss, D. and Hof, C. (2023) Trait overdispersion in dragonflies reveals the role and drivers of competition in community assembly across space and season. *Ecography* (in press), e06918. DOI: 10.1111/ecog.06918.

O'Malley, Z.G., Compson, Z.G., Orlofske, J.M., Baird, D.J., Curry, R.A. and Monk, W.A. (2020) Riparian and in-channel habitat properties linked to dragonfly emergence. *Scientific Reports* 10, e17665. DOI: 10.1038/s41598-020-74429-7.

Oliveira-Junior, J.M.B., Teodósio, M.A. and Juen, L. (2021) Patterns of co-occurrence and body size in dragonflies and damselflies (Insecta: Odonata) in preserved and altered Amazonian streams. *Austral Entomology* 60, 436–450. DOI: 10.1111/aen.12535.

Onishko, V.V., Kosterin, O.E. and Voinov, I.O. (2023) Results of odonatological studies in southern Primorye, Russia, in 2011-2020. *International Dragonfly Fund-Report 177*.

Osborn, R. and Samways, M.J. (1996) Determinants of adult dragonfly patterns at new ponds in South Africa. *Odonatologica* 25, 49–58.

Outomuro, D. and Johansson, F. (2019) Wing morphology and migration status, but not body size, habitat or Rapoport's rule predict range size in North-American dragonflies (Odonata: Libellulidae). *Ecography* 42, 309–320. DOI: 10.1111/ecog.03757.

Padffke, T. and Suhling, F. (2003) Temporal priority and intra-guild predation in temporary waters: an experimental study using Nambian desert dragonflies. *Ecological Entomology* 28, 340–347. DOI: 10.1046/j.1365-2311.2003.00513.x.

Palacino-Rodríguez, F., Brito, J.S., Calvão, L.B., Gonzalez, A.S. and Juen, L. (2020a) In Neotropical savannas, altitude affects the diversity of the Anisoptera but not the Zygoptera (Insecta: Odonata). *Marine and Freshwater Research* 72, 766–773. DOI: 10.1071/MF20182.

Palacino-Rodríguez, F., Palacino, D.A., Mungula-Steyer, R. and Juen, L. (2020b) Effects of seasonality and environmental change on an Andean damselfly *Mesamphiagrion laterale* (Odonata: Coenagrionidae). *Journal of Insect Conservation* 24, 499–511. DOI: 10.1007/s10841-020-00237-z.

Patten, M.A. and Benson, B.R. (2023) A broader flight season for Norway's Odonata across a century and a half. *Oikos* 2023, e09882. DOI: 10.1111/oik.09882.

Paulson, D.R. (2021) Nocturnal roosting of neotropical libellulid dragonflies: perhaps only *Orthemis* roosts in groups. *Notulae odonatologicae* 9, 285–290. DOI: 10.5281/zenodo.4746216.

Paulson, D.R., Landeira-Dabarca, A. and Haave-Audet, E. (2020) First nocturnal roosting aggregations of dragonflies reported from New World tropics. *Notulae odonatologicae* 9, 263–268. DOI: 10.5281/zenodo.4268593.

Pereira de Gouvêa, T.P., Dias de Oliveira, T.M., Ferreira, E.D.F., de Castro Jacques, G., Teofilo-Guedes, G. *et al.* (2022) Response of Odonata communities to dry season in a deciduous forest in the Northern Minas Gerais, Brazil. *EntomoBrasilis* 15, e1020. DOI: 10.12741/ebrasilis.v15.e1020.

Pinkert, S., Brandl, R. and Zeuss, D. (2016) Colour lightness of dragonfly assemblages across North America and Europe. *Ecography* 40, 1110–1117. DOI: 10.1111/ecog.02578.

Pinto, N.S., De Marco, P. and Moura, R.R (2022) Know your enemy: the dragonfly *Erythrodiplax fusca* (Libellulidae) uses eavesdropping to obtain information about potential rivals. *Behavioural Processes* 202, e104741. DOI: 10.1016/j.beproc.2022.104741.

Pither, J. and Taylor, P.D. (1998) An experimental assessment of landscape connectivity. *Oikos* 83, 166–174.

Pollard, J.B. and Berrill, M. (1992) The distribution of dragonfly nymphs across a pH gradient in south-central Ontario lakes. *Canadian Journal of Zoology* 70, 878–885. DOI: 10.1139/z92-125.

Powney, G.D., Cham, S.S.A., Smallshire, D. and Isaac, N.J.B. (2015) Trait correlates of distribution trends in Odonata of Britain and Ireland. *PeerJ* 3, e1410. DOI: 10.7717/peerj.1410.

Price, P.W., Westoby, M. and Rice, B. (1988) Parasite-mediated competition: some predictions and tests. *The American Naturalist* 131, 544–555. DOI: 10.1086/284805.

Pritchard, G. and Kortello, A. (1997) Roosting, perching, and habitat selection in *Argia vivida* Hagen and *Amphiagrion abbreviatum* (Selys) (Odonata: Coenagrionidae), two damselflies inhabiting geothermal springs. *The Canadian Entomologist* 129, 733–743. DOI: 10.4039/Ent129733-4.

Purse, B.V. and Thompson, D.J (2003) Emergence of the damselflies, *Coenagrion mercuriale* and *Ceriagrion tenellum* (Odonata: Coenagrionidae), at

their northern range margins. *European Journal of Entomology* 100, 93–99.
Purse, B.V., Hopkins, G.W., Day, K.J. and Thompson, D.J. (2003) Dispersal characteristics and management of a rare damselfly. *Journal of Applied Ecology* 40, 716–728.
Raczyński, M., Stoks, R., Johansson, F. and Sniegula, S. (2021) Size-mediated priority effects are trait-dependent and consistent across latitudes in a damselfly. *Oikos* 130, 1535–1547. DOI: 10.1111/oik.08353.
Raczyński, M., Stoks, R., Johansson, F., Bartoń, K. and Sniegula, S. (2022) Phenological shifts in a warming world affect physiology and life history in a damselfly. *Insects* 13, e622. DOI: 10.3390/ insects13070622.
Rantala, M.J., Ilmonen, J., Koskimäki, J., Suhonen, J. and Tynkkynen, K. (2004) The macrophyte, *Stratiotes aloides*, protects larvae of dragonfly *Aeshna viridis* against fish predation. *Aquatic Ecology* 38, 77–82. DOI: 10.1023/B:AECO.0000021005.22624.16.
Rehfeldt, G.E. (1990) Anti-predator strategies for oviposition site selection of Pyrrhosoma nymphula (Zygoptera: Odonata). *Oecologia* 85, 233–237.
Remsburg, A.J., Olson, A.C. and Samways, M.J. (2008) Shade alone reduces adult dragonfly (Odonata: Libellulidae) abundance. *Journal of Insect Behavior* 21, 460–468. DOI: 10.1007/s10905-008-9138-z.
Resende, B.O., Ferreira, V.R.S., Brasil, L.S., Calvão, L.B., Mendes, T.P. *et al*. (2021) Impact of environmental changes on the behavioral diversity of the Odonata (Insecta) in the Amazon. *Scientific Reports* 11, e9742. DOI: 10.1038/s41598-021-88999-7.
Richter, O., Suhling, F., Müller, O. and Kern, D. (2008) A model for predicting the emergence of dragonflies in a changing climate. *Freshwater Biology* 53, 1868–1880. DOI: 10.1111/j.1365-2427.2008.02012.x.
Rivas-Torres, A., Sánchez-Guillén, R.A. and Cordero-Rivera, A. (2019) Alternative reproductive strategies in black-winged territorial males of *Paraphlebia zoe* (Odonata, Thaumatoneuridae). *PeerJ* 7, e6489. DOI:10.7717/peerj.6489.
Rivera-De la Parra, L., Sarma, S.S.S. and Nandini, S. (2016) Changes in prey preferences of dragonfly naiads of *Rhionaeschna multicolour* (Hagen, 1861) (Odonata: Aeshnidae) in the presence and the absence of macrophytes. *Aquatic Insects* 37, 241–252. DOI: 10.1080/01650424.2016.1211291.
Roberts, D. (2012) Responses of three species of mosquito larvae to the presence of predatory dragonfly and damselfly larvae. *Entomologia Experimentalis et Applicata* 145, 23–29. DOI: 10.1111/j.1570-7458.2012.01300.x.
Roberts, D. (2017) Mosquito larvae can detect water vibration patterns from nearby predator. *Bulletin of Entomological Research* 107, 499–505. DOI: 10.1017/S)))7485316001140.
Rocha-Ortega, M., Rodríguez, P., Bried, J., Abbott, J. and Córdoba-Aguilar, A. (2020). Why do bugs perish? Range size and local vulnerability traits as surrogates of Odonata extinction risk. *Proceedings of the Royal Society B: Biological Sciences* 287, e20192645. DOI: 10.1098/rspb.2019.2645.
Rocha-Ortega, M., Khelifa, R., Sandall, E.L., Deacon, C., Sánchez-Rivero, X. *et al*. (2023) Linking traits to extinction risk in Odonata. In: Córdoba-Aguilar, A., Beatty, C.D. and Bried, J.T (eds) *Dragonflies and Damselflies: Model Organisms for Ecological and Evolutionary Research* (2nd ed., pp. 343–357). Oxford University Press, Oxford. DOI:10.1093/oso/9780192898623.003.0024.
Rodrigues, Y.K. and Beldade, P. (2020) Thermal plasticity in insects' responses to climate change and to multifactorial environments. *Frontiers in Ecology and Evolution* 8, e271. DOI: 10.3389/fevo.2020.00271.
Rolff, J. and Siva-Jothy, M.T (2004) Selection on insect immunity in the wild. *Proceedings of the Royal Society B* 271, 2157–2160. DOI: 10.1098/rspb.2004.2859.
Rosario, K., Dayaram, A., Marinov, M., Ware, J., Kraberger, S. *et al*. (2012) Diverse circular ssDNA viruses discovered in dragonflies (Odonata: Epiprocta). *Journal of General Virology* 93, 2668–2681. DOI: 10.1099/vir.0.045948-0.
Rundle, S.D., Bilton, D.T., Abbott, J.C. and Foggo, A. (2007) Range size in North American *Enallagma* damselflies correlates with wing size. *Freshwater Biology* 52, 471–477. DOI: 10.1111/j.1365-2427.2006.01712.x.
Rüppell, G., Hilfert-Rüppell, D., Schneider, B. and Dedenbach, H. (2020) On the firing line: interactions between hunting frogs and Odonata. *International Journal of Odonatology* 23, 199–217. DOI: 10.1080/13887890.2020.1733328.
Russell, R.W., May, M.L., Soltesz, K.L. and Fitzpatrick, J.W. (1998) Massive swarm migrations of dragonflies (Odonata) in Eastern North America. *American Midland Naturalist* 140, 325–342.
Saha, P.D. and Marathe, J.M. (2017) Odonata diversity of Jalgaon city and around (Maharashtra). *Indian Journal of Entomology* 79, 191–201. DOI: 10.5958/0974-8172.2017.00039.6.
Samejima, Y. and Tsubaki, Y. (2010) Body temperature and body size affect flight performance a damselfly. *Behavioral Ecology and Sociobiology* 64, 685–692. DOI: 10.1007/s00265-009-0886-3.
Samways, M.J. (1989a) Insects and the disturbance landscape. *Agriculture, Ecosystems and the Environment* 27, 183–194.
Samways, M.J. (1989b) Taxon turnover in Odonata across a 3000 m altitudinal gradient in southern Africa. *Odonatologica* 18, 263–274.
Samways, M.J. (2003) Conservation of an endemic dragonfly fauna in the Seychelles archipelago. *Odonatologica* 32, 177–182.
Samways, M.J. (2008) *Dragonflies and Damselflies of South Africa*. Pensoft, Sophia, Bulgaria.
Samways, M.J. (2010) Impacts of extreme weather and climate change on South African dragonflies. *BioRisk* 5, 73–84. DOI: 10.3897/biorisk.5.843.

Samways, M.J. and Osborn, R. (1998) Divergence in a transoceanic circumtropical dragonfly on a remote island. *Journal of Biogeography* 25, 935–946.
Samways, M.J., Deacon, C., Kietzka, G.J., Pryke, J.S., Vorster, C. and Simaika, J.P. (2020) Value of artificial ponds for aquatic insects in drought-prone southern Africa: a review. *Biodiversity and Conservation* 29, 3131–3150. DOI: 10.1007/s10531-020-02020-7.
Sánchez-Guillén, R.A., Martínez-Zamilpa, S.M.J., Jimés, J.G., Forbes, M.R.L. and Córdoba-Aguilar, A. (2013) Maintenance of polymorphic females: do parasites play a role? *Oecologia* 171, 105–113. DOI: 10.1007/s0042-012-2388-7.
Sang, A. and Teder, T. (2011) Dragonflies cause spatial and temporal heterogeneity in habitat quality for butterflies. *Insect Conservation and Diversity* 4, 257–264. DOI: 10.1111/j.1752-4598.2011.00134.x.
Santolamazza, S., Baquero, E. and Cordero-Rivera, A. (2011) Incidence of *Anagrus obscurus* (Hymenoptera: Mymaridae) egg parasitism on *Calopteryx haemorrhoidalis* and *Platycnemis pennipes* (Odonata: Calopterygidae: Platycnemididae) in Italy. *Entomological Science* 14, 366–369. DOI: 10.1111/j.1479-8298.2011.00454.x.
Scales, J.A. and Butler, M.A. (2016) The relationship between microhabitat use, allometry and functional variation in the eyes of *Megalagrion* damselflies. *Functional Ecology* 30, 356–368. DOI: 10.1111/1365-2435.12479.
Schad, A.N., Kennedy, J.H. and Dick, G.O. (2016) Secondary production and seasonal development of epiphytic *Enallagma civile* Hagen, 1861 (Odonata: Coenagrionidae) in a newly constructed urban wetland floodway ecosystem. *Aquatic Insects* 37, 159–173. DOI: 10.1080/01650424.2016.1191653.
Schindler, M., Fesl, C. and Chovanec, A. (2003) Dragonfly associations (Insecta: Odonata) in relation to habitat variables: a multivariate approach. *Hydrobiologia* 497, 169–180. DOI: 10.1023/A:1025476220081.
Schmidt, E. (1985) Habitat inventorization, characterization and bioindication by a "Representative Spectrum of Odonata Species (RSO)." *Odonatologica* 14, 127–133.
Serrano-Menses, M.A., Córdoba-Aguilar, A., Méndez, V., Layen, S.J. and Székely, T. (2007) Sexual size dimorphism in the American rubyspot: male body size predicts male competition and mating success. *Animal Behaviour* 73, 987–997. DOI: 10.1016/j.anbehav.2006.08.012.
Shapoval, N.A., Shapoval, G.N. and Shapoval, A.P. (2022) New, rare and vagrant damselflies and dragonflies (Insecta: Odonata) in the Kaliningrad Oblast, north-western Russia. *Acta Biologica Sibirica* 8, 261–279. DOI: 10.14258/abs.v8.e17.
Sharma, D., Brahma, S., Saha, N., Kundu, M., Saha, G.K. and Aditya, G. (2016) Association of larval Odonata and hydrophytes in wetlands of West Bengal, India: implications for conservation and monitoring. *Journal of Entomology and Zoology Studies* 4, 35–39.
Shelly, T.E. (1982) Foraging behavior of light- versus shade-seeking adult damselflies in a lowland neotropical forest (Odonata: Zygoptera). *Physiological Zoology* 55, 335–343.
Shelton, J.M., Samways, M.J. and Day, J.A. (2015) Non-native rainbow trout change the structure of benthic communities in headwater streams of the Cape Floristic Region, South Africa. *Hydrobiologia* 745, 1–15. DOI: 10.1007/s10750-014-2067-2).
Simpkin, J.L., Britten, H.B. and Brussard, P.F. (2000) Effects of habitat fragmentation and differing mobility on the population structures of a great basin dragonfly (*Sympetrum corruptum*) and damselfly (*Enallagma carunculatum*). *Western North American Naturalist* 60, 320–332.
Siva-Jothy, M.T. (2000) A mechanistic link between parasite resistance and expression of a sexually selected trait in a damselfly. *Proceedings of the Royal Society B* 267, 2523–2527. DOI: 10.1098/rspb.2000.1315.
Siva-Jothy, M.T. and Plaistow, S.J. (1999) A fitness cost of eugregarine parasitism in a damselfly. *Ecological Entomology* 24, 465–470. DOI: 10.1046/j.1365-2311.1999.00222.x.
Smith, J., Samways, M.J. and Taylor, S. (2007) Assessing riparian quality using two complementary sets of bioindicators. *Biodiversity and Conservation* 16, 2695–2713. DOI: 10.1007/s10531-006-9081-2.
Sniegula, S., Golab, M.J. and Johansson, F. (2017) Cannibalism and activity rate in larval damselflies increase along a latitudinal gradient as a consequence of time constraints. *BMC Evolutionary Ecology* 17, e167. DOI: 10.1186/s12862-017-1010-3.
Sniegula, S., Golab, M.J. and Johansson, F. (2019) Size-mediated priority and temperature effects on intra-cohort competition and cannibalism in a damselfly. *Journal of Animal Ecology* 88, 637–648. DOI: 10.1111/1365-2656.12947.
Sokolova, Y.Y., Kryukova, N.A., Glupov, V.V. and Fuxa, J.R. (2006) *Systenostrema alba* Larsson 1988 (Microsporidia, Thelohaniidae) in the dragonfly *Aeshna viridis* (Odonata, Aeshnidae) from South Siberia: morphology and molecular characterization. *Journal of Eukaryotic Microbiology* 53, 49–57. DOI: 10.1111/j.1550-7408.2005.00075.x.
Sokolovska, N., Rowe, L. and Johansson, F. (2000) Fitness and body size in mature odonates. *Ecological Entomology* 25, 239–248. DOI: 10.1046/j.1365-2311.2000.0025.x.
Söndgerath, D., Rummland, J. and Suhling, F. (2012) Large spatial scale effects of rising temperatures: modelling a dragonfly's life cycle and range throughout Europe. *Insect Conservation and Diversity* 5, 461–469. DOI: 10.1111/j.1752-4598.2011.00182.x.
Srygley, R.B. (2003) Wind drift compensation in migrating dragonflies *Pantala* (Odonata: Libellulidae). *Journal of Insect Behavior* 16, 217–232.

Stewart, S.S. and Vodopich, D.S. (2018) Environmental effects on wing shape and wing size of *Argia sedula* (Odonata: Coenagrionidae). *International Journal of Odonatology* 21, 189–203. DOI: 10.1080/13887890.2018.1523752.

Steytler, N.S. and Samways, M.J. (1995) Biotope selection by adult male dragonflies (Odonata) at an artificial lake created for insect conservation in South Africa. *Biological Conservation* 72, 381–386. DOI: 10.1016/0006-3207(94)00052-R.

Stoks, R. (2001) What causes male-biased sex ratios in mature damselfly populations? *Ecological Entomology* 26, 188–197. DOI: 10.1046/j.1365-2311.2001.00303.x.

Stoks, R. and Córdoba-Aguilar, A. (2012) Evolutionary ecology of Odonata: a complex life cycle perspective. *Annual Review of Entomology* 57, 249–265. DOI: 10.1146/annurev-ento-120710-100557.

Stoks, R. and McPeek, M.A. (2006) A tale of two diversifications: reciprocal habitat shifts to fill ecological space along the pond permanence gradient. *The American Naturalist* 168, S50–S72.

Stoks, R., Swillen, I. and De Block, M. (2012) Behaviour and physiology shape the growth accelerations associated with predation risk, high temperatures and southern latitudes in *Ischnura* damselfly larvae. *Journal of Animal Ecology* 81, 1034–1040. DOI: 10.1111/j.1365-2656.2012.01987.x.

Su, G., Dudley, R., Pan, T., Zheng, M., Peng, L. and Li, Q. (2020) Maximum aerodynamic force production by the wandering glider dragonfly (Pantala flavescens, Libellulidae). *Journal of Experimental Biology* 223, jeb218552. DOI: 10.1242/jeb.218552.

Suárez-Tovar, C.M., Rocha-Ortega, M., González-Voyer, A., González-Tokman, D. and Córdoba-Aguilar, A. (2019) The larger the damselfly, the more likely to be threatened: a sexual selection approach. *Journal of Insect Conservation* 23, 535–545. DOI: 10.1007/s10841-019-00142-0.

Suhling, F. (1996) Interspecific competition and habitat selection by the riverine dragonfly *Onychogomphus uncatus*. *Freshwater Biology* 35, 209–217. DOI: 10.1046/j.1365-2427.1996.00491.x.

Suhling, I. and Suhling, F. (2013) Thermal adaptation affects interactions between a range-expanding and a native odonate species. *Freshwater Biology* 58, 705–714. DOI: 10.1111/fwb.12074.

Suhling, F., Sahlén, G., Kasperski, J. and Gaedecke, D. (2005) Behavioural and life history traits in temporary and perennial waters: comparisons among three pairs of sibling dragonflies. *Oikos* 108, 609–617.

Suhling, F., Martens, A. and Marais, E. (2009) How to enter a desert: patterns of Odonata colonisation of arid Namibia. *International Journal of Odonatology*, 12, 287–308. DOI: 10.1080/13887890.2009.9748346.

Suhling, F., Martens, A. and Suhling, I. (2017) Long-distance dispersal in Odonata: examples from arid Namibia. *Austral Ecology* 42, 544–552. DOI: 10.1111/aec.12472.

Suhonen, J., Hilli-Lukkarinen, M., Korkeamäki, E., Kuitunen, M., Kullas, J. *et al.* (2010) Local extinction of dragonfly and damselfly populations in low- and high-quality habitat patches. *Conservation Biology* 24, 1148–1153. DOI: 10.1111/j.1523-1739.2010.01504.x.

Suhonen, J., Korkeamäki, E., Salmela, J. and Kuitunen, M. (2014) Risk of local extinction of Odonata freshwater generalists and specialists. *Conservation Biology* 28, 783–789. DOI: 10.1111/cobi.12231.

Suhonen, J., Ilvonen, J.J., Korkeamäki, E., Nokkala, C. and Salmela, J. (2022) Using functional traits and phylogeny to understand local extinction risk in dragonflies and damselflies (Odonata). *Ecology and Evolution* 12, e8648. DOI: 10.1002/ece3.8648.

Suutari, E., Rantala, M.J., Salmela, J. and Suhonen, J. (2004) Intraguild predation and interference competition on the endangered *Aeshna viridis*. *Oecologia* 140, 135–139. DOI: 10.1007/s00442-004-1559-6.

Svensson, E.I., Runemark, A., Verzijden, M.N. and Wellenreuther, M. (2014) Sex differences in developmental plasticity and canalization shape population divergence in mate preferences. *Proceedings of the Royal Society B* 281, e20141636. DOI: 10.1098/rspb.2014.1636.

Swaegers, J., Janssens, S.B., Ferreira, S., Watts, P.C., Mergeay, J. *et al.* (2014) Ecological and evolutionary drivers of range size in *Coenagrion* damselflies. *Journal of Evolutionary Biology* 27, 2386–2395. DOI: 10.1111/jeb.12481.

Sysiak, M., Pietrzak, B., Kubiak, M., Bednarska, A. and Mikulski, A. (2023) Chemical cannibalistic cues make damselfly larvae hide rather than hunt. *Scientific Reports* 13, e13556. DOI: 10.1038/s41598-023-40732-2.

Takahashi, Y., Kagawa, K., Svensson, E.I. and Kawata, M. (2014) Evolution of increased phenotypic diversity enhances population performance by reducing sexual harassment in damselflies. *Nature Communications* 5, e4468. DOI: 10.1038/ncomms5468.

Taylor, P.D. (2006) Movement behaviors of a forest odonate in two heterogenous landscapes. In: Cordero-Rivera, A. (ed.) *Forests and Dragonflies* (pp. 225–238). Pensoft, Sophia, Bulgaria.

Taylor, P.D. and Merriam, G. (1995) Wing morphology of a forest damselfly is related to landscape structure. *Oikos* 73, 43–48.

Taylor, P.D. and Merriam, G. (1996) Habitat fragmentation and parasitism of a forest damselfly. *Landscape Ecology* 11, 181–189. DOI: 10.1007/BF02447516.

Thompson, D.J., Hassall, C., Lowe, C.D. and Watts, P.C. (2011) Field estimates of reproductive success in a model insect: behavioural surrogates are poor predictors of fitness. *Ecology Letters* 14, 905–913. DOI: 10.1111/j.1461-0248.2011.01655.x.

Thorp, J.H. and Cothran, (1984) Regulation of freshwater community structure at multiple intensities of dragonfly predation. *Ecology* 65, 1546–1555.

Troast, D., Suhling, F., Jinguji, H., Sahlén, G. and Ware, J. (2016) A global population genetic study of *Pantala flavescens*. *PLoS ONE* 11, e0148949. DOI: 10.1371/journal.pone.0148949.

Tsubaki, Y., Samejima, Y. and Siva-Jothy, M.T. (2010) Damselfly females prefer hot males: higher courtship success in males in sunspots. *Behavioral Ecology and Sociobiology* 64, 1547–1554. DOI: 10.1007/s00265-010-0968-2.

Turgeon, J., Stoks, R., Thum, R.A., Brown, J.M. and McPeek, M.A. (2005) Simultaneous Quaternary radiations of three damselfly clades across the Holarctic. *The American Naturalist* 165, E78–E107.

Uboni, C., Jugovic, J., Tordoni, E., Pizzul, E., Riservato, E. and Bacaro, G. (2020) Dragonfly (Odonata) diversity patterns in mixohaline coastal wetlands. *Estuaries and Coasts* 43, 375–386. DOI: 10.1007/s12237-019-00687-y.

Van Gossum, H., Beatty, C.D., Charlat, S., Waqa, H., Markwell, T. *et al*. (2007) Male rarity and putative sex-role reversal in Fijian damselflies (Odonata). *Journal of Tropical Ecology* 23, 591–598. DOI: 10.1017/S0266467407004373.

Van Huyssteen, P. and Samways, M.J. (2009) Overwintering dragonflies in an African savanna (Anisoptera: Gomphidae, Libellulidae). *Odonatologica* 38, 167–172.

Van Remsburg, A.J. and Turner, M.G. (2009) Aquatic and terrestrial drivers of dragonfly (Odonata) assemblages within and among north-temperate lakes. *Journal of the North American Benthological Society* 28, 44–56. DOI: 10.1899/08-004.1.

Van Schalkwyk, J., Samways, M.J. and Pryke, J.S. (2014) Winter survival by dragonfly adults in the Cape Floristic Region. *International Journal of Odonatology* 17, 17–30. DOI: 10.1080/13887890.2014.880382.

Vander Vorste, R., Stubbington, R., Acuña, V., Bogan, M.T., Bonada, N. *et al*. (2021) Climatic aridity increases temporal nestedness of invertebrate communities in naturally drying rivers. *Ecography* 44, 860–869. DOI: 101111/ecog.05349.

Verberk, W.C.E.P. and Calosi, P. (2012) Oxygen limits heat tolerance and drives heat hardening in the aquatic nymphs of the gill breathing damselfly *Calopteryx virgo* (Linnaeus, 1758). *Journal of Thermal Biology* 37, 224–229. DOI: 10.1016/j.jtherbio.2012.01.004.

Verheyen, J., Temmerman, K., De Block, M. and Stoks, R. (2018) Voltinism-associated differences in winter survival across latitudes: integrating growth, physiology, and food intake. *Oecologia* 186, 919–929. DOI: 10.1007/s00442-018-4079-5.

Vilenica, M. (2017) Ecological traits of dragonflies (Odonata) assemblages along an oligotrophic Dinaric karst hydrosystem. *Annales de Limnologie-International Journal of Limnology* 53, 377–389. DOI: 10.1051/limn/2017019.

Waller, J.T. and Svensson, E.I. (2017) Body size evolution in an old insect order: no evidence for Cope's Rule in spite of fitness benefits of large size. *Evolution* 71, 2178–2193. DOI: 10.1111/evo.13302.

Ware, J., Kohli, M.K., Mendoza, C.M., Troast, D., Jinguji, H. *et al*. (2022) Evidence for widespread gene flow and migration in the Globe Skimmer dragonfly *Pantala flavescens*. *International Journal of Odonatology* 25, 43–55. DOI: 10.48156/1388.2022.1917166.

Watanabe, M., Matsuoka, H. and Taguchi, M. (2004) Habitat selection and population parameters of *Sympetrum infuscatum* (Selys) during sexually mature stages in a cool temperate zone of Japan (Anisoptera: Libellulidae). *Odonatologica* 33, 169–179.

Watts, P.C., Rouquette, J.R., Saccheri, I.J., Kemp, S.J. and Thompson, D.J (2004) Molecular and ecological evidence for small-scale isolation by distance in an endangered damselfly, *Coenagrion mercuriale*. *Molecular Ecology* 13, 2931–2945. DOI: 10.1111/j.1365-294X.2004.02300.x.

Watts, P.C., Saccheri, I.J., Kemp, S.J. and Thompson, D.J. (2006) Population structure and the impact of regional and local habitat isolation upon levels of genetic diversity of the endangered damselfly *Ceriagrion mercuriale* (Odonata: Zygoptera). *Freshwater Biology* 51, 193–205. DOI: 10.1111/j.1365-2427.2005.01478.x.

Watts, P.C., Rousset, F., Saccheri, I.J., Leblois, R. and Thompson, D.J. (2007) Compatible genetic and ecological estimates of dispersal rates in insect (*Coenagrion mercuriale*: Odonata: Zygoptera) populations: analysis of 'neighbourhood size' using a more precise estimator. *Molecular Ecology* 16, 737–751. DOI: 10.1111/j.1365-294X.2006.03184.x.

Watts, P.C., Keat, S. and Thompson, D.J. (2010) Patterns of spatial genetic structure and diversity at the onset of a rapid range expansion: colonisation of the UK by the small red-eyed damselfly *Erythromma viridulum*. *Biological Invasions* 12, 388–3903. DOI: 10.1007/s10530-010-9779-7.

Wellborn, G.A., Skelly, D.K. and Werner, E.E. (1996) Mechanisms creating community structure across a freshwater gradient. *Annual Review of Ecology and Systematics* 27, 337–363.

Westbrook, J.K. and Eyster, R.S. (2017) Atmospheric environment associated with animal flight. In: Chilson, P.B., Frick, W.F., Kelly, J.F. and Liechti, F. (eds.) *Aeroecology* (pp. 13–44). Springer, Cham, Switzerland.

White, E.L., Hunt, P.D., Schlesinger, M.D., Corser, J.D. and deMaynadier, P.G. (2015) Prioritizing Odonata for conservation action in the northeastern USA. *Freshwater Science* 34, 1079–1093. DOI: 10.1086/682287.

Wikelski, M., Moskowitz, D., Adelman, J.S., Cochran, J., Wilcove, D.S. and May M.L. (2006) Simple rules guide dragonfly migration. *Biology Letters* 2, 325–329. DOI: 10.1098/rsbl.2006.0487.

Wildermuth, H. (1993) Habitat selection and oviposition site recognition by the dragonfly *Aeshna juncea* (L.): an experimental approach in natural habitats (Anisoptera: Aeshnidae). *Odonatologica* 22, 27–44.

Wittwer, T., Sahlén, G. and Suhling, F. (2010) Does one community shape another? Dragonflies and fish in Swedish lakes. *Insect Conservation and Diversity* 3, 124–133. DOI: 10.1111/j.1752-4598.2010.00083.x.

Wohlfahrt, B., Mikolajewski, D.J., Joop, G. and Suhling, F. (2006) Are behavioural traits in prey sensitive to the risk imposed by predatory fish? *Freshwater Biology* 51, 76–84. DOI: 10.1111/j.1365-2427.2005.01475.x.

Wonglersak, R., Fenberg, P.B., Langdon, P.G., Brooks, S.J. and Price, B.W. (2020) Temperature-body size responses in insects: a case study of British Odonata. *Ecological Entomology* 45, 795–805. DOI: 10.1111/een.12853.

Worthen, W.B. and Horacek, H.J. (2015) The distribution of dragonfly larvae in a South Carolina stream: relationships with sediment type, body size, and the presence of other larvae. *Journal of Insect Science* 15, e31. DOI: 10.1093/jisesa/iev013.

Zebsa, R., Khelifa, R., Kahalerras, A., Djalal, H. and Houhamdi, M. (2014) Emergence pattern, site selection, and seasonal regulation of *Onychogomphus costae* Selys, 1885 (Odonata: Gomphidae) in northeastern Algeria. *Aquatic Insects* 36, 257–265. DOI: 10.1080/01650424.2015.1087574.

Zebsa, R., Khelifa, R. and Kahalerras, A. (2015) Adults movement pattern and habitat preferences of the Maghribian endemic *Gomphus lucasii* (Odonata: Gomphidae). *Journal of Insect Science* 15, e151. DOI: 10.1093/jisesa/iev128.

Zebsa, R., Mahdjoub, H. and Khelifa, R. (2022) Similar response of a range expanding dragonfly to low- and high-elevation predators. *Diversity* 14, e302. DOI: 10.3390/d14040302.

Zettlemoyer, M.A. and Peterson, M.L. (2021) Does phenological plasticity help or hinder range shifts under climate change? *Frontiers in Ecology and Evolution* 9, e689192. DOI: 10.3389/fevo.2021.689192.

Zeuss, D., Brandl, R., Brändle, M., Rahbek, C. and Brunzel, S. (2014) Global warming favours light-coloured insects in Europe. *Nature Communications* 5, e3874. DOI: 10.1038/ncomms4874.

Zeuss, D., Brunzel, S. and Brandl, R. (2017) Environmental drivers of voltinism and body size in insect assemblages across Europe. *Global Ecology and Biogeography* 26, 154–165. DOI: 10.1111/geb.12525.

Zia, A., Din, A.-U., Azam, I., Munir, A. and Afsheen, S. (2018) Effect of salinity gradients on species composition of Odonata naiads. *Arthropods* 7, 11–25.

5 Stressors on Freshwater Ecosystems and Dragonflies

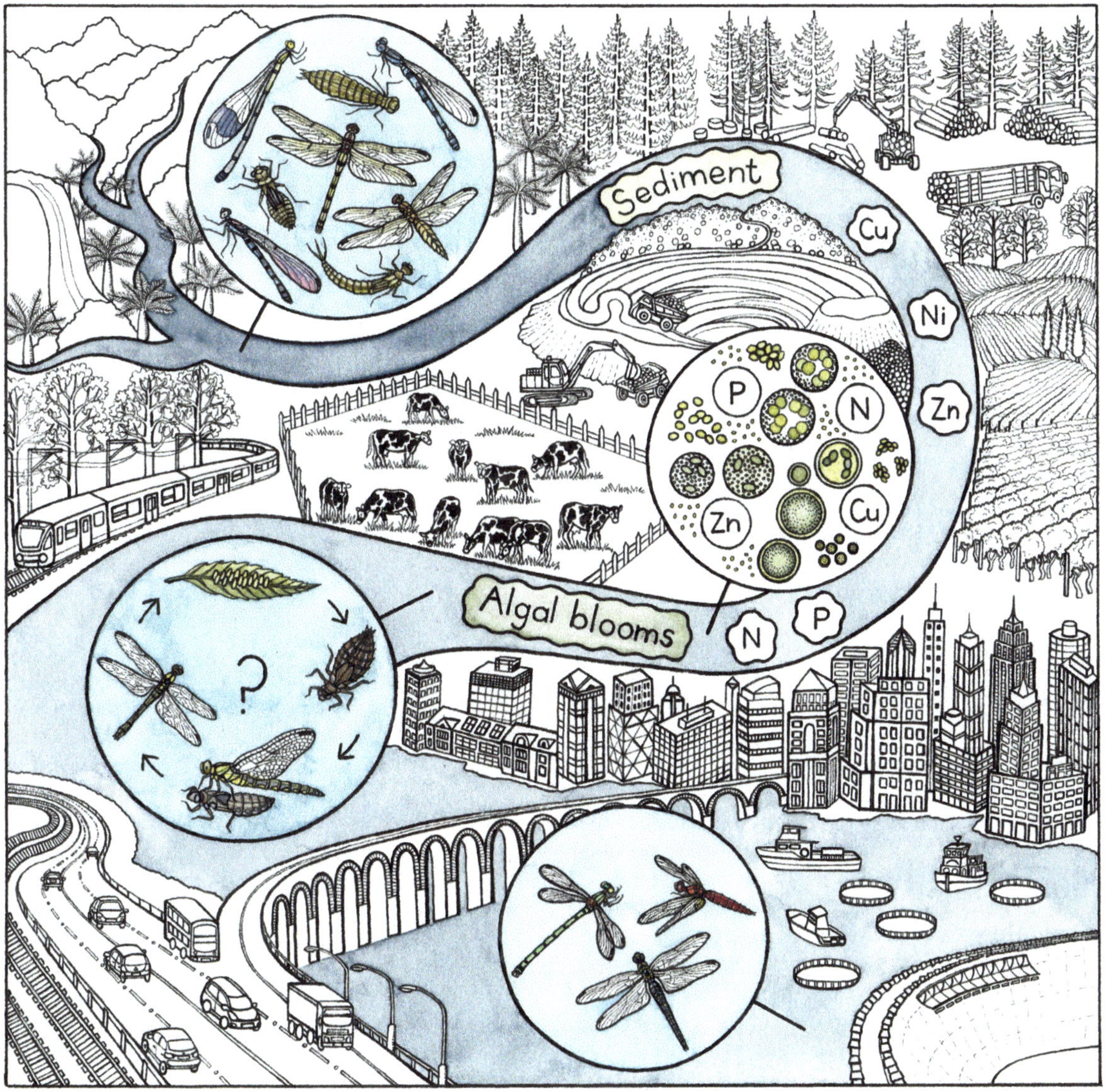

Illustration: Megan Louise Potgieter.

DOI: 10.1079/9781789248395.0005

Abstract

Global freshwaters experience various stressors, including climate change and extreme weather. Healthy headwater streams and their riparian zones are especially important, and home to many dragonfly habitat specialists.

Considerations of spatial scale are essential in the conservation of healthy natural ecosystems. At the large scale, quality catchments are important, while much dragonfly conservation usually occurs at specific locations.

Decline among dragonfly habitat specialists is concerning. In contrast, some habitat-generalist species benefit from landscape transformation.

Stressors are various and many. Some are visually obvious, like tropical forest loss. Others are pervasive and not readily detected, like chemical contaminants which can do harm far away after being conveyed by water movement.

Human impacts on dragonflies may be direct, and/or they may be indirect through changes in food webs. The intensity of multiple stressors, both rural and urban, is becoming increasingly significant. The survivors constitute a homogenized subset of the historical species assemblage, composed of the most tolerant species.

5.1 Stressors on Freshwater Systems

5.1.1 Human impacts on freshwater systems overall

Freshwater ecosystems, along with their margins and riparian zones, are among the most biodiverse on the planet, having high levels of biodiversity and providing irreplaceable services for society (Lynch *et al.*, 2023). Yet these ecosystems are declining on every continent and in every river basin, notwithstanding that some rivers and other water bodies have been remediated in recent years. Currently, about one third of global freshwater discharges pass through agricultural, industrial or urban areas. About one fifth of arable land is now irrigated, with this proportion likely to surpass one third by the mid-twenty-first century. Less than one fifth of the world's pre-industrial freshwater wetlands remain. However, projections indicate that less than one tenth will remain by 2050, with water transfer megaprojects likely to have a major impact. Human use of critical water resources will approach 50% by 2050 (Albert *et al.*, 2021). However, with some firm policy actions, freshwater biodiversity and ecosystem decline can be slowed, and is among the highest of all humanity's priorities for a sustainable future (Fig. 5.1).

Freshwater systems are disadvantaged in various ways. They can be influenced by events way upstream, are receivers of land-use effluents, are subject to human activities on the surrounding land and along the riparian zone, and there are competing interests among many types of human water consumers. Furthermore, freshwater systems also support many irreplaceable endemic species that are unable to escape the various adverse impacts.

Human impacts on freshwater systems can lead to five general stressors: overexploitation, water pollution (agricultural, industrial, urban, pharmaceutical), flow modification, degradation and even destruction of habitats, and invasion by alien species (Dudgeon *et al.*, 2006). While dragonflies on their own are not directly overexploited, aspects of their habitat may be severely depleted. All the other four categories impact dragonflies at one location or another.

Reid *et al.* (2019) updated the list of impacts, which have intensified or are novel. These include changing climate, increase in commerce, infectious diseases, harmful algal blooms, expanding hydropower, emerging contaminants, engineered nanomaterials, microplastic pollution, light and noise pollution, freshwater salination, declining calcium, and cumulative stressors. While some of these may not yet be known to affect dragonflies directly, they may alter the food webs of which they are part.

We also need to consider the relationship between landscape change and its effect on hydrological and ecological processes. Ecohydrological changes mostly concern the amount and quality of in-channel run-off, water quality and soil loss. Landscape topography and its spatial distribution can change rainfall run-off processes and lead to changes in run-off yield. In turn, water quality is intimately related to topographical composition and spatial patterns on high ground versus low ground (Fu *et al.*, 2005), with changes in sediment load along the length of a river (Syvitski *et al.*, 2014), as well as being associated with changes in socio-ecological processes (Cai *et al.*, 2019). Furthermore, for sediment and many nutrients, the effects of adjacent land uses can be detected up to 4 km away from stream flows. This means that the wider landscape must always be considered (Houlahan and Findlay, 2004), as the overall system is connected by dynamic groundwater movement (Ward, 1998).

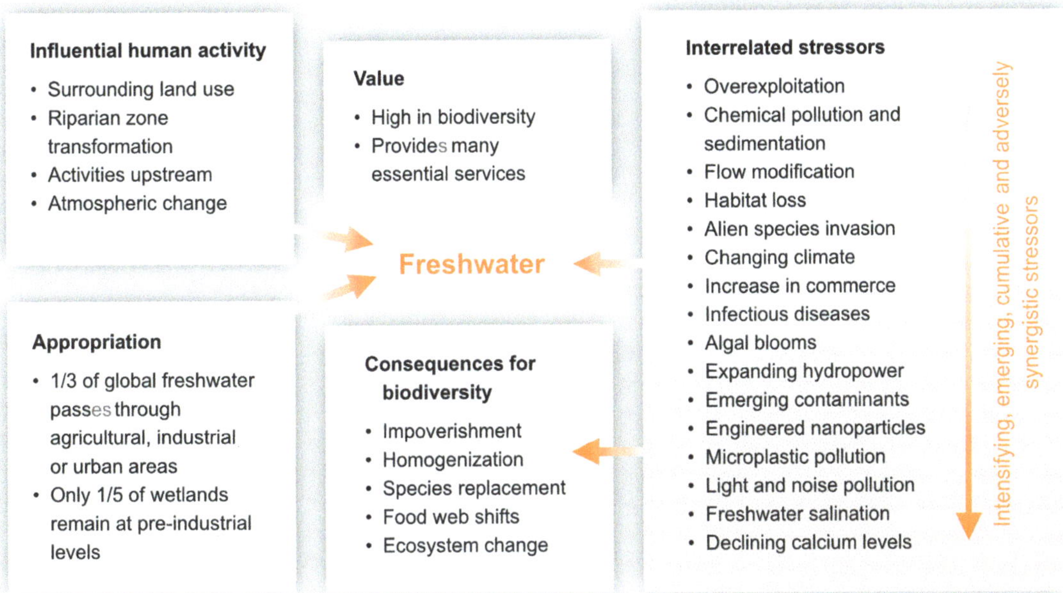

Fig. 5.1. Freshwaters have great value as they are high in biodiversity and provide many essential ecosystem services. However, they have suffered a high level of appropriation, due to significant and influential human activity. Many interrelated stressors have been identified as affecting freshwater biodiversity. Their overall effects are intensifying and coupled with emerging new stressors they are often cumulative and adversely synergistic. The severe consequences for biodiversity and humanity are now becoming clear. (Illustration: Robyn Symons).

Of concern is that there is overall homogenization of freshwater faunas and ecosystem change emanating from the interactive effects of habitat degradation and loss, loss of specialist indigenous species and invasion by alien species (Strayer, 2010). When these processes are considered together, the result is often not so much a loss of species but a replacement of sensitive species (Rahel, 2002; Ma *et al.*, 2023).

Spatial scale and apparency are other important considerations. The loss of a patch of forest through timber extraction is visually clear and may greatly change local dragonfly assemblage composition. In contrast, the effect of polluting chemicals may not be apparent and can be very widespread. There are about 100,000 chemical compounds in daily use that can enter freshwater systems. They are strongly associated with land use in catchments and caused by intensive agriculture and/or urban activities. Yet we know very little about the chemical risks that many of these compounds may have on the biota. A pan-European study of 4000 sites and 223 compounds showed that the most impactful compounds on an invertebrate representative (*Daphnia magna*) by far were pesticides, at 87% of the sites, followed by polycyclic aromatic hydrocarbons, organotin compounds and brominated fire retardants, as well as some other compounds (Malaj *et al.*, 2014). Yet microcrustaceans, for example, can be an important component of a dragonfly's diet (Dudová *et al.*, 2019). Their mortality, as well as that of other prey types, through exposure to pesticides and other compounds, could have a major effect on the dragonfly's food base. This is in addition to any direct ecotoxicological risk to dragonfly larvae.

5.1.2 Stressors on rivers

5.1.2.1 The complexity of river systems

River systems are complex, owing to a wide variety of relationships between spatial pattern and spatial scale, which are often not linear. This can make it difficult to separate human from historical influences (Allan, 2004). River landscapes include the components of the land surface that are influenced by a river, such as the aquatic habitats within the river, the riparian zone and associated marshy areas. Rivers that share similar hydrologic, geomorphic and climatic conditions are known as 'hydrogeomorphic patches' (Kopp and Allen, 2021; Scown and Thomas, 2023). These patches and the functional connectivity between them play a major role in determining the local biodiversity patterns and food webs, of which dragonflies are a part. Human-induced changes to hydrogeomorphic patches can potentially isolate otherwise functionally connected patches (Fig. 5.2).

Of concern is that while rivers are naturally intermittent (parts of them having lower discharge rates and even drying out during parts of the year) due to climatic, hydrologic, geological and geomorphic drivers, they have been exacerbated by human activity and associated climate change in recent years (DelVecchia *et al.*, 2022). Six of the world's largest rivers have become intermittent in their mainstem, and over 400 European rivers have dried earlier and for longer, inevitably having a profound effect on the biota (Datry *et al.*, 2023).

5.1.2.2 Significance of headwater streams

The global stream and river area is estimated to be 485,000–662,000 km^2, making up 0.30–0.56% of the land surface area. While moderately sized rivers make up the greatest global area, the total global headwater stream length, and therefore riparian interface, is dominated by headwater streams (Downing *et al.*, 2012). In turn, these headwater streams contribute about 70% of the water volume, with their chemistry greatly influencing water quality in the lower reaches (Alexander *et al.*, 2007) (Fig. 5.3).

Headwater streams are highly significant hydrologically and ecologically, making it essential that they are assessed quantitatively and qualitatively, and in terms of the risks that they face and their conservation value (Amatulli *et al.*, 2022). Headwater streams are home to many aquatic species and contribute highly and disproportionately to river network biodiversity (Finn *et al.*, 2011). Yet they can differ according to various interrelated environmental variables such as their orientation to the sun, local precipitation patterns and substrate characteristics,

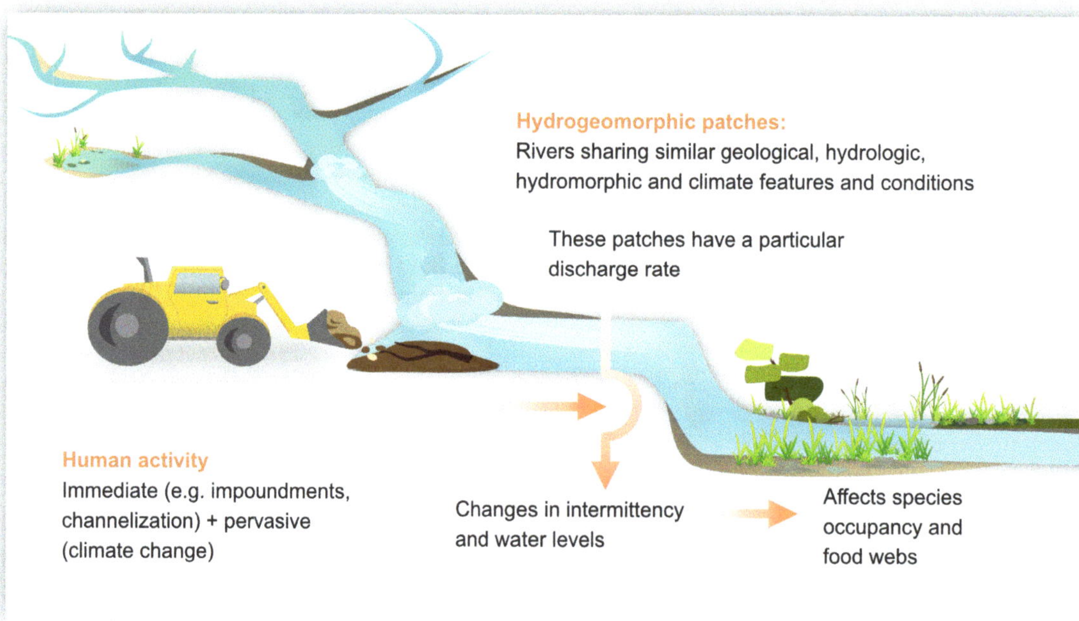

Fig. 5.2. River systems are complex systems at various spatial scales, both of themselves and in relation to the land surface. Rivers that share similar various major features and conditions are known as hydrogeomorphic patches. When human activity impinges on these systems there can be a change in intermittency and in aquatic species composition and interactions. (Illustration: Robyn Symons).

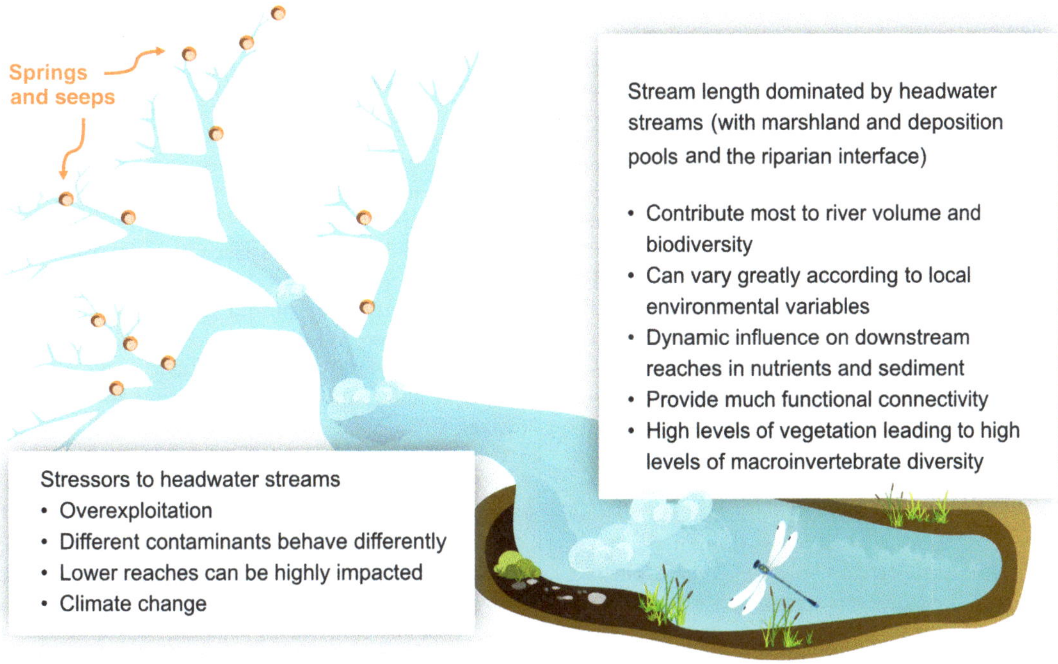

Fig. 5.3. Headwater streams, with their attendant seeps, springs, marshes and deposition pools, are the dominant freshwater system in terms of length, interfacing with the riparian zone, and in contributing to water volume. They are also dynamic and hydromorphologically diverse, so contributing much to ecological heterogeneity and beta diversity. However, they are currently subject to several major stressors. (Illustration: Robyn Symons).

which can define, for example, their macroinvertebrate assemblages (Wang, L. *et al.*, 2022).

Headwater streams are dynamic environments which influence downstream reaches by retaining or conveying sediment and nutrients, providing refuges or habitats for aquatic organisms, and providing functional connectivity both along the water and the riparian zone, which means they can convey stressors such as pollution to downstream reaches (Nguyen *et al.*, 2023). In high-relief topographies, there is a distinct channel often hemmed in by rocky valley walls. In low-relief topographies, while having a distinct channel, streams often inundate parts of the surrounding area. Small streams are also subject to the effects of underlying geology and local climate. In arid areas they may dry out at certain times of the year, often leaving remnant pools. They also initiate springs or seeps and feed into ponds and lakes (Wohl, 2017).

Headwater streams vary greatly in macroinvertebrate diversity between regions and continents. While the streams are often poor in alpha diversity of aquatic macroinvertebrates, they are often high in beta diversity, driven by high levels of physical and vegetation heterogeneity and a wide range of local environmental conditions. There may even be high gamma diversity among catchments (Clarke *et al.* 2008).

Tropical water systems are characterized by rapid temperature gradients, varied flow, natural disturbance regimes and high organic input. These features lead to high habitat heterogeneity, high biological diversity and distinct life-history strategies, often associated with high levels of local endemism. In turn, however, they are highly susceptible to climate change (Encalada *et al.*, 2019).

Tropical mountain streams arise in the warm and wet mountains, descending through dense forest and feeding rivers and floodplains. The high tropical mountains are characterized by high rainfall associated with cloud forest and species-rich biodiversity, with about 2.4% of cloud forest already having been lost globally (Karger *et al.*, 2021). For dragonflies of the Yungas cloud forest in north-western Argentina, and where there are 11 cloud forest dragonfly endemic species, the most species-rich zone is the foothills. However, this is also the most threatened area, with 90% of the foothill rainforest lost to logging and replaced by agriculture (Von Ellenreider, 2009).

The biological value of headwater streams also emphasizes the importance of understanding the behaviour of biological contaminants. Nitrogen and phosphorus loads originate mainly from agricultural fields, affecting headwater stream quality. But the relationship between these pollutants varies relative to spatial scale. At the large spatial scale (>100 km^2) both nitrogen and phosphorus are statistically related to agricultural pressure. However, at the smaller scale and that of headwater streams (<20 km^2), nitrogen and phosphorus behave differently. While landscape composition (percentage of agricultural land use) controls nitrogen exports to freshwater, it is landscape configuration (the spatial arrangement/distribution/organization of the agricultural elements) that determines phosphorus movement. These findings mean that it is essential to monitor and mitigate contaminants at various spatial scales to ensure headwater stream quality (Casquin *et al.*, 2021).

An increase in the amount of agricultural fertilizer exacerbates the adverse effect of water pollution, with associated deterioration of water quality in upper watershed areas. There is also a threshold effect when fertilizer application intensity reaches a certain high level, aggravating the effect of water pollution. In turn, rainfall leads to migration of pollutants across the lower reaches (Xu *et al.*, 2022), while sediment run-off from human activities in densely populated areas can be four times greater than that from natural run-off (Gu *et al.*, 2019).

5.1.3 Stressors on standing waters

The global loss of natural wetlands averages between 54–57%, with losses as high as 87% in some areas since the year 1700. There has been a much faster rate (3.7 times) of wetland loss during the twentieth and early twenty-first centuries, with a loss of 64–71% of wetlands since 1900. Losses have been larger and faster for inland rather than coastal natural wetlands. Although the rate of wetland loss in Europe has slowed, and in North America has remained low since the 1980s, the rate has remained high in Asia, where large-scale and rapid conversion of coastal and inland natural wetlands is continuing. It is unclear whether the investment by national governments in the Ramsar Convention on Wetlands has reduced these rates of loss, suggesting that there is a need to improve knowledge relative to change in wetland areas worldwide, particularly in Africa, the Neotropics and Oceania (Davidson, 2014) (Fig. 5.4).

Between 1984–2015, permanent surface water disappeared from almost 90,000 km^2, roughly the size of Lake Superior in North America. However, 184,000 km^2 appeared elsewhere, mostly through impoundments but also from climate change events such as heavy rainfall and glacier melt. Water loss is more geographically concentrated than water gain, with over 70% of global net water loss in the Middle East and Central Asia, linked to drought, river diversion, damming and unregulated withdrawal. Furthermore, Australia and the USA, as well as other areas, have suffered from long-term droughts (Pekel *et al.*, 2016). Droughts and potential evapotranspiration are set to increase, affecting many parts of the world as climate warming continues. The exact area, extent and timing of them is difficult to predict, despite our best efforts (e.g. Ionita and Nagavciuc, 2021).

Concentration of dissolved oxygen in aquatic systems helps regulate biodiversity, nutrient biogeochemistry, greenhouse gas emissions and drinking water quality, yet is being altered by climate warming and human activity. In lakes, solubility of dissolved oxygen decreases with increasing water temperatures, and oxygen losses in warming lakes can be amplified by enhanced decomposition and stronger thermal stratification. However, oxygen levels may increase from enhanced plant production. In recent years, there has been a decline in dissolved oxygen in both surface and deep-water habitats, suggesting that climate change has altered the physicochemical features of lakes, and more so than in the oceans (Jane *et al.*, 2021). These changes are of concern because there is positive feedback among deoxygenation events that could magnify the effects of climate change on lakes (Lewis *et al.*, 2023).

By 2100, it is estimated that there will be a net area of Ramsar wetland loss of at least 6000 km^2 (about 1%). The number of Ramsar sites with area loss over 10% will increase by 19–65% under low emissions, 148–243% under high emissions and about 16% with a global mean warming of 2°C relative to 1.5°C. It is anticipated that sites most vulnerable to shrinkage are in the Mediterranean, Mexico, Central America and South Africa, highlighting that climate mitigation is essential for future Ramsar wetlands conservation (Xi *et al.*, 2021).

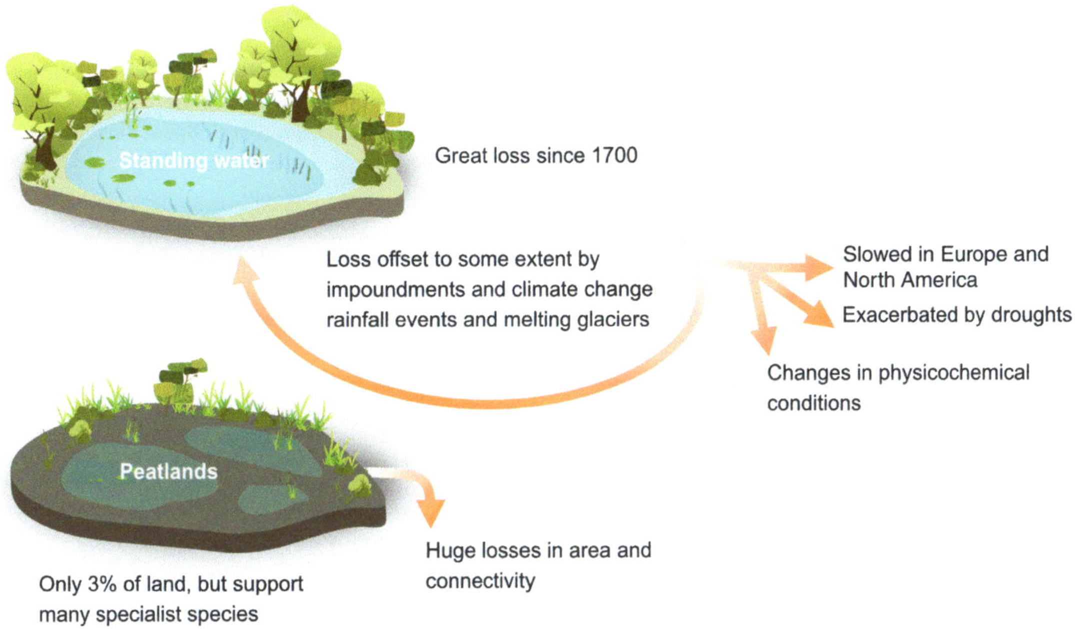

Fig. 5.4. There has been a great loss in the area of global standing water since the year 1700, although the loss has slowed in Europe and North America in recent years. Droughts exacerbate standing water loss in some areas of the world. Decrease in water level and increased water warming is changing physicochemical conditions. The standing water losses are partially offset by impoundments and by events associated with climate change in some areas. The tiny extent of peatlands, typified by many habitat-specialist species, have both decreased in total area and their connectivity with each other. (Illustration: Robyn Symons).

While peatlands cover only about 3% of the land surface, mostly in the northern hemisphere, they are special habitats for many species. In Finland alone, more than half of the peatlands have been drained, mainly for forestry, threatening many species, including dragonflies (Elo *et al.*, 2015). In Switzerland, wetland coverage substantially decreased after 1850, a time when wetland patches were largely connected by a 500 m buffer around all wetland patches enabling dispersal of wetland organisms. Despite extensive wetland loss, the wetland networks remained largely intact until 1950, but then collapsed into small groups of only a few wetland patches. Besides direct loss of wetland habitats, increased habitat fragmentation is now limiting metapopulation dynamics and hindering genetic exchange between populations (Gimmi *et al.*, 2011).

5.1.4 The debate on freshwater insect decline and partial recovery

While van Klink *et al.* (2020) estimated an increase in global freshwater insect biomass and abundance of about 11% over recent decades, this estimate was largely based on a strong bias toward the cooler regions of the northern hemisphere. Jähnig *et al.* (2021) challenge these results, not only because of the geographical bias, but also because sensitive species could be replaced by ecologically more tolerant ones. Furthermore, Desquilbet *et al.* (2020) indicate methodological flaws in the analysis. Nevertheless, at least in central European streams, there has been an overall increase in both abundance and species richness of aquatic insects over the last few decades (Haubrock *et al.*, 2023b). This improvement in aquatic biota is largely due to release from stressors, especially through river rehabilitation, as well as the spread of generalist and vagile taxa and those adapted to warmer conditions, especially at degraded sites (Pilotto *et al.* 2020). At higher taxonomic levels, river macroinvertebrates in the UK over the years 2002–2019 have not fared so well. Although Odonata as a group was included, and was stable, the sampling did not account for any possible shifts in species composition (Powell *et al.*, 2023). At stream sites across forested, grassland, urban and agricultural land types across the USA, a 27-year study indicated macroinvertebrate density declined by 11% and richness increased by 12.2%, while insect density and richness declined by 23.3% and 6.8%, respectively. Furthermore, differences in richness and composition between urban and agricultural versus forested and grassland streams increased over time. Urban and agricultural streams saw a loss of the few disturbance-sensitive taxa they once had while gaining disturbance-tolerant taxa (Rumschlag *et al.*, 2023).

5.2 Important Conceptual Issues for Dragonfly Conservation

5.2.1 Shifting baseline syndrome

In preparation for tackling pervasive and more extreme climatic events we must overcome a perceptual challenge known as the 'shifting baseline syndrome' (Pauly, 1995). In essence, with ongoing environmental deterioration, people's valuing of nature, and even appreciation of its crucial role in our well-being and survival, is continually being eroded with current environmental conditions being accepted without being questioned. What was considered as a normal environment and levels of biodiversity by past human generations is continually giving way with each new generation to a new normal: living in a continually depleted biotic world, and in effect, having dependence on a more precarious environmental state (Fig. 5.5; and also Fig. 7.6).

This shifting baseline must be recognized when monitoring general trends in insect decline. Different results come from natural areas versus those being continuously impacted by habitat loss through increasingly extensive and intensive urban, suburban and agricultural development (Forister *et al.*, 2023).

Addressing this shift in perception of the baseline starts with the young, both at home and in the school, where there needs to be more 'rescuing the extinction of experience' through more contact with wild nature (Samways, 2007). This process involves being in touch with nature first-hand and understanding the dire consequences for the future should everyone not play their part in looking after what we have and being personally involved in restoring ecosystems where possible, which also involves engaging with insects (Simaika and Samways, 2018). Fortunately, we are seeing concern and a call for

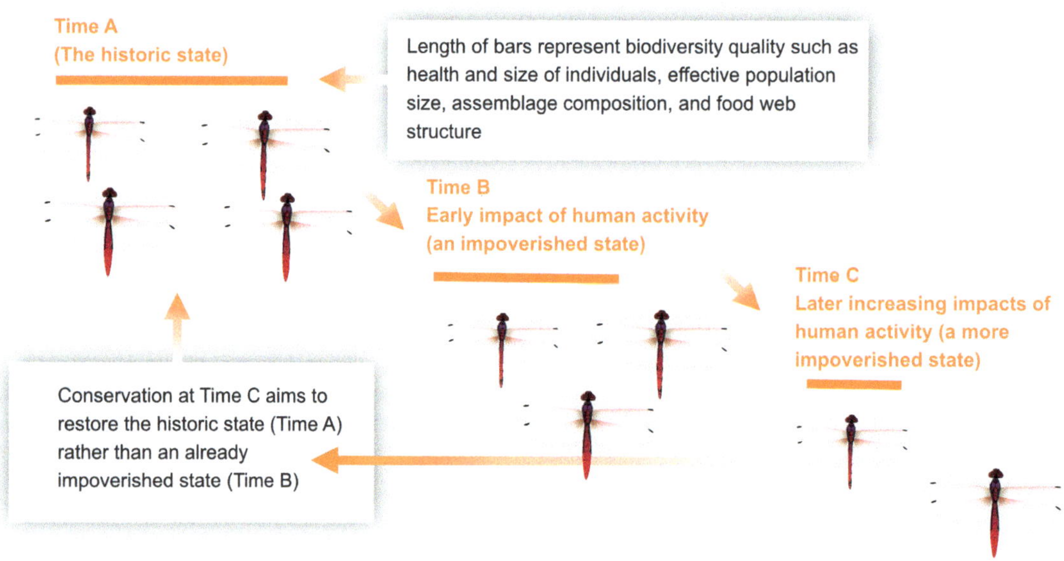

Fig. 5.5. The shifting baseline syndrome. Each human generation is experiencing an increasingly depleted biotic world away from the historic state. Conservation aims to restore the historic state rather than an already partially impoverished state, although the dynamics of dragonfly population shifts must also be considered. (Illustration: Robyn Symons).

action already at the COP (Conference of the Parties) Climate Change Conferences where the younger generation are centrally involved, meaning that there is greater hope for dealing with the shifting baseline syndrome.

An important extension on the points above is that while we may rank and measure threats to changes in current global biodiversity, conservation action mostly takes place at specific localities. Across the world these localities may not only differ greatly in both type and level of threat, but also in the species and ecosystems present. Furthermore, there is the socio-economic context. What this means is that population levels and quality from one location to the next may differ. In other words, the baseline is different. The differences are usually the outcome of human pressure on the one hand and effective conservation on the other, while the ranking of threats depends on the local context and the methods that we use to measure them (Samways, 2020; Bellard *et al.*, 2022).

However, the shifting baseline must be viewed as something dynamic rather than fixed in this rapidly changing world. In the case of dragonflies in the Swiss canton of Basel-Stadt in 2008–2009, ten new species had been recorded since 1980, while five had disappeared since 1980 and another two since 1999. Besides some possible natural dynamic changes in dragonfly population locations, there have been the negative effects of increased direct human pressure and the more pervasive effect of climate change, leading to dragonfly population shifts. In turn, these negative effects have been countered by restoration activities, leading to an overall net increase in local species numbers (Küry and Christ, 2010).

5.2.2 Complexity of predicting future security for dragonflies

Species are adapted to seasonal change in environmental conditions, having been exposed to these regular environmental oscillations over many millennia. However, the degree of seasonal change varies from year to year, often leading to strong weather events. These changes in environmental conditions are now being exacerbated by the increasingly rapid anthropogenic climate change, a major threat for both temperate and tropical insects (Johansson *et al.*, 2020), with associated extreme weather events becoming more widespread (Ripple *et al.*, 2021).

Understanding the effect of various stressors on food webs is critical for conservation action. One of the reasons for this is that adult dragonflies, insectivorous birds and bats share a remarkably similar arthropod food base, yet are phylogenetically distantly related competitors. The implication here is that we can envision a future where the food base decreases due to contamination and/or their mass emergence phenology is disrupted due to climate warming. Repercussions could be population crashes, especially as there is much competition for the same food items (Vesterinen *et al.*, 2020). Although small temperature anomalies result in weak compensatory dynamics, extreme events synchronize dynamics (Evans *et al.*, 2022). Importantly however, climate change models tend to underestimate the effect of extreme weather events on natural terrestrial and agricultural systems and, at least in Europe, also underestimate the adverse effects on water systems, owing to water retention measures (Schewe *et al.*, 2019). However, the current issue is that today's populations are facing increasingly frequent and more extreme weather events than in the immediate past (Harvey *et al.*, 2023).

How populations and communities respond to current environmental change depends on the events and filters that they have experienced in the past. We get a glimpse of this in southern Africa, where there has been a long history of El Niño–Southern Oscillation (ENSO) events with periodic floods, droughts and fires. Some dragonfly species in the region may move to more southern latitudes to avoid adverse environmental conditions. One biogeographically dynamic species, *Aciagrion dondoense*, had never been recorded in South Africa before the year 2000. However, following an intense ENSO wet phase involving Cyclone Eline and intense flooding farther north in Mozambique, it established at some coastal marshes. In 2001, at one South African marsh it became so abundant that it was the dominant species. Then when the ENSO dry phase started and the marsh dried up, it decreased in abundance and had completely disappeared by 2006, and not seen since in the country (Samways and Deacon, 2022).

In the same area, some species, such as the rare but highly mobile *Orthetrum robustum*, are abundant in many shallow pools during ENSO wet phases, retreating to just a few permanent and richly vegetated coastal lakes during droughts. Some other species shift in elevation ranges in different years as suitable stepping-stone habitats become available (Samways and Niba, 2010). Furthermore, the ENSO dry phase is a time of high fire risk. Local endemic dragonfly taxa are well adapted to these natural fire impacts. For example, the four *Syncordulia* species, although confining themselves to their mountain retreats in the highly fire-prone Cape Floristic Region, are strong fliers. On emergence, they move to forage on the dry mountain slope. But when fire threatens, they retreat into still-green vegetation refuges along river margins and among large boulders (Samways and Deacon, 2022) (Fig. 5.6).

Similarly, in the fire-prone Mediterranean region of Spain, despite a devastating forest fire, five dragonfly species survived the fire as larvae: *Boyeria irene*, *Gomphus pulchellus*, *Onychogomphus forcipatus*, *O. uncatus* and *Cordulegaster boltonii*. This was because these species with long life cycles are resilient and adapted to fire disturbance, unlike species with short life cycles (Márquez-Rodríguez *et al.*, 2023).

The issue today is that extreme weather events are becoming more intense and more frequent, leading to a new and ever intense global norm, and likely to leave a long legacy over large geographical areas, especially in the case of drought (Ball-Damerow *et al.*, 2014). For dragonflies, as with all species, it is environmental bottlenecks that determine the fate of population survival in any one place, not the average conditions. Besides, we must also be aware that the degree of population change relative to stressors can vary both over space as well as time (Blüthgen *et al.*, 2022). As environmental legacies and ever-increasing global change are likely to have greatly varying effects at different locations, it means that it is essential to focus on the mechanisms behind the biological response function to understand and predict the effects of future extreme climatic events (van de Pol *et al.*, 2017).

In the case of dragonflies, there are many considerations associated with their individual, population, assemblage and community interrelated responses to climate change. These include development times and patterns, environmental temperature, photoperiod, diapause, phenology, pigmentation, basking behaviour, microhabitat use, activity patterns, wing-whirring, trophic interactions, body size, immune responses, geographic distributions, drought, pollution, landscape connectivity, sea-level rise, evolutionary adaptation, processes at geographic range margins and extinction risk (Hassall and Thompson, 2008). This extensive list indicates the complexity of understanding, let alone predicting, the precise effects of climate change on dragonflies, especially when there are additional environmental factors to consider like spatial scale, topographic heterogeneity and interaction between stressors (Fig. 5.7).

5.2.3 Significance of multiple stressors

A 'stressor' from a conservation perspective is an anthropogenic influence on a range of ecological hierarchies from populations to ecosystem processes (Orr *et al.*, 2020; Jackson *et al.*, 2021). Today, human-induced effects are so widespread and pervasive that we need to think in terms of multiple stressors (Birk *et al.*, 2020). The overarching stressor event of climate warming interacts with many localized stressors, such as landscape transformation and pollution (Wang, M. *et al.*, 2022), with, for example, ponds and shallow lakes requiring strategic and comprehensive assessment (Labat and Usseglio-Polatera, 2023). In turn, local stressors vary in time, duration and intensity. Stressors may be distal (such as a polluting upstream event) or proximal (such as removal of riparian trees) both of which can be interactive (Fig. 5.8).

There are essentially two sets of multiple stressor events. The first includes synergistic interactions, where the effect of two or more stressors is more than the sum of their components. The second refers to antagonistic interactions, where the effect of two or more stressors is less than the sum of their parts. When two or more synergistic interactions are negative on a defined ecological component at a certain time and place, while also including ecological memory (how past stressors influence current and future responses), we refer to their combined effect as being adversely synergistic. The result may be reduction of ecological integrity and ecosystem change, usually with local loss of certain species, though with some species possibly benefitting from the combined stressors. The outcome is often food web simplification and community homogenization, where habitat generalists with a resilient biology are frequently the ones that prevail.

When the overall effects of urbanization and agriculture, with their multiple stressors, are compared, the two types of landscape transformation have a similar effect on dragonfly assemblages. The outcome of the two types of landscape change is that while there may be little change in species richness, the assemblages are homogenized through a shift to a collective of more habitat-generalist species at the expense of the specialist endemics (Kietzka *et al.*, 2018).

Fig. 5.6. In El Niño–Southern Oscillation event-prone southern Africa, the local dragonflies are well-adapted to extreme weather events having experienced them many times in the past. As drought and fire often go together in areas of greatly fluctuating water levels, the local species are highly dispersive habitat generalists with wide geographic ranges. When drought threatens, they move to perennial water bodies, mostly far away, which are often suboptimal for breeding. In contrast, and shown here, where perennial mountain streams are present, drought is not a problem, but fire is. However, the highly adapted endemic species like *Ceratogomphus triceraticus* (top left) and *Syncordulia* species (see Fig. 3.19) soon detect a fire approaching (top right), quickly leave their montane foraging areas, which can suffer major damage (middle left), and head for riparian refuges. However, they avoid riverine areas that are not refuges (middle right). Instead, they relocate to fire refuges that do not burn but are close to the water's edge, which may be treed (bottom left) or composed of low-canopy vegetation (bottom right). These species are not geographically wide dispersers but are nevertheless highly mobile, enabling them to swiftly locate safe refuges. (Photos: Michael Samways; Illustration: Robyn Symons).

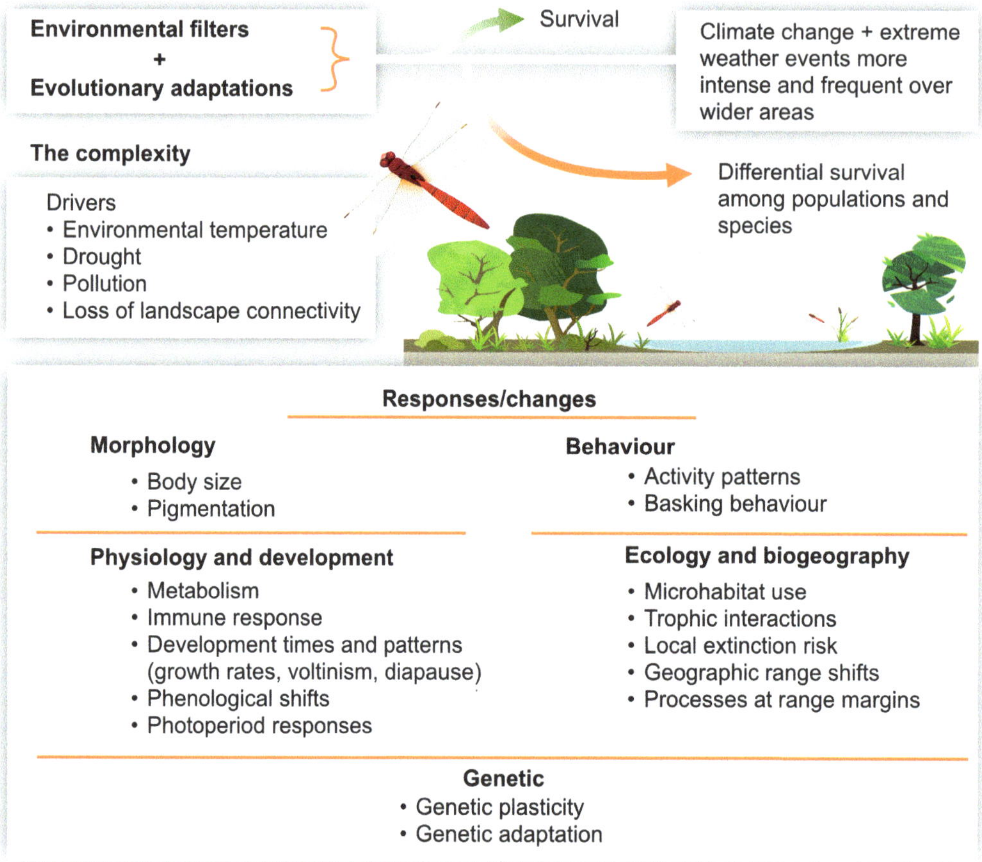

Fig. 5.7. Evolutionary adaptation to successive environmental filters over deep time has led to the survival of the suites of species we see today. However, recent climate change and attendant extreme weather events are causing differential survival of dragonfly populations and species in many areas. The complexity of predicting how each population and species will respond to the various climate change-associated drivers is immense. The responses and/or changes that are likely to occur relate to many aspects of dragonfly biology, including morphology, physiology and development of the larva, adult behaviour, and ecology and biogeography, all underpinned by genetic plasticity and possible genetic adaptation as time passes. (Illustration: Robyn Symons).

Furthermore, species can experience different sets of stressors over time. A study of *Coenagrion hastulatum* in the Netherlands between 1921–2021 indicated that in earlier years, changes in human use of ponds along with increased sulphur and nitrogen deposition were the main threats. In recent years, the main stressor set is now different, being severe droughts and nitrogen deposition. This means that today restoration of groundwater systems and rewetting measures are a conservation priority (Termaat *et al.*, 2023).

The effects of land-use and climate change are affecting many aquatic organisms, including macroinvertebrates, as well as stream metabolism and decomposition processes. Increasing temperature alters the physicochemical conditions, such as oxygen concentration and nutrient dynamics, and the energy base of invertebrate food webs. Impacts on sediment lead first to changes in algal and microbial species composition and productivity, as well as organic matter decomposition rates. These processes can be aggravated by agriculture and urbanization, leading to more variable flow rates with higher pollutant loading, affecting locally adapted species and shifting community composition and function (Taniwaki *et al.*, 2017). Increased sedimentation

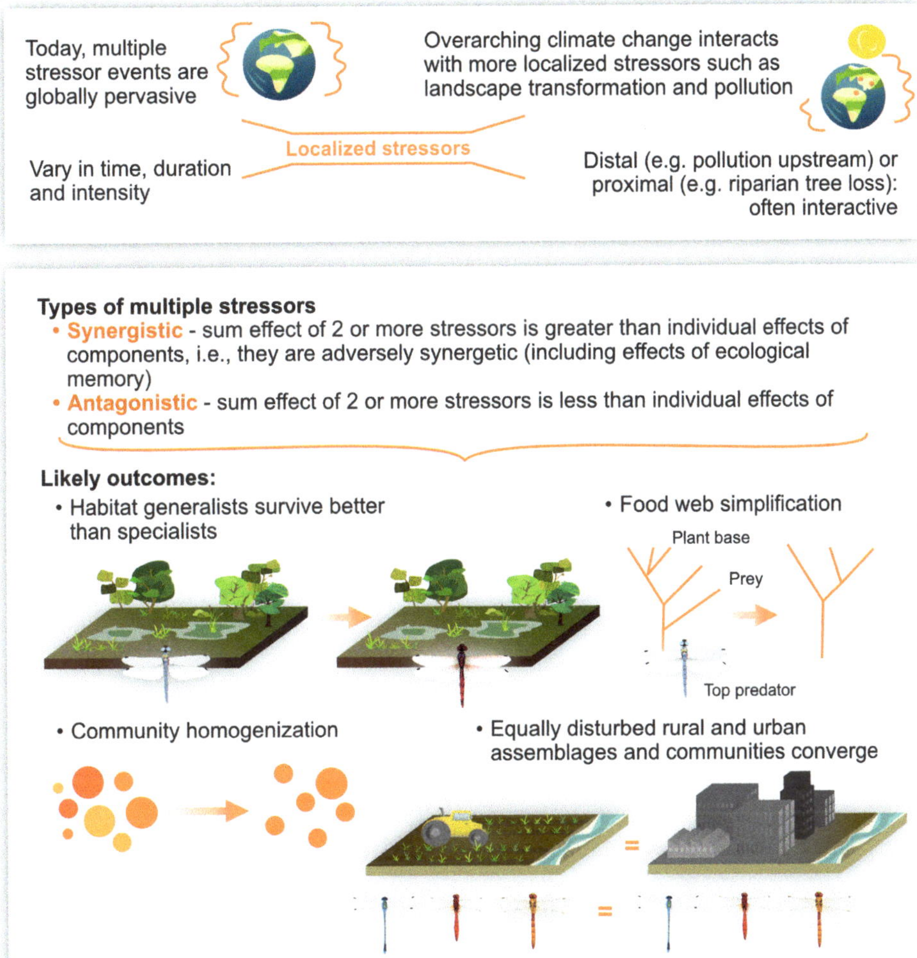

Fig. 5.8. Multiple stressor events are now occurring all around the world. The overarching effect of climate change interacts with many localized stressors which have certain properties. There are two types of multiple stressors. Of greatest concern are those that are adversely synergistic, as they potentially have a major impact on organisms. Some of the stressors may have ecological memory, meaning that while no longer detectable, their effect may still be present. For example, pollutants could still be present in the sediment even though the pollution source may now be under control. Multiple stressors generally have some likely outcomes which are widespread globally, making it important to understand these effects for dragonfly conservation in all parts of the world. (Illustration: Robyn Symons).

can be gradual over time, with, for example, increasing intensity of cereal production and high numbers of livestock enabling fine sediment to accumulate in rivers (Grabowski and Gurnell, 2016).

Across the USA, while anthropogenic activities have a negative effect on streams and rivers, the strongest effect is from natural geoclimatic and geomorphic factors. Nevertheless, sediment characteristics are affected by enhanced flood flows, with basin and riparian agriculture the main cause of streambed instability and an excess of fine sediments. Riparian land transformation in general reduces riparian vegetation cover complexity, and when combined with reduced summer flow levels, are the main determinants of instream habitat simplification (Kaufmann *et al.*, 2022a).

Furthermore, in over 1.2 million km of US streams and rivers, flowing water condition is poor over 22–24% of their length, due to human disturbance of riparian zones, streambed sediment change and loss of riparian vegetation cover, with another 14% due to a decrease in instream habitat complexity. Nationally, stream length in poor condition ranges between 4–42%. Relative risk estimates indicate that streams and rivers with poor sediment ratings, low riparian cover complexity and/or reduced instream habitat cover conditions are 1.4–2.6 times more likely to have fish or macroinvertebrate assemblages in poor condition (Kaufmann *et al.* 2022b).

European rivers are affected by many stressors which affect water quality and hydromorphology, with currently only about 40% of the rivers reaching 'good ecological status' in terms of ecological integrity according to the European Water Framework Directive. Stressors such as extent of urban and agricultural land use in the riparian zone, alteration of mean annual flow and of base flow, total phosphorus load, total nitrogen load and various toxic compounds suggest that overall 39.4% of changes are due to altered hydromorphology (morphology: 23.2%; hydrology: 16.2%), 34.4% to nutrient enrichment and 26.2% by toxic substances. Importantly, over half of the total adverse effect is due to stressor interaction, especially nutrient enrichment and toxic substances frequently and strongly interacting. These findings suggest that management strategies at the catchment scale are necessary to attain a good ecological status for the flowing waters in Europe (Lemm *et al.*, 2021) (Fig. 5.9).

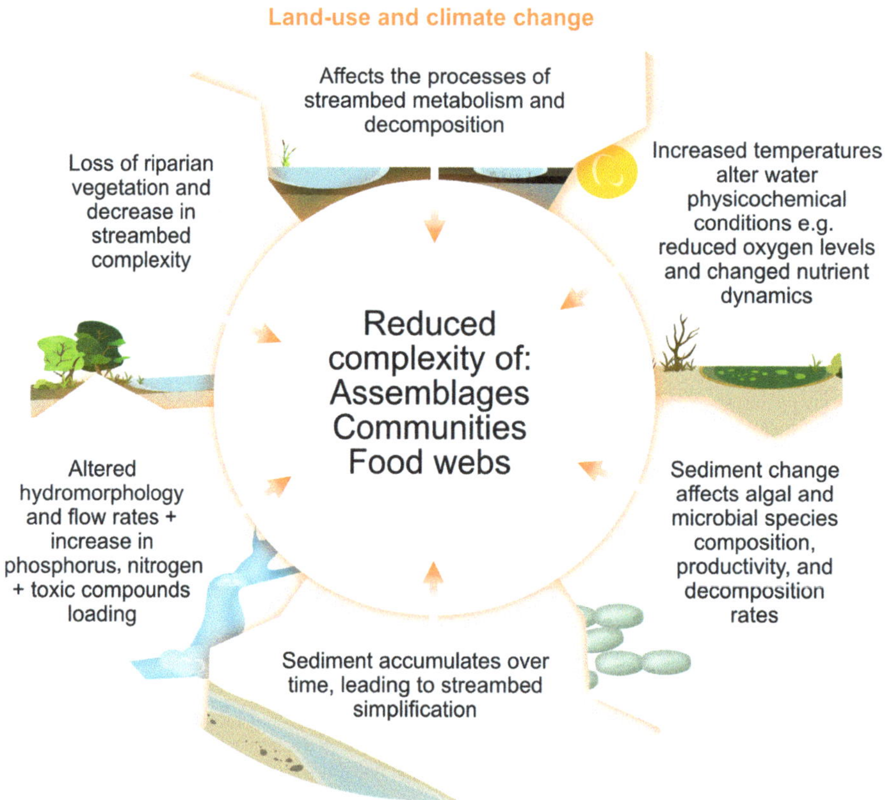

Fig. 5.9. The adversely synergistic effect of land-use and climate change affects the processes of streambed metabolism and decomposition. The adverse drivers include several interrelated variables related to increased temperature, sedimentation, altered hydromorphology and flow rates, and increases in system loading by phosphorus, nitrogen and toxic compounds. These altered states are also often related to loss of riparian vegetation and a decrease in streambed complexity. The outcome on biodiversity of these interrelated stressors is a reduction in complexity of assemblages, communities and food webs. (Illustration: Robyn Symons).

5.2.4 Multiple stressors affect multiple traits

There is an increasingly strong interaction between land transformation and climate change putting ever more pressure on the well-being of biodiversity (Neff *et al.*, 2022). A main spin-off from a warming world is that the overarching and continuous stressor of climate warming is resulting in stronger and more frequent weather events such as droughts, floods and high winds. Water-level drops especially can lead to temporary or even ongoing ecosystem change. For many aquatic insects, these changes can be virtually permanent, with drought, for example, extirpating a remnant isolated population in an arid zone (Fig. 5.10).

One way of viewing the effect of the suite of climate-related stressors is in terms of their different pressures on various organismal traits. There are stressors that have a direct and immediate effect upon the various organisms, such as loss of riparian forest cover. Other stressors have ecological memory (legacy effects) relative to the position of organisms in the food chain. We see this among some material stressors, such as pollutants, which can bioaccumulate along the food chain. Yet other

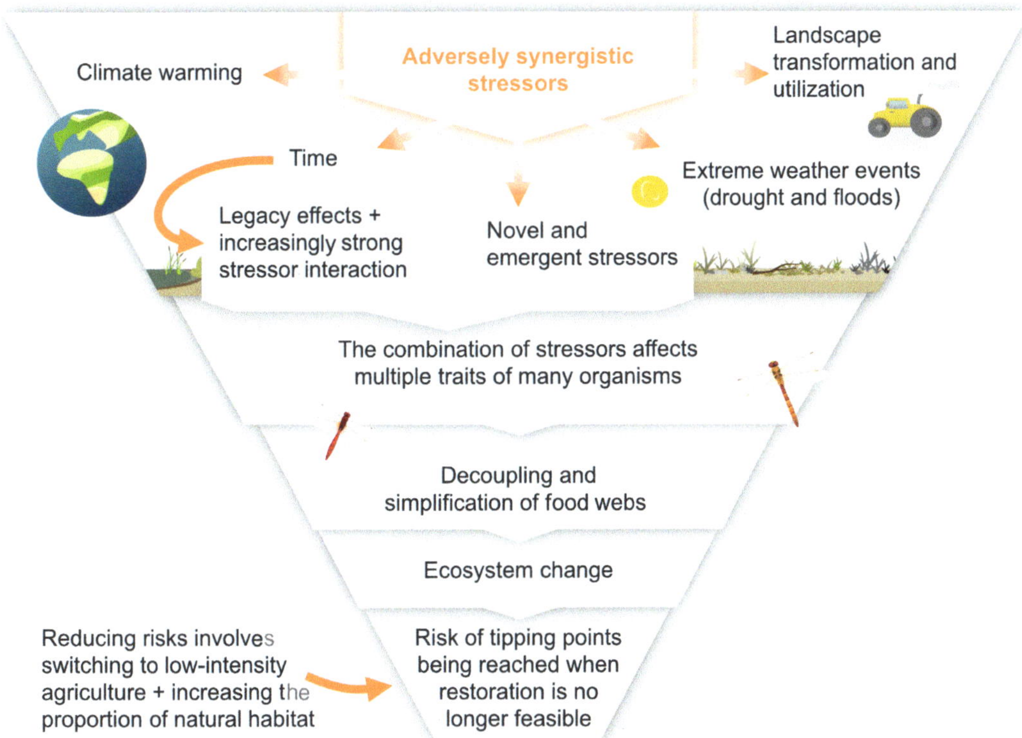

Fig. 5.10. In basic terms, there is a triad of stressors: climate warming, extreme weather events, and landscape transformation and utilization. The interaction between these is usually adversely synergistic. In addition, often the most severe stressors are those that are novel to the various species, and to which there has been no previous evolutionary adaptation. Emergent stressors (those that appear as a threat when they are detected as such) must also be considered. Time is also an important component as there may be legacy effects, such as pollutants bound to sediment, and as time goes by there are often increasingly strong stressor interactions. The combination of all the stressors impinges upon many traits of many organisms, which can decouple and simplify food webs, leading to ecosystem change. The concern today is that there is a risk of tipping points being reached where the changed and simplified food webs cannot be restored. Evidence to date suggests that the risk can be reduced by switching to low-intensity agriculture and increasing the proportion of natural habitat. (Illustration: Robyn Symons).

stressors, such as climate warming, can cause developmental shifts among populations. In turn, extreme weather events can be better understood relative to traits associated with decreasing water levels. A dragonfly may disappear from a pond that is drying up, with no imminent possibility of re-establishment when potential source populations are too far away relative to its powers of dispersal, and that assumes that the rain replenished pond is still an optimal habitat. This means that multiple stressors can impinge on different traits of many organisms in the same place at the same time, decoupling food web structures (Thackeray *et al.* 2016).

The interaction between historical climate warming and intensive agriculture is associated with reductions of almost 50% in insect abundance, and 27% in species richness of insect assemblages relative to those in relatively natural habitats with low rates of historical climate warming, particularly in the tropics. Importantly though, high levels of availability of nearby natural habitat often mitigate reductions in insect abundance and richness associated with low-level agricultural land use under substantial climate warming. Where high levels (75% cover) of natural habitat are available, abundance and richness are reduced by 7% and 5% respectively, compared with reductions of 63% and 61% in areas where only 25% natural habitat is present. Overall, this suggests that insect biodiversity will likely benefit from mitigating climate change through the preservation of natural habitat throughout landscapes, while also reducing the intensity of agriculture (Outhwaite *et al.*, 2022).

The adverse effects of air pollutants and persistent organic pollutants, such as pesticides, are enhanced by a warming climate. The outcome is altered ice and snow melt, food webs, lipid dynamics and organic carbon cycling, resulting in pollutants and pesticides accumulating in water, soil and body tissues. These impacts are aggravated by altered biotransformation of contaminants to more bioactive metabolites and impaired homeostasis, especially affecting species at the edge of their physiological tolerance. Temperature rise also increases volatilization of persistent pollutants and pesticides into the atmosphere, while increased rainfall enhances deposition of airborne pollutants and pesticides through elevated run-off. During drought events these effects are often adversely synergistic with reduced water volumes and increased salinity. Overall, the concern is that as tipping points are reached, restoration becomes increasingly difficult (Noyes *et al.*, 2009).

In arid areas, such as North Africa, wetlands are valuable in supporting species that are rare and/or irreplaceable. When high anthropogenic pressure is combined with increased climate warming, the system will likely have reduced resilience as the combined impacts become too much for some species to survive (Khelifa *et al.*, 2022). A severe drought in the region caused a reduction in dam water depth and discharge into a river, with associated changes in water chemistry. While multiple stressors like climate and human water demands could have extirpated local populations of *Calopteryx exul* in the river, the species is sufficiently responsive to avoid both low- and high-water levels (drought and floods) (Khelifa *et al.*, 2021b).

Mediterranean-type ecosystems are particularly susceptible to drought. When the odonate assemblages in the North African region are compared with those in the southern African region, both are subject to multiple natural and human-induced stressors, likely to be exacerbated by future climate change and human population growth. While dragonflies in North Africa are threatened by a combination of both climatic and human threats, those in southern Africa are mostly threatened by climatic factors, especially drought (Khelifa *et al.*, 2021a).

Even without invoking climate change, there can be legacy effects, lasting many years, even decades. A study of Danish streams showed that groundwater, as well as rainfall bringing down atmospheric pesticides and pollutants, along with surface run-off, can pose a severe ecotoxicological threat to the stream ecosystem. Both contemporary and legacy pesticides, along with their metabolites and impurities, can enter the body of the stream and bind with sediment causing chronic toxicity that can last a long time (McKnight *et al.*, 2015).

Fertilizer is the most important source of nitrogen (about 50%), followed by its presence in rainfall, exported to streams as nitrate (Caraco, 2021). In Italian ponds, there is a decrease in dragonfly species richness with increasing nitrogen concentration. Only *Lestes barbarus*, an ephemeral pond inhabitant, is tolerant of such eutrophic conditions (Carchini *et al.*, 2007).

Stream macroinvertebrates in New Zealand are subject to multiple stressors such as nutrient enrichment and fine sediment input in relation to climate change. There is a complex interplay between the stressors, with benthic macroinvertebrates responding according to the interaction between habitat quality (with or without sediment), resource availability (with or without nutrient enrichment) and a tendency to drift downstream, especially in response to deposited fine sediment, or to emerge early as temperature rises. Presence of sediment and increased temperature together result in a decrease in the body size of organisms (Piggott *et al.*, 2015).

Similar findings come from Europe, where high sediment load also causes benthic macroinvertebrates to drift. In comparison with sediment, the effects of nitrogen and phosphorus are relatively weak, and the interactive effects are complex. While high sediment reduces total macroinvertebrate abundance, this decrease is amplified by phosphorus enrichment at low sediment levels, with the opposite effect at medium levels and little effect at high sediment levels (Davis *et al.*, 2018).

In Sweden, extreme weather events dominate flood risk for planning periods lasting a few decades. However, in the longer term, toward the end of the century, the flood risk is instead dominated by the risk of increased sea-level rise (Hieronymus and Kalén, 2022).

Not surprisingly the study of multiple stressors is central to dragonfly conservation, whether in terms of their populations and place in food webs, or the extent to which their local habitat is secure in the face of ongoing multiple stressor events.

While organisms are adapted to a set of natural gradients, such as pH, and are a baseline for understanding the effects of these gradients on locally adapted communities, it is the appearance of novel stressors that species have not experienced in the past, and to which they have little adaptive capacity, that can pose the greatest threat. Other stressors may become emergent and are currently unknown, or appear insignificant, but might in the future place further pressure on populations and ecosystems.

5.2.5 Spatial effects: Widespread versus local stressors

The spatial scale of freshwater assessments is fundamental to understanding widespread impacts versus local ones (Fig. 5.11). In the Neotropical savanna of Brazil, the most important regional stressors are increased levels of agricultural development, water turbidity and fine sediment on macroinvertebrate diversity. At the local scale, an increased proportion of pasture and nitrogen are the key stressors (Silva *et al.*, 2018).

In a nationwide study of Mexican dragonflies in undisturbed forested areas of all types, 23.4% of all national species are maintained, dropping greatly when trees are removed. Agricultural areas support 15.9%, and urban areas, even the megalopolis of Mexico City, account for 10.5%. The remainder are in mixed or moderately degraded areas, with arid areas supporting few species. Overall, species richness is promoted by a combination of oligotrophic waters, an abundance of macrophytes and good riparian cover. Interestingly though, highest species richness is in intermediate disturbance areas. While these large-scale patterns are apparent, a local focus at smaller spatial scales is more appropriate for conservation action. A reason for this is that species populations with various levels of functional connectivity and site selection traits have different ecologies according to whether they are zygopterans or anisopterans, small- or large-bodied (e.g. *Ischnura ramburii* versus *Remartinia luteipennis*), display male territorial behaviour or not (e.g. *Hetaerina titia* versus *Enallagma novahispaniae*), inhabit running or standing water (*Brechmorhoga tepeaca* versus *Orthemis discolor*), or inhabit open or closed canopies (*Erythemis vesiculosa* versus *Brechmorhoga praecox*) – all having widespread geographical distributions but display higher or lower population levels as local environmental conditions change, leading to species sorting (Cuevas-Yáñez *et al.*, 2017).

The issue of threats relative to spatial scale is an important one. In southern Africa, threats to grassland species are ostensibly mostly from alien tree plantations. Analysis at the coarse spatial scale (1:250,000) concludes that afforestation with alien trees is one of the most severe agents of transformation of indigenous grasslands (Neke and Du Plessis, 2004). However, at the smaller spatial scale, the region supports extensive networks of conservation corridors which provide set-aside natural habitat to such an extent that these conservation areas support dragonfly assemblages in both running and standing waters that are equal in dragonfly diversity to those in neighbouring extensive protected areas (Kietzka *et al.*, 2015, 2021b; Pryke *et al.* 2015).

However, we must be careful not to overgeneralize. The tropical forest volcanic mountains of Mayotte island in the Indian Ocean have small streams trickling down their ancient slopes. Here, alien trees like mango have replaced the indigenous trees in places. This replacement does not deter establishment by dragonfly species, even by the endemic species, as the deep shade cover with sunflecks is

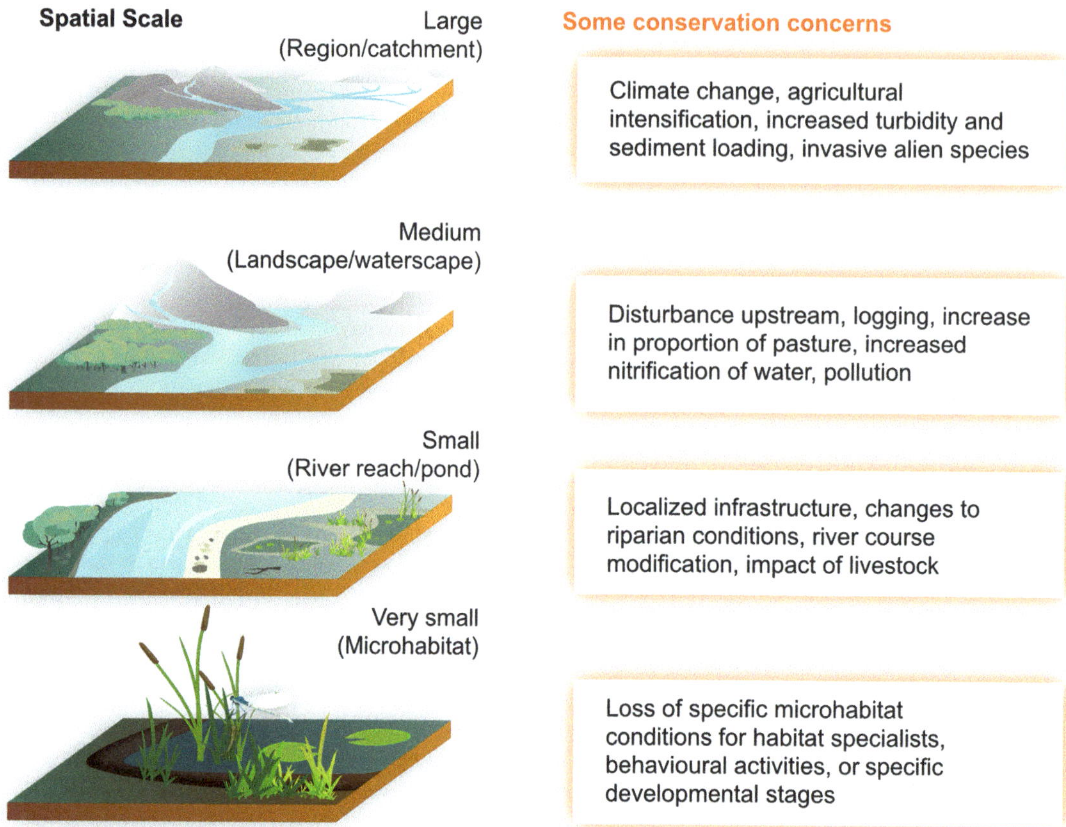

Fig. 5.11. Recognition of the importance of spatial scale is fundamental in ecology and conservation. There are several caveats here, including the species-area effect, where there will be more species the larger the area being considered. Furthermore, large-scale effects can trickle down and compound stressors at the local scale. Shown here are some conservation concerns at various spatial scales. The list is not totally inclusive and will vary from one region or local area to another according to climate, traits of the focal species, and historical and current land-use activities. (Illustration: Robyn Symons).

equivalent to that provided by the native trees (Samways, 2003).

Across Czechia, decreased levels of naturalness due to environmental degradation do not reduce species richness but have a major effect on species composition, especially relating to threatened species. These findings emphasize that natural areas are required if we are to conserve the threatened species (Dolný *et al.*, 2021).

5.2.6 Extinction debt

Adverse stressor impacts on dragonflies are often manifested long after the stressor events. The populations meanwhile have incurred an 'extinction debt'. Furthermore, impacts on one population may later affect other populations, possibly even leading to regional extinction (Fig. 5.12).

A study across Europe on several taxa, including dragonflies, suggests that it is essential to recognize threats in their early stages of impact to address the risk of an extinction debt which may have been accruing over many years, if not decades, following impact (Dullinger *et al.*, 2013). This means that we are reducing populations in a slow but sure way without really appreciating the extent. In the UK, freshwater macroinvertebrates, including dragonflies, showed an overall decline until the mid-1990s, then recovered to pre-1970 levels, likely due to water quality improvements (Outhwaite *et al.*, 2020), offsetting climatic debt (Vaughan and Gotelli, 2019) and extinction debt.

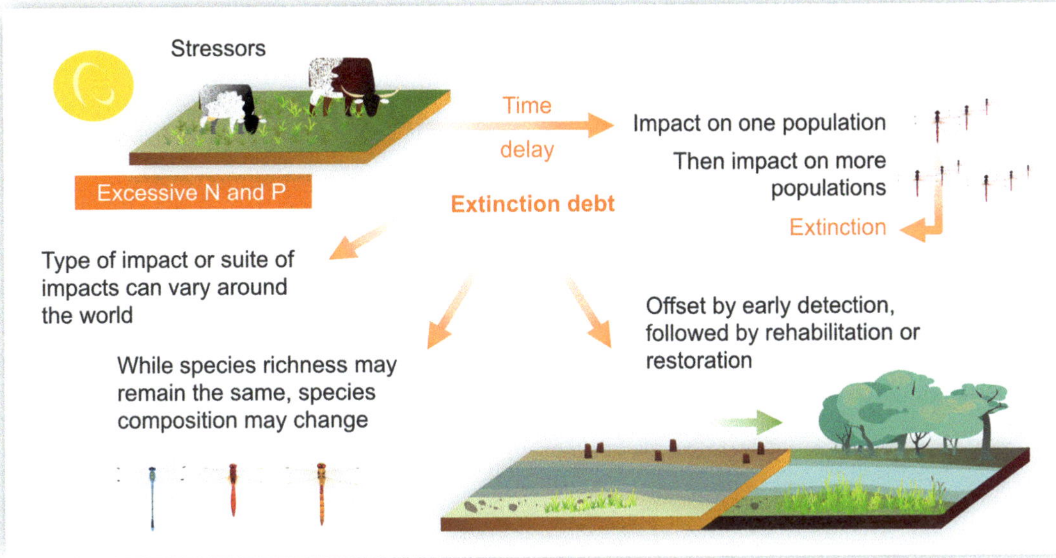

Fig. 5.12. The impacts of stressors on dragonfly populations often occur following a time delay, known as the extinction debt. The impact on one population may be followed by an impact on other populations, which can eventually lead to regional extinction. However, the type of impact can vary greatly in different parts of the world and may not necessarily (according to the type of impact) lead to a decline in species richness, but instead a decline in species composition in any one area due to differential responses by the various species at different locations. This means that it is important to detect the effect of stressors in their early stage to instigate effective rehabilitation or restoration activities. (Illustration: Robyn Symons).

In Britain and Ireland, despite a huge number of contributions from citizen scientists, the effect of any one impact on dragonflies can be hard to ascribe. Nevertheless, the main drivers appear to be habitat loss and degradation through land drainage, afforestation, acidification and lack of appropriate management. While 41% of resident species and/or regular migrants have increased their area of occupancy, this does not necessarily mean that their abundance has remained high. These range changes are also a warning regarding the effects of climate change, especially considering that 11% of dragonfly species have declined in terms of their area of occupancy (Taylor *et al*., 2021).

There is still great concern for freshwater systems in the warm parts of the globe and where human population growth continues to be high. For example, Romero *et al.* (2021) focused on 20 years of data (2000–2019) from the upper Paraná floodplain, Brazil, and found a pervasive decline in abundance of all insect orders, including dragonflies, and families regardless of their ecological function or body size. Drivers of decline include invasive alien fish, water transparency and changes in nitrogen–phosphorus ratios. However, it is often difficult to pin down specific drivers of change in insect diversity as there are so many different types of impact and they often vary from place to place, across realms and across taxa (Pilotto *et al*., 2020; Wagner *et al*., 2021).

In Mexico between 1980–2010, there was an acceleration of land-use change, greatly affecting dragonfly assemblages according to the type and intensity of change, spatial scale and hydrologic region. Overall, dragonflies have been resilient to the land-use change, with species richness higher in transformed areas than natural ones. However, species composition changed, with secondary vegetation helping to slow the loss of species in *Macrothemis*, *Phyllogomphoides* and *Progomphus* (Rocha-Ortega *et al*., 2019).

5.3 Examples of Multiple Stressors on Dragonflies

5.3.1 Various types of multiple stressors

Local impacts can greatly affect the freshwater fauna, including dragonflies, of montane streams and rivers (Fig. 5.13). Dragonflies along the Guadiamar

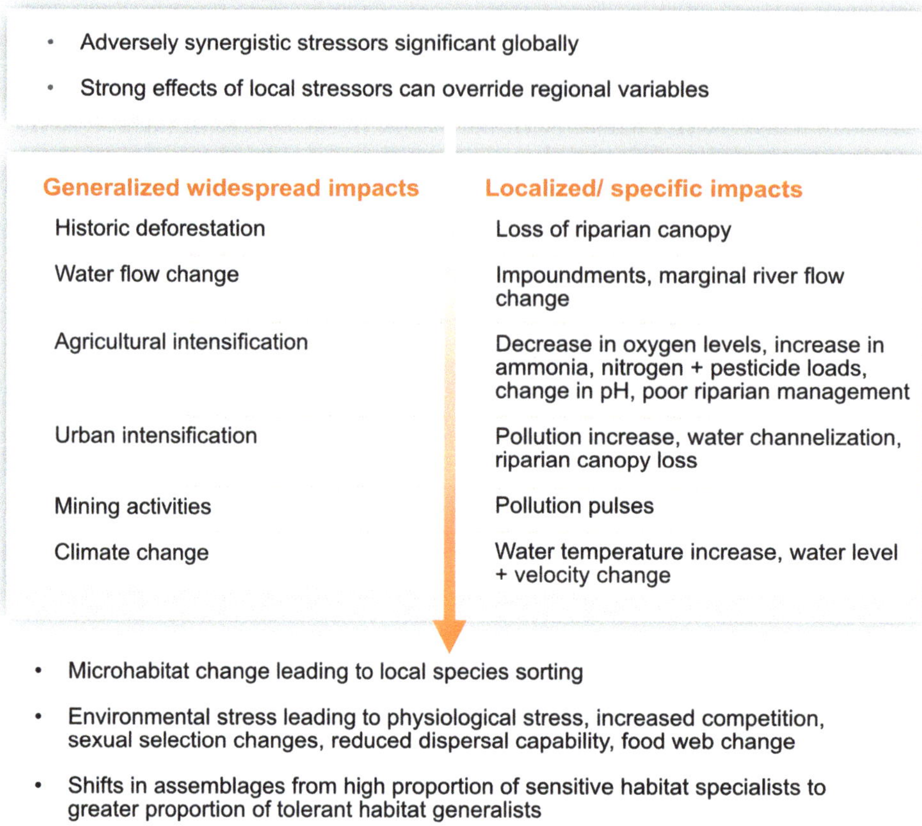

Fig. 5.13. Various types of adversely synergistic stressors affect dragonfly assemblages around the world. Sometimes these stressors are so great that they override regional factors. The stressors may be roughly divided into generalized/widespread impacts and more localized/specific impacts. Effects on dragonflies arising from stressor impacts include fine-scale sorting of species as microhabitats change. These changes affect several physiological and behavioural traits simultaneously, as well as changing food web structures. In naturally forested areas, often assemblages show a shift from ones with a high proportion of sensitive shade-loving habitat specialists to ones with a proportionately higher number of tolerant sun-loving habitat generalists at any one location. (Illustration: Robyn Symons).

river on the Iberian Peninsula have been subject to many human pressures, including historical deforestation, a major mining spill and various agricultural and urban impacts, degrading habitats of both the larvae and adults. These impacts reduce levels of gomphids, corduliids and calopterygids, while favouring species of open, sunny habitats, like certain libellulids and aeshnids, and species tolerant of habitat degradation like *Aeshna mixta* and *Ischnura graellsii* (Ferreras-Romero *et al.*, 2009).

Similarly, along the flowing Wadi Cherf river in Algeria, *I. graellsii*, as well as the localized endemic *Platycnemis subdilatata*, are highly tolerant of disturbance, including pollution from farming activities. Conversely, *Calopteryx exul*, *Coenagrion mercuriale* and *Gomphus lucasii* are highly threatened by human activity (Bouhala *et al.*, 2019). In the same region, odonate species richness in the Wadi Bouaroug river increased from 13 to 21 species from the early 1990s to mid-2010s. During that time, annual water temperatures increased by 1–4°C as river reach and physicochemical conditions changed along with loss of riparian vegetation and canopy cover, leading to an influx of sun-loving species such as *Paragomphus genei*, *Crocothemis erythraea*, *Sympetrum fonscolombii* and *Trithemis* spp. However, over the same time, habitat specialists, like *Coenagrion puella* and *G. lucasii*, decreased (Morghad *et al.*, 2019).

In Amazonian streams, both local variables (state of the local habitat, vegetation cover over the streams and certain physicochemical factors) and regional variables (elevation gradient, bioclimate and forest characteristics) determine the composition of dragonfly assemblages. This suggests that it is important to measure all these variables at various spatial scales to ascertain the effects and impacts of a changing environment on the dragonflies in the region (Oliveira-Junior *et al.*, 2019). However, when modified versus natural streams are compared, environmental conditions relating to the local environment, such as riparian vegetation, physicochemical conditions and human impact, are paramount, and override spatial factors in the region. The local environmental variables sort the odonate assemblages, with fine-scale habitat suitability the main determinant, especially for damselflies. But other factors such as competition, sexual selection, dispersal capability and more cryptic environmental variables may also be playing a role. This is especially the case for the small-bodied zygopterans, which in this region are more habitat specialist than the anisopterans (Oliveira-Junior *et al.*, 2017).

However, this is not to say that all anisopterans are safe in the region. Gomphids in the genus *Phyllocycla* are vulnerable to a whole host of human impacts, with only 55% of suitable habitat remaining in Brazil, mostly in the Amazon, and mining and hydroelectric power activities posing a threat to this stronghold (Araújo *et al.*, 2020).

In the Brazilian Cerrado Maranhense, some species are classified as generalists and others as specialists of untransformed habitat. Overall, habitat generalists are larger in size than species that specialize in intact local environments, with wing width and thorax length together being important in separating the two groups. There is also an interplay between body morphometrics and spatial niche size, with specialists not always having the lowest values, as they can occupy a range of microhabitats within the overall habitat space. However, of concern is that over half of the species are rare relative to the level of fragmentation among local aquatic environments, emphasizing that intact habitat, along with the attendant variety of optional microhabitats, is essential for conserving whole odonate assemblages in the area (Pereira-Moura *et al.*, 2023).

Similar findings come from across the Atlantic Ocean in Ghana, where river flow rate, percentage canopy cover and channel width are the key factors determining dragonfly assemblages. Agriculturally disturbed areas are dominated by sun-loving dragonflies, while the forests support the many habitat specialists (Seidu *et al.*, 2018).

In temperate regions, such as the UK, distribution of dragonflies along river systems relates to both natural changes and management impacts, such as weir construction, management of vegetation overhanging the river and agricultural run-off. Along the River Stour catchment, larval assemblages are directly influenced by marginal flow velocity, biochemical oxygen demand and phosphate concentrations, and indirectly by extent of shade cover and floating vegetation. Adults respond directly to shade cover, land-use type and bank height, and indirectly to biochemical oxygen demand and ammonia concentrations. While some human disturbance, such as weir presence, can benefit species richness, disturbance in general is harmful to the habitat specialists such as *Calopteryx splendens*, *Platycnemis pennipes* and *Libellula fulva*. Overall, the results suggest conservation efforts should focus on these habitat specialists (Hofmann and Mason, 2005). Furthermore, as seen in the case of New Zealand rivers, it is important for managers to address the issue of multiple stressors and the interactions between them (Townsend *et al.*, 2008).

Even dragonfly species with broad ecological tolerance can be sensitive to hydromorphic stressors such as scouring of the riparian vegetation, channel modification and loss of woody debris. Along a lowland river in Croatia, these stressors, along with changes in oxygen saturation, pH, ammonium, water temperature and total nitrogen, are important environmental variables for the larvae. However, while *Calopteryx virgo* and gomphids are associated with higher ammonium and nitrogen levels, the other species are not (Maoduš *et al.*, 2022).

5.3.2 Impact of pollutants

Coal combustion waste exposes dragonfly larvae to complex mixtures of potentially toxic metals and metalloids (Fig. 5.14). The highest concentrations occur in upstream sites close to contaminant sources. Body form and habitat use are significant, especially in response to nickel, barium, strontium, vanadium, beryllium, cadmium and chromium, which accumulate to a greater extent in sprawler and climber-sprawler behavioural types. In contrast, arsenic and selenium accumulate most in burrowers, and although some differential accumulation occurs

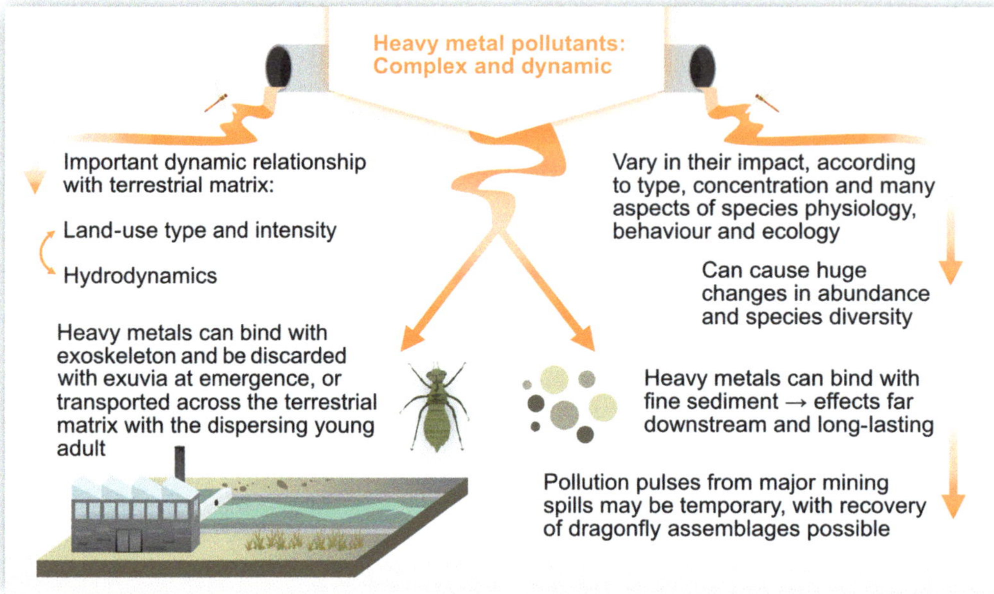

Fig. 5.14. Heavy metal pollution is a significant stressor about which relatively little is known as it is highly complex and dynamic in behaviour. Its impact varies greatly depending on pollutant type, concentration and many aspects of species biology. When severe, this type of pollution can have a major depressing effect on dragonfly abundance and species diversity. In both agricultural and urban environments, there is an important relationship between land-use type and intensity and the hydrodynamic response. As heavy metals can bind with fine sediments, the adverse effect can be felt downstream and can be long-lasting. Within the dragonfly body, heavy metals can combine with the exoskeleton. However, they may be either left behind in the exuvia or are exported into the surrounding matrix by the young dispersing adult. Heavy metal pollution can be ongoing and pervasive but, in the case of an accidental mining spill, can be short-lived, with recovery possible over time as the dragonfly species gradually reassemble, so long as they are still present in the local area and able to recolonize the formerly disturbed site. (Illustration: Robyn Symons).

in sprawlers, it varies greatly between sites (Fletcher *et al.*, 2017).

Dragonfly larvae of standing waters close to an iron and steel factory in Finland were responsive to changes in iron, manganese and cadmium, while accumulating high concentrations of zinc. They accumulated little cadmium, as the larvae free themselves of it quickly (Nummelin *et al.*, 2007).

Chromium ore mining releases considerable sediment. In the Philippines, this activity caused adult dragonfly density to drop to only 10% and species richness to 42% compared to the natural forest, with impact greatest on the endemic forest species (Villanueva, 2010).

In the north-western Ghats, India, times of both monsoon and post-monsoon, along with water pollution, strongly affect zygopteran diversity, whereas anisopteran diversity is only affected by the monsoon. Nevertheless, some damselflies, such as *Ischnura senegalensis*, *Ceriagrion coromandelianum*, *Copera marginipes* and *Heliocypha bisignata* are partially tolerant of urban pollution (Koparde, 2016).

Pollution in Pakistan rivers from agricultural and industrial activity leads to certain heavy metals (lead, iron, zinc, copper) from water and sediment accumulating in dragonfly tissues. However, there is not only great variation between species in the extent to which their bodies accumulate the metals, but also in terms of their body parts (head, thorax, wings, abdomen). Two species, *Neurothemis tullia* and *Orthetrum anceps*, are particularly tolerant of very high levels of the metals (Noor-Ul-Islam *et al.*, 2021). In general, the widespread dragonfly species appear to be relatively tolerant of high levels of heavy metals. Larvae of two libellulids, *Erythemis simplicollis* and *Pachydiplax longipennis*, can tolerate environmentally realistic levels of copper, cadmium and lead in their tissues, especially bound into their exoskeleton, with some of these metals being left behind in exuviae after the adult has emerged (Tollett *et al.*, 2009).

In California, larvae of *Libellula*, *Pachydiplax* and *Erythemis* are closely associated with bottom sediment where they tend to accumulate the highest levels of metals, especially copper and zinc. This is especially the case within artificial wetlands created to contain industrial effluent metal concentrations with regulated limits. Furthermore, the effects are also seen downstream, with larvae in the lower reaches often accumulating as high metal concentrations as those upstream (Fletcher *et al.*, 2020). At the same sites, Fletcher *et al.* (2022) also found that metals can accumulate in both the emergent body and the exuvia, although concentrations vary greatly, especially for vanadium and cadmium. These metals contrast with copper, zinc and manganese, which tend to leave the wetland with the teneral adult. Overall, the situation is complex, varying with species, metal type, accumulated metal concentration, body mass and number of emerging individuals, with all factors contributing to the moderation of the extent to which metals leave the water and are redistributed around the local landscape.

A river in the south-west of the Iberian Peninsula suffered a major mining spill in 1998, with discharge of large quantities of mud and acidic waters with high heavy metal concentrations eliminating the local dragonfly fauna. Recovery began two years later with the appearance of seven former resident species, and interestingly, an additional four species that are widespread across Africa (*Brachythemis leucosticta*, *Orthetrum chrysostigma*, *O. trinacria* and *Trithemis annulata*). However, full recovery of the historical assemblage has been affected by other stressors such as diffuse pollution from local small towns (Ferreras-Romero *et al.*, 2009).

Surrounding land use can make a difference in the extent to which heavy metals accumulate in dragonflies. Accumulation of copper and zinc is greater in larvae from industrial stormwater ponds compared to those draining highways or in open spaces (Karouna-Renier and Sparling, 2001). In the case of sugarcane that receives regular fertilizer application, local stream dragonflies have highly elevated levels of copper and zinc (Corbi *et al.*, 2008).

Many types of pharmaceuticals and endocrine disruptors enter freshwaters globally. The contaminants that bioaccumulate through the aquatic food chain can be transferred to terrestrial ecosystems through the emerging and dispersing adults. Caddisflies, for example, pass on contaminants to both larval and adult dragonflies where there is bioaccumulation and bioconcentration. But the extent of accumulation and concentration varies among odonate life stages (the larvae of some dragonflies show greatest accumulation, while in others it is the adult), type of contaminant and the species involved in the various food linkages. In turn, these contaminants also pose a risk to other terrestrial predators, such as spiders, birds and bats (Previšić *et al.*, 2021) (Fig. 5.15).

Contamination by microplastics is a major emerging problem for human health and aquatic organisms (Sazli *et al.*, 2023). Even small concentrations of polystyrene nanoplastics emanating from both water and sediment accumulate in the tissues of the gomphid *Aphylla williamsoni*, leading to biochemical damage. There is a redox (oxidation and reduction as complementary processes) imbalance inferred from oxidative stress responses and antioxidant defences. There are also neurotoxic effects with the potential to affect nerve and neuromuscular function (Guimarães *et al.*, 2021).

Pollution as manifested by a combination of algae, plastic debris, dumped building materials and/or sewerage is a major driver of change to odonate assemblages along streams in Algeria. These stressors are also adversely synergistic with increased water temperatures and soft sediment. Nevertheless, there is also some tolerance to pollution, with *Ischnura graellsii* and *Platycnemis subdilatata* being the most pollution-tolerant species in this region (Bouhala *et al.*, 2019).

At a stream in Brisbane, Australia, both larval and adult abundances of 36 species dropped substantially just downstream of a sewage treatment plant. Although the adult assemblage is greatly restored farther downstream, larval abundance remains low. The ten stream-specialist species are most affected by effluent. Adults of only one of the stream species are present but in low numbers at the most contaminated site, while two other species are not present downstream from the effluent outflow. Adult and larval dragonfly diversities give a qualitative indication of water quality, with chlorine likely the most important toxicant immediately downstream of the effluent inpouring (Watson *et al.*, 1982).

Treatment of effluent is aimed at restoring water condition. However, a test on *Coenagrion hastulatum* in Sweden shows that wastewater treatment-plant effluent is not equivalent to natural freshwater when some behavioural and physiological traits are analysed. Effluent exposure reduces activity and foraging behaviour but generates a faster escape response. It also changes 14 bodily compounds,

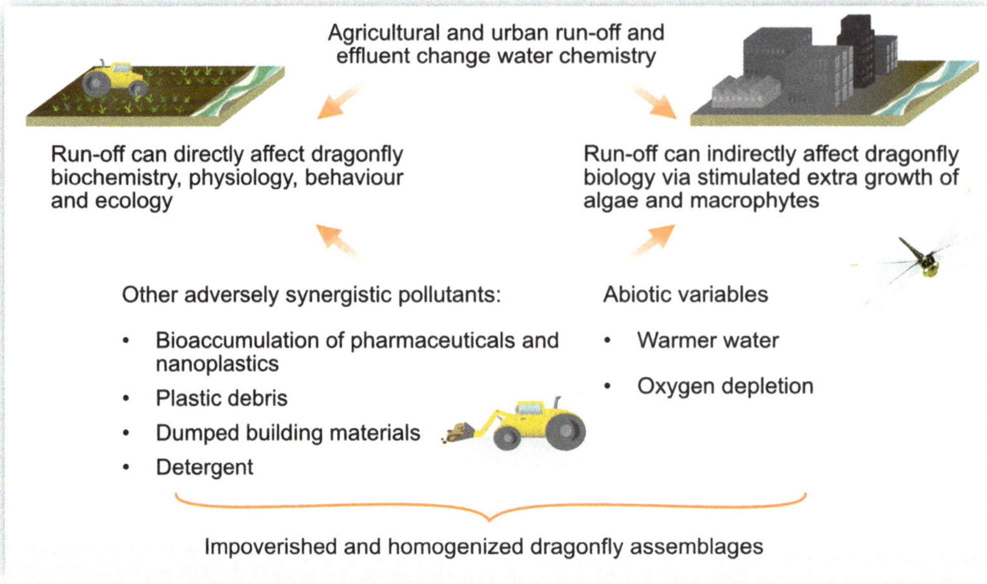

Fig. 5.15. Agricultural and urban pollutant run-off and effluent can be major stressors on certain freshwater systems. The pollutants variously directly affect dragonfly biochemistry, physiology, behaviour and ecology. Where pollutants stimulate plant growth, there can be indirect effects, such as increase in density of algae and macrophytes. These pollutants, besides being adversely synergistic with abiotic variables such as water warming and low oxygen levels, can also be synergistic with other types of pollutants such as pharmaceuticals, nanoplastics, plastic debris, dumped building materials and detergents. The overall outcome is usually impoverished and homogenized dragonfly assemblages. (Illustration: Robyn Symons).

leading to an increase in protein metabolism and oxidative stress (Späth *et al.*, 2022a), while also changing metabolite composition during larval development (Späth *et al.*, 2022b).

Dragonfly assemblages in Croatian streams are heavily impacted by agricultural run-off, especially from sewerage from cattle farms. While some species such as *Ischnura elegans* and *Platycnemis pennipes* tolerate a wide range of polluting conditions, some, such as *Gomphus vulgatissimus* and *Somatochlora meridionalis*, are greatly affected by the impact. Yet two species, *Aeshna cyanea* and *Calopteryx splendens*, increase in abundance with increasing pollution. The main environmental variables driving these odonate changes are chemical oxygen demand, total organic carbon, water temperature and dissolved oxygen concentration. However, there is also an intermediate effect via the aquatic macrophytes which benefit from this particular polluted environment (Vilenica *et al.*, 2021).

Even in reserve areas there can be adverse impacts from human activity. In Gunung Jerai Forest Reserve, Malaysia, larvae in the Libellulidae and Euphaeidae dominate in terms of abundance with increase in human pressure, affecting water chemistry while also increasing levels of sedimentation. However, there is a difference between the two taxa. Libellulid larvae are most affected by water velocity and biochemical oxygen demand, while euphaeid larvae are influenced by changes in water temperature and depth, pH and oxygen demand (Suhaila *et al.*, 2016).

Chemical pollutants coming from domestic use, fruit tree cultivation and various industries in a river in South Africa bioaccumulate in the tissues of gomphid larvae. The highest levels relate to copper, manganese, antimony and zinc (Addo-Bediako and Malakane, 2020).

Hand-washing detergent is used in the small streams of Mayotte island in the Comoros archipelago. This activity locally homogenizes the dragonfly fauna. While widespread African species like *Crocothemis erythraea*, *Palpopleura lucia*, *Orthetrum stemmale* and *Trithemis arteriosa* remain common in the polluted areas, Comoros endemics, such as *Pseudagrion pontogenes* and *Platycnemis agrioides*, are largely absent. The few remaining individuals of *P. pontogenes* have detergent-stained and distorted wings (Samways, 2003).

5.3.3 Impact of pesticides

Pesticides applied on land often redistribute and poison freshwater habitats (Stehle and Schultz, 2015), posing a local extinction risk for many species. Furthermore, there are indications that a planetary boundary has been crossed where chemical pollution is also unsafe for humanity (Persson *et al.*, 2022) (Fig. 5.16).

Neonicotinoid insecticides are among the pesticides that pose greatest risk to insect populations generally, including aquatic species (Van Dijk *et al.*, 2013; Morrissey *et al.*, 2015). These insecticides have a major adverse effect on freshwater food webs, with huge implications for many taxa (Sánchez-Bayo *et al.*, 2016). Environmentally significant concentrations of the neonicotinoid thiacloprid draining into natural ditches affect the survival, growth, moulting and mobility of *Ischnura elegans* in the Netherlands. Emergence is also strongly affected, which is additive to predation pressure upon larvae in the wild. This in turn can affect population levels in the long term and could be one reason for the decline of this species, and perhaps many other odonates, in that country and elsewhere in recent years (Barmentlo *et al.*, 2019).

Other insecticides are also of concern in dragonfly conservation. The systemic broad-spectrum phenylpyrazole insecticide fipronil is applied to nursery boxes of rice seedlings before planting out into paddies. This and other insecticides, such as the neonicotinoid imidacloprid and the ryanoid chlorantraniliprole, have decreased the density of

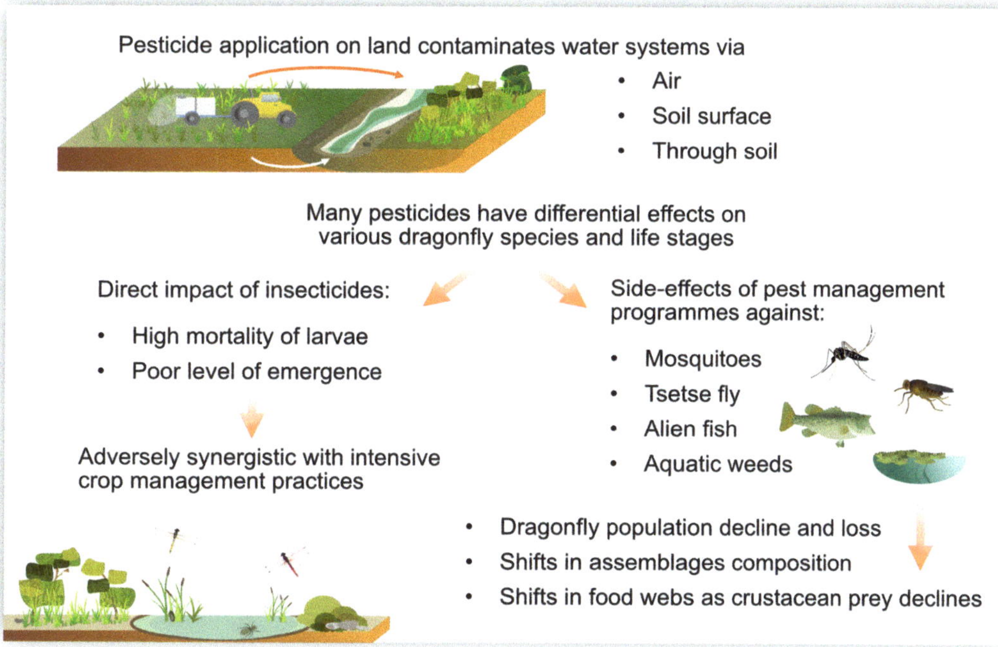

Fig. 5.16. Pesticides applied on land often find their way to local water bodies via the air, across the soil surface and/or through the soil. There are many pesticides, and according to their concentration and even their breakdown products, they can have an adversely differential effect on various dragonfly species and their life stages. There are specific impacts on larvae, especially when various pest management programmes are instigated, including those using insecticides, piscicides and herbicides. Among the most important impacts of insecticides, in particular, are high mortality of larvae and as a cause of poor levels of emergence to become adults. Some of these pesticide impacts can be adversely synergistic with intensive management of crop fields. The outcome is dragonfly population decline and even local population loss. These population changes are also often associated with dragonfly assemblage shifts. As pesticides can impoverish the dragonfly food base, especially that of small crustaceans, there can also be great shifts in food webs. (Illustration: Robyn Symons).

Sympetrum frequens in rice paddies in Japan in recent years. However, there may be some other adversely synergistic stressors also involved, such as mid-summer drainage of paddies and/or harsh tillage (Nakanishi *et al.*, 2020). While the insecticides imidacloprid and fipronil dissipate in the water within a few days, they can have a strong effect on *Sympetrum infuscatum*. Fipronil can eliminate young larvae within nine days of application. Imidacloprid is not so directly impactful but leads to low adult emergence via its effect on susceptible and diminished microcrustacean prey numbers (Jinguji *et al.*, 2013). Furthermore, exposure to imidacloprid, even at low dosage levels, impairs predator-recognition learning in damselfly larvae, having important implications for chemosensory cognition of non-target species in aquatic environments (Wickramasingha *et al.*, 2024).

Species traits are important relative to their exposure to various pesticides used in rice cultivation. The herbicide pentoxazone decreases aquatic macrophyte abundance and has a stronger effect on predators, like dragonflies, living on the submerged plants than on the bottom inhabitants. Different species respond differently to the herbicide, and only one phytophilous species (*Indolestes peregrinus*) and one benthic species (*Orthetrum albistylum*) are significantly affected by it, with three benthic dragonfly species even more abundant in the presence of the herbicide. However, these indirect effects of the herbicide are insignificant in comparison with the direct lethal effects of the insecticide fipronil (Hashimoto *et al.*, 2019).

The carbamate carbaryl and the organophosphate azinphos-methyl both affect the New Zealand endemic damselfly *Xanthocnemis zealandica*. However, the response to the two compounds differs. Carbaryl reduces hatching success, increases the mortality of the younger instars and the number of adults emerging. Azinphos-methyl affects all instars, but especially instar seven (Hardersen and Wratten, 2000).

There has always been concern over the extent to which chemical mosquito control affects the non-target macroinvertebrate fauna. Among the commonly used products is the pyrethroid cypermethrin and the growth regulator diflubenzuron, which at recommended doses can affect both *Ischnura elegans* and its potential *Daphnia magna* prey. Both *I. elegans* eggs and larvae suffer higher than normal mortality when exposed to these insecticides, and larval growth rates are retarded even at low concentrations. *D. magna* is immobilized, leading to death, even at greatly reduced concentrations. Although these insecticides break down in the water within three days, they still leave a legacy of direct mortality and disturbed food webs, as well as breakdown products with unknown effects (Subrero *et al.*, 2019).

Aerial spraying of deltamethrin at doses used to control tsetse fly (*Glossina morsitans*) in Botswana prevents the emergence of several species of dragonflies, even the organic pollution tolerant *Ischnura senegalensis* (Schuran, 2008). This case shows how blanket spraying against a pest can greatly affect non-target dragonflies.

The ketone rotenone is used to control alien fish, but causes the mortality of *Anax imperator* larvae, increasingly so at higher concentrations. Of concern, is that while this species is relatively tolerant of low concentrations, many other species lower in the food web, such as daphnids and mayflies, are highly susceptible, suggesting that use of this pesticide poses risks for the river environment in general (Dalu *et al.*, 2015).

Rotenone has a strong effect on many macroinvertebrates, but so do alien fish like the bass *Micropterus dolomieu* and *M. salmoides*, which are also significant predators of small endemic fish in the lower reaches of mountain streams of the Cape in South Africa. A strategic way forward is to remove the alien fish with the pesticide, and then let the macroinvertebrates recolonize the treatment area from the headwater streams above (Woodford *et al.*, 2013).

5.3.4 Global warming and pesticides

Global warming and pesticide pollution are major adversely synergistic stressors (Fig. 5.17). Exposure of *Ischnura elegans* larvae to insecticides (e.g. chlorpyrifos) while under food stress reduces their ability to deal with heat stress. Cool latitude unstressed individuals are less sensitive to food stress, but sensitive to pesticide stress and high heat exposure, while low latitude individuals are better able to withstand extreme temperatures. Individuals at both latitudes are susceptible to pesticide exposure, but more so in the case of the cool latitude individuals. These findings illustrate that the delayed effects of larval pesticide exposure can spill over to adults, but according to adult heat exposure capabilities. As both warm and cool elevation individuals respond differently to heat and pesticide

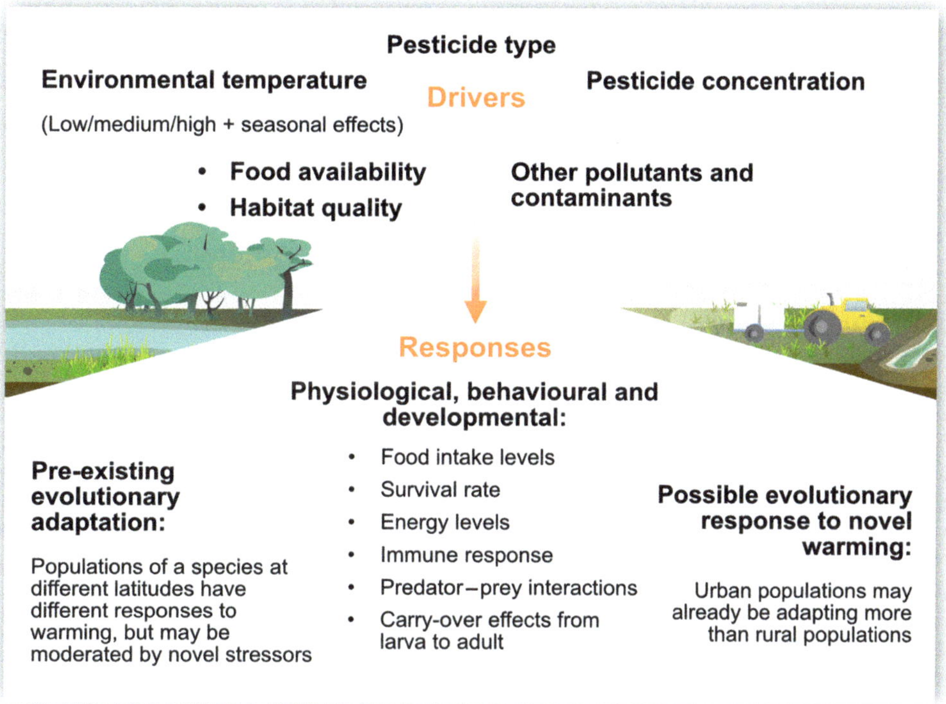

Fig. 5.17. There is a complex relationship between damselfly larvae and novel environmental conditions. Larvae are affected by contamination by pesticides, which can be interactive with other pollutants and contaminants. Their responses are also strongly associated with other factors, especially temperature and food availability, against a background of many factors under the umbrella of habitat quality. The particular larval responses are various, and are mostly a combination related to physiology, behaviour and development. Importantly, there can be carry-over effects from the larva to the adult. There is some pre-existing evolutionary adaptation to warming among populations according to latitude, but the presence of novel stressors can alter responsiveness. There already appears to be some evolutionary adaptation to warming, with urban populations better adapted to a warming environment than rural ones. (Illustration: Robyn Symons).

stressors, earlier evolution along natural thermal gradients is likely playing a role in shaping sensitivity to pesticides. This is some indication that the warm-adapted phenotypes are better able to cope with the two stressors simultaneously, so long as sufficient food is available (Janssens *et al*., 2014).

Additional studies indicate that warming makes chlorpyrifos more toxic to *Ischnura elegans* larvae in terms of its survival, growth rate and food intake. However, local thermal adaptation does not counter the effect of warming on pesticide toxicity. This indicates that gradual thermal evolutionary adaptation will not be effective enough for the species to survive, for example, a local 4°C rise in temperature (Dinh Van *et al*., 2014a). Furthermore, when *I. elegans* is exposed to elevated temperatures, it is more susceptible to a contaminant (zinc) and likely only able to tolerate these changed conditions when there is thermal evolution and/or migration of a warm-adapted genotype (Dinh Van *et al*., 2013).

There are other ramifications associated with the interaction between a warming summer climate and chlorpyrifos in *Ischnura elegans* in relation to its position in the food web. The pesticide reduces the predatory behaviour of the species toward its prey, with an even greater reduction in this behaviour at higher temperatures. Its behavioural response to its own predators is also reduced but does not deteriorate further at higher temperatures. The ecological implication is that wild larvae will suffer reduced food intake and be at more risk of being eaten by their predators. This will be aggravated at higher temperatures when the pesticide is present (Dinh Van *et al*., 2014b).

There are additional effects of climate warming and chlorpyrifos on *Ischnura elegans*. A simulated warmer winter temperature of 8°C leads to higher winter survival and growth rate than at 4°C, a temperature commonly experienced by high latitude populations. The effect is especially pronounced among low-latitude populations, suggesting thermal adaptation. Although chlorpyrifos adversely affects survival, growth rate, lipid content and acetylcholinesterase activity, there is an increase in phenoloxidase activity. Overall, this suggests that sensitivity to pesticides will be greatest when bodily energy reserves are low (Arambourou and Stoks, 2015).

Damselflies, like *Ischnura elegans*, are exposed to a range of temperatures on a daily basis, affecting physiological traits related to oxidative stress and bioenergetics. Responses to chlorpyrifos vary along a gradient of mean temperatures (12–34°C). While most physiological traits are not greatly affected by chlorpyrifos at intermediate temperatures (20–24°C), most of the traits are negatively affected by chlorpyrifos at very high (28–32°C) and to lesser extent at lower temperatures (16°C and below), especially under daily temperature fluctuation. Overall, this means that it is essential to evaluate pesticide toxicity along temperature gradients that are realistic relative to their life in the wild, especially at different latitudes (Verheyen *et al.*, 2023).

When there are interactive effects of temperature and contaminants (carbaryl, chlorpyrifos, esfenvalerate, endosulfan, perfluorooctane sulfonic acid and zinc) on damselflies, some of the contaminants are adversely synergistic. Larval exposure to these contaminants affects levels of fat content, immune response and predator–prey interactions. Importantly however, some of these effects carry through into the adult. Furthermore, these delayed effects are also related in a complex and differential way to increased temperature. These findings have profound significance not just in the case of multiple stressors but also relative to increased vulnerabilities to a changing climate. This is especially so when a current stressor, such as temperature, is directly linked to changes in immune response (Stoks *et al.* 2015).

There are complex and interactive effects when *Enallagma cyathigerum* larvae are exposed to the organophosphate chlorpyrifos and the environmental variables of water temperature and different food levels under simulated natural conditions. Importantly, some of these effects can also carry over into the adult. Chlorpyrifos exposure results in smaller adults, except in individuals reared at high food levels. However, neither food level nor temperature have a significant effect on survival in the absence of chlorpyrifos, but the pesticide reduces survival at high temperatures. Chlorpyrifos causes a 3-days longer development time at low food levels but leads to a 5-days shorter development time at high food levels. Growth rate slows at low food levels, leading to smaller adults. Although seemingly counter-intuitive, low food levels and low temperatures enable adults to better tolerate the cold, a response which is not affected by the pesticide (Janssens and Stoks, 2013). In turn, urban populations of *Coenagrion puella* are more able to tolerate the same pesticide than are rural populations, suggesting some local adaptation (Tüzün *et al.*, 2015).

The level of pesticide impacts depends on the concentration in the local environment. At relatively low environmentally realistic concentrations, pesticides can alter the interaction between behavioural (food intake) and physiological (food assimilation and conversion efficiency). However, the effect of pesticide stress is strongly interactive with predation pressure. Pesticides such as atrazine and endosulfan at environmentally realistic levels cause stress in *Coenagrion puella* but the stress is far less than that imposed by predators such as conspecifics and fish. Moderate pesticide stress elicits compensatory physiological responses, especially when there is an abundance of food. However, there is still an adversely synergistic interaction between the two stressors, especially for atrazine and predators. Yet while endosulfan can increase larval growth rate and food ingestion at low concentrations, this becomes an adverse interaction in the presence of predators (Campero *et al.*, 2007).

5.3.5 Food web changes from chemical contamination

Medical and veterinary antibiotics are not readily biodegradable on entering aquatic systems. The widely used antibiotic norfloxacin can affect the growth rate and behaviour of various organisms at various levels in the food web. This can lead to changes in the direction and strength of trophic cascades, such as the green algae–microcrustacean–larval damselfly food chain (Wan *et al.*, 2020).

There is increasing concern that multiple stressors can affect aquatic food webs by changing the relative abundance of the component organisms. Application in mesocosms of the biocide *Bacillus thuringiensis israelensis* (Bti) for mosquito control can reduce the abundance of non-target chironomid fly larvae by 53–87%. These larvae are the preferred food of the juvenile newts, *Lissotriton* spp. This food web change increases juvenile newt susceptibility to predation by *Aeshna cyanea*, possibly because of physiological stress on the young newts (Allgeier *et al.*, 2019). However, in an earlier experiment, there were no significant direct effects of Bti on either larvae or adults of the libellulid mosquito predator *Erythemis simplicicollis* (Painter *et al.*,1996).

Copper, cadmium and zinc can accumulate in young chinook salmon after eating juvenile chironomids, caddisflies and mayflies (Saiki et al., 2001), and can also accumulate in dragonflies. Copper, cadmium, mercury and manganese can accumulate in the tissues of *Tramea cophisa*, and while not influencing food selectivity, feeding rate decreases after exposure to the contaminants (Lima *et al.*, 2021).

Investigation of how freshwater contamination by plastics affects dragonflies is in its infancy. Besides there being possible direct effects on the physiology of dragonflies, plastics also pose a bioaccumulation risk, as they can contain various elements such as nickel, cadmium and lead. Dragonflies may also be affected by changes to their food chain, whether through alterations to algae or to other invertebrates, with, for example, the morphology of chironomids changing in response to plastics contamination. Furthermore, there are concerns over sediment quality where plastic materials can accumulate (Azevedo-Santos *et al.*, 2021).

Engineered silver nanoparticles have many commercial applications, but potentially pose a risk to freshwater systems through changing interactions between species. *Daphnia magna* avoids being eaten by *Anax junius* larvae through horizontal and vertical movements. While silver nanoparticles alone increase the *Daphnia* population by 185%, in the presence of both nanoparticles and the predator, its population decreases by 91% as its movements are inhibited. This suggests that the use of these nanoparticles can change predator–prey population dynamics (Pokhrel and Dubey, 2012).

5.4 Infrastructure Changes to Rivers

5.4.1 River engineering

5.4.1.1 The significance of dams

With increasing human demands for water coupled with greater environmental uncertainty, impoundment of flowing water for torrent control, storage, abstraction and electrical power generation is an inevitable development. The placement of a barrage across flowing water reduces velocity and disrupts the river continuum, with the effects dependent on river system and barrage type (Zhao *et al.*, 2019) (Fig. 5.18).

The size and construction of dam walls depend on the demands for water, ranging from highly local for a small local human community through to very large for water transfer to high-demand, distant places, especially cities. Often the water levels of reservoirs fluctuate greatly depending on the degree of demand relative to the amount of rainfall in the catchment. These huge changes in dynamics place much pressure on the biological communities both in the water and on the margins and riparian zones, especially in areas of the world with high water demand (Wang *et al.*, 2018).

The number of dams worldwide is massive, with the USA alone having about 2 million of them. 75,000 of these have dam walls of over 2 m in height (Tiemann *et al.*, 2004). Furthermore, there are also many other, smaller impoundments which regulate water. Weirs may be put in place to slow water flow, lessening the risk of local flooding, often of residential areas.

The impacts of water impoundment are physical, chemical, geomorphologic and hydrologic arising from changes to water dynamics. There are essentially four zones associated with an impoundment: the unperturbed upstream river, the impounded still water body, the turbulent outflow zone immediately below the dam wall and the downstream river reach.

Once impoundment has begun, the system directly upstream shifts from a running water system to a flooded standing water one. Thereafter, succession among biological communities depends on the relative constancy of water levels. This is determined by the amount of water recharge and the level of human water demand. When water recharge is low and demand is high, the water level drops. However, when recharge is high and demand is low, the water level increases. Overall, these great fluctuations in water recession versus transgression lead to a riparian zone above the dam wall that is largely depauperate of permanent biological communities when water levels are low.

In contrast, where both water recharge and demand are relatively constant, the water level fluctuates very little. There is then time for colonization by biota of the dam margins, followed by ecological

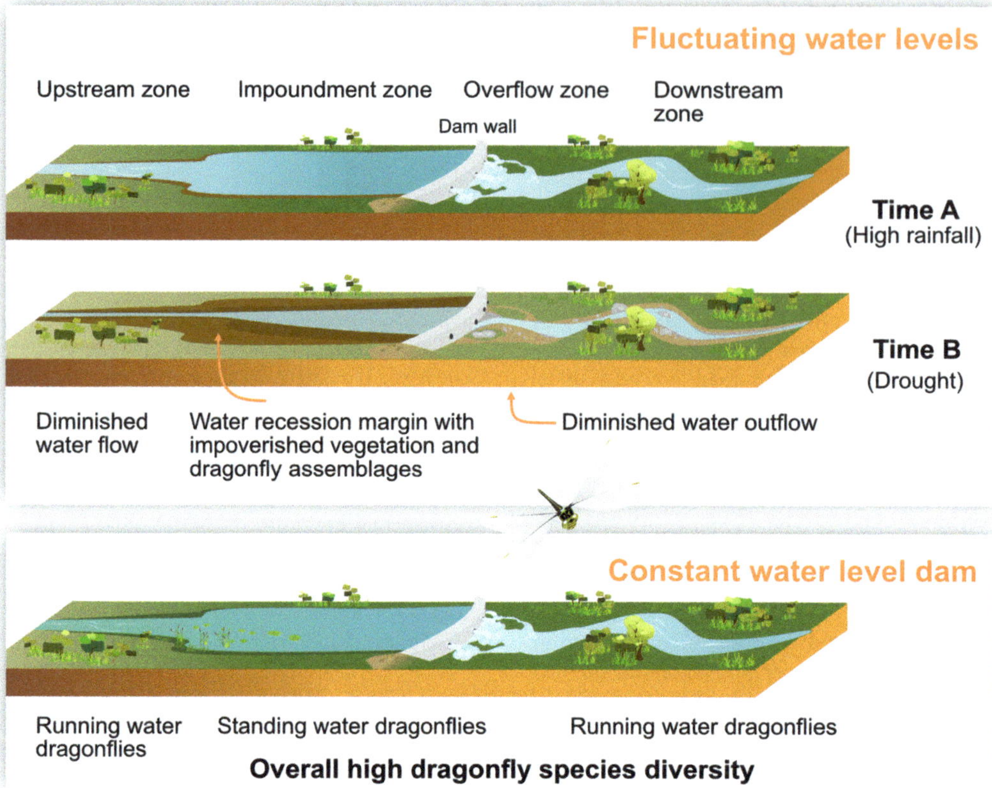

Fig. 5.18. Running water impoundment has various effects on physical, chemical, geomorphologic and hydrologic features, all of which greatly affect the biota. Dragonflies are affected according to the size of the running water system and hence the size of the dam wall. The water levels behind large dams, like these here, may fluctuate greatly. During times of high rainfall water levels are high (Time A) but during times of drought water levels are low (Time B). Above the large dam wall, the changes in water recharge result in oscillating water transgressions (Time A) in times of water plenty and recessions in times of drought (Time B), making the dam margins highly impoverished in vegetation and dragonflies. However, in the case of other dams, where water levels remain relatively constant, appropriate water chemistry and riparian vegetation can encourage high species richness of both plants (aquatic and riparian) and dragonflies. (Illustration: Robyn Symons).

succession leading in time to a rich mix of both running and standing water communities in the area around the dam according to their position relative to the inflowing river.

5.4.1.2 *Effect of large river dams on dragonflies*

Dragonfly diversity is greatly affected by the construction of large river dams. These large dams have a major impact on dragonflies both upstream and downstream. They are often built in dry areas to store water and to control water release for various purposes ranging from hydroelectric power generation through to irrigation. They often fluctuate greatly in water level according to local weather conditions, whether high water levels after heavy rainfall or low water levels in times of drought.

Damming of the Boussellam river in Algeria caused a decrease in vegetation diversity below the dam, and a drop from 17 adult dragonfly species upstream to 14 downstream. However, the total species richness is 23, indicating much change in species composition on both sides of the dam wall, as well as there being a change in abundance. Most of the change is due to zygopterans rather than anisopterans. Water temperature is lower upstream while pH is lower downstream. Oxygen levels are higher downstream, possibly due to the turbulence from the outflowing water from the dam flowing over a rocky substrate with silt deposition in the interstices. This may be a reason why the rare *Onychogomphus uncatus* is present only downstream. Interestingly, the very rare and highly threatened *Calopteryx exul* is present both upstream and downstream, although much less abundant downstream (Chelli *et al.*, 2020).

The severe drought in Algeria 2014–2017 had a major effect on *Calopteryx exul* populations on the Seybouse river. The impact was both direct from the dry conditions and from high demands on the local dam, reducing water depth levels and the degree of water discharge. This led to a change in water chemistry, particularly increases in phosphates, ammonia and sulphates. *Calopteryx exul* is also susceptible to high levels of discharge, increasing water velocity and changing riparian vegetation structure and composition (Khelifa *et al.*, 2021b).

In central Brazil, odonate richness is affected by small hydroelectric dam installation. There are different effects on zygopterans versus anisopterans due to physical and chemical changes, especially waterflow, pH and turbidity, to water bodies. Zygoptera species richness is strongly affected by dissolved oxygen, with richness highest where there are high levels of dissolved oxygen. This is likely the result of greater amounts of riparian vegetation, along with higher water flow and lower water temperature, leading to increased dissolved oxygen. As regards Anisoptera, damming increases local species richness, although with higher levels of dissolved oxygen and water turbidity, anisopteran richness decreases. This response suggests that turbidity and lower levels of dissolved oxygen are associated with higher water temperature and lower water flow. The outcome is that anisopteran species richness is optimal where there are intermediate levels of disturbance (Klein *et al.*, 2018).

From an upstream perspective, a dammed river with relatively constant water levels can benefit dragonflies, with a shift in prevalence from running water species to standing water species. In the Dinaric ecoregion in the western Balkan of south-east Europe, reservoirs created by dammed rivers support a wide range of odonate species but only when there is the right water chemistry along with an abundance of aquatic and riparian vegetation with much spatial heterogeneity. While most species are habitat generalists, the rare and threatened *Lindenia tetraphylla* is also present (Vilenica *et al.*, 2020).

5.4.1.3 Effect of small river dams on dragonflies

The effects of small river and stream dams have a much more local effect than that of large dams of social and industrial importance (Fig. 5.19). The damming of a small perennial river in an undisturbed area of the highly biodiverse Cape Floristic Region, South Africa, caused complex changes in both aquatic and riparian realms. Although natural aquatic biodiversity increases with decreases in elevation, a result of higher levels of food availability and less extreme environmental conditions (Davies and Day, 1988), this elevation gradient is reversed by the presence of a dam. Immediately below the dam wall there is a distinct decrease in larval abundance and diversity of synlestids, coenagrionids, aeshnids, gomphids and libellulids. While there is some recovery downstream over the next 1 km, odonate diversity and abundance does not reach the levels of those in the natural riverine area above the reservoir. The impoverishment in the downstream reaches comes from a combination of interrelated factors. The reduced cover of the vegetation canopy downstream results in reduced organic input into the aquatic system and therefore less resources for dragonfly prey. Associated increases in more homogenous, sunlit conditions, also lead to lower dissolved oxygen levels and pH, resulting in more algal growth than in the upstream microhabitat-diverse riverine areas. In short, the immediate below-dam macroinvertebrate (including odonate) assemblages are only a subset of the more diverse ones upstream of the reservoir (Bredenhand and Samways, 2009). However, a small perennial stream impoundment elsewhere in South Africa, although changing the local dragonfly assemblage in the immediate area from one characteristic of running water to one of still water, results in an immediate switching back to a running water assemblage within a few metres downstream (Samways and Steytler, 1996). The differences in outcomes between these two small running water systems is largely about the size of the impoundment: the larger river impoundment has a much more extensive effect downstream than the small stream impoundment.

Damming of small rivers associated with a major river in the Brazilian subtropical southern Atlantic Forest changes solar incidence, water flow and the temperature of water bodies. When the semi-natural, shaded riverine sites are compared with the sunny standing water sites there is a major shift in odonate

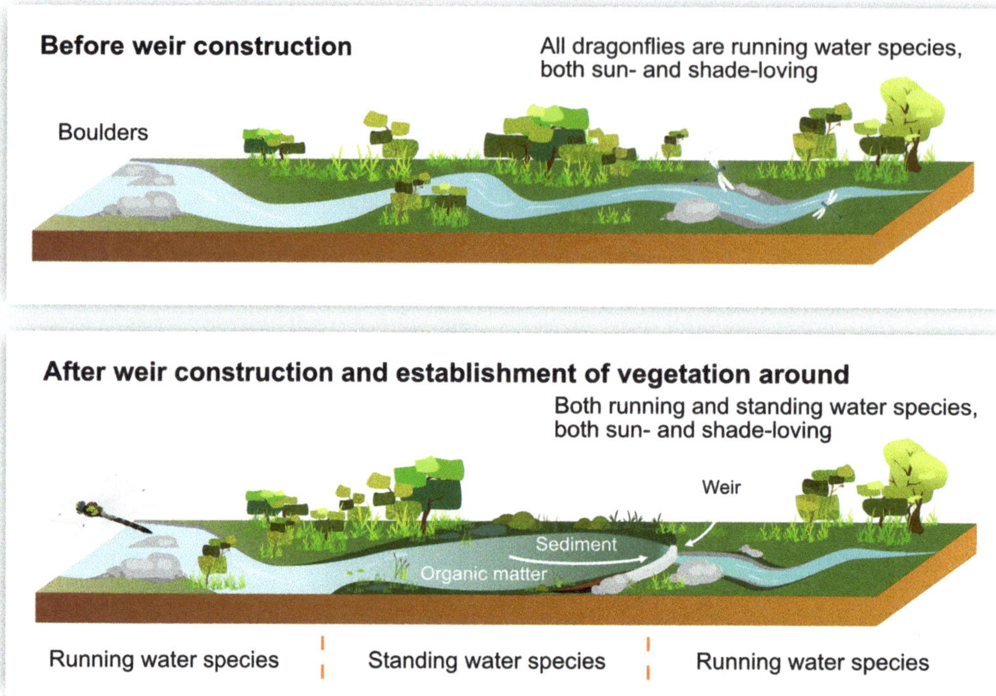

Fig. 5.19. Small impoundments, like weirs, across small rivers are very different in their impact on dragonflies compared to large dams across major rivers, with greatly fluctuating water levels. When the small river is perennial and has various vegetation types and other features, such as boulders, along its length, it has a variety of running water dragonfly species, both sun- and shade-loving. However, after impoundment the dragonfly assemblage increases greatly as standing water species colonize the pond. There is no overall loss of species in the same reach of the river. However, where, before impoundment, sediment and organic debris would drift downstream, they are now blocked by the weir, causing infilling and allowing growth of marshland vegetation. (Illustration: Robyn Symons).

species richness and diversity. Species richness *increases* greatly in the disturbed sites, going to 43 species at the dammed sites from 22 at the semi-natural sites, with only 4 species shared by the two systems. The odonate fauna moves away from largely forest specialists to widespread habitat generalists (Araujo *et al.* 2022).

5.4.1.4 Dragonfly dynamics associated with small impoundments

Very small impoundments across streams and small rivers present some interesting opportunities for dragonflies through a local change in water velocity and marginal changes in water chemistry over time. Beaver Dam in Pennsylvania, USA, was constructed in the 1930s forming a small reservoir of 2.7 ha. Two major storms over the years broke the impoundment, resulting in the lake becoming a wet meadow with a small pond. A total of 22 species were recorded in the immediate area over that time. The succession of dragonflies in this Appalachian locality has been very interesting. While the earlier reservoir attracted typical pond species (*Lestes* spp., *Cordulia shurtleffi*, *Dorocordulia libera*, *Ladona julia*, *Leucorrhinia frigida*, *L. intacta* and *Gomphus borealis*), these have since become rarer. In recent years, some neonatives have seemingly arrived in response to climate warming, including *Archilestes grandis*, *Enallagma divagans*, *E. basidens*, *Libellula flavida*, *L. cyanea*, *L. incesta* and *Sympetrum obstrusum*. *Enallagma anna* arrived in the mid-2000s, but had disappeared by 2014, making its presence uncertain. Typical bog and fen species have become more prominent, such as *Amphiagrion saucium*, *Chromagrion conditum*, *Enallagama hageni*, *Nehalennia grandis*, *Libellula quadrimaculata*, *L. semifasciata*, *Somatochlora tenebrosa*, *S. walshii* and *Leucorrhinia hudsonica*. Species of concern that appear to be increasing include *Enallagma divagans* and *Somatochlora elongata*, while other species of conservation interest seem to be declining, such as *Lestes eurinus*, *S. walshii*, *Cordulia shurtleffi* and *Gomphus borealis* (Schiffer *et al.*, 2015).

However, not all stream barrages provide good opportunities for dragonflies. Wetland ecosystems ('veredas') in the Brazilian Cerrado are hydromorphic soils that retain much rainwater, releasing it slowly as streams and pools during the dry season. Damming of the flow leads to changes in hydrologic regimes, reduced ecological integrity and a decrease in dragonfly diversity. These changes come about not only from modified hydrologic regimes, especially oxidation-reduction potential, but also from cattle trampling the marginal vegetation when they congregate to drink water (Brasil *et al.*, 2021).

In contrast, weirs for reducing high water velocity and increasing water availability in the savanna of the Kruger National Park in South Africa simply cause species switching or provide new habitat for others. The weirs lead to local avoidance by *Olpograstra fuelleborni* and the appearance of *Palpopleura lucia*, for instance, as well as the torrent-water species *Zygonyx natalensis* in association with the weir wall (Stewart and Samways, 1998).

A small (550 m circumference) and shallow (less than 2 m deep) pond was created by impounding a small river in a South African botanical garden specifically for plant and insect conservation. The initial number of dragonfly (all riverine) species was 12, and increased to 26 two years after the weir was constructed, which included the inflowing and outflowing stream as well as the pond (Steytler and Samways, 1995). The pond was then periodically assessed over the subsequent 30 years, during which time the pond silted up and became a wetland with the stream re-establishing along its centre. The pond was dredged in 2017, immediately prior to which the species richness was 17. By 2020, species richness had increased to 30 (Kietzka *et al.*, 2021a), with another 12 species recorded on separate occasions in between (Willis and Samways, 2013). Overall, this illustrates the highly dynamic nature of this assemblage and the rapid colonization or disappearance of dragonfly species as the ecosystem changes over time.

5.4.1.5 *Comparison of impoundment type*

In summary, water impoundments, when very large and subject to great variation in both recharge rates and water demand, without doubt have a major effect on aquatic biota. Such barrages along major rivers are a major stressor, especially when adversely synergistic with other factors such as pollution, excessive sedimentation, water abstraction for irrigation and climate change with attendant extreme weather events. This does not bode well for many indigenous aquatic species as dam walls cannot be torn down.

At the other end of the spectrum are small impoundments or weirs that provide an increasing number of microhabitats that increase, for example, local dragonfly diversity. In effect, they reconcile societal needs (e.g. for mitigating flash flooding and acting as a pool for water during droughts) and those of dragonflies (e.g. providing extra local microhabitats and refuges).

5.4.2 Water diversion and drainage

5.4.2.1 *Soft drainage channels*

Water channels may be constructed to divert water into agricultural fields, especially for crop production in times of reduced rainfall (Fig. 5.20). Rice cultivation often involves damming, redirection of water and forest clearing. In Sikkim Himalayas in northern India, while organic rice paddies have slightly higher dragonfly diversity than natural forest, this open ecosystem is dominated by sun-loving and disturbance-tolerant species, especially those in the Libellulidae. This contrasts with the dragonfly assemblage of the forest, characterized by habitat specialists, especially in the Platystictidae and Chlorocyphidae (Dewan *et al.*, 2019).

Water diversion can be adversely synergistic with other stressors. There have been contrasting changes in abundance of dragonflies in Japan since the year 2000. While some species have remained stable and a few have increased, two species (*Asiagomphus pryeri* and *Sympetrum eroticum*) have declined substantially, seemingly associated with multiple stressors including pesticide application in rice paddies (Kita *et al.*, 2021).

With the loss of so much lowland floodplain habitat in Japan, rice paddies have become a refuge for many still water dragonfly species, including threatened species such as *Lestes japonicus*, *Libellula angelina*, *Sympetrum maculatum* and *S. uniforme*. However, the rice paddies are also subject to various stressors like eutrophication, desiccation, sedimentation, pesticides, successional vegetation changes,

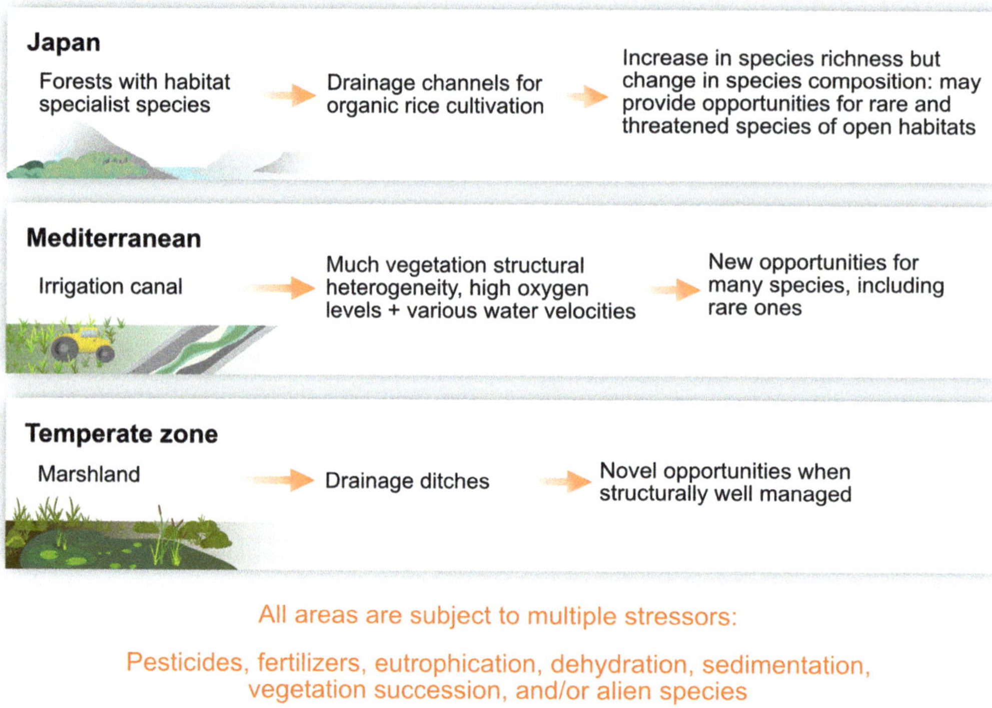

Fig. 5.20. Water diversion or drainage ditches, while largely associated with land conversion for agriculture, can provide opportunities for many dragonfly species, even rare and threatened ones. However, all share the same risk of various multiple stressors. (Illustration: Robyn Symons).

and invasive fish, invertebrates and plants, all impacts which require conservation action. While running water habitats are relatively secure in comparison with these artificial dammed environments, they are still vulnerable to the risk of large dam construction (Kadoya *et al.*, 2009).

A similar pattern of establishment is seen on Lombok island, Indonesia where clearance of forest for agriculture, such as rice cultivation, leads to a shift in the dragonfly assemblage to a disturbance-tolerant sun-loving set of species. While three species of conservation interest (*Pseudagrion pilidorsum declaratum*, *Libellago lineata* and *Gynacantha subinterrupta*) are associated with streams with high-canopy cover, *P. pilidorsum* is also present in the more open irrigated areas (Ilhamdi *et al.* 2021).

A 12.4 km long irrigation canal in Provence, southern France supports over 70% of the dragonfly species known from the local area. This high species richness is largely due to the presence of much structural diversity among the aquatic plants, along with high oxygen levels and various water velocities. The canal supports several rare and threatened species. Among these are exceptionally high densities of *Onychogomphus uncatus* larvae, and larval densities of *Gomphus pulchellus* and *G. vulgatissimus* reaching densities of those found in natural waters (Schridde and Suhling, 1994).

The construction of drainage ditches for land reclamation has accelerated in previous decades, although restoration practices are being implemented (Paul *et al.*, 2023). In low-lying Lincolnshire, UK, drainage ditches have been used to convert native fenland to agricultural land. These ditches, when not subject to agricultural stressors, such as pesticide and fertilizer pollution, are significant habitats for a wide range of macroinvertebrate taxa, including dragonflies (Hill *et al.*, 2016). Such ditches require regular maintenance to ensure their drainage capacity. In the Netherlands, this should be done about every three years to maintain a water depth of 40–80 cm for optimal conditions for dragonfly larvae (Twisk *et al.*, 2000).

5.4.2.2 *Channelization using hard surfaces*

Where water is guided along constructed channels (revetments) made of concrete or gabions, the water margins are not well vegetated, and the water often has high velocity following high rainfall but can be very low in times of little rain (Fig. 5.21). These factors mean that water cannot permeate the surrounding area to create localized wet areas or feed into the groundwater. In dry regions, this channelization can cause localized hydrologic drought. Furthermore, without vegetation cover, there is very little input of organic matter, also contributing to biotic impoverishment of the channel water. The high velocity, lack of vegetation on which to grasp and a dearth of food means that larvae are washed along the channel (Narf, 1985).

Amami Island, Japan has a moist subtropical climate with glossy, broad-leaved, evergreen forest supporting high levels of local endemism among the dragonflies. Adults use the complex structures of the riparian forest as habitats for reproduction, while the larvae inhabit the submerged roots and leaf litter. However, channel enlargement for flood control has involved the removal of the riparian cover and cementing of the banks for river margin protection. The outcome in the channelized zones was reduced shade, disconnection of terrestrial and aquatic habitats, and homogenization of the stream arising from fine sedimentation, leading to severe impoverishment of the dragonfly fauna, with only one or no species at all remaining (Sakai *et al.*, 2019).

5.5 Matrix Changes Associated with Freshwater Systems

5.5.1 Vegetation changes

5.5.1.1 *Tropical forest loss*

Natural forests maintain water quality, with deforestation and forest landscape fragmentation leading to declines in water quality (Fig. 5.22). However, there is much variation in water quality responses depending on the selected water quality variables, watershed properties, forest types, climate and land-use context (Qiu *et al.*, 2023).

Fig. 5.21. Hard surface channels are poor in dragonfly species. In the channel shown here only two habitat generalist species were present, compared to at least eight species in equivalent lengths of neighbouring natural streams. (Photo: Michael Samways).

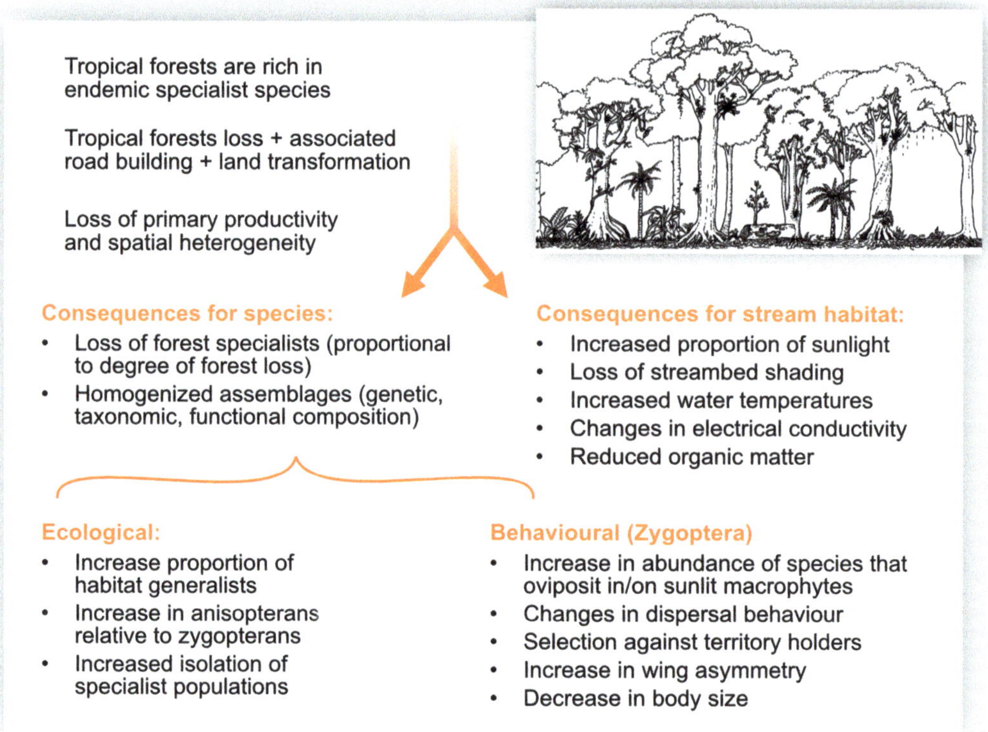

Fig. 5.22. Tropical forests are remarkably rich in many rare and localized endemic dragonfly species that depend on intact primary forest. The huge impact on dragonflies of tropical forest loss across the world is very similar, whatever their location. Tree canopy loss is often adversely synergistic with other stressors. Overall, there is loss of primary productivity and spatial heterogeneity which has huge consequences for stream habitats, as well as for the species themselves. The ecological consequences are largely the same globally. Research in South America has also indicated that there are behavioural consequences as well, mostly for zygopterans. (Illustration: Robyn Symons).

Loss of natural forest cover is especially concerning in the tropics with its high levels of biodiversity coupled with massive deforestation (e.g. Brito *et al.*, 2020). This loss leads to biotic homogenization as local communities genetically, taxonomically and/or functionally converge in their composition. The underlying concern is that high biotic heterogeneity, as measured by biological community change over space and time (beta diversity), is giving way to far less variation in space and/or time leading to homogeneity among aquatic communities.

In Amazonian streams, the intactness of environmental conditions, such as stream shading, bank angle, quantity of wood on the streambed, electrical conductivity and lower water temperature, favours zygopteran species richness over anisopteran species richness. But loss of the natural riparian vegetation structure greatly changes the overall assemblage structure, especially relative to the proportions of the two suborders (Oliveira-Junior *et al.*, 2017). Also in the Amazon, and in two biogeographic regions with distinct levels of endemism, degradation of the riparian zone causes biotic homogenization of the zygopteran assemblages. This process is manifested through rare species being suppressed while the most common species become super dominant (Brasil *et al.*, 2017).

The loss of Amazonian forest cover, while having little effect on taxonomic diversity (species richness and abundance), has a large effect on phylogenetic diversity. While habitat-specialist zygopterans, like *Acanthallagma*, *Hetaerina* (*Mnesarete*), *Aeolagrion*, *Microstigma* and *Mecistogaster* species, occur frequently along the forested streams, they are replaced by more sun-loving habitat generalists, such as *Erythrodiplax*, *Orthemis*, *Acanthagrion* and *Ischnura* species, in the open habitats (Carvalho *et al.*, 2013, 2021).

Another problem associated with deforestation in the region is that populations of forest specialists are becoming increasingly isolated from each other (De Marco Júnior *et al.*, 2015). Deforestation can also occur because of road building in the Amazon. This activity changes odonate species composition, favouring open water anisopterans while disfavouring zygopterans, with local loss of species such as *Dicterias atrosanguinea* and *Chalcopteryx scintillans* that otherwise inhabit forest interiors (Monteiro Júnior *et al.*, 2013). Furthermore, the effects of deforestation in this region can also result in changes to odonate morphology in deforested areas, such as fluctuating wing asymmetry in *Argia tinctipennis* increasing greatly while body size decreases, compared to in naturally forested habitats (Pinto *et al.*, 2012).

In fragmented forests in Indiana, USA, wing length in *Calopteryx maculata* changes according to the spatial scale over which physiological variation is affected by local environmental conditions relative to its dispersal ability. Wing length in this species in this disturbed ecosystem is affected by local conditions at distances of about 1–5 km (Goldner and Holland, 2023).

Behavioural diversity can also be affected by forest loss. In a savanna–forest patch region of Brazil, changes in forest cover alter site suitability for laying eggs. Odonate species with epiphytic oviposition (eggs deposited on surfaces) exhibit an overall decline in richness through loss of forest-associated substrates such as woody debris, leaves and roots. However, species with exophytic oviposition (eggs deposited on a water surface) increase as sunlit side channels become available. Furthermore, epiphytic species that choose macrophytes, as well as endophytic species (eggs deposited inside vegetation), are less affected as more macrophytes associated with more sunlit conditions become available (Rodrigues *et al.*, 2019).

In the Amazon, environmental variables representing stream condition as well as vegetation presence are changed by deforestation. Various odonate traits related to reproductive behaviour, dispersal and territoriality, are responsive to these variables. The removal of riparian vegetation has a marked effect on many Zygoptera species, suggesting that deforestation selects against territorial behaviour and niche specialization in this group, many species of which depend on intact habitats (Resende *et al.*, 2021), leading to local species extirpation and changes in assemblage composition (Juen *et al.*, 2014). This means that to retain the ecological integrity of odonate assemblages, especially the zygopterans, it is critical to retain intact riparian zones (Oliveira-Junior *et al.*, 2017), while maintaining net primary productivity and ecosystem heterogeneity in terms of temperature and rainfall (Brasil *et al.*, 2019).

Besides the local environmental variables being important, there are also the regional ones to consider, such as various climatic zones and the extent of forest cover for maintaining the full complement of species (Oliveira-Junior *et al.*, 2019). However, in this region it is not so critical to keep natural forest intact for retaining anisopteran assemblages, which can benefit from forest clearance through the provision of sunnier habitats and increased electrical conductivity which promotes algal and benthic macroinvertebrate populations. This means that while assemblages may be no poorer in species richness with transformation, there can be great changes in assemblage composition, related not only to forest cover but also to macrophyte abundance (Pereira *et al.*, 2019).

In the drier Cerrado–Caatinga ecotone of the northern Brazilian state of Maranhão, odonate species composition varies according to the level of naturalness versus transformation. Yet this variation does not change species richness or abundance, due largely to local substitution of extirpated habitat-specialist species by generalists in the transformed areas (Veras *et al.*, 2020). Further results from the same region indicate that disturbed streams and vegetation favour anisopterans over zygopterans owing to streams with less canopy cover receiving increased light and heat compared to streams with dense natural cover. The decrease in canopy cover filters odonate species, likely due to specific ecophysiological requirements such as thermoregulation strategies that are dependent on body size, and flight behaviour and oviposition behaviour, both important in the selection of microhabitats. Sites with less canopy cover favour the presence of macrophyte selection by endophytic species, also supporting species with exophytic oviposition that do not require riparian vegetation, as with most anisopterans (Veras *et al.*, 2022).

Much of the tropical Atlantic Forest in the state of Bahia, Brazil has been converted to pasture or cacao culture (cacao trees are planted in the understorey of forested areas), leading to an *increase* in dragonfly species richness. However, the three landscape types have very different species composition. Natural forest supports specialists such as

Heteragrion aurantiacum, *Heliocharis amazona*, *Leptagrion macrurum*, *Kiautagrion acutum*, *Aceratobasis cornicauda* and an unidentified Gomphidae species. In turn, pasture favours open area specialists such as *Ischnura capreolus*, *Erythrodiplax paraguayensis*, *E. fusca* and *Perithemis lais*, while cacao cultivation supports some forest species such as *Forcepcioneura serrabonita*, *Heteragrion consors* and *Perilestes fragilis* (Santos and Rodrigues, 2022a, b).

Costa Rican streams see differences in species richness, species composition, proportion and the number of widespread species between forest and cultivated areas, with only three of the 11 forest species present at streams in agricultural areas. In this geographical region, streams in agricultural areas are not suitable as reproductive habitats for most of the forest species, with deforestation posing a major threat to them (Hofhansl and Schneeweihs, 2008).

Other tropical forest areas have experienced impacts on dragonflies through forest loss, as in Colombia, where 89% of the local endemic species, mostly zygopterans, are associated with primary tropical forest (Bota-Sierra *et al.*, 2021). Sungai Wain Protection Forest, East Kalimantan, Indonesia, now contains only a few fragments of primary rainforest, the bulk having either been transformed to agriculture or is now secondary forest, and subject to fire. These transformations have led to great changes in species composition and functional diversity, especially of habitat specialists, mostly Zygoptera, in proportion to the intensity of the transformation (Dolný *et al.*, 2012). In central Java, Indonesia, ten odonate species are confined to tropical forest, while three species are widespread habitat generalists (*Crocothemis servilia*, *Trithemis aurora* and *Agriocnemis pygmaea*) but restricted to agricultural areas. Two local endemics have different habitat tolerances. While the Java and Bali endemic *Paragomphus reinwardtii* is confined to natural forest, the zygopteran *Heliocypha fenestrata* occupies habitats in both forest and agricultural areas (Astuti *et al.*, 2022). In central Java, while undisturbed natural forest has the highest levels of odonate species richness, tourist sites are also species-rich, with equivalent species diversity, which is encouraging for ecotourism (Nafisah and Soesilohadi, 2021).

On the Sangihe Islands of north Sulawesi, Indonesia, the highest dragonfly diversity is around waterfalls, dams and forests, with the lowest diversity in plantations and agricultural settlement areas. The high diversity areas are characterized by the high relative humidity and low temperatures associated with forested habitats. This contrasts with the low relative humidity and high temperatures in open settlement areas. The Sulawesi endemic *Celebargiolestes orri* and the more widely distributed *Prodasinera autumnalis* are confined to forests with natural levels of canopy cover and good water quality (Koneri *et al.*, 2022).

In Borneo, the highest dragonfly species diversity is in intact primary forest, where 60% of all recorded species occur, with 32% of all species present only in this minimally disturbed ecosystem. In turn, the proportion of habitat specialists decreases along a gradient from intact primary forest, slightly degraded primary forest, secondary forest, heavily degraded forest to deforested areas, with overall forest degradation leading to fewer species, great change in species composition and less species diversity (Dolný *et al.*, 2011).

In the montane (about 1000 m above sea level) rainforest of Papua New Guinea, 40% of the odonate species, mostly rare and possibly localized endemic zygopterans with high beta diversity, inhabit intact streams and rivers with high primary canopy cover, including one phytotelmata specialist, *Podopteryx selysi*. The maintenance of canopy cover and avoidance of other disturbances are essential (Oppel, 2005). Almost all the narrow-range endemic odonates in the Sahyadri Tiger Reserve in the northern Western Ghats, India, occupy natural headwater streams with dense canopy and open areas of water, indicating that it is essential to maintain intact ecosystems for the survival of these species (Koparde *et al.*, 2015).

Similar findings come from Mount Hamiguitan, Mindanao, Philippines, where there are extremely high levels of national endemic species (94% for Zygoptera, especially in Platystictidae and Platycnemididae). Most of these species require dense forest with undisturbed riparian and aquatic vegetation, as well as optimum water temperatures (Villanueva and Mohagan, 2010).

Primary Ghanaian forest is not only characterized by dense canopy cover, but also by low water temperatures and low turbidity. When this ecosystem is disturbed by logging and other human impacts, there is a local loss of forest specialists such as *Umma cincta*, *Chlorocypha selysi* and *Micromacromia zygoptera* in favour of sun-loving habitat generalists (Seidu *et al.*, 2017).

5.5.1.2 Phytotelmata

Phytotelmata are cavities in terrestrial plants that collect water. They may be formed from modified leaves, leaf axils, flowers, stem holes or depressions, open fruits or fallen leaves, with about 50 odonate species known to occupy these microhabitats. Some of the phytotelmata for odonates are in the Bromeliaceae, specifically the tank bromeliads which hold open water in their overlapping leaf bases. Over 12 species of Zygoptera have been reported from Neotropical tank bromeliads, some of which are specialists to this habitat, especially in *Leptagrion*, *Bromeliagrion* (Coenagrionidae), and *Mecistogaster* (Pseudostigmatidae), which prey mostly on mosquito larvae (Frank and Lounibos, 2008).

South American rainforests are home to the very large (19 cm wingspan) giant helicopter damselfly *Megaloprepus caerulatus*, with larvae living in tree holes, especially in primary forest (Fincke, 2006). These species are all threatened by deforestation. However, other helicopter damselflies *Mecistogaster linearis* and *M. ornata* readily colonize secondary forest, while *Mecistogaster modesta*, which inhabits tank bromeliads, is 36 times more abundant in secondary forest, seemingly because it favours more sunlit conditions (Srivastava *et al.*, 2005).

The small and Critically Endangered protoneurid *Roppaneura beckeri* is endemic to the Brazilian Atlantic Forest, breeding only in the terrestrial umbels of *Eryngium floribundum*. It is so localized and has such unusual biological characteristics and evolutionary significance that it is a priority species for conservation, especially in forest fragments in peri-urban environments (Polizeli and Pinto, 2021).

Shade is a major environmental variable for dragonflies in coastal Kenyan forests. While species richness is low in these forested habitats, they support several specialized odonate species, including two inhabitants (*Coryphagrion grandis* and *Hadrothemis scabrifrons*) of water-filled tree holes that require closed-canopy forest and a dense understorey. With loss of trees, there is local loss of the forest specialists, although an increase in species richness is associated with colonization of disturbed sites by sun-loving habitat generalists (Clausnitzer, 2003) (Fig. 5.23).

5.5.1.3 High elevation, subtropical and temperate forest loss

At around 2000 m above sea level in the Himalayas, the anisozygopteran *Epiophlebia laidlawi* and parapatric species *Caliphaea confusa*, *Davidus abberans*, *Chlorogomphus atkinsoni*, *Neallogaster hermionae* and *N. ornata* are potentially at threat from deforestation, road building, and terrace agriculture (Brockhaus and Hartmann, 2009). Forest mountain streams in the Chakrashila Wildlife Sanctuary in western Assam, India, support ten odonate habitat specialists, while degraded forest supports none. The greatest threats are illegal logging, firewood gathering, forest fires and cattle grazing (Choudhury *et al.*, 2020).

Fig. 5.23. A water-filled tree hole in a Kenyan tropical forest (upper left), the breeding habitat of *Coryphagrion grandis* (upper right) and *Hadrothemis scabrifrons* (bottom). (Photos: Viola Clausnitzer).

In the Pirapó river basin in southern Brazil, high abiotic and biotic structural diversity, as well as leaf litter biomass, plus high habitat heterogeneity across the subregion leads to the highest diversity among the larval odonate assemblages. Loss of the riparian vegetation not only changes habitat suitability for certain species, but also reduces leaf litter input into the streams (Souza *et al.*, 2015). Also in southern Brazil, the semi-deciduous Atlantic Forest has been severely degraded with only remnants remaining. These remnants are essential for supporting rare specialists, such as *Heteragrion triangulare* (Renner *et al.*, 2015).

In the Brazilian Cerrado headwater streams, zygopteran larvae require no or minimal disturbance, and are far less tolerant of human disturbance than are anisopteran larvae. These responses are associated with the traits intrinsic to the two groups, and which shape the overall odonate assemblage in response to elevated conductivity and dissolved solids, higher percentage pasture in the catchment, higher percentage exposed soil and higher stream embeddedness. Important zygopteran traits are long bodies and caudal lamellae respiration, great thermal sensitivity and endophytic oviposition. In contrast, the anisopterans with their cylindrical bodies, internal gills, endothermic thermoregulation and in many cases burrowing habits, are more tolerant of human activity (Silva *et al.*, 2021).

In the Eastern Highlands of Zimbabwe, natural forest loss, alongside plantations and mining operations, threatens local dragonfly species, such as the endemic *Pseudagrion inyangae*, outside protected areas (Mafuwe and Moyo, 2020).

In Sweden, the highest species richness of lake-inhabiting dragonfly larvae is in undisturbed boreal forests rich in aquatic plants that provide a wide range of microhabitats. In comparison, logged forest has lower species richness of both plants and dragonflies, especially among species that overwinter as eggs or young larvae. Increased water temperature fluctuations and nutrient leakage into the water are possible reasons for the decrease in dragonfly richness. Univoltine species are not affected as they are less dependent on constant water conditions (Sahlén, 1999).

In Western Australia, one population of the rare and highly localized *Petalura hesperia* has been extirpated by tree clearing for agriculture (Barrett and Williams, 1998) (Fig. 5.24).

5.5.1.4 Transformation of mixed- and low-canopy habitats

There can be similar outcomes, whether for high or low canopies, when land transformation takes place. In both Neotropical forests and grasslands, intensive human land use has a stronger effect on aquatic biodiversity than local climate and environmental variables. In both biomes, species richness and functional diversity of animal and macrophyte assemblages decrease with the increasing cover of agriculture, pasture and urbanization. Human land uses are associated with functional homogenization of both animal and macrophyte assemblages,

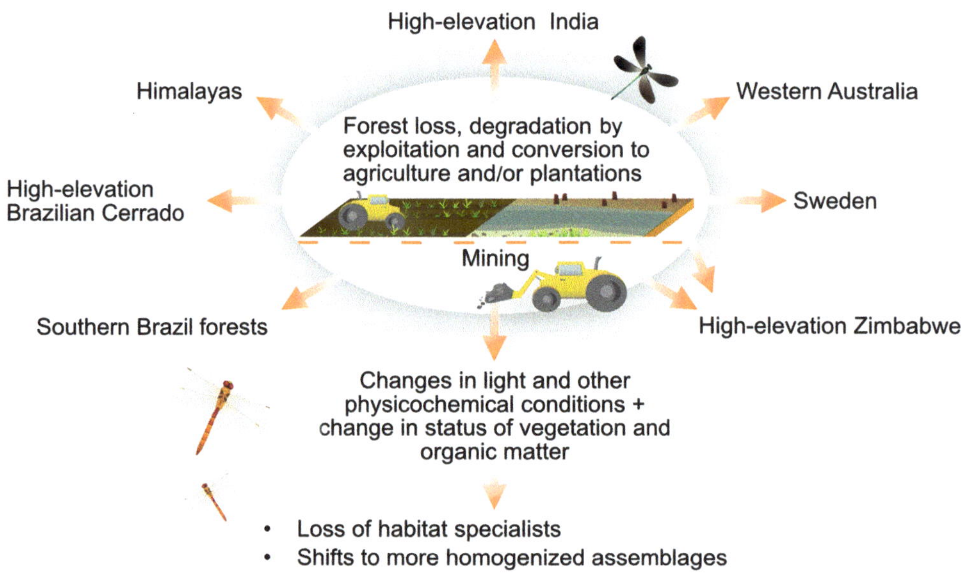

Fig. 5.24. Away from tropical lowlands, we have glimpses of the impact of human activity on dragonflies in tall canopy ecosystems in various parts of the world. While the specific stressors vary from one geographical area to another, the general impact is largely the same. The stressors act via increased sunlight, changing the physicochemical conditions of the water, in particular, and inevitably the vegetation types and structures, as well as causing a decrease in organic matter. The various stressors mostly lead to loss of dragonfly habitat-specialist species in favour of generalists, along with a shift to more homogenized assemblages. (Illustration: Robyn Symons).

reducing animal biomass through direct and indirect pathways mediated by declines in taxonomic and functional diversities (Moi *et al.*, 2023).

While endemic species often conjure great interest arising out of human fascination for rarity, not all range-restricted endemics are highly susceptible to landscape transformation. In Algeria, the percentage of agriculture-free area within the geographical range of the North African endemic *Platycmemis subdilatata* declined from 80% to 26% in recent years. The species has survived this major transition, and better than most other local odonates, by frequently using croplands and grasslands (Khelifa *et al.*, 2021c).

Groundwater-dependent ecosystems and their associated habitat-specialist odonate species are being increasingly threatened globally. Petalurid dragonflies in Australia are one group facing this threat. *Petalura gigantea* (Fig. 4.28) is a long-lived species with larvae that burrow in areas with emergent seepage or where there is a water table that is sufficiently high to saturate the peaty substrate where oviposition and larval burrow establishment takes place. This means that the species is vulnerable to any lowering of the water table through, for example, groundwater abstraction or longwall coal mining, along with the compounding effects of more intense fire regimes as projected for future climate change (Baird and Burgin, 2016).

On the Newnes Plateau in the Blue Mountains of New South Wales, Australia is a scarce peat swamp (mire) ecosystem. However, longwall mining is greatly lowering the water tables that underpin these local mires, posing a key threat to groundwater-dependent biota, including *Petalura gigantea* (Baird and Benson, 2020). Importantly, this mining method is adversely synergistic with subsequent fire impacts, including peat combustion, causing irreversible damage to the mires. Of great concern is that the undermined mires, unlike the non-undermined mires, fail to respond to the onset of good rains, with almost no resprouting of typical and often long-lived, re-sprouter sedgeland and shrub species, leading to local ecosystem collapse and loss of groundwater-dependent biota (Keith *et al.*, 2023) (Fig. 5.25).

The Pampas biome of South America is a mix of subtropical grassland, wetlands, streams and rivers, often with some riparian forest. The streams and rivers bordered by grassland or forest are complex

Fig. 5.25. Groundwater-dependent montane mire on the Newnes Plateau in the Blue Mountains, New South Wales, Australia, showing seepage areas and vigorous resprouting sedgeland vegetation in March 2020, following drought and severe wildfire in December 2019, and then above average rainfall in early 2020 (left). This *Petalura gigantea* peat swamp habitat was not undermined by longwall coal mining and the minimal combustion of peat and post-fire recovery is in stark contrast to the nearby undermined peat swamp in March 2020, which suffered ecosystem collapse (right). (Photos: Ian Baird).

ecosystems, rich in dragonfly species dependent on the type of water body and cover, but sensitive to human interference (Fig. 5.26). Species richness and composition vary among both land cover and aquatic habitat types, while abundance varies only according to land cover. The highest odonate species turnover is in the forest, while grassland is composed of mostly good colonizers. Maintaining high landscape and waterscape heterogeneity is essential, as rivers and streams support 9 unique species, and forest 17, mostly small damselflies. In forest patches, clear cold water surfaces from groundwater, forming small streams or swamps that support particular dragonfly assemblages of conservation priority. However, species richness and diversity decrease when grassland and forest are converted to arable land or degraded by cattle ranching. While these transformed areas attract mostly habitat-generalist species with good dispersal abilities, they also attract some species, such as *Aphylla molossus*, *Orthemis aequilibris* and *Tholymis citrina*, adapted to hot, dry and temporary water conditions (Renner *et al.*, 2018).

Also, in the Pampas, localities that have only a few common species are characteristically associated with a set of rare species, while localities with a preponderance of common species show no clear pattern. The common species are strongly positively associated with water temperature, turbidity, dissolved oxygen and pH, while negatively associated with drier habitats. In contrast, rare species are positively but weakly associated with grassland, but negatively with agriculture, salinity and conductivity. Locations with six or more common species have a reduced number of rare species compared to sites with fewer common species, with the common species seemingly reducing available niche space for less competitive and rarer species. While the original assemblages may have been species-poor with only a few regionally common species, current

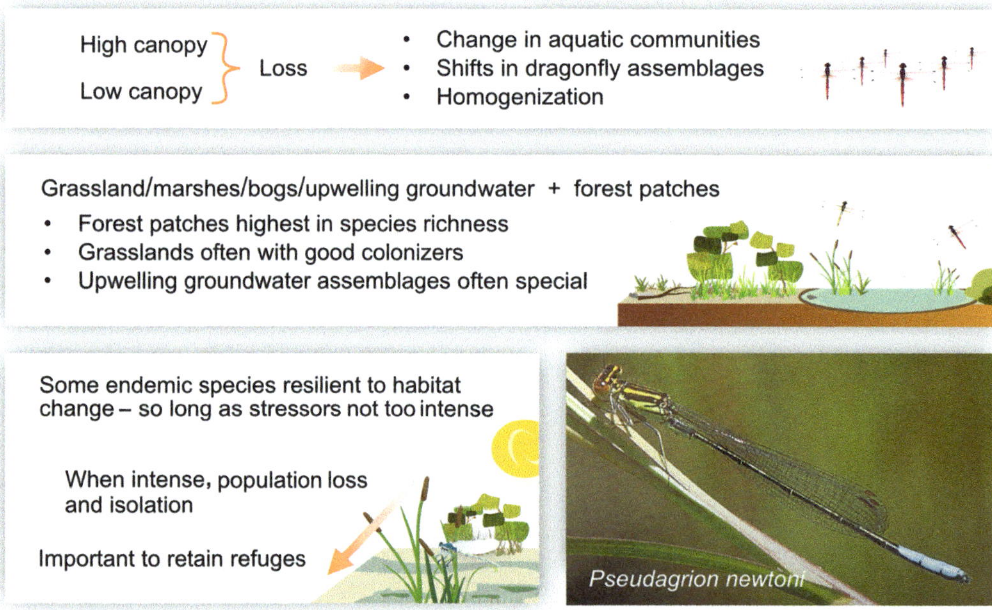

Fig. 5.26. Dragonfly responses to transformation of low-canopy habitats are largely similar to those of changes in high-canopy habitats. While natural forest patches of the Pampas support the most species, the grasslands are dominated by good colonizers. Ground-welling water assemblages are special and require conservation action. Homogenization of assemblages is the most common outcome following agricultural intensification. Endemism does not necessarily equate to habitat sensitivity, so long as habitat transformation is not too intense. Geographic range retraction and isolation of populations is often the result when marshes and peat bogs are agriculturally converted. For many low-canopy species refuges from agricultural intensification are important for their survival. The extremely rare and localized *Pseudagrion newtoni* is a microhabitat specialist that perches only on tall grass stems. It disappears when these special low-canopy conditions are lost. (Photo: Michael Samways; Illustration: Robyn Symons).

transformation has increased the number of common species, negatively affecting the rare species (Renner *et al.*, 2022).

Furthermore, landscape modification of the Pampas through agricultural intensification leads to taxonomic homogenization of odonate assemblages in lakes, streams and temporary wetlands. Cropland and natural grassland sites differ in water chemistry and odonate species composition, with higher variation in composition in grassland than in cropland. There is interplay between agricultural and grassland land uses, as well as in the variation in composition of odonate assemblages. Maintaining a mix of grassland and cropland in fields adjacent to water bodies can limit the negative effects of agricultural intensification (Pires *et al.*, 2022).

Although *Lestes dryas* is widespread in the Holarctic, it is much rarer and more localized in England than it was in former decades. The reason for its decline has been loss of ancient fen marshlands to agricultural reclamation. Besides habitat loss, it has also suffered from increasing population isolation, precluding its establishment in better habitat in times of drought (Moore, 1980). Loss of bog and mire habitats, mostly from draining, is the main reason for dragonfly declines associated with these habitats in the Netherlands (van Strien and van Grunsven, 2023).

In the eastern Mediterranean (Croatia) long-term moderate land and river transformation can enhance land cover heterogeneity, increasing dragonfly species diversity in intermittent rivers. However, as dragonfly assemblages can become functionally more homogenous in response to human transformation, it does not mean that these sites have more conservation value than natural ones. Nevertheless, these transformed rivers, with the right aquatic and riparian habitats, are significant for improving local dragonfly diversity, including some threatened species such as *Cordulegaster heros* (Vilenica *et al.*, 2022). Incidentally, the most northern population of this species, on the southern slopes of the Carpathian Mountains in Czechia, currently has a stable population (Holuša and Holušová, 2022).

5.5.1.5 Complexity of multiple stressors in the Mediterranean region

The effect of multiple stressors must be understood to assess the real threats to the dragonfly fauna. However, these multiple effects can be complex. For example, some ponds in the Akfadou massif forest, north-east Algeria are subject to stressors such as impacts from cattle overgrazing and tourist visitation at the pond margins, leading to loss of vegetation and development of a mud layer. Despite these disturbances, the ponds have relatively stable dragonfly assemblages. Another pond, however, has been invaded by the lakeside bulrush *Schoenoplectus lacustris*, greatly reducing the amount of open water. Yet the pond has high levels of abundance and species richness, but reduced species evenness through domination by Lestidae and Coenagrionidae, with 80% of the total population made up of just three species: *Ischnura graellsii*, *Lestes virens* and *Chalcolestes viridis* (Chelli and Moulaï, 2019).

Montenegro in the eastern Mediterranean is home to several dragonfly populations of conservation concern, including *Lestes parvidens*, *Caliaeschna microstigma*, *Lindenia tetraphylla*, *Gomphus schneiderii*, *Cordulegaster heros*, *C. bidentata*, *Epitheca bimaculata* and *Selysiothemis nigra*. However, the increasing demand for water to support the growing number of tourists along the Adriatic coast, and for agricultural irrigation, along with the increased frequency and duration of hot and dry periods, have together resulted in the drying out of streams, impacting the local odonate assemblages (De Knijf *et al.*, 2013).

5.5.2 Stressors from other human activities

5.5.2.1 Impact of plantations

Eucalyptus plantations in the Cerrado biome of Minas Gerais, Brazil reduce odonate species richness and abundance compared to natural areas, and more so for zygopterans than anisopterans. The main reasons for the decreases are degraded riparian zones, and a decrease in the diversity of aquatic vegetation and organic debris along pond margins (Borges *et al.*, 2021).

Oil palm (*Elaeis guineensis*) plantations have become widespread in warm forested areas, with natural vegetation buffers instigated in some places to help protect biodiversity. A meta-analysis of plantations lacking buffers shows that among the various studies, primary or disturbed forests host between 44% and 19% fewer stream taxa, respectively, according to the study. In contrast, in plantations with buffers the differences overall range between 0% and 24%, respectively. However, stream community composition differs between forests and plantations despite the presence of riparian buffers, due to agrochemicals and altered environmental conditions such as temperature changes, worsened water conditions, microhabitat loss, and food and shelter depletion. Yet abundance does not differ among land uses, as increases in generalist species offset population declines among vulnerable forest specialists. This means that while buffers help freshwater biodiversity to some degree, there is no substitute for retaining primary natural forest for maximal conservation of tropical forest freshwater biodiversity (Rojas-Castillo *et al.*, 2023).

As regards dragonflies, oil palm plantations with 30-metre forest buffers adjacent to Amazonian streams affect the water physicochemical properties and habitat quality of the streams. Streams surrounded by more oil palms, as well as human infrastructure such as roads, have higher pH values, and less woody material than natural areas. These changes lead to a change in the adult odonate assemblage, largely through replacement of forest habitat-specialist species by non-forest specialists (Carvalho *et al.*, 2018). Forested streams in the region have more wood in the streambed compared to vegetation-buffered oil palm sites, which have increased proportions of fine sediment and higher water temperatures. Alpha diversity of zygopteran and anisopteran larvae is higher in buffered oil palm areas than in forested areas, although zygopteran biotic heterogeneity is higher at oil palm sites but not for anisopterans, suggesting that this buffer-managed land transformation counteracts habitat simplification and faunal homogenization (Mendes *et al.*, 2019).

There is a difference in morphological diversity among dragonfly larvae in oil palm versus forest areas in the eastern Amazon. There is a significant relationship between genus abundance and environmental variables, as well as between genera and morphological traits. Changes to, or loss of, riparian vegetation increases sediment load, decreasing the amount of suitable plant substrate available, leading to simplification of morphological traits in the odonate assemblage (Mendes *et al.*, 2020).

In Peninsular Malaysia, small-scale oil palm agriculture has riparian corridors of artificial waterways associated with natural peat swamp and mangrove forest. Odonate species richness increases with vegetation cover, water temperature and dissolved oxygen, but decreases with water level, pH, and total dissolved solids. Species composition is influenced by dissolved oxygen and pH. These changes likely come about from application of fertilizers and chemical pesticides used to clear natural vegetation to make way for more palm oil trees (Ishak *et al.*, 2021).

In Sumatra, Indonesia, oil palm plantations support 41 odonate species. However, assemblages in ditches vary between core and roadside edge habitats. Although heavy rainfall fills ditches and promotes larval development, planting of understorey vegetation provides little improvement (Luke *et al.*, 2020).

In southern Brazil, alien *Pinus elliotti* and indigenous *Araucaria angustifolia* plantation trees are established among mixed wet forest, a component of the highly threatened Atlantic Forest biodiversity hotspot. Among a total of 42 species, pine forests have the highest species richness, all habitat generalists in low abundance. This suggests that the alien element has given rise to new species assemblages composed of species adapted to the novel habitat of pine afforestation. However, habitat specialists are largely confined to natural forest. *Araucaria* plantations have an intermediate species composition. Despite the differences in habitat preference between generalist and specialist species, alien plantation habitats may provide stepping-stone habitats enabling habitat colonization (Renner *et al.*, 2016a). Furthermore, odonate species compose 14 functional groups across various water types, indicating that the functional composition varies between natural forest and especially pine plantations. Natural forest favours specialist traits, suggesting that natural forest conversion to tree plantations, especially alien ones, leads to a shift toward less specialized dragonfly assemblages with changed functional group composition (Dalzochio *et al.*, 2017). Among three types of aquatic environments (artificial lakes, streams/rivers, swamps) species richness is similar, although species composition differs. Lake assemblages differ substantially from those in the other habitats, with artificial lake species likely originating from other degraded areas nearby, suggesting that the remnants of the Atlantic Forest have already been greatly transformed (Renner *et al.*, 2016b).

As is so often the case, stressors compound, as in the Eastern Highlands, Zimbabwe, a biodiversity hotspot, where there is much higher dragonfly diversity in streams in protected versus transformed areas. The compound stressors of alien tree plantations, mining activities, vegetation clearing and tourist impact appear to be the main drivers reducing dragonfly diversity and abundance (Mafuwe and Moyo, 2020; Mafuwe *et al.*, 2021).

In Wales, UK, *Cordulegaster boltonii* requires well-developed marginal stream habitats for life-cycle completion. However, the habitats are degraded by conifer plantation trees (mostly *Picea sitchenisis*) abutting Welsh moorland streams, locally excluding the dragonfly species (Ormerod *et al.*,1990).

5.5.2.2 Impact of monocropping

The adverse effect of intensively managed matrices around ponds can affect the insect fauna within them via changes in the composition and structure of the macrophytes. This is because of a strong association between the insects and the particular plants in the water system determining the structure and function of the local communities (Stamenković *et al.*, 2023).

In São Paulo state, Brazil, sugarcane is extensive, although the associated streams and ponds are not depauperate in dragonfly species (Ferreira-Peruquetti and Fonseca-Gessner, 2003). In South Africa, artificial ponds in sugarcane have significantly lower pH and dissolved oxygen compared to the other pond types and have a higher cover of macrophytes compared to those in a neighbouring protected area. This results in sugarcane, plantation (*Acacia* and *Eucalyptus*) and protected area ponds having similar dragonfly species richness and diversity, although complementary in species composition. Among all three pond types, there is a strong relationship between species composition and vegetation cover, representing a combination of high-quality sunny perching sites, and a preferred habitat for many still water species. Sugarcane and protected area ponds share 22 out of the 33 total species shared among the three pond types. Abundance of both anisopterans and zygopterans increases greatly with an increase in grass/reed cover, especially in the sugarcane ponds and irrespective of their size. However, two species, *Pseudagrion citricola* and *Pinheyschna subpupillata*, are present only in protected area ponds (Briggs *et al.*, 2019) (Fig. 5.27).

5.5.2.3 Impacts of invasive alien aquatic plants

Invasion by floating vegetation, such as water hyacinth (*Pontederia crassipes*), changes water hydrodynamics,

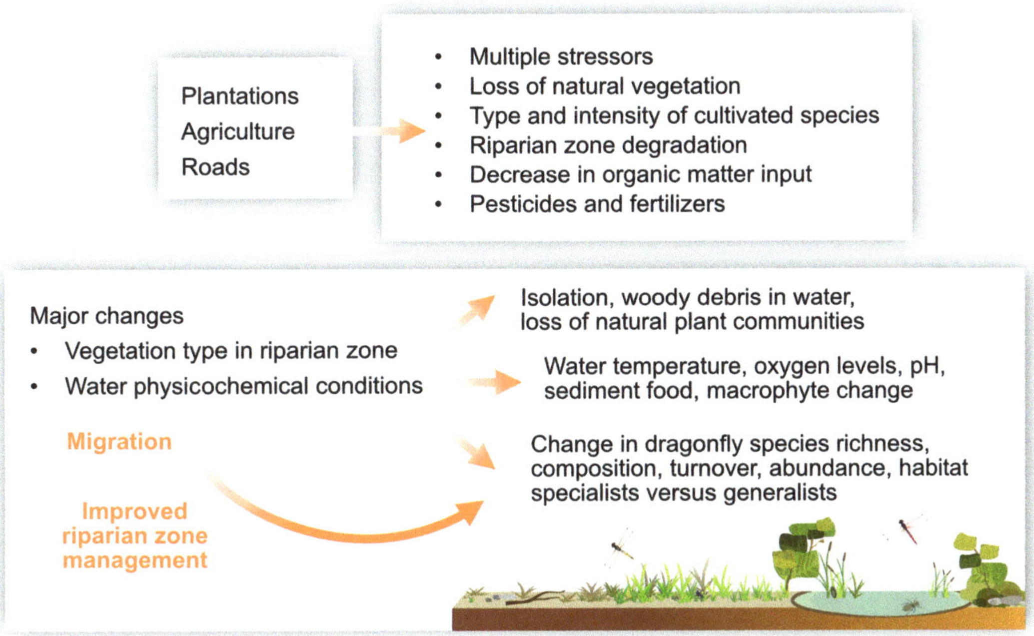

Fig. 5.27. Plantations, agriculture and associated roads result in multiple stressors on dragonflies in various water ecosystems. The main changes are to land cover in the riparian zone and altered physicochemical variables in the water. These changed conditions affect many aspects of dragonfly ecology. However, improvements to both the matrix and especially the riparian zone can go a long way to retaining intact dragonfly assemblages. (Illustration: Robyn Symons).

overwhelms native aquatic macrophytes and can be a nuisance to humans. However, water weeds also have benefits, not least for absorbing polluting compounds, and when harvested, for improving soil fertility (Kleinschroth *et al.*, 2021).

A weir across the Letaba River in the Kruger National Park attracted water hyacinth which then became an attractive habitat for six dragonfly species that were not present elsewhere along the river (Stewart and Samways, 1998). However, this is not to suggest that alien macrophytes like water hyacinth should be encouraged. When slow moving but invaded reaches of the Nseleni River, South Africa, were compared to uninvaded reaches, many aquatic macroinvertebrates were much less abundant in the invaded reaches, with decreases in larval abundance of synlestids, lestids, coenagrionids, aeshnids, gomphids and libellulids (Coetzee *et al.*, 2014).

Floating *Salvinia minima* invaded Lake Kariba in southern Africa after the large Kariba hydroelectric dam was built, leading to establishment followed by a great population increase of *Pseudagrion nubicum* (Balinsky, 1967). Floating *Pistia stratiotes* and the submerged *Ceratophyllum demersum* also invaded the new reservoir. Anisopteran larvae (*Orthetrum* sp., *Acisoma* sp., *Trithemis* sp., *Thylomis tillarga*, *Brachythemis leucosticta* and other libellulids) increased greatly on *P. stratiotes*. In contrast, zygopteran larvae (mostly *Pseudagrion* sp., but also '*Enallagma*' sp. and *Ischnura senegalensis*) increased on *C. demersum* (Petr, 1968).

The free-floating aquatic plant *Salvinia molesta* can be highly invasive, threatening aquatic insect site colonization, assemblage structure and ability to complete life cycles through its changing of the local biotope as it forms increasingly dense mats. Different mat coverages of 25%, 50% and 100% all reduce dissolved oxygen, pH and light availability in the water. Without the weed, only *Sympetrum* sp. larvae are present, making up 7% of the macroinvertebrate fauna. However, in the presence of the weed at 25% cover, dragonflies increase in percentage presence with an associated assemblage change: *Anax* sp. (5%), *Erythemis* sp. (24%) and *Sympetrum* sp. (9%). When the weed cover increases to 50% and 100%, only *Erythemis* sp. is present at 53% of the total macroinvertebrate fauna. These figures illustrate how dragonfly proportions can change greatly as water weed invasion progresses (Wahl *et al.*, 2021). *Salvinia molesta* can be controlled biologically using the weevil *Cyrtobagous salviniae*, with success being measured by the recovery of the macroinvertebrate assemblage (Motitsoe *et al.*, 2020).

The invasive alien giant reed *Arundo donax* locally reduces populations of *Ischnura genei* (a Mediterranean island endemic threatened by habitat loss) and *Crocothemis erythraea* in Malta (Balzan, 2012). In Korea, invasive alien reeds like *Phragmites communis*, *P. japonica* and *Typha orientalis* impoverish local odonate assemblages (Lee *et al.*, 2020).

5.5.2.4 Impacts of invasive alien terrestrial plants

Most South African dragonflies require sunlit conditions, an open-canopy environment to which they are adapted. When these conditions are not met, they can decline greatly. This happens when their natural habitats become invaded by fast-growing alien trees, especially *Acacia*, *Pinus* and *Eucalyptus*. Dense canopies of alien trees have two effects: eliminating favoured sunny microhabitats of flying adults (Remsburg *et al.*, 2008), and inhibiting the growth of aquatic and riparian indigenous vegetation used for perching, resting and oviposition. The outcome is substantial impoverishment and homogenization of the dragonfly assemblages, with the ancient specialist species being especially affected. The remaining habitat supports only a diminished assemblage composed of species that can tolerate the novel shady conditions, mostly by making use of the remaining gaps in the alien tree cover or microhabitats that are still present in-channel (Samways and Sharratt, 2010).

The South African dragonfly fauna is overall much more heliophilic than, for example, the Amazonian fauna, and greatly affected by the shading out of natural habitats, in particular by *Acacia mearnsii*. However, a nationwide programme to remove these riparian aliens meant that a countrywide search could begin to find rare and localized species that had not been seen for many years, even in their type localities. Three species of great concern were rediscovered: *Spesbona angusta* after 83 years of not being observed, *Pseudagrion newtoni* after 41 years and *Proischnura polychromatica* after 30 years. They were all rediscovered at locations where invasive alien trees had been removed. Fortunately for these species, small remnant source populations must have been present somewhere in the vicinity. But had this alien tree clearance programme not been undertaken there would have been a real risk of extinction (Samways *et al.*, 2005) (Fig. 5.28).

5.5.2.5 *Impact of invasive alien aquatic animals*

As we saw in Chapter 4, fish are significant predators of dragonfly larvae and many odonate species have morphological and behavioural adaptations to escape indigenous fish predators, honed by many years of predator–prey interactions and evolution. Artificial ponds lacking fish can have greater dragonfly species richness than those with fish, suggesting that some dragonfly species are highly susceptible to fish predation (Maynou *et al.*, 2017). However, in the case of alien fish, this is a novel impact that theoretically could go either in the way of direct predation on dragonfly larvae or benefit a dragonfly larva through large alien fish eating smaller fish predators. Mostly, fish are generalist opportunist feeders that change their diet as they develop from hatchling through to adult. This means that by and large alien fish are most likely to be harmful to dragonfly larval populations. However, very little is known about the extent to which this is the case from a global perspective. There are also differences among fish species relative to their feeding behaviour. Trout feed largely on the surface, even taking flying damselflies close to the water surface, while carp can graze on aquatic plants to such a degree that the aquatic habitat is degraded. Furthermore, there are other possible predators which could compound the situation, as well as impacts from other stressors, both abiotic and biotic, which together could potentially have a major impact on some dragonfly species.

One example of direct fish impact is the introduction of various poeciliid fish species into the Hawaiian Islands. These predators are strongly implicated as having reduced the abundance of, and even causing local extirpation of, some populations of the endemic *Megalagrion* damselflies (*M. leptodemas*, *M. oceanicum* and *M. xanthomelas*) (Englund, 1999). In Canada, out of 22 stormwater retention ponds, the two containing alien carp *Cyprinus carpio* were the only ones that supported no odonate larvae, despite larvae being common in the other 20 ponds (Ackerman and Galloway, 2003).

Fig. 5.28. A river reach that was formerly rich in local endemic dragonfly species in the Cape Floristic Region, South Africa that has now been totally overwhelmed by invasive alien *Acacia* trees. Here the river (marked by a blue line) is not even visible and devoid of any of the endemic species. (Photo: Michael Samways).

Certain frogs are predators of some odonate larvae or adults and have a suppressing effect on their populations. Of note, is that the only odonate species listed as Extinct on the IUCN Red List of Threatened Species is *Sympetrum dilatatum*, endemic to the mid-Atlantic St. Helena island. It was likely eliminated by the invasive alien frog *Strongylopus grayii* (https://www.iucnredlist.org/species/21226/193512121 accessed 19 March 2024).

In Italy, the invasive crayfish *Procambaras clarkii* strongly reduces several anisopteran and zygopteran species in streams and wetlands (Siesa *et al.*, 2014). In Germany, the invasive crayfish *Pontastacus leptodactylus* is adversely synergistic with wastewater pollution, altered hydrology, the impact of fine organic material and dense riparian alder tree cover, leading to an impoverished odonate fauna (Ott, 2019). In the same country, there is also differential sensitivity among the various dragonfly species to the various invasive crayfish species (Ott, 2023).

The young stages of the invasive zebra mussel *Dreissena polymorpha* in North America can colonize the bodies of dragonfly larvae. This behaviour has a strong but differential effect on dragonfly larvae according to their lifestyle in a Michigan lake. The large-sized sprawlers *Hagenius brevistylus* and *Didymops transversa* that live on top of the substrate are the most susceptible to mussel colonization. The attachment has a strong effect on the larval survivorship and mobility of the penultimate and final instars of *H. brevistylus* in particular. The shallow burrower *Epitheca princeps* and the two deeper burrowers *Dromogomphus spinosus* and *Progomphus obscurus* are far less susceptible (Fincke *et al.*, 2009).

With increasing temperatures in central Europe, there has been a shift in the proportion of native versus alien benthic macroinvertebrates between 2000–2018. While overall assemblage species richness increased by 12.3% and abundance by 14.9%, alien species richness increased 5-fold and abundance 40-fold, especially among the warm-adapted species. Elevation and stream size were identified as the main drivers of this change (Haubrock *et al.*, 2023a) (Fig. 5.29).

5.5.2.6 *Impact of fish farming*

In central Europe, the ecological value of some ponds is decreased by intensive fish farming. Besides fish predation pressure, there is also macrophyte reduction, high nutrient loads and turbidity in these production ponds. Different assemblages respond to these stressors in various ways. Odonates are sensitive to the quality of the aquatic environment and overall landscape structure, while also accurately reflecting differences in pond management intensity. In contrast, amphibians respond only to landscape structure, and diving beetles to the quality of the aquatic environment, while aquatic Heteroptera are insensitive to management and habitat quality. With the exception of oviposition strategies, there are no functional traits that provide adaptive advantages or disadvantages relative to management intensity. In sum, fish farming imposes intense pressure on the system, homogenizing the structure of assemblages. This means that it would be difficult to implement management measures that would be of equal benefit to all species across all taxonomic groups. The key management response is to maintain variability among ponds and their surroundings. Nevertheless, fish-farming ponds are really only secondary habitats for aquatic biodiversity which are not an equal alternative to natural wetlands (Harabiš *et al.*, 2023).

5.5.2.7 *Impact of domestic livestock*

Intensive and extensive livestock farming can have a huge impact on water systems and their dragonflies, often in concert with other types of landscape transformation, in many parts of the world. These include the subtropical grasslands (Pampas) of South America (Renner *et al.*, 2019).

Livestock seeking water can damage vegetation around pond margins and increase water turbidity. When this livestock activity is too intense there is a reduction in dragonfly diversity, as found in Italy (Carchini *et al.*, 2005) and in macroinvertebrates in general in Belgium (Declerck *et al.*, 2006). However, when livestock stocking rates are low or intermediate, macroinvertebrate diversity is maintained (Céréghino *et al.*, 2008). Including all macroinvertebrates in such assessments is important, as dragonfly prey must also be retained.

Trampling during dry seasons in semi-arid regions leads to loss of surface vegetation biomass which inhibits plant regeneration in the following wet season, and increases soil loss rate (Dunne *et al.*, 2011). In the Brazilian Cerrado, natural streams have much more intact habitats than pastures. Yet there is higher abundance but differences in dragonfly species composition in streams bordered by pasture relative to natural areas. Transformation to pasture adjacent to streams is also associated with an

Fig. 5.29. Invasive alien species have a variable effect on dragonflies. Alien aquatic plants, combined with the filling of a new dam, can result in colonization by dragonflies, followed by an increase in their population levels. However, along rivers, aquatic weeds can cause local population declines and shifts in assemblages relative to weed density. Alien fish have caused major declines, while an introduced frog has led to total extinction of one island dragonfly species. Alien crayfish are also causing declines in dragonfly abundance, while alien mussels reduce the survivorship and mobility of certain species. (Illustration: Robyn Symons).

increase in the overall body size of dragonfly species (Calvão *et al.*, 2018).

Reduction in abundance of *Enallagma* larvae was a predictable and accurate indicator of vegetation loss by cattle grazing around prairie potholes in Canada. Significantly, it was the height and density of the vegetation rather than plant species composition that was important. Adults were also associated with taller, denser vegetation but away from the water's edge where grazing had not reduced the structure of marginal vegetation. Cattle grazing here did not directly reduce water quality, with the damselflies only affected by vegetation change. Furthermore, high damselfly density was also associated with overall aquatic macroinvertebrate species richness. The management implication is that tall, emergent vegetation representing high-quality damselfly larval habitat is an indicator overall of good wetland quality (Foote and Hornung, 2005).

In Rajasthan, India, livestock pressure is a threat to wetland species, even in reserves (Singh and Hermans, 2019). Similarly, in the alpine Cansiglio Forest reserve in north-eastern Italy, ponds support 21 odonate species. Cattle grazing negatively affects the assemblages, reducing both species richness and abundance, largely favouring only the habitat generalists. When grazing is intensive, pond margins are damaged and pond drying is exacerbated. Yet at the other extreme, when ponds are abandoned, shrubs invade the entire pond surface, overall suggesting moderate levels of grazing are optimal for maintaining dragonfly diversity (Uboni *et al.*, 2021).

5.5.2.8 *Human consumption of dragonflies*

People are also predators of dragonflies in certain parts of the world, especially for protein intake, the content of which can vary between 40–65%. In China, larvae of several species are often eaten, especially *Lestes praemorsa*, *Gomphus cuneatus* and *Crocothemis servilia*. In the Neotropics, there is a preference for aeshnid species, while in Laos individual adult *Anax guttatus* are eaten (Williams *et al.*, 2021).

5.5.2.9 *Changes in odonate behaviour in response to landscape transformation*

Behavioural responses by individuals of a species to human-induced landscape transformation are an important dimension of the effects of stresses on populations, especially when they occupy novel environments. Occupation of these new conditions may involve genetic plasticity and change in response to new selection pressures. Where pressures are intense and the genome sufficiently accommodating, there can be changes in genetic architecture with associated shifts in behavioural syndromes (Karlsson Green *et al.*, 2016). However, these trait shifts can be complex as not all behavioural traits will be perfectly synchronized, possibly leading to a break in stable evolutionary correlations between the involved behavioural traits.

In Brazil, the calopterygid *Hetaerina* (*Mnesarete*) *pudica* shows behavioural shifts in transformed streams where cattle enter the water to drink. In the transformed areas, males invest more time and energy in chase and patrol flights, reflecting greater investment in fight behaviour and in searches for mates. This investment in shifted energy expenditure may redirect energy that could be invested in reproduction, or alternatively, it could be interpreted as males spending more time looking for mates (Guillermo and Juen, 2021).

5.6 Effects of Urbanization on Dragonflies

5.6.1 The process of urbanization

Cities are characterized by having a high human density of more than 620 individuals per square kilometre, where 75% or more of the population are not engaged in rural activities. They also have a high proportion of impervious hardscape, tall buildings and many communication networks of power lines, roads and railway lines. Cities are also typically a little warmer than the surrounding rural and natural areas, a phenomenon known as the heat island effect.

Both cities and towns often continue to grow outward, with infilling by other infrastructure as part of this process of expansion. Today, more people are urban rather than rural. Furthermore, urban and rural areas are connected through flows of water, food, goods, energy and people, making the effect of towns and cities often felt beyond their boundaries (Grimm *et al.*, 2008). In turn, a town is a built-up area larger than a village and smaller than a city, but overall with a lower skyline than a city, and usually a higher proportion of green space relative to hardscape.

The growth of a town into a city is associated with a decreasing proportion of people working in agriculture, sometimes with urban centres merging through conurbation, urban sprawl, infilling and densification. These processes lead to less habitat being available for wildlife such as insects, with insect diversity and abundance both decreasing with ongoing densification (New, 2015). Often, biotic homogenization occurs as the habitat specialists begin to decrease and disappear locally in favour of generalist species, with urbanization sometimes even overriding the effect of river regulation in terms of impacts on benthic macroinvertebrates (Calderon *et al.*, 2023). Some insect species, often alien, thrive in dense urban environments, and feed on human organic products. However, dragonflies are not among this group as they require a well-vegetated landscape with an abundance of free-flying prey. The main threats facing dragonflies are loss of habitat, both aquatic and riparian, pollution and high traffic volume (Fig. 5.30).

Towns and cities in the past often developed on a river. Besides the river supplying water, in the era before mechanized transport, it was especially valuable for transporting goods in and out of the urban centre. This development has also meant that natural water flow, substrate type and quality, water chemistry and pollution, as well as loss of riparian zones has greatly affected dragonfly populations. However, in some small cities and towns, where the river and its riparian zone have remained in a near-natural state, there may be a substantial dragonfly fauna. Similarly, artificial ponds, which in some areas may be many centuries old and formerly a necessity for livestock, may also support a significant dragonfly assemblage. However, there is no universal effect of urbanization on insect assemblages across the world, as the geographical, ecosociological and infrastructural contexts vary so much (Ferzoco *et al.*, 2023). Furthermore, in the case of

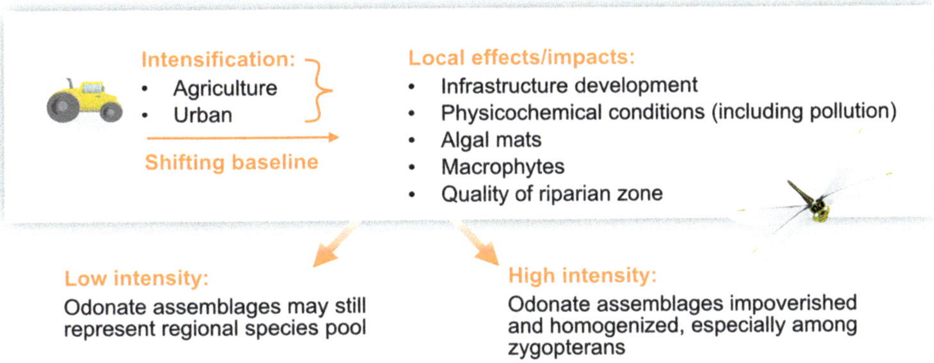

Fig. 5.30. Agricultural and urban impacts often have similar local effects on dragonfly assemblages, according to the shifting baseline (the result of legacies from past impacts) and the level of intensification. The environmental changes and stressors involved are various and may be adversely synergistic. There is a great difference between the effect of low- and high-intensity stressors. A suite of local low-intensity stressors may have little impact on local dragonfly assemblages, while local high-intensity stressors may impoverish them, usually at the expense of habitat specialists, especially among zygopterans. (Illustration: Robyn Symons).

urban ponds, biological interactions and abiotic water conditions override urban effects on the terrestrial matrix and on aquatic insect diversity levels. There are also different dependences and demands on terrestrial and/or aquatic habitats by different taxonomic groups and variation among their colonization behaviours (Keinath *et al.*, 2023).

5.6.2 The rural–urban gradient

Spatial scale associated with urbanization is an important consideration in urban dragonfly conservation. Local factors can be the main drivers of resident dragonfly diversity. Urban ponds are particularly good for illustrating this point as they are essentially islands and readily comparable. In and around the megacity of Paris, France, odonate assemblage composition along the rural–urban gradient is affected most by the density of buildings around any one pond, irrespective of where that pond is situated along the gradient. High densities of buildings around ponds reduce odonate assemblage diversity. However, moderate to high densities of macrophytes promote resident odonate diversity. The conclusion here is that certain fine-scale urban effects can override the general coarse-scale effects of urbanization (Jeanmougin *et al.*, 2014).

Odonate species can respond differently to urbanization according to whether the system is running or standing water, as in a rural–urban gradient at Louisville, Kentucky, USA. For standing waters, urban and rural sites support similar anisopteran species richness, although zygopteran species richness is much lower at urban than rural sites. In the case of odonate assemblage composition, it is the effect of urban development, closeness of the urban centre and level of algal coverage that drives changes. For running water, the main drivers of assemblage composition are water temperature and distance from the urban centre. The change in assemblage composition, as well as the type of ecosystem, can be related to the relative dispersal capability and habitat specificity of the various species, with zygopterans overall more habitat specific than anisopterans (Prescott and Eason, 2018).

Grid cell analysis across Germany on odonates associated with all water types found that 92.6% of all 81 central European dragonfly species occurred in cities, as well as 85.7% of 14 species of conservation concern. Assemblages of city species were subsets of the regional species pool, while city species richness increased with regional species numbers. Some cities even hosted the complete regional species pool. There are distinct land-use-related dragonfly assemblages that follow two gradients, one being the degree of transformation and the other a gradient from open land to forest. Urban landscapes in the rural areas are characterized by distinct sets of species assemblages, all species of which are opportunistic and prefer warm environmental conditions. The urban areas also have higher alpha diversity than intensive agricultural landscapes.

However, similar values of gamma diversity across all land-use categories suggest a countrywide homogenized dragonfly fauna, likely caused by land use in the past. Overall, while urban landscapes maintain higher species diversity than agricultural landscapes, they modify species composition (Goertzen and Suhling, 2019). Nevertheless, good-quality habitat in some cities in the country can support certain rare species of conservation concern, such as *Aeshna viridis* and *Ophiogomphus cecilia* (Goertzen and Suhling, 2015).

The rural–urban gradient also needs to be viewed from a species trait perspective. Among wetlands along rural–urban gradients on Rhode Island, USA, anisopteran species richness, evenness and diversity do not differ. However, there are changes in species composition. This is the result of several species showing a strong relationship with the level of urbanization, with these habitat generalists being most common at urban sites and those with fish. This contrasts with some of the rarer species which favour the rural sites. So, while urban wetlands can host many anisopteran species, rural wetlands are important for the habitat-specialist species (Lubertazzi and Ginsberg, 2010).

Villalobos-Jiménez *et al.* (2016) reviewed the response of stream-inhabiting dragonflies to rural–urban gradients. They found that 21 out of 31 studies on the topic reported lower diversity in built-up city cores, with 12 of the cases related to pollutant concentrations in the water, as well as to other co-occurring stressors such as vegetation loss and urban development. Often there is an increase in abundance of species in the Libellulidae but a decrease in those in the Gomphidae. However, sensitivity versus tolerance matters not so much at the family level but rather at that of specific species. Furthermore, there are often carry-over effects, such as from female adult to the larva, or the persistence of impacts between life stages from larva to adult. Overall, there is a decrease in dragonfly diversity, less so from the effect of urbanization per se, but rather owing to the intensive management associated with urban environments.

Among species-rich stream odonate assemblages in the municipality of Manaus, in the Amazon region of Brazil, species richness declines with increasing urbanization, especially in the case of damselflies which prefer less-disturbed habitats. In contrast, among anisopterans, species richness is highest in intermediate habitats along the rural–urban gradient (Monteiro Júnior *et al.*, 2015). In and around Belém, also in the Amazon region of Brazil, urbanization leads to the loss of K-strategists (species with slow growth, long development time, large body size) in favour of *r*-strategists (species with fast growth and small biomass), with habitat specialists rather than habitat generalists most affected. Higher urban temperatures also benefit some of the *r*-strategists, which increase in abundance with urban transformation (Brito *et al.*, 2021).

In central Mexico, zygopterans are affected more by urbanization than are anisopterans. There is higher zygopteran species richness and abundance at sites with low versus high urban development, although no such differences among the anisopterans along the gradient. However, there is a positive statistical relationship between anisopteran species richness, as well as their abundance, and human waste products percentage and macrophyte cover, although there is also a negative relationship with dissolved solids in the water (Suárez-Tovar *et al.*, 2022).

There is much variation in odonate response to urbanization according to geographical location and local context. For example, the larval odonate fauna of Georges River, Sydney, Australia, is neither affected by urban development nor degradation of water quality with increased levels of urbanization (Tippler *et al.*, 2018). In contrast, in South Carolina, USA, small habitats isolated in urban areas are depauperate, supporting only nested subsets of species compared to larger natural areas (Worthen and Chamlee, 2020). In the same geographical area, at the small, plot scale, odonate larval abundance and diversity increases with increasing amounts of dead debris, possibly providing refuge from predators. In turn, adult abundance and diversity is related to the amount of aquatic and shoreline vegetation used as perches. But at large spatial scales, odonate diversity is related more to natural changes in habitat quality than to urbanization. Zygopteran diversity increases downstream of urbanization, while anisopteran diversity is greatest in sunny, open habitats with fields, wetlands and open water (Worthen *et al.*, 2021). Along Texas streams, anisopteran larvae species in rural versus urban streams are similar, with a complex set of environmental variables determining their diversity and distribution, along with the type of benthic habitat also playing an important role (Bowles and Kleinsasser, 2022).

In the UK, the most species-rich urban ponds in terms of macroinvertebrate assemblages are those with complex macrophyte stands and floating vegetation, with low nutrient concentrations and embedded in some green space. The most impoverished ponds are those in highly urban settings and with hard-engineered edges, heavy shading and nutrient-rich waters (Thornhill *et al.*, 2017). Furthermore, pond density in the UK has been decreasing. Over the years 1904–2009, the number of ponds in the country declined by 82% with pond density overall decreasing from 7.1 km^2 to 1.3 km^2. There was also a concurrent increase in distance between ponds by up to 49%, indicating increasing pond isolation with fewer neighbouring ponds (Thornhill *et al.*, 2018).

Across 19 sites in central Mexico, the relationship between human waste, water chemistry, macrophyte cover and odonate diversity indicates that there is higher damselfly species richness and abundance in sites with low rather than high levels of urbanization. However, there are no differences in anisopteran assemblages along the gradient. There is a positive relationship for both anisopteran species richness and abundance, and waste percentage and macrophyte cover, but a negative relationship with dissolved solids in the water. Overall, the odonate assemblages are fairly tolerant to urbanization, with anisopterans being more tolerant of the urbanization process than zygopterans (Suárez-Tovar *et al.*, 2022).

There is a contrasting effect of urbanization among coastal cities of the dune, pond and lagoon ecosystem of Rio Grande do Sul, Brazil where anisopteran species richness, composition and abundance are affected by the intensity of urbanization. Unusually, this is not the case for zygopterans, which in this non-forested environment appear to be dominated by habitat generalists, apart from *Telebasis corallina* which is not tolerant of urbanized areas. This species and the anisopterans *Erythemis credula*, and an unidentified *Erythrodiplax* species, are indicative of non-urbanized areas (Sganzerla *et al.*, 2021).

When artificial ponds are compared with natural ones in the Anza-Borrego Desert State Park, California, the dragonfly assemblages associated with natural oases, semi-developed ponds (artificial ponds with natural vegetation) and developed ponds (on golf courses) are distinctly different, driven partly by aquatic vegetation but also the level of development. The natural springs support certain specialist species, such as *Argia vivida*, while the developed ponds support widespread opportunistic species, especially in the genera *Enallagma*, *Ischnura* and *Sympetrum* (Garfield *et al.*, 2019).

In the city of Pune, Maharashtra, India, where dragonfly diversity is driven by the intensity of land use, river species richness is highest at moderately urbanized sites. The urban centre is so built up and the rural areas are so heavily impacted by logging, channelization and sewerage effluent that the dragonfly fauna is severely impoverished (Jere *et al.*, 2020). In the same city, industrial pollution is also highly impoverishing to the local dragonfly assemblages (Saha *et al.*, 2018). Nevertheless, a very small (350 m^2), highly polluted site (Khadki) with degradable and non-degradable waste, in the centre of Pune, has still been able to support 17 odonate species (Saha and Gaikwad, 2015), assuming ecological relaxation will not take place in due course. Future local species loss is plausible, as along two river basins in the same state, there has been rapid industrialization and urbanization. The outcome has been habitat deterioration to such an extent that at certain monitored sites there has been a 31% loss of dragonfly species over the 50-year period prior to 2009–2010 (Kulkarni and Subramanian, 2013).

A global review of the traits of sensitive versus tolerant odonate species in all types of water bodies suggests that habitat preference is not linked with sensitivity to urbanization. However, larger species with weak sexual dimorphism are the most sensitive species. Furthermore, anisopteran species with yellow to red rather than dull coloration tended to be less sensitive to urban areas, while for zygopteran species, territorial species tended to be less sensitive to urban areas (Suárez-Tovar *et al.*, 2023b). In Mexico, the damselflies *Hetaerina americana* and *H. vulnerata*, which have red wingspots and are territorial, showed great tolerance of urban conditions and seemingly are under no environmental stress (Suárez-Tovar *et al.*, 2023a).

Odonate individuals are more successful at catching prey in more urbanized local environments. This may be due to the overall more simplified structure of urban vegetation, changes in odonate flight behaviour or an increase in acceptable fly prey items (Suárez-Tovar *et al.*, 2024).

5.6.3 Refuges for dragonflies in urban environments

Sometimes, it is fine-scale effects that determine dragonfly presence in certain urban ecosystems (Fig. 5.31). The city of Cape Town in the Cape Floristic Region biodiversity hotspot in South Africa has lost 97% of its wetlands through infilling and urban development. The low-lying wetlands are under severe threat from many stressors arising

Fig. 5.31. As the effects of urban stressors are largely local, it means that high-quality refuges can play an important role in supporting local odonate assemblages. Table Mountain in the Cape Floristic Region (CFR) biodiversity hotspot is a sky island with perennial streams fed by both rainfall and condensation. It supports many odonate species, including CFR endemics, despite being surrounded by sprawling urbanization. While large sky island refuges like this may be able to support species in the long term, small urban refuges subject to intense stressors may not. (Photo: Michael Samways).

from the process of urbanization, and are depauperate in aquatic biodiversity, driven by a combination of local environmental variables such as artificially high nutrient loads, wetland transformation and invasive alien trees (Deacon *et al*., 2021b). Yet the centre of the city is Table Mountain, an important and large mountain refuge for endemic biodiversity, that is home to several highly endemic dragonflies. The headwater streams, arising from condensation as well as rainfall on the mountain, are today fully protected. The only dragonfly loss has been the very rare and highly threatened *Orthetrum rubens*, with the mountain being its type locality. However, it has not been seen there since the 1930s. It is largely urban sprawl that prevents it from recolonizing the mountain (Deacon and Samways 2021a, 2021b).

Furthermore, the 78 ha Lekki Conservation Centre on the outskirts of Lagos, Nigeria is a forest swamp and savanna grassland enclave which supports a relatively rich dragonfly assemblage. While most species are widespread African species, it is home to some West African swamp forest species such as *Copera sikassoensis* and *Ceriagrion rubellocerinum* (Ojonugwa *et al*., 2020).

There are other instances where fine-scale assessment in urban areas has revealed urban refuges. In an urbanized area of Michigan, USA, besides there being an abundance of habitat-generalist species, there are populations of habitat specialists such as *Cordulegaster obliqua* in small forest streams, *Stylurus spiniceps* and *Macromia taeniolata* in large rivers, and *Amphiagrion saucium* and *Nannothemis bella* in acidic wetlands. Furthermore, certain species of conservation concern in the region are also present, including *Stylurus notatus*, *S. plagiatus*, *Hetaerina titia*, *Rhionaeschna mutata* and *C. obliqua* (Craves and O'Brien, 2013).

In dry regions, dragonflies are compromised by the scarcity and unreliability of water supplies. However, in Texas, USA, urban reservoirs are more dragonfly species-rich than either natural pools or those surrounded by grasslands or croplands. Of the 33 species recorded, seven are exclusive to urban ponds. The reason for this is that the urban ponds are fed by consistently available urban run-off and hold water for longer, even during droughts, making them refuges in a more stressful matrix (Husband and McIntyre, 2021).

In and around Manaus, in the Amazon region of Brazil, stream dragonfly diversity is directly linked to the level of degradation, with assemblages distinctly different and more diverse than those in both intermediate and degraded habitats, as forest specialists are without suitable habitat (Monteiro Júnior *et al.*, 2014). Similarly, in South Africa, dragonfly assemblage composition changes with the level of urbanization, with the sun-loving habitat generalists replacing the habitat specialists. However, retaining wooded riparian strips along rivers enables national endemics like *Chlorolestes tessellatus* and *Allocnemis leucosticta* to persist, even in an urban environment (Samways and Steytler, 1996).

5.6.4 Effect of urbanization on dragonfly biology

Urbanization also causes changes in the physiology of certain species. The growth of the city of Tehuixtla, Mexico has led to a reduction of terrestrial vegetation and an increase in wastewater effluent in the streams. The damselfly *Hetaerina americana* responds to these environmental changes by showing a drop in abundance of both larvae and adults. Physiological effects among adults include a drop in lipid levels and muscle biomass. Egg survival also drops (Córdoba-Aguilar and Rocha-Ortega, 2019).

Although some odonate species are such habitat specialists and so highly phylogenetically constrained that urbanization is taking a toll on their populations, there are some signs that evolutionary adaptation is already taking place in some more genetically resilient species in response to urbanization. For example, urban populations of *Coenagrion puella* are better able to tolerate heatwaves than are rural populations, due to the larvae suffering less from a depletion of energy availability among immune-challenged individuals (Tüzün and Stoks, 2021). Among males of this same species there are strong interactive effects of hatching period and level of urbanization on metamorphic traits, with both treatments strongly affecting lifetime mating success. Although both treatments influence lifetime mating success, only the hatching period effect is mediated by metamorphic traits, which in turn is related to the level of urbanization (Tüzün and Stoks, 2018).

In London, UK, while there are some small responses by certain dragonflies to an urban heat island, the effects are negligible on emergence patterns compared to those of climate change, owing to their aquatic habitats buffering microclimatic conditions which regulate the urban heat island effect. This effect has implications for urban design and management (Villalobos-Jiménez and Hassall, 2017).

Ischnura elegans is widespread and abundant in ponds throughout much of Europe, and able to tolerate many stressors. One manifestation of its tolerance of urbanization is that its wing morphology is unaffected by the urban environment. However, other flight-related traits, such as flight endurance, flight performance, fat content and dispersal abilities might also be involved in its wide habitat tolerance (Villalobos-Jiménez and Hassall, 2019).

Interestingly, *Ischnura elegans*, in an urban environment, displays 'synapse organization' among single nucleotide polymorphisms (SNPs), accumulated by the accretion of different genes associated with candidate SNPs in urbanized areas. This emphasizes the highly polygenic nature of adaptation, where the genetic response is achieved through a subtle adjustment of allele frequencies in different genes that contribute to adaptive phenotypes. Overall, a polygenic adaptive response in the nervous system to urbanization specifically implicates genes involved in synapse organization (Babik *et al.*, 2023).

5.6.5 Other impacts associated with infrastructure development

5.6.5.1 Dragonfly roadkill

Roads with little traffic are not a significant barrier to most flying odonates. However, moving vehicles are a mortality factor. In western Romania, of the 555 invertebrates killed by traffic, especially in forested areas, less than 2% were dragonflies, with much higher percentages for many other insect taxa (Popovici *et al.*, 2018). Yet in India, dragonfly deaths in national parks and suburban shrubland were 61% of the total for insects (Rao and Girish, 2007). Traffic volumes and speed play a role, but so does landscape context, as well as exactly how insect mortalities are measured and compared.

An investigation of four roads with different traffic volumes in Illinois, USA gave the mean number of dragonfly individuals killed as between 2–35 individuals/km/day. Although *Plathemis lydia* and *Libellula luctuosa* represented, respectively, only 14% and 31% of the number of dragonflies observed, they made up more than 70% of the deaths. Their low-level flight is a major factor in this mortality. Yet *Tramea lacerata* also flies low, but its flight agility enables it to dodge moving vehicles better than the other two species (Soluk *et al.*, 2011).

Mackinac Bay, Michigan, USA is an extensive coastal wetland bordered by a highway that separates a marsh occupied by dragonflies defending territories and breeding from the adjacent forest where they forage and rest. Daily mortality from traffic varies greatly, ranging between 10–256 individual deaths per kilometre, with 60%+ being males. The effects of roads which skirt wetlands on dragonflies is concerning as dragonfly populations are often small and restricted to wetland habitats which are already suffering multiple stressors (Riffell,1999).

Road type and the ecological context can have variable effects on adult dragonflies, some positive and others negative. Access roads to Russian forests are foraging areas for many dragonfly species as well as shelter zones from predators that hunt in open areas. Furthermore, dragonflies disperse along roads and clearings cut into the forest, leading to elevated local species richness. They also display modified behaviour with respect to daily activity patterns, choice of night roosts, territorial behaviour, etc. However, paved roads with heavy, high-speed traffic passing can be a major mortality factor, and of concern for rare and threatened species. Furthermore, environmental conditions along forest roads deteriorate in the autumn when air temperatures remain below 15°C and rain puddles accumulate on dirt roads, causing dragonflies to become trapped in these puddles and die (Timofeev, 2016).

5.6.5.2 Boating pressure on dragonflies

The various impacts from recreational boating include a decrease in water quality, damage and disturbance to plants and other organisms, and the moving around of invasive alien species. Recreational boating can increase wave frequency and magnitude (increasing shear stress) and alter wave direction relative to wind-generated waves. The type and speed of boats have an effect, as does water depth, shoreline slope and the structural complexity of aquatic vegetation, all of which affect boat-generated waves. Odonate assemblage composition of adults and exuviae change through a combination of factors associated with local habitat structure, landscape configuration and boating pressure. However, adults and larvae (sampled as exuviae) are influenced by different factors. In the case of adult species composition, landscape configuration has the greatest effect, followed by habitat structure, and boating pressure. For the young stages, landscape configuration and habitat structure are most important, with boating an added stressor (Hall *et al.*, 2015).

5.6.5.3 Effect of bridges and their abutments

Bridges and their supporting structures across rivers (abutments) locally shade the river, change river behaviour and affect microhabitats in the immediate vicinity. These drivers cause a local shift in the composition of the macroinvertebrate fauna (Yu *et al.*, 2016), including its functional diversity (Gál *et al.*, 2023).

Three bridges across a small river in Germany did not completely fragment a *Calopteryx splendens* population habitat, with 13% of all recaptured individuals successfully passing under at least one bridge. However, in the case of one wide but low bridge more than 70% of the approaching individuals turned back (Schutte *et al.*, 1997).

5.6.5.4 Effect of artificial light on dragonflies

Artificial lights at night change the species composition of riparian predators and scavengers, increasing the aquatic-to-terrestrial subsidy flux, which could cascade through the riparian food web. As artificial light affects processes that link adjacent ecosystems, it is possibly significant for odonates, especially via their prey and at emergence, with *Ischnura elegans* and *Lestes sponsa* caught in emergence traps in Germany (Manfrin *et al.*, 2017). Although, in another study, flight intercept traps adjacent to artificial lighting did not catch any odonates in Germany (Perkin *et al.*, 2014), artificial lighting can be particularly significant in tropical forests where crepuscular and nocturnally active flying adults are frequently caught in entomological light traps. This needs to be quantified as these forests face increasing pressure from illuminated infrastructure development.

5.6.5.5 *Possible effect of electromagnetic radiation on dragonflies*

Electromagnetic radiation (EMR) from human communication infrastructure affects the behaviour of many organisms, including insects (Balmori, 2015; Thill *et al.*, 2023). Behavioural changes include disturbance to swarming and navigation in bees, colony disruption, orientation and locomotion in ants, magnetoreception in cockroaches and fly reproduction. Ecologically, EMR variously affects abundance and composition of wild pollinators, with the potential to affect food webs. However, to date there appears to be no definitive information on the effect of EMR on dragonfly behaviour or ecology.

5.6.5.6 *Risk of ecological traps for dragonflies*

Ecological traps are preferred habitats that confer lower fitness than another available habitat. At the ecosystem level, some altered wetlands may provide essential or valuable habitat while others could decrease both local and regional population persistence, and ultimately lead to local extinctions (Sievers *et al.* 2018).

Certain artificial shiny dark surfaces are attractive to some flying dragonflies, especially libellulids. The bodywork of parked dark-coloured cars and windscreens often attracts dragonflies (Watson, 1992; Stevani *et al.*, 2000; Burrial and Ocharan, 2003; Günther, 2003; Wildermuth and Horváth, 2005), leading to oviposition on the cars in some cases. As there are so many parked cars around the world at any one time, this is a major effect, although the extent to which this may reduce dragonfly populations is unknown.

Other attractive surfaces include shiny black gravestones (Horváth *et al.*, 2007) and crude oil surfaces (Horváth *et al.*, 1998). The reason for these various behavioural misjudgements is that dragonflies are attracted to the intense horizontal polarization of reflected visible and ultraviolet light, which can be stronger than that from the water surface. This ecological trapping has been described as polarized light pollution (Horváth *et al.*, 2009) and can vary according to the physiological condition of individuals, potentially having strong eco-evolutionary effects (Ensaldo-Cárdenas *et al.*, 2021).

Ecological traps can also be significant in rural settings. Among the shallow ephemeral wetlands of the Great Plains of North America (playas) there is not a steady increase in odonate richness with wetland area, but rather there is a threshold response to area in the size range 8–18 ha, in large part because larger wetlands are deeper and have less emergent vegetation than shallow ones. This means that in larger pools, perching and oviposition sites are restricted. However, as sedimentation from agricultural activity decreases with water depth, allowing plant growth, the threshold effect is not as strong in cropland as in natural grassland wetlands. The result is that the increased emergent vegetation gives a false signal to adults seeking a territory or oviposition site relative to natural pools, with these disturbed pools acting as ecological traps as the pool dries out (Hernandez *et al.*, 2006).

In Italy, some modified rivers dry out at certain times of the year. These systems attract adults of certain dragonfly species during the wet phase, leading to development of the young stages. However, with the onset of the wet phase, these systems can be ecological traps to which the adults are attracted but with unsuitable conditions for development of the larvae (Hardersen, 2008).

Fishponds in Czechia are ecological traps for *Sympetrum depressiusculum* and *S. sanguineum*. These ponds, while attracting ovipositing individuals, are largely unsuitable for completion of the life cycle, due to several factors including pond depth, and fish disturbance and predation (Šigutová *et al.*, 2015).

5.7 Interspecific Interactions among Dragonflies

5.7.1 Introgressive hybridization

Heterospecific pairing among sympatric odonates is not uncommon (Bick and Bick, 1981; Miller and Fincke, 2004), and likely often undetected, especially when morphically similar species are engaged in the activity (Dey and Pal, 2022). However, pairings of *Lestes sponsa* in the UK have been observed even with species of other genera (Pickess, 2016). Sometimes these pairings lead to interspecific hybrids, as, for example, the coenagrionids *Paracercion sieboldii* and *P. melanotum* in Japan (Okude and Futahashi, 2020).

As we saw in Chapter 4, many dragonflies are currently showing range shifts in response to a rapidly changing global climate. Any one individual has in effect three choices: to stay put and tolerate the changing conditions according to the fixed and plastic traits it possesses, to move to a more optimal habitat, or to stay and possibly die. This shuffling of the fauna, and at a slower pace also the flora, often involves new associations between species, or more significantly, between genotypes, which may or may not have had previous contact. Both the receiving genotype, which could be under some stress associated with the ongoing changing climate and associated extreme weather events, and the advancing genotype then experience a novel relationship. This novelty has potentially various outcomes. There could be increased competition, a change in predator–prey relationships, and among closely related species, chances of hybridization.

Introgressive hybridization (introgression) is of significance for dragonflies. It occurs when there is transfer of genetic material from one gene pool into another through repeated backcrossing of an interspecific hybrid with one of its parents leading to a variably complex mixture of genes in the offspring. This process appears to have been significant among odonates in the deep past, resulting in reticulate phylogenies with net-like species relationships. As these genetic exchanges can become fixed over time, they are important even for defining higher-level taxa over long periods of time. For example, about 30% of the genome of the Anisozygoptera is due to introgression. The evolutionary benefit of introgression lies in it being an important source of novel variation leading to speciation, diversification, adaptation to new environments and shifts in geographic boundaries (Suvarov *et al.*, 2022).

Current rapid climate change is seeing many new zones of geographic overlap between formerly geographically isolated taxa. Among insects in general, these new associations can affect local species persistence, and at times can increase or decrease local insect species diversity. Among odonates in the current era, introgression is a significant factor. Some taxa, for example, in *Ischnura* and *Mnais*, appear to be particularly engaging of introgression (Sánchez-Guillén *et al.*, 2016). While it is still early days in the discovery of modern introgression events, it seems that it is much rarer in anisopterans than zygopterans, perhaps due to the copulatory apparatus being more specific, preventing interspecific mating, despite this group's higher mobility.

Introgression has a profound effect on the way we do conservation. First, it highlights difficulties with the biological species concept, that is, one good biological species is not supposed to interbreed with another. Furthermore, introgression may be considered as a source of novel biodiversity, while contrastingly, a cause of biodiversity loss. Conservation arguably should not so much focus on the biological species concept but rather address conservation units based on the impact on biodiversity when taxa are lost. This implies that some conservation programmes should aim at protecting biodiversity as a dynamic evolutionary process (Quilodrán *et al.*, 2020).

Having said that, we cannot ignore protection of highly threatened species. *Ischnura gemina* is a threatened species in the San Francisco Bay area of California. One of the concerns is that this species is currently at risk from displacement and/or extinction by the more widespread *I. denticollis*, aggravated by this species also benefitting from *I. gemina* genes enabling it to better adapt to new environmental conditions (Sánchez-Guillén *et al.*, 2014).

Ischnura elegans and *I. graellsii* are closely related and share many morphological, genetic and phenotypic traits, as well as a preference for similar habitats, leading to some hybridization between the two species. *I. elegans* has greatly expanded its geographic range in Spain in the last several decades. Early hybridization in the late nineteenth century subsequently led to this species replacing *I. graellsii* while modifying its original niche to become more like the niche of *I.graellsii*. Hybridization is ongoing with bidirectional backcrosses through transfer of genetic material with a slight lowering of overall genetic differentiation where the species are in close contact, irrespective of when they first came into contact. The different genetic mixes are explained by reproductive barriers in hybrid zones of various strengths (the strongest being in the oldest hybrid zone), continuous recolonization events, closeness to the arrival site and time since colonization (Sánchez-Guillén *et al.*, 2022).

5.7.2 Potential interspecific competition in a changing world

Model predictions of the effect of climate warming on the European *Gomphus vulgatissimus* suggest that it would require less development time resulting in its spread northward. At the southern end of its range, the species may experience increased competition (Söndgerath *et al.*, 2012).

Siblings *Sympetrum striolatum* and *S. vulgatum* are common and widespread in central Europe, and while *S. striolatum* is showing strong positive population trends, *S. vulgatum* is declining. Their hatching rates decrease under drought conditions, especially in *S. vulgatum*. *S. vulgatum* varies greatly in development between clutches, while the eggs of *S. striolatum* develop much faster and hatch highly synchronously giving it a temporal advantage leading to a size-advantage over *S. vulgatum*. However, the *S. vulgatum* larva grows faster. Nevertheless, the current downward population trends are partly due to lower survival of *S. vulgatum* under drought, as opposed to the phenological size benefits of *S. striolatum*. Overall, this means that the population dynamics of the two species can be explained by climatically-induced changes in their interactions (Hogreve and Suhling, 2022).

While there has been much debate around neonatives possibly outcompeting already well-established local species, there is little evidence to support the argument one way or the other. However, some interesting results have started to emerge. The damselfly *Erythromma viridulum* has in recent years expanded its range in the UK, mostly because it has colonized areas with good-quality habitat. Investigations on its possible negative or positive effects on already established odonate populations indicate that anisopteran species generally persist rather well in the presence of the newcomer. Zygopterans also show little change in persistence, although there are some possible negative effects on its congener *E. nijas* and on *Platycnemis pennipes* (Cranston *et al.*, 2023) (Fig. 5.32).

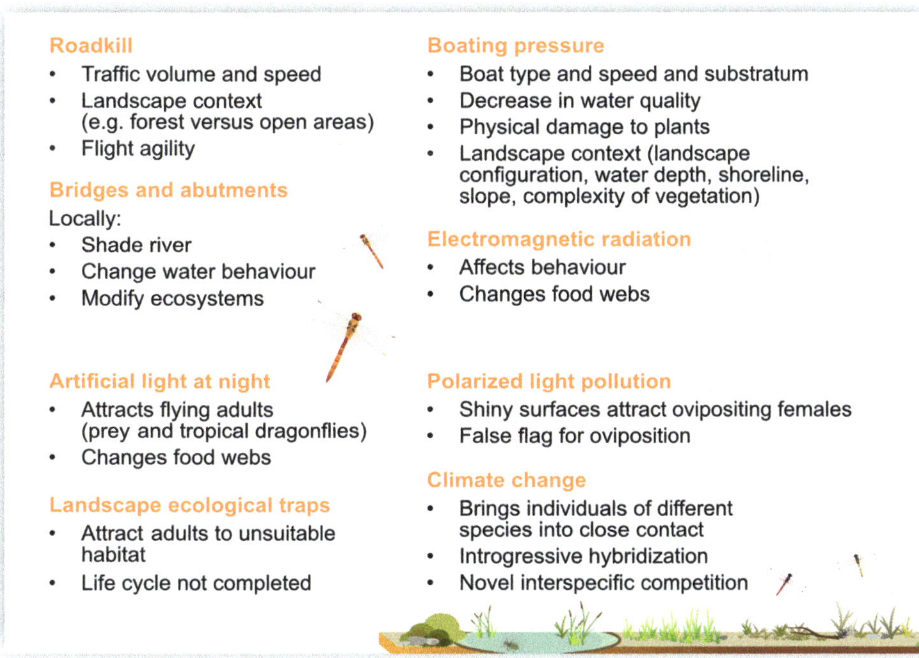

Fig. 5.32. There are some additional stressors not considered earlier that affect dragonfly individuals and populations at the local level, yet are widespread stressors. These include the human movement impacts of roadkill and boating pressure, bridges and their supporting structures, the potential effect of electromagnetic radiation, artificial light at night, polarized light pollution from shiny surfaces, landscape ecological traps, climate change, increasing introgressive hybridization and novel competition events. Each stressor has particular qualifiers resulting in various outcomes. (Illustration: Robyn Symons).

References

Ackerman, J. and Galloway, T.D. (2003) Odonata larvae in urban retention ponds in Winnipeg, Manitoba, Canada. *Proceedings of the Entomological Society of Manitoba* 59, 5–15.

Addo-Bediako, A. and Malakane, K. (2020) Preliminary assessment of chemical elements in sediments and larvae of Gomphidae (Odonata) from the Blyde River of the Olifants River system, South Africa. *International Journal of Environmental Research and Public Health* 17, e8135. DOI: 10.3390/ijerph17218135.

Albert, J.S., Destouni, G., Duke-Sylvester, S.M., Magurran, A.E., Oberdorff, T. *et al.* (2021) Scientists' warning to humanity on the freshwater biodiversity crisis. *Ambio* 50, 85–94. DOI: 10.1007/s13280-020-01318-8.

Alexander, R.B., Boyer, E.W., Smith, R.A., Schwarz, G.E. and Moore, R.B. (2007) The role of headwater streams in downstream water quality. *Journal of the American Water Resources Association* 43, 41–59. DOI: 10.1111/j.1752-1688.2007.00005.x.

Allan, J.D. (2004) Landscapes and riverscapes: the influence of land use on stream ecosystems. *Annual Review of Ecology and Systematics* 35, 257–284. DOI: 10.1146/annurev.ecolsys.35.1202022.110122.

Allgeier, S., Friedrich, A. and Brühl, C.A. (2019) Mosquito control based on *Bacillus thuringiensis israelensis* (Bti) interrupts artificial wetland food chains. *Science of the Total Environment* 686, 1173–1184. DOI: 10.1016/j.scitotenv.2019.05.358.

Amatulli, G., Marquez, J.G., Sethi, T., Kiesel, J., Grigoropoulou, A. *et al.* (2022) Hydrography90m: a new high-resolution global hydrographic dataset. *Earth System Science Data* 14, 4525–4550. DOI: 10.5194/essd-14-4525-2022.

Arambourou, H. and Stoks, R. (2015) Warmer winters modulate life history and energy storage but do not affect sensitivity to a widespread pesticide in an aquatic insect. *Aquatic Toxicology* 167, 38–45. DOI: 10.1016/j.aquatox.2015.07.018.

Araujo, B.R., Pinto, Â.P. and Padial, A.A. (2022) Influence of landscape homogenization due to river damming on dragonfly (Odonata) community structuring in a subtropical forest in the southern Atlantic Forest. *Ecohydrology* 15, e2419. DOI: 10.1002/eco.2419.

Araújo, M.F.A., De Marco, P., Juen, L. and Tôrres, N.M. (2020) Vulnerability of *Phyllocycla* species (Odonata: Gomphidae) to current and planned anthropic activities by the Brazilian Government. *Neotropical Entomology* 49, 24–32. DOI: 10.1007/s13744-019-00714-4.

Astuti, A., Nayasilana, I.N., Sugiyarto and Budiharjo, A. (2022) Community structure of dragonflies (Odonata) in Gunung Bromo's Forest Area with Special Purpose (FASP), Karanganyar, Central Java, Indonesia. *Biodiversitas* 23, 2493–2501. DOI: 10.13057/biodiv/d230529.

Azevedo-Santos, V.M., Brito, M.F.G., Manoel, P.S., Perroca, J.F., Rodrigues-Filho, J.L. *et al.* (2021) Plastic pollution: a focus on freshwater biodiversity. *Ambio* 50, 1313–1324. DOI: 10.1007/s13280-020-01496-5.

Babik, W., Dudek, K., Marszałek, M., Palomar, G., Antunes, B. and Sniegula, S. (2023) The genomic response to urbanization in the damselfly *Ischnura elegans*. *Evolutionary Applications* 16, 1805–1818. DOI: 10.1111/eva.13603.

Baird, I.R.C. and Benson, D. (2020) Serious impacts of longwall coalmining on endangered Newnes Plateau Shrub Swamps, exposed by the December 2019 bushfires. *Australasian Plant Conservation* 29, 12–15.

Baird, I.R.C. and Burgin, S. (2016) Conservation of a groundwater-dependent mire-dwelling dragonfly: implications of multiple threatening processes. *Journal of Insect Conservation* 20, 165–178. DOI: 10.1007/s10841-016-9852-3.

Balinsky, B.I. (1967) On some intrinsic and environmental factors controlling the distribution of dragonflies (Odonata), with redescription and a new name for a little known species. *Journal of the Entomological Society of Southern Africa* 29, 3–22.

Ball-Damerow, J.E., M'Gonigle, L.K. and Resh, V.H. (2014) Local and regional factors influencing assemblages of dragonflies and damselflies (Odonata) in California and Nevada. *Journal of Insect Conservation* 18, 1027–1036. DOI: 10.1007/s10841-014-9709-6.

Balmori, A. (2015) Anthropogenic radiofrequency electromagnetic fields as an emerging threat to wildlife orientation. *Science of the Total Environment* 518–519, 58–60. DOI: 10.1016/j.scitotenv.2015.02.077.

Balzan, M.V. (2012) Associations of dragonflies (Odonata) to habitat variables within the Maltese Islands: a spatiotemporal approach. *Journal of Insect Science* 12, e87. DOI: 10.1673/031.012.8701.

Barmentlo, S.H., Vriend, L.M., van Grunsven R.H.A. and Vijver, M.G. (2019) Environmental levels of neonicotinoids reduce prey consumption, mobility and emergence of the damselfly *Ischnura elegans*. *Journal of Applied Ecology* 56, 2034–2044. DOI: 10.1111/1365-2664.13459.

Barrett, M.D. and Williams, M.R. (1998) Distribution of the Western Petalura dragonfly *Petalura hesperia* Watson in Western Australia. *Pacific Conservation Biology* 4, 149–154.

Bellard, C., Marino, C. and Courchamp, F. (2022) Ranking threats to biodiversity and why it doesn't matter. *Nature Communications* 13, e2616. DOI: 10.1038/s41467-022-30339-y.

Bick, G.H. and Bick, J.C. (1981) Heterospecific pairing among Odonata. *Odonatologica* 10, 259–270.

Birk, S., Chapman, D., Carvalho, L., Spears, B.M., Andersen, H.E. *et al.* (2020) Impacts of multiple stressors on freshwater biota across spatial scales and ecosystems. *Nature Ecology and Evolution* 4, 1060–1068. DOI: 10.1038/s41559-020-1216-4.

Blüthgen, N., Staab, M., Achury, R. and Weisser, W.W. (2022) Unravelling insect declines: can space replace time? *Biology Letters* 18, e20210666. DOI: 10.1098/rsbl.2021.0666.

Borges, L.R., Barbosa, M.S., Carneiro, M.A.A. and Santos, J.C. (2021) Habitat integrity drives Odonata diversity in *Eucalyptus*-dominated landscape. *Environmental Monitoring and Assessment* 193, e12. DOI: 10.1007/s10661-020-08740-1.

Bota-Sierra, C.A., Flórez-V, C., Escobar, F., Sandoval-H., J., Novelo-Gutiérrez, R. *et al.* (2021) The importance of tropical mountain forests for the conservation of dragonfly biodiversity: a case from the Colombian Western Andes. *International Journal of Odonatology* 24, 233–247. DOI: 10.23797/2159-6719_24_18.

Bouhala, Z., Khemissa, C., Márquez-Rodríguez, J. Ferreras-Romero, M., Samraoui, F. and Samraoui, B. (2019) Ecological correlates of odonate assemblages of a Mediterranean stream, Wadi Cherf, northeastern Algeria: implications for conservation. *International Journal of Odonatology* 22, 181–197. DOI: 10.1080/13887890.2019.1688199.

Bowles, D.E. and Kleinsasser, L.J. (2022) Environmental determinates of distribution for dragonfly nymphs (Odonata: Anisoptera) in urban and non-urban east Texas streams, USA. *Hydrobiology* 1, 76–88. DOI: 10.3390/hydrobiology 1010006.

Brasil, L.S., Vieira, T.B., de Oliveira-Junior, J.M.B., Dias-Silva, K. and Juen, L. (2017) Elements of metacommunity structure in Amazonian Zygoptera among streams under different spatial scales and environmental conditions. *Ecology and Evolution* 7, 3190–3200. DOI: 10.1002/ece3.2849.

Brasil, L.S., Silverio, D.V., Cabette, H.S.R., Batista, J.D., Vieira, T.B. *et al.* (2019) Net primary productivity and seasonality of temperature and precipitation are predictors of the species richness of the damselflies in the Amazon. *Basic and Applied Ecology* 35, 45–53. DOI: 10.1016/j.baae.2019.01.001.

Brasil, L.S., Ferreira, V.R.S., Resende, B.O., Juen, L., Batista, J.D. *et al.* (2021) Dams change beta diversity of aquatic communities in the veredas of the Brazilian Cerrado. *Frontiers in Ecology and Evolution* 9, e612642. DOI: 10.3389/fevo.2021.612642.

Bredenhand, E. and Samways, M.J. (2009) Impact of a dam on benthic macroinvertebrates in a small river in a biodiversity hotspot: Cape Floristic Region, South Africa. *Journal of Insect Conservation* 13, 297–307 DOI: 10.1007/s10841-008-9173-2.

Briggs, A.J., Pryke, J.S., Samways, M.J. and Conglong, D.E. (2019) Complementarity among dragonflies across a pondscape in a rural landscape mosaic. *Insect Conservation and Diversity* 12, 241–250. DOI: 10.1111/icad.12339.

Brito, J.G., Roque, F.O., Martins, R.T., Nessiman, J.L., Oliveira, V.C. *et al.* (2020) Small forest losses degrade stream macroinvertebrate assemblages in the eastern Brazilian Amazon. *Biological Conservation* 241, e108263. DOI: 10.1016/j.biocon.2019.108263.

Brito, J.P., Carvalho, F.G. and Juen, L. (2021) Response of the zygopteran community (Odonata: Insecta) to change in environmental integrity driven by urbanization in eastern Amazonian streams. *Ecologies* 2, 150–163. DOI: 10.3390/ ecologies2010008.

Brockhaus, T. and Hartmann, A. (2009) New records of *Epiophlebia laidlawi* Tillyard in Bhutan, with notes on its biology, ecology, distribution, zoogeography and threat status (Anisozygoptera: Epiophlebiidae). *Odonatologica* 38, 203–215.

Burrial, A. T. and Ocharan, F. J. (2003) ¿Coches como hábitat para libélulas? Algunos machos de *Crocothemis erythraea* creen que sí. *Boletin de la Sociedad Entomologia Aragonesa* 32, 214–215.

Cai, D., Fraedrich, K., Guan, Y., Guo, S. and Zhang, C. (2019) Urbanization and climate change: insights from eco-hydrological diagnostics. *Science of the Total Environment* 647, 29–36. DOI: 10.1016/j.scitotenv.2018.07.319.

Calderon, M.R., Almeida, C.A., Jofré, M.B., González, S.P. and Miserendino, M.L. (2023) Flow regulation by dams impacts more than land use on water quality and benthic communities in high-gradient streams in a semi-arid region. *Science of the Total Environment* 881, e163468. DOI: 10.1016/j.scitotenv.2023.163468.

Calvão, L.B., Juen, L., Oliveira Junior, J.M.B., Batista, J.D. and De Marco Júnior, P. (2018) Land use modifies Odonata diversity in streams of the Brazilian Cerrado. *Journal of Insect Conservation* 22, 675–685. DOI: 10.1007/s10841-018-0093-5.

Campero, M., Slos, S., Ollevier, F. and Stoks, R. (2007) Sublethal pesticide concentrations and predation jointly shape life history: behavioral and physiological mechanisms. *Ecological Applications* 17, 2111–2122. DOI: 10.1890/07-0442.1.

Caraco, N. (2021) A tribute to tributaries: river studies elucidate links between human activity and nutrient export across a broad range of watersheds. *Ambio* 50, 742–744. DOI: 10.1007/s13280-020-01462-1.

Carchini, G., Solimini, A.G. and Ruggiero, A. (2005) Habitat characteristics and odonate diversity in mountain ponds of central Italy. *Aquatic Conservation: Marine and Freshwater Ecosystems* 15, 573–581. DOI: 10.1002/aqc.741.

Carchini, G., Della Bella, V., Solimini, A.G. and Bazzanti, M. (2007) Relationships between the presence of odonate species and environmental characteristics of lowland ponds of central Italy. *Annales de Limnologie-International Journal of Limnology* 43, 81–87.

Carvalho, F.G., Pinto, N.S., Oliveira Júnior, J.M.B. and Juen, L. (2013) Effects of marginal vegetation removal on Odonata communities. *Acta Limnologica Brasiliensia* 25, 1–10. DOI: 10.1590/S2179-975X2013005000013.

Carvalho, F.G., Oliveira Roque, F., Barbosa, L., Assis Montag, L.F. and Juen, L. (2018) Oil palm plantation is not a suitable environment for most forest specialist species of Odonata in Amazonia. *Animal Conservation* 21, 526–533. DOI: 10.1111/acv.12427.

Carvalho, F.G., Duarte, L., Nakamura, G., Seger, G.D.S. and Juen, L. (2021) Changes of phylogenetic and taxonomic diversity of Odonata (Insecta) in response to land use in Amazonia. *Forests* 12, e1061. DOI: 10.3390/f12081061.

Casquin, A., Dupas, R., Gu, S., Couic, E, Gruau, G. and Durand, P. (2021) The influence of landscape spatial configuration on nitrogen and phosphorus exports in agricultural catchments. *Landscape Ecology* 36, 3383–3399. DOI: 10.1007/s10980-021-01308-5.

Céréghino, R., Ruggiero, A., Marty, P. and Angélibert, S. (2008) Biodiversity and distribution patterns of freshwater invertebrates in farm ponds of a south-western French agricultural landscape. *Hydrobiologia* 597, 43–51. DOI: 10.1007/s10750-007-9219-6.

Chelli, A.M. and Moulaï, R. (2019) Diversity and ecological diagnosis of dragonflies of high-mountain temporary ponds in the Akfadou Massif Forest (Algeria). *Zoology and Ecology* 29, 28–37. DOI: 10.1080/21658005.2019.1.4.

Chelli, A., Moulaï, R. and Djemai, A. (2020) Does the Tichi Haf dam construction affect dragonfly and damselfly (Odonata: Insecta) assemblages of the Boussellam watercourse (central north Algeria)? A preliminary study. *Zoology and Ecology* 30, 38–47. DOI: 10.35513/21658005.2020.1.5.

Choudhury, K., Chakravarty, S. and Saikiya, M.K. (2020) Diversity and habitat preference of odonate in Chakrashila Wildlife Sanctuary, Western Assam, India. *International Journal of Advanced Research* 8, 1132–1140. DOI: 10.2174/IJAR01/12110.

Clarke, A., Mac Nally, R., Bond, N. and Lake, P.S. (2008) Macroinvertebrate diversity in headwater streams: a review. *Freshwater Biology* 53, 1707–1721. DOI: 10.1111/j.1365-2427.2008.02041.x.

Clausnitzer, V. (2003) Dragonfly communities in coastal habitats of Kenya: indication of biotope quality and the need of conservation measures. *Biodiversity and Conservation* 12, 333–356.

Coetzee, J.A., Jones, R.W. and Hill, M.P. (2014) Water hyacinth, *Eichhornia crassipes* (Pontederiaceae), reduces benthic macroinvertebrate diversity in a protected subtropical lake in South Africa. *Biodiversity and Conservation* 23, 1319–1330. DOI: 10.1007/s10531-014-0667-9.

Corbi, J.J., Trivinho-Strixino, S. and dos Santos, A. (2008) Environmental evaluation of metals in sediments and dragonflies due to sugar cane cultivation in neotropical streams. *Water, Air, and Soil Pollution* 195, 325–333. DOI: 10.1007/s11270-008-9749-1.

Córdoba-Aguilar, A. and Rocha-Ortega, M. (2019) Damselfly (Odonata: Calopterygidae) population decline in an urbanizing watershed. *Journal of Insect Science* 19, 1–6. DOI: 10.1093/jisesa/iez063.

Cranston, J., Isaac, N.J.B. and Early, R. (2023) Associations between a range-shifting damselfly (*Erythromma viridulum*) and the UK's resident Odonata suggest habitat sharing is more important than antagonism. *Insect Conservation and Diversity* 16, 416–426. DOI: 10.1111/icad.12630.

Craves, J.A. and O'Brien, D.S. (2013) The Odonata of Wayne County, MI: inspiration for renewed monitoring of urban areas. *Northeastern Naturalist* 20, 341–362.

Cuevas-Yáñez, K., Benítez, M. and Córdoba-Aguilar, A. (2017) Large-scale human environmental intervention is related to a richness reduction in Mexican odonates. *Revista Mexicana de Biodiversidad* 88, 664–673. DOI: 10.1016/j.rmb.2017.06.008.

Dalu, T, Wasserman, R.J., Jordaan, M., Froneman, W.P. and Weyl, O.L.F. (2015) An assessment of the effect of rotenone on selected non-target aquatic fauna. *PLoS ONE* 10, e0142140. DOI: 10.1371/journal.pone.0142140.

Dalzochio, M.S., Périco, E. Renner, S. and Sahlén, G. (2017) Effect of tree plantations on the functional composition of Odonata species in the highlands of southern Brazil. *Hydrobiologia* 808, 283–300. DOI: 10.1007/s10750-017-3431-9.

Datry, T., Truchy, A., Olden, J.D., Busch, M.H., Stubbington, R. *et al.* (2023) Causes, responses, and implications of anthropogenic versus natural flow intermittence in river networks. *BioScience* 73, 9–22. DOI: 10.1093/biosci/biac09.

Davidson, N.C. (2014) How much wetland has the world lost? Long-term and recent trends in global wetland area. *Marine and Freshwater Research* 65, 934–941. DOI: 10.1071/MF14173.

Davies, B. and Day, J. (1988) *Vanishing Waters*. University of Cape Town Press, Cape Town, South Africa.

Davis, S.J., hUallacháin, D.Ó, Mellander, P.-E., Kelly, A.-M., Matthaei, *et al.* (2018) Multiple-stressor effects of sediment, phosphorus and nitrogen on stream macroinvertebrate communities. *Science of the Total Environment* 637–638, 577–589. DOI: 10.1016/scitotenv.2018.05.052.

De Knijf, G., Vanappelghem, C. and Demolder, H. (2013) Odonata from Montenegro, with notes on taxonomy, regional diversity and conservation. *Odonatologica* 42, 1–29.

De Marco Júnior, P., Nóbrega, C.C., Souza, R.A. and Neiss, U.G. (2015) Modeling the distribution of a rare Amazonian odonate in relation to future deforestation. *Freshwater Science* 34, 1123–1132. DOI: 10.1086/682707.

Deacon, C. and Samways, M.J. (2021a) A review of the impacts and opportunities for African urban dragonflies. *Insects* 12, e190. DOI: 10.3390/insects12030190.

Deacon, C. and Samways, M.J. (2021b) Urban threats and conservation measures relating to aquatic arthropods on the iconic Table Mountain, South Africa: a review. *Basic and Applied Ecology* 56, 192–212. DOI: 10.1016/j.baae. 2021.07.009.

Declerck, S., De Bie, T., Ercken, D., Hampel, H., Schrijvers, S. *et al.* (2006) Ecological characteristics of small farmland ponds: associations with land use practices at multiple spatial scales. *Biological Conservation* 131, 523–532. DOI: 10.1016/j.biocon.2006.02.024.

DelVecchia, A.G., Shanafield, M., Zimmer, M.A., Busch, M.H., Krabbenhoft, C.A. *et al.* (2022) Reconceptualizing the hyporheic zone for nonperennial rivers. *Freshwater Science* 41, 167–182. DOI: 10.1086/720071.

Desquilbet, M., Gaume, L., Grippa, M., Céréghino, R., Humbert, J.-F. *et al.* (2020) Comment on "Meta-analysis reveals declines in terrestrial but increases in freshwater insect abundances". *Science* 370 (6523), 1–2. DOI: 10.1126/science.abd8947.

Dewan, S., Darnal, N., Acharya, B.K., Subramanian, K.A., Chettri, B. and Jins, V.J. (2019) Effectiveness of organic terrace rice cultivation in conservation of odonates in Sikkim, Eastern Himalaya, India, *International Journal of Odonatology* 22, 207–222. DOI: 10.1080/13887890.2019.1700175.

Dey, R. and Pal, A. (2022) Heterospecific pairing between male *Agriocnemis kalinga* Nair & Subramanian, 2014, and female *A. pygmaea* Rambur, 1842 (Odonata: Coenagrionidae). *Notulae odonatologicae* 9, 241–245. DOI: 10.5281/zenodo.4268585.

Dinh Van, K., Janssens, L., Debecker, S., De Jonge, M., Lambret, P. *et al.* (2013) Susceptibility to a metal under global warming is shaped by thermal adaptation along a latitudinal gradient. *Global Change Biology* 19, 2625–2633. DOI: 10.1111/gcb.12243.

Dinh Van, K., Janssens, L., Debecker, S. and Stoks, R. (2014a) Temperature- and latitude-specific individual growth rates shape the vulnerability of damselfly larvae to a widespread pesticide. *Journal of Applied Ecology* 51, 919–928. DOI: 10.1111/1365-2664.12269.

Dinh Van, K., Janssens, L., Debecker, S. and Stoks, R. (2014b) Warming increases chlorpyrifos effects on predator but not anti-predator behaviours. *Aquatic Toxicology* 152, 215–221. DOI: 10.1016/j.aquatox.2014.04.011.

Dolný, A., Bárta, D., Lhota, S., Rusdianto, and Drozd, P. (2011) Dragonflies (Odonata) in the Bornean rain forest as indicators of changes in biodiversity resulting from forest modification and destruction. *Tropical Zoology* 24, 63–86.

Dolný, A., Harabiš, F., Bárta, D., Lhota, S. and Drozd, P. (2012) Aquatic insects indicate terrestrial habitat degradation: changes in taxonomical structure and functional diversity of dragonflies in tropical rainforest of East Kalimantan. *Tropical Zoology* 25, 141–157. DOI: 10.1080/03946975.2012.717480.

Dolný, A., Ožana, S., Burda, M., and Harabiš, F. (2021) Effects of landscape patterns and their changes to species richness, species composition, and the conservation value of odonates (Insecta). *Insects* 12, e478. DOI: 10.3390/insects12060478.

Downing, J.A., Cole, J.J., Duarte, C.M., Middelburg, J.J., Melack, J.M. *et al.* (2012) Global abundance and size distribution of streams and rivers. *Inland Waters* 2, 229–236. DOI: 10.5268/IW-2.4.502.

Dudgeon, D., Arthington, A.H., Gessner, M.O., Kawabata, Z.-I., Knowler, D.J. *et al.* (2006) Freshwater biodiversity: importance, threats, status and conservation challenges. *Biological Reviews* 81, 163–182. DOI: 10.1017/S1464793105006950.

Dudová, P., Boukal, D.S. and Klecka, J. (2019) Prey selectivity and the effect of diet on growth and development of a dragonfly, *Sympetrum sanguineum*. *PeerJ* 7, e7881 DOI: 10.7717/peerj.7881.

Dullinger, S., Essl, F., Rabitsch, W., Erbe, K.-H., Gingrich, S. *et al.* (2013) Europe's other debt crisis caused by the long legacy of future extinctions. *PNAS* 110, 7342–7347. DOI: 10.1073/pnas.1216303110.

Dunne, T., Western, D. and Dietrich, W.E. (2011) Effects of cattle trampling on vegetation, infiltration, and erosion in a tropical rangeland. *Journal of Arid Environments* 75, 58–69. DOI: 10.1016/j.aridenv.2010.09.001.

Elo, M., Penttinen, J. and Kotiaho, J.S. (2015) The effect of peatland drainage and restoration on Odonata species richness and abundance. *BMC Ecology* 15, e11. DOI: 10.1186/s12898-015-0042-z.

Encalada, A.C., Flecker, A.S., LeRoy Poff, N., Suárez, E., Herrera-R, G.A. *et al.* (2019) A global perspective on tropical montane rivers. *Science* 365, 1124–1129. DOI: 10.1126/science.aax1682.

Englund, R.A. (1999) The impact of introduced poeciliid fish and Odonata on the endemic *Megalagrion* (Odonata) damselflies of Oahu Island, Hawaii. *Journal of Insect Conservation* 3, 225–243.

Ensaldo-Cárdenas, A.S., Rocha-Ortega, M., Schneider, D., Robertson, B.A. and Córdoba-Aguilar, A. (2021) Ultraviolet polarized light and individual condition drive habitat selection in tropical damselflies and dragonflies. *Animal Behaviour* 180, 229–238. DOI: 10.1016/j.anbehav.2021.04.019.

Evans, L.C., Melero, Y., Schmucki, R., Boersch-Supan, P.H., Brotons, L. *et al.* (2022) Bioclimatic context of species' populations determines community stability. *Global Ecology and Biogeography* 31, 1542–1555. DOI: 10.1111/geb.13527.

Ferreira-Peruquetti, P.S. and Fonseca-Gessner, A.A. (2003) Odonata communities on natural areas of Cerrado and monoculture of northeastern São Paulo State, Brazil: relationship between land use and richness. *Revista Brasiliera de Zoologia* 20, 219–224.

Ferreras-Romero, M., Márquez-Rodríguez, J. and Ruiz-García, A. (2009) Implications of anthropogenic disturbance factors on the Odonata assemblage in a Mediterranean fluvial system. *International Journal of Odonatology* 12, 413–428.

Ferzoco, I.M.C., Murray-Stoker, K.M., Hasan, L.S., Javier, C.M., Modi, V. *et al.* (2023) Freshwater insect communities in urban environments around the globe: a review of the state of the field. *Frontiers in Ecology and Evolution* 11, e1174166. DOI: 10.3389/fevo.2023.1174166.

Fincke, O.M. (2006) Use of forest and tree species, and dispersal by giant damselflies (Pseudostigmatidae): their prospects in fragmented forests. In: Cordero Rivera, A. (ed.) *Forests and Dragonflies*. Pensoft, Sophia, Bulgaria, pp. 103–125.

Fincke, O.M., Santiago, D., Hickner, S. and Bienek, R. (2009) Susceptibility of larval dragonflies to zebra mussel colonization and its effect on larval movement and survivorship. *Hydrobiologia* 624, 71–79. DOI: 10.1007/s10750-008-9667-7.

Finn, D.S., Bonada, N., Múrria, C. and Hughes, J.M. (2011) Small but mighty: headwaters are vital to stream network biodiversity at two levels of organization. *Journal of the North American Benthological Society* 30, 963–980. DOI: 10.1899/11-012.1.

Fletcher, D.E., Lindell, A.H., Stillings, G.K., Blas, S.A. and McArthur, J.V. (2017) Trace element accumulation in lotic dragonfly nymphs: genus matters. *PLoS ONE* 12, e0172016. DOI: 10.1371/journal.pone.0172016.

Fletcher, D.E., Lindell, A.H., Stankus, P.T., Fletcher, N.D., Lindell, B.E. and McArthur, J.V. (2020) Metal accumulation in dragonfly nymphs and crayfish as indicators of constructed wetland effectiveness. *Environmental Pollution* 256, e113387. DOI: 10.1016/j.envpol.2019.113387.

Fletcher, D.E., Lindell, A.H., Stankus, P.T., Fulghum, C.M. and Spivey, E.A. (2022) Species- and element-specific patterns of metal flux from contaminated wetlands versus metals shed with exuviae in emerging dragonflies. *Environmental Pollution* 300, e118976. DOI: 10.1016/j.envpol.2022. 118976.

Foote, A.L. and Hornung, C.L.R. (2005) Odonates as biological indicators of grazing effects on Canadian prairie wetlands. *Ecological Entomology* 30, 273–283.

Forister, M.L., Black, S.H., Elphick, C.S., Grames, E.M., Halsch, C.A. *et al.* (2023) Missing the bigger picture: why insect monitoring programs are limited in their ability to document the effects of habitat loss. *Conservation Letters* 16, e12951. DOI:10.1111/conl.12951.

Frank, J.H. and Lounibos, L.P. (2008) Insects and allies associated with bromeliads: a review. *Terrestrial Arthropod Reviews* 1, 125–153. DOI: 10.1163/187498308X414742.

Fu, B.-J., Zhao, W.-W., Chen, L.-D., Liu, Z.-F. and Lü, Y.-H. (2005) Eco-hydrological effects of landscape pattern change. *Landscape and Ecological Engineering* 1, 25–32. DOI: 10.1007/s11355-005-0001-5.

Gál, B., Weiperth, A., Farkas, J. and Schmera, D. (2023) Road crossings change functional diversity and trait composition of stream-dwelling macroinvertebrate assemblages. *Scientific Reports* 13, e20698. DOI: 10.1038/s41598-023-47975-z.

Garfield, E., Kang, T., Kolding, I. and Mayo, J. (2019) Odonate communities across a desert development gradient. *California Ecology and Conservation Research* Fall 2019, 1–9. DOI: 21973/N35W97.

Gimmi, U., Lachat, T. and Bürgi, M. (2011) Reconstructing the collapse of wetland networks in the Swiss lowlands 1850–2000. *Landscape Ecology* 26, 1071–1083. DOI: 10.1007/s10980-011-9633-z.

Goertzen, D. and Suhling, F. (2015) Central European cities maintain substantial dragonfly species richness – a chance for biodiversity conservation? *Insect Conservation and Diversity* 8, 238–246. DOI: 10.1111/icad.12102.

Goertzen, D. and Suhling, F. (2019) Urbanization versus other land use: diverging effects on dragonfly communities in Germany. *Diversity and Distributions* 25, 38–47. DOI: 10.1111/ddi.12820.

Goldner, J.T. and Holland, J.D. (2023) Wing morphology of a damselfly exhibits local variation in response to forest fragmentation. *Oecologia* 202, 369–380. DOI. 10.1007/s00442-023-05396-9.

Grabowski, R.C. and Gurnell, A.M. (2016) Diagnosing problems of fine sediment delivery and transfer in a lowland catchment. *Aquatic Sciences* 78, 95–106. DOI: 10.1007/s00027-015-0426-3.

Grimm, N.B., Faeth, S.H., Golubiewski, N.E., Redman, C.L., Wu, J. *et al.* (2008) Global change and the ecology of cities. *Science* 319, 756–760. DOI: 10.1126/science.1150195.

Gu, C., Mu, X., Gao, P., Zhao, G. and Sun, W. (2019) Changes in runoff and sediment load in the three parts of the Yellow River basin, in response to climate change and human activities. *Hydrological Processes* 33, 585–601. DOI: 10.1002/hyp.13345.

Guillermo, R. and Juen, L. (2021) Odonate ethodiversity as a bioindicator of anthropogenic impact. *International Journal of Odonatology* 24, 149–157. DOI: 10.23797/2159-6719_24_11.

Guimarães, A.T.B., Rodrigues, A.S.L., Pereira, P.S., Silva, F.G. and Malafia, G. (2021) Toxicity of polystyrene nanoplastics in dragonfly larvae: an insight on how these pollutants can affect benthic macroinvertebrates. *Science of the Total Environment* 752, e141936. DOI: 10.1016/j.scitotenv.2020.141936.

Günther, A. (2003) Eiablage von *Sympetrum vulgatum* auf ein parkendes Auto (Odonata: Libellulidae). *Libellula* 22, 19–23.

Hall, A.M., Mccauley, S.J. and Fortin, M.-J. (2015) Recreational boating, landscape configuration, and local habitat structure as drivers of odonate community composition in an island setting. *Insect Conservation and Diversity* 8, 31–42. DOI: 10.1111/icad.12080.

Harabiš, F., Hronková, J., Holer, T. and Šípková, H. (2023) Selective effect of fish farming management on freshwater diversity. *Biodiversity and Conservation* 32, 735–753. DOI: 10.1007/s10531-022-02521-7.

Hardersen, S. (2008) Dragonfly (Odonata) communities at three lotic sites with different hydrological characteristics. *Italian Journal of Zoology* 75, 271–283. DOI: 10.1080/11250000801925227.

Hardersen, S. and Wratten, S.D. (2000) Sensitivity of aquatic life stages of *Xanthocnemis zealandica* (Odonata: Zygoptera) to azinphos-methyl and carbaryl. *New Zealand Journal of Marine and Freshwater Research* 34, 117–123. DOI: 10.1080/00288330.2000.9516919.
Harvey, J.A., Tougeron, K., Gols, R., Heinen, R., Abarca, M. *et al.* (2023) Scientists' warning on climate change and insects. *Ecological Monographs* 93, e1553. DOI: 10.1002/ecm.1553.
Hashimoto, K., Eguchi, Y., Oishi, H., Tazunoki, Y., Tokuda, M. *et al.* (2019) Effects of a herbicide on paddy predatory insects depend on their microhabitat use and an insecticide application. *Ecological Applications* 29, e01945. DOI: 10.1002/eap.1945.
Hassall, C. and Thompson, D.J. (2008) The effects of environmental warming on Odonata: a review. *International Journal of Odonatology* 11, 131–153.
Haubrock, P.J., Cuthbert, R.N. and Haase, P. (2023a) Long-term trends and drivers of biological invasions in Central European streams. *Science of the Total Environment* 876, e162817. DOI: 10.1016/j.scitotenv.2023.162817.
Haubrock, P.J., Pilotto, F. and Haase, P. (2023b) Multidecadal data indicate increase of aquatic insects in Central European streams. *Science of the Total Environment* 879, e163017. DOI: 10.1016/j.scitotenv.2023.163017.
Hernandez, K.M., Reece, B.A. and McIntyre, N.E. (2006) Effects of anthropogenic land use on Odonata in playas of the southern high plains. *Western North American Naturalist* 66, 273–278.
Hieronymus, M. and Kalén, O. (2022) Should Swedish sea level planners worry more about mean sea level rise or sea level extremes? *Ambio* 51, 2325–2332. DOI: 10.1007/s13280-022-01748-6.
Hill, M.J., Chadd, R.P., Morris, N., Swaine, J.D. and Wood, P.J. (2016) Aquatic macroinvertebrate biodiversity associated with artificial agricultural drainage ditches. *Hydrobiologia* 776, 249–260. DOI: 10.1007/s10750-016-2757-z.
Hofhansl, F.P. and Schneeweihs, S. (2008) Banderillas: effects of deforestation on dragonflies (Insecta, Odonata) in the Pacific lowland of Costa Rica. *Stapfia 88, zugleich Kataloge der oberösterreichischen Landesmuseen Neue Serie* 80, 237–247.
Hofmann, T.A. and Mason, C.F (2005) Habitat characteristics and the distribution of Odonata in a lowland river catchment in eastern England. *Hydrobiologia* 539, 137–147. DOI: 10.1007/s10750-004-3916-1.
Hogreve, J. and Suhling, F. (2022) Development of two common dragonfly species with diverging occupancy trends. *Journal of Insect Conservation* 26, 571–581. DOI: 10.1007/s10841-022-00396-1.
Holuša, O. and Holušová, K. (2022) Population density and abundance of the northernmost population of *Cordulegaster heros* (Anisoptera: Cordulegastridae) in Europe (Czech Republic) with notes on its biogeographical range. *Diversity* 4, e854. DOI: 10.3390/d14100854.
Horváth, G., Bernáth, B. and Molnár, G. (1998) Dragonflies find crude oil visually more attractive than water: multiple-choice experiments on dragonfly polarotaxis. *Naturwissenschaften* 85, 292–297.
Horváth, G., Malik, P., Kriska, G. and Wildermuth, H. (2007) Ecological traps for dragonflies in a cemetery: the attraction of *Sympetrum* species (Odonata: Libellulidae) by horizontally polarizing black gravestones. *Freshwater Biology* 52, 1700–1709. DOI: 10.1111/j.1365-2427.2007.01798.x.
Horváth, G., Kriska, G., Malik, P. and Robertson, B. (2009). Polarized light pollution: a new kind of ecological photopollution. *Frontiers in Ecology and the Environment*, 7, 317–325. DOI: 10.1890/080129.
Houlahan, J.E. and Findlay, C.S. (2004) Estimating the 'critical' distance which adjacent land-use degrades wetland water and sediment quality. *Landscape Ecology* 19, 677–690. DOI: 10.1023/B:LAND.0000042912.87067.35.
Husband, D.M. and McIntyre, N.E. (2021) Urban areas create refugia for odonates in a semi-arid region. *Insects* 12, e431. DOI: 10.3390/insects12050431.
Ilhamdi, M.L., Idrus, A.A., Santoso, D., Hadiprayitno, G. and Syazali, M. (2021) Species richness and conservation priority of dragonflies in the Suranadi Ecotourism Area, Lombok, Indonesia. *Biodiversitas* 22, 1846–1852. DOI: 10.13057/biodiv/d220430.
Ionita, M. and Nagavciuc, V. (2021) Changes in drought features at the European level over the last 120 years. *Natural Hazards and Earth System Sciences* 21, 1685–1701. DOI: 10.5194/nhess-21-1685-2021.
Ishak, M., Norhisham, A.R., Thomas, S.M., Nurhidayu, S., Ghazali, A. and Azhar, B. (2021) Physicochemical properties as driver of Odonata diversity in oil palm waterways. *Frontiers in Forests and Global Change* 4, e613064. DOI: 10.3389/ffgc.2021.613064.
Jackson, M.C., Pawar, S. and Woodward, G. (2021) The temporal dynamics of multiple stressor effects: from individuals to ecosystems. *Trends in Ecology & Evolution* 36, 402–410. DOI: 10.1016/j.tree.2021.01.005.
Jähnig, S.C., Baranov, V., Altermatt, F., Cranston, P., Friedrichs-Manthey, M. *et al.* (2021) Revisiting global trends in freshwater insect biodiversity. *WIREs Water* 8, e1506. DOI: 10.1002/wat2.1506.
Jane, S.F., Hansen, G.J.A., Kraemer, B.M., Leavitt, P.R., Mincer, J.L. *et al.* (2021) Widespread deoxygenation of temperate lakes. *Nature* 594, 66–70. DOI: 10.1038/s41586-021-03550-y.
Janssens, L. and Stoks, R. (2013) Fitness effects of chlorpyrifos in the damselfly *Enallagma cyathigerum* strongly depend upon temperature and food level and can bridge metamorphosis. *PLoS ONE* 8, e68107. DOI: 10.1371/journal.pone.0068107.
Janssens, L., Dinh Van, K. and Stoks, R. (2014) Extreme temperatures in the adult stage shape delayed effects of larval pesticide stress: a comparison between latitudes. *Aquatic Toxicology* 148, 74–82. DOI: 10.1016/j.aquatox.2014.01.02.

Jeanmougin, M., Leprieur, F., Loïs, G. and Clergeau, P. (2014) Fine-scale urbanization affects Odonata species diversity in ponds of a megacity (Paris, France). *Acta Oecologica* 59, 26–34. DOI: 10.1016/j.actao.2014.05.008.

Jere, A., Darshetkar, A., Patwardhan, A. and Koparde, P. (2020) Assessing the response of odonates (dragonflies and damselflies) to a tropical urbanization gradient. *Journal of Urban Ecology* 6, 1–7. DOI: 10.1093/jue/juaa029.

Jinguji, H., Thuyet, D.Q., Uéda, T. and Watanabe, H. (2013) Effect of imidacloprid and fipronil pesticide application on *Sympetrum infuscatum* (Libellulidae: Odonata) larvae and adults. *Paddy and Water Environment* 11, 277–284. DOI 10.1007/s10333-012-0317-3.

Johansson, F., Orizaola, G. and Nilsson-Örtman, V. (2020) Temperate insects with narrow seasonal activity periods can be as vulnerable to climate change as tropical insect species. *Scientific Reports* 10, e8822. DOI: 10.1038/s41598-020-65608-7.

Juen, L., de Oliveira-Júnior, J.M.B., Shimano, Y., Mendes, T.P. and Cabette, H.S.R. (2014) Composition and richness of Odonata (Insecta) in streams with different levels of conservation in a Cerrado-Amazonian forest ecotone. *Acta Amazonica* 44, 223–233. DOI: 10.1590/S0044-59672014000200008.

Kadoya, T., Suda, S.-i. and Washitani, I. (2009) Dragonfly crisis in Japan: a likely consequence of recent agricultural degradation. *Biological Conservation* 142, 1899–1905. DOI: 10.1016/j.biocon.2009.02.033.

Karger, D.N., Kessler, M., Lehnert, M. and Jetz, W. (2021) Limited protection and ongoing loss of tropical cloud forest biodiversity and ecosystems worldwide. *Nature Ecology and Evolution* 5, 854–862. DOI: 10.1038/s41559-021-01450-y.

Karlsson Green, K., Eroukhmanoff, S., Harris, S., Pettersen, L.B. and Svensson, E.I. (2016) Rapid changes in genetic architecture of behavioural syndromes following colonization of a novel environment. *Journal of Evolutionary Biology* 29, 144–152. DOI: 10.1111/jeb.12769.

Karouna-Renier, N.K. and Sparling, D.W. (2001) Relationships between ambient geochemistry, watershed land-use and trace metal concentrations in aquatic invertebrates living in stormwater treatment ponds. *Environmental Pollution* 112, 183–192.

Kaufmann, P.R., Hughes, R.M., Paulsen, S.G., Peck, D.V., Seeliger, C.W. *et al.* (2022a) Physical habitat in conterminous US streams and rivers, Part 1: geoclimatic controls and anthropogenic alteration. *Ecological Indicators* 141, e109046. DOI: 10.1016/j.ecolind.2022.109046.

Kaufmann, P.R., Hughes, R.M., Paulsen, S.G., Peck, D.V., Seeliger, C.W. *et al.* (2022b) Physical habitat in conterminous US streams and rivers, Part 2: a quantitative assessment of habitat condition. *Ecological Indicators* 141, e109047. DOI: 10.1016/j.ecolind.2022.109047.

Keinath, S., Onandia, G., Griesbaum, F. and Rödel, M.-O. (2023) Effects of urbanization, biotic and abiotic factors on aquatic insect diversity in urban ponds. *Frontiers in Ecology and Evolution* 11, e1121400. DOI: 10.3389/fevo.2023.1121400.

Keith, D.A., Benson, D.H., Baird, I.R.C., Watts, L., Simpson, C.C. *et al.* (2023) Effects of interactions between anthropogenic stressors and recurring perturbations on ecosystem resilience and collapse. *Conservation Biology* 37, e13995. DOI: 10.1111/cobi.13995.

Khelifa, R., Deacon, C., Mahdjoub, H., Suhling, F., Simaika, J.P. and Samways, M.J. (2021a) Dragonfly conservation in the increasingly stressed African Mediterranean-type ecosystems. *Frontiers in Environmental Science* 9, e660163. DOI: 10.3389/fenvs.2021.660163.

Khelifa, R., Mahdjoub, H., Baaloudj, A., Cannings, R.A. and Samways, M.J. (2021b) Effects of both climate change and human water demand on a highly threatened damselfly. *Scientific Reports* 11, e7725. DOI: 10.1038/s41598-021-86383-z.

Khelifa, R., Mahdjoub, H., Baaloudj, A., Cannings, R.A. and Samways, M.J. (2021c) Remarkable population resilience in a North African endemic damselfly in the face of rapid agricultural transformation. *Insects* 12, e353. DOI: 10.3390/insects12040353.

Khelifa, R., Mahdjoub, H. and Samways, M.J. (2022) Combined climatic and anthropogenic stress threaten resilience of important wetlands in an arid region. *Science of the Total Environment* 806, e150806. DOI: 10.1016/j.scitotenv.2021.150806.

Kietzka, G.J., Pryke, J.S. and Samways, M.J. (2015) Landscape ecological networks are successful in supporting a diverse dragonfly assemblage. *Insect Conservation and Diversity* 8, 229–237. DOI: 10.1111/icad.12099.

Kietzka, G.J., Pryke, J.S. and Samways, M.J. (2018) Comparative effects of urban and agricultural land transformation on Odonata assemblages in a biodiversity hotspot. *Basic and Applied Ecology* 33, 89–98. DOI: 10.1016/j.baae.2018.08.008.

Kietzka, G.J., Pryke, J.S., Gaigher, R. and Samways, M.J. (2021a) 32 years of essential management to retain value of an urban dragonfly awareness pond. *Urban Ecosystems* 24, 1295–1304. DOI: 10.1007/s11252-021-01115-5.

Kietzka, G.J., Pryke, J.S., Gaigher, R. and Samways, M.J. (2021b) Webs of well-designed conservation corridors maintain river ecosystem integrity and biodiversity in plantation mosaics. *Biological Conservation* 254, e108965. DOI: 10.1016/j.biocon.2021.108965.

Kita, A., Nakahara, M. and Tokuda, M. (2021) Changes in Odonata abundance between 2000 and 2015–2016 in Saga Plain, northern Kyushu, Japan. *Journal of Insect Conservation* 24, 575–583. DOI: 0.1007/s10841-020-00226-2.

Klein, C.E, Pinto, N.S., Spigoloni, Z.A.V., Bergamini, F.M., de Melo, F.R. *et al.* (2018) The influence of small hydroelectric power plants on the richness and composition of Odonata species in the Brazilian savanna. *International Journal of Odonatology* 21, 33–44. DOI:10.1080/13887890.2017.1419884.

Kleinschroth, F., Winton, R.S., Calamita, E., Niggemann, F., Botter, M. *et al.* (2021) Living with floating vegetation invasions. *Ambio* 50, 25–137. DOI: 10.1007/s13280-020-01360-6.

Koneri, R., Nangoy, M.J. and Siahaan, P. (2022) Species diversity of dragonflies on the Sangihe Islands, North Sulawesi, Indonesia. *Applied Ecology and Environmental Research* 20, 1763–1780. DOI: 10.15666/aeer/2002_17631780.

Koparde, P. (2016) Damsels in distress – seasons, habitat structure and water pollution changes damselfly diversity and assemblages in urban wetlands. *Animal Biology* 66, 305–319. DOI: 10.1163/15707563-00002506.

Koparde, P., Mhaske, P. and Patwardhan, A. (2015) Habitat correlates of Odonata species diversity in the northern Western Ghats, India. *Odonatologica* 44, 21–43.

Kopp, D. and Allen, D. (2021) Scaling spatial pattern in river networks: the effects of spatial extent, grain size and thematic resolution. *Landscape Ecology* 36, 2781–2794. DOI: 10.1007/s10980-021-01270-2.

Kulkarni, A.S. and Subramanian, K.A. (2013) Habitat and seasonal distribution of Odonata (Insecta) of Mula and Mutha river basins, Maharashtra, India. *Journal of Threatened Taxa* 5, 4084–4095.

Küry, D. and Christ, J. (2010) Odonata fauna and conservation in the canton Basel-Stadt (NW Switzerland). *Mitteilungen der Naturforschenden Gesellschaften beider Basel* 12, 105–118.

Labat, F. and Usseglio-Polatera, P. (2023) A new bioassessment multimetric index (BECOME) and diagnostic tool (BECOMEd) for small standing waters. *Ecological Indicators* 154, e110831. DOI: 10.1016/j.ecolind.2023.110831.

Lee, S.-D., Bae, S.-H. and Lee, G.-G. (2020) Understanding the impact of environmental changes on the number of species and populations of Odonata after creating a constructed wetland. *Korean Journal of Environment and Ecology* 34, 515–529. DOI: 10.13047/KJEE.2020.34.6.515.

Lemm, J.U., Venohr, M., Globevnik, L., Stefanidis, K., Panagopoulos, Y. *et al.* (2021) Multiple stressors determine river ecological status at the European scale: towards an integrated understanding of river status deterioration. *Global Change Biology* 27, 1962–1975. DOI: 10.1111/gcb.15504.

Lewis, A.S.L., Lau, M.P., Jane, S.F., Rose, K.C., Be'eri-Shlevin, Y. *et al.* (2023) Anoxia begets anoxia: a positive feedback to the deoxygenation of temperate lakes. *Global Change Biology* 30, e17046. DOI: 10.1111/gcb.17046.

Lima, J.C.S., Moreira, R.A., Neto, A.J.G, Andrade, D.P., Freitas, E.C. *et al.* (2021) Metal toxicity can affect dragonfly nymphs and ostracods predation rates and food selectivity: ecological implications on food webs. *Water, Air and Soil Pollution* 232, e288. DOI: 10.1007/s11270-021-05248-3.

Lubertazzi, M.A.A. and Ginsberg, H.S. (2010) Emerging dragonfly diversity at small Rhode Island (U.S.A.) wetlands along an urbanization gradient. *Urban Ecosystems* 13, 517–533. DOI: 10.1007/s11252-010-0133-8.

Luke, S.H., Advento, A.D., Dow, R.A., Aryawan, A.A.K., Barclay, H. *et al.* (2020) Complexity within an oil palm monoculture: the effects of habitat variability and rainfall on adult dragonfly (Odonata) communities. *Biotropica* 52, 366–379. DOI: 10.1111/btp.12749.

Lynch, A.J., Cooke, S.J., Arthington, A.H., Baigun, C. Bossenbroek, L. *et al.* (2023) People need freshwater biodiversity, *WIRES Water* 10, e1633. DOI: 10.1002/wat2.1633.

Ma, Y., Yu, Z., Jia, S. Wu, N., Yin, K. *et al.* (2023) Multiple anthropogenic stressors influence the taxonomic and functional homogenization of macroinvertebrate communities on the mainstream of an urban-agricultural river in China. *Journal of Environmental Management* 341, e118017. DOI: 10.1016/j.jenvman.2023.118017.

Mafuwe, K. and Moyo, S. (2020) Dragonfly (Odonata) community structure in the Eastern Highlands Biodiversity Hotspot of Zimbabwe: potential threats of land use changes on freshwater invertebrates. *International Journal of Odonatology* 23, 291–304. DOI: 10.1080/13887890.2020.1768156.

Mafuwe, K., Tambara, E., Matutu, F.F., Maforimbo, C., Tsamba, J. *et al.* (2021) Community assembly of adult odonates in lacustrine systems of an understudied world heritage site of south-eastern Zimbabwe. *International Journal of Odonatology* 24, 122–138. DOI: 10.2397/2159-6719_24_9.

Malaj, E., von der Ohe, P.C., Grote, M., Kühne, R., Mondy, C.P. *et al.* (2014) Organic chemicals jeopardize the health of freshwater ecosystems on the continental scale. *PNAS* 111, 9549–9554. DOI: 10.1073/pnas.1321082111.

Manfrin, A., Singer, G., Larsen, S., Weiß, N., van Grunsven, R.H.A. *et al.* (2017) Artificial light at night affects organism flux across ecosystem boundaries and drives community structure in the recipient ecosystem. *Frontiers in Environmental Science* 5, e61. DOI: 10.3389/fenvs.2017.00061.

Maoduš, I.V., Pozojević, I., Vilenica, M. and Mihaljević, Z. (2022) Longitudinal dynamics of Odonata assemblages in an anthropogenically disturbed lotic system. *International Journal of Limnology* 58, e7. DOI: 10.1051/limn/2022005.

Márquez-Rodríguez. J., Samraoui, B. and Ferreras-Romero, M. (2023) Effect of forest fires on a Mediterranean Odonata assemblage. *International Journal of Odonatology* 26, 27–35. DOI: 10.48156/1388.2023.1917199.

Maynou, X., Martín, R. and Aranda, D. (2017) The role of small secondary biotopes in a highly fragmented landscape as habitat and connectivity providers for dragonflies (Insecta: Odonata). *Journal of Insect Conservation* 21, 517–530. DOI: 10.1007/s10841-017-9992-0.

McKnight, U.S., Rasmussen, J.J., Kronvang, B., Binning, P.J. and Bjerg, P.I. (2015) Sources, occurrence and predicted aquatic impact of legacy and contemporary pesticides in streams. *Environmental Pollution* 200, 64–76. DOI: 1016/j.envpol.2015.02.015.

Mendes, T.P., Benone, N.L. and Juen, L. (2019) To what extent can oil palm plantations in the Amazon support assemblages of Odonata larvae? *Insect Conservation and Diversity* 12, 448–458. DOI: 10.1111/icad.12357.

Mendes, T.P., Amado, L.L., Ribeiro, R.A.B. and Juen, L. (2020) Morphological diversity of Odonata larvae (Insecta) and abiotic variables in oil palm plantation areas in the Eastern Amazon. *Hydrobiologia* 847, 161–175. DOI: 10.1007/s10750-019-04079-y.

Miller, N.M. and Fincke, M.O. (2004) Mistakes in sexual recognition among sympatric Zygoptera vary with time of day and colour morphism (Odonata: Coenagrionidae). *International Journal of Odonatology* 7, 471–491. DOI: 10.1080/13887890.2004.9748233.

Moi, D.A., Barrios, M., Tesitore, G., Burwood, M., Romero, G.Q. *et al.* (2023) Human land-uses homogenize stream assemblages and reduce animal biomass production. *Journal of Animal Ecology* 92, 1176–1189. DOI: 10.1111/1365-2656.13924.

Monteiro Júnior, C.S., Couceiro, S.R.M., Hamada, N. and Juen, L. (2013) Effect of vegetation removal for road building on richness and composition of Odonata communities in Amazonia, Brazil. *International Journal of Odonatology* 16, 135–144. DOI: 10.1080/13887890.2013.764798.

Monteiro Júnior, C.S., Juen, L. and Hamada, N. (2014) Effects of urbanization on stream habitats and associated adult dragonfly and damselfly communities in central Brazilian Amazonia. *Landscape and Urban Planning* 127, 28–40. DOI: 10.1016/j.landurbplan.2014.03.006.

Monteiro Júnior, C.S., Juen, L. and Hamada, N. (2015) Analysis of urban impacts on aquatic habitats in the central Amazon basin: adult odonates as bioindicators of environmental quality. *Ecological Indicators* 48, 303–311. DOI: 10.1016/j.ecolind.2014.08.021.

Moore, N.W. (1980) *Lestes dryas* Kirby: a declining species of dragonfly (Odonata) in need of conservation: notes on its status and habitat in England and Ireland. *Biological Conservation* 17, 143–148.

Morghad, F., Samraoui, F., Touati, L. and Samraoui, B. (2019) The times they are a changin': impact of land-use shift and climate warming on the odonate community of a Mediterranean stream over a 25-year period. *Vie et Milieu: Life and Environment* 69, 25–33.

Morrissey, C.A., Mineau, P., Devries, J.H., Sanchez-Bayo, F., Liess, M. *et al.* (2015) Neonicotinoid contamination of global surface waters and associated risk to aquatic invertebrates: a review. *Environment International* 74, 291–303. DOI: 10.1016/j.envint.2014.10.024.

Motitsoe, S.N., Coetzee, J.A., Hill, J.M. and Hill, M.P. (2020) Biological control of *Salvinia molesta* (D.S. Mitchell) drives aquatic ecosystem recovery. *Diversity* 12, e204. DOI: 10.3390/d12050204.

Nafisah, N.A. and Soesilohadi, R.C.H. (2021) Community structure of dragonfly (Ord: Odonata) in natural forest and tourist sites Petungkriyono Forest, Central Java, Indonesia. *Journal of Tropical Biodiversity and Biotechnology* 6, e67328. DOI: 10.22146/jtbb.67328.

Nakanishi, K., Uéda, T., Yokomizo, H. and Hayashi, T.I. (2020) Effects of systemic insecticides on the population dynamics of the dragonfly Sympetrum frequens in Japan: statistical analyses using field census data from 2009 to 2016. *Science of the Total Environment* 703, e134499. DOI: 10.1016/j.scitotenv.2019.134499.

Narf, R.P. (1985). Aquatic insect colonization and substrate changes in a relocated stream segment. *The Great Lakes Entomologist* 18, 84–92.

Neff, F, Korner-Nievergelt, F., Rey, E., Albrecht, M., Bollmann, K. *et al.* (2022) Different roles of concurring climate and regional land-use changes in past 40 years' insect trends. *Nature Communications* 13, e7611. DOI: 10.1038/s41467-022-35223-3.

Neke, K.S. and Du Plessis, M.A. (2004) The threat of transformation: quantifying the vulnerability of grasslands in South Africa. *Conservation Biology* 18, 466–477. DOI: 10.1111/j.1523-1739.2004.00157.x.

New, T.R. (2015) *Insect Conservation and Urban Environments*. Springer, Switzerland.

Nguyen, H.H., Kiesel, J., Peters, K., Hering, D., Sinclair, J.S. and Haase, P. (2023) Stream macroinvertebrate community metrics consistently respond to a spatiotemporal disturbance gradient but composition is more context-dependent. *Landscape Ecology* 38, 3133–3151. DOI: 10.1007/s10980-023-01769-w.

Noor-Ul-Islam, H., Khan, K., Zia, S.A., Naeem, M. and Shams, W.A. (2021) Heavy metals accumulation in dragonflies (Odonata) and their habitats in District Swabi, Khyber Pakhtunkhwa, Pakistan: assessing dragonfly bionomics in the region. *Bulletin of Environmental Contamination and Toxicology* 107, 838–847. DOI: 10.1007/s00128-021-03338-w.

Noyes, P.D., McElwee, M.K., Miller, H.D., Clark, B.W., Van Tiem, L.A. *et al.* (2009) The toxicology of climate change: environmental contaminants in a warming world. *Environment International* 35, 971–986. DOI: 10.1016/j.envint.2009.02.006.

Nummelin, M., Lodenius, M., Tulisalo, E., Hirvonen, H. and Alanko, T. (2007) Predatory insects as bioindicators of heavy metal pollution. *Environmental Pollution* 145, 339–349. DOI: 10.1016/j.envpol.2006.03.002.

Ojonugwa, E., Williams, A.B. and Kehinde, K.A. (2020) An inventory of Odonata in relation to water quality of

Lekki Conservation Centre, Nigeria. *British Journal of Entomology and Natural History* 33, 2905.1–2905.16.

Okude, G. and Futahashi, R. (2020) Interspecific hybrid between *Paracercion sieboldii* and *P. melanotum* from Japan (Odonata: Coenagrionidae). *Notulae odonatologicae* 9, 204–208. DOI: 10.5281/zenodo.3823261.

Oliveira-Junior, J.M.B., De Marco Junior, P., Dias-Silva, K., Leitão, R.P., Leal, C.G. *et al.* (2017) Effects of human disturbance and riparian conditions on Odonata (Insecta) assemblages in eastern Amazon basin streams. *Limnologica* 66, 31–39. DOI: 10.1016/j.limno.2017.04.007.

Oliveira-Junior, J.M.B, Dias-Silva, K., Teodósio, M.A. and Juen, L. (2019) The response of neotropical dragonflies (Insecta: Odonata) to local and regional abiotic factors in small streams of the Amazon. *Insects* 10, e446. DOI: 10.3390/insects10120446.

Oppel, S. (2005) Habitat associations of an Odonata community in a lower montane rainforest in Papua New Guinea. *International Journal of Odonatology* 8, 243–257. DOI: 10.1080/13887890.2005.9748256.

Ormerod, S.J., Weatherley, N.S. and Merrett, W.J. (1990) The influence of conifer plantations on the distribution of the Golden Ringed Dragonfly *Cordulegaster boltoni* (Odonata) in upland Wales. *Biological Conservation* 53, 241–251.

Orr, J.A., Vinebrooke, R.D., Jackson, M.C., Kroeker, K.J., Kordas, R.L. *et al.* (2020) Towards a unified study of multiple stressors: divisions and common goals across research disciplines. *Proceedings of the Royal Society B* 287, e20200421. DOI: 10.1098/rspb.2020.0421.

Ott, J. (2019) Dragonflies as indicators of the restoration of the river Oster: results of the third success-control-study (Insecta: Odonata). *Libellula Supplement* 15, 147–162.

Ott, J. (2023) Sensibilität von Libellen gegenüber invasiven Krebsen. *Naturschutz und Landschaftsplanung* 55, 30–36. DOI: 10.1399/NuL.2021.01.01.

Outhwaite, C.L., Gregory, R.D.,Chandler, R.E., Collen, B. and Isaac, N.J.B. (2020) Complex long-term biodiversity change among invertebrates, bryophytes and lichens. *Nature Ecology and Evolution* 4, 384–392. DOI: 10.1038/s41559-020-1111-z.

Outhwaite, C.L., McCann, P. and Newbold, T. (2022) Agriculture and climate change are reshaping insect biodiversity worldwide. *Nature* 605, 97–102. DOI: 10.1038/s41586-022-04644-x.

Painter, M.K., Tennessen, K.J. and Richardson, T.D. (1996) Effects of repeated applications of *Bacillus thuringiensis israelensis* on the mosquito predator *Erythemis simplicicollis* (Odonata: Libellulidae) from hatching to final instar. *Environmental Entomology* 25, 184–191.

Paul, S.S., Hasselquist, E.M., Jarefjäll, A. and Ågren, A.M. (2023) Virtual landscape-scale restoration of altered channels helps us understand the extent of impacts to guide future ecosystem management. *Ambio* 52, 182–194. DOI: 10.1007/s13280-022-01770-8.

Pauly, D. (1995) Anecdotes and the shifting baseline syndrome of fisheries. *Trends in Ecology and Evolution* 10, 430.

Pekel, J.-F., Cottam, A., Gorelick, N. and Belward, A.S. (2016) High-resolution mapping of global surface water and its long-term changes. *Nature* 540, 418–422. DOI: 10.1038/nature20584.

Pereira, D.F.G., Oliveira-Junior, J.M.B. and Juen, L. (2019) Environmental changes promote larger species of Odonata (Insecta) in Amazonian streams. *Ecological Indicators* 98, 179–192. DOI: 10.1016/j.ecolind.2018.09.020.

Pereira-Moura, L., Veras, D.S., de Carvalho, F.G., Juen, L., and Marques Couceiro, S.R. (2023) Habitat specificity and morphology-main filters for the distribution of Odonata in the Cerrado Maranhense, Brazil. *Aquatic Ecology* 57, 443–458. DOI: 10.1007/s10452-023-10021-1.

Perkin, E.K., Hölker, F. and Tockner, K. (2014) The effects of artificial lighting on adult aquatic and terrestrial insects. *Freshwater Biology* 59, 368–377. DOI:10.1111/fwb.12270.

Persson, L., Carney Almroth, B.M., Collins, C.D., Cornell, S., de Wit, C.A. *et al.* (2022) Outside the safe operating space of the planetary boundary for novel entities. *Environmental Science and Technology* 56, 1510–1521. DOI: 10.1021/acs.est.1c04158.

Petr, T. (1968) Population changes in aquatic invertebrates living on two water plants in a tropical man-made lake. *Hydrobiologia* 32, 449–484. DOI: 10.1007/BF00155401.

Pickess, B.P. (2016) *Lestes sponsa* (Hansemann) (Emerald Damselfly) and mixed couplings. *Journal of the British Dragonfly Society* 32(1), 39–43.

Piggott, J.J., Townsend, C.R. and Matthaei, C.D. (2015) Climate warming and agricultural stressors interact to determine stream macroinvertebrate community dynamics. *Global Change Biology* 21, 1887–1906. DOI: 10.1111/gcb.12861.

Pilotto, F., Kühn, I., Adrian, R., Alber, R., Alignier, A. *et al.* (2020) Meta-analysis of multidecadal biodiversity trends in Europe. *Nature Communications* 11, e3486. DOI: 10.1038/s41467-020-17171-y.

Pinto, N.S., Juen, L., Cabette, H.S.R. and De Marco Jr, P. (2012) Fluctuating asymmetry and wing size of *Argia tinctipennis* Selys (Zygoptera: Coenagrionidae) in relation to riparian forest preservation status. *Neotropical Entomology* 41, 178–185. DOI: 10.1007/s13744-012-0029-9.

Pires, M.M., Sahlén, G. and Périco, E. (2022) Agricultural land use affects the heterogeneity of Odonata communities in the Brazilian Pampa. *Journal of Insect Conservation* 26, 503–514. DOI: 10.1007/s10841-021-00349-0.

Pokhrel, L.R. and Dubey, B. (2012) Potential impact of low-concentration silver nanoparticles on predator-prey

interactions between predatory dragonfly nymphs and *Daphnia magna* as a prey. *Environmental Science and Technology* 46, 7755–7762. DOI: 10.1021/es204055c.
Polizeli, L. and Pinto, Â.P. (2021) Taxonomy and conservation concerns of the critically endangered *Roppaneura beckeri*, a phytotelm-breeding damselfly in the southern Brazilian Atlantic Forest. *Bulletin of Insectology* 74, 91–101.
Popovici, P.-V., Bondar, A., Bodog, D.-E., Vicaş, P.-T. and Cupşa, D. (2018) Road mortality on two secondary roads near Abrămuţ locality, western Romania: effects of year period and road surrounding habitats. *Muzeul Olteniei Craiova. Oltenia. Studii şi comunicări. Ştiinţele Naturii* 34, 103–110.
Powell, K.E., Oliver, T.H., Johns, T., González-Suárez, M., England, J. and Roy, D.B. (2023) Abundance trends for river macroinvertebrates vary across taxa, trophic group and river typology. *Global Change Biology* 29, 1282–1295. DOI: 10.1111/gcb.16549.
Prescott, V.A. and Eason, P.K. (2018) Lentic and lotic odonate communities and the factors that influence them in urban versus rural landscapes. *Urban Ecosystems* 21, 737–750. DOI: 10.1007/s11252-018-0752-z.
Previšić, A., Vilenica, M., Vučković, N., Petrović, M. and Rožman, M. (2021) Aquatic insects transfer pharmaceuticals and endocrine disruptors from aquatic to terrestrial ecosystems. *Environmental Science and Technology* 55, 3736–3743. DOI: 10.1021/acs.est.0c07609.
Pryke, J.S., Samways, M.J. and De Saedeleer, K. (2015) An ecological network is as good as a major protected area for conserving dragonflies. *Biological Conservation* 191, 537–545. DOI: 10.1016/j.biocon.2015.07.036.
Qiu, M., Wei, X., Hou, Y., Spencer, S.A. and Hui, J. (2023) Forest cover, landscape patterns, and water quality: a meta-analysis. *Landscape Ecology* 38, 877–901. DOI: 10.1007/s10980-023-01593-2.
Quilodrán, C.S., Montoya-Burgos, J.I. and Currat, M. (2020) Harmonizing hybridization dissonance in conservation. *Communications Biology* 3, e391. DOI: 10.1038/s42003-020-1116-9.
Rahel, F.J. (2002) Homogenization of freshwater faunas. *Annual Review of Ecology and Systematics* 33, 291–315. DOI: 10.1146/annurev.ecolsys.33.010802.150429.
Rao, R.S.P and Girish, M.K.S. (2007) Road kills: assessing insect casualties using flagship taxon. *Current Science* 92, 830–837.
Reid, A.J., Carlson, A.K., Creed, I.F., Eliason, E.J., Gell, P.A. *et al.* (2019) Emergent threats and persistent conservation challenges for freshwater biodiversity. *Biological Reviews* 94, 849–873. DOI: 10.1111/brv.12480.
Remsburg, A.J., Olson, A.C. and Samways, M.J. (2008) Shade alone reduces adult dragonfly (Odonata: Libellulidae) abundance. *Journal of Insect Behavior* 21, 460–468. DOI: 10.1007/s10905-008-9138-z.
Renner, S., Périco, E., Sahlén, G., Santos, D.M. and Consatti, G. (2015) Dragonflies (Odonata) from the Taquari River valley region, Rio Grande do Sul, Brazil. *Check List* 11, e1740. DOI: 10.15560/11.5.1740.
Renner, S., Périco, E. and Sahlén, G. (2016a) Effects of exotic tree plantations on the richness of dragonflies (Odonata) in Atlantic Forest, Rio Grande do Sul, Brazil. *International Journal of Odonatology* 19, 207–219. DOI: 10.1080/13887890.2016.1251343.
Renner, S., Périco, E. and Sahlén, G. (2016b) Man-made lakes form species-rich dragonfly communities in the Brazilian Atlantic Forest (Odonata). *Odonatologica* 45, 135–154. DOI: 10.5281/zenodo.163441.
Renner, S., Périco, E., Dalzochio, M.S. and Sahlén, G. (2018) Water body type and land cover shape the dragonfly communities (Odonata) in the Pampa biome, Rio Grande do Sul, Brazil. *Journal of Insect Conservation* 22, 113–125. DOI: 10.1007/s10841-017-0042-8.
Renner, S., Périco, E., Dalzochio, M.S. and Sahlén, G. (2019) Ecoregions within the Brazilian Pampa biome reflected in Odonata species assemblages. *Austral Ecology* 44, 461–472. DOI: 10.1111/aec.12680.
Renner, S., Périco, E., Dalzochio, M.S. and Sahlén, G. (2022) The balance of common vs. rare: a study of dragonfly (Insecta: Odonata) assemblages in the Brazilian Pampa biome. *Neotropical Biodiversity* 8, 188–199. DOI: 10.1080/23766808.2022.2071405.
Resende, B.O., Ferreira, V.R.S., Brasil, L.S., Calvão, L.B., Mendes, T.P. *et al.* (2021) Impact of environmental changes on the behavioral diversity of the Odonata (Insecta) in the Amazon. *Scientific Reports* 11, e9742. DOI: 10.1038/s41598-021-88999-7.
Riffell, S. K. (1999) Road mortality of dragonflies (Odonata) in a Great Lakes coastal wetland. *The Great Lakes Entomologist* 32, 63–73.
Ripple, W.J., Wolf, C., Newsome, T.M., Gregg, J.W., Lenton, T.M. *et al.* (2021) World scientists' warning of a climate emergency 2021. *BioScience* 71, 894–898. DOI: 10.1093/biosci/biab079.
Rocha-Ortega, M., Rodríguez, P. and Córdoba-Aguilar, A. (2019). Can dragonfly and damselfly communities be used as bioindicators of land use intensification? *Ecological Indicators* 107, e105553. DOI: 10.1016/j.ecolind.2019.105553.
Rodrigues, M.E., De Oliveira Roque, F., Guillermo-Ferreira, R., Saito, V.S. and Samways, M.J. (2019) Egg-laying traits reflect shifts in dragonfly assemblages in response to different amount of tropical forest cover. *Insect Conservation and Diversity* 12, 231–240. DOI: 10.1111/icad.12319.
Rojas-Castillo, O., Kepfer Rojas, S., Juen, L., Fogaça de Assis Montag, L., Carvalho, F.G. *et al.* (2023) Meta-analysis contrasting freshwater biodiversity in forests and oil palm plantations with and without riparian buffers. *Conservation Biology* 38, e14172, DOI: 10.1111/cobi.1172.

Romero, G.Q., Moi, D.A., Nash, L.N., Antiqueira, P.A.P., Mormul, R.P. and Kratina, P. (2021) Pervasive decline of subtropical aquatic insects over 20 years driven by water transparency, non-native fish and stoichiometric imbalance. *Biology Letters* 17, e20210137. DOI: 10.1098/rsbl.2021.0137.

Rumschlag, S.L., Mahon, M.B., Jones, D.K., Battaglin, W., Behrens, J. *et al.* (2023) Density declines, richness increases, and composition shifts in stream macroinvertebrates. *Science Advances* 9, eadf4896.

Saha, P.D. and Gaikwad, S.M. (2015) Odonata assemblage at a small marshy land in Khadki (Pune city): an assessment. *Journal of Entomology and Zoology Studies* 3, 53–64.

Saha, P.D., Rahane, S. and Gaikwad, S.M. (2018) Alpha diversity and species assemblage of Odonata from different habitats in Pune-Chindchwad Municipal Corporation (PCMC) Area, Pune, Maharashtra. *International Conference on Frontiers in Life and Earth Science* 5, 184–191.

Sahlén, G. (1999) The impact of forestry on dragonfly diversity in central Sweden. *International Journal of Odonatology* 2, 177–186. DOI: 10.1080/13887890.1999.9748128.

Saiki, M.K., Martin, B.A., Thompson, L.D. and Welsh, D. (2001) Copper, cadmium, and zinc concentrations in juvenile chinook salmon and selected fish-forage organisms (aquatic insects) in the upper Sacromento River, California. *Water, Air, and Soil Pollution* 132, 127–139.

Sakai, M., Suda, S.-i, Okeda, T., Nomura, R., and Washitani, I. (2019) The importance of riparian subtropical lucidophyllous forest to odonate conservation. *Aquatic Conservation: Marine and Freshwater Ecosystems* 29, 682–692. DOI: 10.1002/aqc.3099.

Samways, M.J. (2003) Threats to the tropical island dragonfly fauna (Odonata) of Mayotte, Comoro archipelago. *Biodiversity and Conservation*, 12, 1785–1792. DOI: 10.1023/A:1024120505611.

Samways, M.J. (2007) Rescuing the extinction of experience. *Biodiversity and Conservation* 16, 1995–1997. DOI: 10.1007/s10531-006-9144-4.

Samways, M.J. (2020) *Insect Conservation: A Global Synthesis*. CAB International, Wallingford, UK.

Samways, M.J. and Deacon, C. (2022) Extinction reprieve for the ancient and imperiled dragonflies at the southern tip of Africa. In: DellaSala, D.A., Goldstein, M.I. (eds) *Imperiled: The Encyclopedia of Conservation*, vol. 1. Elsevier, Amsterdam, Netherlands, pp. 471–484. DOI: 10.1016/B978-0-12-821139-7.00047-7.

Samways, M.J. and Niba, A.S. (2010) Climate and elevational range of a South African dragonfly assemblage. *BioRisk* 5, 85–107. DOI: 10.3897/biorisk.5.844

Samways, M.J. and Sharratt, N.J. (2010) Recovery of endemic dragonflies after removal of invasive alien trees. *Conservation Biology* 24, 267–277. DOI: 10.1111/j.1523-1739.2009.01427.x.

Samways, M.J. and Steytler, N.S. (1996) Dragonfly (Odonata) distribution patterns in urban and forest landscapes, and recommendations for riparian management. *Biological Conservation* 78, 279–288. DOI: 10.1016/S0006-3207(96)00032-8.

Samways, M.J., Taylor, S. and Tarboton, W. (2005) Extinction reprieve following alien removal. *Conservation Biology* 19, 1329–1330. DOI: 10.1111/j.1523-1739.2005.00176.x.

Sánchez-Bayo. F., Goka, K. and Hayasaka, D. (2016) Contamination of the aquatic environment with neonicotinoids and its implication for ecosystems. *Frontiers in Environmental Science* 4, e71. DOI: 10.3389/fenvs.2016.00071.

Sánchez-Guillén, R.A., Muñoz, J., Hafernik, J., Tierney, M. Rodriguez-Tapia, G. and Córdoba-Aguilar, A. (2014) Hybridization rate and climate change: are endangered species at risk? *Journal of Insect Conservation* 18, 295–305. DOI: 10.1007/s10841-014-9637-5.

Sánchez-Guillén, R.A., Córdoba-Aguilar, A., Hansson, B., Ott, J. and Wellenreuther, M. (2016) Evolutionary consequences of climate-induced range shifts in insects. *Biological Reviews* 91, 1050–1064. DOI: 10.1111/brv.12204.

Sánchez-Guillén, R.A., Arce-Valdés, L.R., Swaegers, J., Chauhan, P., Cávez-Rios, J. *et al.* (2022) Variable genomic patterns of hybridization in an expanding hybrid zone of damselflies. *Prensa: Authorea* 3 *(2022).*

Santos, L.R. and Rodrigues, M.E. (2022a) Dragonflies (Odonata) in cocoa growing areas in the Atlantic Forest: taxonomic diversity and relationships with environmental and spatial variables. *Diversity* 14, e919. DOI: 10.3390/ d14110919.

Santos, L.R. and Rodrigues, M.E. (2022b) Land uses for pasture and cacao cultivation modify the Odonata assemblages in Atlantic Forest areas. *Diversity* 14, e672. DOI: 10.3390/d14080672.

Sazli, D., Nassouhi, D., Ergönül, M.B. and Atasagun, S. (2023) A comprehensive review on microplastic pollution in aquatic ecosystems and their effects on aquatic biota. *Aquatic Sciences and Engineering* 38, 12–46. DOI: 10.26650/ASE20221186783.

Schewe, J., Gosling, S.N., Reyer, C., Zhao, F., Ciais, F. *et al.* (2019) State-of-the-art global models underestimate impacts from climate extremes. *Nature Communications* 10, e1005. DOI: 10.1038/s41467-019-08745-6.

Schiffer, C.N., Leppo, B. and White, III, H.B. (2015) Odonata of Beaver Dam, Huntingdon County, Pennsylvania: a record of faunal succession in a changing habitat. *Argia* 27(1), 12–23.

Schridde, P. and Suhling, F. (1994) Larval dragonfly communities in different habitats of a Mediterranean running water system. *Advances in Odonatology* 6, 89–100.

Schuran, E. (2008) The impact of deltamethrin on larval development of dragonflies (Odonata) of the Okavango Delta. Diploma Thesis, Anhalt University (AH), Germany.

Schutte, G., Reich, M., and Plachter, H. (1997) Mobility of the rheobiont damselfly *Calopteryx splendens* (Harris) in fragmented habitats (Zygoptera: Calypterygidae). *Odonatologica* 26, 317–327.
Scown, M.W. and Thomas, M.C. (2023) The discontinuum of river networks: the importance of geomorphic boundaries. *Landscape Ecology* 38, 1307–1319. DOI: 10.1007/s10980-023-01626-w.
Seidu, I., Danquah, E., Nsor, C.A., Kwarteng, D.A. and Lancaster, L.T. (2017) Odonata community structure and patterns of land use in the Atewa Range Forest Reserve, Eastern Region (Ghana). *International Journal of Odonatology* 20, 173–189. DOI: 10.1080/13887890.2017.1369179.
Seidu, I., Nsor, C.A., Danquah, E. and Lancaster, L.T. (2018) Odonata assemblages along an anthropogenic disturbance gradient in Ghana's Eastern Region. *Odonatologica* 47, 73–100. DOI: 10.5281/zenodo1239947.
Sganzerla, C., Dalzochio, M.S., Prass, G.S. and Périco, E. (2021) Effects of urbanization on the fauna of Odonata on the coast of southern Brazil. *Biota Neotropica* 21, e20201122. DOI: 10.1590/16760611-BN-2020-1122.
Siesa, M.E., Padoa-Schioppa, E., Ott, J., De Bernardi, F. and Ficetola, G.F. (2014) Assessing the consequences of biological invasions on species with complex life cycles: impact of the alien crayfish *Procambarus clarkii* on Odonata. *Ecological Indicators* 46, 70–77. DOI: 10.1016/j.ecolind.2014.05.036.
Sievers, M., Hale, R., Parris, K.M. and Swearer, S.E. (2018) Impacts of human-induced environmental change in wetlands on aquatic animals. *Biological Reviews* 93, 529–554. DOI: 10.1111/brv.12358.
Šigutová, H., Šigut, M. and Dolný, A. (2015) Intensive fish ponds as ecological traps for dragonflies: an imminent threat to the endangered species *Sympetrum depressiusculum* (Odonata: Libellulidae). *Journal of Insect Conservation* 19, 961–974. DOI: 10.1007/s10841-015-9813-2.
Silva, D.R.O., Herlihy, A.T., Hughes, R.M., Macedo, D.R. and Callisto, M. (2018) Assessing the extent and relative risk of aquatic stressors on stream macroinvertebrate assemblages in the neotropical savanna. *Science of the Total Environment* 633, 179–188. DOI: 10.1016/j.scitotenv.2018.03.017.
Silva, L.F.R., Castro, D.M.P., Juen, L., Callisto, M., Hughes, R.M. and Hermes, M.G. (2021) Functional responses of Odonata larvae to human disturbances in neotropical savanna headwater streams. *Ecological Indicators* 133, e108367. DOI: 10.1016/j.ecolind.2021.108367.
Simaika, J.P. and Samways, M.J. (2018) Insect conservation psychology. *Journal of Insect Conservation* 22, 635–642. DOI: 10.1007/s10841-018-0047-y.
Singh, D. and Hermans, J. (2019) Odonata observations in the Bandh Baretha region, Rajasthan, India. *Brachytron* 20, 38–48.
Soluk, D.A., Zercher, D.S. and Worthington, A.M. (2011) Influence of roadways on patterns of mortality and flight behavior of adult dragonflies near wetland areas. *Biological Conservation* 144, 1638–1643. DOI: 10.1016/j.biocon.2011.02.015.
Söndgerath, D., Rummland, J. and Suhling, F. (2012) Large spatial scale effects of rising temperatures: modelling a dragonfly's life cycle and range throughout Europe. *Insect Conservation and Diversity* 5, 461–469. DOI: 10.1111/j.1752-4598.2011.00182.x.
Souza, A.M., Fogaça, F.N.O., Cunico, A.M. and Higuti, J. (2015) Does the habitat structure control the distribution and diversity of the Odontofauna? *Brazilian Journal of Biology* 75, 598–606. DOI: 10.1590/1519-6984.18213.
Späth, J., Fick, J., McCallum, E., Cerveny, D., Nording, M.L. and Brodin, T. (2022a) Wastewater effluent affects behaviour and metabolomic endpoints in damselfly larvae. *Scientific Reports* 12, e6830. DOI: 10.1038/s41598-022-10805-9.
Späth, J., Brodin, T., McCallum, E., Cerveny, D., Fick, J. and Nording, M.L. (2022b) Metabolomics reveals changes in metabolite profiles due to growth and metamorphosis during the ontogeny of the northern damselfly. *Journal of Insect Physiology* 136, e104341. DOI: 10.1016/j.jinsphys.2021.104341.
Srivastava, D.S., Melnychuk, M.C. and Ngai, J.T. (2005) Landscape variation in the larval density of a bromeliad-dwelling zygopteran, *Mecistogaster modesta* (Odonata: Pseudostigmatidae). *International Journal of Odonatology* 8, 67–79. DOI: 10.1080/13887890.2005.9748244.
Stamenković, O., Piperac, M.S., Milošević, D., Čerba, D., Cvijanović, D. *et al.* (2023) Multiple anthropogenic pressures and local environmental gradients in ponds governing the taxonomic and functional diversity of epiphytic macroinvertebrates. *Hydrobiologia* 851, 45–65. DOI: 10.1007/s10750-023-05311-6.
Stehle, S. and Schultz, R. (2015) Agricultural insecticides threaten surface waters at the global scale. *PNAS* 112, 5750–5755. DOI: 10.1073/pnas.1500232112.
Stevani, C.V., Porto, J.S., Trindade, D.J. and Bechara, E.J.H. (2000) Automotive clearcoat damage due to oviposition of dragonflies. *Journal of Applied Polymer Science* 75, 1632–1639.
Stewart, D.A.B. and Samways, M.J. (1998) Conserving dragonfly (Odonata) assemblages relative to river dynamics in an African savanna game reserve. *Conservation Biology* 12, 683–692. DOI: 10.1111/j.1523-1739.1998.96465.x.
Steytler, N.S. and Samways, M.J. (1995) Biotope selection by adult male dragonflies (Odonata) at an artificial lake created for insect conservation in South Africa. *Biological Conservation* 72, 381–386. DOI: 10.1016/0006-3207(94)00052-R.
Stoks, R., Debecker, S., Dinh Van, K. and Janssens, L. (2015) Integrating ecology and evolution in aquatic

toxicology: insights from damselflies. *Freshwater Science* 34, 1032–1039. DOI: 10.1086/682571.
Strayer, D.L. (2010) Alien species in fresh waters: ecological effects, interactions with other stressors, and prospects for the future. *Freshwater Biology* 55 (Suppl. 1), 152–174. DOI:10.1111/j.1365-2427.2009.02380.x.
Suárez-Tovar, C.M., Castillo-Pérez, E.U., Sandoval-García, I.A., Schondube, J.E., Cano-Santana, Z. and Córdoba-Aguilar, A. (2022) Resilient dragons: exploring Odonata communities in an urbanization gradient. *Ecological Indicators* 141, e109134. DOI: 10.1016/j.ecolind.2022.109134.
Suárez-Tovar, C.M., Rocha-Ortega, M. and Córdoba-Aguilar, A. (2023a) Is body condition of Mexican rubyspot (Odonata: Zygoptera) associated with urbanization? *Journal of Insect Conservation* 27, 961–969. DOI: 10.1007/s10841-023-00512-9.
Suárez-Tovar, C.M., Rocha-Ortega, M., Juen, L. and Córdoba-Aguilar, A. (2023b) From the forest to the city: the persistence of dragonflies and damselflies in the urban jungle. *Biodiversity and Conservation* 33, 91–113. DOI: 10.1007/s10531-023-02733-5.
Suárez-Tovar, C.M., Pereira, J.L.S., Rocha, T., Rivera-Duarte, J.D., Juen, L. and Córdoba-Aguilar, A. (2024) Fierce city hunters: more effective predation of dragonflies and damselflies in urbanized areas. *Animal Behaviour* 208, 51–58. DOI: 10.1016/j.anbehav.2023.12.002.
Subrero, E., Sforzini, S., Viarengo, A. and Cucco, M. (2019) Exposure to anti-mosquito insecticides utilized in rice fields affects survival of two non-target species, *Ischnura elegans* and *Daphnia magna*. *Paddy and Water Environment* 17, 1–11. DOI: 10.1007/s10333-018-0678-3.
Suhaila, A.H., Che Salmah, M.R. and Nurul Huda, A. (2016) Composition and distribution of Odonata larvae and its relationship with physicochemical water quality in northern peninsular Malaysia. *Malaysian Journal of Science* 35, 198–209. DOI: 10.22452/mjs.vol35no2.9.
Suvarov, A., Scornavacca, C., Fujimoto, M.S., Bodily, P., Clement, M. *et al.* (2022) Deep ancestral introgression shapes evolutionary history of dragonflies and damselflies. *Systematic Biology* 71, 526–546. DOI: 10.1093/sysbio/syab063.
Syvitski, J.P.M., Cohen, S., Kettner, A.J. and Brakenbridge, G.R. (2014) How important and different are tropical rivers? An overview. *Geomorphology* 227, 5–17. DOI: 10.1016/j.geomorph.2014.02.029.
Taniwaki, R.H., Piggott, J.J., Ferraz, S.F.B. and Matthaei, C.D. (2017) Climate change and multiple stressors in small tropical streams. *Hydrobiologia* 793, 41–53. DOI: 10.1007/s10750-016-2907-3.
Taylor, P., Smallshire, D.J., Parr, A.J., Brooks, S.J., Cham, S.A. *et al.* (2021) *State of Dragonflies in Britain and Ireland*. British Dragonfly Society, Old Weston, Huntingdon, UK.
Termaat, T., Ketelaar, R., van Kleef, H.H., Verberk, W.C.E.P., van Grunsven, R.H.A. and Wallis De Vries, M.F. (2023) Spearhead blues: how threats to the damselfly *Coenagrion hastulatum* changed over time. *Journal of Insect Conservation* (in press). DOI: 10.1007/s10841-023-00537-0.
Thackeray, S.J., Henrys, P.A., Hemming, D., Bell, J.R., Botham, M.S. *et al.* (2016) Phenological sensitivity to climate across taxa and trophic levels. *Nature* 535, 241–245. DOI:10.1038/nature18608.
Thill, A., Cammaerts, M.-C. and Balmori, A. (2023) Biological effects of electromagnetic fields on insects: a systematic review and meta-analysis. *Reviews in Environmental Health* (in press). DOI: 10.1515/reveh-2023-0072.
Thornhill, I., Batty, L., Death, R.G., Friberg, N.R. and Ledger, M.E. (2017) Local and landscape scale determinants of macroinvertebrate assemblages and their conservation value in ponds across an urban land-use gradient. *Biodiversity and Conservation* 26, 1065–1086. DOI: 10.1007/s10531-016-1286-4.
Thornhill, I., Batty, L., Hewitt, M., Friberg, N.R. and Ledger, M.E. (2018) The application of graph theory and percolation analysis for assessing change in the spatial configuration of pond networks. *Urban Ecosystems* 21, 213–225. DOI: 10.1007/s11252-017-0724-8.
Tiemann, J.S., Gillette, D.P., Wildhaber, M.L. and Edds, D.R. (2004) Effects of lowhead dams on riffle-dwelling fishes and macroinvertebrates in a Midwestern river. *Transactions of the American Fisheries Society* 133, 705–717.
Timofeev, A.M. (2016) The ecology and behavioral features of dragonflies (Insecta, Odonata) in the Forest-Steppe Zone of Central Chernozem Region. *Russian Journal of Ecology* 47, 501–507. DOI: 10.1134/S106741361605012X.
Tippler, C., Wright, I.A., Davies, P.J. and Evans, C.R. (2018) Are Odonata nymph adversely affected by impaired water quality in urban streams. *Austral Ecology* 43, 890–902. DOI: 10.1111/aec.12630.
Tollett, V.D., Benvenutti, E.L., Deer, L.A. and Rice, T.M. (2009) Differential toxicity to Cd, Pb, and Cu in dragonfly larvae (Insecta: Odonata). *Archives of Environmental Contamination and Toxicology* 56, 77–84. DOI: 10.1007/s00244-008-9170-1.
Townsend, C.R., Uhlmann, S.S. and Matthaei, C.D. (2008) Individual and combined responses of stream ecosystems to multiple stressors. *Journal of Applied Ecology* 45, 1810–1819. DOI: 10.1111/j.1365-2664.2008.01548.x.
Tüzün, N. and Stoks, R. (2018) Pathways to fitness: carry-over effects of late hatching and urbanisation on lifetime mating success. *Oikos* 127, 949–959. DOI: 10.1111/oik.05033.
Tüzün, N. and Stoks, R. (2021) Lower bioenergetic costs but similar immune responsiveness under a heat wave in urban compared to rural damselflies. *Evolutionary Applications* 14, 24–35. DOI: 10.1111/eva.13041.

Tüzün, N., Debecker, S., Op de Beeck, L. and Stoks, R. (2015) Urbanisation shapes behavioural responses to a pesticide. *Aquatic Toxicology* 163, 81–88. DOI: 10.1016/j.aquatox.2015.04.002.

Twisk, W., Noordervliet, M.A.W. and ter Keurs, W.J. (2000) Effects of ditch management on caddisfly, dragonfly and amphibian larvae in intensively farmed peat areas. *Aquatic Ecology* 34, 397–411.

Uboni, C., Borsato, V. and Bacaro, G. (2021) Odonate fauna assemblages in the "Cansiglio Forest" (Insecta: Odonata). *Rendiconti Lincei. Scienze Fisiche e Naturali* 32, 899–910. DOI: 10.1007/s12210-021-01029-6.

van de Pol, M., Jenouvrier, S., Cornelissen, J.H.C. and Visser, M.E. (2017) Behavioural, ecological and evolutionary responses to extreme climatic events: challenges and directions. *Philosophical Transactions of the Royal Society B* 372, e20160134. DOI: 10.1098/rstb.2016.0134.

Van Dijk, T.C., Van Staalduinen, M.A. and Van der Sluijs, J.P. (2013) Macro-invertebrate decline in surface water polluted with imidacloprid. *PLoS ONE* 8, e62374. DOI: 10.1371/journal.pone.0062374.

van Klink, R., Bowler, D.E., Gongalsky, K.B., Swengel, A.B., Gentile, A. and Chase, J.M. (2020) Meta-analysis reveals declines in terrestrial but increases in freshwater insect abundances. *Science* 368 (6489), 417–420. DOI: 10.1126/science.aax9931.

van Strien, A.J. and van Grunsven, R.H.A. (2023) In the past 100 years dragonflies declined and recovered by habitat restoration and climate change. *Biological Conservation* 277, e109865. DOI: 10.1016/j.biocon.2022.109865.

Vaughan, I.P. and Gotelli, N.J. (2019) Water quality improvements offset the climatic debt for stream macroinvertebrates over twenty years. *Nature Communications* 10, e1956. DOI: 10.1038/s41467-019-09736-3.

Veras, D.S., Lustosa, G.S., Moura, L.P., Ferreira, M.F.R. and Juen, L. (2020) Differences in land use modify Odonata assemblages in the Cerrado-Caatinga ecotone. *Acta Limnologica Brasiliensia* 32, e15. DOI: 10.1590/S2179-975X7119.

Veras, D.S., Pinto, N.S., Calvão, L., Lustosa, G.S., Azevêdo, C.A.S. and Juen, L. (2022) Environmental thresholds of dragonflies and damselflies from a Cerrado-Caatinga ecotone. *Environmental Monitoring and Assessment* 194, e614. DOI: 10.1007/s10661-022-10310-6.

Verheyen, J., Cuypers, K. and Stoks, R. (2023) Adverse effects of the pesticide chlorpyrifos on the physiology of a damselfly only occur at the cold and hot extremes of a temperature gradient. *Environmental Pollution* 326, e121438. DOI: 10.1016/j.envpol.2023.121438.

Vesterinen, E.J., Kaunisto, K.M. and Lilley, T.M. (2020) A global class reunion with multiple groups feasting on the declining insect smorgasbord. *Scientific Reports* 10, e16595. DOI: 10.1038/s41598-020-73609-9.

Vilenica, M., Pozojević, I., Vučković, N. and Mihaljević, Z. (2020) How suitable are man-made water bodies as habitats for Odonata? *Knowledge and Management of Aquatic Ecosystems* 421, e3. DOI: 10.1051/kmae/2020008.

Vilenica, M., Kerovec, M., Pozojević, I., and Mihaljević, Z. (2021) Odonata assemblages in anthropogenically impacted lotic habitats. *Journal of Limnology* 80, 17–25. DOI: 10.4081/jlimnol.2020.1968.

Vilenica, M., Rebrina, F., Ružanović, L., Gulin, V. and Brigić, A. (2022) Odonata assemblages as a tool to assess the conservation value of intermittent rivers in the Mediterranean. *Insects* 13, e584. DOI: 10.3390/insects13070584.

Villalobos-Jiménez, G. and Hassall, C. (2017) Effects of the urban heat island on the phenology of Odonata in London, UK. *International Journal of Biometeorology* 61, 1337–1346. DOI 10.1007/s00484-017-1311-7.

Villalobos-Jiménez, G. and Hassall, C. (2019) Wing shape patterns among urban, suburban, and rural populations of *Ischnura elegans* (Odonata: Coenagrionidae). *International Journal of Odonatology* 22, 37–49. DOI: 10.1080/13887890.2018.1564074.

Villalobos-Jiménez, G., Dunn, A.M. and Hassall, C. (2016) Dragonflies and damselflies (Odonata) in urban ecosystems: a review. *European Journal of Entomology* 113, 217–232. DOI: 10.14411/eje.2016.027.

Villanueva, R.J.T. (2010) Adult Odonata community in Dinagat Island, the Philippines: impact of chromium ore mining on density and species composition. *Odonalologica* 39, 133–140.

Villanueva, J.R. and Mohagan, A.B. (2010) Diversity and status of Odonata across vegetation types in Mt. Hamiguitan Wildlife Sanctuary, Davao Oriental. *Asian Journal of Biodiversity* 1, 25–35.

Von Ellenreider, N. (2009) Odonata of the Argentine Yungas cloud forest: distribution patterns and conservation status. *Odonatologica* 38, 39–53.

Wagner, D.L., Grames, E.M., Forister, M.L., Berenbaum, M.R. and Stopak, D. (2021) Insect decline in the Anthropocene: death by a thousand cuts. *PNAS* 118, e2023989118. DOI: 10.1073/pnas.2023989118.

Wahl, C., Diaz, R. and Kaller, M. (2021) Invasive floating fern limits aerial colonization and alters community structure of aquatic insects. *Wetlands* 41, e60. DOI: 10.1007/s13157-021-01457-y.

Wan, L., Long, Y., Hui, J., Zhang, H., Hou, Z. *et al.* (2020) Effect of norfloxacin on algae-cladoceran grazer-larval damselfly food chains: algal-morphology-mediated trophic cascades. *Chemosphere* 256, e127166. DOI: 10.1016/j.chemosphere.2020.127166.

Wang, F., Maberly, S.C., Wang, B. and Liang, X. (2018): Effects of dams on riverine biogeochemical cycling and ecology. *Inland Waters* 8, 130–140. DOI: 10.1080/20442041.2018.1469335.

Wang, L., Lv, X., Li, J., Tan, L., Rizo, E.Z. and Han, B.P. (2022) Species diversity and community composition of macroinvertebrates in headwater streams of two subtropical neighboring lowland basins. *Diversity* 14, e402. DOI: 10.3390/d14050402.

Wang, M., Chen, L., Wu, L., Zhang, L., Xie, H. and Shen, Z. (2022) Review of nonpoint source pollution models: current status and future direction. *Water* 14, e3217. DOI: 10.3390/w14203217.

Ward, J.V. (1998) Riverine landscapes: biodiversity patterns, disturbance regimes, and aquatic conservation. *Biological Conservation* 83, 269–278.

Watson, J.A.L. (1992) Oviposition by exophytic dragonflies on vehicles. *Notulae odonatologicae* 3, 137.

Watson, J.A.L., Arthington, A.H. and Conrick, D.L. (1982) Effect of sewage effluent on dragonflies (Odonata) of Bulimba Creek, Brisbane. Australian *Journal of Marine and Freshwater Research* 33, 517–528.

Wickramasingha, P.D., Morrissey, C.A., Phillips, I.D., Crane, A.L., Ferrari, M.C.O and Chivers, D.P. (2024) Exposure to the insecticide, imidacloprid, impairs predator-recognition learning in damselfly larvae. *Environmental Pollution* 342, e123085. DOI: 10.1016/j.envpol.2023.123085.

Wildermuth, H. and Horváth, G. (2005) Visual deception of a male *Libellula depressa* by the shiny surface of a parked car (Odonata: Libellulidae). *International Journal Odonatology* 8, 97–105.

Williams, D.D., Williams, S.S. and van Huis, A. (2021) Can we farm aquatic insects for human food or livestock feed? *Journal of Insects as Food and Feed* 7, 121–127. DOI: 10.3920/JIFF2021.x002.

Willis, C.K. and Samways, M.J. (2013) *Dragonfly & Damselfly Trail Guide: KwaZulu-Natal National Botanical Garden*. South African National Biodiversity Institute, Pretoria, South Africa.

Wohl, E. (2017) The significance of small streams. *Frontiers of Earth Science* 11, 447–456. DOI: 10.1007/s11707-017-0647-y.

Woodford, D.J., Barber-James, H.M., Bellingan, T.A., Day, J.A., de Moor, F.C. *et al.* (2013) Immediate impact of piscicide operations on a Cape Floristic Region aquatic insect assemblage: a lesser of two evils? *Journal of Insect Conservation* 17, 959–973. DOI: 10.1007/s10841-013-9578-4.

Worthen, W.B. and Chamlee, M.G. (2020) Determinants of adult odonate community structure at several spatial scales: effects of habitat type and landscape context. *International Journal of Odonatology* 23, 365–379. DOI: 10.1080/13887890.2020.1796831.

Worthen, W.B., Fravel, R.K. and Horne, C.P. (2021) Downstream changes in odonate (Insecta: Odonata) communities along a suburban to urban gradient: untangling natural and anthropogenic effects. *Insects* 12, e201. DOI: 10.3390/insects12030201.

Xi, Y., Peng, S., Ciais, P. and Chen, Y. (2021) Future impacts of climate change on inland Ramsar wetlands. *Nature Climate Change* 11, 45–51. DOI: 10.1038/s41558-020-00942-2.

Xu, H., Tan, X., Liang, J., Cui, Y. and Gao, Q. (2022) Impact of agricultural non-point source pollution on river water quality: evidence from China. *Frontiers in Ecology and Evolution* 10, e858822. DOI: 10.3389/fevo.2022.858822.

Yu, Z., Wang, H., Wang, R., He, T., Cao, Q. *et al.* (2016) The effects of bridge abutments on the benthic macroinvertebrate community. *Polish Journal of Environmental Studies* 25, 1–7. DOI: 10.15244/pjoes/61879.

Zhao, W., Guo, W., Zhao, L., Li, Q., Cao, X. and Tang, X. (2019) Influence of different types of small hydropower stations on macroinvertebrate communities in the Changjiang River Basin, China. *Water* 11, e1892. DOI:10.3390/w11091892.

6 Dragonfly Conservation Action

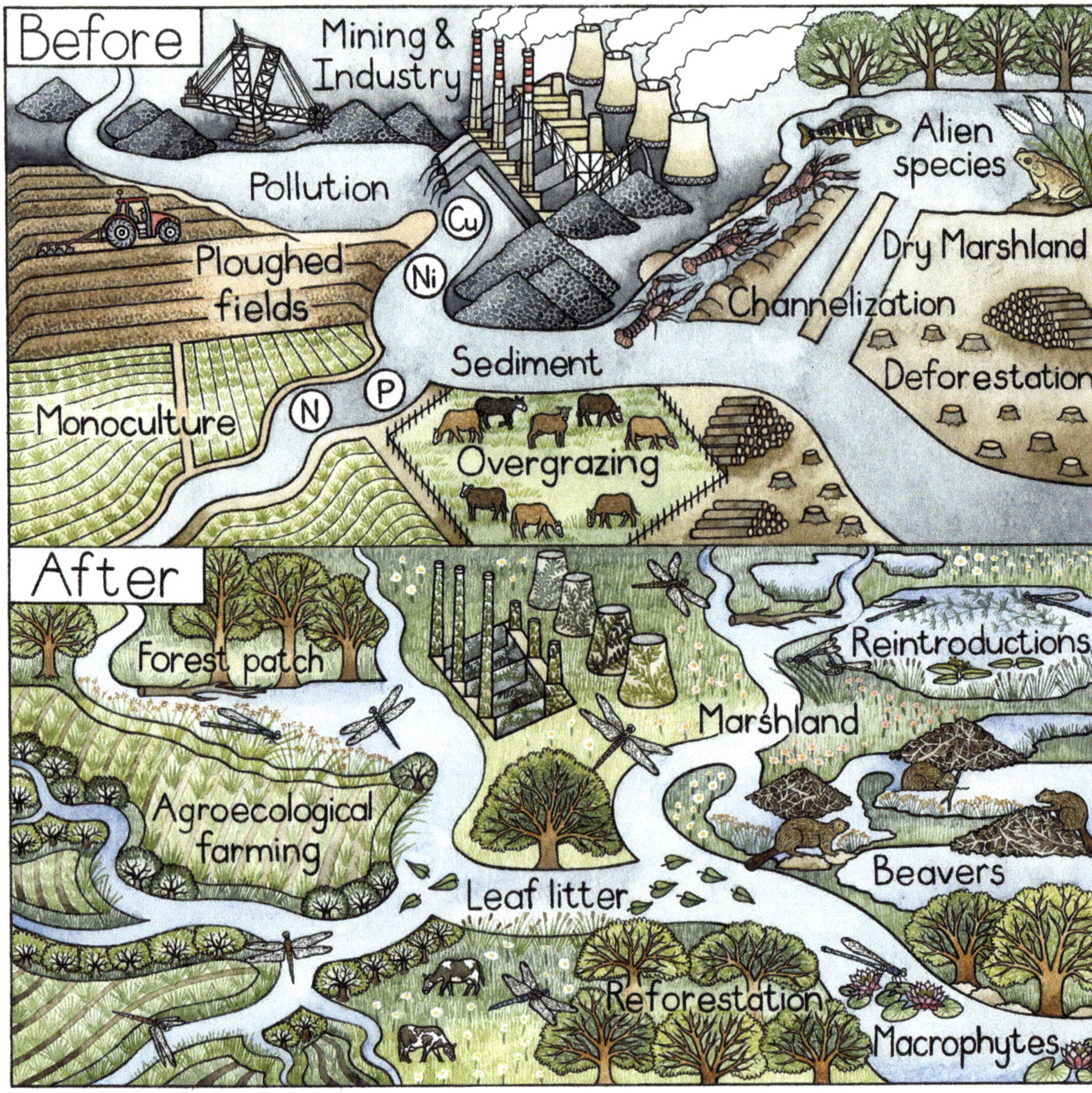

Illustration: Megan Louise Potgieter.

DOI: 10.1079/9781789248395.0006

Abstract

Effective stressor mitigation and ecosystem restoration aims to reinstate the historical condition of high ecological integrity.

Protected areas are critical for conservation of many habitat-specialist dragonflies. Their value is enhanced by the instigation of adjacent networks or large-scale conservation corridors and regreening.

Dragonflies, as sentinels of both aquatic and terrestrial realms, help ensure the recovery of ecosystems at large spatial scales. Running water catchment-level conservation is at the heart of the recovery process. For standing waters, central are high levels of landscape and pondscape heterogeneity and functional connectivity.

Artificial ponds benefit many dragonfly species, both rural and urban, though favouring habitat generalists over specialists and increasing their area of occupancy. All healthy freshwaters require the protection of ecotones and banks. Cleaner water along with the restoration of riparian zones and ancient peatland bogs are greatly helping threatened dragonflies.

Genetic analyses are central for preserving and enabling genetic diversity among populations.

6.1 Calls for More and Better Freshwater Protection

With changes to freshwater dynamics and quality worldwide, there are now calls for urgent action. As the quality and quantity of freshwater determines the fate of both humans and biodiversity, long-term freshwater protection is required. This is against the background that current approaches to freshwater protection are inadequate due in part to freshwater ecosystems often being secondary or unaddressed components of area-based terrestrial protection. From a global perspective, freshwater conservation is largely inadequately addressed when terrestrial measures are used. As freshwater-dependent species are declining faster than marine and terrestrial species, and with increasing pressures on human communities in many drought- and flood-prone areas of the world, approaches to protect freshwaters must balance the needs of both people and nature (Higgins *et al.*, 2021; Lynch *et al.*, 2023).

These concerns stimulated Tickner *et al.* (2020) to present an Emergency Recovery Plan to reduce freshwater biodiversity loss. Priority actions include: 1) Improving environmental flows (e.g. better water management for power generation, more flood risk reduction, better storing and delivery of water for agricultural, industrial and domestic uses, changes in quantity, timing and variability of water flows and levels); 2) Improving water quality; 3) Protecting and restoring important and critical habitats; 4) Managing the exploitation of freshwater ecosystem resources; 5) Preventing and controlling invasive alien species; and 6) Safeguarding and restoring river connectivity.

Dealing with these issues requires practical and realistic solutions (Oberdorff, 2022) that include the consideration of five general domains: data infrastructure, ecology, management, social ecology and monitoring (Maasri *et al.*, 2022), all feeding into aquatic insect conservation (Samways *et al.*, 2020). Furthermore, the increasingly severe effects of extreme weather and unusually low or high rainfall events are bringing into sharp focus the urgency of freshwater biodiversity protection and conservation (Fig. 6.1).

6.2 Principles of Freshwater Restoration and Management

6.2.1 Ecosystem restoration

6.2.1.1 Basic principles of restoration

Restoration ecology is the scientific activity of enabling the recovery of a disturbed or greatly altered ecosystem. The putting of restoration ecology into practice is ecological restoration. One of the main aims of restoration is to bring back the functional and compositional aspects of the historic (original, natural) ecosystem (Raimundo *et al.*, 2018). Generally, the overall term 'restoration' is used.

There are some challenges for restoration, which lie in four interrelated realms. First, as biodiversity is so complex, it is not possible to know all the original aspects of biodiversity relative to the huge number of interactions and dynamics that were once taking place between organisms in food webs. Important traits, like mobility relative to shifts among these interactions and the dynamics of environmental conditions, must be considered relative to changes in the past. Second, knowing which historic site or state to use as a model for a restoration project presents many challenges (Fuller *et al.*, 2017). Third, often there is not an intact and extensive reference site left to act as a baseline for current restoration activities. Fourth, as ecosystems are dynamic over time and change according to legacy and prevailing conditions, a decision must be made on which historic state in the past the restoration

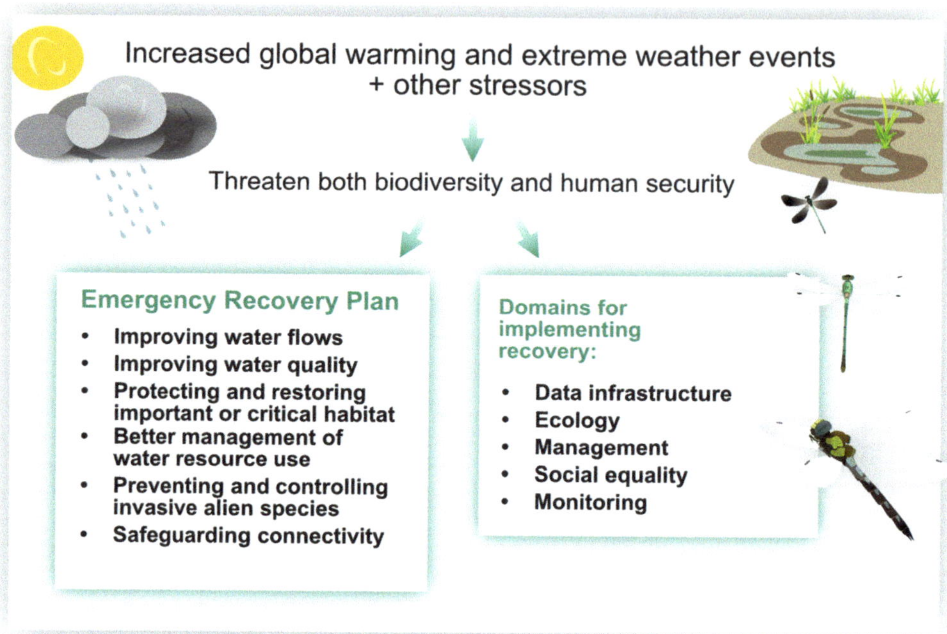

Fig. 6.1. Increasingly strong global warming and extreme weather events, plus other localized stressors, threaten both biodiversity and human security, which are interconnected. This has led to the drawing up of an Emergency Recovery Plan (Tickner *et al.*, 2020) for freshwater, focusing on six key issues that need to be addressed. These actions rest in five domains to be considered if we are to make progress with freshwater conservation. (Illustration: Robyn Symons).

activity is aiming to mimic (Gibbons *et al.*, 2008). The legacy conditions and events include, for example, natural events such as glaciations, droughts, fires or floods, as well as more recent human-induced perturbations such as persistent insecticide input or the impact of invasive alien species. Furthermore, we do not know what local or more widespread extinction events have occurred in the past.

In view of these challenges, there has been a broadening of the view on restoration targets. Rather than a strict focus on the recreation of a particular historic ecosystem, a progressive approach in these rapidly changing times is to embrace a range of possible and predictable restoration outcomes (Brudvig, 2017). Arguably, this perspective is acceptable in these current times where novel landscapes are becoming increasingly commonplace and can have substantial conservation value. This approach must also feed into ensuring resilient landscapes and waterscapes that can withstand ever-increasing environmental impact (Samways, 2020).

Palmer (2016) identifies some themes that are fundamental to restoration ecology: 1) History shapes ecosystems, and determines what is achievable in terms of restoration; 2) Genetic and biological diversity reduces ecological vulnerability, and is an important consideration relative to local extinction, invasion behaviour, effects of disturbance and great fluctuations in the rates and direction of ecological processes; 3) Environmental heterogeneity at a wide range of spatial and temporal scales can enhance diversity, species coexistence, population and species persistence, and historical ecosystem function; 4) Spatial scale, in turn, feeds into landscape context in the sense that the density and arrangement of habitat patches have a strong effect on the recovery of ecological patterns and processes; 5) Ecological systems, such as those dominated by alien species, can persist in altered or degraded states that can be difficult to reverse; 6) Environmental filters and species interactions together determine restoration outcomes, with restoration activities overcoming the constraints posed by these factors; and 7) Rarely do ecological systems exhibit equilibrium dynamics, meaning that many restoration trajectories and alternative states are possible.

Palmer (2016) then articulates emerging themes in restoration ecology, which enhance ecosystem resilience as an important restoration goal and are based on: 1) Species traits are a significant factor that reflect responses to changing environmental conditions, while recognizing their importance provides insight into how changes in species composition and relative abundances may affect ecosystem-level processes and restoration outcomes; 2) Although species may show spatial distribution shifts in their responses to changes in the environment, these shifts can be modified by biotic interactions, dispersal, demographic processes, etc.; 3) There are strong feedback loops between plants and other members of the community which influence ecosystem structure, composition, function and restoration; and 4) Climate change and extreme weather events are causing instability in both historical and restored ecosystems, making restoration trajectories difficult to predict.

Not surprisingly, these complexities mean that the restoration trajectory is not necessarily a simple linear process. In the case of river macroinvertebrate assemblages, these can vary greatly from year to year during the recovery period, as well as there being variation among the different microhabitat types, even at the same sites (Bae *et al.*, 2023) (Fig. 6.2).

6.2.1.2 Assessment in ecological restoration

Predicting the outcomes of a restoration project is rarely easy, as there is much variation among restoration approaches, efforts and implementation. Furthermore, making sound predictions must consider variation in species composition and spatial patterns in both the disturbed state prior to restoration and what restored state transpires in practice. To assess the two states requires assessments using sets of measures such as the abiotic and biotic conditions of the habitat, the taxonomic composition and diversity, and the functional and phylogenetic diversity of the focal organisms to rate the extent of recovery or restoration (Brudvig *et al.*, 2017).

Opportunities for restoration also depend on the type and degree of damage incurred. These include soil compaction, local changes in water regimes, presence or not of pollutants and ongoing threats from invasive alien organisms. This means that a thorough assessment of the system's state is required to be clear on exactly what is the starting

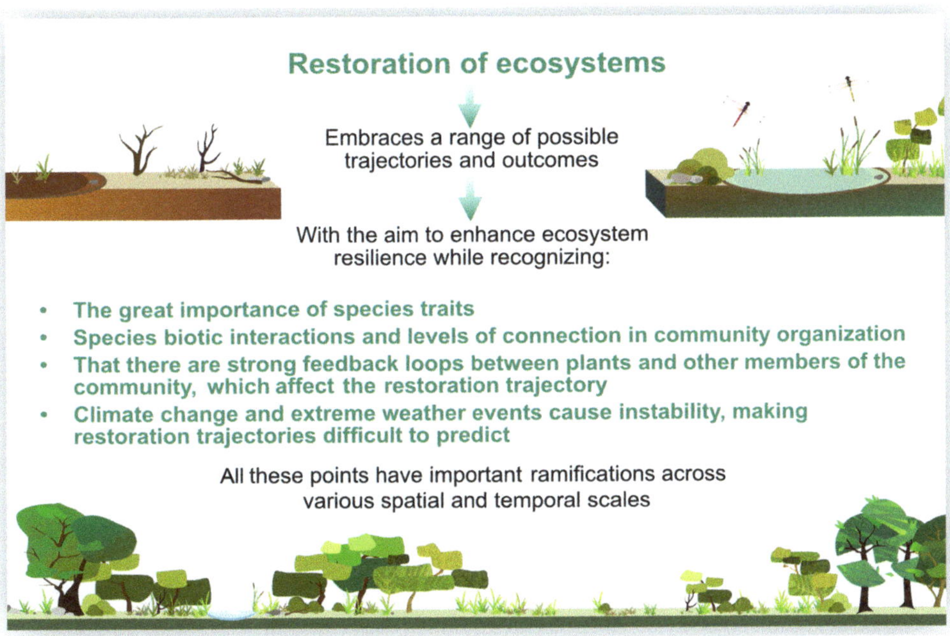

Fig. 6.2. Ecosystem restoration is not a simple linear process, but rather embraces a range of trajectories and possible outcomes, with the aim of enhancing ecosystem resilience during times of stress. The approach recognizes four important points, which have ramifications across various spatial and temporal scales. (Illustration: Robyn Symons).

point. This is in effect a triage assessment that determines the 'restorability' of the focal system, i.e. its potential for truly being restored (Samways, 2000). For rivers, for example, this initial approach can be developed into systematic conservation planning aimed at establishing networks of protected areas that represent the full range of species and ecosystems. This approach is based on three key attributes of a catchment: irreplaceability, condition and vulnerability (Linke *et al.*, 2007).

Furthermore, a system's inherent restorative capability also depends on its ability to recover from severe natural environmental disturbances, such as storms and drought (Schmitz, 2012). In turn, this means including important ecological factors, which for odonate larvae, for example, can vary throughout the year (Mendes *et al.*, 2018).

Beyond identifying the focal species and baseline condition for restoration of a particular system, these factors include: 1) Determination of the extent of passive versus managed restoration, and which kind of additive active restoration measures are required; 2) Determination of the extent to which there will be establishment of the focal species and assembly of focal communities; and 3) Determination of the estimated timescale necessary for these processes to play out. The restoration activity must also formulate the questions, goals and testable hypotheses relative to both ecological theory and past successes (Török and Helm, 2017), as well as an assessment of the functional traits of the focal organisms (Ostertag *et al.*, 2015) (Fig. 6.3).

6.2.1.3 Managing the recovery process

While passive restoration (the natural recovery of a system) is often the most expedient approach, it may be necessary to engage intervention ecology, where some sort of management activity is required to alter the restoration trajectory according to the initial targets. This is especially the case where the aim is to restore maximal naturalness and high ecological integrity. Sometimes however, where damage is both high and extensive, plant species, for example, of a different provenance from the focal site may have to be included as they may have special characteristics with which to improve the highly degraded state. Management restoration where one of the most important goals is creating the right ecological conditions for reintroduction of animals that were formally extirpated is known as 'rewilding'. This approach is usually at a large spatial scale over extensive landscapes.

It is essential that the relative success of any restoration project is assessed (for dragonflies, summarized by Harabiš *et al.* (2023)). These assessments and ongoing monitoring are important to highlight how well a restoration project is progressing. Such

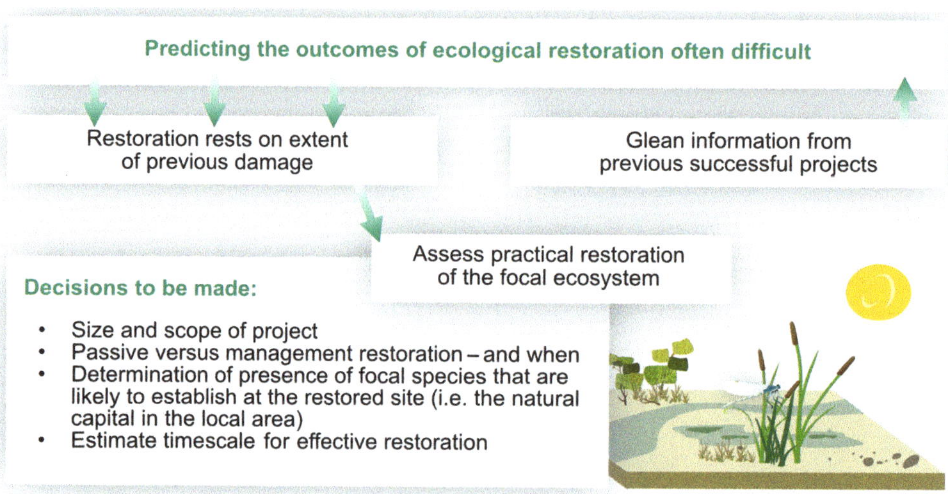

Fig. 6.3. Predicting the outcomes of ecological restoration is rarely certain, although some insight can be gleaned from previous successes, when available. A significant starting point is to identify the extent of the damage to the focal ecosystem. It is then important to assess the practicality of any restoration project, which inevitably means that there are some fundamental decisions to be made before embarking on the exercise. (Illustration: Robyn Symons).

activities may not be across the whole restoration site, but often at specific locations to ensure greater landscape heterogeneity and ecological resilience in the face of ongoing environmental and climate change.

Sometimes, unforeseen events like extreme flooding, drought or fire reset the succession trajectory. These events combined with climate-induced range changes can alter the odonate assemblages, for the better for some species, while for the worse for others. At Beaver Dam Pond, Pennsylvania, USA, and following changes in beaver activity, species like *Libellula flavida*, *L. cyanea*, *L. incesta*, *Archilestes grandis*, *Enallagma divagans*, *E. basidens* and *Sympetrum obtrusum* appeared recently, seemingly in response to climate change (Shiffer *et al.*, 2015).

Once begun, the restoration project then allows the community to go through ecological succession. This may require intervention management in the form of establishment of certain riparian and/or aquatic plants, and then managing succession to ensure habitat heterogeneity through selective removal of certain highly competitive plant individuals. Slowing plant succession can be achieved by removing certain trees to reduce shading of the oxygen-producing in-water macrophytes (Carmo *et al.*, 2023). Macrophyte growth resulting from shade removal can improve macroinvertebrate levels, as in UK ponds. Nevertheless, a wider geographical perspective is also required through maintenance of a range of ponds at different successional stages to support regional macroinvertebrate diversity (Sayer *et al.*, 2012).

In Germany, increasing levels of direct sunlight increase sedge and macrophyte cover, improving dragonfly species richness and diversity. In turn, removal of dead wood, organic sludge, trees and shrubs from the water, as well as removal of woody vegetation and alien herbs from the riparian zone, also encourages dragonfly recovery (Janssen *et al.* 2018).

In terms of restoration of dragonfly habitat, there are two interrelated approaches: restoring ecosystems overall versus restoration of specific species or species groups. At the ecosystem scale, restoration largely focuses on the plants and recovery of botanical naturalness alongside improvement in the plant structural cover and water physicochemical conditions. The assumption often is that dragonfly individuals will then establish passively. Importantly however, the arrival of individuals at the recovering site is contingent upon there being a source population from which these individuals can disperse and colonize. Furthermore, there must be no ecological traps in the area undergoing restoration, which could otherwise subvert ecological recovery.

In the case of species-level restoration, the prime objective is to create the specific habitat conditions of the focal species (or group of species). This means ensuring that all conditions, resources, and opportunities are available at the right time for all the stages in the life cycle. Species recovery may be at a historical site where the species has declined following disturbance, transformation and/or contamination. In the current era of intense climate change, restoration may be at a new climatically suitable site that requires some habitat improvement with a good chance of success (Fig. 6.4).

6.2.2 Related approaches to restoration ecology

While restoration aims to bring back historic ecological integrity and resilience, there are related approaches that are less demanding, and which put human needs and aspirations to the fore. These approaches include ‘ecological landscaping’, ‘rehabilitation’, ‘reclamation’ and ‘greening’, all of which aim principally to improve upon adverse changes that formerly greatly changed landscapes and waterscapes (Fig. 6.5).

Ecological landscaping is often undertaken in urban areas where botanical gardens, parks, housing estates or business parks mostly use indigenous plant species to improve the aesthetics of an area for maintaining human well-being while appreciating the provenance of the species. It often includes the introduction of water bodies which can improve the area of occupancy of certain species with, for example, green roof landscaping on the tops of buildings (Dvorak, 2023).

Wastewater can be treated through ecological landscaping using ‘water filter gardens’. These are landscaped and well-vegetated lakes that are attractive to both wildlife and humans and clean the water by natural filtration. This approach avoids extensive use of hard infrastructure and ecologically insensitive chemical treatment. This advance has been pioneered, for example, by Niteroi City Park, Rio de Janeiro, Brazil.

Rehabilitation and reclamation are often undertaken with the aim of improving, for example, rivers and mine sites with the intent to repair damage rather than specifically restoring the historic condition. In turn, greening relates to ecological

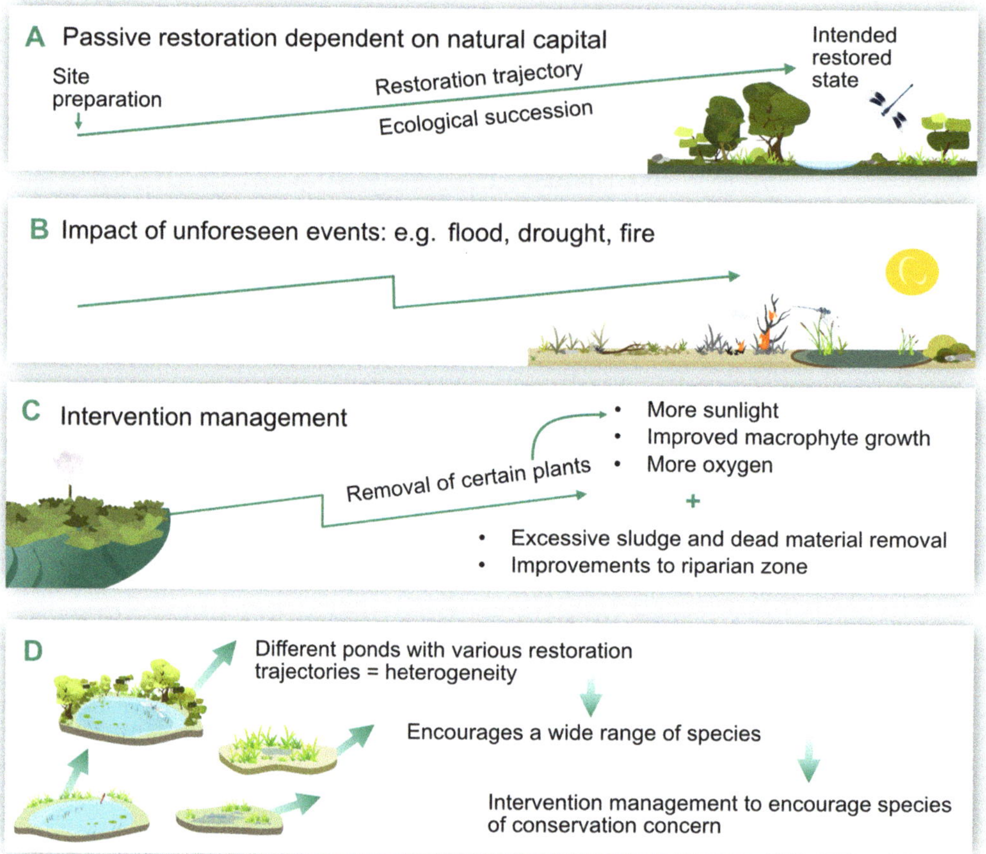

Fig. 6.4. Passive restoration is about letting the chosen site regenerate naturally, the success of which depends on the level of presence of natural capital (species of natural provenance). Once the site has been prepared, it will, principally through plant community succession (ecological succession), move toward the intended restored state (A). However, unforeseen events, such as floods, drought and fire may subvert the recovery trajectory (B). Furthermore, the site may not recover as intended, usually because certain plants, such as alien ones, develop at a fast rate, changing the trajectory away from the intended one. The recovery trajectory then needs intervention management. Mostly, this involves the removal of certain large plant individuals that are excessively shading out the site. More sunlight then reaches the water surface, encouraging growth of macrophytes, leading to higher oxygen levels. Other interventions may also be required, such as sludge and excessive dead wood removal from the water. Improvement of the riparian zone must be considered, as water and adjacent land improvement go together (C). In the case of ponds, the succession trajectory may differ from one pond to the next. This may happen naturally because of various factors such as the local hydromorphology. Intervention management may also be used, with the aim of creating as much landscape/pondscape heterogeneity as possible for the maximum number of opportunities for a wide range of dragonfly species. In some cases, specific management activities may be employed to benefit certain species, usually those of conservation concern (D). (Illustration: Robyn Symons).

engineering, where indigenous or non-invasive alien vegetation, along with physical structures, such as gabions and wall or log barriers, are used to prevent topsoil from shifting, for example, to reduce erosion and prevent the siltation of water courses. As with restoration, these less 'purist' approaches usually lead to ecological succession taking its course, often resulting in a gradual increase in local biodiversity, mostly of more habitat-generalist species.

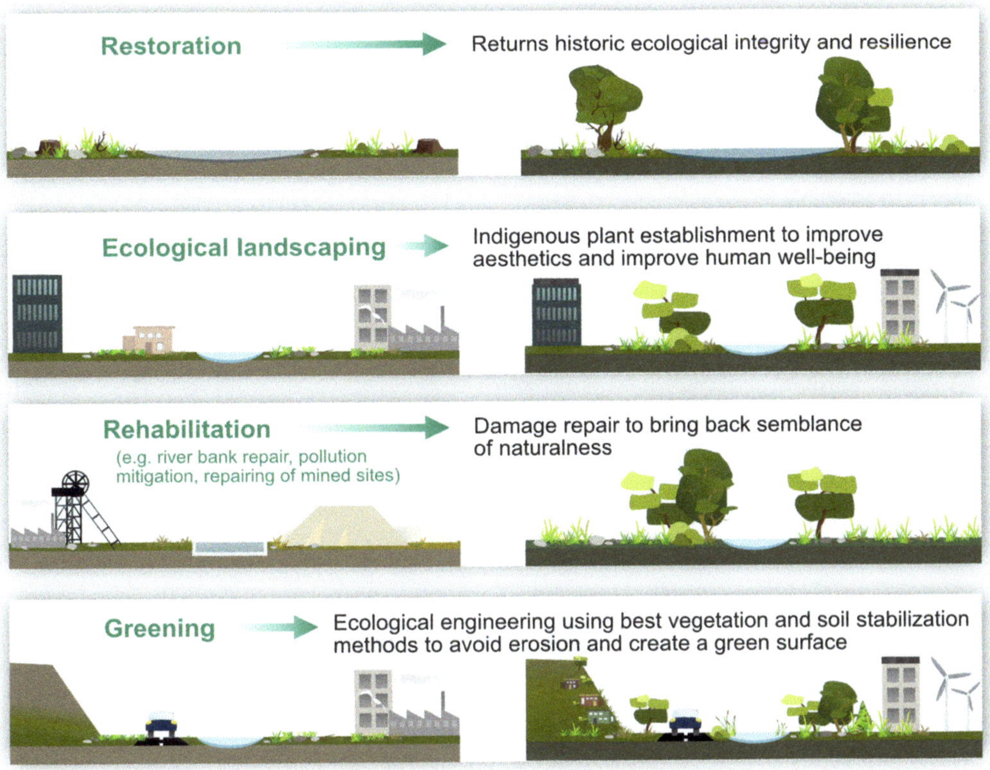

Fig. 6.5. Restoration aims to restore the historic state of ecological integrity and resilience, which may be passive or with some sort of intervention management. Other approaches to landscape/waterscape recovery, while considering nature, largely put humans first. Ecological landscaping uses indigenous plants to improve landscape/waterscape aesthetics based on human values while in effect, respecting the provenance of the species. In turn, rehabilitation is about ecosystem damage repair after intense human activity, with the aim of bringing back some degree of naturalness. Greening (sometimes called regreening) is an ecological engineering approach that, for example, establishes soil and vegetation as stabilization measures, or ecologically improves wetlands to purify freshwater flows. These are not hard and fast categories, and in some areas a combined approach is used. However, all have the common denominator that time is necessary for the full benefits to be felt. (Illustration: Robyn Symons).

6.2.3 Reconciliation ecology

Today, there are many novel ecosystems across the world. These are very important in dragonfly conservation. Among these novel ecosystems are modified running waters, reservoirs and artificial ponds, all having a terrestrial matrix which must also be considered. Each of these novel waterscape/landscape features have their own species sets, communities and species interactions. In many cases, when designed to be biodiversity-friendly, they can play a major role in sustaining dragonfly populations across a region, either as water oases in a dry area or as enablers of metapopulation dynamics. This new reality of maintaining and even improving novel ecosystems for high-grade biodiversity and ecological resilience is known as reconciliation ecology (Moyle, 2014) (Fig. 6.6).

6.3 Extending the Effectiveness of Protected Areas

6.3.1 The value of protected areas

A rule of thumb is that proclaimed terrestrial protected areas should cover at least 10% of a terrestrial biome, but this amount varies from one region of the world to another, ranging from 2% to over 12% (Brooks *et al.*, 2004). Currently, the area of protected land including freshwater covers about

Fig. 6.6. Reconciliation ecology recognizes that novel ecosystems created by humans can play a major role in local conservation. Such ecosystems, when well-designed and managed, can play a major role in supporting local dragonfly populations by providing extra habitat and maintaining metapopulation dynamics, mostly of habitat-generalist species. Seen here is a pond created on a golf course in the Cape Winelands Biosphere Reserve in the Cape Floristic Region, South Africa, with an eco-friendly matrix. (Photo: Michael Samways).

15% of the globe (Chauvenet *et al.*, 2017). But not all reserves are fully protected (Leroux *et al.*, 2010), with an emphasis to improve the quality of many of them (Barnes *et al.*, 2018). The intention is to have 17% of the land surface conserved, while also being ecologically representative and well managed (Di Minin and Toivonen, 2015). Many nation states have improved their protected area coverage, mostly through adding small reserves (Kuempel *et al.*, 2016).

The huge value of protected areas for structural, compositional and functional biodiversity, including that of insects, is undeniable, though much more needs to be done in terms of planning and management (Chowdhury *et al.*, 2023a) and how connectivity among protected areas is measured (Hughes *et al.*, 2023). Furthermore, it is essential that traditional biodiversity conservation strategies should be combined with strategies purposely developed in response to a warming climate and more frequent and intense weather events. Recommendations for improving the effectiveness of protected areas fall into five main categories: 1) Ensure sufficient physical and functional connectivity; 2) Protect climate refuges, especially those important for habitat-specialist species (in particular those of cold water in a warming world (Manfrin *et al.*, 2023)); 3) Protect a few large rather than many small areas; 4) Protect areas that are predicted to become important for biodiversity in the future; and 5) Complement permanently protected areas with those of interim status until such time as these new areas are formally designated as fully protected by law (Ranius *et al.*, 2023). These approaches are important, as most dragonfly species, as with most other insect orders, are currently poorly represented by protected areas around the world (Chowdhury *et al.*, 2023b).

However, regarding point 3 above in the case of dragonflies, the conservation of small areas can also be important, even if they are stepping-stone habitats and notwithstanding that good-quality novel habitats can also be important. Furthermore, in the case of point 5, complementing protected areas need not just be temporary. There are several means of doing this, not least by directly connecting protected areas with large-scale networks of conservation corridors across adjacent novel landscapes. The point here is that the effectiveness of protected areas can be extended in various ways.

6.3.2 Concerns related to protected areas

The major concern is that protected areas are sometimes prone to catastrophic events such as drought, fire or floods, while also subject to the effect of climate warming, meaning that they may not be optimally sited in the future to conserve current biodiversity as they are meant to. This means that protected areas should not be seen and managed as islands. Efforts should be made to functionally connect them. However, as only about 10% of protected areas are functionally connected at various distances of up to 100 km between them, much more needs to be done to improve linkages (Saura *et al.*, 2017).

Nevertheless, as many dragonfly species expand their geographical ranges poleward across transformed landscapes in response to climate warming, most are disproportionately colonizing protected areas as they advance. This means that networks of protected areas are stepping-stones of suitable breeding sites, enabling geographic range shifts (Chapter 3).

As many species remain in the protected areas that they colonize, these areas are essential to biodiversity conservation both now and into the future. The challenge is to balance retaining current species with colonization by new species (Thomas *et al.*, 2012). Connectivity among protected areas enables metapopulation dynamics, encouraging organisms to maintain genetic viability, and in some cases, helping them to locally survive climate change. Some dragonflies in Britain have already naturally established protected areas in response to a warming world (Gillingham *et al.*, 2015). In contrast, in the European Alps, where there is ongoing glacier retreat, there is rearrangement of water courses leading to novel patterns of river networks. Although these changes will provide new high-elevation refuges, they may not be sited in existing protected areas (Wilkes *et al.*, 2023).

An essential consideration when planning for expansion and improved viability of protected area networks is to combine adequate representation of both threatened species and ecosystems (Polak *et al.*, 2015). This also involves inclusion of sites of special significance for their irreplaceable biota (Butchart *et al.*, 2012).

Sometimes simply having larger protected and/or well-managed areas for maintaining the local dragonfly fauna is effective, notwithstanding climate change and other unforeseen adverse events. The larger the size of a protected area, the better, so long as increased size equates to representation of good-quality habitat.

In the Pune metropolitan area in India, odonate abundance and species richness increases with an increase in the size of parks, the presence of large and perennial water sources, and an abundance of aquatic vegetation. However, good park management practices are also essential for promoting species richness and diversity, supplemented with the planting of indigenous trees and maintenance of pollution-free water bodies (Saha and Gaikwad, 2014).

One concern is that globally more than half of protected land areas and two-thirds of the number of protected units (using various modelled climate change scenarios) are at risk of climate connectivity failure. This raises doubts as to the extent to which many species will be able to undergo successful climate-induced range shifts among protected areas. The issue is that protected areas will not be sufficiently functionally connected for many species to shift in a warming climate. When species disappear from protected areas and there is not enough immigration of species suited to the warming climate, many protected areas may lose species under ongoing climate change. Recent pledges to conserve 30% of the planet by 2030 emphasize the need for innovative land management strategies that enable species to undergo range shifts (Parks *et al.*, 2023) (Fig. 6.7).

6.3.3 Examples of protected areas for dragonflies

In Argentina, 70% of the national dragonfly species are present in protected areas (Muzón and Von Ellenreider, 1999), although some of them are at risk given ongoing climate change (Nava-Bolaños *et al.*, 2021). The Bejaia wetlands in north-eastern Algeria, which include Wadi Soummam and Lake Mezaia, as well as high mountain forest ponds in Gouraya National Park, together support 33 dragonfly species,

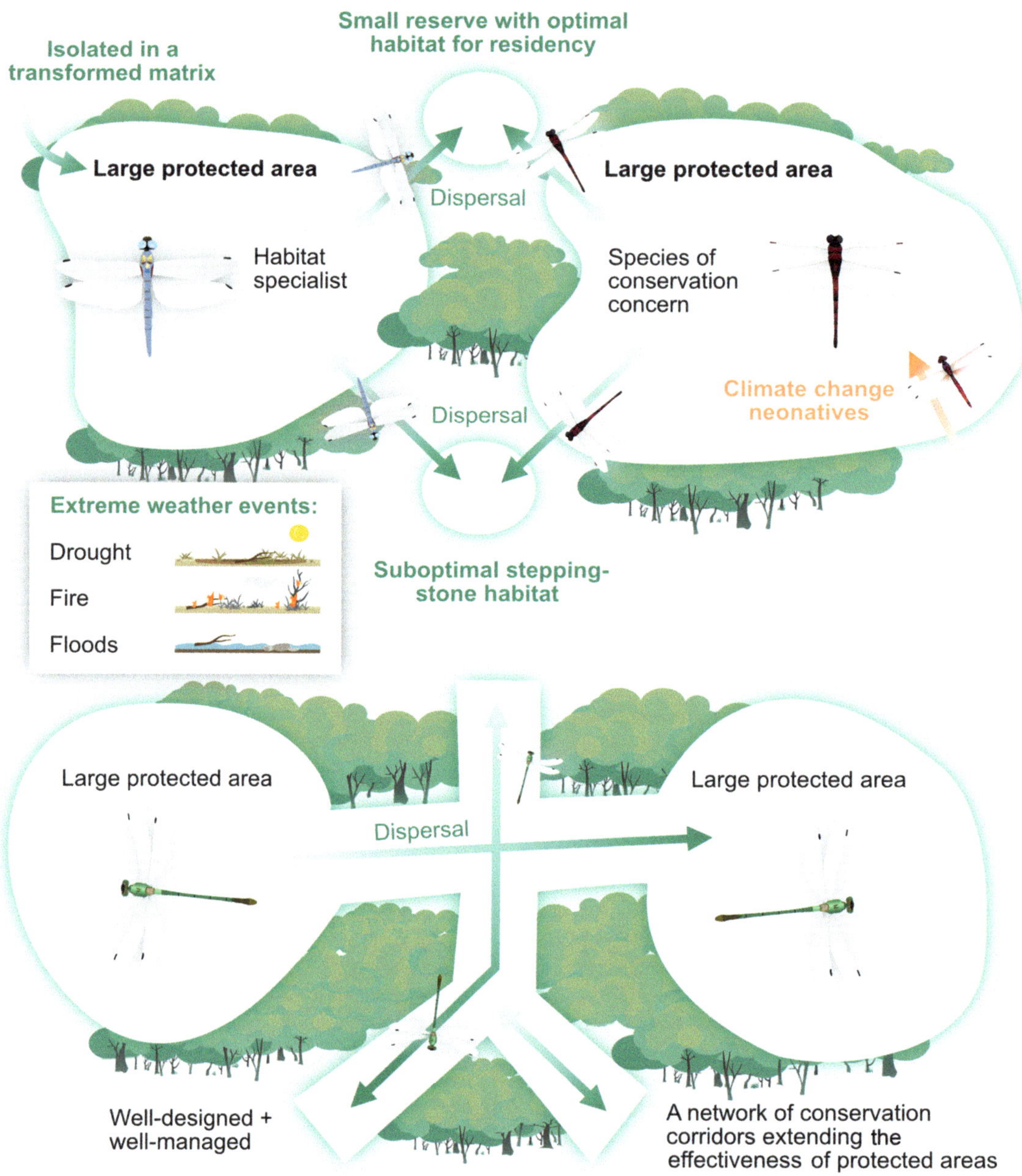

Fig. 6.7. The value of formally protected areas lies in their ability to safeguard many habitat specialists and species of conservation concern. However, protected areas are not without challenges, such as being fixed entities in the face of climate change, subject to extreme, even catastrophic, weather events, and being physically isolated in a transformed matrix. While large areas are preferred, for dragonflies, some small reserves can also be valuable as residency areas, especially when there is functional connectivity that enables metapopulation dynamics and hence gene flow. Small reserves may also function as stepping-stone habitats for dispersing individuals between larger formally protected areas. Interestingly, some opportunistic climate change neonatives are already colonizing some protected areas. Protected areas may also be complemented by instigation of networks of conservation corridors, allowing free movement across the wider landscape. These corridors must be well-designed and well-managed. (Illustration: Robyn Symons).

52% of the species known from Algeria, including two Maghrebian endemic taxa, *Platycnemis subdilatata* and *Enallagma deserti* (Chelli and Moulaï, 2019). Another example is the Kole wetlands in the state of Kerala, India which cover 13,632 ha yet support one quarter (44 species in all, including *Agriocnemis keralensis*, endemic to the Western Ghats, and *Libellago indica*, endemic to peninsular India) of all the odonate species in the state (38,863 km^2, 285 times the size of the wetlands) (Chandran *et al.*, 2021). Similarly, the Kattampally wetland in the Western Ghats is the richest wetland in Kerala in terms of Odonata species diversity. This protected area also supports some local endemics, such as *Ceriagrion chromothorax*, *Caconeura* cf. *risi*, *Pseudagrion indicum* and *Platylestes kirani* (Rodrigues *et al.*, 2022).

An example of an important coastal protected area for dragonflies is the 3280 km^2 iSimangaliso Wetland Park in the Maputaland-Pondoland-Albany global biodiversity hotspot, South Africa. A total of 86 species have so far been recorded from the park, about half the number of nationally recorded species and distinct subspecies, yet the park is only 0.27% the size by area of the country. The taxa include 12 national endemics, and the 7 nationally red-listed taxa *Aciagrion dondoense*, *Agriocnemis gratiosa*, *A. ruberrima ruberrima*, *Pseudagrion coeleste umsingaziense*, *Gynacantha villosa*, *Diplacodes pumila* and *Urothemis luciana* (Hart *et al.*, 2014).

While dragonflies as a group have been instrumental along with other biodiversity in the setting aside of protected areas, there are few reserves dedicated specifically for dragonfly conservation. However, through motivation and financial support from R. Mackenzie Dodds, Europe's first dragonfly reserve was established at Ashton Water, Northamptonshire, UK. This was followed by the establishment of the The Dragonfly Centre at Wicken Fen, Cambridgeshire, UK (Ashworth, 2021). A wildlife and dragonfly reserve (Serra de São José) of 4758 ha was established in south-eastern Minas Gerais, Brazil for dragonfly protection. It has a wide variety of habitats at 1000–1430 m above sea level, ranging from springs, small streams and rivers through to temporary and permanent standing water, mostly in grassland. Importantly, some of the flowing waters have gallery forest. The reserve supports 128 odonate species (49 zygopterans and 79 anisopterans), 15.5% of the species recorded from Brazil. The richest habitats are natural or semi-natural ponds with relatively low levels (less than 30%) of aquatic macrophyte cover (Bedê *et al.*, 2015).

An important protected area in Italy is the Val Grande National Park situated between the Alps and the Insubria region, with high rainfall and complex orography. To date, 25 odonate species have been recorded there, representing 26% of the Italian fauna, 36% of Piedmont region and 58% of Verbano-Cusio-Ossola province. The park supports populations of the boreo-alpine species *Somatochlora alpestris* and *Sympetrum danae*, which are scattered and isolated in the Italian Alps, and the endemic European *Cordulegaster bidentata*. It also supports riverine odonate taxa which are concentrated in south-western Europe where they are threatened by droughts and freshwater exploitation (Pompilio and Mosini, 2021).

While streams in the Amazônia National Park, Brazil have higher habitat integrity than streams outside the protected area, there is little difference in species richness and abundance. Only zygopteran species composition differs between both areas, with habitat integrity and forest cover of some streams outside the national park being similar to those inside, supporting forest-specialist species. Furthermore, the national park is a source for species outside the park (Brito *et al.*, 2023).

6.3.4 Nuances of protected areas for dragonfly conservation

We cannot assume that all protected areas have equal value for maintaining special dragonfly species or assemblages. The enormous 19,485 km^2 Kruger National Park, South Africa, was instigated essentially to conserve large savanna mammals in an El Niño–Southern Oscillation (ENSO) prone area, naturally subject to floods and droughts. Odonatologically, this is an area with a highly varying and unpredictable climate not conducive to the development of local endemism. Among the 51 species from three major rivers (from north to south: Olifants, Sabie, Crocodile), none have even national endemic status. Large water flow fluctuations have resulted in high species turnover, with most species in reaches with low to intermediate natural disturbance, and assemblages separating according to the vegetation and physicochemical conditions of each river system. Aquatic macrophytes promote dragonfly species richness, including two invasive aliens (*Pistia stratiotes* and *Pontederia crassipes*) which provide attractive oviposition sites for certain zygopterans. Small weirs (to supply water sources for vertebrates in times of drought) also encourage certain dragonfly species. While they are mostly species associated with standing water, the waterfall species *Zygonyx natalensis* also benefits greatly from this disturbance factor when the weirs overflow (Stewart and Samways,1998).

Study of the Sabie River in the Kruger National Park, with its wide range of habitat types, including forested locations, sand bars, flood zones and riverine pools, shows that individual habitat types play a major role in boosting dragonfly numbers. This emphasizes that spatial heterogeneity within a major protected area greatly increases species richness (Clark and Samways, 1996).

In Africa, there are hotspots of dragonfly diversity at the continental scale. However, it is the fine spatial scale that is significant for dragonfly conservation, as there can be prioritization for dragonfly conservation at realistic logistical levels, both within and outside hotspots. Furthermore, by operating at the fine-scale for a large geographical area, there is the potential for local conservation managers to consider inclusion of the priority areas that are not formally protected, and that are adjacent to existing reserve networks. Where this is not possible, or where areas fall within the mosaic of different landscape uses, these areas can be monitored for future possible threats allowing for proactive dragonfly conservation actions (Simaika *et al.*, 2013).

The structural similarity between large rivers and lakes, along with differences among basins, likely indicates historical events in colonization that have shaped dragonfly assemblages (Juen *et al.*, 2007). Similar findings come from the Cape Floristic Region in South Africa where even small and closely spaced montane river catchments have different dragonfly assemblages, indicating very long-term isolation of individual catchments (Simaika and Samways, 2011). These findings emphasize that geographic representation of habitats is important for conserving the full local complement of species.

6.3.5 Physically extending the effectiveness of protected areas

6.3.5.1 Networks of large-scale conservation corridors

Protected areas in effect can be made larger by having networks of large-scale conservation corridors emanating from their borders into the surrounding working landscape, such as timber plantations (Pryke and Samways, 2012). The conservation corridors may incorporate natural streams and rivers running within them, as well as various types of standing water.

In the Maputaland-Pondoland-Albany biodiversity hotspot, South Africa, networks of large conservation corridors across timber plantation mosaics are adjacent to, and connected with, certain major protected areas. When dragonflies in the corridors and protected areas are compared, their species richness, diversity and abundance are largely the same in the networks as in the protected areas. The drivers are also largely the same. Dragonfly species richness, assemblage composition and abundance levels are positively correlated with corridor width, with both landscape (vegetation structure) and water variables (wide streams, low water turbidity, substrate type, shallow water) being similar in both the protected area and corridor systems (Kietzka *et al.*, 2021b). However, when these corridors are wide, they can even complement adjacent protected areas in terms of their dragonfly assemblages (Van Schalkwyk *et al.*, 2023).

Interestingly, human disturbance in these networks of conservation corridors has less effect than natural environmental variables on odonate species richness and composition. This emphasizes the importance of recognizing appropriate local-scale variables for conserving natural heterogeneity both in the water and on land. Water quality management in these production landscape conservation corridors is a top priority. This priority together with timber trees being situated at least 30 m (and mostly much more) from the water's edge, are the main reasons why there is equivalence of the dragonfly fauna in the two systems and why conservation corridors effectively extend the functional reach of the protected areas (Kietzka *et al.*, 2015).

At low elevations in this region, large indigenous mammals such as elephants and hippos are present, creating mud wallows. When these wallows, as well as ponds and small marshes in the networks of conservation corridors are compared with those in adjacent protected areas, the networks share 74% of the 47 dragonfly species, while supporting the same number of range-restricted species. Five species are particular to the conservation corridors and seven species to the protected areas. A reason for the difference between these results for still waters and those for running waters is due to heterogeneity in vegetation diversity and physicochemical conditions among the different standing water types, affecting both richness and diversity. Wallows are the poorest in terms of species richness due to the continual disturbance by the mammals. However, the wallows, ponds and small marshes are largely complementary in species composition. High species turnover is a manifestation of these different standing water bodies across the wider landscape (Pryke *et al.*, 2015) (Fig. 6.8).

6.3.5.2 Significance of the pondscape

Vegetation cover in both natural and artificial ponds is central for maintaining species richness and composition in networks of conservation corridors, which increase the area of occupancy for 84% of the dragonfly species. Natural and artificial ponds are complementary in their dragonfly species diversity. However, many other environmental variables are also necessary to encourage the full suite of local diversity and for enabling pondscape resilience in this ENSO event area. The message here is that maintaining a diverse and heterogeneous pondscape is important for conserving local dragonfly diversity in this biodiverse area, with good-quality artificial ponds having considerable conservation value (Deacon *et al.*, 2018). This is also the case in the Neotropics, where good-quality riparian zones are essential for conserving benthic macroinvertebrates in general (Linares *et al.*, 2023).

Landscape graphs (graph theory) provide an assessment of functional connectivity across the landscape/waterscape based on the movement of species, with connectivity values identifying the most important habitat patches and corridors for maintaining or restoring landscape connectivity.

Fig. 6.8. Conservation corridors of indigenous grassland among planted timber trees, and adjacent to protected areas, conserve headwater streams and their riparian zones (top left). Where the land surface is flat, wetlands are conserved (top right), and in some places, extensive wetlands form (bottom left). At lower elevations where megaherbivores are present (hippos here), there can be much disturbance to the water and the vegetation, resulting in a naturally impoverished but disturbance-tolerant dragonfly assemblage (bottom right). (Photos: Michael Samways).

However, results vary greatly depending on the graph construction method. This means that to evaluate the predictive performance of each model for pondscape functional connectivity, the validation process should be conducted using different biological datasets, such as frogs and dragonflies (Godet and Clauzel, 2021) (Fig. 6.9).

6.3.5.3 Biosphere reserves

The *Man and the Biosphere Programme* (United Nations Educational, Scientific and Cultural Organization) is an important international initiative that instigates biosphere reserves. Worldwide, there are currently 738 biosphere reserves in 134 countries. Each biosphere reserve consists of three zones: 1) A central core zone where there is full biodiversity preservation; 2) A buffer zone around the core with only agroecological farming as well as natural areas; and 3) An outer, transition zone where there is more extensive agroecological farming. This zoning approach constrains stressors, by avoiding conversion to conventional agriculture, while limiting agroecological activities and infrastructure development to specific areas.

Biosphere zoning aims to harmonize biodiversity conservation and human activities while also providing opportunities for much of the biodiversity to expand beyond the protected area (the core zone) and into the surrounding matrix. In short, biosphere reserves promote a balanced relationship between conservation and sustainable development by enabling functional connectivity among the various landscapes, promoting high levels of arthropod conservation.

In South Africa, the Kogelberg Biosphere Reserve in the heart of the Cape Floristic Region biodiversity hotspot is rich in localized endemic dragonfly species. Of the 38 species in the reserve, six taxa (*Chlorolestes conspicuus*, *C. umbratus*, *Ecchlorolestes peringueyi*, *Pseudagrion furcigerum*, *Syncordulia venator* and *Orthetrum julia capicola*) are endemic to the Cape Floristic Region, and another five are national endemics, the rest being a combination of regionally rare or common species. Four species are red-listed as Near Threatened (1 species) or Vulnerable (3 species), emphasizing that the biosphere reserve is a stronghold and a microhotspot in terms of supporting endemic and threatened species. Furthermore, rather surprisingly, the total

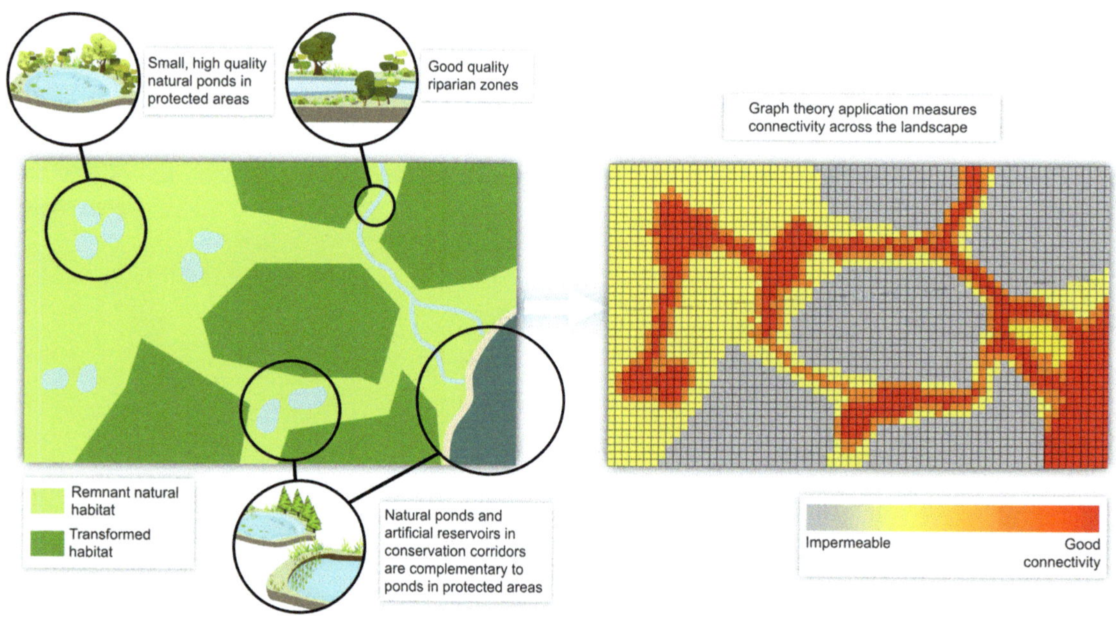

Fig. 6.9. Natural and artificial ponds are highly complementary in transformed landscapes. Pondscape resilience can further be improved through maintaining heterogeneity at the landscape scale and by ensuring a well-connected terrestrial matrix among natural and artificial ponds. To better identify critical pathways of functional connectivity, graph theory can be used to inform conservation action. (Illustration: Charl Deacon).

abundance of the endemic and red-listed species is greater than that of the widespread habitat generalists. The core zone is by far the most important with 33 species. In turn, the buffer zone supports 20 species, and the transition zone 14 species, but with 8% of species exclusive to the buffer zone and 5% exclusive to the transition zone, none of which are endemic or threatened red-listed species, the importance of the core zone is emphasized (Grant and Samways, 2011) (Fig. 6.10).

6.3.5.4 The relative value of the matrices around protected areas

Comparing dragonfly diversity in a protected area versus its surroundings in the Brazilian Atlantic Forest hotspot indicates both a greater total species richness, and greater anisopteran richness in the surrounding modified matrix. While species composition differs less between the sites inside the protected area than in the surroundings, some species inside the protected area contribute greatly to species diversity and beta diversity, so maintaining regional dragonfly diversity. The emphasis here is that while the protected area surrounding matrix is not without value for many dragonfly species, protected areas are vital for supporting threatened species and forest specialists such as those developing in phytotelmata (*Leptagrion macrurum* and *L. acutum*) (Ribeiro, C. *et al.*, 2022).

6.3.6 Urban refuges for dragonflies

In some parts of the world, there are urban green spaces of considerable conservation value for dragonflies. The 194 ha Kolkata Ecopark, India, is one of the last remnants of Rajarat grassland, with a heterogeneous mix of grassland, wetland, bushes and trees. To date, 34 odonate species have been recorded there, more than any other local urban parks or gardens. Some of the species are locally rare, with two, *Pseudagrion rubriceps* and *P. decorum* being very rare (Samanta *et al.*, 2022). Similarly in the vicinity of Tagum City, Philippines, where relative humidity and ambient temperature affects the abundance of certain species, a botanical garden is rich and diverse in dragonfly taxa, despite only having a small ditch (Vitor *et al.*, 2022).

One concern with refuge areas in an intensely modified matrix, and often increasingly so, is that without functional connectivity, there is a risk of ecological relaxation. This may be caused by, for

Fig. 6.10. The Kogelberg Biosphere Reserve, South Africa, with a view from the core zone out across the buffer zone to the transition zone. (Illustration: Julia van Schalkwyk).

example, extreme weather events, a sudden new threat, and/or genetic impoverishment. To tackle this challenge the Durban Metropolitan Open Space System (D'MOSS), South Africa set aside 80,000 ha to help protect species in the Maputaland-Pondoland-Albany global biodiversity hotspot. Its action plan includes protection of natural inland wetlands limiting the need for additional costly stormwater ponds in urban areas. The project also removes invasive alien vegetation to restore natural hydrological function. This project has value for biodiversity, including dragonflies, through protection of both aquatic habitats and riparian zones. Among the activities has been conversion of agricultural areas such as abandoned sugarcane fields, where former irrigation reserves have become dragonfly-rich ponds (Samways, 2008a). To date the project is home to 70 dragonfly species, including three national endemics. Some species, such as *Phaon iridipennis*, has even penetrated deep into the urban area. Overall, this project indicates that land and water restoration through ecosystem-based adaptation can have exceptional value for dragonflies (Deacon and Samways, 2021) (Fig. 6.11).

Fig. 6.11. A restored artificial pond that was once an irrigation pond for sugarcane farming but is now part of the Durban Metropolitan Open Space System, South Africa (top). This pond is rich in macrophyte heterogeneity and also in dragonfly species, including some tropical/subtropical species at the southern margin of their geographic range. From left to right *Lestes tridens*, *Urothemis edwardsii*, and *U. assignata*. (Photos: Michael Samways).

6.3.7 Value of ecotones between running and standing waters

6.3.7.1 *Small-scale ecotones*

Ecotones between any two ecosystems can have elevated levels of biodiversity as they draw upon species from both adjacent ecosystems as well as supporting certain ecosystem specialists. An ecotone between an artificial lake and its outflowing stream in South Africa supports species from both adjacent ecosystems, with overlapping macroinvertebrate assemblages, including those of dragonflies. One reason for this is that the ecotone is composed of a wide range of microhabitats reflected on either side of the ecotone. Furthermore, the lake is disturbed by small waves from windiness, while the undulating stream has edges composed of sheltered slow-flowing refuges as well as regions of faster-moving water which have an effect on the macroinvertebrate fauna, like that of the lake waves. The overall outcome is that the aquatic ecotone between the lake and the stream is not at the mouth of the stream but along the edges of both systems. In turn, the centres of the lake and the stream, although aquatically connected, are quite different in their macroinvertebrates and made up of separated micropatches with sharp boundaries, faunistically connected through the assemblages of the edge ecotone. The ecotone is in effect a semi-permeable membrane to the drifting stream fauna with movement back again apparently mostly in the air, according to a patchy metapopulation model, with both the river continuum and boundary concepts being applicable to this dual system. The ecotone does not display any edge effects, and so accords more with a variegation than a fragmentation model. The conservation implication here is that such ecotones provide some resilience to the aquatic environment overall and should be zoned as a conservation priority (Samways and Stewart, 1997).

At ecotones between feeder streams and lakes in Ohio, USA riparian canopy cover, relative algal densities and emergent vegetation all play a role in determining dragonfly assemblage composition. Though canopy cover explains 55% and 75% of abundance and species richness variation, respectively, across the three habitat types, the relationships are strongest in ecotones. Ecotones have very similar assemblages to each other, while their collective assemblage is a subset of the species occurring in the lake. The few stream species are all present in the ecotone (Cunningham-Minnick *et al.*, 2019) (Fig. 6.12).

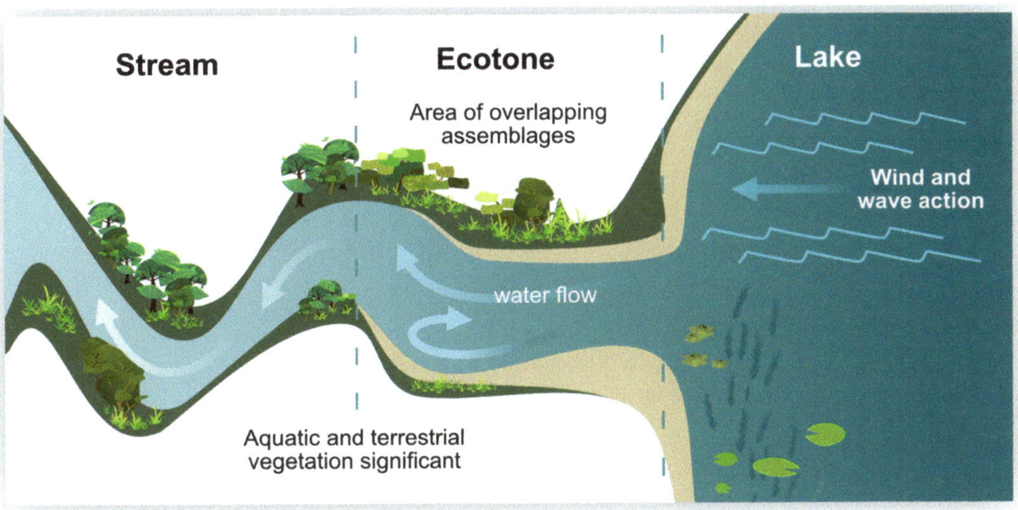

Fig. 6.12. Ecotones between flowing and standing water systems may support a characteristic macroinvertebrate and dragonfly fauna from both systems. Assemblage composition is driven by abiotic variables, such as water velocity in the stream and wind action on the lake, and biotic variables, such as aquatic and terrestrial vegetation cover. Largely, these ecotones fit the variegation model of species sorting. The conservation value of ecotones lies in their augmentation of local ecological resilience resulting from a wide range of microhabitats. (Illustration: Robyn Symons).

6.4 Specific Restoration Considerations for Running Waters

6.4.1 Whole catchment approaches

The macronutrients phosphorus, nitrogen and carbon, along with light energy, are the basis of primary productivity that sustains freshwater systems. Headwater streams are the uppermost portions of a catchment, typically representing 60–80% of the total stream/river length and draining 70–80% of the total catchment area. They play a fundamental role in the biogeochemical connectivity between the aquatic and terrestrial realms, while controlling the direction of supply of macronutrients to the downstream river networks as they converge with decreasing elevation (Jarvie *et al.*, 2018). Furthermore, headwater streams and their networks are highly significant for aquatic macroinvertebrate diversity (Sitati *et al.*, 2024) (Fig. 6.13).

The small size of headwater streams means that they are especially responsive to both natural and anthropogenic disturbances, including debris flows, changes in vegetative cover, sediment input and in organic matter input. Surface run-off into headwater streams may occur only during storm events or snowmelt. In warm regions though, where there is dense and persistent cloudiness, condensation may also feed headwater streams (MacDonald and Coe, 2007).

There are several positive effects of headwaters on downstream biodiversity, including offering

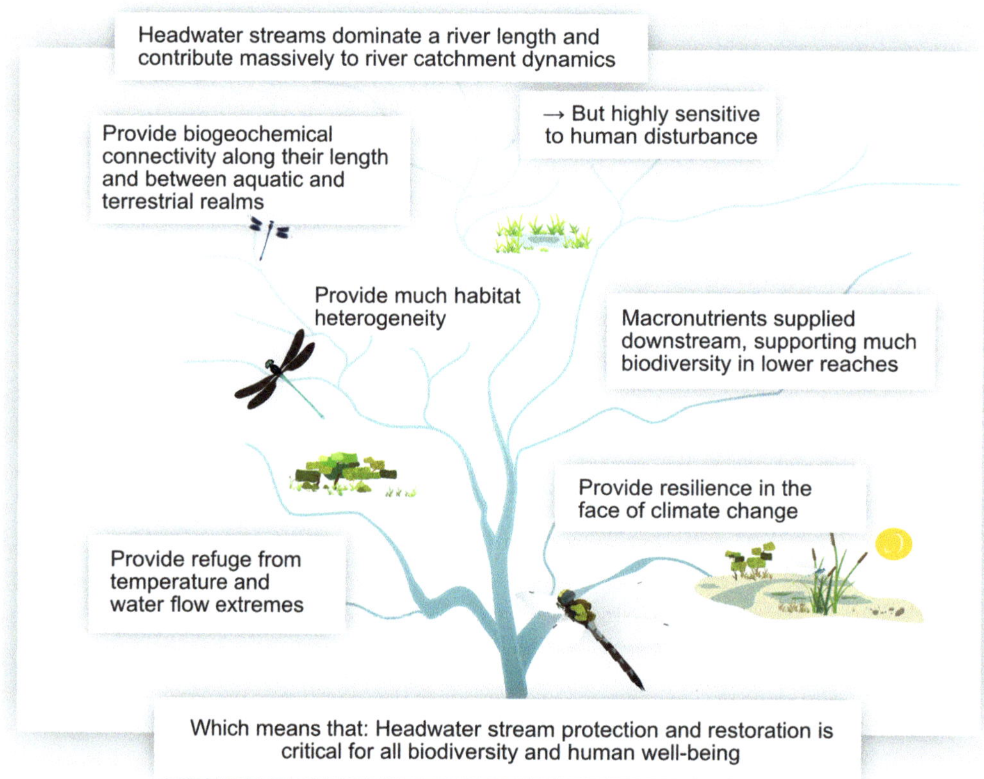

Fig. 6.13. The conservation value of headwater streams cannot be overemphasized, especially as they are so sensitive to human disturbance. Besides contributing to most of a river's length and water catchment capability, they provide biogeochemical connectivity both along their lengths, and between aquatic and terrestrial realms. They also supply essential macronutrients to downstream reaches. Headwaters are often highly heterogeneous, providing a wide range of natural habitats for a disproportionately high number of species. Furthermore, they provide refuge from temperature and water flow extremes, as well as ecological resilience in the face of climate change. The conservation message is loud and clear: headwater streams require the strictest protection and restoration where necessary for the benefit of a huge range of species and for human well-being. (Illustration: Robyn Symons).

refuge from temperature and flow extremes, competitors, predators and introduced species. Headwater streams also support populations of colonists, are a rich source of food items and create movement corridors up and down catchments. Degradation and loss of headwaters and their functional connectivity to ecosystems downstream threaten the biological integrity of entire river networks (Meyer *et al.*, 2007).

In the Amazon, both regional and local environmental changes are important predictors of odonate larval assemblages in headwater streams. Importantly, environmental conditions that maintain resources for larval development within the stream channel must be considered, as also must vegetation integrity for the adult. Knowing these conditions is a baseline for restoration where there has been deterioration through human activity (Ribeiro *et al.*, 2009).

To address the challenge of freshwater biota conservation, assessments are required to determine what areas to restore and what species to consider, while recognizing that in many catchments multiple habitat types should be restored. The implication is that biodiversity safeguards are required through policy and implemented in practice. Doing this avoids undermining the biophysical relationships that provide ecosystem resilience to climate change. However, conservation action involves integrated large-scale science that has a positive effect across biome boundaries (Temperton *et al.*, 2019). This is relative to the fact that there is clear evidence, from Brazil, for example, that protecting and restoring forests and their headwater streams is of critical importance for all freshwater biodiversity (Santos *et al.*, 2020).

A detailed pan-European review reveals the complexity of river restoration with its highly variable effects, along with the necessity for clearly set objectives and planning procedures. The most positive effects of river restoration have been on terrestrial and semi-aquatic organism groups such as dragonflies. Ecologically sensitive river widening projects have been significant in this process. River widening involves removing bank and bank fixation, flattening of riverbanks, and widening of the river cross sections, all of which are aimed at increasing the wetted area. Positive responses by the biota are primarily through increased abundance of organisms, but there is very little indication that overall biodiversity is increased. Specific traits rather than species richness or total abundance benefit most from restoration interventions. Habitat restoration also affects trophic structure, primarily by increasing the breadth of resources assimilated by consumers. Overall, it is essential to consider the whole catchment so that recolonization of restored habitats is made possible (Friberg *et al.*, 2016; Kupilas *et al.*, 2016).

An assessment of Dutch river restoration indicates that while there has been a great increase in stream restoration efforts, largely motivated by environmental legislation, proper monitoring of the effectiveness of the restoration measures is often lacking. Furthermore, there is a mismatch between restoration goals and restoration measures undertaken. Measures are still mainly focused on hydromorphological techniques, with biological goals under-represented. Additionally, restoration practices take place mainly at small spatial scales, despite the widely recognized relevance of tackling multiple stressors acting over large spatial scales toward stream recovery. These findings suggest that it is now timely to instigate monitoring programmes over long time-periods to ascertain whether the restoration measures taken have led to the desired results. There also needs to be diagnosis of the dominant stressors and planning for restoration measures at the appropriate spatial scale of these stressors, which is generally the catchment scale (Oliveira *et al.*, 2020).

6.4.2 River dynamics and species traits

Biodiversity typically varies along environmental gradients, with species diversity, functional diversity and species turnover often associated with greater environmental heterogeneity arising out of an increase in the number of niches becoming available. Along a river, heterogeneous environments offer a greater range of conditions, such as water pH, flow velocity and variability of resources, including coarse particulate organic matter and establishment of various plant species. However, the extent to which aquatic organisms establish along the river and sort themselves depends on the range of functional traits among them. Then there is the overlay of environmental dynamics, such as variable precipitation and temperature, as well as the extent of human disturbance and the degree of invasion by alien species (Fig. 6.14).

Among streams in the Brazilian Amazon, different types of anthropogenic disturbances affect the relationship between environmental heterogeneity

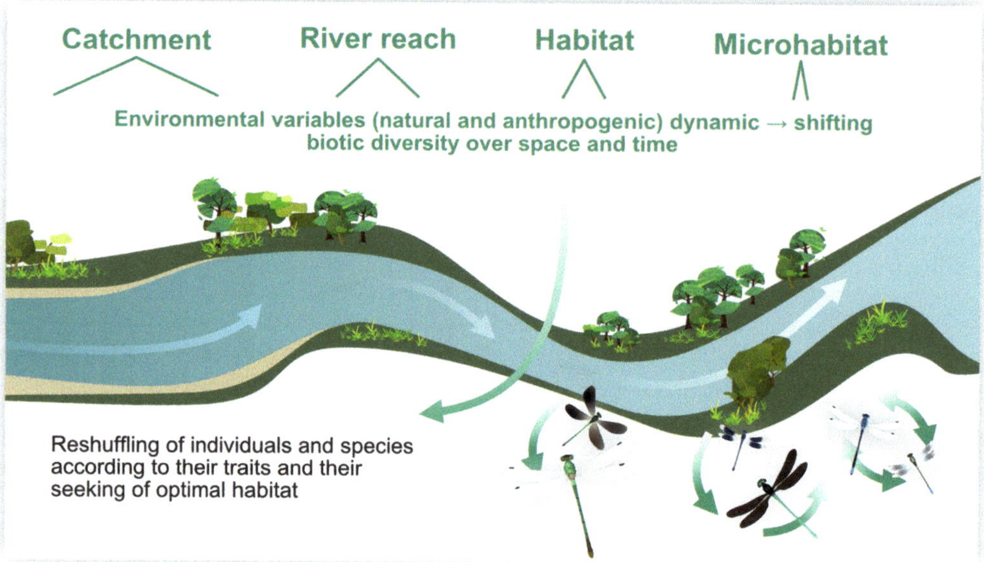

Fig. 6.14. A river is a highly dynamic system in terms of both water and the terrestrial interface. Conservation action is cognizant of the spatial scale of these dynamics, from catchment down to microhabitat (symbolized here by wedge width). The environmental variables, both natural and anthropogenic, vary in their effects on the biota over both space and time. This means that biotic diversity is heterogeneous while also continually shifting. The way in which communities shift in composition is in accordance with the suite of traits that the component species possess, and the ability of individuals among those species to select, establish, feed and breed in optimal habitats and microhabitats that are forever changing. (Illustration: Robyn Symons).

and dragonfly species composition according to species traits (Calvão *et al.*, 2023). In turn, among streams at multiple spatial scales in the Brazilian savanna, taxonomic and functional alpha diversities of aquatic insects shift at the local scale. The upshot for ecosystem management is that increased heterogeneity must consider the relevant factors for the focal organisms at a given spatiotemporal scale. Maintaining spatial water flow diversity effectively increases local alpha diversity, whereas increases in water depth variety potentially increases local taxonomic and functional alpha diversity in highly disturbed streams.

There are various factors that influence the biodiversity–heterogeneity relationship, including the absence of species in the regional species pool to colonize various parts of the river, ongoing human stressors on the organisms, focus on the wrong spatial scale and misreading the relationship between heterogeneity relative to the traits of the organisms. This means that it is essential to establish which aspects of heterogeneity are important for the focal organisms, such as riverbank height and water depth, as well as the historical and current levels of human disturbance in the Brazilian savanna (Agra *et al.*, 2021).

6.4.3 Shifting baseline

Another consideration for ecological restoration is the shifting baseline syndrome (see Chapter 5, Section 5.2) (Fig. 6.15). An example of how there can be great changes in ecosystems is seen in the case of the Yellow River, China. There have been major historical shifts in the sediment transport regime. When the two main drivers of sediment transport, climate change and food production, are considered over time, there have been historical climatic changes (e.g. the Medieval Warm Period) resulting in changes in sediment transport. However, in the past, when anthropogenic stresses dominated, there was a regime shift, leading to a tipping point which caused sediment transport to no longer be resilient to climate drivers. The example of the Yellow River is useful for understanding cascading effects within various socio-ecological systems. As there have been

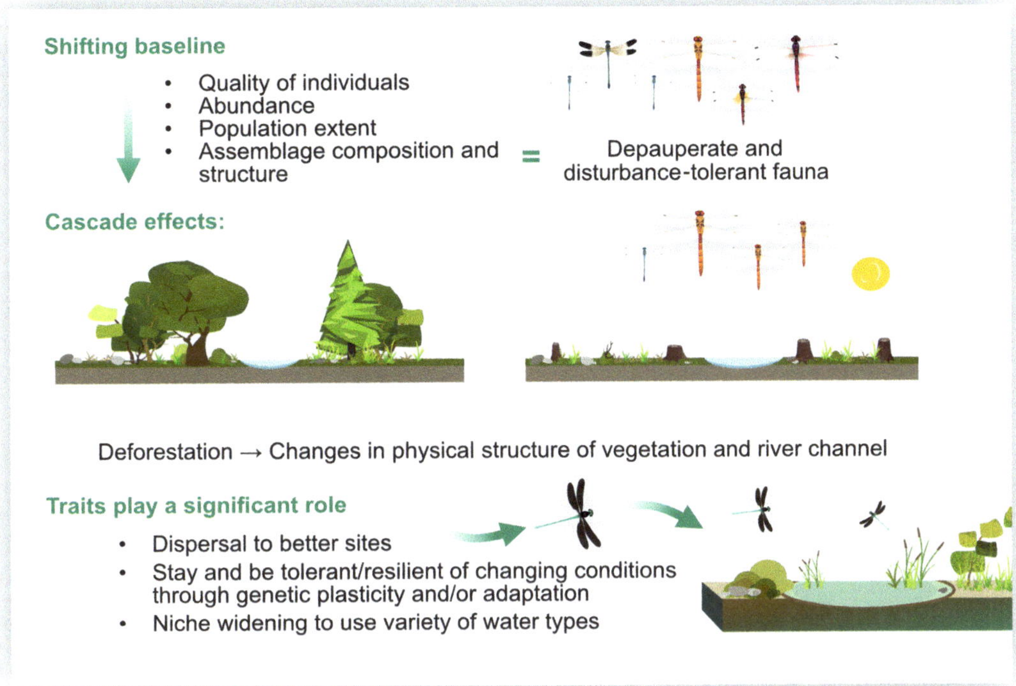

Fig. 6.15. Owing to the prior effects of stressors, an assemblage may already be depauperate, i.e. the baseline has shifted. This means that although the site may be the subject of restoration, its successful restoration must be measured against an undisturbed reference site. The shifting baseline manifests as quality of individuals, changes in abundance, physical extent of population, and/or assemblage composition and structure, all resulting in a depauperate odonate fauna composed of disturbance-tolerant and resilient species. Such shifts may come about, for example, through the cascading effects of deforestation leading to physical changes to the vegetation and water channel. Traits play a major role in these faunal shifts. A species may disperse and establish in good habitat elsewhere. Another species may stay and become tolerant and resilient through genetic plasticity and/or adaptation. When this occurs, there is niche widening, enabling the occupation of various water types. There may also be a shift in the prey type consumed. (Illustration: Robyn Symons).

similar changes globally, it is crucial to understand the relationship between regime shifts and their drivers/impacts when developing resilience management strategies within river basin systems (Song *et al.*, 2020).

There is possibly evidence of a shifting baseline already taking place among odonates in Mexico. An index of vulnerability for odonates uses species traits and ecological niche modelling to relate changes in vegetation cover to the habitat preferences of each species and their sensitivity to habitat deterioration. While the most sensitive species are in areas with no land-use change, odonates overall do not show a clear preference for conserved versus perturbed land-use areas. A few sensitive species are resilient, occurring in many land-use types, suggesting that they now may be more generalist than previously recorded. Furthermore, 35 species are not found in any of the best conserved areas, implying that these species have instead adapted to seemingly unfavourable habitats. For example, *Enallagma semicirculare* and *Gynacantha nervosa*, which appear to have been more widely distributed in the past, are now less common yet occupy many land-use types. Also, 59 species present in optimal habitat are also present in less optimal habitats, suggesting some genetic plasticity to land-use change has taken place by remaining and adapting. This niche widening is manifested by use of both running and standing water or shifts in their flight period. Seemingly, their dietary requirements are still being met, perhaps by changing their prey species (Rodríguez-Tapia *et al.*, 2020).

Along the Turvo Sujo river below the town of Viçosa in upland central Brazil, the odonate assemblage was relatively insensitive to local conditions such as the physical characteristics of riparian vegetation and channel structure, as well as dissolved oxygen levels. This was apparently due to an already depauperate odonate assemblage that has survived in the system after deforestation 40–50 years earlier (Silva *et al.*, 2010).

6.4.4 Downstream effects

Headwater streams are an important consideration when restoring river reaches farther downstream. For example, three rivers in Ireland support 154 macroinvertebrate taxa, with headwater streams having the highest number of unique species, 24 in total, making up 17% of the total macroinvertebrate taxon richness of all rivers. Headwaters also support many species also present farther downstream (38% of species), emphasizing their potential for acting as sources to downstream sites should human pressure cause local extinctions. This means that headwaters are not only important for catchment biodiversity but may also be critical source habitats for the restoration of downstream sites (Callanan *et al.*, 2014).

In the Amazon, floodplain wetlands are functionally connected with both the surrounding landscape and upstream into the catchment. Integrating terrestrial and aquatic conservation frameworks is crucial for safeguarding the ecosystem functioning of complex floodplain–river systems through maintenance of functional connectivity. These improvements to the whole system are key for guaranteeing the long-term persistence and resilience of these highly dynamic ecosystems (Reis *et al.*, 2019).

In 1998, the Guadiamar river in south-western Spain suffered a major toxic mining waste spill in its middle reaches. The river prior to that had already been subject to other stressors related to land and river transformation, impoverishing the odonate fauna. Ten years after the spill, the dragonfly assemblage in the affected area was again like that of the unaffected upper reaches, with some species being abundant, such as *Erythromma lindenii*, *Ischnura graellsii* and two *Platycnemis* species, as well as attracting widespread anisopterans like *Anax imperator*, *Crocothemis erythraea*, *Orthetrum chrysostigma* and *Sympetrum striolatum*. Interestingly, and by way of comparison, the lower reaches of the river that were also affected by the mining spill but suffered from many years of permanent and diffuse urban and agricultural pollution, supported poor dragonfly assemblages with only tolerant species like *I. graellsii*, *Aeshna mixta*, *Brachythemis leucosticta* and *Trithemis annulata* being present (Ferreras-Romero *et al.*, 2009).

6.4.5 Relative recovery of dragonflies in running versus standing waters

Dragonflies can be resilient and are able to quickly recover when their habitats are restored, so long as source populations are nearby. Water and physical habitat improvement in the Netherlands has benefitted running water species more than those of standing waters (van Strien and van Grunsven, 2023). Recovery has led to a better conservation status for many of the species. Furthermore, southern species showed more positive trends than northern species, suggesting that climate change has contributed greatly (Termaat *et al.*, 2015). Van Grunsven *et al.* (2020) update the data to 2019, and conclude that the Dutch dragonfly fauna is highly dynamic with winners and losers over the last 30 years. Unexpectedly, critical species such as *Leucorrhinia caudalis* have come back, while some very common species, like the moorland *Lestes sponsa* and *Sympetrum danae*, have declined greatly. Overall, there are differences in trends of species from different habitats. Improved water quality in running waters has led to the recovery of species in this water type. However, a major driver of changes in the fauna is climate change, seen by an increase in thermophilic species. Furthermore, climate change is also likely causing a decline in moorland species that prefer lower temperatures. The adverse impacts will not only be through changes in temperature, but also because of more weather extremes, especially droughts (Fig. 6.16).

6.4.6 Importance of restoring vegetation

6.4.6.1 Aquatic macrophytes

The macrophyte community (aquatic macrophytes) is a key element in freshwater ecosystems, especially those that are shallow. In these systems, local species richness and composition are strongly influenced by site-level physical heterogeneity and hydrological connectivity and, in general, more so than the type of land use and nutrient loads (Bolpagni *et al.*, 2020). However, macrophytes

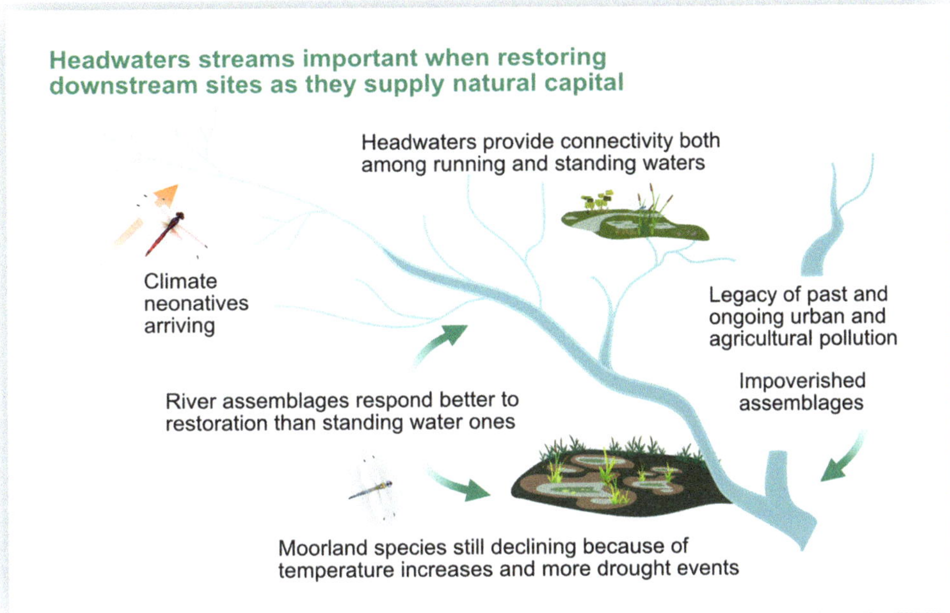

Fig. 6.16. In Europe, there are winners and losers among dragonflies relative to restoration in a rapidly changing world. A baseline consideration is that headwater streams provide connectivity among all water types, running and standing. These streams are also important for providing natural capital (in terms of remaining species) to previously degraded and now restored downstream sites. Overall, odonate assemblages in rivers have responded better to restoration than those in standing waters. Climate neonatives are also appearing, adding to the assemblages, especially those from southern rather than northern species. However, moorland species continue to be under pressure as temperatures rise and droughts become more likely. Legacy issues must also be considered, with diffuse urban and agricultural pollution inhibiting dragonfly assemblage recovery. (Illustration: Robyn Symons).

change the physicochemical environment, and are important in nutrient cycling, produce biomass used by herbivores and detritivores, and provide habitat for many other organisms, leading to effects at higher trophic levels (Thomaz, 2023), including on dragonflies (Fig. 6.17).

In turn, and especially important for all stages in the dragonfly life cycle, the composition of riparian vegetation affects macrophytes. The species richness of macrophytes increases with low-growing vegetation, important for macrophyte establishment in streams. Together, low-growing vegetation cover and herbaceous cover improves macrophyte diversity, with increased dominance of low-growing herbs (by, for example, removal of woody riparian plants that otherwise create too much shade) favouring pioneering macrophytes (Carmo *et al.*, 2023). In short, it is easy to see how dragonfly larval and adult assemblage composition and abundance are changed both directly and indirectly by any shifts in the riparian plant community.

In Croatian intermittent streams and rivers, macrophyte abundance, water electrical conductivity and water velocity operating together are the most significant determinants of odonate assemblage composition. Macrophyte-rich sites promote higher abundance as well as greater taxonomic and functional diversity compared to the macrophyte-poor sites. Assemblages at macrophyte-rich sites are characterized by larger body size and a greater proportion of species preferring standing over temporary running water conditions. Rich aquatic plant assemblages also determine the various patterns of larval development and drought resilience strategies. In contrast, macrophyte-poor sites are preferred by running water species, with larval development all year round and with no specific drought-resisting strategies (Vilenica *et al.*, 2022b).

In Vienna, Austria, a hydroelectric power plant along a 21-kilometre length of shoreline associated with the Danube river formerly had steep embankments, but was restructured by creating shallow

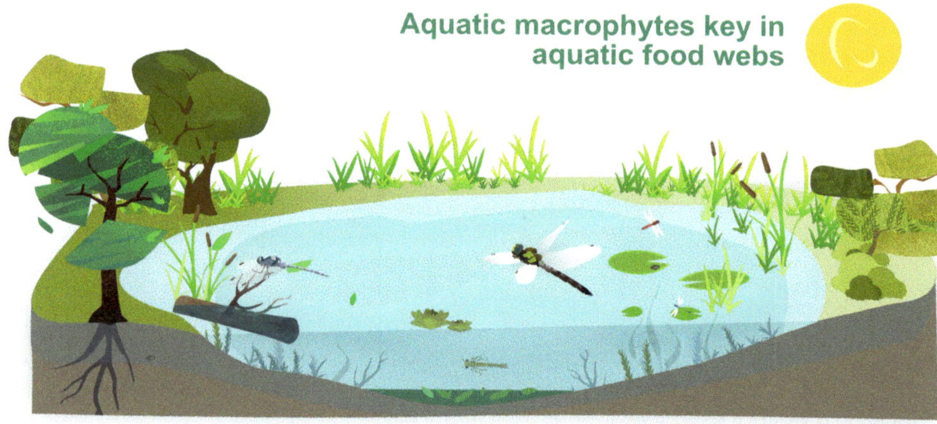

Fig. 6.17. Indigenous aquatic macrophytes of all types are central to the restoration of watery ecosystems. They are the main primary producers converting the sun's energy into available energy for the whole aquatic food web. Aquatic macrophytes respond to abiotic variables relating to water velocity, depth and physicochemistry, while they produce oxygen, food and structures for the dependant animals. The various macrophyte species together produce a whole variety of microhabitats and much spatial heterogeneity in the aquatic realm, resulting in high levels of biodiversity. They are affected by the vegetation of the riparian zone. Leaf and twig fall from trees along the margin provide nutrients, while fallen tree trunks in the water create local micro-environments. However, the living riparian trees, when tall, may shade the water and inhibit aquatic macrophyte growth. In turn, the various macrophytes, emergent, floating and submerged, also interact, creating spatial heterogeneity. The establishment of ecological integrity among macrophytes is fundamental to dragonfly and freshwater conservation. (Illustration: Robyn Symons).

water areas, gravel banks, small permanent backwaters and temporary waters. After four years, the rehabilitated area was colonized by 24 dragonfly species (31% of all Austrian species). The development of macrophyte communities was instrumental in enabling dragonflies to establish. The first colonizing species were the floodplain species *Ischnura elegans*, *I. pumilio* and *Sympetrum striolatum* making up over 90% of the individuals, followed in later successional stages by *Platycnemis pennipes*, *Sympetrum fonscolombii*, *Coenagrion puella*, *Anax imperator* and *Orthetrum cancellatum* (Chovanec *et al.*, 2000, 2002).

When construction of two canals in Vienna was followed by restoration management involving establishment of a historic-type plant community along with high-quality water, there was rapid colonization by odonates. At one canal over five years, species richness went from 1 to 26 species, and at the other canal over four years, from 15 to 29 species (Chovanec and Raab, 1997).

The invasive water plants *Pistia stratiotes* and/or *Pontederia crassipes* support populations of various odonate species both in Florida, USA, and other warmer parts of the world (Lounibos *et al.*, 1990). However, to date these two plant species appear to support only an homogenized dragonfly fauna and do not support any highly threatened odonate species outside the plants' natural provenance.

6.4.6.2 Riparian vegetation

In Pará state, Brazil, when conventional logging (to within 10 m of the river margin) and reduced-impact logging (polycyclic logging where any one area is only logged again after 30 years of regeneration according to Forest Stewardship Certification) are compared to historically natural areas, there are few differences in odonate abundance and species richness in the three areas. However, there are great differences in species composition between conventional logging, reduced-impact logging and historically natural areas. Conventional logging not only changes the degree of vegetation canopy but also aspects of the stream channel, reducing water quality, dissolved oxygen concentration and stream microclimate. As conventional logging leaves some vegetation patches intact, some species like *Fyglia amazonica*, *Psaironeura tenuissima* and *Heteragrion silvarum* prevail. Species turnover also plays a role, as indicated by each of the three areas having some unique species (Calvão *et al.*, 2016).

In the same region, species turnover is the main component of beta diversity associated with indigenous forest, reduced-impact logging or conventional logging. This again suggests that management based on reduced-impact logging could be adopted to guarantee socio-economic benefits while also minimizing the impacts of logging on local dragonfly diversity (Mendes *et al.*, 2021).

Odonate larvae assemblages in the Atlantic Forest streams of southern Brazil show great changes in composition according to substrate type and river reaches with different riparian vegetation widths. Of note is that assemblage structure varies relative to organic (litter) substrate in reaches with narrower riparian vegetation strip widths compared to inorganic (stones and gravel) substrates in reaches with wider riparian vegetation strips. These findings suggest different management strategies are required regarding riparian width retention, and according to the in-stream conditions (Pires *et al.*, 2020; Graziano *et al.*, 2022).

In the central savanna Cerrado region of Brazil, there is a weak threshold of 54% of native forest necessary to sustain the full zygopteran assemblage. Furthermore, while legally there is a requirement that riparian buffers of 30 m on both sides of a watercourse must be retained, a width of 250 m is more realistic to support the full zygopteran fauna (Rodrigues *et al.*, 2016).

In the coastal forests of Kenya, the effect of disturbance on dragonfly assemblages is mainly related to the degree of shade from the forest canopy. With increasing loss of the forest canopy, species richness *increases*, although composition changes greatly, with forest specialists replaced by common habitat-generalist species. There is spatial heterogeneity among the forest species, with forested spring sources and swamp forest, for example, having distinct dragonfly assemblages composed of habitat-specialist species that are highly sensitive to habitat disturbance. With increasing disturbance of all habitat types (stream, spring source or swamp forest), species turnover decreases and the assemblages become homogenized. However, some species of forest streams that are more than 2 m wide show a higher tolerance to habitat disturbance and may occur where the forest canopy has been locally removed. The sun/shade mosaic along the forest streams is a natural phenomenon caused by tree-fall gaps due to the activities of large mammals or flooding. Certain species, such as *Coryphagrion grandis*, *Gynacantha usambarica* and *Thermochoria jeanneli*, are forest-habitat specialists confined to this habitat type and are at risk of local extirpation when forests are destroyed. Nevertheless, a strip of natural riparian vegetation of at least 10 m along each side of streams and rivers in a cultivated landscape is recommended to reduce population isolation. These measures are also effective for dragonfly conservation through promoting landscape heterogeneity, as well as decreasing the destructive impact of flooding and drought (Clausnitzer, 2003).

In South Africa, while much of the odonate fauna is sun-loving, there are certain species that are typical of riparian zones fringed with native trees. These trees often allow sunflecks to reach the water and ground surfaces, promoting growth of a wide range of understorey macrophytes. This vegetation complexity makes up a riparian margin of 20–30 m separating the watercourse from various disturbances such as urban development and alien tree plantation blocks. The historically vegetated riparian zone enables establishment of species like *Pseudagrion hageni* as well as the national endemics *Chlorolestes tessellatus* and *Allocnemis leucosticta* (Samways and Steytler, 1996). Similar findings come from the Bumbu river, Democratic Republic of Congo, where maintaining heterogeneous vegetation along riverbanks supports a diverse dragonfly assemblage characterized by species such as *Chalcostephia flavifrons*, *Ceriagrion corallinum*, *Lestes virgatus*, *L. ictericus* and *L. tridens* (Sisa *et al.*, 2022).

Bearing in mind that dragonflies are part of complex food webs and require prey, fringing trees at wetlands can help freshwater organisms in general to survive droughts through provision of refuges for the dormant stages. The surface temperatures of dry sediment are cooler beneath fringing trees, and the sediment has higher organic matter, holds more water and is less dense than open water sediment (Strachan *et al.*, 2016).

Restoration of riparian zones can take some time to achieve. Dissolved nutrient concentrations often increase directly after initial logging of the trees and can remain elevated for at least five years. Nevertheless, there is a successional progression in nutrient release and uptake in riparian zones, followed by a recovery period and full return to predisturbance conditions. However, full ecosystem recovery may take in the order of 10–20 years (Wood *et al.*, 2022) (Fig. 6.18).

Fig. 6.18. The conservation of riparian zones is fundamental to freshwater conservation. These zones are the buffer between aquatic and terrestrial realms, while providing functional connectivity along rivers. Maintaining riparian zone stability ensures the continuance of natural ecosystem processes at the interface between water and land. Logging of a forested riparian zone is usually disastrous, and so these zones are often protected by law. However, what riparian width should be conserved depends on many factors that range from ecosystem type through to land-use type and intensity, as well as what is acceptable socio-ecologically and in terms of commercial interests. The point is that any riparian buffer zone is better than none, and when present, the wider the better. But landscape planners need hard figures, usually with riparian zones proclaimed as 20–30 m wide. This figure though, falls short of the 250 m required in Amazonian forest for maintaining the full natural complement of odonates, but would probably contribute greatly to maintaining Kenyan forest specialists. (Illustration: Robyn Symons).

In Hungary, sport fishing has raised concerns over preserving the ecological integrity of the banks of floodplain water bodies. Reconciling sport fishing and maintenance of intact local dragonfly assemblages is achieved by ensuring that the permanent fishing stands do not exceed 8 m in width, and separation of fishing stands by at least 12 m of intact riparian sections (Müller *et al.*, 2003).

6.4.7 The debate on riparian buffer zone width and multi-taxon conservation

Importantly, dragonfly conservation must be integrated with conservation of other taxa across entire catchments, such as fish and other invertebrates, that are sensitive even to low levels of forest loss (Brito *et al.*, 2020). Furthermore, as pointed out by Leal *et al.* (2020), conservation planning in Brazil should be based on both aquatic and terrestrial protection which together have much greater value than either one, but with insignificant loss in terrestrial preservation. These perspectives led Martins *et al.* (2021) to conclude that: 1) Logging should be prohibited in riparian linear protected areas that are less than 100 m wide on each side of headwater streams. Linear protected areas should also be maintained in a network of catchments across all biomes and as many landscape types as possible; 2) A rigorous monitoring programme be implemented that uses standardized methods to assess and regulate land use better; and 3) Conservation planning areas must consider aquatic biota as well as terrestrial biota.

In the western Amazon, riparian forest is crucial for maintaining the ecological stability of the streams, while also essential for retaining zygopteran populations (Brito *et al.*, 2021). A Brazilian nationwide study established thresholds of native vegetation cover loss. Thresholds varied greatly among biomes, buffer sizes and biological sets, ranging between 0.5–77.4% for fish, 2.9–37.0% for aquatic invertebrates and 3.8–43.2% for a subset of aquatic invertebrates (Ephemeroptera, Plecoptera, Trichoptera (as a group: EPT), Odonata and Diptera). Thresholds were lower for smaller riparian buffers (50 m and 100 m) than larger ones (200 m and 500 m), indicating that land use should be distanced from streams. Furthermore, thresholds were lower at lower percentages of riparian vegetation loss in the small buffers. Significantly, thresholds were critically lower for invertebrates, with a reduction of only 6.5% of native vegetation cover within a 50-metre riparian buffer enough to have a significant impact them. The great variability in biodiversity responses to loss of native riparian vegetation suggests caution in the use of a single riparian width for conservation actions or policy decisions nationwide. As conservation practice requires managerially realistic recommendations, it is suggested that maintaining at least 50-metre-wide riparian buffers on either side of streams would contribute greatly to protecting local freshwater habitats. However, putting in place even wider riparian buffers (about 100 m wide on both banks) would be beneficial to a range of freshwater taxa (Dala-Corte *et al.*, 2020). Furthermore, retaining remnant forest fragments in the region can improve EPT diversity in a transformed and open landscape that can only support a reduced level of diversity (Faria *et al.*, 2023).

6.4.8 Matrix effects

6.4.8.1 Plantation practices

Amazonian oil palm plantations have buffer zones with a maximum width of 30 m on each bank inside oil palm areas, composed of secondary forest. These streams are rich in genera of odonate larvae owing to the relatively high levels of stream and bank heterogeneity and are richer in genera than natural forest. Biotic heterogeneity in terms of genera turnover of Zygoptera larvae is high in the streams owing to highly variable responses to the environmental heterogeneity, with selection of foraging sites by adults likely determining larval distribution (Mendes *et al.*, 2019).

In Brazil, the cabruca agroecological approach of planting cacao under a native tree canopy greatly benefits both larval and adult odonates. Important environmental variables for both life stages include water conductivity, watercourse channel width and dissolved oxygen, with spatial factors, such as canopy cover and intactness, important predictors of adults. Cabruca areas maintain high levels of odonate diversity, including forest specialists and those sensitive to landscape transformation (Santos and Rodrigues, 2022a, b) (Fig. 6.19).

6.4.8.2 Restoring the wooded matrix

In northern Minnesota, USA, when old-growth forest, mature secondary forest and recently clear-cut areas are compared in terms of their dragonfly fauna in streams, ponds, lakes and bogs, recently logged areas have the lowest species richness and diversity. The highest species richness and diversity is in the old-growth primary forest, where patches of well-established trees also increase beta diversity and therefore species heterogeneity. Logging has such an adverse effect on the dragonfly assemblages that full recovery takes several decades. Functional connectivity among remaining primary forest is essential and may be provided in part by the 300 m riparian old-growth forest that fringes the Mississippi headwaters (Rith-Najarian, 1998).

By 1870, Hong Kong had been extensively deforested. However, following extensive assisted and natural reforestation, today it is again largely covered by trees with pockets of intense urbanization. This reforestation has enabled many dragonfly

Fig. 6.19. The 'cabruca' agroecological approach (left) combines the production of cacao (centre) with stream intactness (right), supporting an odonate assemblage with high ecological integrity. (Photos: Leandro Juen).

species, that presumably were once there but went extinct locally due to deforestation, to return to the area, especially from mainland China. This recovery suggests that secondary forest can be highly effective in supporting many species, although some forest specialists likely will take many more years to return (Samways, 2008b) (Fig. 6.20). Measuring the recovery of dragonfly fauna goes hand in hand with measuring forest recovery from a botanical point of view (Bressane *et al.*, 2023).

In East Africa, *Coryphagrion grandis* and *Hadrothemis scabrifrons* inhabit tree-hole phytotelmata as larvae. Their habitat is seemingly in short supply as 65% of all phytotelmata surveyed in different coastal forests in Kenya were occupied by larvae of one or both species. This also emphasizes the value of mature forest for maintaining forest specialists like these species in the region (Clausnitzer and Lindeboom, 2002) (see Fig. 5.23).

During a river's recovery time, in-stream processes and hydrologic connectivity between stream channel and subsurface zones may influence nitrogen processing. This means restoration aimed at increasing hydrologic connectivity in riparian zones may enhance denitrification rates by increasing soil organic carbon availability and altering hydrologic flow paths. Because geology, geomorphology, riparian soils, hydrologic flows and nitrogen loads all influence denitrification rates, they must all be considered in the restoration project, which by necessity must consider bank reshaping and stabilization, channel reconstruction, riparian revegetation and other physical factors (Mayer *et al.*, 2022).

6.4.8.3 Restoring riparian zones through removal of invasive alien plants

Invasive alien plants can be a serious problem for dragonflies in some parts of the world. Invasive alien trees are particularly threatening in watercourses in certain dry climates, especially Mediterranean-type climates with hot dry summers and cool wet winters. Alien trees, especially in the genera *Acacia*, *Eucalyptus* and *Pinus*, can take hold and spread rapidly along watercourses, shading out local vegetation on the banks and at running water margins. Streams and small rivers are especially vulnerable owing to their small size, and risk of

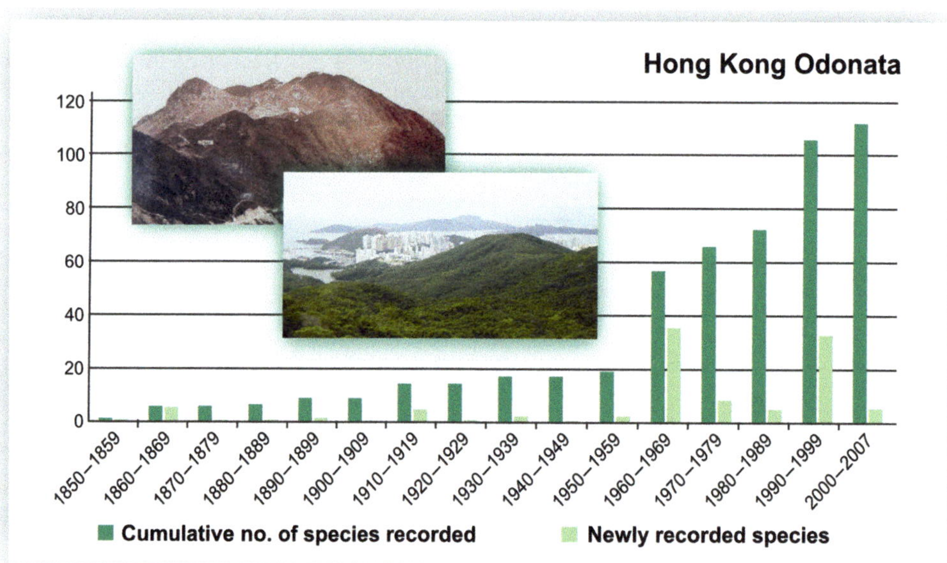

Fig. 6.20. By 1870, Hong Kong had been extensively deforested (top image), but, unlike many tropical areas, has undergone considerable assisted and natural reforestation, so that today it is largely green, with pockets of intense urbanization (bottom image). This regreening has meant that many dragonfly species that presumably once were there, but went extinct locally due to deforestation, have been able to return to the area (bar chart). (Top image and graph courtesy of Keith Wilson from presentation at the 17th International Symposium of Odonatology, Hong Kong, 2006. Bottom image from Samways (2008b), courtesy of Oxford University Press).

having their entire water surface covered (see Fig.5.28). They often also establish as a strip along the water's edge with no sunlit gaps as might be the case with indigenous vegetation. There is a direct effect on dragonfly adults through a drop in direct sunlight levels, while also losing oviposition sites as local low-canopy vegetation is shaded out. Removal of these alien trees is imperative for the local freshwater fauna in general, but particularly so for odonates.

South Africa, a largely dry country with a mostly sun-loving odonate fauna, along with high levels of endemism in the mountainous and high plateau regions, is highly vulnerable to invasion by alien trees. Their removal has been a national priority, not specifically for dragonflies, but to reinstate scarce natural water flows, while also providing employment among local communities. When invasive alien trees are removed, dragonflies can recover well.

At sites across the Cape Floristic Region, of the 23 species sampled, over 30% of the recovered species are endemics: the Cape endemics *Chlorolestes umbratus*, *Elattoneura frenulata*, *Pseudagrion furcigerum* and *Orthetrum julia capicola*, and the national endemics *Allocnemis leucosticta*, *Pseudagrion draconis* and *Pineyschna subpupillata* (Samways and Sharratt, 2010). Other notable endemics and near-endemics that have recovered locally through removal of invasive alien trees include *Chlorolestes apricans*, *C. conspicuus*, *C. fasciatus*, *Spesbona angusta*, *Metacnemis angusta*, *Pseudagrion newtoni*, *Proischnura polychromatica*, *Syncodulia gracilis*, *S. legator*, *S. serendipator*, *S. venator* and *Ceratogomphus triceraticus*, a substantial number of very rare, and in most cases, threatened species (Samways *et al.*, 2005; Samways and Simaika, 2016).

However, removal of alien trees on the iconic Table Mountain, Cape Town, never saw the recovery of the highly localized *Orthetrum rubens*, at this its type locality (Samways and Deacon, 2022). The reason for this is that there are no source populations in the vicinity and there is a hot, dry plain, also transformed by urban development, between Table Mountain and the nearest mountain populations.

Besides these endemic species, populations of many other dragonfly species, both nationally rare and common in the country, have recovered following alien tree removal. This includes the savanna areas which are without localized endemics owing to the historical uncertainty associated with local climate (Samways and Grant, 2006; Magoba and Samways, 2010; Modiba *et al.*, 2017) (Fig. 6.21).

Additionally, invasive alien trees, like *Acacia mearnsii*, in South Africa can increase nitrogen runoff from riparian zones (Petersen *et al.*, 2020). Bearing in mind that this tree species can greatly disadvantage the dragonfly assemblage through shading out of natural habitat, there is yet another reason for removing these alien trees from riparian zones. The various life stages of zygopterans in particular can help monitor the restoration, as adults are differentially sensitive to various in-water and on-margin vegetation types, while the larvae respond to stream physicochemical conditions.

There are, however, nuances associated with certain alien trees. For example, the alien *Quercus robur* in South Africa does not appear to inhibit the national endemic *Chlorolestes tessellatus* nor the near-endemic *C. elegans*, possibly because of sunflecks penetrating the canopy and to which individuals are attracted. In turn, the endemic *Pseudagrion vaalense*, especially the females, use individual alien weeping willow trees (*Salix babylonica*) for perching, with indigenous trees scarce in its grassland habitat. In turn, nationally endemic *Proischnura rotundipennis* is highly tolerant of low-growing American bramble (*Rubus cuneifolius*) which readily invades its marshy habitat, and endemic *Africallagma sapphirinum* occupies artificial lakes fringed with the locally invasive grass *Hyparrhinia hirta*.

6.5 Recovery and Restoration of Standing Waters

6.5.1 Significance of standing waters

Standing waters are of major ecological significance for supporting high levels of biodiversity and are receiving much conservation attention (Keddy, 2023). The global area of natural lakes and ponds (304 million of them) is about 4.2 million km^2. A large proportion of water coverage is dominated by the millions of water bodies smaller than 1 km^2. In turn, the water coverage of large, engineered reservoirs is about 260,000 km^2, while the water behind localized farm impoundments covers between 0.1–6% of farm area or 77,000 km^2 globally (Downing *et al.*, 2006). Revised estimates using satellite imagery suggest that there are about 117 million standing water bodies each with a surface area greater than 0.2 ha. Combined, the surface area covered by standing water bodies is 3.7% of Earth's non-glaciated land area, dominated by large

Fig. 6.21. This site, on top of Table Mountain, Cape Town, South Africa, was previously afforested with alien pine trees which were removed about three years prior to this photo. A remnant stump is visible here (top left). The recovery of aquatic macrophytes was fast, with mats soon establishing in the height of the dry summer when water levels are low (top right). The natural fynbos vegetation also soon recovered (bottom left). Odonate assemblage recovery was astonishing, with the Dragonfly Biotic Index (DBI) (see Chapter 7) site score increasing sixfold (i.e. a Dragonfly Recovery Score (DRS) of 600%) after removal of the alien trees. The high-scoring recovery was due largely to the reappearance of many high-scoring endemic species. In the hot savanna, odonate assemblages are much more species-rich, but composed of low-scoring widespread habitat generalists. While removal of alien trees increased the DBI site score, the DRS was only 114%, not because of the low effectiveness of restoration, but simply because the dragonfly fauna was entirely composed of Least Concern species. (Photos: Michael Samways; Illustration: Charl Deacon).

and intermediate-sized lakes in terms of total surface area (Verpoorter *et al.*, 2014).

Importantly, both natural and artificial ponds contribute substantially to carbon sequestration. Carbon burial rates are higher in ponds with macrophytes, fish and high nitrogen : phosphorus loads (Holgerson *et al.*, 2024).

6.5.2 Standing water-body connectivity and dynamics

Lakes often have high levels of biodiversity but differ greatly according to interrelated factors like underlying geology, elevation, topography, isolation, drainage and floodplain presence. They can be considered as metasystems associated with the flow of matter and chemicals (both nutrients and pollutants) that flow across the waterscape. Their position in the landscape affects the extent to which they accumulate chemicals and species, including alien ones. Highly connected lakes at low elevations in the landscape accumulate these components more than high elevation lakes. Inevitably, this means that lake restoration should consider these dynamics at multiple spatial scales (Heino *et al.*, 2020).

Standing water bodies, especially smaller ones such as ponds (standing water bodies with an area of less than 2 ha) can vary considerably in size according to changes in season, degree of weather events and climate, and may form multipond systems (Chen et al., 2020). In the UK, where there is a high density of ponds, there are great changes in both their numbers and wetted areas between seasons and years, driven by variations in regional rainfall. The variation in hydroperiod and extent of wet areas means that different ponds often have very different plant communities and other associated biodiversity. Ponds under different land uses respond to rainfall over different timescales suggesting that pond types may vary in sensitivity to climate change. High rainfall extremes benefit the pondscape by inundating sites and disrupting the normally adverse effects of intensive land use. Overall, this suggests that threats to temperate ponds from climate change will be a complex interplay between actual variations in regional weather coupled with land use in the matrix around each pond (Jeffries, 2016) (Fig. 6.22).

6.5.3 Heterogeneity and standing water-body recovery

Across an area, such as an extensive bog, the various macroinvertebrate species often display high species turnover while having distinct local distribution patterns. These patterns may be associated, for example, with interrelated differences in nutrient concentrations, dissolved oxygen, water depth, hydroperiod, and vegetation type and structure. These factors indicate that restoration should focus not necessarily on increasing the size of one habitat

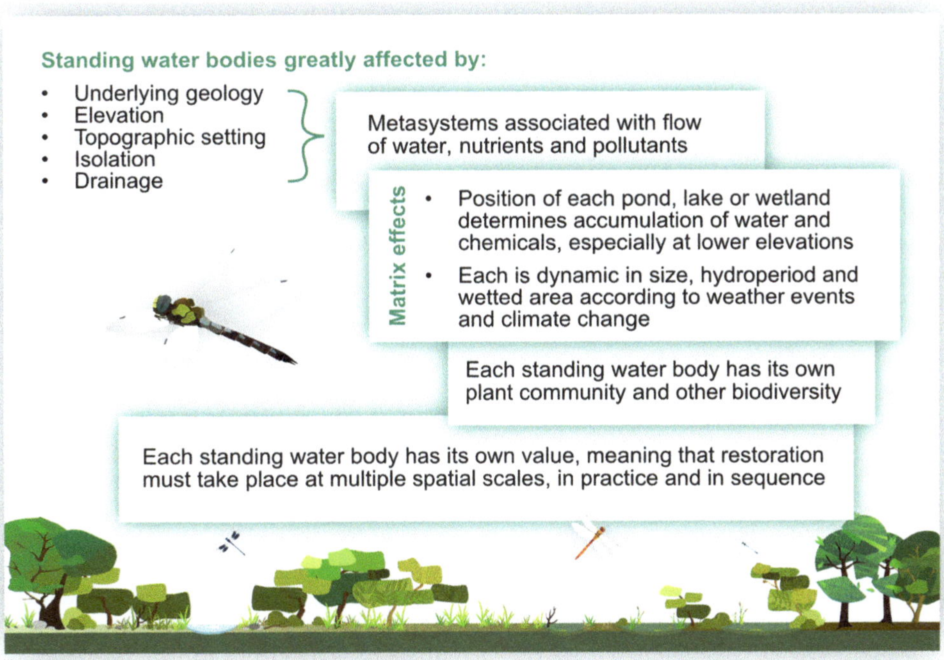

Fig. 6.22. Standing water bodies are greatly affected by five important underlying factors. This makes them metasystems associated with the flow of water, nutrients and pollutants. The position on the landscape of each pond, lake or wetland determines the accumulation of these three components, and strongly so at lower elevations. Each is dynamic in size, hydroperiod and wetted area according to weather events, especially downpours or drought, and increasingly so with ongoing climate change. Each is also subject to matrix effects, both natural and anthropogenic. All these effects together lead to each standing water body having its own plant community and other associated species. In effect, this means that each standing water body has its own value, suggesting that restoration must take place at multiple spatial scales. As conservation action cannot take place everywhere at once, restoration is usually done in sequence. (Illustration: Robyn Symons).

type but rather on improving habitat heterogeneity. Focus must then be on the processes that underpin this heterogeneity, such as variation in hydrologic conditions emanating from differences in seepage, water infiltration and flow.

Risks come when large areas are rewetted all at once or where there is some sort of large disturbance, such as the building of a causeway. The restoration process requires that no temporary or permanent water types are lost. At the same time, larger spatial conservation areas may need to be considered. This may be in the form of improving management outside reserve areas or extending conservation efforts beyond the areas of restoration focus.

Improvement measures may include, for instance, restoring regional hydrology and increasing water infiltration by highly selective logging. Meanwhile, within the reserve, restoration should be temporally and/or spatially graded, one area at a time. This allows species to recover or redistribute in response to the changes implemented. All the while, monitoring should take place to adapt, or even change management type and intensity, with the intention of aiming for high heterogeneity as well as high functional connectivity across the watery landscape (Verberk *et al.*, 2006).

Restoration of wetlands in its broadest sense has two goals: increasing landscape biodiversity while maintaining their functioning as nutrient traps from land to freshwaters. Both goals are dependent on a set of physical, chemical and biological features. A combination of the wetland features, such as shallowness, large surface area and high shoreline complexity, support high levels of biodiversity across taxa, including macrophytes and macroinvertebrates, while also improving nitrogen retention. In contrast, small and deep wetlands are more efficient at phosphorus retention but have less value for biodiversity. This means that when designing new wetlands or restoring degraded ones, area, depth and shoreline complexity have fundamental effects on nutrient retention and biodiversity, and yet at times these effects can be conflicting (Hansson *et al.*, 2005).

A restoration study of the odonate assemblages in a prairie pothole region of Iowa, USA illustrates that homogenous stands of shoreline vegetation are depauperate in terms of both species richness and abundance compared to stands of shoreline vegetation with diverse vertical structure, while the diverse stands also favour species of local conservation concern, i.e. those at risk and uncommon. The number of occurrences of species of conservation concern is four times greater in mixed than in monotypic vegetation. Overall, the monotypic vegetation largely supports very high occurrences of generalist odonate species such as *Ischnura verticalis*, *Lestes unguiculatus*, *Erythemis simplicicollis*, *Libellula pulchella* and *Pachydiplax longipennis*. Habitat specialists restricted to the mixed vegetation include *Amphiagrion saucium*, *Enallagma apsersum*, *Ischnura hastata*, *Aeshna constricta*, *Celithemis elisa*, *Leucorrhinia intacta*, *Sympetrum corruptum* and *S. vicinum* (Mabry and Dettmann, 2010).

In Germany, urban dragonfly species richness depends on habitat heterogeneity and availability, which can decrease in the heart of many cities leading to a homogenized subset of habitat-generalist species, including *Ischnura elegans*. However, some regionally rare species may benefit from urban ponds, with species like *Lestes barbarus* associated with temporary ponds and *Ischnura pumilio* with new ponds. While the regional species pool determines which species are locally present (i.e. they are part of the natural capital), those that do colonize urban ponds are mostly generalists. The implication for dragonfly conservation is that conservation planning in urban areas should not focus on a few rare species but rather on improvement of habitat structure, which includes habitat heterogeneity (i.e. aquatic and riparian vegetation diversity) as well as clean water, even among small ponds. Some anthropogenic disturbance is not necessarily harmful to these assemblages, especially when pond management is ecological rather than unnatural. A further point is that there should be a variety of pond types in terms of water level and hydroperiod, as well as ecological succession stages, to maintain high dragonfly diversity and assemblage resilience across a wide area (Goertzen and Suhling, 2013).

In Spain, ponds in Castilla–La Mancha botanical garden were compared for taxonomic and functional diversity with nearby natural lakes and an urban pond. Compared to natural lakes, the odonate assemblages were less species-rich and their functional diversity was lower in the botanical garden ponds, consisting mostly of habitat generalists. Two reasons for the differences were that the natural lakes supported running water species in the feeder streams and rivers entering them, while also not having the same level of alien fish predation as in the botanical garden ponds. The urban pond was depauperate as it lacked vegetation and had fluctuating water levels (Pinilla-Rosa *et al.*, 2023).

6.5.4 Peatland and bog restoration

Peats and bogs are often ancient and support some rare and threatened habitat-specialist dragonflies. Odonate diversity in Belarus differs according to distance to the water body, wind speed and bog water level, with the highest diversity along lakeshores. There are distinct assemblages associated with lagg zones, lake shores and hollows, although assemblages were similar in pine bogs and open bogs. Differences among assemblages are mainly driven by cold-adapted, highly specialized species, such as *Leucorrhinia albifrons*, *L. dubia*, *Sympetrum danae*, and two abundant generalist species *S. sanguineum* and *S. vulgatum*. As peat bog specialists are among the most threatened of Palaearctic insects, it means that relatively intact Belarusian peat bogs are important conservation refuges for specialists in this type of ecosystem (Sushko, 2021).

In Germany, bogs are important habitats for dragonfly conservation. When restoration of rewetted industrial peat cuts, with or without former agricultural use, are compared to rural peat cuts, restored bogs are as diverse in overall species richness as rural peat cuts. All dragonfly species in the area can recolonize at least some of the nutrient-poor restored bogs. However, the high nutrient content caused by former agricultural use in some bogs prevents the characteristic dragonfly fauna of the bogs from returning. Nevertheless, they are important secondary habitat, especially for transition-bog species. Overall, these findings suggest that restoration measures that create a network of small oligo- to mesotrophic water bodies are the best way forward for maintaining local dragonfly diversity (Krieger *et al.*, 2019).

A military training area in Czechia provides opportunities for dragonfly conservation. Among 48 species in standing water bodies, richness was similar in these areas compared to the surrounding landscape, although there were slightly more species of conservation value in the training area. Habitat type did not influence species richness, though there was a distinct difference in the conservation value of different habitat types. Peat bogs and extensive ponds had the highest conservation value, while artificial pools and farm ponds had the lowest. Species such as *Coenagrion hastulatum*, *Lestes dryas* and *Sympetrum danae* were associated with late successional habitats, especially peatlands which were mostly inside the training area. In contrast, species such as *L. virens* and *S. striolatum* were associated with early successional and temporary habitats, mostly outside the training area (Harabiš and Dolný, 2018).

In Japan, to determine the trajectory of wetland site recovery a species nestedness approach was used. Species potentially capable of inhabiting the restoration site are flagged. Based on detailed knowledge of their habitat requirements, types of habitat, such as bogs and marshes, ponds, or slow-moving water, can be selected for preferential restoration. The methodology involves using species nestedness in a presence–absence matrix to predict which species will recover at the restoration site in accordance with the regional 'rank order of species prevalence' (i.e. a spectrum ranging from being common through to being rare in the region). The method successfully predicted which species over time will likely establish in the various wetland habitat types and highlights which species are so rare and threatened that they become candidates for reintroduction (Kadoya *et al.*, 2008).

6.5.5 Importance of buffers and natural margins

Dragonfly individuals use the margins around ponds and lakes to mature, forage, roost and in some cases, mate. Around a 185-hectare lake in Mississippi, USA, odonates were present and equally abundant 10–160 m from the water's edge, with mature males outnumbering females and immatures at the 10–40 m distance, whereas it was the reverse at the 130–160 m distance. These findings indicate that wide buffer zones around wetlands may be essential to protect odonate assemblages associated with ponds, especially females and sexually immature adults (Bried and Ervin, 2006).

The presence of tall sedges at water margins and submerged macrophytes are the main drivers for significantly increasing dragonfly diversity and species composition complexity at restored ponds in south-western Germany. Restoration measures were solely mechanical involving removal of deadwood, organic sludge, trees and shrubs from the water, and removal of woody vegetation and alien herbs from the riparian zone. This activity led to increased light levels and higher macrophyte cover, such as that of *Chara* sp. and *Potamogeton gramineus* in the water and fast-growing tall sedges in restored ponds. Two zygopteran and one anisopteran species are mainly associated with restored ponds, with *Lestes sponsa* and *Sympetrum vulgatum* occurring exclusively in restored ponds, and only 6% of *Coenagrion puella* individuals in unrestored ponds (Janssen *et al.*, 2018).

Important factors associated with emergent macrophytes relative to dragonflies are the amount and type of cover relative to season in subtropical South Africa. Dragonfly species richness overall increases with sedge cover by up to 70% and bulrush cover by up to 40%, but then decreases above that level. Dragonfly species richness responded to sedge and tall grass/reed cover in spring, but bulrush and sedge cover late summer. Species composition also changes with season, especially in relation to surface macrophytes and bulrush in spring but only surface macrophytes in late summer (Briggs *et al.*, 2019a).

In Taiwan, park ponds under intense human management (for tourism and education) and natural ponds without any management are richer in dragonfly species than ponds used for fish farming, with the natural ponds the richest of all. These natural ponds are also rich in macrophytes and support well-vegetated margins raising the level of microhabitat heterogeneity. In ponds with such healthy margins, dragonfly richness was not affected by surrounding land-use type (Chien *et al.*, 2019).

Among UK ponds, zygopterans more so than anisopterans, rely on both habitat quality, such as pond vegetation and water transparency, coupled with good-quality buffers along pond perimeters. In this case, the buffers need only be 2 m wide. This, however, is likely a minimum width, with buffers 25–30 m wide probably more beneficial for both dragonflies and other freshwater biodiversity, especially as some species like *Coenagrion puella*, *Ischnura elegans* and *Enallagma cyathigerum* spend time away from water, only returning to breed (Raebel *et al.*, 2012b).

Also in the UK, ponds dominated by floating and submerged macrophytes and having the greatest water transparency, support the highest dragonfly abundance and species richness, with associated partial to complete buffer strips around their perimeters. Ponds low in the amount of aquatic vegetation are turbid, yielding no signs of breeding even when they are buffered. The recommendation is that agri-environment schemes should support buffer strip creation and management (Raebel *et al.*, 2021a).

In a Spanish semi-arid Mediterranean area, restoration of traditional water infrastructures encourages increase of aquatic vegetation cover, while appropriate and regular management of riparian vegetation restores natural water conditions that enable the recovery of odonate diversity and abundance, as opposed to a simple increase in species richness. Key restoration activities include the removal of the alien reed *Arudo donax* and ensuring the right quantity and variety of aquatic macrophytes such as *Chara* sp. and *Cladophora* sp. Associated restoration activities include maintenance of the correct water flow and ensuring low concentrations of nutrients and low electrolytic conductivity, as well as regular control of sediment and nutrient accumulation, alongside stabilization of hydrological conditions to prevent extreme water flows. Finally, limiting tourist and recreational access, and designating restricted biodiversity conservation zones to reduce anthropogenic pressure, also contribute to the recovery of dragonflies and other threatened taxa (Monzó and Verdú, 2022) (Fig. 6.23).

6.5.6 Ecosystem change associated with stream impoundments

6.5.6.1 Effects of small impoundments across streams

In the European Alps, mountain impoundments to store water are being built to overcome a lack of snow during winter. However, compared with natural ponds, dragonfly species richness in these reservoirs is lower. The species present in these reservoirs constitute a subset (33% by species numbers) of the regional dragonfly fauna, with no species unique to the reservoirs. Nevertheless, these new water bodies likely may assist certain species in their shift toward higher elevations by providing stepping-stone habitats. However, these new habitats need to be of good quality, including a wide variety of microhabitats, be protected from pollution, and function hydrologically like natural ponds, otherwise they may become ecological traps (Fait *et al.*, 2020).

A fine-scale study of odonate microhabitats at an impounded stream creating a dragonfly conservation pond in South Africa showed that 23 odonate species can occupy a rural pond of just 400 m^2. Among the occupiers were two national endemics *Chlorolestes tessellatus* and *Allocnemis leucosticta*, both requiring slow-flowing water and dappled shade. The main drivers were aquatic and emergent macrophytes, as well as various levels of shade from the riparian vegetation to small differences in water velocity. The conservation message here is

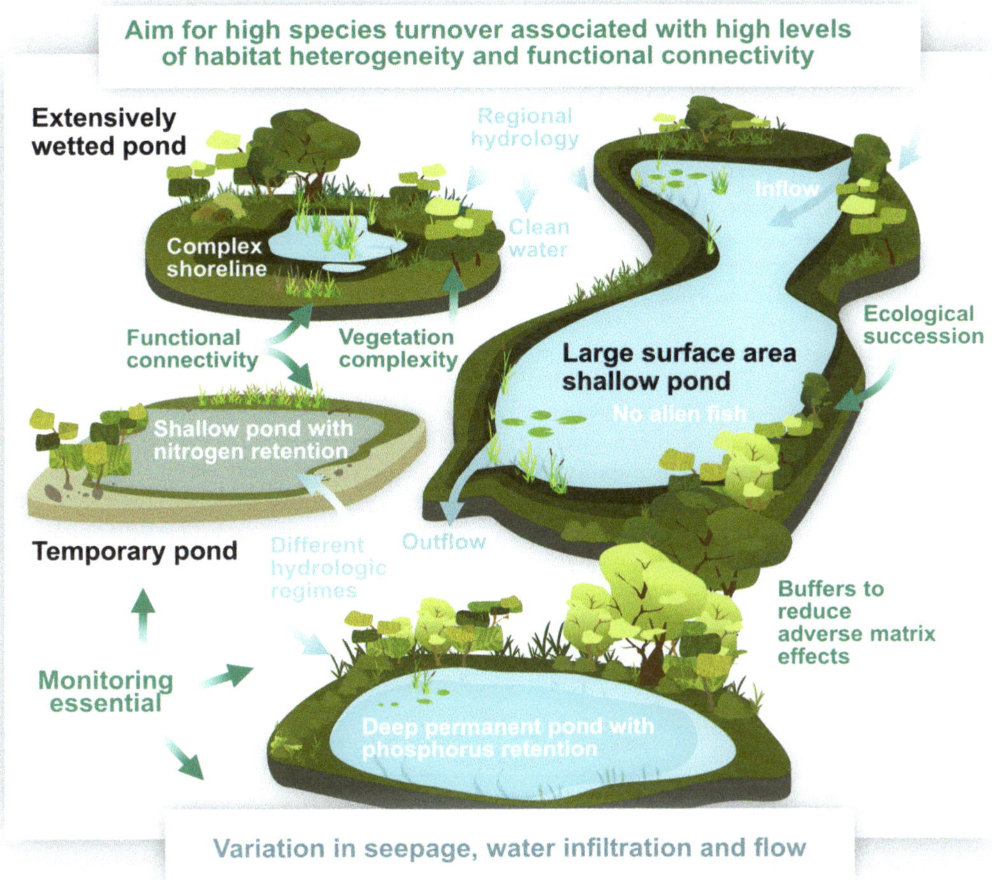

Fig. 6.23. In any one local area, conservation and restoration aims for high spatial species turnover associated with high levels of habitat heterogeneity. Natural population dynamics and species sorting then occurs when there is high functional connectivity within and among the ponds. Maintaining regional hydrology is important, which includes always having clean water. Shown here are some of the various restoration and management approaches used to conserve dynamic standing waterscapes. Monitoring is essential to establish the level of success and to point to any intervention management that may be required. (Illustration: Robyn Symons).

that on a base of clean water and provision of a wide range of microhabitats, species from the matrix will come (Samways *et al.* 1996a).

6.5.6.2 Managing impounded streams for encouraging dragonfly awareness

An important implication for dragonfly conservation is that the restoration of wetlands should aim for shallowness and much heterogeneity, while also being aware that ecological succession may gradually fill up the water body with silt and plants.

In South Africa, a stream in the KwaZulu–Natal National Botanical Garden (Pietermaritzbug), which curved around the base of a wooded hill, was impounded to create an artificial lake (550 m circumference, 2.4 ha and no more than 2 m deep) (Kingfisher Lake) specifically for rare water plant conservation and to further enhance a dragonfly awareness trail (Suh and Samways, 2001, 2005). It was decided to site a trail at this botanical garden as it was accessible to many urban people while also being overall extremely rich in dragonfly species, owing to its great variety of microhabitats and supporting both higher and lower elevation species, all of which are highly visible to human visitors. The artificial lake also attracted many standing water species that would otherwise not be present. It was also part of an initiative by the South African National Biodiversity Institute to encourage awareness of dragonflies present in botanical gardens (Willis and Samways, 2011). The trail consists of various loops with 'view points' that have species information boards. Leaflets were made to provide further information, but popularity demanded that a more comprehensive booklet be made (Willis and Samways, 2013) (Fig. 6.24).

When this artificial lake was instigated, the aim was to maximize habitat heterogeneity. Factors like sunlight versus shade, various degrees of still versus flowing water, and a wide range of plant species and structures were a priority. Within months of the lake filling with water, dragonfly species richness increased from 12 to 26, which included habitat-specialist endemics like *Chlorolestes tessellatus* and *Pseudagrion hageni hageni*. Only one species, the common and widespread *Paragomphus cognatus* was lost to inundation, although it remained in typical abundance immediately above and within 80 m of the impoundment downstream (Steytler and Samways, 1995).

The lake was studied over a 32-year period since inception. During that time, there was vegetation overgrowth and natural infilling by inorganic and organic material, followed by overgrowth by emergent macrophytes. Nevertheless, the stream continued to cut a path through this overgrowth close to its original course. This led to a return to a largely running water dragonfly fauna with a few standing water species remaining in the marshy overgrowth area.

Re-establishment of the lake according to its intended aim of creating microhabitat heterogeneity for both plants and local dragonflies, then required management intervention in the form of dredging, removal of alien plants and selective planting of indigenous marginal vegetation, alongside maintenance of the inflow and outflow regimes to sustain the intended complement of dragonfly species, which was achieved (Kietzka *et al.*, 2021a) (Fig. 6.25).

Shallow ponds in the Netherlands were restored by dredging, reducing fish biomass, softening the

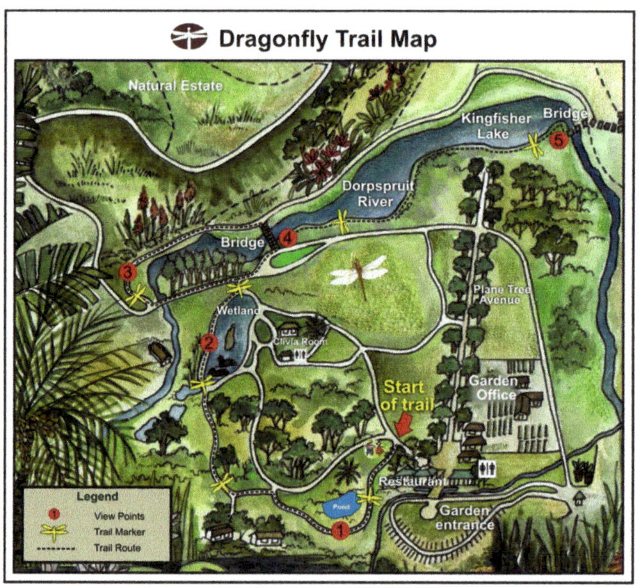

Fig. 6.24. The Dragonfly Trail Map for the KwaZulu–Natal National Botanical Garden, Pietermaritzburg, South Africa, with five recommended stopping-off points (each a 'View point') with dragonfly species information boards, two of which are beside the artificially created Kingfisher Lake (top right). The small weir is labelled 'Bridge' and has a walkway for visitors to cross. (Illustration: Linette Ferreira and reproduced with permission from the South African National Biodiversity Institute).

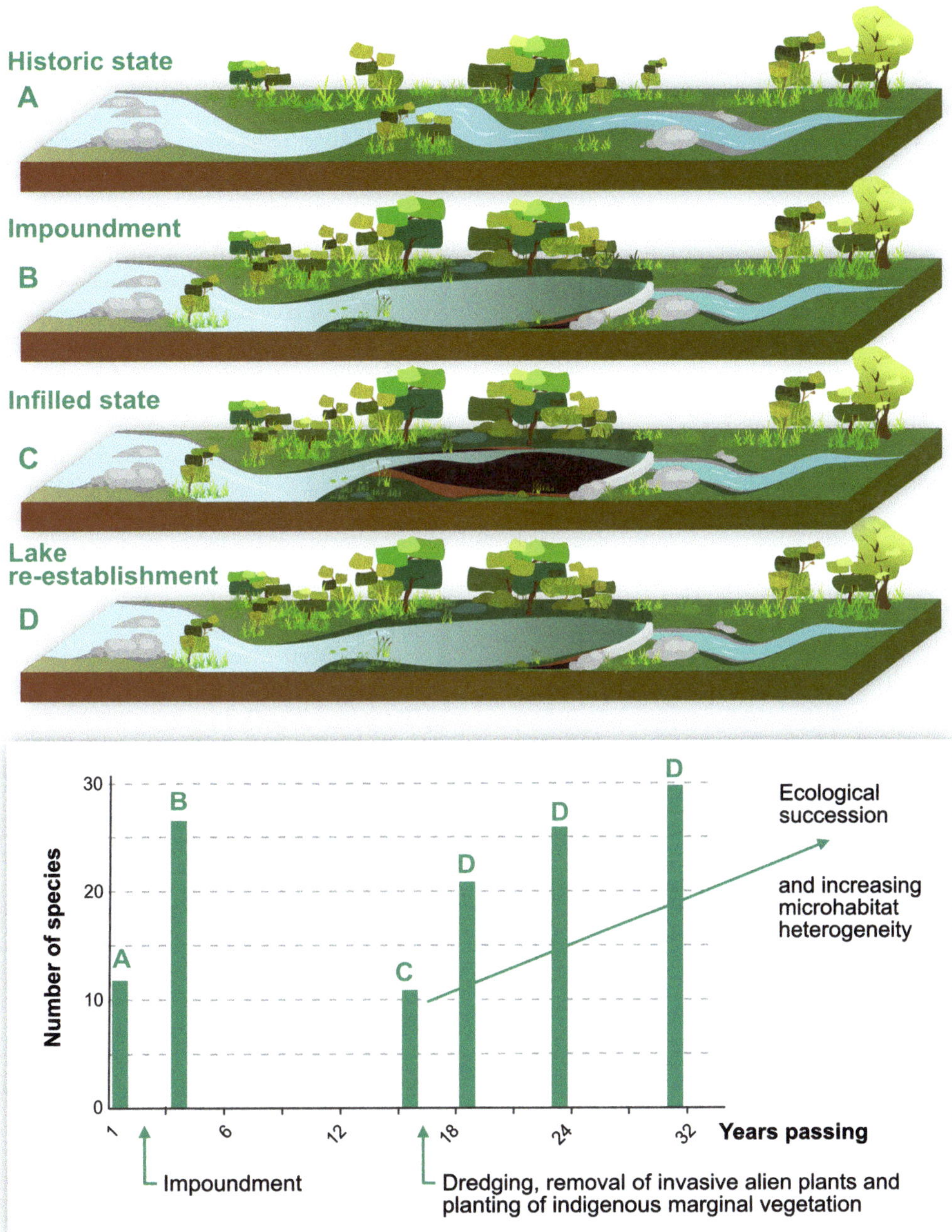

Fig. 6.25. One of the main reasons for this stream impoundment was to create the shallow Kingfisher Lake to locally increase the variety of microhabitats and attract a wider range of dragonfly species. This was part of an initiative in the KwaZulu–Natal National Botanical Garden to enhance dragonfly conservation awareness among visitors to the garden. The historic state of the stream (A) supported 12 species, mostly running water species. After impoundment with a small weir, leading to the formation of the lake (B), within a few months, the number increased to 26 species. As the years passed, the number of species dropped considerably, a result of silt and organic matter accumulation, followed by overgrowth of pioneer marshy vegetation. Meanwhile, the stream still cut a course through the overgrowth close to its original path (C). The lake was then dredged, cleared of invasive alien plants and landscaped with indigenous vegetation. From that point onward, and with ecological succession, the number of species increased considerably (D). (Illustration: Robyn Symons).

banks, planting of macrophytes and importantly by reducing external nutrient load, especially from stormwater run-off (Lürling *et al.*, 2023).

6.5.6.3 Restoration of beaver wetlands

Beavers (*Castor canadensis* in North America and *C. fiber* in Eurasia) are important ecosystem engineers that dam streams and small rivers, as well as gnawing down smaller diameter trees, especially within 10 m of the water's edge (Trentanovi *et al.*, 2023). Their activity in Canada has a profound effect on local ecosystems by decreasing water velocity where dams are constructed, increasing the volume-to-surface ratio, and by digging their 200–300 m long foraging channels, they increase wetland perimeters by 575%. This activity increases ecosystem complexity, landscape heterogeneity and functional connectivity (Hood and Larson, 2015).

A group of ponds in Virginia, USA once had the most diverse, abundant and biogeographically interesting Odonata fauna, including the most boreal and state-rare species of any tributary in the area. However, a long-term study in the early 2000s, showed that beavers had abandoned the local area, possibly due to reduced availability of live hardwood trees, greatly reducing the number of odonates present in a few small remaining ponds (Roble *et al.*, 2009).

As regards invertebrates, there are two important perspectives: how beaver dams change stream invertebrate communities, and the creation of wetland habitats within and beyond the boundary of the streams. Beta diversity among invertebrates of beaver wetlands in the USA is nearly twice as high as in any other individual successional stage, with each stage having a particular taxonomic composition. This means that beavers have some ability to stabilize beta and regional diversity in the face of wetland loss and climate change (Bush *et al.*, 2019). This is encouraging for dragonflies, especially as beavers on both continents are being reintroduced to parts of their former geographic ranges. Dragonflies and their food base, both in the water and on land, are being secured along with rewilding with beavers.

Leucorrhinia glacialis and *L. hudsonica*, although common in Canada and the north-eastern USA, are rare in central Appalachia, USA, where they are isolated at the extreme southern edge of their geographic range. The Appalachian wetlands are permanent, acidic and fishless, and the dragonflies rarely disperse far from their natal ponds. Most of the wetlands containing these species were created by beavers, with some ponds rapidly shrinking due to a lack of dam maintenance by the beavers. The isolation and rarity of habitats supporting both *L. glacialis* and *L. hudsonica* in the area, along with their poor dispersal abilities and apparent reliance on beaver activity, suggests an uncertain future for these locally rare dragonfly species. Protection of these rare habitats and populations, along with possible dragonfly translocation into these dams along with beaver management, may help sustain them (Liberto and Studinski, 2021).

In Scotland, the reintroduction of the European beaver with its water engineering capabilities has resulted in positive ecological effects, including on degraded freshwater systems and on macroinvertebrates in the region. Construction of features that may prove difficult to replicate by conventional methods of habitat creation suggests that in the future, beavers could contribute more widely to the restoration of freshwater habitats. This though, assumes that suitable strategies are in place that harmonize beaver conservation with human aspirations for land use (Law *et al.*, 2016).

Following introduction of the beaver into Argyll, Scotland in 2009, it has since colonized some new sites, while abandoning others. The effects of the beaver on local ecosystems are a complex interaction of weather, water levels, vegetation dynamics and flow dynamics, all of which vary over time. Over a few years, the effect of beaver activity on *Brachytron pratense* have been neutral to possibly slightly negative. As regards other odonate species, there have been no changes in the number of species present at localities with beaver activity, although *Aeshna cyanea* has colonized other sites in the area most likely due to climate change (Batty, 2017). Reintroduction of the beaver into Cornwall, UK, has led to increased landscape and waterscape heterogeneity now supporting 17 species of odonates (Jones and Harris, 2022).

Following the loss of beavers in Sweden, it was reintroduced between 1922 and 1939, since when it has increased substantially. In terms of macroinvertebrates overall, there is no difference in diversity above and below beaver dams, with effects similar over a wide geographical area. However, pond age and size has had a large effect on species diversity. Young ponds are the most diverse in comparison with medium-aged ponds. Small ponds are significantly more diverse than medium-sized and large ponds (Johansson, 2014).

Other findings in Sweden show that in their natural range, beavers create ponds which may seem to be like other shallow wetlands. However, they differ subtly in their physical characteristics and disturbance effects. Beaver wetlands have specific assemblages that differ from other wetlands and are not simply a subset of other assemblages. This is especially so for plants and also for insects such as water beetles. These points indicate that beaver wetland restoration provides a highly feasible, valuable and complementary way forward to conserve local wetlands (Law *et al.*, 2019).

After 10–15 years, beaver reintroduction into the Hürtgenwald woodland area, Germany, saw a greatly elevated local odonate species richness. Beaver wetlands supported 29 species, dams abandoned by beavers within the last 1–3 years supported 14 species, and ponds without beavers supported only 4 species. Some beaver wetland species are locally rare, including *Ceriagrion tenellum*, *Cordulegaster boltonii*, *Orthetrum coerulescens*, *Leucorrhinia pectoralis* and *L. rubicunda* (Schloemer *et al.*, 2013) (Fig. 6.26).

6.5.7 Standing water-body size and waterscape heterogeneity

6.5.7.1 Comparison of ponds and lakes

According to the species-area hypothesis, larger water bodies would be expected to support the most species. Furthermore, small water bodies are often more iso-

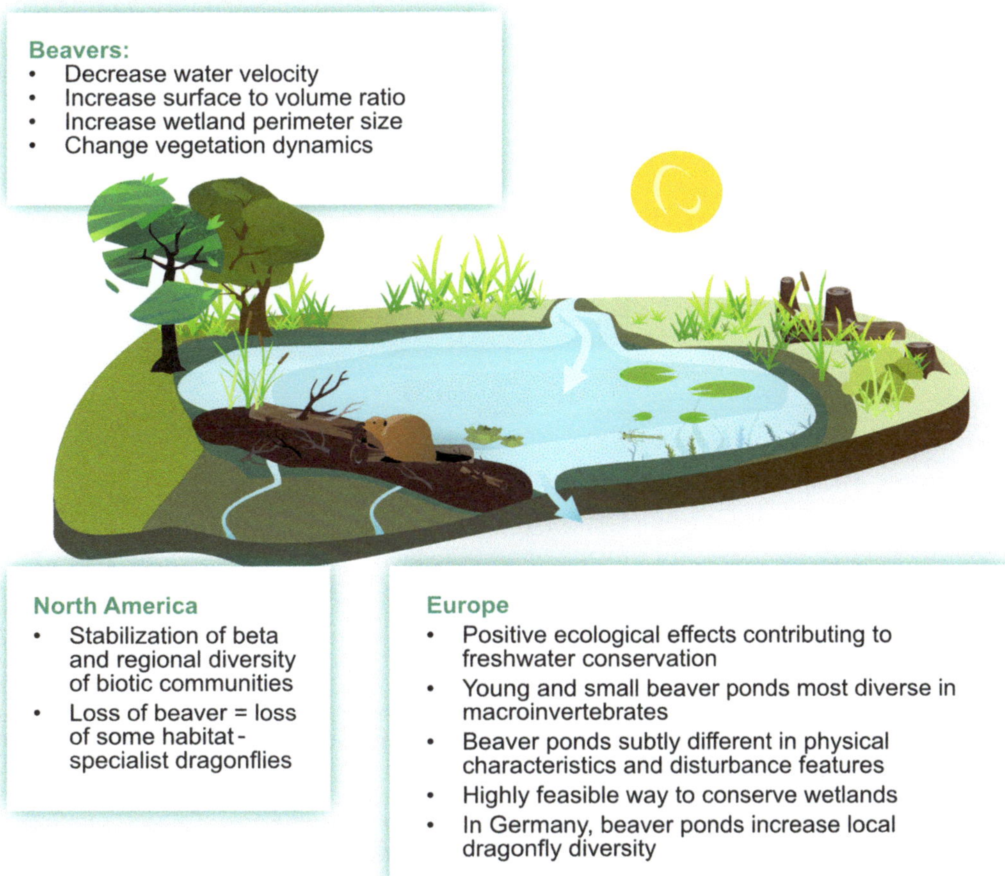

Fig. 6.26. The two beaver species, one in North America and the other in Europe, are important ecological engineers that have major positive local effects on water and vegetation dynamics. Evidence so far indicates that beaver rewilding, especially in Europe, is beneficial for biotic, macroinvertebrate and dragonfly diversity. Overall, beavers are good for freshwater and dragonfly conservation and restoration. (Illustration: Robyn Symons).

lated, are difficult to colonize and are expected to have higher extinction rates, all of which suggest smaller water bodies will have low local diversity. But the species-area hypothesis does not necessarily stand up in the case of standing water bodies.

Among European mountain lakes related to a size of about 2 ha there are ecological thresholds in aquatic communities. Yet, the question of ponds versus lakes is not simply about size and isolation, as in the case of many terrestrial ecosystems, but relates to great differences in abiotic variables. The most obvious case is where the water body edge is shallow and often warm, whereas the centre may be deep with characteristic thermal water layers. Furthermore, smaller ponds regionally support the highest plant and macroinvertebrate diversity relative to other water body types. This may be due to factors like higher beta diversity, greater variety of pond types over small spatial scales and small catchment areas with a wide range of physicochemical conditions across a region, promoting higher biodiversity at the regional scale.

In high mountain areas in Europe, species turnover between standing water bodies is driven by greater environmental variability and degree of pond isolation. High elevation ponds do not support greater regional diversity, nor do they have a higher number of unique taxa compared to lakes, but nevertheless the high among-site diversity of ponds relative to lakes emphasizes the relevance of ponds for regional diversity (Hamerlík *et al.*, 2014).

6.5.7.2 *Effects of water-body size*

The debate around pond size needs much focus as it is central to dragonfly conservation (Fig. 6.27).

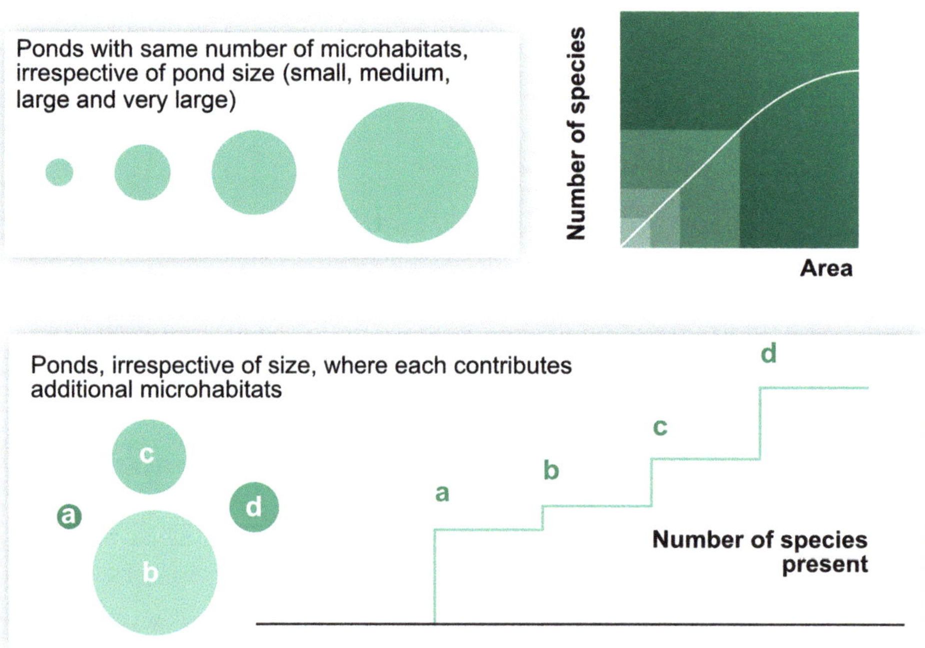

Fig. 6.27. Typically, among terrestrial systems, there is an increase in the number of species present the larger the area. This increase is due to more microhabitats being included, with each one able to support different species, overall leading to the species accumulation curve climbing. However, there is a finite number of microhabitats and hence species that can be supported, and so the species accumulation curve tends to flatten out (top right). Ponds are slightly different, with size on its own not necessarily contributing more species where there are the same number of microhabitats per pond (top left). But as ponds can vary greatly in number and types of microhabitat, irrespective of pond size, some small ponds can be richer in species than some larger ponds. When all these different ponds are considered together, there is an increase in the number of species present as each pond is considered in turn, but the increase is in steps, according to which pond is next included (bottom). (Illustration: Robyn Symons).

Large ponds (less than 2 ha) and particularly lakes (more than 2 ha) usually have zones related to water depth, ranging from the shallow littoral zone to the deep benthic zone when the focus is on the substratum. Several studies in different countries and different contexts have found that an increase in pond size leads to higher odonate diversity (Osborn and Samways, 1996; Oertli *et al.*, 2002; Kadoya *et al.*, 2004; Ruggiero *et al.*, 2008; De Marco *et al.*, 2014; Jooste *et al.*, 2020) while others have found that it does not (Carchini *et al.*, 2007), and occasionally leads to a decrease in species diversity (Goertzen and Suhling, 2013; Chien *et al.*, 2019). These different findings do not necessarily contradict each other as the standing water ecosystem is both complex and varies greatly from one geographical area or context to the next, and each pond can have very different features.

There are several issues here. First, pond size can matter for certain species, but varies among taxa. We must view this against odonates requiring prey and having predators. Second, most odonate larvae require some sort of botanical substrate, although burrowing gomphids, for example, are an exception. We see this at artificial lakes in South Africa where scuba observations show that odonate larvae are confined to the plant-rich littoral zone and do not live at depths greater than 1.2 m where there is only sediment and some algae (Samways *et al.*,1996b). Third, increasing water depth toward the centre of a pond or lake often increases the open water zone (where light might not penetrate: the limnetic zone) relative to the littoral zone which is well lit and plant rich. Fourth, even when a pond is shallow, there may be great differences in physicochemical and biotic conditions with various degrees of biological homogeneity versus heterogeneity of the water body. Fifth, adult odonates require terrestrial areas where they can emerge as exuviae, mature, perch for foraging, rest at night, mate and disperse. All species require this terrestrial realm for at least one of these behaviours. Sixth, there is much more chance that a small pond will be more sheltered from high winds and storms than a large lake in the same terrain. However, small ponds in dry geographical regions are more susceptible to drought unless they have a reliable perennial water inflow.

In the hot lowlands of Zululand, South Africa there is an interplay between pond size and type of matrix. While odonate species richness is similar among various ponds, abundance is highest at large ponds regardless of natural terrestrial patch size. In turn, zygopteran functional richness and diversity is driven by pond and natural patch size, suggesting that they are sensitive to landscape fragmentation. In contrast, anisopterans are more resilient to fragmentation, selecting suitable habitats mostly on water chemistry and vegetation characteristics. Overall, large ponds are occupied by different odonate assemblages compared to small ponds, with occupancy strongly associated with mobility traits. This emphasizes that ponds in both small and large natural terrestrial patches have conservation value. From a conservation point of view, a pondscape that represents various pond sizes and terrestrial contexts is important for maintaining the potential odonate diversity in the area. Furthermore, adjacent natural terrestrial habitats also connect nearby aquatic habitats, enabling odonates to move across the landscape in response to natural and artificial drivers (Deacon *et al.*, 2024).

There is another reason why pond size is important, and that relates to practical aspects of conservation. Among Swiss standing water bodies with a mean size of 8817 m^2, dragonfly species richness increases with pond size. However, it does not mean that all ponds must be as big as possible to promote dragonfly diversity. The reason for this is that there are confounding factors like pond depth, vegetation type, structure and density, physicochemical conditions of the water, level of microhabitat heterogeneity, substrate type, and so on. Maintaining the small Swiss ponds, promotes species richness and/or richness of threatened pond species in the region, which on the face of it is contrary to what would be expected from biogeographic theory. In general, pond conservation, whether protection, restoration and/or management, is about promoting conservation of all ponds, regardless of their area. As creation of small ponds is a relatively easy task, the fruitful way forward for all interested parties, whether farmers, local authorities, nature conservation organizations or schools, is to promote and care for a wide variety of ponds, and ideally ponds should not be too far apart (Oertli *et al.*, 2002).

6.5.7.3 Subsidence ponds

Under natural conditions it is not unusual for sinkholes to form because of land collapse. In Virginia, USA, acidic wetlands are formed from localized dissolution and collapse of karst terrain (dolomite and limestone) followed by inflow of alluvial material from nearby mountain slopes. These sinkhole ponds and marshes can be remarkably rich in dragonfly species, including isolated populations of *Celithemis martha* and *Nehalennia integricollis* (Roble, 1999).

6.5.7.4 *Extreme weather events and pond dragonflies*

The Cape Floristic Region at the southern tip of Africa has a Mediterranean-type climate (cool wet winters and hot dry summers) and its ponds, whether natural or artificial, have greatly fluctuating water levels, and more so than in many tropical or temperate climates. However, in this region, there appears to be a long history of adaptation by dragonflies to naturally extreme seasonal weather conditions and fluctuating water levels, with drought events also affecting water physicochemical conditions.

In the Cape Floristic Region, some larvae are most abundant in ponds with high conductivity, such as *Crocothemis* and *Ischnura*. Most other genera occupy ponds with relatively low conductivity, leading to shifts in larval dragonfly species composition. Along the pH gradient, *Crocothemis* larvae prefer alkaline conditions, while *Ceratogomphus* and *Trithemis* are in ponds with close to neutral pH. Other genera, such as *Africallagma*, *Ischnura* and *Anax*, occur across the whole pH gradient.

Biotic factors are also important. Both adults and larvae increase in species richness with higher emergent marginal vegetation cover. In turn, adult species richness, but not abundance, increases with conductivity, percentage of submerged vegetation cover and percentage of emergent vegetation cover, although it is highest where there is less than 20% emergent vegetation cover. In turn, larval species richness increases slightly with increasing percentage of reed cover, although their abundance decreases with an increase in percentage of reed cover and water temperature, but then increases with increasing conductivity. Overall, the Cape Floristic Region dragonflies are well-adapted to water level fluctuation in this drought-prone region. The conservation message here for conserving a rich dragonfly assemblage is to retain a wide range of ponds with different climate-related dynamics. The range includes both constant or fluctuating water levels, and where there is diverse aquatic and marginal vegetation, as well as environmental gradients related to elevation and topographic influences (Jooste *et al.*, 2020) (Fig. 6.28).

In countries like South Africa where there are great differences in pond types owing to huge differences in local climate (including seasonality of rainfall, condensation events and extreme weather events) and topography, dragonflies inhabit a wide range of water bodies. Some species are specialists relative to greatly changing weather events. *Bradinopyga cornuta* breeds in pools in large boulders that have been hollowed out by small rotating boulders over very long periods of time when the river is in spate (Fig. 6.29). *Pantala flavescens* occupies very small temporary pools following rainfall events. *Hemistigma albipunctum* larvae will leave a habitat where the water level drops to the substrate to emerge as adults as a marsh dries out. At the other extreme, are swamp specialists like *Zyxomma atlanticum* that occupies swamp forest channels that draw down to a series of pools in the dry season. In the same habitat, *Gynacantha usambarica* requires this drop in water level to lay its eggs in the mud.

6.5.8 Natural pond complementarity with artificial ponds

6.5.8.1 *Artificial ponds as complementary enhancers of dragonfly assemblages*

For many years it has been known that artificial ponds, especially those used for local water retention for farming, can play a major role in supporting dragonfly populations. This is especially so when still water species are abundant across a region and the landscape is dominated by fast flowing and seasonal rivers and where ponds are well-established and rich in aquatic and marginal vegetation. In short, good-quality farm ponds can act as local nature reserves.

This is the case between 600–1800 m above sea level (all at the same latitude) in the Drakensberg mountains in South Africa where artificial ponds (farm ponds) at these elevations provide sheltered and attractive habitats that are otherwise lacking. Four Zygoptera (*Elattoneura glauca*, *Ischnura senegalensis*, *Pseudagrion hageni* and *P. massaicum*) (out of 17 in the area) and seven Anisoptera (*Acisoma inflatum*, *Pineyschna subpupillata*, *Brachythemis leucosticta*, *Ceratogomphus pictus*, *Crocothemis erythraea*, *Orthetrum hintzi* and *O. trinacria*) (out of 23 in the area) were locally present only in the farm ponds.

Fig. 6.28. In the Mediterranean-type climate of the Cape Floristic Region, South Africa, ponds and other wetland-type habitats fluctuate greatly between the cool wet winters and hot dry summers. However, the amount of fluctuation depends on various factors, especially water replenishment by perennial feeder streams coming from the mountains. The local dragonfly fauna is well-adapted to the shifting hydrological, physicochemical and vegetation dynamics. The conservation message is to conserve as many natural standing water bodies as possible, especially perennial streams and ponds where local endemic species are sentinels. Shown here is a perennial feeder stream and pond with two of its sentinel endemic species, *Chlorolestes umbratus* (left) and *Elattoneura frenulata* (right). (Photos: Michael Samways).

Fig. 6.29. Some dragonflies are habitat specialists but that does not necessarily mean that they have a small geographic distribution or that they are threatened. For example, *Bradinopyga cornuta* is an extreme habitat specialist that breeds in rock-scoured potholes along fast-moving savanna rivers in southern and eastern Africa. (Photos: Michael Samways).

However, elevation influences the attractiveness of the ponds versus natural habitats. At all elevations, most or all species occurred at least in artificial ponds. However, it was only at the higher elevations (1501–1800 m) that all the species occurred in at least natural ponds. At middle elevations (901–1200 m), 78% occurred only in natural ponds. The emphasis here is that while the artificial ponds at the high elevations (1501–1800 m) add little to the local dragonfly fauna, at lower and intermediate elevations (601–1500 m), well-established and well-vegetated ponds play a major role in increasing the regional area of occupancy for many species (Samways, 1989) (Fig. 6.30).

In the South African Drakensberg at 640–1100 m above sea level, a set of relatively small and newly created artificial ponds attracted dragonfly assemblages where composition was strongly determined by certain abiotic and biotic environmental variables. Importantly, these variables overruled any assembly rules associated with interspecific competition. However, larger ponds were not more species-rich than smaller ponds because factors such as water quality, vegetation type and microhabitat diversity, acting together, overrode pond size. Species richness was greatest at shallow, well-vegetated ponds with clear, oxygenated water. Sunlight/shade and marginal/submerged vegetation gradients were the main drivers of assembly patterns at the ponds. In turn, assemblage patterns among the ponds were influenced by different variables representing different ecological successional stages (Osborn and Samways, 1996).

Importantly, the many artificial ponds (about 0.5 million in South Africa), needed for human water security, have greatly changed the local distribution patterns of surface freshwater across the region, increasing the area of occupancy for many dragonfly species, mostly highly vagile habitat generalists (Samways, 1989). Furthermore, some national endemics, such as *Africallagma sapphirinum*, breed in some of the larger reservoirs. In times of drought, some artificial ponds decrease to little more than a series of large turbid puddles, with, for example, *Pseudagrion commoniae* still remaining. Overall, dragonflies have benefitted from the artificial ponds and reservoirs, enabling population resilience during droughts (Samways *et al.*, 2020) (Fig. 6.31).

A highly transformed area in western Spain has compromised dragonfly diversity. However, farm ponds in the area support dragonflies from various other water types (rivers, streams, and reservoirs and other farm ponds) in an area where permanent water bodies are scarce. Overall, the highest dragonfly diversity and abundance is in farm ponds, which can also function as stepping-stone habitats for individuals seeking more optimal habitats (Paz *et al.*, 2021).

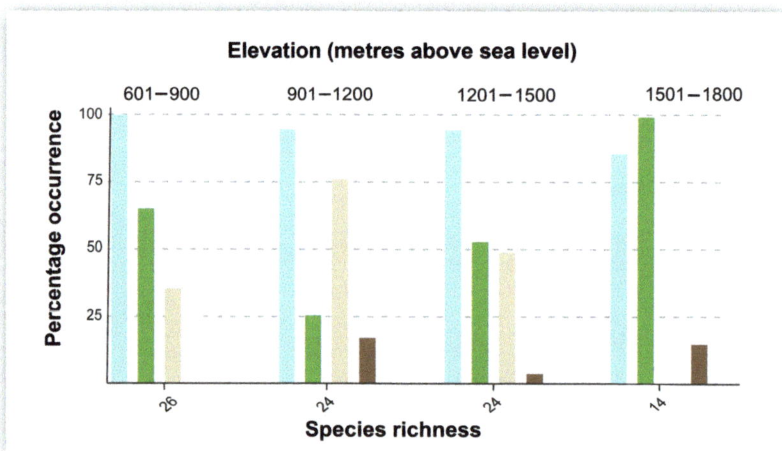

Fig. 6.30. Artificial ponds (here farm dams) play a very important role in supporting local odonate assemblages. However, the extent to which they do so depends on elevation. Given here is the odonate species richness at four different elevation zones in the Drakensberg mountains, South Africa. Here, blue bars represent percentage species occurrence in farm dams, green in natural ponds, light brown in farm dams but not natural ponds and dark brown in natural ponds but not farm dams. At all three lower elevation zones, farm dams have higher species occurrences than in natural ponds. The height of the blue bars across all elevations emphasizes the value of farm dams for increasing the area of occupancy for many species. The light brown bars at the two middle elevations emphasize the complementary value of farm dams relative to natural ponds. (Based on Samways (1989); Illustration: Robyn Symons).

Fig. 6.31. It is well known that many opportunistic odonate species readily colonize certain artificial water bodies. However, some highly localized species may also do so. This reservoir at about 1000 m above sea level supports a strong resident population of the highly localized South African endemic *Africallagma sapphirinum* along its margins. (Photos: Michael Samways).

Various artificial water bodies, such as small stream impoundments, flooded quarries, flooded gravel pits and flooded sandpits in Slovakia supported 44 odonate species, including several red-listed ones, such as *Epitheca bimaculata*, *Leucorrhinia caudalis* and *L. pectoralis*. Although species richness was greatest at impoundments, these largely comprised widespread habitat generalists, with the rarer and threatened species mostly in the flooded pits (Balázs *et al.*, 2022).

Dragonflies in drought-prone areas are known for their rapid development, an adaptation for seasonal aquatic habitats. They also often remain at natal ponds as the subsequent winter rains are likely (Suhling *et al.*, 2009). In very arid areas, artificial ponds used for irrigation and domestic water supply can be local areas of high dragonfly diversity. This is the case in arid central Iran where artificial water bodies support not so much a homogenized dragonfly fauna as an arid region-adapted one, with some characteristic species in the genera *Lestes* and *Sympecma*. These water bodies also show no increase in anisopterans over zygopterans. As these water bodies are largely on private land, they are inadvertent dragonfly reserves where the species benefit from human water impoundment schemes (Barzoki *et al.*, 2020).

The Cape Floristic Region global biodiversity hotspot, South Africa, while rich in endemic dragonfly species, has few natural ponds yet many artificial ponds, mostly used by local farms for irrigation. Although the artificial ponds support relatively few rare or threatened dragonfly species, they have increased the area of occupancy and population sizes of a lot of generalist species, many from nearby natural deposition pools or from unknown local sources. However, some local (*Chlorolestes conspicuus*, *Elattoneura frenulata*, *Pseudagrion furcigerum*, *Orthetrum julia capicola*) and national (*Pseudagrion draconis*, *Pinheyschna subpupillata*) endemic species were also recorded in the artificial ponds. Although having small biogeographic ranges, they are locally common species. Overall dragonfly assemblages, as well as anisopteran and zygopteran assemblages separately, in artificial ponds differed from assemblages in deposition pools, indicating that artificial ponds are to some extent novel ecosystems with arguably hybrid assemblages. Habitat type, elevation and temperature were significant drivers in structuring overall species assemblages. The environmental drivers for anisopteran and zygopteran species composition differed. For anisopterans, riparian vegetation and the level of landscape connectivity was important, while temperature was not. For zygopterans, river catchment, habitat type and water temperature were most important. Artificial ponds likely act as stepping-stone habitats across the fragmented landscape, so long as they are managed in such a way as to have perennial and constant water levels along with high levels of vegetation complexity and heterogeneity to conserve all species, both habitat generalists and some specialists (Simaika *et al.*, 2016).

In South Carolina, USA, streams and rivers have a nested subset of dragonfly species and fewer individuals than lakes. Disturbed barren land and grasslands reduce species diversity. Yet at large scales, transformed landscapes had mixed effects. Small, isolated habitats in urban areas had a depauperate and nested subset of species compared to larger natural areas. Importantly however, odonate abundance and diversity was highest at artificial lakes and ponds, indicating the value of these artificial water bodies for supporting local odonate diversity (Worthen and Chamlee, 2020).

Across three mountain ranges in Czechia, *Somatochlora alpestris* now breeds in artificial habitats of peat pools created by the movement of heavy machinery on now abandoned forest roads. Some of the pools have been colonized in high numbers, with tens of larvae of different instars and exuviae, as well as an abundance of adults. Successful colonization of the species was largely due to closeness of source sites, optimal habitat conditions and reduction in frequency of disturbance (Kašák *et al.*, 2023).

Niche packing of ponds can also be viewed across different ponds. Among a suite of temporary and permanent ponds in the Pampas biome, Brazil, where there is a total of 82 species distributed across 73 permanent and temporary lakes and ponds, species occupancy ranges between 1–54 per water body. About 70% of the species occurred in less than 10% of the water bodies, with many species being rare, and often making up the majority in many of the assemblages, with no difference in species richness between temporal and permanent water bodies (Renner *et al.*, 2020).

6.5.8.2 Managing ponds

A group of 37 ponds in south-western France supported 40% of regional dragonfly species richness, including some rare species. The ponds were variable in terms of what they were used for, including cattle watering, irrigation, duck farming, aesthetic reasons, bait farming or were abandoned, and were surrounded by a wide range of landscape types which did not affect species richness. Ducks, however, did reduce dragonfly species richness. The conclusion was that each species in turn had a particular preference, or not, for a particular set of conditions present in each pond (Ruggiero *et al.*, 2008).

In the Barcelona area of Spain, where rivers are in poor condition, small ponds greatly enhance local dragonfly diversity. The creation and management of ponds and the restoration of running and still water habitats in short river stretches are a possible cost-effective approach to increase general dragonfly species richness and abundance. Although ponds are not optimal habitats for riverine species, they could provide stepping-stone or interim habitats. In turn, good pond management will safeguard habitat heterogeneity, especially by having a range of pond types. A wide range of odonate species can also be encouraged through the presence of macrophyte-dominated early and mid-successional stages to avoid ecological succession through to dry habitats (Maynou *et al.*, 2017).

A large (40 km shoreline) water storage reservoir in the highlands of Poland supports about 26% of the nation's total recorded odonates. Important features include a shallow littoral zone with much vegetation heterogeneity both in the water and along the banks. This is a good example of where human requirements go hand in hand with maintenance of dragonfly diversity. However, there is always room for managerial tweaking of the habitats, especially in support of rare and threatened species, such as monitoring plant ecological succession and adjusting for optimal botanical diversity. With the increasing demand for water storage (as well as flood amelioration), new reservoirs can use these findings for optimal design for both humans and dragonflies (Buczyński *et al.*, 2022).

A group of constructed agricultural wetlands in Finland host 17 habitat-generalist odonate species, making up about 53% of the local still water species and around 40% of all species occurring in the region. Odonate abundance was positively associated with the aquatic vegetation structural diversity, while species richness increased with floating macrophytes and water transparency. This suggests that improving heterogeneity of aquatic vegetation should be enhanced when planning and managing wetlands for high dragonfly diversity. While protecting natural wetlands is of primary importance, constructed wetlands have an important role to play in maintaining wetland diversity, and must be an important consideration for policy makers (Huikkonen *et al.*, 2020).

Similar findings come from a fragment of the threatened Seasonal Deciduous Forest (Atlantic Forest biome) in Rio Grande do Sul, Brazil. The number of dragonfly species (20 in streams and 45 in farm ponds) equals about 26% of the known species in the province. This dragonfly diversity is associated with high plant diversity and many wetlands in the region act as source habitats (Pires *et al.*, 2019).

While there are no natural standing water habitats in Papuk Nature Park, Croatia, the presence of many artificial ponds, such as fishponds, gravel pits and reservoirs offer the opportunity for occupation by many standing water species. Old artificial ponds in the area are especially important habitats, especially as over the years the environmental conditions in these systems have become much closer to natural lake habitats or swamps, although they should be managed to avoid deterioration (Vilenica *et al.*, 2022a) (Fig. 6.32).

Certain odonate species require specific management practices. In southern France, the rare and threatened *Lestes macrostigma* inhabits brackish still waters. Conservation of the species involves promoting larval survival, through preserving flood marshes and ponds from late autumn (November) to late spring (June). This approach should be coupled with maintaining water levels of 30 cm or more during spring, which leads to increased adult size at emergence, and therefore improved reproductive success. Additionally, to improve functional connectivity, there should be instigation of networks of ponds to provide stepping-stone habitats

Fig. 6.32. Examples of well restored and managed standing water bodies in Papuk Nature Park, Croatia. (Photos: Marina Vilenica).

across the local region. These ponds would provide oviposition sites for adults, maturation sites for larvae, and they need only be small (200–1000 m^2) (Lambret *et al.*, 2023).

6.5.8.3 Improving fishponds away from being ecological traps

Central European intensive fishponds are ecological traps for two species of libellulid, *Sympetrum depressiusculum* (see Fig. 4.24) and *S. sanguineum*. Although the ponds attract ovipositing pairs of both species, they are largely unsuitable for dragonfly breeding through a combination of factors such as pond depth and fish disturbance and predation. This is concerning in the case of S. *depressiusculum* which is a threatened species in Europe. However, the pressure on this species can be mitigated by synchronizing the time and place of pondside grass mowing and leaving high-quality uncut hay meadows away from the ponds as source habitats (Šigutová *et al.*, 2015). However, this also means not neglecting the aquatic environment which must also be improved, so that there is complementarity between both the aquatic growing environment and appropriate management of the terrestrial foraging landscape. This is especially important late in the season for anisopterans where patches of suitable terrestrial habitat are critical (Hykel *et al.*, 2016). This is not so much the case for zygopterans, as they die off late in the season when the dragonflies are still active (Harabiš, 2017) (Fig. 6.33).

Fig. 6.33. *Sympetrum depressiusculum* is a threatened species in central Europe and attracted to fishponds which are ecological traps for it (top left). The trapping effect was mitigated by periodic draining of a source pond (top right) to promote the growth of aquatic vegetation to mimic conditions similar to those in the natural habitats of *S. depressiusculum* (bottom left). Management of the trap habitats through mowing of the surrounding bank vegetation during times of oviposition prevented *S. depressiusculum* from ovipositing into trap habitats (bottom right), promoting colonization of the source pond by the species. (Photos: Hana Šigutová (top left and right); Aleš Dolný (bottom left and right)).

6.5.8.4 *Complementary artificial ponds relative to landscape context*

In France, ponds in different landscape contexts (forest, urban and agricultural fields), support odonate assemblages according to local characteristics, with some species present in only forest or field ponds, and urban ponds lacking some species. Odonate assemblages are driven mainly by vegetation, macroinvertebrate prey availability and pond size. The presence and abundance of larvae are related to aquatic vegetation, which varies among ponds. Overall, it is vegetation and its heterogeneity within and around ponds, shaped by human activity type, that determines local odonate assemblage diversity (Le Gall *et al.*, 2018).

Across a protected area–sugarcane–forestry mosaic in South Africa, odonate species richness is similar in various artificial ponds. However, assemblage composition differs in response to vegetation cover, with the different pond contexts complementary, and ponds in the transformed areas (sugarcane and timber production areas) contributing greatly to regional dragonfly diversity. While species composition is responsive to many variables, only sugarcane ponds are significantly different from the other pond types, and their composition is driven mostly by pond characteristics such as vegetation cover and dissolved oxygen. As the area is at an intermediate elevation between the coast (dominated by thicket/savanna vegetation) and mountains (dominated by grasslands), species from both elevations occupy many of the ponds, indicating assemblage crossover. However, certain pond types are preferred by certain species. Of note is that coastal species prefer sugarcane ponds, with all nine of the predominantly coastal species, including *Tramea limbata*, *Urothemis assignata* and *Rhyothemis semihyalina*, present only at sugarcane ponds. As sugarcane is largely a coastal crop, and these coastal species associate with the local reedy habitats in natural ponds and marshes in the coastal region, they find this tall graminaceous crop close to their natural preferences (Briggs *et al.* 2019b).

A switch from conventional to environmentally friendly rice cultivation in Japan involves foregoing use of insecticides although still using a herbicide and a fungicide. The change in cultivation practice increased the abundance of emerging anisopterans (*Sympetrum frequens* and *S. infuscatum*) 87-fold, and zygopteran adults (*Mortonagrion selenion*, *Ischnura asiatica*, *Indolestes peregrinus* and *Lestes temporalis*) 9.9-fold. However, the benefits of switching to environmentally friendly rice cultivation to benefit odonates must also involve assurance that all other conditions in the landscape are suitable for them to complete their life cycles (Baba *et al.*, 2019). These conditions include specific aquatic and riparian vegetation, including trees, close to paddies (Hamasaki *et al.*, 2011).

In northern Brazil, pond dragonfly abundance, species richness and composition were all adversely affected in the most disturbed landscapes. Greatest abundance and highest richness were in preserved natural ponds. Importantly, the best-preserved natural ponds supported a unique set of species, emphasizing the importance of conservation of these natural ponds (Santos *et al.*, 2021).

Garden ponds in the UK can act as refuges for certain aquatic insect taxa, but in comparison with field ponds, perform poorly in the case of most taxa. However, for dragonflies, garden ponds can serve as refuges for five odonate species which occurred extensively in both field and garden ponds, with vegetation diversity a key driver. *Ischnura elegans* was the most abundant zygopteran within garden and field ponds (Hill and Wood, 2014).

6.5.8.5 *Value of stormwater ponds*

Stormwater ponds are constructed to control high levels of water flow over hard surfaces during rain downpours, to retain road pollutants such as trace metals and hydrocarbons, and to harness toxic spills in case of a road accident.

Stormwater ponds associated with a road highway in southern France had a high concentration of copper and zinc in the top sediment layer and herbicides in the water column. Nevertheless, dragonfly species were still present, but species richness was higher in ponds with a natural bottom than in ones with an artificial bottom (plastic membrane). Two species, *Ischnura pumilio* and *Erythromma viridulum*, that are uncommon in the region were present in the ponds with a natural bottom (Scher and Thièry, 2005).

Stormwater ponds along a road highway in an agricultural and semi-natural environment in Czechia are smaller, not so eutrophic, less shaded by trees, have less fish and are less functionally connected than a variety of other nearby rural ponds.

They also have higher plant diversity and pH and are more densely overgrown with riparian vegetation. Compared to the other ponds, the stormwater ponds have higher odonate species richness, more species of conservation concern, higher species abundance and more pioneer species that prefer early succession habitats. However, high diversity of both aquatic and marginal vegetation, along with woody riparian zones arising from conditions associated with late ecological succession, encourage dragonfly species that prefer these conditions. The relative isolation of the stormwater ponds, along with their more diverse environmental conditions, leads to each of these ponds having different assemblages as well as higher species turnover among the ponds compared to rural ponds. Overall, these highway stormwater ponds greatly enhance and support local dragonfly diversity (Šigutová *et al.*, 2022) (Fig. 6.34).

A total of 31 odonate species, comprising 61% of the species in an urban area in the province of Uppland in Sweden, were recorded in stormwater ponds, a large component of the regional odonate species pool. The nestedness component of beta diversity was higher than the turnover component, and mainly the result of increased pond area and total vegetation cover in the water and on the banks. Important specific vegetation variables are the percentage of cover of floating algae scums, emergent aquatic macrophytes and tall shore vegetation (Johansson *et al.*, 2019).

In Canada, stormwater ponds had fewer Anisoptera species and lower abundance, although more Zygoptera species than other ponds. Some stormwater ponds had similar odonate assemblages as in natural ponds. The drivers of abundance and species composition were largely macrophyte composition, while also strongly linked to the presence of specific wetland plant species (Perron and Pick, 2020).

Stormwater ponds can vary in their degree of permanency ranging from temporarily containing water, through semi-perennial when they drop to small pools of water, to perennial ponds. Temporary ponds in Germany, with their warmer microclimates and short hydroperiod differ from the other

Fig. 6.34. Species composition of rural control ponds (top left) and stormwater ponds (top right) differ significantly. Stormwater ponds display higher species turnover. However, taxonomic diversity does not differ between the pond types. Indicator species of stormwater ponds are *Sympetrum striolatum* (bottom left), *Ischnura pumilio* (bottom centre) and *Enallagma cyathigerum* (bottom right), all pioneer species which prefer early successional habitats. No indicator species were detected in the surrounding rural ponds, whose assemblages were composed of widespread habitat generalists. (Photos and scans: Aleš Dolný).

two pond types in terms of their dragonfly diversity. Interestingly, species richness and exuviae density of locally threatened species, such as *Libellula depressa* and *Ischnura pumilio*, were highest in the temporary ponds while lowest in the perennial ponds. These ponds play an important role in dragonfly conservation as they are of high quality and are generally well-maintained (Holtmann *et al.*, 2018, 2019).

6.5.8.6 Restored mining ponds

In central Europe's industrial region, Upper Silesia, abandoned mining shafts have collapsed, resulting in the formation of subsidence ponds. They are heterogeneous, displaying various stages of ecological succession. Fifty odonate species have been recorded from these ponds, including rare habitat specialists such as *Orthetrum coerulescens* and *Leucorrhinia pectoralis*. These ponds are in effect refuges, as low-elevation bogs are now very rare and largely confined to higher elevations (Dolný and Harabiš, 2012).

In the same region, using different approaches to restoration across a wide range of post-coal mining ponds resulted in a rich dragonfly assemblage of 32 species, including eight threatened ones. The key lay in having a variety of pond types that differed in their vegetation, bottom substrate, water shading and in the surrounding terrestrial matrix (Harabiš *et al.*, 2013).

Also, in the same region, further details throw more light on understanding the value of mining ponds. High-value post-mining pondscapes, in terms of dragonfly diversity, depend on the presence of a variety of vegetation successional stages, as well as exclusion of harmful technical reclamation activities which can otherwise lead to poor habitat conditions and loss of sensitive and habitat-specialist species. Though some populations do occupy pools shortly after reclamation, this is followed by a decline in assemblage composition complexity over time. One issue with these reclamation ponds is that they do not overall provide sufficient early succession stages or suitable conditions for pioneer dragonfly species. Furthermore, populations of habitat generalists are mostly stable, but this is not the case for habitat specialists. However, the specialists can still colonize novel aquatic habitats in early successional stages but only when source populations are nearby. Furthermore, there are high local extinction rates among thermophilic Mediterranean species, seemingly due to them being marginal and on the edge of their geographic ranges. Overall, management of post-mining ponds should aim for high habitat heterogeneity through the creation of mixed moderate disturbance and successional habitats (Harabiš and Dolný, 2015a). Doing this can help the recovery of certain range-retreating species such as *Sympecma paedisca* in central Europe (Harabiš, 2023).

In the Carpathian Mountains, Slovakia, dragonfly assemblages associated with four standing water types (standing water ponds, fishponds (used for farming fish), mining ponds (depressions left after excavating sand and gravel), and marshland (natural shallow depressions filled by ground and surface water)) were assessed for their odonate assemblage composition. Important drivers of dragonfly diversity are bank characteristics and littoral vegetation, method and intensity of site management, stage of ecological succession of aquatic habitats, and mountain and submontane landscape features. The highest species richness is associated with habitats with a wide range of different microhabitats, especially in mining ponds. These ponds are characterized, for example, by *Lestes dryas*, *L. virens*, *L. barbarus*, *Chalcolestes viridis*, *Libellula depressa*, *Pyrrhosoma nymphula*, *Erythromma viridulum*, *Sympetrum striolatum* and *Ischnura elegans*. Inflowing streams to these ponds are occupied by species such as *Calopteryx splendens*, *Gomphus vulgatissimus* and *Onychogomphus forcipatus*. In contrast, fishponds are characterized by the habitat generalists such as *Platycnemis pennipes*, *Ischnura elegans* and *L. depressa*. While the most degraded ponds provide habitat for a relatively low number of species with broad ecological tolerance, rare and specialist species require high-quality ponds in which to reproduce (Petrovičová *et al.*, 2021).

Among a group of adjacent mining ponds (after gravel and sand extraction) in Poland, 42 dragonfly species were present, with 35 recorded as resident. One third of all species are confined to single ponds. Similarity indices for species presence were all below 53%, indicating great variation among dragonfly assemblages in the various ponds. These differences reflect the precise habitat selectivity of the various species, related to pond area and plant diversity which explained 49% of the total variance associated with pond occupation. Many of the species are of conservation concern, suggesting that these sites should be retained rather than be infilled (Bobrek, 2021).

Lake remediation in France through liming (to counteract post-mining acidification) can reduce odonate diversity and larval numbers in comparison with naturally recovered lakes with no intervention liming, as well as subtly affecting adult assemblages. Lakes left to recover naturally supported 22 species as adults and 18 species as emerging larvae, with *Lecorrhinia albifrons* exclusive to naturally recovered ponds. Limed ponds supported 19 species as adults and 16 as emerging larvae, with only *Coenagrion lindenii* exclusive to limed ponds. Overall, this suggests that post-mining, lakes are best left to recover naturally without any intervention liming, despite the adverse effects of initial acidification (D'Amico *et al.*, 2004).

6.5.8.7 *Golf course ponds*

Golf courses, while of dubious merit in terms of water consumption, can play an important role where ponds and streams are a component of the rough. They could be considered as nodes in peri-urban areas. Around Stockholm, Sweden, several odonate species are present on golf courses, including two species which are only locally present at golf course ponds, including the red-listed *Leucorrhinia pectoralis* and *Enallagma cyathigerum*, both with preferred aquatic plant communities (Colding *et al.*, 2009). In South Africa, one of the strongest and most southerly populations of *Tetrathemis polleni* occurs at well-treed ponds on Pennington Gold Course in southern KwaZulu-Natal (Fig. 6.35).

6.5.9 Speed of recovery of dragonfly assemblages following conservation action

The speed of recovery of dragonfly assemblages can be remarkably rapid assuming source populations are present nearby. A lake in Buenos Aires, Argentina, was given focused restoration involving physical cleaning out of discarded items, planting of native herbs and trees, planting of grass, adding of mulch, cessation of pruning of vegetation and maintenance of indigenous vegetation including *Pontederia crassipes*, which forms dense mats known as 'embalsados'. Within two years the lake had accumulated 81% of the regional odonate diversity compared to 21% for a nearby unrestored lake (Lozano *et al.*, 2022).

However, among rewetted peatlands in Finland recovery of the full potential dragonfly assemblage will likely take more than three years, even with source populations nearby. Nevertheless, restoration of these peatlands in terms of recovery of abundance and species richness was very effective and holds much promise for peatland dragonfly conservation when effective restoration measures are put in place (Elo *et al.*, 2015).

Drainage and afforestation of peatland causes extensive habitat degradation and species loss in central Germany, where restoration has focused on stabilizing habitats and local water tables. The result was that habitat conditions improved over an 18-year period. During that time, characteristic peatland vegetation recovered slowly once the improved water-holding capacity reached a point where peat began to accumulate. In contrast, dragonflies colonized the restored small water bodies within three years. However, their diversity then decreased a little because of fewer and shallower ponds with the passing of time. While many deep and diverse ponds promote more heterogeneous conditions, certain dragonfly species have more specific pond requirements. For example, *Somatochlora arctica* needs shallow pools in *Sphagnum* mats, while *Aeshna juncea* prefers deeper ponds. In sum, though dragonflies readily colonize restored ponds, a variety of bog pond types are required to support the full range of characteristic peatland dragonflies (Strobl *et al.*, 2020).

While vegetation succession at newly created ponds is important, among more mature ponds, intermediate stages of succession may produce the highest dragonfly species richness. Later successional stages may become too shaded as bushes and trees mature along the banks, shading the water, while the pond may also infill with vegetation debris (Kolar *et al.*, 2021).

In France, there is no difference in species richness between restored and newly created ponds, as most species colonize the ponds in the first year, although there is a difference in colonization patterns between Anisoptera and Zygoptera. Assemblage composition is related to the landscape matrix of the ponds, as well as the time since pond creation or restoration. Abundances are positively related to pond connectivity, especially among zygopterans. While anisopteran larvae are not affected much by the vegetation, the endophytic oviposition of zygopterans means that they were more sensitive to vegetation type. Overall, these findings confirm that odonates are ready colonizers, so long as individuals from source populations

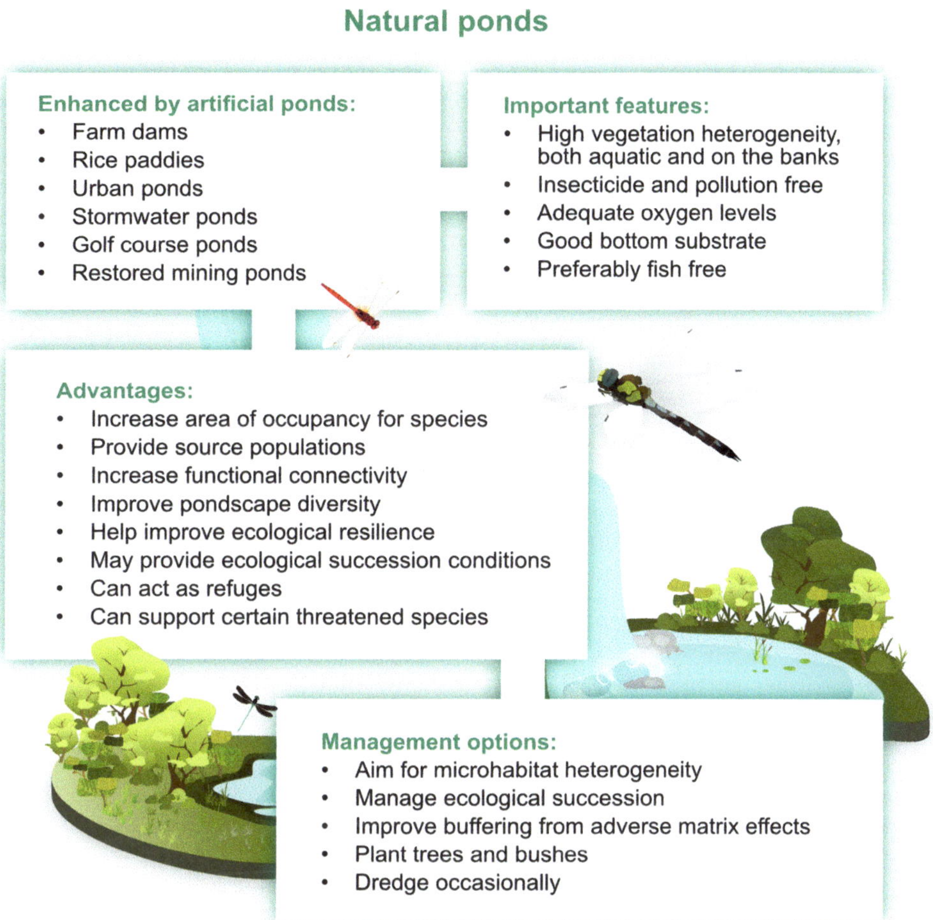

Fig. 6.35. Maintaining natural ponds in tip-top condition is essential. But this is not to overlook the additional value of a range of artificial pond types, particularly for enhancing the overall pondscape viability. High-quality artificial ponds of all sorts have some important common features, which are not dissimilar to those of natural ponds. The advantages of good-quality artificial ponds are several. However, some may require certain management procedures to ensure that their value for dragonflies and other biota is maintained. (Illustration: Robyn Symons).

can reach the newly suitable habitat. Furthermore, creation of new ponds is as effective as pond restoration for maintaining local odonate diversity (Minot *et al.*, 2021).

In Japan, groups of parents, teachers, children and conservation activists are building ponds on school grounds specifically for dragonflies, also helping teach awareness of odonates. There are now many hundreds of ponds like this in many parts of the country, which are appropriate for urban situations because many species are capable of locating suitable habitat on their own and surviving. However, the larvae of smaller species sometimes need to be brought to new ponds to establish populations (Primack *et al.*, 2000).

In Virginia, USA, newly constructed, fishless ponds suggest that dragonfly assemblages at new ponds can be species-rich. However, low physiological production in newly constructed ponds indicates that these ponds may not yet have enough prey available for naturally high levels of physiological condition and abundance, which may take some time to establish (Braccia *et al.*, 2007).

Within three years of creation, a new, ecologically managed pond in Vienna, Austria had attracted about 17 dragonfly species, with natural and rapid colonization by the submerged macrophyte *Potamogeton perfoliatus* being an important enabling factor (Chovanec, 1994).

Parts of a 400-hectare wetland in Greene County, Ohio, USA, that had previously been drained for agriculture and peat harvesting was restored, to improve groundwater levels. Of the 54 Odonata species recorded from the county, 36 occur in the wetland and 26 were recorded specifically at restored sites within five years of restoration. While these new habitats were not specifically constructed to attract odonates, various successful design principles became clear. First, the wetlands were made to resemble natural habitats by creating a very gently changing topography over existing topography, while avoiding the building of water containment structures. Second, natural hydrologic conditions were restored, involving provision of a range of water depths and fluctuating hydroperiods to establish high plant diversity. Third, habitat diversity was improved by establishing a diversity of plant species from seed, greenhouse stock or locally collected stock. Fourth, all the sites were located close to existing wetlands to encourage recruitment of a wide range of biodiversity. Although the restored sites are open, each is situated on low ground and near trees or shrubs that can serve as windbreaks or perching sites (Roush and Amon, 2003).

There can be rapid colonization of new ponds from across the sea. The small (26 ha) Cousine Island, Seychelles was formerly waterless. As part of general island restoration, small (a few square metres) and shallow (a few centimetres) ponds were constructed for giant tortoises to bathe. Within just over two years, *Ceriagrion glabrum*, *Diplacodes trivialis*, *Orthetrum stemmale wrightii*, *Pantala flavescens* and *Tramea limbata* were breeding on the island after having crossed at least 2.5 km of open sea (Samways, 1998).

6.5.10 Effect of large herbivores on dragonflies

Africa still retains many native megaherbivores, some of which interact with dragonflies after a long historic association. *Brachythemis leucosticta* adults follow large game animals, catching flying insects flushed out of the grass. However, this interaction between game and dragonflies is strongest adjacent to water bodies.

In South Africa, herds of up to 600 strong of the indigenous Cape buffalo can have a huge temporary impact through trampling of riparian vegetation, reducing populations of *Ischnura senegalensis*, *Ceriagrion glabrum*, *Nesciothemis farinosa*, *Diplacodes lefebvrii* and *Acisoma* sp. (Stewart and Samways, 1998).

Elephants require much water and are often at the water's edge, or they enter the water where they trample vegetation and the substratum. Along a meandering river with a vegetation-rich margin in Botswana, dragonfly abundance varies according to the level of elephant disturbance. At extremely high and medium levels of impact, dragonfly abundance is lowest, but surprisingly, at high levels of impact it is highest. However, only 7 out of 31 species occur at any one site, indicating that different species tolerate different levels of impact. Nevertheless, the species (16 in total) that occur in very high or extremely high impact areas, also occupy low impact areas, suggesting considerable habitat tolerance, with only *Ischnura senegalensis* and *Brachythemis leucosticta* most abundant in the most disturbed areas. In contrast, *Ceriagrion katamborae*, *Pseudagrion deningi*, *Agriocnemis victoria*, *Orthetrum icteromelas*, *Trithemis monardi*, *T. stictica*, *Rhyothemis fenestrina* and *Aethriamanta rezia* are very sensitive to elephant impact. Overall, it is the variety of habitat types, including various levels of elephant impact, that supports the full complement of local species (Samways and Grant, 2008) (Fig. 6.36).

6.6 Dragonfly Species-level Recovery and Conservation

6.6.1 Principles

6.6.1.1 *General principles of species conservation*

Around the world, parties to the Convention on Biological Diversity are working on a post-2020 Global Biodiversity Framework to tackle the ongoing loss of biodiversity. Among the goals is to get countries to reduce the extinction rate at least tenfold and to halve the risk of species extinctions across all taxonomic and functional groups. While there has been a focus on area-based conservation and threat mitigation, through, for example, habitat restoration, there also needs to be targeted recovery actions. These include reintroductions and breeding site provision to prevent species extinction. Furthermore, there needs

Fig. 6.36. The African elephant is still abundant in many areas in southern Africa, such as the Okavango swamps (top). Elephants regularly wade in the water where they trample the plants and substrate, with some species like *Ischnura senegalensis* (bottom left) and *Brachythemis leucosticta* (bottom right) most abundant in these highly impacted areas. (Photos: Michael Samways).

to be active management actions that enable the recovery and conservation of species and their genetic diversity (Bolam *et al.*, 2022).

To preserve the greatest number of species, there is debate whether we should give priority to cost-effective actions or to focus solely on the most threatened species. However, a decision analysis framework shows that allocating resources solely to the most threatened species will not minimize the number of extinctions in the long term. The reason for this is that the threatened species approach does not account for the risk faced by less threatened species with the potential of going extinct in the future. It is principally a short-term goal, often the perspective of conservation organizations (Wilson *et al.*, 2011).

Translating these findings into dragonfly conservation implies that a primary focus for the long term would best be attention to the higher spatial levels of ecosystems, landscapes and even whole catchments to conserve all dragonfly faunas. However, this does not preclude effective species-level activities (Byers *et al.*, 2022). Besides, sometimes the start of decline of certain species may indicate the onset of some chronic event which may adversely affect many species in the end (Fig. 6.37).

6.6.1.2 *Significance of maintaining effective population size*

We focus here on populations of single dragonfly species in relation to the various abiotic and biotic

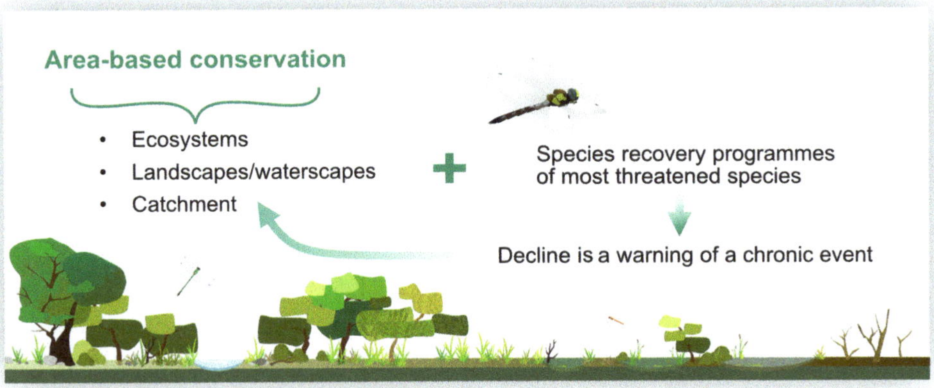

Fig. 6.37. Area-based conservation can be supplemented with species recovery programmes, which often focus on the most threatened species. The value of this approach is that the decline of certain species may be a warning sign of a chronic event, which feeds back into appreciating the importance of area-based conservation while giving certain focal species special attention. (Illustration: Robyn Symons).

components of their environment. These components include interactions between the focal species and other organisms. We highlight what conservation measures have succeeded, or not, to ensure the persistence of local populations. For dragonflies, these other organisms include prey, predators, competitors and parasites, as well as plants, with which individuals of the population interact. It is not just the presence or absence of these other organisms that is important, but also their abundance. We ask which, and in what quantity, these other organisms, as well as the prevailing physicochemical conditions, maximize the chances of survival. It also means determining what can be done to increase species effective population size: a population that is contributing the most to the next generation through more breeding individuals. Effective population size, alongside local movement and migration rates, are important evolutionary parameters that affect population survival and determine the relative influence of natural selection and genetic drift (Watts *et al.*, 2007).

In practice, increasing effective population size is done by removing or mitigating stressors as well as encouraging healthy, genetically diverse sets of breeding individuals to ensure population continuance. A stark reminder is that species extinction is often through gradual loss of individuals from a population until finally there is the death of the last individual.

But then any one population of a species also lives in a community and an ecosystem, which means that to conserve a species we need to conserve its habitat within those systems, which is the sum of the resources needed to sustain it in the long term. As dragonflies are amphibiotic, this implies that we must conserve both the aquatic and terrestrial realms. It also means that we must be acutely aware that ecosystems vary over time, especially relative to season and extreme weather events, as well as ongoing climate warming (Fig. 6.38).

The value of protected areas is substantial for conserving certain dragonfly species, especially in warm mountainous areas. *Hypolestes* is a zygopteran genus endemic to the Greater Antilles in the Caribbean, inhabiting streams in forests. Two of the species in this genus are currently considered as highly threatened through loss of their habitat: *H. trinitatis* from Cuba (Vulnerable) and *H. clara* from Jamaica (Endangered). The third species of the genus is *H. hatuey* from the island of Hispaniola (Haiti and Dominican Republic) and is assessed as Near Threatened. All three species have more than 10% of their extent of occurrence in protected areas, which may or may not be sufficient to conserve them in view of ongoing forest loss and a changing climate (Torres-Cambas *et al.*, 2016).

6.6.1.3 *The species concept and genetics*

The term 'species' from a basic conservation perspective, is the whole population of similar individuals that can all potentially mate and leave

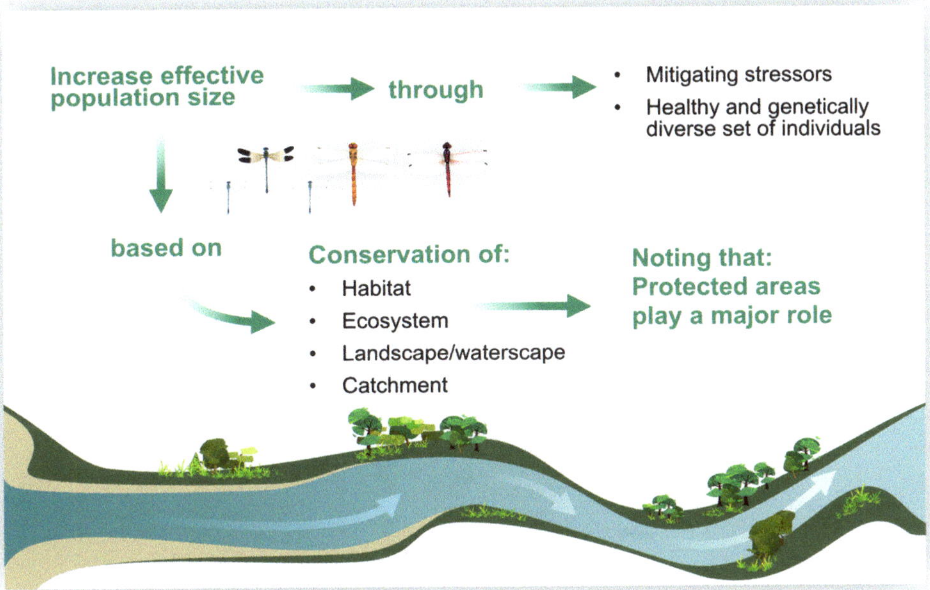

Fig. 6.38. The basic tenet of species-level conservation is to increase effective population size through mitigating stressors and encouraging a healthy and genetically diverse set of individuals. This approach is based on conservation of the hierarchy area-based perspectives, while recognizing that protected areas can play a major role for ensuring population persistence, especially of habitat-specialist species. (Illustration: Robyn Symons).

viable offspring. In turn, species conservation is the maintenance of a *variety* of potentially interbreeding individuals and their collective genetic diversity.

Two or more populations of a species may be geographically isolated, and although still potentially able to interbreed, are not able to do that as there is perhaps a physical barrier separating them. Individuals that constitute any one of the various populations may have a set of morphological features, behaviours and physiologies (i.e. traits; Chapter 4) that are slightly different from those in another population. The difference is based on small genetic differences, though they are still able to mate and breed with other individuals from the other populations. However, different populations of any one species may have a slightly different niche from those of another in a widely separated geographical location, especially when the individuals are of low mobility. These physically separated populations with slightly different characteristics are known as evolutionarily significant units or by the expression 'molecular operational taxonomic units' (mOTUs). These are not to be confused with 'morphological operational taxonomic units' (MOTUs), species that are morphologically identified (morphotyped) but without necessarily having a scientific name (Fig. 6.39).

Different mOTUs may also be subject to different types and levels of stressors, either because threats vary from one place to another and/or because a particular genetic constitution means poor ability of individuals to survive any one or multiple stressors. This is the case for *Coenagrion puella*, where the North African mOTU is highly threatened and deserving of urgent conservation focus relative to the European one (Ferreira *et al.*, 2016).

A related issue, and important in dragonfly conservation, is misidentification of certain species until genetic studies provide clarity. *Orthetrum nitidinerve* is a western Mediterranean endemic known from 17 breeding populations in Sardinia and Sicily, Italy, as well as Algeria (El Bouhissi *et al.*, 2022). Records from continental Italy are due to misidentification. The species has declined greatly in Sicily, though several large populations remain in Sardinia. The species is threatened by agriculture and grazing intensification or agricultural

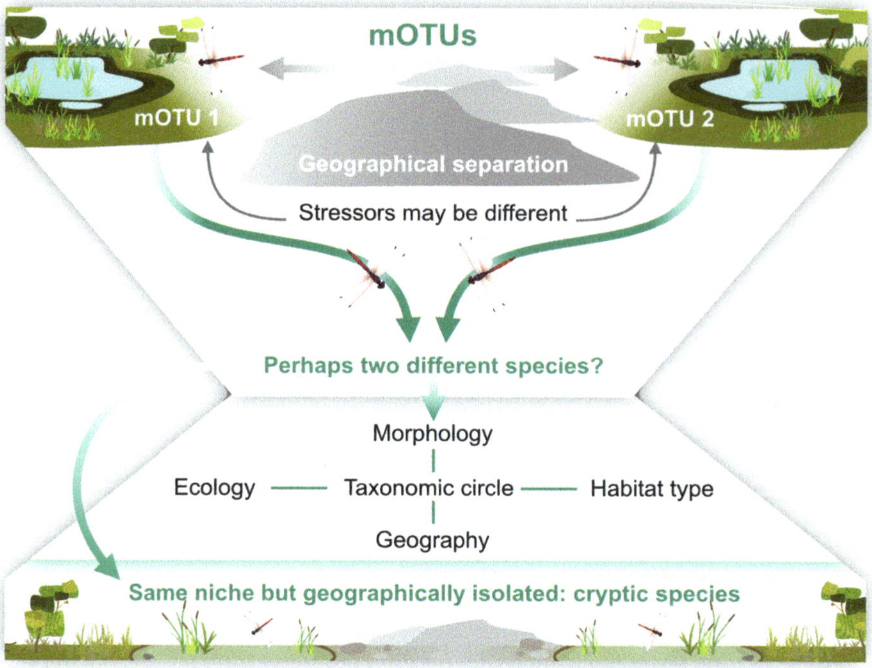

Fig. 6.39. Molecular operational taxonomic units (mOTUs) are genetically distinct populations of the same species, often geographically separated. Each may be subject to different stressors. Research on mOTUs may suggest that different populations should be raised to the rank of different species. This may be confirmed by the taxonomic circle of slight differences in morphology, ecology, habitat preference and geographical distribution. In other cases, different mOTUs may share the same niche but are geographically separated, constituting separate, morphologically very similar cryptic species. (Illustration: Robyn Symons).

abandonment, as well as drought and desiccation from water overexploitation and climate change (Assandri *et al.*, 2020).

Sometimes genetic analyses can identify phylogenetically distinct populations of conservation significance. On Taiwan, the '*Ischnura aurora*-like' damselfly has been confirmed as an isolated population of the widespread Asian species *I. rubilio*. It is a unique island phenotype that has declined greatly in recent years, with only one population remaining and, as a result of the genetic studies, is now receiving conservation attention (Hu and Futahashi, 2023).

Detailed genetic analysis may raise suspicions that perhaps genetic divergence is sufficient to warrant raising a mOTU to the rank of species. For example, the widespread and common species *Nannophya pygmaea* has four mOTUs in East Asia. The genetic divergence between these mOTUs is high enough that perhaps among them are two species (Low *et al.*, 2016).

When genetic divergence is used alongside other data on morphology, ecology, habitat type and geography, together they are known as a 'taxonomic circle'. This may lead to sufficient confidence in raising mOTUs to the rank of separate species. This was done in southern Africa for a large sample of ostensible *Trithemis stictica*, leading to the discovery of two new species. This was based not only on genetic isolation, but also on ecological niche shift, fixed size differences and most likely reproductive isolation (Damm *et al.*, 2010). Similar findings come from North America for the nominate species *Hetaerina americana* (Vega-Sánchez *et al.*, 2024).

When dragonfly genomes are examined in detail, one ostensible species can turn out to be more than one species, with all individuals apparently having the same niche. These are known as 'cryptic species', which are sometimes difficult to separate on morphology alone. Genetic analyses revealed that the charismatic Central American genus *Megaloprepus* is composed of more species than the well-known *M. caerulatus*. Following genetic analysis, a new species, *M. diaboli* was discovered, along with a further two former subspecies now also included in the genus: *M. latipennis* and *M. brevistigma*. These four species occupy the same ecological niche, and all share high sensitivity to heat. They also have low dispersal capabilities across open areas, and when this behaviour is seen against the historical fragmentation of Neotropical old-growth forests, the result has been small distributional ranges and geographic isolation. This means that the conservation status of the four species should be re-evaluated relative to regional and local endemism (Feindt and Hadrys, 2022).

A further aspect of the value of genetic studies is that they can confirm a species presence at a site of interest. Morphological and phylogenetic evidence identified two female individuals collected in Oceana County, Michigan, USA, as *Somatochlora hineana*, confirming the presence of this threatened species at a new disjunct location lacking some previously thought important habitat features, such as underlying geology that is typical of other *S. hineana* populations (Craves *et al.*, 2022).

After a dragonfly species has passed through the evolutionary filters of the past, individuals in a population must be able to survive current globally rapidly changing conditions. Furthermore, there must be enough individuals to overcome small population effects like finding optimal habitat or a mate and avoiding inbreeding effects. There must also be enough genetic diversity and plasticity for the population to survive genetic bottlenecks, which otherwise might lead to decreased fitness (i.e. inbreeding depression) and increased risks of extinction, as in the case of Critically Endangered *Libellula angelina* in Korea (Kim *et al.*, 2020) (Fig. 6.40).

One aim of species population-level conservation is to decrease the gap between actual average level of fitness and the maximal possible fitness. In other words, to ensure maximal adaptation while reducing maladaptation, a process known as 'reducing genetic load'. Populations with low genetic diversity are often considered to be at greater risk of decline and extinction than populations with high genetic diversity. However, the risk to populations needs to be viewed from historic geomorphic, dispersal trait and current types of landscape change perspectives. In turn, conservation action aims to improve functional

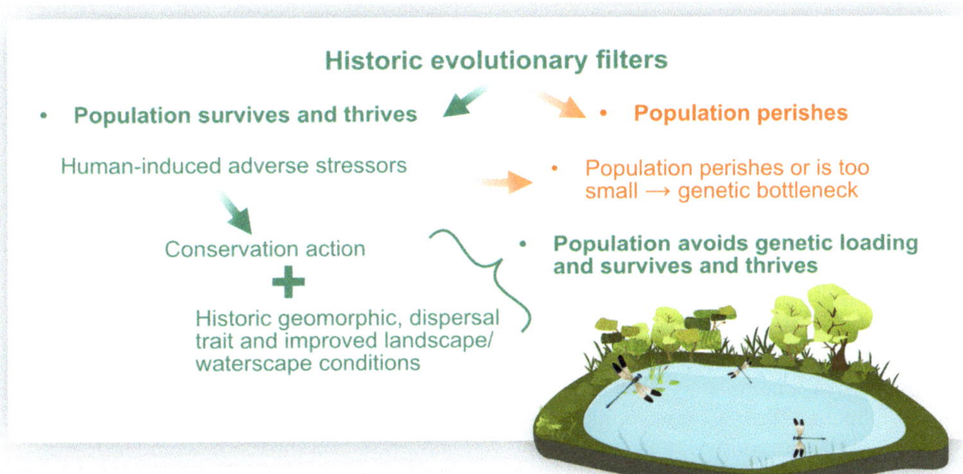

Fig. 6.40. Natural pressures in the past have placed evolutionary filters on species, with them now having to face new stressors from human activity. Population persistence depends on there being enough individuals to overcome small population effects ranging from finding optimal habitat or a mate through to avoiding inbreeding effects like having enough genetic diversity and plasticity for healthy population persistence. If not, there will be inbreeding depression. Conservation aims to overcome genetic bottlenecks through ensuring maximal adaptation while minimizing maladaptation, i.e. reducing genetic load. The success of this action is set against the background of a favourable geomorphic, dispersal trait and improved landscape/waterscape conditions. (Illustration: Robyn Symons).

connectivity and landscape/waterscape conditions for effective population levels.

The Hawaiian endemic *Megalagrion* species (see Fig. 3.15) and island populations give some insight into the significance of the interplay between these perspectives. *M. xanthomelas* and *M. pacificum* share many haplotypes among populations, driven by high dispersal ability. However, there is less haplotype sharing among populations of two other species, *M. calliphya* and *M. hawaiiense*, even within the larger Maui Nui super island, suggesting that these two species do not disperse well across land or water. Their phylogenies also reflect the geological history of the islands. In terms of genetic health, there is high diversity in most populations of *M. calliphya*, *M. hawaiiense*, and Hawai'i island populations of *M. xanthomelas*. However, there is low diversity in genetically bottlenecked or relict populations, including those on O'ahu and Maui of *M. xanthomelas* and *M. pacificum*. Surprisingly, there is low genetic diversity in O'ahu Ko'olau and Hawai'i Onomea populations of *M. hawaiiense*, the Moloka'i population of *M. pacificum*, and the West Maui population of *M. calliphya*, suggesting that conservation should focus on these populations to maximize their long-term genetic health (Jordan *et al.*, 2007).

6.6.1.4 Recognizing different effects of stressors on populations

Knowledge of the impact that various stressors have on a system, and by extension, on the individuals, has another facet. When significant adverse stressors last a long time, the chances of population improvement diminish over time, leading eventually to local population extinction. Where a stressor is simply a pulse, such as a mining spill, over time there may be recolonization from source habitats nearby and according to the dispersal traits of those individuals. An example is part of the Guadiamar river, Spain, where a catastrophic mining spill devastated the dragonfly fauna, yet many of the species' populations recovered, as source populations were locally present. However, ongoing alteration of the river system by pollution and/or forest degradation continued to pose a major problem for long-term survival of the populations (Ferreras-Romero *et al.*, 2009).

A further issue is that during times of environmental change and/or type and level of stressor, there may be a restriction of resource availability, i.e. a resource bottleneck. For example, an event associated with climate change (e.g. low water levels) can lead to diminished or changed resources. However, when a competitor or predator is not favoured, there may be a lifting of the resource bottleneck on a population of a particular species. Usually though, it is the resource specialists, such as those requiring a specific microhabitat or prey, that are most disadvantaged.

6.6.1.5 Importance of adaptive capacity

Adaptive capacity is a combination of a species' genetic characteristics, behavioural and physiological disposition, and traits like life-history type and dispersal/colonization abilities for short-term survival and long-term evolutionary potential. However, this adaptive capacity is rarely fully realized. The reason for this is that at any one time and place there is a set of conditions, both abiotic and biotic, that constrain population size and viability, and therefore the effective population size. In short, a species' realized adaptive capacity is its ability to survive environmental change.

In the face of ongoing stressors, especially multiple stressors, the chances of fully realizing adaptive capacity are becoming less than would have been the case in the historic state. Species are now confronted with many more challenges to long-term survival than in the past. However, insect species vary in the degree to which they can respond, positively or negatively, to current environmental changes, with some even benefitting from certain human activity, as often seen in the Neotropics when anisopterans replace zygopterans following forest loss without necessarily any change in overall species richness.

Some dragonfly species are constrained by certain specialisms. This may, for example, be a specific egg-laying site, as in the case of *Lestes macrostigma*, which inhabits brackish waters and where oviposition is associated only with the rush *Bolboschoenus maritimus* and three other similarly structured plant species (Matushkina and Lambret, 2011). However, there must also be caution when associating a dragonfly species with certain plant species. *Aeshna viridis* is often considered as being associated with the reed-like *Stratiotes aloides*. Though in Sweden, only half the ponds occupied by this plant had *A. viridis* present, despite the dragonfly's good dispersal capabilities. It seems the dragonfly will also occupy ponds rich in plant species and with good ecological status, important when considering its conservation (Andersen *et al.*, 2016).

Understanding a dragonfly's realized adaptive capacity relative to its recovery is seen in threatened *Calopteryx exul*. High levels of both occupancy and colonization depend on the presence of the water plant *Potamogeton nodosus*. Local extinction risk increases with habitat disturbance, while chances of survival improve when *P. nodosus* is present. Restoration involves augmentation of *P. nodosus* patches, the total number of individuals, the number of reproductive events and philopatry increase, leading to an overall increase in the damselfly's conservation status (Khelifa and Mellal, 2017).

6.6.1.6 Significance of the realized niche

The fundamental niche of a species is all the environmental conditions in which it can survive, reproduce and leave viable offspring (i.e, conditions that approach its adaptive capacity). However, in the real world there are various constraints on a species, such as competition, predation, disease and certain environmental conditions, that lead to it only being able to survive in certain areas and under specific conditions, which together constitute its realized niche. Defining the realized niche of a species is difficult as there is variation among all the many environmental variables that impinge on a species over both space and time, and on the traits and phenotypic plasticity present among individuals making up a focal population.

Regional studies can help define the realized niche of a species. In the case of threatened *Leucorrhinia pectoralis* across Estonian landscapes that have undergone various degrees of transformation or restoration in recent years, lake size and the proportion of short riparian vegetation determine whether the species is present or absent across the various landscape types. The species is more specific in restored landscapes, with large effects from other environmental variables. In sum, the realized niche of the species in the unrestored landscapes is constrained by the present-day transformed habitats. The message here is that if a species' realized niche is derived from local distribution patterns without incorporating landscape history, it can lead to an erroneous assessment of the realized niche. Furthermore, landscape restoration can provide information on a species' habitat dependencies before habitat degradation occurred, so long as restoration mitigation reflects the full range of former landscape characteristics (Iversen *et al.*, 2016).

6.6.1.7 Restoration considerations for species-level conservation

The fully restored state of a habitat is in effect a substitute for the historic state. That historic state in any one area may be difficult to determine where extensive transformation has occurred. Furthermore, some shifting baseline effects may have occurred of which we now have no knowledge (Chapter 5, Section 5.2). This emphasizes the value of having strong populations of any one dragonfly species, and preferably at many locations. This is essential, for example, when translocation of individuals is to take place as part of a reintroduction project (see Section 6.6.3).

Insect species which have low genetic diversity, low dispersal–colonization capacity, special life-history requirements and have specialized traits are the ones likely to suffer most from human-induced adverse stressors. The conservation response is to integrate estimates of population size, demographic dynamics and behavioural traits, to determine the opportunities versus challenges for maintaining populations in a rapidly changing world.

There is, however, an important caveat here. Use of the word 'habitat specialist' must not assume that all habitat specialists are at great risk of local extinction. The term is relative to the type of human transformation and the traits of the focal dragonfly species. *Pantala flavescens*, for example, is arguably a habitat specialist of small often temporary pools in warm climates, yet its amazing flight ability enables it to occupy a whole range of small water pools created by human activity, ranging from vehicle ruts to agricultural excavations, and even bird baths. This is very different from, say, a highly specialist damselfly species, such as *Drepanosticta berlandi* in Indonesia, which occupies heavily shaded streams in tropical rainforest and is finely adapted to this relatively unchanging forest environment (Zulhariadi *et al.*, 2022).

Furthermore, the preferred habitat of an odonate species may vary across its geographic range. *Coenagrion mercuriale*, a protected species in Europe, has slightly different preferences for the proportions of emergent and of submerged vegetation across the continent (Rouquette and Thompson, 2005) (Fig. 6.41).

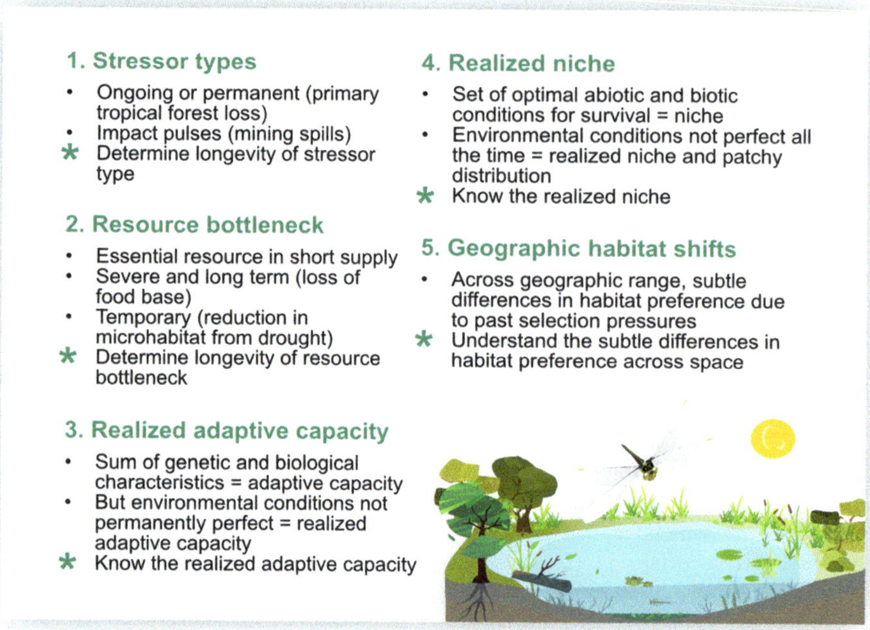

Fig. 6.41. Some important considerations when conserving dragonfly effective population size: 1. Stressors may be long-term and chronic, such as loss of primary tropical forest for a habitat-specialist species. Other stressors may be short term, such as a mining spill. Understanding the reversibility of stressors is important in habitat restoration. 2. A resource bottleneck occurs when an essential resource needed for survival is in short supply. It may be severe and long term, such as a loss of the food base from water poisoning by a persistent contaminant, or it may be temporary, such as a decline in optimal microhabitat as water levels drop during drought. Knowing the potential longevity of such a bottleneck is essential. 3. The adaptive capacity of a species is the sum of its genetic and biological features for maximal survival in a particular area. However, this adaptive capacity is rarely fully realized as at any one place and time it may not have the ability to survive all the current abiotic and biotic environmental conditions, resulting in constraints on effective population size. Knowing the realized adaptive capacity of a species is essential for the conservation of effective population size. 4. Similarly, while the niche of a species is the set of abiotic and biotic environmental conditions enabling a species to survive, there may be certain subtle yet significant conditions that are not optimal owing to small yet important shifts in environmental conditions over space and time. This is the realized niche of a species, often resulting in a patchy distribution. Knowing the exact niche of a species is essential to its conservation. 5. Different selection pressures in the past in various geographic areas may lead to geographically separated populations of a species having slightly different habitat preferences. These differences must be understood according to the location of the focal population. (Illustration: Robyn Symons).

6.6.2 Conservation of specific dragonfly species

The extremely rare and localized *Coenagrion hylas* has very limited distribution in Europe, living in small, isolated populations, while also being a strict habitat specialist. Yet the species has good dispersal capabilities of up to 30 km, while the adult is also long-living (average 12 days, maximum 40 days). Populations show no signs of inbreeding. Altogether, these traits suggest that the creation of stepping-stone habitats is a highly viable management option to sustain, or best-case scenario to improve, its current population status (Landmann *et al.*, 2021).

In north-western Spain, the genetics of *Macromia splendens* illustrate that it is highly philopatric, due not to its lack of flight capability, but because of its strict ecological requirements for local forests. This suggests that linking local forest canopies is important for maintaining the populations of this species in the area (Amorín *et al.*, 2009).

Some species of anisopterans require very specific habitat conditions and present conservation challenges owing to this requirement coupled with their specific traits. For example, *Rhionaeschna mutata* requires small, heavily vegetated, acidic, fish-free ponds with a wooded riparian edge and with sphagnum moss present. Furthermore, the species is locally rare and confined to transient or ephemeral ponds. It appears to have been subject to multiple human stressors, especially the connecting of formerly isolated ponds enabling local fish invasion, so leading to increased turbidity of ponds and changes in the food web, as well as the loss of trees across the landscape, and perhaps even more stress from climate change and extreme weather events. Conservation of these increasingly rare habitats is a priority for the species (Schilling *et al.*, 2019).

Another rare and charismatic species, *Epiophlebia laidlawi*, is threatened by deforestation, water abstraction, erosion and cattle grazing in its Himalayan broadleaf and subtropical pine forest habitat. The best protection is conservation of specific habitats in vast protected areas. This has been partly done in Nepal. Here, water and wildlife conservation areas protect the springs and headwaters of important river catchments through prohibition of forest clearance and reforestation programmes to provide increased soil surface stabilization in formerly logged areas (Brockhaus and Hartmann, 2009).

Cordulegaster boltonii in the UK is an inhabitant of upland rivers and streams. It is adversely affected by conifer plantation forestry. The clearing of plantation trees to both sides of the streams along a width of 10 m or 10–50 m did not see a recovery of the species. This was due to inadequate recovery of its scarce bank habitat: undercut banks on the submerged parts of *Juncus* and *Molinia* buried in organic detritus (Ormerod *et al.*, 1990).

Sometimes human-disturbed habitats have great value for certain odonates. Small streams and rivulets in agricultural landscapes are threatened habitats throughout Europe. A population of *Coenagrion ornatum* was found in the Elbe valley, Czechia after it was thought to be regionally extinct. Further small populations were subsequently found, also in other river basins. Interestingly, it prefers highly altered slow velocity and silty stream sites. Ineffective management unfortunately extirpated the species instead of strengthening the existing population through a misunderstanding of its habitat preferences. It is now known that *C. ornatum* requires a diversity of macrophytes in small, well-warmed sections of channels without being shaded out by woody vegetation, and streams without gravel or concrete substrates. Ongoing use of channels, particularly for drainage, helps protect the species, so long as overgrowing vegetation is removed, and the stream bed is improved by establishing buffers or overflows in the form of development of rich vegetation along the watercourses (Harabiš and Dolný, 2015b) (Fig. 6.42).

6.6.3 Dragonfly reintroductions

Coenagrion mercuriale as a species is assessed as Near Threatened on the IUCN Red List, although

Fig. 6.42. An example of a locally rare and threatened damselfly species, *Coenagrion ornatum* (inset), with a specific habitat preference, and being supported by rural (left) and even urban streams (right). (Photos and scans: Aleš Dolný).

many of its populations across its geographic range in Europe and North Africa are either extinct or highly threatened (La Porta and Goretti, 2019). In Morocco, for example, nearly 45% of the populations are extinct in Tunisia and only a single population remains. Formerly, the species occupied areas that became subject to intense levels of human activity and it is now mainly restricted to high elevation areas, with remaining populations threatened by increasing demand for water, agricultural intensification and land conversion (Ferreira *et al.*, 2015). However, some new populations are being discovered as in north-western Algeria (El Bouhissi *et al.*, 2022).

In the UK, *Coenagrion mercuriale* occurs mainly in two habitat types in southern England. At a stronghold in the New Forest, it occurs in narrow streams or boggy runnels that emanate from base-rich substrates, while elsewhere it also occurs in the ditch systems around watery meadows in the flood plains of two chalk rivers. Overall, the species requires a habitat where the channel substrate is mostly silt and has wide underwater ledges (berms), as well as in-channel emergent dicot vegetation and bankside monocot vegetation. Trees have a negative impact on its density (Rouquette and Thompson, 2005).

The species was reintroduced to a recipient site in south-western England from the New Forest donor site about 140 km farther east, and where the species is locally abundant. The initial introduction was of 57 mature individuals, and then after improvement in the water supply by removal of birch trees and cattle brought in to graze the recipient site, as well as impounding a stream to reduce flow velocity, the species established. A further 400 females and 100 males were reintroduced two years later, along with ponies to help with grazing. The critical features of a suitable recipient site were no trees, grazing by livestock and the removal of accumulated silt at impoundments to ensure its population persistence. The reintroduction at the time was considered a success (Thompson 2010; Thompson *et al.*, 2015). In Algeria, where the species is severely threatened, the height of the vegetation for the larva for adult emergence is also important, increasing sharply up to 1 m tall, then tailing off above that (Mellal *et al.*, 2018).

When the *Coenagrion mercuriale* UK reintroduction is compared to that of *Ischnura gemina* in the San Francisco Bay area of the USA, there are some similarities in the importance of identifying exactly which aspects of the reception site are critical to success. These considerations led to *I. gemina* successfully establishing. However, the project ultimately failed due to lack of adequate follow-up management of the recipient site. The site required clearing of some of the aquatic vegetation, but not too much to have an adverse effect on the larval microhabitat or the bottom sediment. It also required the removal of the woody, shading vegetation on the banks to mimic historical natural disturbance (Hannon and Hafernik, 2007). Comparing the reintroductions of the two species emphasizes that it is vital to have follow-up management following the reintroduction of a donor set of individuals.

Urothemis edwardsii (Fig. 6.11), while common in many areas in Africa and beyond, is listed as Critically Endangered on the Mediterranean IUCN Red List (Riservato *et al.*, 2009). In northern Africa, it is restricted to just a few remaining water bodies in Algeria. The restoration activities combined introduction of both final instar larvae (up to 10% of the donor population each year) and transfer of eggs. Using final instar larvae was crucial because the level of mortality for this stage is low compared to that of eggs. However, the transfer of eggs was also important as they likely contributed to the increase in population size while also reducing overharvesting of the donor larvae population. Restoration success was also due to a great boost in population size of the recipient population through occupation of new microhabitats. Important too, was that there was no evidence of overharvesting of the donor population, making the project sustainable. Another point is that males wander up to 1.2 km away from the water and require a wide buffer around water bodies. However, some threats remain, including water pumping and livestock grazing at emergence sites. The potential for low genetic diversity and the possibility of future inbreeding depression may possibly be eased a little now that the metapopulation size has increased (Khelifa *et al.*, 2016). A further important finding is that there must be sufficient vegetation height for the exuviae (Affef, 2019). Follow-up studies until 2021 indicate that the local populations remain viable, but colonization of other potential sites is limited by distances over 1.2 km, human activities and infrastructures, and invasion by European carp *Cyprinus carpio*, notwithstanding that climate change and drought continue to threaten *U. edwardsii* in northern Africa (Zouaimia *et al.*, 2022) (Fig. 6.43).

Fig. 6.43. *Urothemis edwardsii* was successfully translocated as larvae and eggs across lakes with precisely the right habitat in El Kala National Park, Algeria. The success of the project, both in terms of establishment at the recipient site and any possible impact on the donor site, involved surveying adults (left) and particularly exuviae (right). (Photos: Rassim Khelifa).

Sometimes dragonfly conservation focuses on the protection of specific populations of widespread species. *Leucorrhinia dubia* declined in the UK greatly in previous decades. However, an important population remains at Chartley Moss Nature Reserve in central England. Its genome reveals that this population at least is most closely related to populations in Switzerland and France. This suggests that this species colonized Britain from south-western Europe after the retreat of ice following the Last Glacial Maximum (Johansson, 2017). The species inhabits acidic bog pools with floating sphagnum moss. A site was identified (Foulshaw Moss in Cumbria, UK) as a recipient site for the species from a remaining location, also in Cumbria, for the species. Reintroduction involved translocation of larvae of the species between 2010–2014 (Clarke, 2014). The result was excellent, and by the end of 2016, about 2000 exuviae were recorded, indicating successful breeding at the new site.

In Czechia, *Leucorrhinia dubia* was reintroduced to artificially created bog pools. Eighty last instar larvae were translocated in 2001. By 2015–2017, over 100 exuviae and 75 adults were recorded, by which time it was the dominant species, with no change in composition and abundance of the local dragonfly assemblage. Importantly, the translocation showed that just a small number of last instar larvae can lead to establishment of a self-sustaining and genetically viable population, so long as there is high-quality habitat available and active collaboration with nature conservation authorities, both of which are vital for success. Genetic analysis showed that the translocated population might in the future serve as a source of genetic variation for the original population should it be depleted (Dolný *et al.*, 2018).

6.7 Dragonflies As Biological Control Agents

Dragonflies supply many ecosystem services related to the condition and quality of both freshwaters and the local terrestrial environment. Among other activities, they also export nutrients from the aquatic environment to the terrestrial one, and are important predators, potentially of certain pest species (Simaika and Samways, 2008; May, 2019).

Dragonflies have been investigated as biological control (biocontrol) agents for reducing pest population levels, especially those of mosquitoes in tropical environments. Both realms (water and air) have been considered, with adult dragonflies as predators and potential biocontrol agents of the adult stage of mosquitoes, and larval dragonflies as predators and potential biocontrol agents of mosquito larvae. However, dragonflies, both adults and larvae, are largely opportunistic feeders. As larvae, dragonflies can also be cannibalistic, and are prey themselves, especially for fish. These traits do not make for typically ideal candidates as biocontrol agents, which usually have most potential when they are host-specific, density-dependent and free of enemies.

Furthermore, studies in the laboratory or in mesocosms, which may show that mosquito larvae can be controlled under such artificial conditions, do not necessarily translate into demonstrable reduction of mosquito populations in the wild. This is not to say that the results of some constrained studies do not contribute to suppression of certain pests, but that dragonflies in these studies must be able to prove themselves as reliable biocontrol agents. One case illustrating the value of laboratory studies comes from Pakistan. *Pantala flavescens* among several odonate species is an efficient predator of *Culex quinquefasciatus* (a vector of encephalitis and West Nile fever) third instar larvae and is highly resistant to the increasing level of dissolved ammonium and nitrate (Ilahi *et al.*, 2019).

Biocontrol agents come in three categories associated with their ecological and biogeographical context. The first category is natural control, which involves harnessing the predatory ability of potential biocontrol agents to control pests, often by manipulating the local landscape, a major component of agroecology. The second category is augmentative control (biocontrol by augmentation of numbers of individuals), where a potential agent is artificially boosted in numbers, following mass-rearing, for example, to control a pest. The third category, classical biological control, is the introduction of a foreign natural enemy to control a foreign pest, often with both having the same provenance. This last category can be discounted in the case of dragonflies, as they are generalist predators and would not get clearance to be introduced into foreign lands, and also, they are not at all easy to rear in huge numbers in an insectary as required for release in the field. It is the first two categories that we will focus on here.

6.7.1 Natural vector control and food webs

While there is no doubt that many odonate species can reduce some local populations of medically important pest species, the question is the extent to which dragonflies have genuine public health value. The issue is clearly articulated by Barry and Roberts (2014). Unconstrained populations of *Crocothemis erythraea* and *Ischnura evansi* reduce populations of the mosquito *Culex quinquefasciatus* (a vector of various eucaryote, nematode and viral diseases) by up to 75% over five days under semi-natural conditions. However, in the presence of alarm chemicals from the larger predator *Anax imperator*, the predation efficiency of *C. erythraea* is reduced by about half. *A. imperator* also preys directly on mosquito larvae, with the late instars choosing the largest available prey when these alternatives are available and so this predator may no longer be a major source of mosquito mortality under natural conditions. Although final instar individuals of *I. evansi* are unaffected by predator kairomones, the point is that natural environments contain multiple predators of different ages and size classes. This leads to complex direct and indirect interactions between predator guild members that may further reduce the overall effect on their prey, although presence of alternative prey can result in dietary shifts among the predators. In sum, the reticulate nature of these food webs poses limitations on the value of odonates as natural control agents.

However, odonates overall can play a role in the general suppression of certain disease vectors. In Mexico, although 2.4% of consumption of two species of *Aedes* mosquitoes is a low level, the fact that there are 12 odonate species locally present means that it is possible that odonates as a group may play a relevant role as mosquito vector regulators (Córdoba-Aguilar *et al.*, 2021), especially relative to landscape transformation and climate change (Rengifo-Correa *et al.*, 2022). This may also be the case in India where several odonate species are associated with urban small stormwater tanks (Gajbe, 2021).

Odonate predators (certain *Mecistogaster*, *Megaloprepus*, *Gynacantha* species) depress abundance of various species of mosquitoes both in water-filled tree holes and artificial pots in Panamanian tropical forests. Large odonates depress mosquito abundance better than smaller ones. Despite generic differences in growth rates and size of final instars, odonate genera do not differ in their ability to depress mosquito abundance. Tree-hole odonates inhabit mature and secondary forests in wet and dry lowlands of the Neotropics, Africa and Asia. As they are often the largest and most common predators in tree holes and smaller phytotelmata, which may also support disease vectors such as *Aedes albopictus*, odonates may possibly contribute to disease suppression in these warm environments (Fincke *et al.*, 1997).

6.7.2 Augmentative pest control using dragonfly species

A classic example of pest control is by *Crocothemis servilia* used against *Aedes aegypti* (a vector of yellow

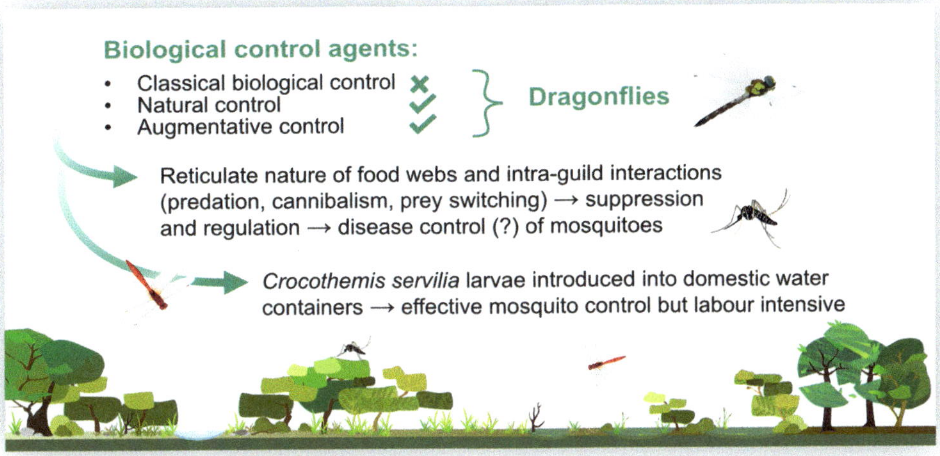

Fig. 6.44. Biological control agents come in three main use categories: classical biological control, natural control and augmentative control. The first can be discounted in the case of possible dragonfly agents, as this approach is not acceptable either in terms of policy or logistics. Dragonflies in the wild, while suppressing and to some extent regulating disease vectors such as mosquitoes, have yet to prove themselves as effective in disease control either for humans or other animals. Although augmentative control of mosquito larvae using the introduced larvae of *Crocothemis servilia* into the confined environment of domestic water containers is effective, it is labour intensive, inhibiting its use. (Illustration: Robyn Symons).

fever and other diseases) at the onset of the monsoon season in Myanmar. Most pre-adult *A. aegypti* occur in domestic water storage containers. Laboratory reared three-week old larvae of *C. servilia* were introduced into these containers. The result was a great drop in pre-adult mosquito population levels, and a drop in the adult population after six weeks. Success came about from confinement of both the prey and predator in the containers, with the dragonfly a voracious predator of mosquito larvae. There was no issue here with the reticulate nature of food webs as is normally the case among dragonflies and their prey. However, this specific control method against *A. aegypti* is labour intensive and depends on the enthusiasm of the local inhabitants for its success (Sebastian *et al.*, 1990) (Fig. 6.44).

References

Affef, B. (2019) Emergence ecology of the Critically Endangered *Urothemis edwardsii* in a new colonized site in El Kala National Park (Algeria). *Zoology and Ecology* 29, 140–145. DOI: 10.35513/21658005.2019.2.10.

Agra, J., Ligeiro, R., Heino, J., Macedo, D.R., Castro, D.M.P. *et al*. (2021) Anthropogenic disturbances alter the relationships between environmental heterogeneity and biodiversity of stream insects. *Ecological Indicators* 121, e107079. DOI: 10.1016/j.ecolind.2020.107079.

Amorín, M.A., Vila, M. and Cordero Rivera, A. (2009) Population genetic structure of two threatened dragonfly species (Odonata: Anisoptera) as revealed by RAPD analysis. In: Habel, J.C. and Assmann, T (eds) *Relict Species: Phylogeography and Conservation Biology*. Springer-Verlag, Berlin and Heidelberg, pp. 295–308. DOI: 10.1007/978-3-540-92160-8_17.

Andersen, E., Nilsson, B. and Sahlén, G. (2016) Survival possibilities of the dragonfly *Aeshna viridis* (Insecta, Odonata) in southern Sweden predicted from dispersal possibilities. *Journal of Insect Conservation* 20, 179–188. DOI: 10.1007/s10841-016-9850-5.

Ashworth, A. (2021) *Dragonfly: A Novel and Poetics (Poetics only)*. PhD thesis, Edge Hill University, Lancashire, UK.

Assandri, G., Bazzi, G., Maggioni, D., Galimberti, A. and Kunz, B (2020) Distribution, autecology, genetic characterization, and conservation of the Western Mediterranean endemic dragonfly *Orthetrum nitidinerve* (Selys, 1841): insights from Italy. *International Journal of Odonatology* 23, 405–422. DOI: 10.1080/13887890.2020.1828194.

Baba, Y.G., Kusumoto, Y. and Tanaka, K. (2019) Positive effect of environmentally friendly farming on paddy field odonate assemblages at a small landscape scale. *Journal of Insect Conservation* 23, 467–474. DOI: 10.1007/s10841-019-00132-2.

Bae, M.-J., Hwang, Y., Ham, S.-N., Kim, S.-Y. and Kim, E.J. (2023) Community recovery of benthic macroinvertebrates in a stream influenced by mining activity: importance of microhabitat monitoring. *Environmental Research* 234, e116499. DOI: 10.1016/j.envres.2023.116499.

Balázs, A., Šipoš, J., Matúšová, Z., Hamerlík, L., Novikmec, M. and Svitok, M. (2022) Comparison of conservation values among man-made aquatic habitats using Odonata communities in Slovakia. *Biologia* 77, 2549–2561. DOI:10.1007/s11756-022-01129-0.

Barnes, M.D., Glew, L., Wyborn, C. and Craigie, I.D. (2018) Prevent perverse outcomes from global protected area policy. *Nature Ecology and Evolution* 2, 759–762. DOI: 10.1038/s41559-018-0501-y.

Barry, M.J. and Roberts, D.M. (2014) Indirect interactions limit the efficiency of Odonata as natural control agents for mosquito larvae. *Journal of Insect Behavior* 27, 626–638. DOI: 10.1007/s10905-014-9455-3.

Barzoki, Z.E., Ebrahimi, M. and Sadeghi, S. (2020) Odonata diversity and species assemblages in the Northwest Central Plateau of Iran. *Journal of Insect Conservation* 24, 459–471. DOI: 10.1007/s10841-019-00211-4.

Batty, P. (2017) Odonata monitoring for the Scottish Beaver Trial 2009–2014 with notes on the post-trial period. *Journal of the British Dragonfly Society* 33, 79–103.

Bedê, L.C., Machado, A.B.M., Piper, W. and Magalhães de Souza, M. (2015) Odonata of the Serra de São José – Brazil's first Wildlife Reserve aimed at the conservation of dragonflies. *Notulae odonatologicae* 8, 117–155.

Bobrek, R. (2021) Post-mining ponds in the Sandomierz Forest (SE Poland) as an important site for the conservation of a species-rich odonate assemblage. *Acta zoologica cracoviensia* 64, 159–168. DOI: 10.3409/azc.64.02.

Bolam, F.C., Ahumada, J., Akçakaya, H.R., Brooks, T.M., Elliott, W. *et al.*, (2022) Over half of threatened species require targeted recovery actions to avert human-induced extinction. *Frontiers in Ecology and the Environment* 21, 64–70. DOI: 10.1002/fee.2537.

Bolpagni, R., Laini, A., Buldrini, F., Ziccardi, G., Soana, E. *et al.* (2020) Habitat morphology and connectivity better predict hydrophyte and wetland plant richness than land-use intensity in overexploited watersheds: evidence from the Po plain (northern Italy). *Landscape Ecology* 35, 1827–1839. DOI: 10.1007/s10980-020-01060-2.

Braccia, A., Voshell Jr, J.R. and Christman, V.D. (2007) The Odonata of newly constructed ponds with life history and production of dominant species. *Aquatic Insects* 29, 115–130. DOI: 10.1080/01650420701329279.

Bressane, A., Siminski, A., Gurjon Gomes, I., Peres Melo, C., Scofano da Rosa, G.C. *et al.* (2023) Prioritization of key indicators for the classification of successional stages in regenerating subtropical Atlantic forest, Southern Brazil: a proposal based on multivariate order statistics. *Environment Systems and Decisions* 43, 232–241. DOI: 10.1007/s10669-022-09881-z.

Bried, J.T. and Ervin, G.N. (2006) Abundance patterns along a wetland buffer. *Wetlands* 26, 878–883.

Briggs, A., Pryke, J.S., Samways, M.J. and Conlong, D.E. (2019a) Macrophytes promote insect conservation in artificial ponds. *Aquatic Conservation: Marine and Freshwater Ecosystems* 29, 1190–1201. DOI: 10.1002/aqc.3157.

Briggs, A.J., Pryke, J.S., Samways, M.J. and Conlong, D.E. (2019b) Complementarity among dragonflies across a pondscape in a rural landscape mosaic. *Insect Conservation and Diversity* 12, 241–250. DOI: 10.1111/icad.12339.

Brito, J.G., Roque, F.O., Martins, R.T., Nessiman, J.L., Oliveira, V.C. *et al.* (2020) Small forest losses degrade stream macroinvertebrate assemblages in the eastern Brazilian Amazon. *Biological Conservation* 241, e108263. DOI: 10.1016/j.biocon.2019.108263.

Brito, J., Calvão, L., Cunha, E., Maioli, L., Barbirato, M. *et al.* (2021) Environmental variables affect the diversity of adult damselflies (Odonata: Zygoptera) in western Amazonia. *International Journal of Odonatology* 24, 108–121. DOI: 10.23797/2159-6719_24_8.

Brito, J.S., Silva, E.C., Ferreira, V.R.S., Bastos, R.C., Cruz, G.M. *et al.* (2023) The importance of national parks in maintaining the habitat integrity and diversity of Odonata species in Amazonian streams. *Journal of Insect Conservation* (in press). DOI: 10.1007/s10841-023-00543-2.

Brockhaus, T. and Hartmann, A. (2009) New records of *Epiophlebia laidlawi* Tillyard in Bhutan, with notes on its biology, ecology, distribution, zoogeography and threat status (Anisozygoptera: Epiophlebiidae). *Odonatologica* 38, 203–215.

Brooks, T.M., Bakarr, M.I., Boucher, T., Da Fonseca, G.A.B., Hilton-Taylor, C. *et al.* (2004) Coverage provided by the Global Protected-Area System: is it enough? *BioScience* 54, 1081–1091. DOI: 10.1641/0006-3568(2004)054[1081:CPBTGP]2.0.CO;2.

Brudvig, L.A. (2017) Toward prediction in the restoration of biodiversity. *Journal of Applied Ecology* 54, 1013–1017. DOI: 10.1111/1365-2664.12940.

Brudvig, L.A., Barak, R.S., Bauer, J.T., Caughlin, T.T., Laughlin, D.C. *et al.* (2017) Interpreting variation to advance predictive restoration science. *Journal of Applied Ecology* 54, 1018–1027. DOI: 10.1111/1365-2664.12938.

Buczyński, P., Walczyk, K., Tańczuk, A., Buczyńska, E., Bojar, P. and Góral, N. (2022) Structural and physical factors as predictors of the species distribution and diversity of dragonflies and damselflies (Odonata) in an upland storage reservoir. *Folia Biologica (Kraków)* 70, 67–78. DOI:10.3409/fb_70-2.08.

Bush, A., Compson, Z.G., Monk, W.A., Porter, T.M., Steeves, R. *et al*. (2019) Studying ecosystems with DNA metabarcoding: lessons from biomonitoring of aquatic macroinvertebrates. *Frontiers in Ecology and Evolution* 7, e434. DOI: 10.3389/fevo.2019.00434.

Butchart, S.H.M., Scharlemann, J.P.W., Evans, M.I., Quader, S., Aricò, S. *et al*. (2012) Protecting important sites for biodiversity contributes to meeting global conservation targets. *PLoS ONE* 7, e32529. DOI: 10.1371/journal.pone.0032529.

Byers, O., Copsey, J., Lees, C., Miller, P. and Traylor-Holzer, K. (2022) Reversing the decline in threatened species through effective conservation planning. *Diversity* 14, e754. DOI: 10.3390/d14090754.

Callanan, M., Baars, J.-R. and Kelly-Quinn, M. (2014) Macroinvertebrate communities of Irish headwater streams: contribution to catchment biodiversity. *Biology and Environment: Proceedings of the Royal Irish Academy* 114, 143–162. DOI: 10.3318/ BIOE.2014.30.

Calvão, L.B., Noueira, D.S., Montag, L.F.A., Lopes, M.A. and Juen, L. (2016) Are Odonata communities impacted by conventional or reduced impact logging? *Forest Ecology and Management* 382, 143–150. DOI: 10.1016/j.foreco.2016.10.013.

Calvão, L.B., Brito, J.S., Ferreira, D., Cunha, E.J., Oliveira-Junior, J.M.B. and Juen, L. (2023) Effects of the loss of forest cover on odonate communities in eastern Amazonia. *Journal of Insect Conservation* 27, 205–218. DOI: 10.1007/s10841-022-00444-w.

Carchini, G., Della Bella, V., Solimini, A.G. and Bazzanti, M. (2007) Relationships between the presence of odonate species and environmental characteristics of lowland ponds of central Italy. *Annales de Limnologie-International Journal of Limnology* 43, 81–87.

Carmo, R.S., Fares, A.L.B.L., Pereira, G.F.M. and Michelan, T.S. (2023) Does the structure of riparian vegetation affect the diversity of macrophytes in eastern Amazonian streams? *Biologia* 78, 79–89. DOI: 10.1007/s11756-022-01181-w.

Chandran, A.V., Jose, S.K. and Gopalan, S.V. (2021) Dragonflies and damselflies (Insecta: Odonata) of the Kole Wetlands, central Kerala, India. *Journal of Threatened Taxa* 13, 17963–17971. DOI: 10.11609/ jott.5885.13.3. 17963-17971.

Chauvenet, A.L.M., Kuempel, C.D., McGowan, J., Beger, M. and Possingham, H.P. (2017) Methods for calculating protection equality for conservation planning. *PLoS ONE* 12, e0171591. DOI: 10.1371/journal. pone.0171591.

Chelli, A. and Moulaï, R. (2019) Ecological characterization of the odonatofauna in lotic and lentic waters of northeast Algeria. *Annales de la Société entomologique de France (N.S.)* 55, 430–445. DOI: 10.1080/00379271.2019.1660215.

Chen, W., Nover, D., Yen, H., Xia, Y., He, B. *et al*. (2020) Exploring the multiscale hydrologic regulation of multipond systems in a humid agricultural catchment. *Water Research* 184, e115987. DOI: 10.1016/j. watres.2020.115987.

Chien, W.C., Li, M.-H. and Li, H.-J. (2019) Appropriate management practices help enhance odonate species richness of small ponds in peri-urban landscapes. *Urban Ecosystems* 22, 757–768. DOI: 10.1007/s11252-019-00859-5.

Chovanec, A. (1994) Man-made wetlands in urban recreational areas: a habitat for endangered species? *Landscape and Urban Planning* 29, 43–54.

Chovanec, A. and Raab, R. (1997) Dragonflies (Insecta, Odonata) and the ecological status of newly created wetlands: examples for long-term bioindication programmes. *Limnologica* 27, 381–392.

Chovanec, A., Schiemer, F., Cabela, A., Gressler, S., Grötzer, C. *et al*. (2000) Constructed inshore zones in river corridors through urban areas-the Danube in Vienna: preliminary results. *Regulated Rivers: Research and Management* 16, 175–187.

Chovanec, A., Schiemer, F., Waidbacher, H. and Spolwind, R. (2002) Rehabilitation of a heavily modified river section of the Danube in Vienna (Austria): biological assessment of landscape linkages on different scales. *International Review of Hydrobiology* 87, 183–195.

Chowdhury, S., Jennions, M.D., Zalucki, M.P., Maron, M., Watson, J.E.M. and Fuller, R.A. (2023a) Protected areas and the future of inset conservation. *Trends in Ecology and Evolution* 38, 85–95. DOI: 10.1016/j. tree.2022.09.004.

Chowdhury, S., Zalucki, M.P., Hanson, J.O., Tiatragul, S., Green, D. *et al*. (2023b) Three-quarters of insect species are insufficiently represented by protected areas. *One Earth* 6, 1–8. DOI: 10.1016/j.oneear. 2022.12.003.

Clark, T.E. and Samways, M.J. (1996) Dragonflies (Odonata) as indicators of biotope quality in the Kruger National Park, South Africa. *Journal of Applied Ecology* 33, 1001–1012.

Clarke, D. (2014) The White-faced Darter (*Leucorrhinia dubia* Vander Linden) re-introduction project in Cumbria. *Journal of the British Dragonfly Society* 30, 54–78.

Clausnitzer, V. (2003) Dragonfly communities in coastal habitats of Kenya: indication of biotope quality and the need of conservation measures. *Biodiversity and Conservation* 12, 333–356.

Clausnitzer, V. and Lindeboom, M. (2002) Natural history and description of the dendrolimnetic larva of *Coryphagrion grandis* (Odonata). *International Journal of Odonatology* 5, 29–44. DOI: 10.1080/13887890.2002.9748175.

Colding, J., Lundberg, J., Lundberg, S. and Andersson, E. (2009) Golf courses and wetland fauna. *Ecological Applications* 19, 1481–1491. DOI: 10.1890/07-2092.1.

Córdoba-Aguilar, A., Miguel-Rodríguez, M.S., Rocha-Ortega, M., Lanz-Mendoza, H., Cime-Castillo, J. and Benelli, G. (2021) Adult damselflies as possible regulators of mosquito populations in urban areas. *Pest Management Science* 77, 4274–4287. DOI: 10.1002/ps.6496.

Craves, J.A., Cognato, A.I., O'Brien, D. and Mahoney, M.J. (2022) A new locality and unexpected haplotypes of the federally-endangered Hine's Emerald dragony, *Somatochlora hineana* (Odonata: Corduliidae). *Bulletin of American Odonatology* 13, 7–17.

Cunningham-Minnick, M.J., Meyer, T.B. and Crist, T.O. (2019) Shifts in dragonfly community structure across aquatic ecotones. *International Journal of Odonatology* 22, 121–133. DOI: 10.1080/13887890.2019.1615006.

D'Amico, F., Darblade, S., Avignon, S., Blanc-Manel, S. and Ormerod, S.J. (2004) Odonates as indicators of shallow lake restoration by liming: comparing adult and larval responses. *Restoration Ecology* 12, 439–446.

Dala-Corte, R.B., Melo, A.S., Siqueira, T., Bini, L.M., Martins, R.T. *et al.* (2020) Thresholds of freshwater biodiversity in response to riparian vegetation loss in the Neotropical region. *Journal of Applied Ecology* 57, 1391–1402. DOI: 10.1111/1365-2664.13657.

Damm, S., Schierwater, B. and Hadrys, H. (2010) An integrative approach to species discovery in odonates: from character-based DNA barcoding to ecology. *Molecular Ecology* 19, 3881–3893. DOI: 10.1111/j.1365-294X.2010.04720.x.

De Marco Jr, P., Nogueira, D.S., Correa, C.C., Vieira, T.B., Silva, K.D. *et al.* (2014) Patterns in the organization of Cerrado pond biodiversity in Brazilian pasture landscapes. *Hydrobiologia* 723, 87–101. DOI: 10.1007/s10750-013-1695-2.

Deacon, C. and Samways, M.J. (2021) A review of the impacts and opportunities for African urban dragonflies. *Insects* 12, e190. DOI: 10.3390/insects12030190.

Deacon, C., Samways, M.J. and Pryke, J.S. (2018) Artificial reservoirs complement natural ponds to improve pondscape resilience in conservation corridors in a biodiversity hotspot. *PLoS ONE* 13, e0204148. DOI: 10.1371/journal. pone.0204148.

Deacon, C., Samways, M.J. and Pryke, J.S. (2024) Interplay between pond size and matrix extent drives odonate diversity patterns in a fragmented landscape. *Biodiversity and Conservation* 32, 4767–4785. DOI: 10.1007/s10531-023-02726-4.

Di Minin, E. and Toivonen, T. (2015) Global protected area expansion: creating more than paper parks. *BioScience* 65, 637–638. DOI: 10.1093/biosci/biv064.

Dolný, A. and Harabiš, F. (2012) Underground mining can contribute to freshwater biodiversity conservation: allogenic succession forms suitable habitat for dragonflies. *Biological Conservation* 145, 109–117. DOI: 10.1016/j.biocon.2011.10.020.

Dolný, A., Šigutová, H., Ožana, S. and Choleva, L. (2018) How difficult is it to reintroduce a dragonfly? Fifteen years monitoring *Leucorrhinia dubia* at the receiving site. *Biological Conservation* 218, 110–117. DOI: 10.1016/j.biocon.2017.12.011.

Downing, J.A., Prairie, Y.T., Cole, J.J., Duarte, C.M., Tranvik, L.J. *et al.* (2006) The global abundance and size distribution of lakes, ponds, and impoundments. *Limnology and Oceanography* 51, 2388–2397.

Dvorak, B. (2023) Designing green roofs for dragonflies. *Living Architecture Monitor* Summer 2023. 1–6.

El Bouhissi, M., Chaib, S., Houhamdi, M. and Khelifa, R. (2022) Checklist of Odonata of Sidi Bel Abbes, Northwest Algeria. *Hydrobiology*, 1, 433–439. DOI: 10.3390/hydrobiology1040025.

Elo, M., Penttinen, J. and Kotiaho, J.S. (2015) The effect of peatland drainage and restoration on Odonata species richness and abundance. *BMC Ecology* 15, e11. DOI: 10.1186/s12898-015-0042-z.

Fait, P., Demierre, E., Ilg, C. and Oertli, B. (2020) Small mountain reservoirs in the Alps: new habitats for alpine freshwater biodiversity? *Aquatic Conservation: Marine and Freshwater Ecosystems* 30, 617–630. DOI: 10.1002/aqc.3306.

Faria, A.P.J., Ligeiro, R., Calvão, L.B., Giam, X., Leibold, M.A. and Juen, L. (2023) Land use types determine environmental heterogeneity and aquatic insect diversity in Amazonian streams. *Hydrobiologia* 851, 281–289. DOI: 10.1007/s10750-023-05190-x.

Feindt, W. and Hadrys, H. (2022) The damselfly genus *Megaloprepus* (Odonata: Pseudostigmatidae): revalidation and delimitation of species-level taxa including the description of one new species. *Zootaxa* 5115 (4), 487–510. DOI: 10.11646/zootaxa.5115.4.2.

Ferreira, S., Martínez-Freiría, F., Boudot, J.-P., Haissoufi, M.E., Bennas, N. *et al.* (2015) Local extinctions and range contraction of the endangered *Coenagrion mercuriale* in North Africa. *International Journal of Odonatology* 18, 137–152. DOI: 10.1080/13887890.2015.1017846.

Ferreira, S., Boudot, J.-P., El Haissoufi, M., Alves, P.C., Thompson, D.J. *et al.* (2016) Genetic distinctiveness of the damselfly *Coenagrion puella* in North Africa: an overlooked and endangered taxon. *Conservation Genetics* 17, 985–991. DOI: 10.1007/s10592-016-0826-5.

Ferreras-Romero, M., Márquez-Rodríguez, J. and Ruiz-García, A. (2009) Implications of anthropogenic disturbance factors on the Odonata assemblage in a Mediterranean fluvial system. *International Journal of Odonatology* 12, 413–428.

Fincke, O.M., Yanoviak, S.P. and Hanschu, R.D. (1997) Predation by odonates depresses mosquito abundance in water-filled tree holes in Panama. *Oecologia* 112, 244–253.

Friberg, N., Angelopoulos, N.V., Buijse, A.D., Cowx, I.G., Kail, J. *et al.* (2016) Effective river restoration in the

21st Century: from trial and error to novel evidence-based approaches. *Advances in Ecological Research* 55, 535–611. DOI: 10.1016/bs.aecr.2016.08.010.

Fuller, R.J., Williamson, T., Barnes, G. and Dolman, P.M. (2017) Human activities and biodiversity opportunities in pre-industrial cultural landscapes: relevance to conservation. *Journal of Applied Ecology* 54, 459–469. DOI: 10.1111/1365-2664.12762.

Gajbe, P.U. (2021) Impact of a small artificial water source on the diversity of odonates (Insecta: Odonata) in an urban landscape. *Arthropods* 10, 60–65.

Gibbons, P., Briggs, S.V., Ayers, D.A., Doyle, S., Seddon, J. *et al.* (2008) Rapidly quantifying reference conditions in modified landscapes. *Biological Conservation* 141, 2483–2493. DOI: 10.1016/j.biocon.2008.07.009.

Gillingham, P.K., Alison, J., Roy, D.B., Fox, R. and Thomas, C.D. (2015) High abundances of species in protected areas in parts of their geographic distributions colonized during a recent period of climate change. *Conservation Letters* 8, 97–106. DOI: 10.1111/conl.12118.

Godet, C. and Clauzel, C. (2021) Comparison of landscape graph modelling methods for analysing pond network connectivity. *Landscape Ecology* 36, 735–748. DOI: 10.1007/s10980-020-01164-9.

Goertzen, D. and Suhling, F. (2013) Promoting dragonfly diversity in cities: major determinants and implications for urban pond design. *Journal of Insect Conservation* 17, 399–409. DOI: 10.1007/s10841-012-9522-z.

Grant, P.B.C. and Samways, M.J. (2011) Micro-hotspot determination and buffer zone value for Odonata in a globally significant biosphere reserve. *Biological Conservation* 144, 772–781. DOI: 10.1016/j.biocon.2010.11.008.

Graziano, M.P., Deguire, A.K. and Surasinghe, T.D. (2022) Riparian buffers as a critical landscape feature: insights for riverscape conservation and policy renovations. *Diversity* 14, e172. DOI: 10.3390/d14030172.

Hamasaki, K., Yamanaka, T., Tanaka, K., Nakatani, Y., Iwasaki, N. and Sprague, D.S. (2011) Environmental characteristics accounting for Odonate assemblages in rural reservoir ponds in Japan. *Japan Agricultural Research Quarterly* 45, 187–196.

Hamerlík, L., Svitok, M., Novikmec, M., Očadlík, M. and Bitušík, P. (2014) Local, among-site, and regional diversity patterns of benthic macroinvertebrates in high altitude waterbodies: do ponds differ from lakes? *Hydrobiologia* 723, 41–52. DOI: 10.1007/s10750-013-1621-7.

Hannon, E.R. and Hafernik, J.E. (2007) Reintroduction of the rare damselfly *Ischnura gemina* (Odonata: Coenagrionidae) into an urban California park. *Journal of Insect Conservation* 11, 141–149. DOI: 10.1007/s10841-006-9027-8.

Hansson, L.-A., Brönmark, C., Nilsson, P.A. and Åbjörnsson, K. (2005) Conflicting demands on wetland ecosystem services: nutrient retention, biodiversity or both? *Freshwater Biology* 50, 705–714. DOI:10.1111/j.1365-2427.2005.01352.x.

Harabiš, F. (2017) Does the management of surrounding terrestrial habitats increase the tendency of odonates to leave aquatic habitats? *Biodiversity and Conservation* 26, 2155–2167. DOI: 10.1007/s10531-017-1350-8.

Harabiš, F. (2023) Post-mining areas as the last area for the expansion of the declining Siberian Winter damselfly (Odonata: Lestidae). *Journal of Insect Conservation* 27, 707–715. DOI: 10.1007/s10841-023-00491-x.

Harabiš, F. and Dolný, A. (2015a) Odonates need natural disturbances: how human-induced dynamics affect the diversity of dragonfly assemblages. *Freshwater Science* 34, 1050–1057. DOI: 10.1086/682234.

Harabiš, F. and Dolný, A. (2015b) Necessity for the conservation of drainage systems as last refugia for threatened damselfly species, *Coenagrion ornatum*. *Insect Conservation and Diversity* 8, 143–151. DOI: 10.1111/icad.12093.

Harabiš, F. and Dolný, A. (2018) Military training areas as refuges for threatened dragonfly species: effect of spatial isolation and military activity. *Biological Conservation* 217, 28–35. DOI: 10.1016/j.biocon.2017.10.021.

Harabiš, F., Tichanek, F. and Tropek, R. (2013) Dragonflies of freshwater pools in lignite spoil heaps: restoration management, habitat structure and conservation value. *Ecological Engineering* 55, 51–61. DOI: 10.1016/j.ecoleng.2013.02.007.

Harabiš, F., Simaika, J.P., Dolný, A., Luke, S.H., Elo, M. *et al.* (2023) Odonata as focal taxa for ecological restoration. In: Córdoba-Aguilar, A., Beatty, C.D. and Bried, J.T. (eds) *Dragonflies and Damselflies: Model Organisms for Ecological and Evolutionary Research*. 2nd ed. Oxford University Press, Oxford, pp. 401–412. DOI:10.1093/oso/9780192898623.003.0028.

Hart, L.A., Bowker, M.B., Tarboton, W. and Downs, C.T. (2014) Species composition, distribution and habitat types of Odonata in the iSimangaliso Wetland Park, KwaZulu-Natal, South Africa and the associated conservation implications. *PLoS ONE* 9, e92588. DOI:10.1371/journal.pone.0092588.

Heino, J., Alahuhta, J., Bini, L.M., Cai, Y, Heiskanen, A.-S. *et al.* (2020) Lakes in the era of global change: moving beyond single-lake thinking in maintaining biodiversity and ecosystem services. *Biological Reviews* 96, 89–106. DOI: 10.1111/brv.12647.

Higgins, J., Zablocki, J., Newsock, A., Krolopp, A., Tabas, P. and Salama, M. (2021) Durable freshwater protection: a framework for establishing and maintaining long-term protection for freshwater ecosystems and the values they sustain. *Sustainability* 13, e1950. DOI: 10.3390/su13041950.

Hill, M.J. and Wood, P.J. (2014) The macroinvertebrate biodiversity and conservation value of garden and field ponds along a rural-urban gradient. *Fundamental and Applied Limnology* 185, 107–119. DOI: 10.1127/fal/2014/0612.

Holgerson, M.A., Ray, N.E. and Russ, C. (2024) High rates of carbon burial linked to autochthonous production in artificial ponds. *Limnology and Oceanography Letters* 9, 43–51. DOI: 10.1002/lol2.10351.

Holtmann, L., Juchem, M., Brüggeshemke, J., Möhlmeyer, A. and Fartmann, T. (2018) Stormwater ponds promote dragonfly (Odonata) species richness and density in urban areas. *Ecological Engineering* 118, 1–11. DOI: 10.1016/j.ecoleng.2017.12.028.

Holtmann, L., Brüggeshemke, J., Juchem, M. and Fartmann, T. (2019) Odonatae assemblages of urban stormwater ponds: the conservation value depends on pond type. *Journal of Insect Conservation* 23, 123–132. DOI: 10.1007/s10841-018-00121-x.

Hood, G.A. and Larson, D.G. (2015) Ecological engineering and aquatic connectivity: a new perspective from beaver-modified wetlands. *Freshwater Biology* 60, 198–208. DOI: 10.1111/fwb.12487.

Hu, F.-S. and Futahashi, R. (2023) Molecular phylogenetic analysis and its impact on the conservation of *Ischnura rubilio* Selys, 1876 (Odonata: Coenagrionidae) in Taiwan. *International Journal of Odonatology* 26, 63–73. DOI: 10.48156/1388.2023.1917031.

Hughes, J., Lucet, V., Barrett, G., Moran, S., Manseau, M. *et al.* (2023) Comparison and parallel implementation of alternative moving-window metrics of the connectivity of protected areas across large landscapes. *Landscape Ecology* 38, 1411–1430. DOI: 10.1007/s10980-023-01619-9.

Huikkonen, I.-M., Helle, I. and Elo, M. (2020) Heterogenic aquatic vegetation promotes abundance and species richness of Odonata (Insecta) in constructed agricultural wetlands. *Insect Conservation and Diversity* 13, 374–383. DOI: 10.1111/icad.12396.

Hykel, M., Harabiš, F. and Dolný, A. (2016) Assessment of the quality of the terrestrial habitat of the threatened dragonfly, *Sympetrum depressiusculum* (Odonata: Libellulidae). *European Journal of Entomology* 113, 476–481. DOI: 10.14411/eje.2016.062.

Ilahi, I., Yousafzai, A.M., Attaullah, M., Haq, T.U., Ali, H. *et al.* (2019) The role of odonate nymphs in ecofriendly control of mosquitoes and sensitivity of odonate nymphs to inorganic nutrient pollutants. *Applied Ecology and Environmental Research* 17, 6171–6188. DOI: 10.15666/aeer/1703_61716188.

Iversen, L.L., Rannap, R., Briggs, L and Sand-Jensen, K. (2016) Variable history of land use reduces the relationship to specific habitat requirements of a threatened aquatic insect. *Population Ecology* 58, 155–164. DOI 10.1007/s10144-015-0516-z.

Janssen, A., Hunger, H., Konold, W., Pufal, G. and Staab, M. (2018) Simple pond restoration measures increase dragonfly (Insecta: Odonata) diversity. *Biodiversity and Conservation* 27, 2311–2328. DOI: 10.1007/s10531-018-1539-5.

Jarvie, H.P., Smith, D.R., Norton, L.R., Edwards, F.K., Bowes, M.J. *et al.* (2018) Phosphorus and nitrogen limitation and impairment of headwater streams relative to rivers in Great Britain: a national perspective. *Science of the Total Environment* 621, 849–862. DOI: 10.1016/j.scitotenv.2017.11.128.

Jeffries, M.J. (2016) Flood, drought and the inter-annual variation to the number and size of ponds and small wetlands in an English lowland landscape over three years of weather extremes. *Hydrobiologia* 768, 255–272. DOI: 10.1007/s10750-015-2554-0.

Johansson, A. (2014) Effects of beaver dams on benthic macroinvertebrates. MSc thesis, Uppsala University, Sweden.

Johansson, F. (2017) Population structure of *Leucorrhinia dubia* (Vander Linden) the White-faced Darter in Europe with special reference to the population at Chartley Moss. *Journal of the British Dragonfly Society* 33, 73–78.

Johansson, F., Bini, L.M., Coiffard, P., Svanbäck, R., Wester, J. and Heino, J. (2019) Environmental variables drive differences in the beta diversity of dragonfly assemblages among urban stormwater ponds. *Ecological Indicators* 106, e105529. DOI: 10.1016/j.ecolind.2019.105529.

Jones, C. and Harris, J. (2022) Beavers, trees and the freshwater environment. *Wood Wise: Tree and Woodland Conservation*. Spring 2022, 28–31.

Jooste, M.L., Samways, M.J. and Deacon, C. (2020) Fluctuating pond water levels and aquatic insect persistence in a drought-prone Mediterranean-type climate. *Hydrobiologia* 847, 1315–1326. DOI: 10.1007/s10750-020-04186-1.

Jordan, S., Barruet, E. and Olaf, M. (2007) Blue *hawaiiense* and beyond: conservation genetics and comparative phylogeography of four Hawaiian *Megalagrion* damselfly species (Odonata: Coenagrionidae). In: N.L. Evenhuis and J.M. Fitzsimons (eds) *Biology of Hawaiian Streams and Estuaries*. Bishop Museum Bulletin in Cultural and Environmental Studies 3, Hawai'i, pp. 247–260.

Juen, L., Cabette, H.S.R., and De Marco Jr., P. (2007) Odonate assemblage structure in relation to basin and aquatic habitat structure in Pantanal wetlands. *Hydrobiologia* 579, 125–134 DOI 10.1007/s10750-006-0395-6.

Kadoya, T., Suda, S.-i. and Washitani, I. (2004) Dragonfly species richness on man-made ponds: effects of pond size and pond age on newly established assemblages. *Ecological Research* 19, 461–467. DOI: 10.1111/j.1440-1703.2004.00659.x.

Kadoya, T., Suda, S-i., Nushihiro, J. and Washitani, I. (2008) Procedure for predicting the trajectory of species recovery based on the nested species pool information: dragonflies in a wetland restoration site as a

case study. *Restoration Ecology* 16, 397–406. DOI: 10.1111/j.1526-100X.2007.00314.x.
Kašák, J., Holuša, O. and Mazalová, M. (2023) Artificial habitat – a chance for survival of a rare montane dragonfly (Odonata): case study on an alpine emerald (*Somatochlora alpestris*). *Journal of Insect Conservation* 27, 315–321. DOI: 10.1007/s10841-023-00457-z.
Keddy, P.A. (2023) *Wetland Ecology: Principles and Conservation*, 3rd *edn*. Cambridge University Press, Cambridge.
Khelifa, R. and Mellal, M.K. (2017) Host-plant-based restoration as a potential tool to improve conservation status of odonate specialists. *Insect Conservation and Diversity* 10, 151–160. DOI: 10.1111/icad.12212.
Khelifa, R., Mellal, M.K., Zouaimia, A., Amari, H., Zebsa, R. *et al*. (2016) On the restoration of the last relict population of a dragonfly *Urothemis edwardsii* Selys (Libellulidae: Odonata) in the Mediterranean. *Journal of Insect Conservation* 20, 797–805. DOI: 10.1007/s10841-016-9911-9.
Kietzka, G.J., Pryke, J.S. and Samways, M.J. (2015) Landscape ecological networks are successful in supporting a diverse dragonfly assemblage. *Insect Conservation and Diversity* 8, 229–237. DOI: 10.1111/icad.12099.
Kietzka, G.J., Pryke, J.S., Gaigher, R. and Samways, M.J. (2021a) 32 years of essential management to retain value of an urban dragonfly awareness pond. *Urban Ecosystems* 24, 1295–1304. DOI: 10.1007/s11252-021-01115-5.
Kietzka, G.J., Pryke, J.S., Gaigher, R. and Samways, M.J. (2021b) Webs of well-designed conservation corridors maintain river ecosystem integrity and biodiversity in plantation mosaics. *Biological Conservation* 254, e108965. DOI: 10.1016/j.biocon.2021.108965.
Kim, M.J., Kim, J.S., Kim, S.-S. and Kim, I. (2020) Development and validation of microsatellite markers for an endangered dragonfly, *Libellula angelina* (Odonata: Libellulidae), with notes on population structures and genetic diversity. *International Journal of Odonatology* 23, 93–102. DOI: 10.1080/13887890.2019.1701573.
Kolar, V., Vlašánek, P. and Boukal, D.S. (2021) The influence of successional stage on local odonate communities in man-made standing waters. *Ecological Engineering* 173, e106440. DOI: 10.1016/j.ecoleng.2021.106440.
Krieger, A., Fartmann, T. and Poniatowski, D. (2019) Restoration of raised bogs: land-use history determines the composition of dragonfly assemblages. *Biological Conservation* 237, 291–298. DOI: 10.1016/j.biocon.2019.06.032.
Kuempel, C.D., Chauvenet, A.L.M. and Possingham, H.P. (2016) Equitable representation of ecoregions is slowly improving despite strategic planning shortfalls. *Conservation Letters* 9, 422–428. DOI: 10.1111/conl.12298.
Kupilas, B., Friberg, N., McKie, B.G., Jochmann, M.A., Lorenz, A.W. and Hering, D. (2016) River restoration and the trophic structure of benthic invertebrate communities across 16 European restoration projects. *Hydrobiologia* 769, 105–120. DOI: 10.1007/s10750-015-2569-6.
La Porta, G. and Goretti, E. (2019) Investigation on the declining Southern Damselfly (*Coenagrion mercuriale*, Odonata) in a Mediterranean population: survival rate and population size. *Journal of Insect Conservation* 23, 667–675. DOI: 10.1007/s10841-019-00160-y.
Lambret, P., Jeanmougin, M. and Stoks, R. (2023) Factors driving larval abundance and adult size of the threatened *Lestes macrostigma* (Odonata): keys for water management and habitat restoration. *Journal of Insect Conservation* 27, 389–402. DOI: 10.1007/s10841-023-00461-3.
Landmann, M., Schlick-Steiner, B.C., Steiner, F.M. and Landmann, A. (2021) Connectivity within isolation: dispersal, population genetics, and conservation of the rarest European damselfly. *Insect Conservation and Diversity* 14, 800–813. DOI: 10.1111/icad.12516.
Law, A., McLean, F. and Willby, N.J. (2016) Habitat engineering by beaver benefits aquatic biodiversity and ecosystem processes in agricultural streams. *Journal of Applied Ecology* 61, 486–499. DOI: 10.1111/fwb.12721.
Law, A., Levanoni, O., Foster, G., Ecke, F. and Willby, N.J. (2019) Are beavers a solution to the freshwater biodiversity crisis? *Diversity and Distributions* 25, 1763–1772. DOI: 10.1111/ddi.12978.
Le Gall, M., Fournier, M., Chaput-Bardy, A. and Husté, A. (2018) Determinant landscape-scale factors on pond odonate assemblages. *Freshwater Biology* 63, 306–317. DOI: 10.1111/fwb.13065.
Leal, C.G., Lennox, G.D., Ferraz, S.F.B., Ferreira, J., Gardner, T.A. *et al*. (2020) Integrated terrestrial-freshwater planning doubles conservation of tropical aquatic species. *Science* 370 (6512), 117–121. DOI: 10.1126/science.aba7580.
Leroux, S.J., Krawchuk, M.A., Schmiegelow, F., Cumming, S.G., Lisgo, K. *et al*. (2010) Global protected areas and IUCN designations: do the categories match the conditions? *Biological Conservation* 143, 609–616. DOI: 10.1016/j.biocon.2009.11.018.
Liberto, B. and Studinski, J. (2021) Abundance and habitat associations of disjunct and regionally rare populations of *Leucorrhinia glacialis* and *L. hudsonica* in the Appalachian Mountains (Odonata: Libellulidae). *Odonatologica* 50, 187–202. DOI: 10.5281/zenodo.5703200.
Linares, M.S., dos Santos, L.B., Callisto, M. and Santos, J.C. (2023) Physical habitat condition as a key tool to

maintain freshwater biodiversity in neotropical artificial ponds, *Water Biology and Security* 2, e100189. DOI: 10.1016/j.watbs.2023.100189.

Linke, S., Pressey, R.L., Bailey, R.C. and Norris, R.H. (2007) Management options for river conservation planning: condition and conservation re-visited. *Freshwater Biology* 52, 918–938. DOI: 10.1111/j.1365-2427.2006.01690.x.

Lounibos, L.P., Escher, R.L., Dewald, L.B., Nishimura, N. and Larson, V.L. (1990) Odonata associated with water lettuce (*Pistia stratiotes* L.) in south Florida. *Odonatologica* 19, 359–366.

Low, V.L., Sofian-Azirun, M. and Norma-Rashid, Y. (2016) Playing hide-and-seek with the tiny dragonfly: DNA barcoding discriminates multiple lineages of *Nannophya pygmaea* in Asia. *Journal of Insect Conservation* 20, 339–343. DOI 10.1007/s10841-016-9860-3.

Lozano, F., Palacio, A., Ramos, L.S., Granato, L., Drozd, A. and Muzón, J. (2022) Recovery of local dragonfly diversity following restoration of an artificial lake in an urban area near Buenos Aires. *Basic and Applied Ecology* 58, 88–97. DOI: 10.1016/j.baae.2021.11.006.

Lürling, M., van Oosterhout, F., Mucci, M. and Waajen, G. (2023) Combination of measures to restore eutrophic urban ponds in the Netherlands. *Water* 15, e3599. DOI: 10.3390/w15203599.

Lynch, A.J., Cooke, S.J., Arthington, A.H., Baigun, C. Bossenbroek, L. *et al.* (2023) People need freshwater biodiversity. *WIRES Water* 10, e1633. DOI: 10.1002/wat2.1633.

Maasri, A., Jähnig, S.C., Adamescu, M.C., Adrian, R., Baigun, C. *et al.* (2022) A global agenda for advancing freshwater biodiversity research. *Ecology Letters* 25, 255–263. DOI: 10.1111/ele.13931.

Mabry, C. and Dettman, C. (2010) Odonata richness and abundance in relation to vegetation structure in restored and native wetlands of the Prairie Pothole Region, USA. *Ecological Restoration* 28, 475–484. DOI: 10.3368/er.28.4.475.

MacDonald, L.H. and Coe, D. (2007) Influence of headwater streams on downstream reaches in forested areas. *Forest Science* 53, 148–168. DOI: 10.1093/forestscience/53.2.148.

Magoba, R.N. and Samways, M.J. (2010) Recovery of benthic macroinvertebrate and adult dragonfly assemblages in response to large scale removal of riparian invasive alien trees. *Journal of Insect Conservation* 14, 627–636. DOI: 10.1007/s10841-010-9291-5.

Manfrin, A., Pilotto, F., Larsen, S., Tonkin, J.D., Lorenz, A.W. *et al.* (2023) Taxonomic and functional reorganization in Central European stream macroinvertebrate communities over 25 years. *Science of the Total Environment* 889, e164278. DOI: 10.1016/ j.scitotenv.2023.164278.

Martins, R.T., Brito, J., Dias-Silva, K., Leal, C.G., Leitão, R.P. *et al.* (2021) Low forest-loss thresholds threaten Amazonian fish and macroinvertebrate assemblage integrity. *Ecological Indicators* 127, e107773. DOI: 10.1016/j.ecolind.2021.107773.

Matushkina, N.A. and Lambret, P.H. (2011) Ovipositor morphology and egg laying behaviour in the dragonfly *Lestes macrostigma* (Zygoptera: Lestidae). *International Journal of Odonatology* 14, 69–82. DOI: 10.1080/13887890.2011.568190.

May, M.L. (2019) Odonata: Who they are and what they have done for us lately: classification and ecosystem services of dragonflies. *Insects* 10, e62. DOI:10.3390/insects10030062.

Mayer, P.M., Pennino, M.J., Newcomer-Johnson, T.A. and Kaushal, S.S. (2022) Long-term assessment of floodplain reconnection as a stream restoration approach for managing nitrogen in ground and surface waters. *Urban Ecosystems* 25, 879–907. DOI: 10.1007/s11252-021-01199-z.

Maynou, X., Martín, R. and Aranda, D. (2017) The role of small secondary biotopes in a highly fragmented landscape as habitat and connectivity providers for dragonflies (Insecta: Odonata). *Journal of Insect Conservation* 21, 517–530. DOI: 10.1007/s10841-017-9992-0.

Mellal, M.K., Zebsa, R., Bensouilah, M., Houhamdi, M. and Khelifa, R. (2018) Aspects of the emergence ecology of the regionally endangered *Coenagrion mercuriale* (Odonata: Coenagrionidae) in Northeast Algeria. *Zoology and Ecology* 28, 224–230. DOI: 10.1080/21658005.2018.1504405.

Mendes, T.P., Luiza-Andrade, A., Cabette, H.S.R. and Juen, L. (2018) How does environmental variation affect the distribution of dragonfly larvae (Odonata) in the Amazon-Cerrado transition zone in Central Brazil? *Neotropical Entomology* 47, 37–45. DOI: 10.1007/s13744-017-0506-2.

Mendes, T.P., Benone, N.L. and Juen, L. (2019) To what extent can oil palm plantations in the Amazon support assemblages of Odonata larvae? *Insect Conservation and Diversity* 12, 448–458. DOI: 10.1111/icad.12357.

Mendes, T.P., Montag, L.F.A., Alvarado, S.T. and Juen, L. (2021) Assessing habitat quality on alpha and beta diversity of Odonata larvae (Insect) in logging areas in Amazon forest. *Hydrobiologia* 848, 1147–1161. DOI: 10.1007/s10750-021-04524-x.

Meyer, J.L., Strayer, D.L., Wallace, J.B., Eggert, S.L., Helfman, G.S. and Leonard, N.E. (2007) The contribution of headwater streams to biodiversity in river networks. *Journal of the American Water Resources Association* 43, 86–103. DOI: 10.1111/j.1752-1688.2007.00008.x.

Minot, M., Aubert, M. and Husté, A. (2021) Pond creation and restoration: patterns of odonate colonization and community dynamics. *Biodiversity and Conservation* 30, 4379–4399. DOI: 10.1007/s10531-021-02312-6.

Modiba, R.V., Joseph, G.S., Seymour, C.L., Fouché, P. and Foord, S.H. (2017) Restoration of riparian systems through clearing of invasive plant species improves functional diversity of Odonate assemblages. *Biological Conservation* 214, 46–54. DOI: 10.1016/j.biocon.2017.07.031.

Monzó, J. and Verdú, J.R. (2022) Effects of restoration and management of Mediterranean traditional water systems on Odonata alpha diversity: a long-term monitoring survey. *Biodiversity and Conservation* 31, 227–243. DOI: 10.1007/s10531-021-02331-3.

Moyle, P.B. (2014) Novel aquatic ecosystems: the new reality for streams in California and other Mediterranean climate regions. *River Research and Applications* 30, 1335–1344. DOI: 10.1002/rra.2079.

Müller, Z., Jakab, T., Tóth, A., Dévai, G., Szállassy, N. *et al.* (2003) Effect of sports fisherman activities on dragonfly assemblages on a Hungarian river floodplain. *Biodiversity and Conservation* 12, 167–179.

Muzón, J. and Von Ellenreider, N. (1999) Status and distribution of Odonata (Insecta) within natural protected areas in Argentina. *Biogeographica* 75, 119–128.

Nava-Bolaños, A., Vrech, D.E., Peretti, A.V. and Córdoba-Aguilar, A. (2021) Argentinian odonates (dragonflies and damselflies): current and future distribution and discussion of their conservation. *Journal of Threatened Taxa* 13, 19448–19465. DOI: 10.11609/jott.7166.13.11.19448-19465.

Oberdorff, T. (2022) Time for decisive actions to protect freshwater ecosystems from global changes. *Knowledge and Management of Aquatic Ecosystems* 423, e19. DOI: 10.1051/kmae/2022017.

Oertli, B., Joye, D.A., Castella, E., Juge, R., Cambin, D. and Lachavanne, J.-B. (2002) Does size matter? The relationship between pond area and biodiversity. *Biological Conservation* 104, 59–70.

Oliveira, P.C.R., van der Geest, H.G., Kraak, M.H.S., Westveer, J.J., Verdonschot, R.C.M. and Verdonschot, P.F.M. (2020) Over forty years of lowland stream restoration: lessons learned? *Journal of Environmental Management* 264, e110417. DOI: 10.1016/j.jenvman.2020.110417.

Ormerod, S.J., Weatherley, N.S. and Merrett, W.J. (1990) The influence of conifer plantations on the distribution of the Golden Ringed Dragonfly *Cordulegaster boltoni* (Odonata) in upland Wales. *Biological Conservation* 53, 241–251.

Osborn, R. and Samways, M.J. (1996) Determinants of adult dragonfly patterns at new ponds in South Africa. *Odonatologica* 25, 49–58.

Ostertag, R., Warman, L., Cordell, S. and Vitousek, P.M. (2015) Using plant functional traits to restore Hawaiian rainforest. *Journal of Applied Ecology* 52, 805–809. DOI: 10.1111/1365-2664.12413.

Palmer, M.A. (2016) Persistent and emergent themes in the linkage of theory to restoration practice. In: Palmer, M.A., Zedler, J.B. and Falk, D.A. (eds) *Foundations of Restoration Ecology*, 2nd ed. Island Press, Washington DC, pp. 517–531. DOI: 10.5822/978-1-61091-698-1_18.

Parks, S.A., Holsinger, L.M., Abatzoglou, J.T., Littlefield, C.E. and Zeller, K.A. (2023) Protected areas not likely to serve as steppingstones for species undergoing climate-induced range shifts. *Global Change Biology* 29, 2681–2696. DOI: 10.1111/gcb.16629.

Paz, V., Baños-Picón, L., Rosas-Ramos, N., Tobajas, E., Tormos, J. and Asís, J.D. (2021) The role of artificial ponds in maintaining dragonfly populations in an intensified farmland landscape. a case of study in Zamora, Spain. *Journal of Natural History* 54, 2439–2454. DOI: 10.1080/00222933.2020.1850901.

Perron, M.A.C. and Pick, F.R. (2020) Stormwater ponds as habitat for Odonata in urban areas: the importance of obligate wetland plant species. *Biodiversity and Conservation* 29, 913–931. DOI: 10.1007/s10531-019-01917-2.

Petersen, C.R., Jovanovic, N.Z. and Grenfell, M.C. (2020) The effectiveness of riparian zones in mitigating water quality impacts in an agriculturally dominated river system in South Africa. *African Journal of Aquatic Science* 45, 336–349. DOI: 10.2989/16085914.2019.1685451.

Petrovičová, K., Langraf, V., David, S., Krumpálová, Z. and Schlarmannová, J. (2021) Distinct Odonata assemblage variations in lentic reservoirs in Slovakia (Central Europe). *Biologia* 76, 3727–3741. DOI: 10.1007/s11756-021-00864-0.

Pinilla-Rosa, M., García-Saúco, G., Santiago, A., Ferrandis, P. and Méndez, M. (2023) Can botanic gardens serve as refuges for taxonomic and functional diversity of Odonata? The case of the botanic garden of Castilla–La Mancha (Spain). *Limnology* 24, 37–50. DOI: 10.1007/s10201-022-00704-3.

Pires, M.M., Kotzian, C.B., Dalzochio, M.S., Sganzerla, C., Prass, G. and Périco, E. (2019) Diversity of Odonata (Insecta) in Seasonal Deciduous Forest fragments in southern Brazil (state of Rio Grande do Sul), with a new record for the state and comments on the seasonal distribution of the species. *Biota Neotropica* 19, e20190769. DOI: 10.1590/1676-0611-BN-2019-0769.

Pires, M.M., Müller, N.F.D., Stenert, C. and Maltchik, L. (2020) Influence of different riparian vegetation widths and substrate types on the communities of larval Odonata (Insecta) in southern Brazilian streams. *Acta Limnologica Brasiliensia* 32, e301. DOI: 10.1590/S2179-975X2520.

Polak, T., Watson, J.E.M., Fuller, R.A., Joseph, L.N., Martin, T.G. *et al.* (2015) Efficient expansion of global protected areas requires simultaneous planning for species and ecosystems. *Royal Society Open Science* 2, e150107. DOI: 10.1098/rsos.150107.

Pompilio, L. and Mosini, A. (2021) The Odonate fauna of Val Grande National Park (Insecta: Odonata). *Fragmenta entomologica* 53, 25–42. DOI: 10.13133/2284-4880/434.

Primack, R., Kobori, H. and Mori, S. (2000) Dragonfly pond restoration promotes conservation awareness in Japan. *Conservation Biology* 14, 1553–1554.

Pryke, J.S. and Samways, M.J. (2012) Ecological networks act as extensions of protected areas for arthropod biodiversity conservation. *Journal of Applied Ecology* 49, 591–600. DOI: 10.1111/j.1365-2664.2012.02142.x.

Pryke, J.S., Samways, M.J. and De Saedeleer, K. (2015) An ecological network is as good as a major protected area for conserving dragonflies. *Biological Conservation* 191, 537–545. DOI: 10.1016/j.biocon.2015.07.036.

Raebel, E.M., Merckx, T., Feber, R.E., Riordan, P., Macdonald, D.W. and Thompson, D.J. (2012a) Identifying high-quality pond habitats for Odonata in lowland England: implications for agri-environment schemes. *Insect Conservation and Diversity* 5, 422–432. DOI: 10.1111/j.1752-4598.2011.00178.x.

Raebel, E.M., Merckx, T., Feber, R.E., Riordan, P., Thompson, D.J. and Macdonald, D.W. (2012b) Multi-scale effects of farmland management on dragonfly and damselfly assemblages of farmland ponds. *Agriculture, Ecosystems and Environment* 161, 80–87. DOI: 10.1016/j.agee.2012.07.015.

Raimundo, R.L.G., Guimarães Jr, P.R. and Evans, D.M. (2018) Adaptive networks for restoration ecology. *Trends in Ecology and Evolution* 33, 664–675. DOI: 10.1016/j.tree.2018.06.002.

Ranius, T., Widenfalk, L.A., Seedre, M., Lindman, L., Felton, A. *et al.* (2023) Protected area designation and management in a world of climate change: a review of recommendations. *Ambio* 52, 68–80. DOI: 10.1007/s13280-022-01779-z.

Reis, V., Hermoso, V., Hamilton S.K., Bunn, S.E. and Linke, S. (2019) Conservation planning for river-wetland mosaics: a flexible spatial approach to integrate floodplain and upstream catchment connectivity. *Biological Conservation* 236, 356–365. DOI: 10.1016/j.biocon.2019.05.042.

Rengifo-Correa, L., Rocha-Ortega, M. and Córdoba-Aguilar, A. (2022) Modeling mosquitoes and their potential odonate predators under different land uses. *EcoHealth* 19, 417–426. DOI: 10.1007/s10393-022-01600-z.

Renner, S., Dalzochio, M.S., Périco, E., Sahlén, G. and Suhonen, J. (2020) Odonate species occupancy frequency distribution and abundance–occupancy relationship patterns in temporal and permanent water bodies in a subtropical area. *Ecology and Evolution* 10, 7525–7536. DOI: 10.1002/ece3.6478.

Ribeiro, C., Rodrigues, M.E., Sahlén, G. and de Oliveira Roque, F. (2022) Dragonflies within and outside a protected area: a comparison revealing the role of well-preserved atlantic forests in the preservation of critically endangered, phytotelmatous species. *Journal of Insect Conservation* 26, 271–282. DOI: 0.1007/s10841-022-00385-4.

Ribeiro, R.A.B., Juen, L. and Brasil Riservato, L.S. (2009) Habitat conditions in streams influence Odonata larval assemblages in the eastern Amazon. *International Journal of Odonatology* 25, 22–30. DOI: 10.48156/1388.2022.1917160

Riservato, E., Boudot, J.-P., Ferreira, S., Jović, M., Kalkman, V.J. *et al.* (2009) *The Status and Distribution of Dragonflies of the Mediterranean Basin*. IUCN, Gland, Switzerland and Malaga, Spain.

Rith-Najarian, J.C. (1998) The influence of forest vegetation variables on the distribution and diversity of dragonflies in a northern Minnesota forest landscape: a preliminary study (Anisoptera). *Odonatologica* 27, 335–351.

Roble, S.M. (1999) Dragonflies and damselflies (Odonata) of the Shenandoah Valley sinkhole pond system and vicinity, Augusta County, Virginia. *Banisteria* No.13, 101–127.

Roble, S.M., Carle, F.L. and Flint Jr, O.S. (2009) Dragonflies and damselflies (Odonata) of the Laurel Fork Recreation Area, George Washington National Forest, Highland County, Virginia: possible evidence for climate change. In: Roble, S.M. and Mitchell, J.C. (eds) *A Lifetime of Contributions to Myriapodology and the Natural History of Virginia: A Festschrift in Honor of Richard L. Hoffman's 80th Birthday*. Virginia Museum of Natural History Special Publication No. 16, Martinsville, VA, USA, pp. 365–399.

Rodrigues, M.E., de Oliveira Roque, F., Quintero, J.M.O., de Castro Pena, J.C., de Sousa, D.C. and De Marco Junior, P. (2016) Nonlinear responses in damselfly community along a gradient of habitat loss in a savanna landscape. *Biological Conservation* 194, 113–120. DOI: 10.1016/j.biocon.2015.12.001.

Rodrigues, M., Nayakkan, A., Nair, V.P., Rowther B.E. and Roshnath, R. (2022) Odonata of Kattampally wetland, Kerala state, India. *Notulae odonatologicae* 9, 441–450. DOI: 10.5281/nodo.v9i9.a7.

Rodríguez-Tapia, G., Rocha-Ortega, M. and Córdoba-Auilar, A. (2020) An index to estimate the vulnerability of damselflies and dragonflies (Insecta: Odonata) to land use changes using niche modelling. *Aquatic Insects* 41, 254–272. DOI: 10.1080/01650424.2020.1781191.

Rouquette, J.R. and Thompson, D.J. (2005) Habitat associations of the endangered damselfly, *Coenagrion mercuriale*, in a water meadow ditch system in southern England. *Biological Conservation* 123, 225–235. DOI: 10.1016/j.biocon.2004.11.011.

Roush, S.A. and Amon, J.P. (2003) Repopulation of restored wetland habitat by Odonata (dragonflies and damselflies). *Ecological Restoration* 21, 174–179.

Ruggiero, A., Céréghino, R., Figuerola, J., Marty, P. and Angélibert, S. (2008) Farm ponds make a contribution to the biodiversity of aquatic insects in a French agricultural landscape. *Comptes Rendus Biologies* 331, 298–308. DOI: 10.1016/j.crvi.2008.01.009.

Saha, P.D. and Gaikwad, S.M. (2014) Diversity and abundance of Odonata in parks and gardens of Pune city. *Journal of Entomology and Zoology Studies* 2, 308–316.

Samanta, T., Chatterjee, L., Roy, A.B., Sinha, S. and Besra, S. (2022) Importance of Ecopark, Kolkata in the context of sustainability, compared to Rajarhat grassland, as a habitat for Odonata (dragonflies and damselflies) diversity. *World News of Natural Sciences* 44, 165–175.

Samways, M.J. (1989) Farm dams as nature reserves for dragonflies (Odonata) at various altitudes in the Natal Drakensberg Mountains, South Africa. *Biological Conservation* 48, 181–187.

Samways, M.J. (1998) Establishment of resident Odonata populations on the formerly waterless Cousine island, Seychelles: an Island Biogeography Theory (IBT) perspective. *Odonatologica* 27, 253–258.

Samways, M.J. (2000) A conceptual model of restoration based on experiences from three remote oceanic islands. *Biodiversity and Conservation* 9, 1073–1083. DOI: 10.1023/A:1008974702634.

Samways, M.J. (2008a) *Dragonflies and Damselflies of South Africa*. Pensoft, Sophia, Bulgaria.

Samways, M.J. (2008b) Dragonflies as focal organisms in contemporary conservation biology. In: Córdoba-Aguilar, A. (ed.) *Dragonflies and Damselflies: Model Organisms for Ecological and Evolutionary Research*. Oxford University Press, Oxford, pp. 97–108.

Samways, M.J. (2020) *Insect Conservation: A Global Synthesis*. CAB International, Wallingford, UK.

Samways, M.J. and Steytler, N.S. (1996) Dragonfly (Odonata) distribution patterns in urban and forest landscapes, and recommendations for riparian management. *Biological Conservation* 78, 279–288. DOI: 10.1016/S0006-3207(96)00032-8.

Samways, M.J. and Stewart, D.A.B. (1997) An aquatic ecotone and its significance in conservation. *Biodiversity and Conservation* 6, 1429–1444.

Samways, M.J. and Grant, P.B.C. (2006) Regional response of Odonata to river systems impacted by and cleared of invasive alien trees. *Odonatologica* 35, 297–303.

Samways, M.J. and Grant, P.B.C. (2008) Elephant impact on dragonflies. *Journal of Insect Conservation* 12, 493–498. DOI: 10.1007/s10841-007-9089-2.

Samways, M.J. and Sharratt, N.J. (2010) Recovery of endemic dragonflies after removal of invasive alien trees. *Conservation Biology* 24, 267–277. DOI: 10.1111/j.1523-1739.2009.01427.x.

Samways, M.J. and Simaika, J.P. (2016) *Manual of Freshwater Assessment: Dragonfly Biotic Index*. South African National Biodiversity Institute, Pretoria, South Africa.

Samways, M.J. and Deacon, C. (2022) Extinction reprieve for the ancient and imperiled dragonflies at the southern tip of Africa. In: DellaSala, D.A. and Goldstein, M.I. (eds), *Imperiled: The Encyclopedia of Conservation*, vol. 1. Elsevier, Amsterdam, pp. 471–484. DOI: 10.1016/B978-0-12-821139-7.00047-7.

Samways, M.J., Caldwell, P.M. and Osborn, R. (1996a) Spatial patterns of dragonflies (Odonata) as indicators for design of a conservation pond. *Odonatologica* 25, 157–166.

Samways, M.J., Osborn, R. and van Heerden, I. (1996b) Distribution of benthic invertebrates at different depths in a shallow reservoir in the KwaZulu-Natal Midlands. *Koedoe* 39, 69–76.

Samways, M.J., Taylor, S. and Tarboton, W. (2005) Extinction reprieve following alien removal. *Conservation Biology* 19, 1329–1330. DOI: 10.1111/j.1523-1739.2005.00176.x.

Samways, M.J., Deacon, C., Kietzka, G.J., Pryke, J.S., Vorster, C. and Simaika, J.P. (2020) Value of artificial ponds for aquatic insects in drought-prone southern Africa: a review. *Biodiversity and Conservation* 29, 3131–3150. DOI: 10.1007/s10531-020-02020-7.

Santos, E.P., Wagner, H.H., Ferraz, S.F.B and Siqueira, T. (2020) Interactive persistent effects of past land-cover and its trajectory on tropical freshwater biodiversity. *Journal of Applied Ecology* 57, 2149–2158. DOI: 10.1111/1365-2664.13717.

Santos, F., Nicasio, K., Silva, K., Martins, J., Périco, E. *et al.* (2021) Can artificial ponds retain dragonfly (Insecta: Odonata) biodiversity? A preliminary study in the Brazilian Amazon. *Austral Entomology* 60, 698-706. DOI: 10.1111/aen.12574.

Santos, L.R. and Rodrigues, M.E. (2022a) Land uses for pasture and cacao cultivation modify the Odonata assemblages in Atlantic Forest areas. *Diversity* 14, e672. DOI: 10.3390/d14080672.

Santos, L.R. and Rodrigues, M.E. (2022b) Dragonflies (Odonata) in cocoa growing areas in the Atlantic Forest: taxonomic diversity and relationships with environmental and spatial variables. *Diversity* 14, e919. DOI: 10.3390/ d14110919.

Saura, S., Bastin, L., Battistella, L., Mandrici, A. and Dubois, G. (2017) Protected areas in the world's ecoregions: how well connected are they? *Ecological Indicators* 76, 144–158. DOI: 10.1016/j.ecolind.2016.12.047.

Sayer, C., Andrews, K., Shilland, E., Edmonds, N., Edmonds-Brown, R. *et al.* (2012) The role of pond management for biodiversity conservation in an agricultural landscape. *Aquatic Conservation: Marine and Freshwater Ecosystems* 22, 626–638. DOI: 10.1002/aqc.2254.

Scher, O. and Thièry, A. (2005) Odonata, Amphibia and environmental characteristics in motorway stormwater retention ponds (Southern France). *Hydrobiologia* 551, 237–251. DOI: 10.1007/s10750-005-4464-z.

Schilling, E.G., Lawrenz, R. and Kundel, H. (2019) A review of the reproductive habitat preferences and conservation challenges of a rare, transient, and ecologically restricted darner dragonfly: *Rhionaeschna mutata.*

International Journal of Odonatology 22, 1–9, DOI: 10.1080/13887890.2018.1554513.
Schloemer, S., Dalbeck, L. and Hamm, A. (2013) Biodiversity in beaverponds. Unpublished Conference Paper, University of Bonn.
Schmitz, O.J. (2012) Restoration of ailing wetlands. *PLoS Biology* 10, e1001248. DOI: 10.1371/journal.pbio.1001248.
Sebastian, A., Sein, M.M., Thu, M.M. and Corbet, P.S. (1990) Suppression of *Aedes aegypti* (Diptera: Culicidae) using augmentative release of dragonfly larvae (Odonata: Libellulidae) with community participation in Yangon, Myanmar. *Bulletin of entomological Research* 80, 223–232.
Shiffer, C.N., Leppo, B. and White, H.B. III (2015) Odonata of Beaver Dam, Huntingdon County, Pennsylvania: a record of faunal succession in a changing habitat. *Argia* 27, 12–23.
Šigutová, H., Šigut, M. and Dolný, A. (2015) Intensive fish ponds as ecological traps for dragonflies: an imminent threat to the endangered species *Sympetrum depressiusculum* (Odonata: Libellulidae). *Journal of Insect Conservation* 19, 961–974. DOI: 10.1007/s10841-015-9813-2.
Šigutová, H., Valušák, P. and Dolný, A. (2022) Highway stormwater ponds as islands of Odonata diversity in an agricultural landscape. *Science of the Total Environment* 837, e155774. DOI: 10.1016/j.scitotenv.2022.155774.
Silva, D. de p., De Marco, P. and Resende, D.C. (2010) Adult odonate abundance and community assemblage measures of stream ecological integrity: a case study. *Ecological Indicators* 10, 744–752. DOI: 10.1016/j.ecolind.2009.12.004.
Simaika, J.P. and Samways, M.J. (2008) Valuing dragonflies as service providers. In: Córdoba-Aguilar, A. (ed.) *Dragonflies: Model Organisms for Ecological and Evolutionary Research.* Oxford, University Press, Oxford, pp. 109–123.
Simaika, J.P. and Samways, M.J. (2011) Comparative assessment of indices of freshwater habitat conditions using different invertebrate taxon sets. *Ecological Indicators* 11, 370–378. DOI: 10.1016/j.ecolind.2010.06.005.
Simaika, J.P., Samways, M.J., Kipping, J., Suhling, F., Dijkstra, K.-D. B. *et al.* (2013) Continental-scale conservation prioritization of African dragonflies. *Biological Conservation* 157, 245–254. DOI: 10.1016/j.biocon.2012.08.039.
Simaika, J.P., Samways, M.J. and Frenzel, P.P. (2016) Artificial ponds increase local dragonfly diversity in a global biodiversity hotspot. *Biodiversity and Conservation* 25, 1921–1935. DOI 10.1007/s10531-016-1168-9.
Sisa, E.M., Kamb, J.-C.T., Owema, V.K., Mutambel, D.H. and Bunda, N.P.M. (2022) Structure of Odonata populations in the riparian strips of the Bumbu River watershed in Kinshasa/RD Congo. *International Journal of Science and Research Archive* 6, 28–39. DOI: 10.30574/ijsra.2022.6.1.0095.
Sitati, A., Yegon, M.J., Masese, F.O. and Graf, W. (2024) Ecological importance of low-order streams to macroinvertebrate community composition in Afromontane headwater streams. *Environmental and Sustainability Indicators* 21, e100330. DOI: 10.1016/j.indic.2023.100330.
Song, S., Wang, S., Fu, B., Liu, Y., Wang, K. *et al.* (2020) Sediment transport under increasing anthropogenic stress: regime shifts within the Yellow River, China. *Ambio* 49, 2015–2025. DOI: 10.1007/s13280-020-01350-8.
Stewart, D.A.B. and Samways, M.J. (1998) Conserving dragonfly (Odonata) assemblages relative to river dynamics in an African savanna game reserve. *Conservation Biology* 12, 683–692. DOI: 10.1111/j.1523-1739.1998.96465.x.
Steytler, N.S. and Samways, M.J. (1995) Biotope selection by adult male dragonflies (Odonata) at an artificial lake created for insect conservation in South Africa. *Biological Conservation* 72, 381–386. DOI: 10.1016/0006-3207(94)00052-R.
Strachan, S.R., Chester, E.T. and Robson, B.J (2016) Fringing trees may provide a refuge from prolonged drying for urban wetland invertebrates. *Urban Ecosystems* 19, 1213–1230. DOI: 10.1007/s11252-016-0548-y.
Strobl, K., Moning, C. and Kollmann, J. (2020) Positive trends in plant, dragonfly, and butterfly diversity of rewetted montane peatlands. *Restoration Ecology* 28, 796–806. DOI: 10.1111/rec.12957.
Suh, A.N. and Samways, M.J. (2001) Development of a dragonfly awareness trail in an African botanical garden. *Biological Conservation* 100, 345–353. DOI: 10.1016/S0006-3207(01)00038-6.
Suh, A.N. and Samways, M.J. (2005) Significance of temporal changes when designing a reservoir for conservation of dragonfly diversity. *Biodiversity and Conservation* 14, 165–178. DOI: 10.1007/s10531-005-3652-5.
Suhling, F., Martens, A. and Marais, E. (2009) How to enter a desert: patterns of Odonata colonisation of arid Namibia. *International Journal of Odonatology*, 12, 287–308. DOI: 10.1080/13887890.2009.9748346.
Sushko, G. (2021) Spatial variation in assemblages of Odonata (Insecta) within habitat gradients in large, pristine peat bogs in Belarus. *Biologia* 76, 575–583. DOI: 10.2478/s11756-020-00558-z.
Temperton, V.M., Buchmann, N., Buisson, E., Durigan, G., Kazmierczak, Ł. *et al.* (2019) Step back from the forest and step up to the Bonn Challenge: how a broad ecological perspective can promote successful landscape restoration. *Restoration Ecology* 27, 705–719. DOI: 10.1111/rec.12989.
Termaat, T., van Grunsven, R.H.A., Plate, C.L. and van Strien, A.J. (2015) Strong recovery of dragonflies in

recent decades in The Netherlands. *Freshwater Science* 34, 1094–1104. DOI: 10.1086/682669.

Thomas, C.D., Gillingham, P.K., Bradbury, R.B., Roy, D.B., Anderson, B.J. *et al.* (2012) Protected areas facilitate species' range expansions. *Proceedings of the National Academy of Sciences of the USA* 109, 14063–14068. DOI: 10.1073/pnas.1210251109.

Thomaz, S.M. (2023) Ecosystem services provided by freshwater macrophytes. *Hydrobiologia* 850, 2757–2777. DOI: 10.1007/s10750-021-04739-y.

Thompson, D.J. (2010) Re-introduction of the Southern Damselfly to Venn Ottery Common, Devon, UK. In: Soorae, P.S. (ed.) *Global Re-introduction Perspectives: Additional Case-studies from around the Globe*. IUCN/SSC Re-introduction Specialist Group, Abu Dhabi, UAE, pp. 29–32.

Thompson, D.J., Thompson, A.L., Kerry, L. and Gotham, P. (2015) Reintroduction and establishment of *Coenagrion mercuriale* (Charpentier) (Southern Damselfly) on Venn Ottery Common, Devon. *Journal of the British Dragonfly Society* 31, 14–23.

Tickner, D., Opperman, J.J., Abell, R., Acreman, M., Arthington, A.H. *et al.* (2020) Bending the curve of global freshwater biodiversity loss: an emergency recovery plan. *BioScience* 70, 330–342. DOI: 10.1093/biosci/biaa002.

Török, P. and Helm, A. (2017) Ecological theory provides strong support for habitat restoration. *Biological Conservation* 206, 85–91. DOI: 10.1016/j.biocon.2016.12.024.

Torres-Cambas, Y., Cabana-Ortero, M., Lorenzo-Carballa, M.O. and Cordero-Rivera, A. (2016) Conservation status and protection of three Antillean endemic damselflies. *Journal of Insect Conservation* 20, 277–284. DOI 10.1007/s10841-016-9862-1.

Trentanovi, G., Viviano, A., Mazza, G., Busignani, L., Magherini, E. *et al.* (2023) Riparian forests throwback at the Eurasian beaver era: a woody vegetation assessment for Mediterranean regions. *Biodiversity and Conservation* 32, 4259–4274. DOI: 10.1007/s10531-023-02687-8.

Van Grunsven, R., Bos, G. and Poot, M. (2020) Strong changes in Dutch dragonfly fauna. *Agrion* 24, 134–138.

Van Schalkwyk, J., Kietzka, G.J., Pryke, J.S., Gaigher, R. and Samways, M.J. (2023) Enhancing semi-aquatic species representativeness beyond protected areas: dragonflies in networks of conservation corridors. *Biodiversity and Conservation* 32, 3991–4005. DOI: 10.1007/s10531-023-02678-9.

van Strien, A.J. and van Grunsven, R.H.A. (2023) In the past 100 years dragonflies declined and recovered by habitat restoration and climate change. *Biological Conservation* 277, e109865. DOI: 10.1016/j.biocon.2022.109865.

Vega-Sánchez, Y.M., Oyama, K., Mendoza-Cuenca, L.F., Gaytán-Legaria, R. and González-Rodríguez, A. (2024) Genomic differentiation and niche divergence in the *Hetaerina americana* (Odonata) cryptic species complex. *Molecular Ecology* 33, e17207. DOI: 10.1111/mec.17207.

Verberk, W.C.E.P., van Duinen, G.A., Brock, A.M.T., Leuven, R.S.E.W., Siepel, H. *et al.* (2006) Importance of landscape heterogeneity for the conservation of aquatic macroinvertebrate diversity in bog landscapes. *Journal for Nature Conservation* 14, 78–90. DOI: 10.1016/j.jnc.2005.11.002.

Verpoorter, C., Kutser, T., Seekell, D.A. and Tranvik, L.J. (2014) A global inventory of lakes based on high-resolution satellite imagery. *Geophysical Research Letters* 41, 6396–6402. DOI: 10.1002/2014GL06064.

Vilenica, M., Katar, M., Koren, T. and Štih Koren, A. (2022a) Dragonfly fauna (Insecta: Odonata) of Papuk Nature Park, Croatia. *Natura Croatica: Periodicum Musei Historiae Naturalis Croatici* 31, 351–364. DOI: 10.20302/NC.2022.31.25.

Vilenica, M., Rebrina, F., Kepčija, R.M., Šegota, V., Rumišek, M. *et al.* (2022b) Aquatic macrophyte vegetation promotes taxonomic and functional diversity of Odonata assemblages in intermittent karst rivers in the Mediterranean. *Diversity* 14, e31. DOI: 10.3390/d14010031.

Vitor, K.C.A., Pagarigan, E.O., Tagoon, M.D.T., Garcia, M.M. and Bautista, M.G. (2022) Distribution and species richness of adult Odonata in urban wetlands in Tagum City, Mindanao, Philippines. *Philippine Journal of Science* 151 (6A), 2173–2182.

Watts, P.C., Saccheri, I.J., Kemp, S.J. and Thompson, D.J. (2007) Effective population sizes and migration rates in fragmented populations of an endangered insect (*Coenagrion mercuriale*: Odonata). *Journal of Applied Ecology* 76, 790–800. DOI: 10.1111/j.1365-2656.2007.01249.x.

Wilkes, M.A., Carrivick, J.L., Castella, E., Ilg, C., Cauvy-Fraunié, S. *et al.* (2023) Glacier retreat reorganizes river habitats leaving refugia for Alpine invertebrate biodiversity poorly protected. *Nature Ecology and Evolution* 7, 841–851. DOI: 10.1038/s41559-023-02061-5.

Willis, C.K. and Samways, M.J. (2011) *Water Dancers of South Africa's National Botanical Gardens*. South African National Biodiversity Institute, Pretoria, South Africa.

Willis, C.K. and Samways, M.J. (2013) *Dragonfly & Damselfly Trail Guide: KwaZulu-Natal National Botanical Garden*. South African National Biodiversity Institute, Pretoria, South Africa.

Wilson, H.B., Joseph, L.N., Moore, A.L. and Possingham, H.P. (2011) When should we save the most endangered species? *Ecology Letters* 14, 886–890. DOI: 10.1111/j.1461-0248.2011.01652.x.

Wood, K.L., Kaushal, S.S., Vidon, P.G., Mayer, P.M. and Galella, J.G (2022) Tree trade-offs in stream restoration:

impacts on riparian groundwater quality. *Urban Ecosystems* 25, 773–795. DOI: 10.1007/s11252-021-01182-8.

Worthen, W.B. and Chamlee, M.G. (2020) Determinants of adult odonate community structure at several spatial scales: effects of habitat type and landscape context. *International Journal of Odonatology* 23, 365–379. DOI: 10.1080/13887890.2020.1796831.

Zouaimia, A., Zebsa, R., Bensakhri, Z., Youcefi, A., Bensouilah, S. *et al*. (2022) Update on the geographic distribution of the critically endangered *Urothemis edwardsii* (Selys, 1849) (Odonata: Libellulidae) in northeastern Algeria. *Annales de la Société entomologique de France (N.S.)* 58, 366–372. DOI: 10.1080/00379271.2022.2099972.

Zulhariadi, M., Irawan, R.D., Zulfaeda, A., Hidayani, N. and Irawan. F. (2022) Dragonflies diversity and land cover changes in the Batubolong River, West Lombok District. *Biotropia* 29, 112–123. DOI: 10.11598/btb.2022.29.2.1637.

7 Freshwater Assessment and Monitoring Using Dragonflies

Illustration: Megan Louise Potgieter.

DOI: 10.1079/9781789248395.0007

Abstract

Assessment is about measuring the state of something, while monitoring measures its change. Freshwater bodies and their associated dragonfly assemblages have a natural historic state. While dynamic, it is the benchmark for measuring change, for worse or better.

As both dragonflies and water bodies, running and standing, are intimately connected with the terrestrial matrix, these amphibiotic insects have great merit for measuring the state of the local watery and land environment. In turn, the conservation of dragonflies depends on the environmental conditions being optimal for them.

Much progress is being made to assess, map and monitor freshwater types, their transformations and their improvement using traditional taxonomic measures alongside new satellite and genome technology, and novel analytical methods.

These advances are also indicating how, where and in what way conservation action should be instigated. These range from global trends in species and ecosystem change, through to effective improvements in the local environment.

7.1 Assessment and Monitoring Principles

7.1.1 Basic principles of assessment and monitoring in natural systems

7.1.1.1 Assessment as compared to monitoring

The natural environment is dynamic and forever changing, while also varying from one locality to the next. Assessment is the determination of a state or status of a focal subject or system at one time at a defined spatial scale (grain) or specified location(s). However, the larger the spatial scale or the greater the inclusion of more subjects, systems and/or states, the more challenging it is to undertake a complete and accurate assessment.

In turn, monitoring is ongoing reassessment over time to detect any changes that may have occurred in the state or status of a focal subject or system at a defined spatial scale or specified location(s). However, the larger the spatial scale or the greater the inclusion of more subjects, systems and/or states, and the longer the time over which monitoring is undertaken, the more challenging it is to sustain accurate reassessments on a regular basis (Fig. 7.1).

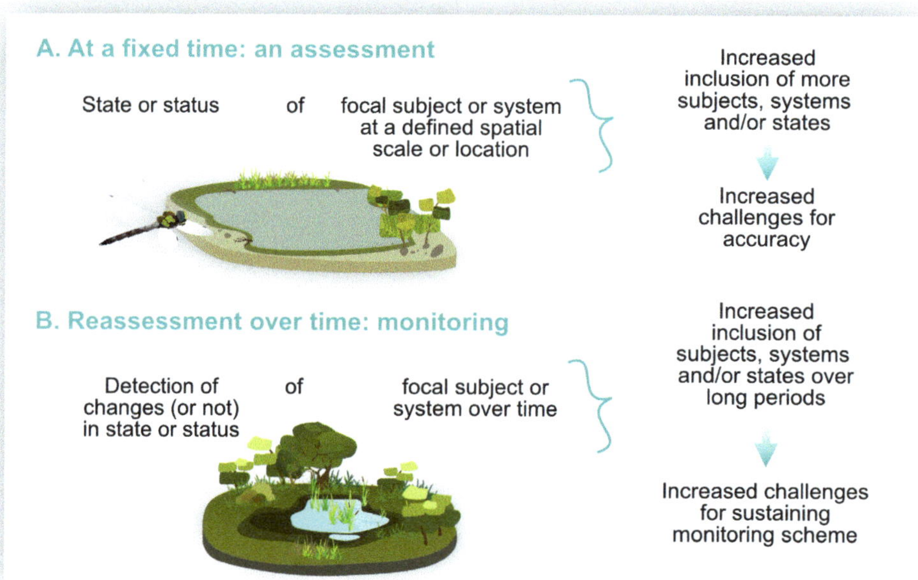

Fig. 7.1. The principles of assessment (A) versus monitoring (B), along with the challenges that both face. (Illustration: Robyn Symons).

7.1.1.2 Significance of spatial scale

In a defined spatial area, events taking place in that space are also influenced by events taking place in the matrix around it. This is because there is flow of material items (from molecules to individuals), energy and information across the permeable boundaries.

The external influence is also hierarchical. A dragonfly individual is present at any one time in a small space, its microhabitat. That microspace has a community of organisms living within it, and many individuals from various species interacting with each other. Furthermore, many individuals of several species may also move among microhabitats (metacommunity dynamics). In turn, the communities in the clusters of microhabitats form part of a larger ecosystem, such as a stream or a pond, positioned in a still larger region, a component of the world.

The small-scale areas are subject to specific short-term abiotic changes such as daily temperature variations and longer-term seasonal changes that affect the local systems on a regular basis. These in turn, are affected by larger events such as climate warming and regional extreme weather events. These events then affect the state of river catchments or wetlands, as well as all the ecosystems and communities within them.

7.1.1.3 Assessment of dragonfly populations and species

In dragonfly population and species conservation, we assess where individuals are positioned at a particular time and at what levels they are present. In the case of a dragonfly population, we may need to know how individuals are distributed across the landscape/waterscape. Developmental stage (larva, exuvia [though technically a developmental manifestation rather than a stage] or adult), sex ratio and perhaps even the physiological condition of the individuals are also considered. Their genetic status, such as whether there is any inbreeding depression, as well as the proportion of morphs, viability of metapopulations or effective population size are other important features. For localized single-species conservation, our goal may be, for example, to determine the viability of the component populations.

7.1.1.4 Multispecies assessments

At the multispecies level, that of dragonfly assemblages, we may want to assess their species richness, which is simply the number of species present at a specified locality. But because there might be species replacement or turnover, it is usually much more valuable to record assemblage composition (the identified specific species that are present). Species diversity (species composition combined with the relative abundance (evenness) of the various species) provides even more insight. However, 'species diversity' is often used instead of assemblage composition to state the mix of species without reference to their relative abundance at a specified location. While species diversity can be condensed into a single numerical index value, it can also be unpacked into a rank-abundance curve of named species (Fig. 7.2).

The number of species sampled in a small ecological unit, such as a community, site, transect or plot, is the unit's alpha diversity. Confusingly, although 'diversity' is given as an epithet, the term does not account for species abundance, just species richness or assemblage composition (the identified number of species present).

When two or more ecological units are compared, the change in the species (or other taxonomic unit) present in the different sampled units is beta diversity. It is composed of two facets: species replacement (species turnover) and shared species present in the different sampled units. The latter component can also be viewed as species nestedness, where species' losses or gains from one sampled unit to the next, are measured. For more than two sampled units, the overlap is known as zeta diversity (Fig. 7.3).

In turn, measuring changes in species in ecological units across a large area, such as a subregion or region, is known as gamma diversity. However, as conservation action is mostly undertaken by socio-political organizations, assessments are influenced by boundaries put in place by law, for example, for a farm, municipality, protected area, county, state/province or a country.

In the case of communities, we may be interested in the extent to which certain dragonfly species are contributing to interaction networks, such as food webs. At the larger spatial scale, the goal may be to assess ecosystem function and the extent to which dragonflies are reflecting ecosystem resilience. Alternatively, the focus may be, for example, the comparison of different ecosystems, landscapes or waterscapes, transformed versus natural.

7.1.1.5 Abiotic environmental variables

As dragonflies are amphibiotic, we must first consider the abiotic variables, including in the freshwater

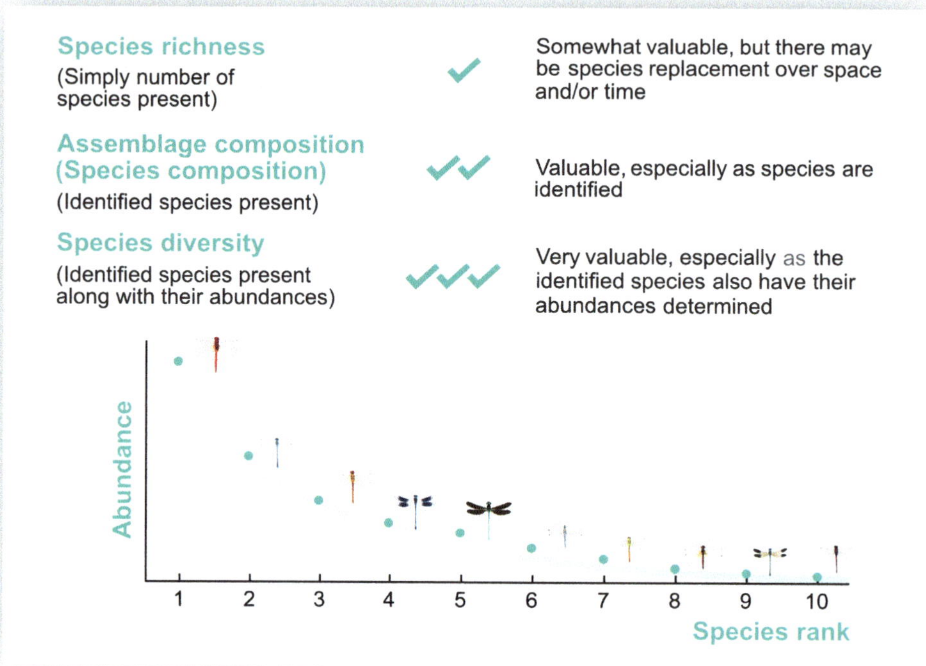

Fig. 7.2. There are three basic ways of recording the species present in a sampled ecological unit, with increasing value for conservation purposes (represented by an increased number of ticks). Species diversity can be described using a single index, but is best visualized with a rank abundance curve, as shown here for an assemblage of ten species. (Illustration: Robyn Symons).

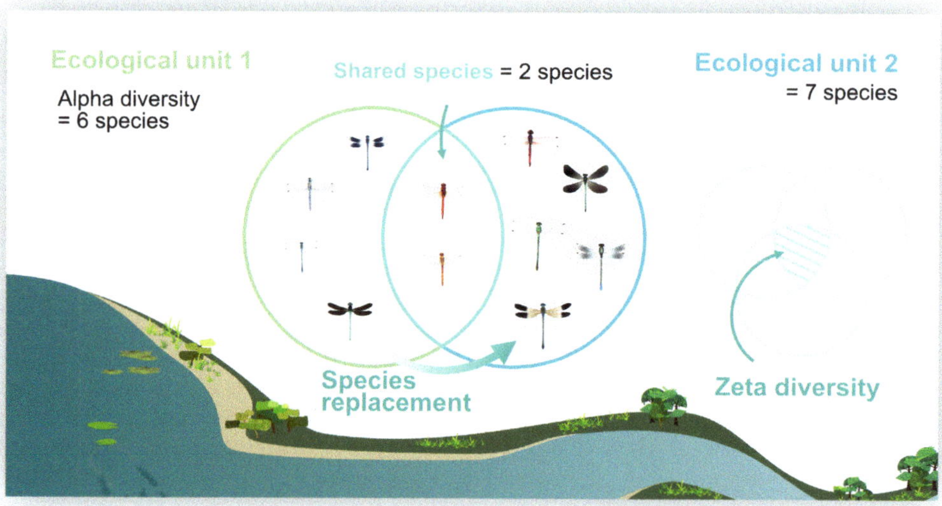

Fig. 7.3. Alpha diversity is the number of sampled species (or another taxonomic unit) present in one ecological unit. In turn, different ecological units often have different species richness and composition. The similarities/differences are beta diversity, composed of: a) shared species in the ecological units (two species here); and b) species replacement (species turnover), with four species in ecological unit 1 being replaced by five species in unit 2. The overlap of species among more than two ecological units is zeta diversity. (Illustration: Robyn Symons).

realm, factors such as water temperature, velocity, turbidity and other physicochemical conditions. Depending on the focal system, these factors may be optimal, or they may not be, with the possibility of the presence of contaminants or pollutants. In the aerial realm of the adult, factors such as temperature, shade levels and wind velocity are among the important considerations when assessing the quality of the riparian zone. At larger spatial scales, the terrestrial matrix around a water body and the effects emanating from upstream in the catchment are also important.

Furthermore, there are events with severe impacts and longevity that need to be considered, such as floods, drought, fire, sea-level rise and in some parts of the world even volcanic eruptions. Besides being physical, these events can change, for example, water chemistry, while they must also be seen against ancient events that have determined current geology and topography, as well as having led to the current biodiversity through historic climate filtering.

7.1.1.6 *Biotic environmental variables*

There are also biotic variables to consider. For dragonfly populations, important factors include the presence, quantity and quality of host plants for the egg, larva and adult, which may vary slightly across a region. Maintaining an effective population means that there must also be sufficient prey, and low numbers of pathogens, parasites, predators and competitors. In the case of dragonflies, some of these challenges come from alien species, especially from invading fish and trees.

At large spatial scales, changes in atmospheric or hydrological conditions, which can exert their influence from great distances, can cause shifts in water condition and in the plant community, affecting dragonfly local abundance and assemblage composition. For example, high rainfall high up in a river catchment area can change the nature of marshes at lower elevations.

7.1.2 Principles of assessment in novel systems

Overlying spatiotemporal natural dynamics are the stressors and impacts from human presence and activity. While some of these are spatially explicit, such as alien riparian tree cover, others are more widespread in their influence, such as excessive nitrogen deposition and pesticide drift in either air or water or both. Many of these stressors are not only multiple, but also subject to the pervasive effect of climate change along with extreme weather events (see Chapter 5).

One challenging issue is that our capacity to design evidence-based freshwater conservation plans is restricted by biodiversity shortfalls, such as incomplete knowledge on taxonomy, geographic distributions, traits and phylogenies of freshwater organisms, as well as datasets being different for geographically separated freshwater systems. This means that there is a need for increased knowledge, especially for functional and phylogenetic features of freshwater communities. In practical terms, this involves combining datasets from different sources through careful data harmonization. Second, more information is required on natural history observations on species habitats and life histories, essential for multitrait databases. Third, fully resolved phylogenies are required for determining the evolutionary relationships of freshwater organisms. When these three hurdles are cleared, there can be a more meaningful comparison of freshwater communities across continents, as well as the capability for mapping functionally important ecosystems and identifying evolutionarily significant areas for conservation (García-Girón *et al.*, 2023). Arguably, dragonflies are an invertebrate group that is leading the way in this regard.

For Amazonian odonates, great progress has been made with the development of the Amazonian Odonata Trait Bank (AMO-TB). The main aim of the trait bank is to produce an initial database with 68 functional traits of 218 species among the over 500 species that occur in the Brazilian Amazon. Information was obtained on behaviour, habit/habitat (of larvae and adults), thermoregulation and geographic distribution. Details of 22 morphological traits were also included. These details led to the production of a functional matrix and identified various functional patterns for the species, as well as determining the relationship between the various trait categories. The ecological and conservation value of this trait bank lies in selection of key traits that represent a set of functional variables that can be used to reduce sampling effort. This is a valuable living and developing resource for a region of the world where biodiversity is under great threat, and possibly an effective working model for other parts of the world (Santos Ferreira *et al.*, 2023).

Against the background that wetlands are characterized by hydrology (e.g. hydroperiod, mean water depth), the presence of hydric soils and the biotic communities present, especially type and extent of aquatic vegetation. Fennessy *et al.* (2007) identified five general considerations that need to be addressed when adapting existing methods or developing new methods to assess wetland ecological condition. These are: 1) defining the extent of the area being assessed; 2) wetland type; 3) approaches to scoring; 4) consideration of highly valued wetland types or features; and 5) procedures for validation with adequate and reliable ecological data. Ideally, the final approach should produce a single integrative score that is rapid yet not costly to undertake. Similar principles apply to running waters.

At large spatial scales, there is often great merit in making comparative assessments across wide areas, to disentangle general effects from local ones. Loss of natural vegetation cover has the same adverse effect on locally adapted species whether in Borneo or Brazil. Pesticides have largely the same adverse effect wherever in the world they are applied. However, local effects may or may not have wide effects depending on type or level of intensity. These include the construction of dams, type and intensity of farming, proportion of hardscape to greenscape in cities, impact of invasive alien organisms, etc. On the positive side, having protected areas, and extensions of them in the form of conservation corridors, along with agroecological approaches such as maintaining landscape heterogeneity, good-quality riparian zones, an abundance of well-managed ponds and greener cities, are all conservation actions that are contributing to effective dragonfly conservation.

Protected areas are important for conserving wetlands, but their effectiveness can be constrained by human activity. Assessing the effect of stressors at large spatial scales can be difficult. Across China, however, when a combination of area and pattern-based landscape indices (Landuse Dynamic Index and Fragmentation Dynamic Index) are combined into an Area-Pattern Quadrant Analysis system for assessing wetland quality, they provide an easy-to-use early warning indicator of wetland landscape change relative to underlying natural and human drivers, as well as providing knowledge-based management guidance by using these standardized and rapid assessment techniques (Mu *et al.*, 2021).

It is often difficult to ascertain why a dragonfly species is not present in a suitable intact natural habitat (Finch *et al.*, 2006). The reason is often because of some past adverse legacy of nature. However, when the species is present in its natural habitat but is absent from a nearby human-impacted site, such as a natural forest clearance or a dredged wetland, the impact on the species is measurable. But there may also be other adverse legacy effects that compound the situation, such as a major pollution event or a breakdown of metapopulation dynamics. Sometimes, these stressors can be teased apart by comparing other localities where the species is present and at various densities. However, population density measurements require caution, as insect populations, including those of dragonflies, can be notoriously variable over time.

All the above are important fundamental considerations when undertaking assessments and are important when undertaking restoration. An important start is to ensure that all the historic natural capital (i.e. ecological integrity) is locally present for colonization and establishment at a restored site. Then there must be an awareness of the potential for new threats or stressors, and even a combination of them. Additionally, there must be an awareness of the spatial context, the degree of change taking place to the terrestrial matrix around the focal freshwater area, and which is significant for dragonfly species to mature or forage away from the water.

7.1.3 Principles of mapping for conservation

7.1.3.1 Some basic principles

Mapping is the plotting and visual representation of items or systems across geographical space, such as individuals, populations, species, assemblages, habitats, biotopes, ponds, landscapes, catchments, biomes or socio-political boundaries. Mapping may also be tailored according to conservation goals. In general terms, apart from individuals and socio-political boundaries, the systems often have diffuse boundaries and can be highly dynamic in nature.

Mapping species is based on the recording and plotting of where individuals were found to occur at one place at one time. Extra spatial and temporal records are then added to the database, enabling an assessment of area of occupancy (AOO) and the extent of occurrence (EOO) of a species occurring across specified areas. Supplementary genetic data may also help determine the health and viability of a population. The data should ideally encompass the absence as well as the presence of individuals, which helps predictive distribution modelling (where a species is likely to occur given the current spatial data across a wide geographical area) and habitat suitability modelling (where a species is likely to occur from the known characteristics of its habitat at the local level). Habitat suitability modelling at large spatial scales can help determine the favorability of conditions for long-term survival of a specified species, or a set of species.

Mapping of optimal habitat can place focus on where further field searches should be done, or perhaps where a species might be reintroduced. Satellite remote sensing, combined with machine learning to produce super resolution images, can help refine these searches. This is especially so when traditional approaches that use land use and land cover data (which focus mostly on the structural properties of landscapes) are complemented with functional properties related to specific ecological processes (Riva and Nielsen, 2021). Designing functional analyses of these data can provide important and complementary information to traditional structural analyses. In turn, dates of species records can contribute to an understanding of the extent to which a species' geographic range has shifted, or is shifting, or not, in relation to various impacts, including climate change (Fig. 7.4).

7.1.3.2 *Significance of spatial scaling*

The science of spatial scaling is complex owing to landscapes and waterscapes being patchy and heterogeneous. Advancement of the topic depends on enabling the prediction and inference of quantities being measured from one spatial scale to another. However, identifying these scaling relationships in landscape ecology is difficult. Besides these variables requiring identification for a particular area or phenomenon, they must also be universal and must apply across space (and/or time) to provide general insights into landscape/waterscape function over large areas.

To understand scaling, we first need to define certain terms and concepts. Traditionally, the term 'scale' of a measurement has employed 'extent' (overall area) and 'grain' (size of the individual units) at which a species' distribution, assemblage, community, ecosystem or landscape/waterscape is studied. Scale has been used to describe the relative size or extent of something and has dimension(s)

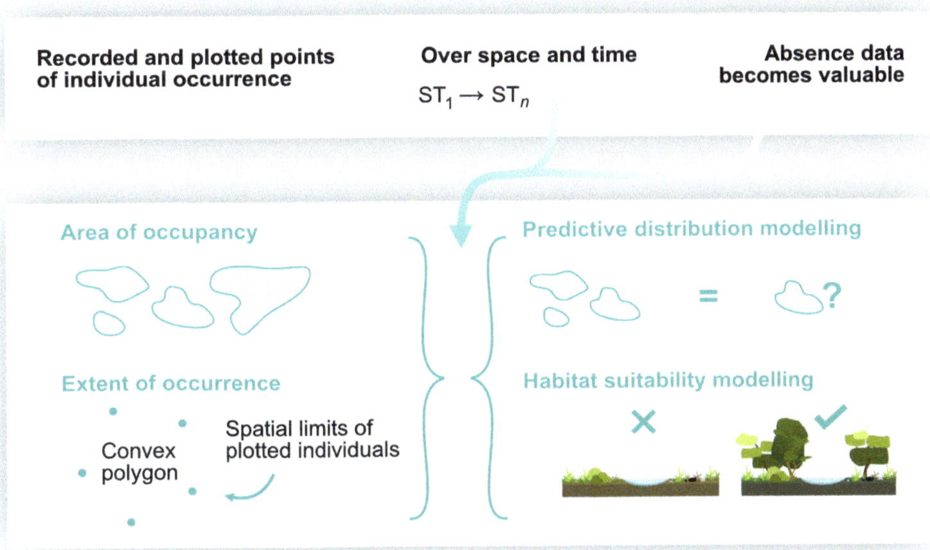

Fig. 7.4. Species mapping is a visualization based on the presence of recorded and map-plotted individuals in different places, with additional plots added over time. Doing this enables the determination of spatial factors such as the area of occupancy and the extent of occurrence of a species. As more plots become available, absence data (no individuals recorded) also becomes valuable. These baseline data then enable predictive distribution modelling and habitat suitability modelling to hone searches for the discovery of more individuals. (Illustration: Robyn Symons).

such as 30 m (the scale or resolution of Landsat imagery) or 2 km (the practical scale unit used in species area of occupancy (AOO) studies). However, any inferences about scale dependency are constrained by the two measures of extent and grain. Furthermore, extent and grain set the bounds of our ability to detect patterns. However, scale is better conceptualized as a dimensionless ratio of the upper to lower limits of a phenomenon in space or time and applied as the distance between two points on a space–time diagram. While extent and grain are both important, it is their ratio that is even more valuable for extending results into applications of scaling theory. From a physical or practical point of view, the extent-to-grain ratio turns dimensional quantities into dimensionless ones so that values can be compared on a relative scale (Frazier, 2023) (Fig. 7.5).

In turn, there is value for landscape ecology and conservation in embracing the concept of 'scope'. While scale is the relative size or extent of something, scope is a ratio of the range (extent) of a measurement relative to the resolution. Scope can also be applied to the number of measurement steps according to the step size or the distance between two points on a space–time diagram (Frazier, 2023).

7.1.3.3 The importance of spatial scale in mapping

Spatial scale is a fundamental principle of mapping, based on the size of the area chosen for recording the presence of a species – the grain size. Any one individual, when it has an exact geographic coordinate, tells us much about its optimal habitat, especially when many individual records are mapped together. Larger spatial scales (grain sizes) of recording, while presenting less information on habitat, are a more practical means of determining the geographic distribution of a species. This is especially so when assessments are repeated over time (reassessment/monitoring), for example, when searching for any geographic range shifts taking place. This is often done using 10 km x 10 km grid cells. At this large scale, recording is about determining cell occupancy by a specific species or set of species at national or greater spatial scales. In short, with the same recording effort, although smaller-scale recording

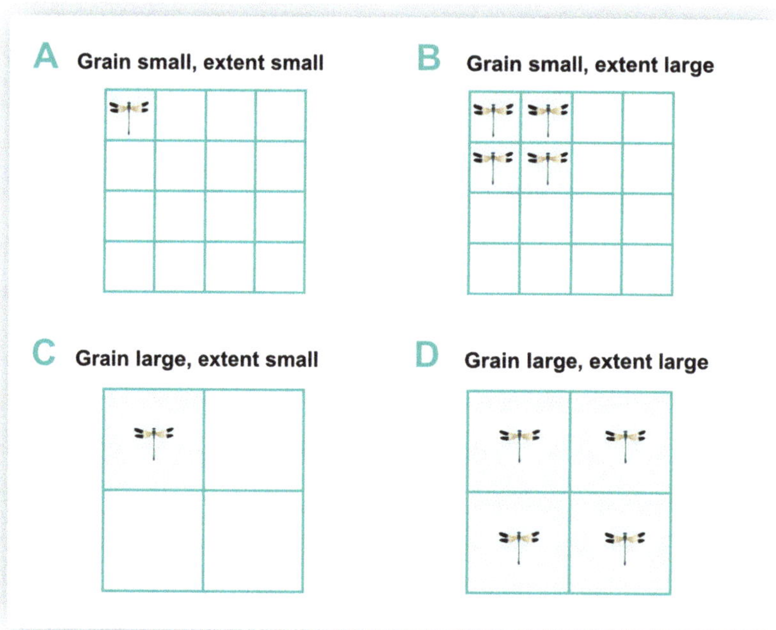

Fig. 7.5. Grain is the size of the area of sampling (the grain size is the same in A and B, but both are different from C and D). Extent is the area (number of squares here) where individuals occur (extent of the species is four times higher both in B than A, and in D than C). The smaller the grain size the more accurate the spatial distribution mapping, but more recording effort is required. The converse is the case with an increase in grain size. (Illustration: Robyn Symons).

provides focused information on precise area of occupancy, larger-scale recording provides strategic information on extent of occurrence (EOO).

Recording all species present at various localities covering an increasingly large area is valuable for constructing species-area curves (i.e. establishing the increase in species richness at increasing spatial scales). This perspective enables determination of the number of species that would be lost when habitat losses, for example, through indigenous forest clearance, occur across an area. However, this can be challenging as sampling may miss species with small geographical ranges, important in lesser-known, species-rich areas of the world. Large-scale features like topography and catchments can play a major role in the determination of species-area curves, as do small-scale factors like unusual ecosystems and refuges. All factors may in turn have significant conservation implications, as certain habitat-specialist species may be present only in small parts of the overall area, and may have already been extirpated (Matthews *et al.*, 2014).

Measurement of species richness can vary at sites as time passes, driven by various factors such as changes in water condition, strong weather events and plant community succession. Odonate species-richness shifts have been noted in standing waters (Dolný *et al.*, 2021) and running waters (Vilenica and Mihaljević, 2022), and both types of systems in the same local area (Kietzka *et al.*, 2021a).

An important corollary here regarding assessment is to always be aware of the possibility of a shifting baseline (see Chapter 5, Section 5.2). This means being clear on how environmental stressors, both natural and anthropogenic, did or might have had an adverse effect in the past. This addresses the question of whether one is measuring the state or status of a focal subject or system that has changed (or is changing) and has become impoverished or homogenized in terms of conditions away from, or not, the natural historic state (Fig. 7.6).

7.1.4 Principles of monitoring for conservation

7.1.4.1 Fundamentals of monitoring

Monitoring is the reassessment of a population, assemblage, community, ecosystem, landscape, region, country or the world over time. It may be at regular intervals when it is feasible and appropriate to do so. As with spatial recording and assessment, temporal monitoring cannot be done on everything over extended time-periods, meaning that the aims,

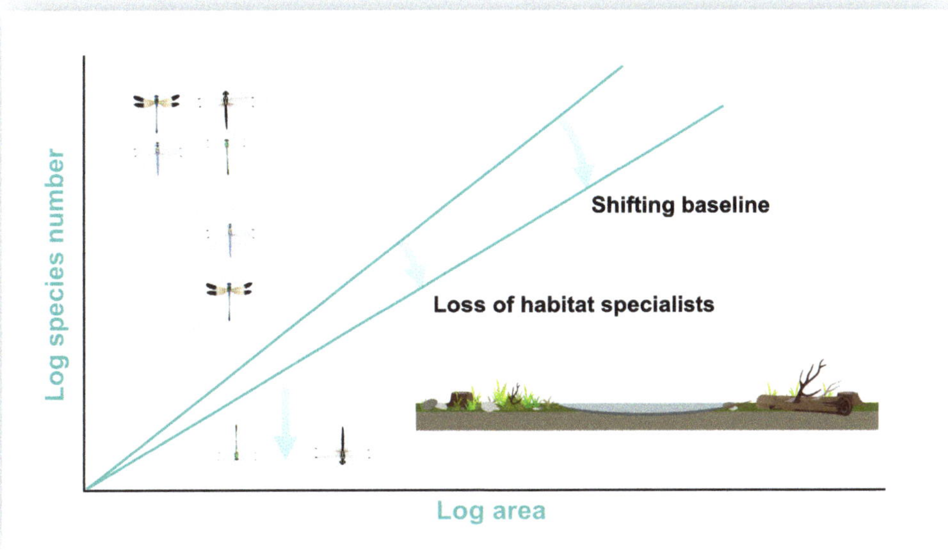

Fig. 7.6. While species richness increases with area, this curve can theoretically tilt downward with a shifting baseline, especially when habitat-specialist species are lost. (Illustration: Robyn Symons).

methodologies and required outcomes have to be carefully considered. For adult dragonflies, monitoring must be done at the right time of day, under good weather conditions, at the appropriate season, and using the right age classes for the task in hand and relative to geographic location.

Monitoring schemes have the most value when they illuminate different aspects of diversity. Besides recording changes in alpha diversity (the number of species present at a focal locality), it is important from a conservation point of view to record changes in temporal beta diversity (i.e. relative changes in species composition of different assemblages over time). In turn, the recording of change in species composition when combined change in evenness (i.e. together they can be used to express diversity), allows comprehensive and insightful information to be obtained.

7.1.4.2 National and regional monitoring schemes

National and regional monitoring schemes are valuable for assessing the presence versus the absence of species. In some cases, the abundance of certain species provides additional information for ascertaining the conservation status of these species and populations. This approach is well-advanced in some parts of the world where dragonfly data have been collected over many years, often with the help of volunteer recorders noting the presence of identified species in 10 x 10 km^2 grid cells/grain size (and in some cases smaller cells) in certain countries or regions. These monitoring schemes have also led to the production of atlases and conservation status reports. They also enable monitoring of the effectiveness of protected areas versus transformed areas.

7.1.4.3 Monitoring freshwater systems

Monitoring of freshwater systems using standard methodology has taken place for many years in various parts of the world. This response stems from freshwater systems being rich in biodiversity and the supply of essential services, yet they suffer from great human impact. Monitoring has involved the recording of changes in physicochemical conditions in freshwaters, especially rivers, to determine the extent of pollution and perturbation whether by sediment or by artificial chemicals such as pesticides or pollutants, or reductions in environmental variables like dissolved oxygen levels and flow rate, all of which can compromise the ecological integrity of a freshwater system. Various aquatic macroinvertebrates, including dragonflies, possess a range of ecophysiological traits. This means that some may tolerate, while others may not, particular levels of stressors on the system. Furthermore, even within species there may be differences among individuals in their response to the presence of stressors, having important implications for rare and threatened dragonfly species (Fig. 7.7).

Lindenmeyer *et al.* (2022) investigated government-instigated monitoring schemes in Australia and asked why some have been successful and others not. They conclude that to be successful, environmental improvement schemes must be sure of eight 'nevers': 1) Never start a new environmental management initiative without being committed to a monitoring scheme; 2) Never start a monitoring scheme without clear questions and goals; (3) Never implement a monitoring scheme without first having a good experimental design; 4) Never ignore matching the purpose and objectives of a monitoring scheme with scheme design; 5) Never change the way monitoring is done without ensuring that new approaches and methods can be calibrated with the old ones; 6) Never try to monitor everything; 7) Never collect data without proper planning, curating and reporting on it; and 8) Never start a monitoring scheme without securing the necessary resources required.

Peñas *et al.* (2023) propose ways to standardize approaches for monitoring freshwater quality and biodiversity status as a means of addressing these 'nevers'. Overall, they suggest that greater effort should be made to construct a coordinated strategy for monitoring freshwater ecosystems at national, continental and global scales. This also involves embracing new technology. Such a strategy should involve a shared baseline of biophysical and biogeochemical variables for each habitat type on agreed protocols that are implemented across regions and administrative borders. They also emphasize that certain freshwater ecosystems, especially mires, need much more focus in this era of rapid climate change.

At the spatial scale of a river catchment in China, where increased human land use leads to biotic homogenization, we need to recognize that aspects of diversity have distinct responses to stressors according to whether taxonomy or function is used to measure changes in ecological communities. While environmental filtering determines taxonomic beta diversity, changes in functional beta diversity

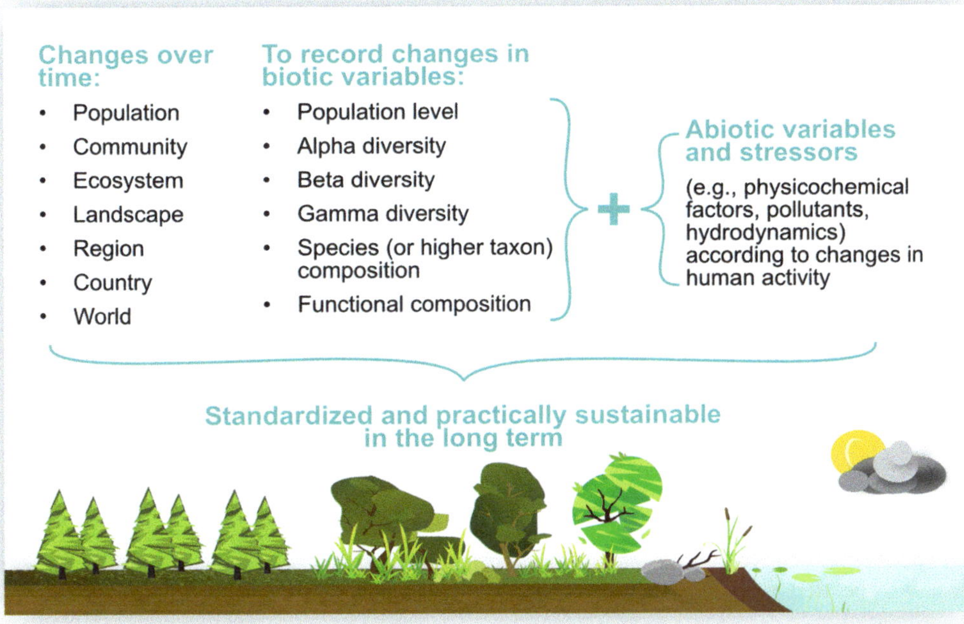

Fig. 7.7. Monitoring freshwater systems is the ongoing reassessment of their condition over time relative to a particular population or at a specified spatial scale. The aim is clearly articulated to record certain biotic variables, while simultaneously recording significant abiotic variables. Monitoring schemes must be well-planned to be practically sustainable in the long term. (Illustration: Robyn Symons).

are driven by the joint effects of stressors and spatial processes. These in turn change over time depending on the degree of water discharge relative to rainfall events in different river reaches. This means that it is important to determine continuous changes in biodiversity along the length of a river in the long term to make informed decisions regarding nutrient input control and any expansion of land to better address river ecological degradation and promote healthy and sustainable development (Ma *et al.*, 2023).

With increasing concerns for the welfare of biodiversity, and as monitoring of freshwater invertebrates can involve the capture and subsequent death of many individuals, there is a call to move to less destructive methods. This includes, for example, the employment of environmental DNA (eDNA) methods (Koperski, 2023).

7.1.5 Principles of surrogate use for conservation

7.1.5.1 Reasons for using surrogates

As it is often so challenging to monitor many insect species all at once, it is often necessary to use surrogates. These are species or species sets that stand in for a larger group of species for assessment or monitoring in conservation planning and implementation. A surrogate is in effect a proxy for a target group.

There are various types of surrogates. An ecological surrogate is a process or element (species, ecosystem or abiotic factors) that represents another aspect of an ecological system or area, known as the 'target'. Under this umbrella term is an indicator surrogate that provides information about ecological systems. For example, measuring selected components of, say, dragonfly assemblage diversity to represent the relative naturalness of an ecosystem, such as rare habitat specialists of natural bogs.

In turn, a management surrogate is a species or group of species that represents another aspect of a system so as to inform effective conservation management. For example, using dragonfly diversity to determine the success or not of a restoration project. Rating a project as successful would be when the restored assemblage, community and ecosystem matches that of a historically intact equivalent.

Environmental surrogates are natural features of the landscape that stand in for a set of species, often a set of rare dragonflies (Kietzka *et al.*, 2023). These surrogates are mostly essential components of the typical macrohabitats for a range of species. For example, having certain large trees, producing both leaf and twig litter as well as logs, may be essential for a set of both rare and common dragonflies to be present or return following restoration. Such features, of course, may take time to establish when restoration is undertaken. Other features can have a more instantaneous effect, such as the use of boulders placed in a stream to manage local stream flow velocity (Fig. 7.8)

7.1.5.2 *Importance of surrogate representativeness*

It is important that the surrogate species or set of species is a good representative of another. However, it is necessary to be alert to certain species requiring special attention because, for example, they are rare and threatened habitat specialists, making it essential to include them. In the case of a restoration project, it can only be considered as completely successful when the habitat specialists have returned along with the habitat generalists. This emphasizes that species composition and turnover are much better measures than simple species richness.

Sometimes, a single dragonfly species may be as effective as a surrogate set of species, although it must be verified that this is the case. Among dragonflies, however, use of a wider range of species is usually necessary, such as a group of habitat specialists that represent a wide range of microhabitats in the ecosystem.

7.1.6 Principles of bioindication for conservation

7.1.6.1 *Dragonflies as bioindicators*

Bioindicators are used to indicate the state or status of diversity or the environment, especially to reveal the level of stress in the system or alternatively the recovery from stress once conservation actions have been undertaken. Dragonflies, as a group, are good bioindicators. The reason for this is that within an assemblage, the various species show different responses to changes in water and riparian condition. Furthermore, some species are more dispersive than others, and some are more tolerant of a certain stressor or set of stressors than other species. The least dispersive and most intolerant species may die out locally with increasing stressor pressure. However, while any species is tolerant of one sort of stressor, it may be highly susceptible to another type of stressor.

The effect of one stressor may be clear, as with the loss of forest trees on certain sets of tropical forest-inhabiting zygopterans. Other stressors may be difficult to detect. This is either because the effects are subtle, such as the long-term effects of a pesticide or pollutant on the larva, or because the stressor pulsed and has since disappeared, as with a mining spill.

Many ostensible bioindicators are arguably better described as responders to, or detectors of, changing conditions. However, responders/detectors can be effective validators of restoration success, indicating that the natural state of the system has returned. This renewed state though, must be monitored after restoration to ensure ecological integrity is maintained. For example, it may be necessary to engage some sort of post-restoration management to correct the trajectory of the system being restored. In the case of dragonflies, this also means that restoration should extend to the terrestrial matrix around a water body.

7.1.6.2 *Testing bioindicators*

Technically dragonfly bioindicators, like all bioindicators, should be selected using a set of postulates, as is done in medicine when verifying the effect of a pathogen. The postulates have a series of steps. First, the historic state is determined with measures of abundance and diversity in the assemblage. This is the reference condition. Second, when the historic state is perturbed or deteriorates in a specific and known way, the response of the various species, according to their various individual traits, is recorded in terms of changed abundance and the diversity of the affected assemblage. Third, when the disturbance or perturbation is removed or resolved, and so long as all species are present in the regional pool (i.e. the natural capital is present), there is a return to the historic assemblage in terms of abundance and diversity of the assemblage. The problem is that many named bioindicators do not undergo testing using this set of postulates, meaning that they really are only presumed bioindicators, or when viewed from another angle, are simply responders to the state of a system (Fig. 7.9).

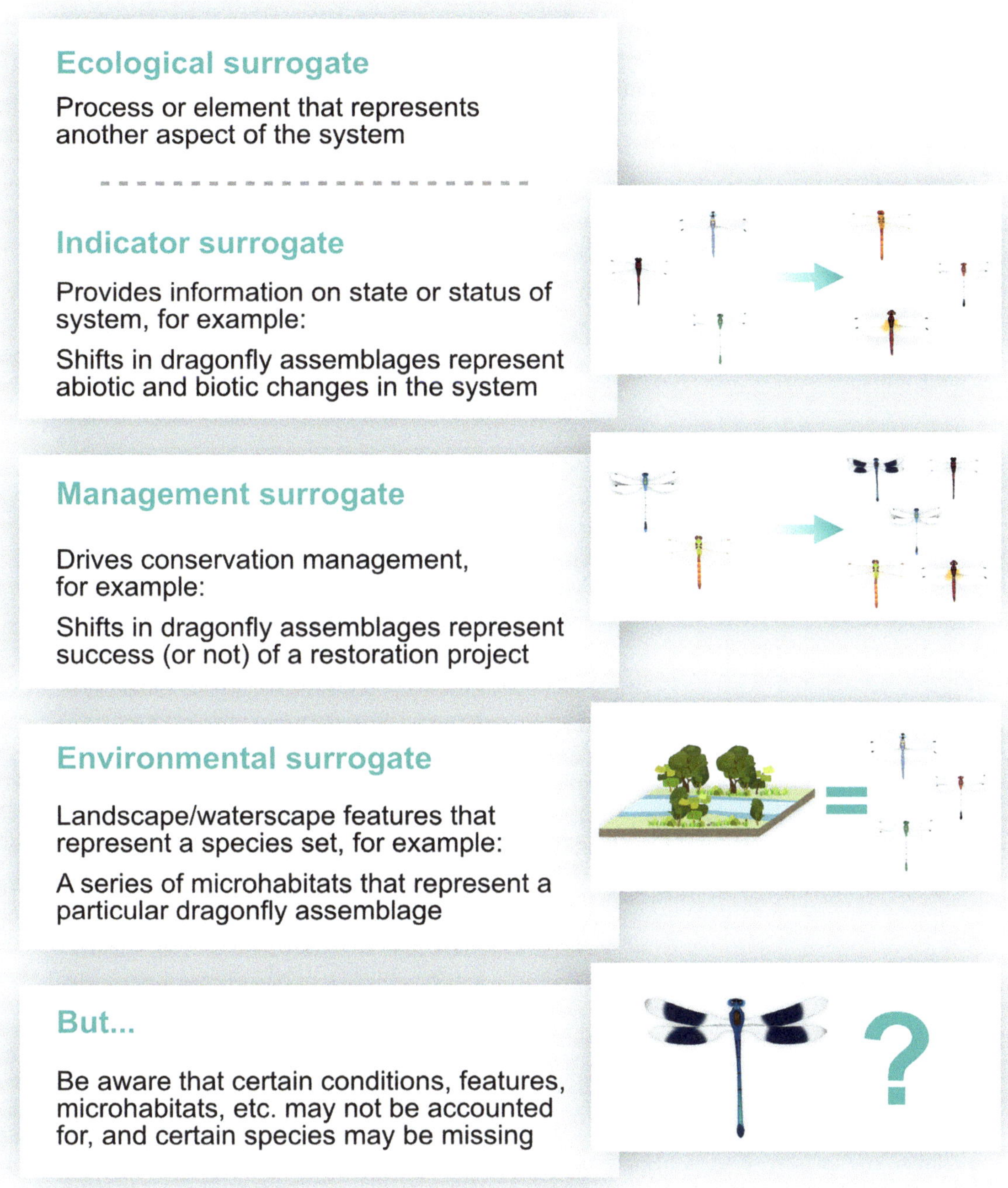

Fig. 7.8. Surrogacy, where something is a proxy for something else, aims to make assessment tractable in view of nature's great complexity. There are three basic categories of surrogates, notwithstanding that dragonfly assemblages are in effect a surrogate for much other biodiversity. Under 'ecological surrogate' is the subcategory 'indicator surrogate'. The other two main categories are 'management surrogate' and 'environmental surrogate'. No surrogate is a perfect umbrella for all other biota sets. A surrogate is not an end in itself as certain conditions, microhabitats, habitat specialists, etc. may have been overlooked in the assessment. (Illustration: Robyn Symons).

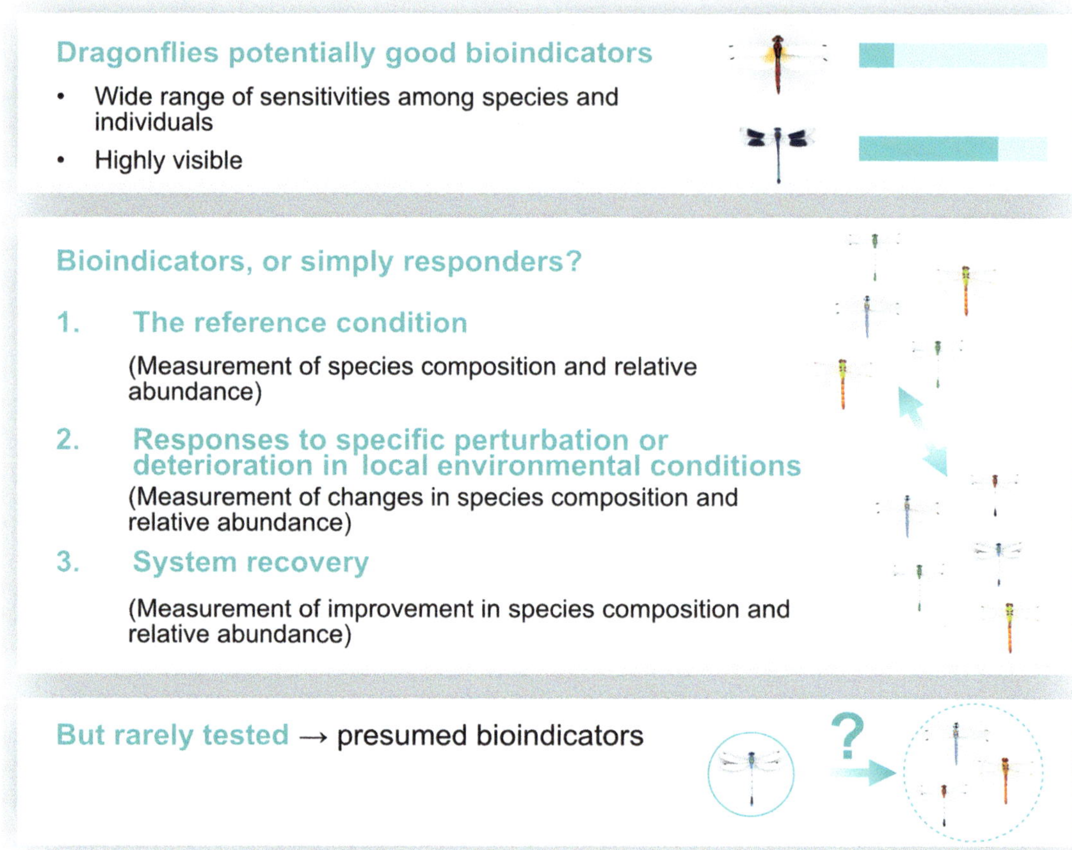

Fig. 7.9. Dragonflies are ostensibly good bioindicators, as their assemblages often display a wide range of sensitivities among the various species and individuals. In practical terms, they are also highly visible organisms. But bioindication value technically requires testing using three steps (1–3). However, as this process is rarely performed, it means that the 'bioindicator' species are only presumed bioindicators. (Illustration: Robyn Symons).

7.2 Freshwater Ecosystem Assessment and Monitoring

7.2.1 Importance of freshwater assessment and monitoring

The importance of freshwater assessment and monitoring is underscored when considering the case of India. The country, besides being rich in odonate species, is facing high water stress of about 66.5% relative to total available water resources, exacerbated by further increases in human population. In terms of water withdrawal, agriculture (90.4%) leads, followed by the municipal (7.4%) and industrial (2.2%) sectors. Furthermore, the gap between wastewater generation and treatment is widening, increasing the incidence of waterborne diseases and heavy metal contamination in the soil. In response, there is an urgent need to assess and monitor socio-ecological systems based on the amount of wastewater treated and its recycling metrics for water sustainability (Biswas *et al*., 2022).

Monitoring the health of aquatic stream communities determines the local effects of various environmental variables on ecosystems and communities, as well as the effects downstream of harmful stressors. This means that it is important to consider both larger- and local-scale measures of biodiversity status and regional gradients in environmental variables and human activity for determining biodiversity levels across catchments. One way to assess the differential effects of both natural events and human activity on a series of stream communities is to consider the regional scale in relation to differences at local spatial scales. However, this is a complex process, though made tractable by using a generalized framework of alpha, beta and zeta diversity, which among Californian stream macroinvertebrates account for 72% of among-stream variation in the mean health of communities shaped largely by the various species finding their optimal niches (Simons *et al.*, 2019).

Local abiotic and biotic environmental variables shape the diversity of natural communities and their distribution, with human activity then reshaping the patterns and functions. Understanding relationships between aquatic macroinvertebrate assemblages and environmental predictors requires multiscale analysis between freshwater systems, riparian zones and the regional landscape, which has been mostly done using species richness, and species and higher-level taxa composition and diversity, as well as the abundance or density of individuals. Aquatic insects, including dragonflies, have played a major role in the assessments and monitoring. However, the use of higher-level taxa is not always efficient at measuring the loss of ecosystem services and extinction of phylogenetic lineages, at least in temperate freshwater systems. Both macroinvertebrates and Ephemeroptera/Plecoptera/Trichoptera (EPT; together popularly known as 'riverflies') have been widely used to assess and monitor running waters, often with an interplay between EPT species richness and abundance (e.g. Bacca *et al.*, 2023; Rivera-Pérez *et al.*, 2023; Santos Ferreira *et al.*, 2023), with dragonflies deserving more attention (Brasil *et al.*, 2020b).

7.2.2 Dragonflies in freshwater assessment and monitoring

Despite the long history of assessing and monitoring freshwaters, few of the traditional monitoring methods provide a global perspective on the loss of biodiversity and its conservation. One reason for this is that in the past much research focused on local issues. However, some methods, especially using macroinvertebrates, have converged in terms of recognizing the value of this group for measuring the quality of freshwaters and their riparian zones and margins. New approaches are being developed that have potential. These include: 1) Consideration of less-studied taxonomic groups; 2) Use of standardized methods across regions allowing global comparisons; and 3) The measurement of change over multiple time points (Jackson *et al.*, 2016).

Regarding the first point, dragonflies are a taxonomic group that has rapidly gained increased relevance in recent years. This is largely due to their prominence in the human psyche, relative ease of identification, relative ease of assessment and monitoring, and a wide range of traits among suites of species in the group. They also display considerable cross correlation with other taxonomic and functional groups or are complementary with them. This is due in part to them being among the top predators in many freshwater systems.

These perspectives feed into the second point: that of global validity. Dragonflies are instantly recognizable and have risen greatly in prominence globally. They are especially significant in biodiversity-rich areas, as well as temperate areas. In short, dragonflies have become excellent sentinels of freshwater quality, whether in terms of deterioration or improvement. Being a relatively easy group to assess and monitor in practice, and with comparatively few phylogenetic constraints, their use can be standardized globally. This though, is not to ignore some special traits of species specific to local areas. These features encourage assessment and monitoring to the benefit of conservation both locally and globally, and so address point 3 above.

Besides great advances in the knowledge of dragonfly taxonomy, phylogeny, distribution, ecology, behaviour, responses to environmental change and conservation, huge strides are also being made in exploring these traits using a whole range of technologies from the molecular at the microscale to remote sensing at the macroscale (Stephenson *et al.*, 2020; Rydhmer *et al.*, 2024). These advances are enabling much better assessment and monitoring for local, regional/national, continental and global scales for freshwater, and dragonfly, protection and conservation in this rapidly changing world. Honing of traditional methods (summarized in Šigutová *et al.* (2023)) as well as the development of new ones are now explored in more detail.

7.2.3 Dragonflies as a component of macroinvertebrate indices

7.2.3.1 Macroinvertebrates for assessing and comparing freshwaters

Various taxa have been used to assess habitat quality or for comparing various types of running or standing water bodies. Most include aquatic plants, and many consider marginal and/or riparian vegetation as well. Rare species of various taxa are often an important component. Physicochemical conditions are usually included. Examples of the studies that include odonates (mostly larvae) are Godreau *et al.* (1999), Painter (1999), Chadd and Extence (2004), Nicolet *et al.* (2004), Williams *et al.* (2004), Solimini *et al.* (2008) and Menetrey *et al.* (2011).

Refinement of macrobenthos indices that use higher-level taxa continues in various parts of the world, especially to select macroinvertebrate subsets that are amenable to quantitative analysis based on comparisons with historic reference sites (e.g. Silva *et al.*, 2017). However, it is important to assess the tolerance of macroinvertebrate taxa before inclusion in biotic indices for improved performance as discriminators of multiple stressors. This was shown in Tanzania, where generalized macroinvertebrate indices performed poorly and failed to discriminate between stressor gradients in a river where there was severe loss of natural vegetation of various types, which had been replaced by grazing and agricultural activities (Masese *et al.*, 2023).

7.2.3.2 Macroinvertebrate rapid assessment protocols

Globally there are knowledge gaps regarding responses by macroinvertebrates to human landscape/waterscape transformation and ongoing climate change. The main stressors to freshwater systems include habitat degradation and loss, hydrological alteration, overexploitation, invasive species, pollution and the multiple effects of climate change. Recommendations that enhance conservation planning and management at multiple spatial scales for integrated catchment management include: the formulation of regulatory measures, an understanding of the determinants of macroinvertebrate diversity both across multiple scales and among taxonomic groups, and the improving collaboration between researchers and conservation practitioners. The integrated use of macroinvertebrate biodiversity information for monitoring of freshwater systems can improve ecosystem management, facilitated by involving conservation psychology, marketing and use of the media and the Internet (Sundar *et al.*, 2019). Macroinvertebrate information can be integrated with data on macrophytes and fish to provide a more comprehensive assessment and prioritization of river networks (Sály *et al.*, 2020).

However, while macroinvertebrate-based assessments and monitoring are important tools for freshwater monitoring and management, their practical application differs from one region to another. There are also some challenges in their implementation globally, which stem from issues such as lack of taxonomists and biostatisticians, high levels of endemism among taxa, especially in the tropics and on certain islands, and limited institutional infrastructures (Ndatimana *et al.*, 2023).

Benthic macroinvertebrates have been widely used to rapidly assess the ecological condition and quality of running waters and to develop assessment indices that summarize the results (Dallas, 2021). While they may differ among regions, the principle is the same. The relative proportion and abundance of the higher-level taxa (mostly at the family level) are gathered using aquatic nets swept across the riverbed that has both bare and vegetated substrata. This approach is based on different taxa and functional groups often possessing different sensitivities to the river condition. It is the change in relative proportion of the various taxa to altered environmental conditions, away from the baseline natural condition, that is important. The sampling protocol produces an index and generates three metrics: the total score (the total of the sensitivity weightings of various taxa recorded at a site), the number of taxa recorded, and the average score per taxon (the total score divided by the number of taxa recorded at a site). These rapid assessment protocols are most useful for determining the extent of major physicochemical changes in rivers (e.g. Ganguly *et al.*, 2018). However, when a set of different macroinvertebrate sampling protocols are used together, they can reflect changes in the extent of landscape degradation (Chirwa and Chilima, 2017).

These macroinvertebrate indices usually include dragonfly larvae, and do not directly reflect the condition of the terrestrial environment of the adult. Nevertheless, the indices may indirectly reflect the condition of the river margin or riparian zone through a change in the proportions of the larvae from the different odonate families. The main problem with these rapid assessment protocols is that the metrics are coarse and only give, for example, a very basic assessment of differences in waterflow and hydroperiod characteristics, and overall have poor discriminatory power between subtle differences in water condition (e.g. the Okavango delta: Dallas and Mosepele, 2020; the length of an East African river: Masese *et al.*, 2023). The most serious issue with these family-level assessments is that they overlook great differences in the sensitivities among the genera and species within any one family. This is the case for odonates, as well as for mayflies and caddisflies, resulting in poor assessment of the impact of a stressor, such as dam construction (Dusabe *et al.*, 2022).

When a rapid macroinvertebrate sampling protocol is compared and standardized (using average score per taxon) across a range of variably disturbed rivers, the species-level odonate sampling protocol (for example, the Dragonfly Biotic Index; see section 7.4.1.6) is far more sensitive. While the two protocols are correlated, the average dissimilarity between sites using macroinvertebrates is only 49% compared to 69% for odonate species assemblages (Simaika and Samways, 2011).

Similarly, in the Ozark Highlands of the USA, a combination of odonate larvae and resident adults were strongly associated with the structure and type of riparian vegetation, while benthic macroinvertebrates were associated with these riparian variables to a lesser degree. The additional environmental variables, such as physicochemical condition of the water, explained only minimal variation in both assemblages. The conclusion is that odonates alone are valuable for assessing and monitoring the riparian structure and condition associated with these streams (Cheri and Finn, 2023) (Fig. 7.10).

7.2.3.3 Assessing freshwaters and their terrestrial matrix

Although conservation action often focuses on terrestrial biodiversity, there is much to be gained when both the freshwater and terrestrial realms are considered together. When more than 1500 Amazonian terrestrial and freshwater species (including Odonata) were sampled, conservation prioritization based on only terrestrial species gave an average of only 22% of the freshwater benefits achieved through freshwater-focused conservation. However, using integrated cross-realm planning, freshwater benefits increased by up to 600% with only a 1% reduction in terrestrial benefits. This means that conservation actions which incorporate both realms, aquatic and terrestrial, can improve the status of freshwater species without compromising terrestrial conservation goals (Leal *et al.*, 2020).

Across a wide range of coastal wetlands in the Great Lakes coastal region of the USA, an assessment of the responsiveness of 66 candidate indicators of agriculture, urban development and point source contaminants, and characterized at multiple spatial scales (100, 500, 1000, and 5000 m buffers, and whole watersheds) led to shortlisting of certain indicators of environmental change. Some fish and *Aeshna* species were two of the strongest indicators of urban development effects, while fish overall were the most important group at all spatial scales to illustrate the effect of agricultural and urban development. Importantly, it is critical to first assess the influence of potentially important non-stressor covariables before assessing the strength of indicator responses to disturbance, while also using an appropriate spatial scale to characterize the various types of disturbance and stress (Brazner *et al.*, 2007).

Andean wetlands in arid climates are isolated ecosystems that are vulnerable to environmental stressors such as pollution and climate change. Using sweep netting across wetland vegetation to sample various insect groups to species level, including odonates, and analysing the data using a wetland community index (based on species distribution modelling and an invertebrate vulnerability index) indicated that odonates are the most sensitive and vulnerable of all the groups (Scheibler *et al.*, 2022).

When macroinvertebrates at the higher, family taxonomic level are compared with species-level odonate (Zygoptera and Anisoptera) assessments, both metrics give similar results for large-scale overall river conditions. However, at the small spatial scale of parts of rivers, odonates are highly sensitive to riparian vegetation, and much more so than macroinvertebrate higher taxa. Furthermore, odonate species are more sensitive to vegetation structure than they are to vegetation composition. Landscape context is also important, with odonate assemblages at point localities being affected by the neighbouring dominant habitat type along the riparian corridor. Both benthic macroinvertebrates and adult odonates are congruent in providing both large spatial scale (river system) and small spatial scale (point locality) information on the impact of stressors such as the effect of riparian invasive alien trees. However, as adult odonate species assemblages are much easier to sample than are macroinvertebrates and are more sensitive to disturbance at both large and small spatial scales, they are the preferred indicator group for rapid assessment of river condition and riparian quality (Smith *et al.*, 2007) (Fig. 7.11).

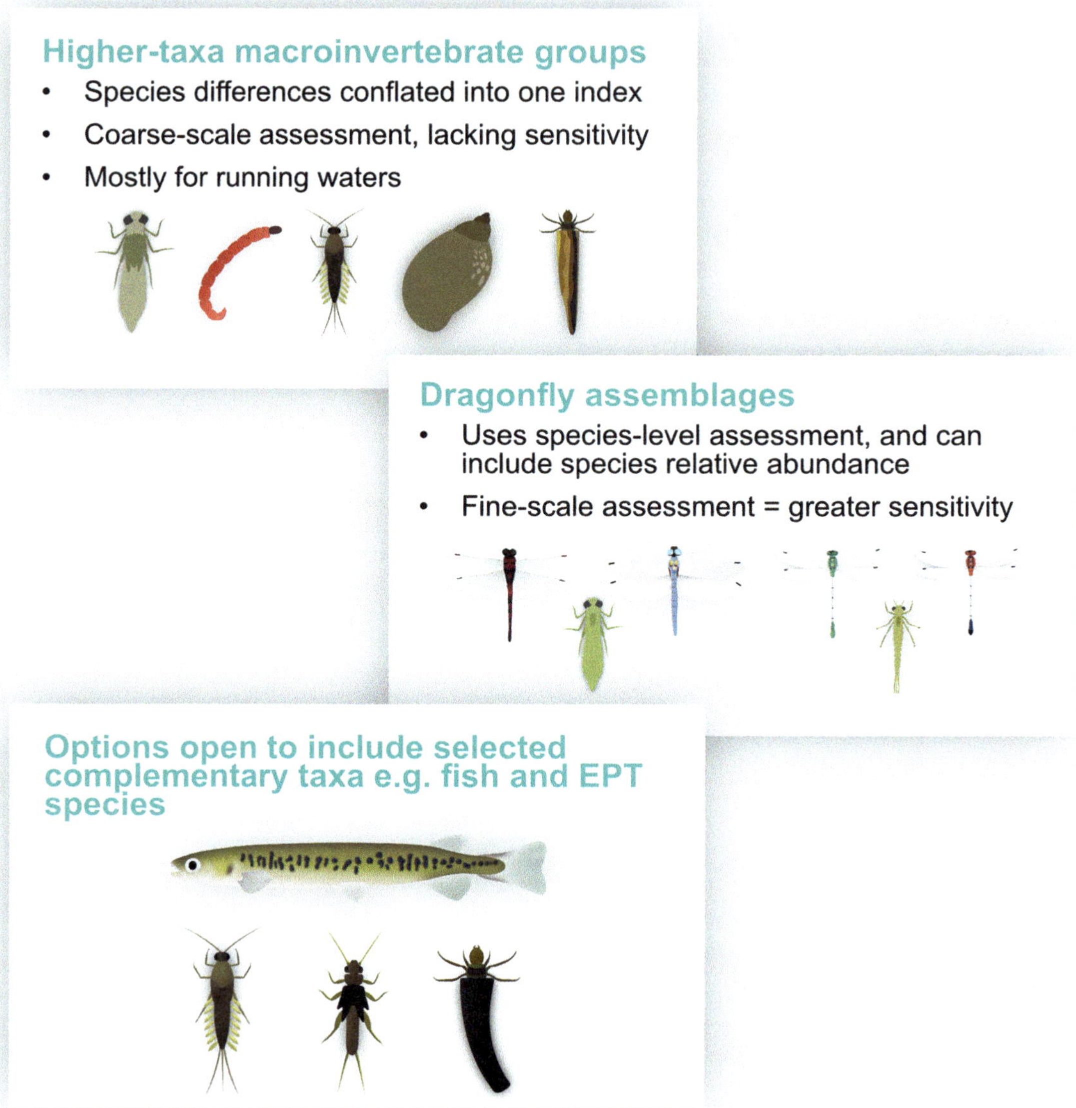

Fig. 7.10. Freshwater assessment using higher-taxa macroinvertebrate groups (including dragonflies at the family or order level) being a coarse approach, can lack sensitivity to significant changes in a water system. Macroinvertebrate indices are mostly applicable to running waters. In contrast, assessment using dragonflies uses more sensitive species-level assessment, and sometimes also relative abundance. Dragonflies can be used to assess both running and standing waters. Using dragonflies does not preclude using other sensitive and complementary taxa such as fish and Ephemeroptera/Plecoptera/Trichoptera (EPT) species. (Illustration: Robyn Symons).

7.2.4 Significance of deep history for dragonfly conservation

Understanding the deep history of dragonflies within a region is fundamental to their conservation, as it gives perspective on the environmental filters that the assemblages have been through. Zygopteran species in Amazonian streams were assessed in terms of their response to three sets of

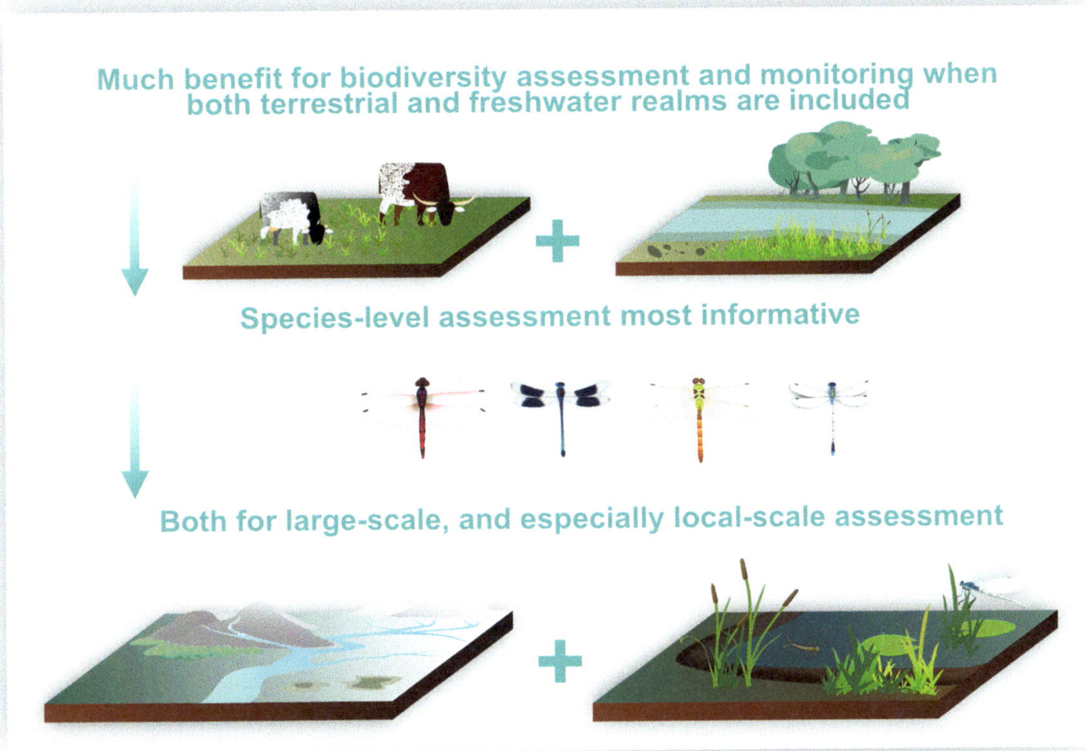

Fig. 7.11. There is much to be gained for biodiversity assessment and conservation when both terrestrial and freshwater realms are included. In turn, species-level assessments are most meaningful, and are effective for both large-scale and especially for fine-scale assessment. (Illustration: Robyn Symons).

predictor variables: environmental variables, areas of endemism (biogeographic) and spatial filters. Species turnover explained 99.4% of the beta diversity, with environmental and biogeographic variables together best explaining these patterns. Overall, there is high gamma diversity, but low alpha diversity. This apparent paradox is explained by high turnover among sites within the landscape, and the diversity patterns being related to both historical biogeographic factors and the spatial structuring of environmental conditions. In terms of conservation, the high turnover and beta diversity, along with their correspondence with areas of endemism, means that conservation of the diversity of these Amazonian assemblages is contingent upon the establishment of a network of large-scale protected areas throughout the major Amazonian interfluves. Ideally, these should be in areas with optimal environmental conditions for the greatest number of species to reduce extinction risk (Brasil *et al.*, 2018).

The Iberá Depression (14,000 km^2) in north-east Argentina is one of the largest wetlands in South America. It is often seen as one unit but is actually composed of three subbasins. Among 128 odonate species, the Iberá Depression is not a single functional unit, with its subbasins more related to external basins than to each other. The reason for this is that the odonates largely respond to the availability of certain wetland types, such as ponds, streams, rivers, swamps, etc., or to specific physical characteristics, such as the type of sediment, the availability of oxygen, etc., instead of to large-scale categorization (del Palacio *et al.*, 2022). The implication for dragonfly conservation is that it is essential to maintain as much intact freshwater heterogeneity as possible.

7.2.5 Developing aerial technologies and analyses for freshwater assessment and monitoring

7.2.5.1 *Remote sensing*

Stream ecological condition is usually measured using both abiotic and biotic variables representing features of the catchment, stream channel and riparian zone. In turn, physicochemical measures of water quality are used to indicate the environmental conditions of the water, catchment land use and land cover quality to identify stressors. Physical habitat structure measures are also used to indicate the structural change of the vegetation. Together, these metrics are used to characterize the associated communities and assemblages.

Remote sensing of catchment land use and land cover provides relatively low-cost environmental information at large spatial scales, which seemingly should minimize the need for fieldwork. However, using a sensing index (the Normalized Difference Vegetation Index (NDVI)) that measures vegetation structure and difference on the ground is insufficient for predicting stream water quality or biotope structure. This suggests that remote sensing indicators should only be used when it is impossible to measure actual water quality and biotope structure in the field (Dias-Silva *et al.*, 2021). Similarly, at a restored lake near Buenos Aeries, Argentina, NDVI did not reflect the observed changes in the structural complexity of the vegetation and therefore could not be used on its own to illustrate dragonfly assemblage recovery (Lozano *et al.*, 2022).

Satellite remote sensing is now contributing to measures of plant alpha diversity. However, measuring beta diversity is more difficult, requiring ground verification (Rocchini *et al.*, 2016), especially for determining habitat quality and life-cycle events (Leitão and Santos, 2019). Nevertheless, it can be used for determining plant community structure and degradation, such as use of quantitative measures derived from terrestrial laser scanning in single-scan mode to describe the structural complexity, canopy openness, foliage height diversity and understory complexity (Seidel *et al.*, 2021).

Automated habitat monitoring to assess changes in optimal habitat for a species can be done annually using the Google Earth Engine cloud-based spatial analysis platform and a dynamic data repository. This approach holds great promise, enabling managers to track the impacts of recent disturbances and to adjust plans to meet goals across increasingly dynamic landscapes (Shirk *et al.*, 2023).

In turn, thermal imaging was valuable in the search for new potential microhabitats for the highly threatened damselfly *Spesbona angusta* (Deacon and Samways, 2017) (Fig. 7.12).

7.2.5.2 *Use of drones*

Drones are another useful methodology for providing spatial and volumetric data to obtain new insights into microclimate modelling. The affordability and ease-of-use of drones have led to them becoming a valuable technological advance for researchers and conservationists in assessing environmental parameters such as topography and canopy height. Data capture and processing methods are sufficiently established to produce data for microclimate models. For finer parameters, such as leaf size, techniques are still novel but show promise.

For most environmental variables, combining spatial landscape characterization from drone data and added data from laboratory and field studies is a useful way to create inputs at relevant spatiotemporal scales, and an opportunity for quantifying landscape structure and heterogeneity at fine resolution (Duffy *et al.*, 2021).

Drones are useful for identifying insect habitats (Habel *et al.*, 2016). When equipped with a 7 X telephoto lens they can also be used to record even small damselflies like *Nannophya pygmaea* in a Japanese wetland that is typically difficult to access by foot or boat. The drone is flown at a low speed at an altitude of 4–5 m to capture sufficiently high-resolution images. This non-invasive and easy-to-use method has much to offer odonate assessment and monitoring, and in turn, their conservation (Niwa and Hirata, 2022).

While drones are valuable tools for providing aerial photographs, in entomological studies, sampling of specimens is often required. By attaching a net bag to a drone and sweeping vegetation, adult and larval insects can be obtained. Drone netting is proving to be a versatile means of insect sampling, particularly in inaccessible terrains. It is time- and cost-effective, minimally invasive and adaptable for many research tasks while producing a catch like that of hand netting. Furthermore, caught specimens stay alive and can be released if not needed (Löcken *et al.*, 2020). Interestingly, dragonfly flight behaviour is being used to improve drone flight capability (Chahl *et al.*, 2021).

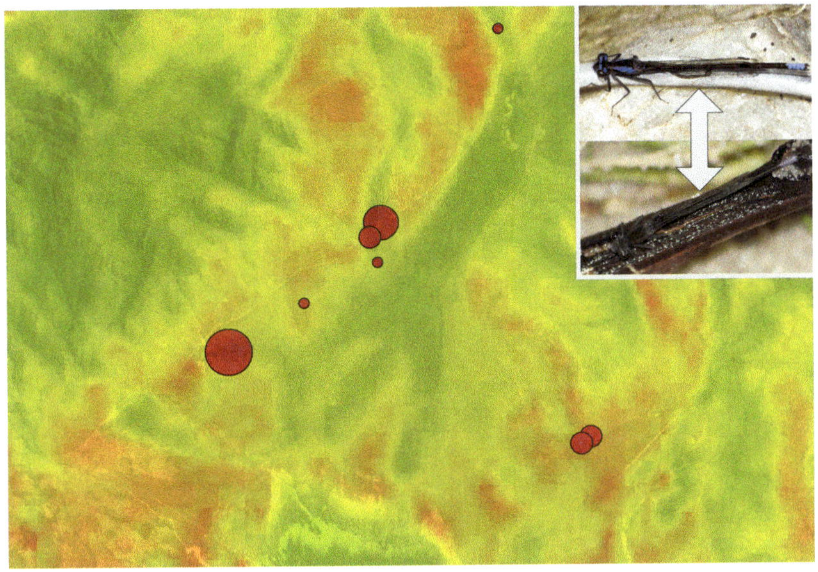

Fig. 7.12. This thermal image (30 m spatial resolution) is part of a marshland in the Cape Floristic Region, South Africa covering about 25 km^2. The scale ranges from green (low solar irradiation) to red (high solar irradiation). The red circles indicate proportional abundance of the extremely rare and threatened *Spesbona angusta* (insert), which displays rapid and reversible body colour change (blue↔brown) in both sexes in response to rapid temperature change, when, for example, a cloud passes over (insert: male, blue in bright light and brown in low light). The larger circles represent more individuals recorded. Its preferred microhabitat is at low to intermediate surface temperatures, shown by the margins between green and orange. (Illustration: Charl Deacon).

Lake research and assessment is being improved using two complementary measurement systems: a floating craft drone and a tethered helium-filled balloon, observed from an accompanying boat while in operation. The system permits fully autonomous operation, with motion controlled by speed settings and waypoints, as well as obstacle detection. On-board instrumentation is connected to a central hub for data storage, with real-time assessment of measurements from the accompanying boat. These measurements are complemented by mesoscale imaging from an instrument package attached to the balloon, recording thermal and red-green-blue imagery and then transmitting these data in real time to a ground station, where surface thermal variations of the lake are displayed (Barry *et al.*, 2019).

7.2.5.3 Camera trapping

Modified terrestrial camera traps have the potential for assessing the relative density of perching dragonflies (Yoshioka *et al.*, 2020). For the multi-habitat user *Sympetrum infuscatum*, which frequently perches and shows relatively short-distance dispersal, its relative population density can be assessed along line-transects. When mature adult dragonflies and exuviae are compared across rice paddies, detection frequency of adults using camera traps in autumn correlates with the density of mature adults recorded at the same time. In turn, for camera-detection frequency in autumn and exuviae in early summer, there is a correlation between the camera-detection frequency of mature adults and exuviae density the following year (Yoshioka *et al.*, 2023).

7.2.5.4 Improving analyses

As freshwater systems are affected by many environmental stressors across multiple spatial scales, it is important to unravel what these stressors are before restoration can be fully effective. One approach for doing this is to use Bayesian networks. They focus on the occurrence of events and predict the likelihood of what caused those events. Bayesian macroinvertebrate models are used for scenario analyses and point to certain types of remedial action. For European lowland streams, recovery can come from improvement in channel velocity and enhancing coarse organic matter (de Vries *et al.*, 2021).

As biodiversity monitoring methods can be highly demanding on time and resources, advances in computer vision combined with deep learning are providing potential new solutions for this major challenge. Cameras and other sensors can effectively, continuously and non-invasively record insects across diurnal and seasonal cycles. Furthermore, these approaches can be used to recognize the physical appearance of specimens and can capture the images by automated imaging. Following training on these data, deep learning models can estimate insect abundance, biomass and diversity, and can quantify variation in phenotypic traits, behaviour and species interactions. Deep learning approaches can be applied to large data sets to obtain ecological information and feed into conservation action by validating image-based taxonomic identification, generating sufficient training data, assisting in the development of reference databases and helping find solutions to integrate deep learning and molecular tools (Høye *et al.*, 2021; van Klink *et al.*, 2022).

Remote sensing has also been used to determine managerial legacies that are significant, for example, in grasshopper conservation (Theron *et al.*, 2022a), and for determining the value of conservation corridors for these insects (Theron *et al.*, 2022b). When also involving machine learning to produce super resolution images, it can establish levels of threats from invasive alien plants to grasshoppers associated with stream margins (Theron *et al.*, 2022c).

7.3 Dragonfly-based Indices of Freshwater and Margin Quality

7.3.1 Principles of sampling dragonfly assemblages

Fundamental sampling methods for odonates and their various life stages are given in Cezário *et al.* (2021) and Resende *et al.* (2020), while approaches and methods for insect conservation in general are given in Samways *et al.* (2010).

At the outset of employing indices to assess freshwater quality, there are some basic factors to consider. When choosing the best index to address a specific conservation question, it is important to first bear in mind that the different life stages differ in mobility, the realm in which they live (aquatic versus terrestrial), and that they are subject to past evolutionary processes and current ecological constraints. There are also practical issues to consider, such as the level of taxonomic and ecological knowledge of the focal species and their traits, the relative value of the various life stages, and field logistical issues like what type of field sampling can be undertaken under difficult field conditions. Last but by no means least, the requirements and qualities of the index, as well as its statistical robustness under various situations or constraints, also need to be considered (Rosset *et al.* 2013).

Inventorying assemblages is essential for determining the value of different sites relative to the environmental variables driving the assemblages. Citizen science plays a major role in this activity, especially when extended to threatened species lists, national red lists and databases on publications, before moving to more sophisticated and effective assessment leading to conservation action (Böhm and Collen 2015).

Importantly, good inventories are needed for conservation monitoring to proceed, especially for determining the conservation status of the focal species. Furthermore, this process can pinpoint species with certain traits and sensitivities which may be surrogates for the whole local assemblage (Robinson *et al.*, 2016). However, there needs to be a decision whether to inventory well in a small defined area, or less comprehensively over a wide area, given the same resources and time available. The choice depends on the conservation question being posed and the urgency for an answer without sacrificing accuracy. There are also data gaps, with, for example, ten new national records of odonates recently added to the Colombian national inventory (Cano-Cobos *et al.*, 2023). Furthermore, surveys need to be updated, as odonate species presences change, especially due to climate change, as in the case of the Lario and Brianza area, Italy (Bazzi *et al.*, 2023).

Often, it is necessary to determine the extent to which sampling is adequate. Species accumulation curves, which plot the addition of newly sampled species with each sampling occasion, are useful for doing this. When all species are summarily accounted for, and no further species are being recorded, the curve flattens out. However, this asymptote is rarely reached despite much sampling, simply because dragonflies ebb and flow across the landscape, often in response to changing weather conditions. Furthermore, some populations die out locally while others become newly established, the dynamics of which are becoming more pronounced with landscape and climate change. In turn, it is important to recognize that there may have been species replacement of, for example, habitat specialists by generalists or a species is simply moving through and not resident. Although there are few invasive alien odonate species, the appearance of a neonative or of a species recovering within its former range would have conservation significance.

7.3.2 Developing dragonfly databases

While reference samples from inventorying constitute basic biodiversity informatics, simple databases of collections and observations are being developed into detailed, interactive and flexible systems of information management, modelling and analysis (Kindsvater *et al.* 2018), important for insect conservation in general (Johnson, 2017).

In turn, ideally, we require species identifications as we cannot inventory nameless species. Having scientific names is essential, and though not an end in and of itself (Wheeler and Miller, 2017), may be considered the beginning of a knowledge base for biodiversity informatics and ecology, and ultimately fundamental to species diversity assessment and conservation (Thomson *et al.*, 2018). A global database, the Global EPTO Database, covering records of Ephemeroptera, Plecoptera, Trichoptera and Odonata, has been launched. For each species record, the drainage basin and subcatchment based on the Hydrography90m dataset, along with the elevation and ecoregion, is captured (Grigoropoulou *et al.*, 2023).

If it is not possible to use named species, then morphospecies (morphological operational taxonomic units (MOTUs)) can be used, so long as they are deposited in a permanent voucher collection where taxonomic verification can take place by a taxonomic expert sometime in the future (e.g. Deitsch and O'Grady, 2022) and they are digitized for the safety of the information that they contain (De Almeida *et al.*, 2021).

Well-preserved MOTUs, while of lesser immediate value than identified specimens, can have their morphology studied in detail and their DNA extracted, assuming good preservation. These MOTUs can be labelled (morphotyped) $Sp_1 \ldots Sp_n$ and characterized morphologically without assigning actual species identification. Odonates assigned in this way have been used, for example, for rapid assessment of the ecological integrity of South American rainforests (Juen and Feest, 2019). MOTUs and/or rapid genetic assessment may be used together in the future, for example, by using cyberinfrastructure that gathers a wide array of character states and genetics (Fig. 7.13).

From a conservation perspective, however, there remains the question of whether some of the inventoried morphotyped specimens are a group of specimens found to be a cryptic species complex, usually following in-depth genetic analysis. Also, we need to determine phenotypic variation, arising out of complex interactions between genotype and environment and resulting in a set of phenotypes within a population or species (phenome) (Houle, 2010), enabling, for example, how dragonflies are responding to stress.

There are also innovations associated with morphological identification. The characters of internal soft tissue can be serially sectioned and technologies for image capturing, such as confocal laser scanning microscopy and three-dimensional electron microscopy reconstruction, can be used to standardize and test new and useful phenomics (Marshall, 2018).

The fine details of morphology and behaviour seen only in living individuals are regularly used in dragonfly assessments. Characteristics seen in life which fade on death, include the colour of the pterostigmas and eyes. Subtle flight movements and other behaviors are also useful in field identification, and these can be recorded in the field, along with specific microhabitat.

There can also be issues with field identification. These may be because of cryptic species or simply because two or more dragonfly species cannot be reliably distinguished in the field. In terms of conservation assessment, the confusion may be highly significant, as when a rare and threatened species is morphologically like a common one. At other times confusion may not matter so much. Adults of *Trithemis dorsalis* and *T. furva* cannot be reliably distinguished in the field even with close-focus binoculars. However, they have different habitat preferences and are mostly geographically separated. There are two approaches to this situation. Either a sample can be physically net-collected and identified with a hand lens to determine which species is present or, alternatively, in locations where they do co-occur, they can be listed as a species pair, as they have the same conservation score for conservation assessment purposes (Samways and Simaika, 2016).

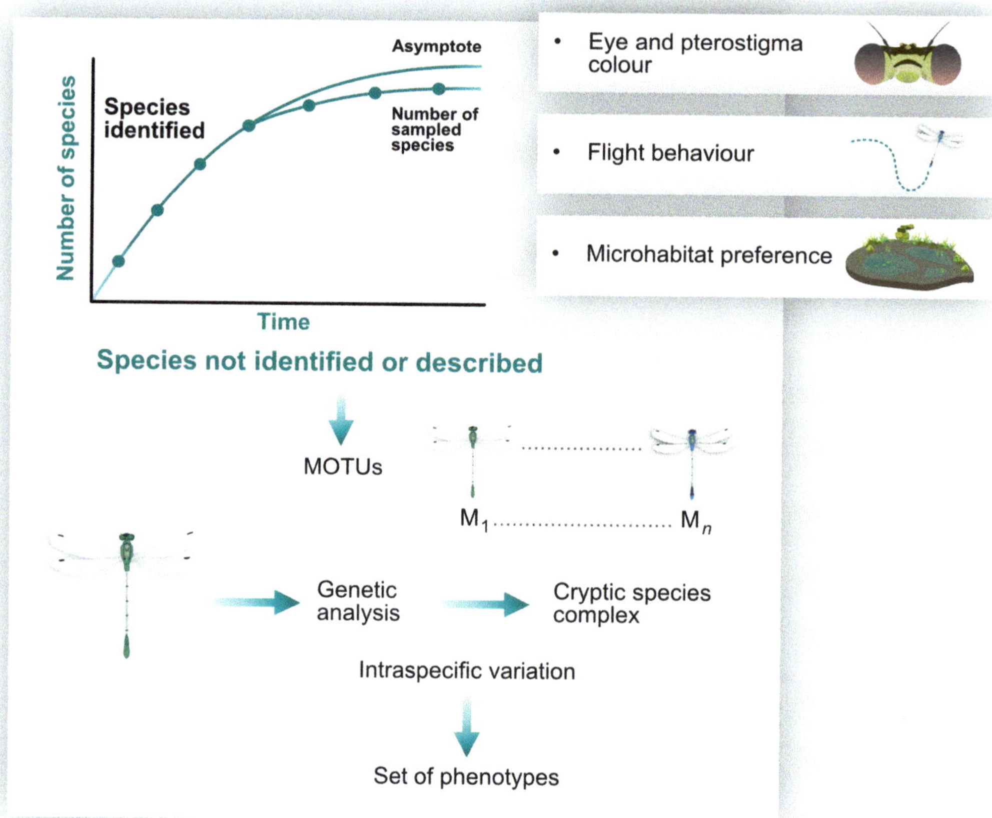

Fig. 7.13. More dragonfly species are usually recorded at any one site the greater the length of time of the survey. This is visually represented as a species accumulation curve, which in practice rarely reaches an asymptote as sampling is finite in time and space, and yet in reality, owing to the dynamics of populations, some new species may appear but others may disappear at the sampled site. Important features, for identifying living dragonflies in the field, besides body morphology and colour patterning, are eye and pterostigma colour, flight behaviour and microhabitat preference. In some parts of the world, species cannot be identified or are not scientifically described but are clearly recognizable. These are morphological operational taxonomic units (MOTUs). Furthermore, and following genetic analysis, one ostensible species may turn out to be a complex of different species (i.e. a set of cryptic species). Additionally, even within a species there can be morphological and other trait variations sometimes making precise identification difficult. (Illustration: Robyn Symons).

7.3.2.1 Value of complete genomes

Insect genomics can be used to assess insect population trends and identify threatened species, as well as help determine the reason for declines and inform conservation strategies. The availability of high-quality reference genomes enables population genomics to address issues like information on past demographic fluctuations, recent population declines (or not) (Tolman *et al.*, 2023), measurement of genetic load levels, and the analysis of genetic adaptation to stressors, as well as delineating molecular operational taxonomic units (mOTUs) and cryptic species (Webster *et al.*, 2023).

There are already several odonate species where reference genomes are available including that of the North American *Hetaerina americana*. This nominate species has a wide distribution across North America and inhabits many locations from sea level to over 2000 m above sea level across many ecoregions with differing water chemistries and temperatures, as well as levels of landscape transformation, although with limited dispersal ability overland (Grether *et al.*, 2023). There is a strong possibility that this nominate species is a cryptic species complex (Vega-Sánchez *et al.*, 2019), with one southern lineage having been raised to the species level (*H. calverti*) (Vega-Sánchez *et al.*, 2020). Furthermore, *H. americana* is one of the species that has declined in California and Nevada in recent years (Ball-Damerow *et al.*, 2014).

Knowing the complete genome of nominate species *Hetaerina americana* enables more focused conservation attention on genetic connectivity and identification of specific barriers to gene flow. Furthermore, this genomic knowledge helps determine levels of connectivity among the riverine habitats, both for the various mOTUs of this species and other species, while also identifying vulnerable aquatic species and local ecosystems such as desert springs (Grether *et al.*, 2023).

7.3.3 Challenges with mapping dragonfly distributions

Mapping the distribution of a dragonfly species at any one time does not always mean that this is the current geographic distribution of a species. Some species move across the landscape, with some even moving across a region. Some others are migratory or partially migratory, with their movements strongly dependent on prevailing weather conditions (Chapter 4, Section 4.3).

The opal slim damselfly *Aciagrion dondoense* was unknown in South Africa prior to the year 2000, when it suddenly appeared. It seemingly moved south from Mozambique owing to a cyclone and massive floods that year. It established in South Africa, even becoming the locally dominant damselfly species in a certain marsh the following year. However, by 2006 it had disappeared from the country, a result of the drying out of its habitat in the next phase of an El Niño–Southern Oscillation (ENSO) cycle (Fig. 3.10). Today, maps show it as being present in South Africa, yet it has not been recorded in the country since (Samways and Simaika, 2016). This doesn't mean, however, that it may not be recorded in the future, especially with the increased likelihood of further extreme weather events.

7.3.4 Significance of large-scale and long-term mapping

7.3.4.1 Mapping at the large scale

Mapping at continental and global scales is valuable for determining the distribution, levels of endemism and threats to dragonflies, for example, across Africa (Clausnitzer *et al.*, 2012). This continental-scale approach has also identified priority areas for dragonfly conservation that largely coincide with biodiversity hotspots (Fig. 7.14). About 72% of the dragonfly species are present in protected areas, covering almost 11% of the African continental landscape (Simaika *et al.*, 2013). These assessments were made possible by taxonomically verified point records providing great accuracy to the mapping (Simaika and Samways, 2010), especially for protected area selection (Simaika and Samways, 2009b). Undertaking mapping at these large scales has value for freshwater insects in general, as protected areas, for example, in Brazil, are biased toward terrestrial systems, with these insects currently declining precipitously (Azevedo-Santos *et al.*, 2019).

While the conservation of whole catchments is recognized as important for dragonfly conservation, there are many mapping methods available, with advantages and disadvantages. Ideally, the extent of occurrence (EOO) should be defined as the sum of the smallest hydrological units of known, inferred or projected resident point occurrences for estimating levels of threat to a taxon. The best unit for dragonflies is the quaternary catchment, and much better than minimum convex polygons, especially for conservation and management (Simaika and Samways, 2010).

Simple presence/absence data (incidence data) is valuable for understanding not only large-scale distributions of odonates, but also small-scale habitat conditions (Finch *et al.*, 2006). However, over large geographic areas, gathering of these data may take a long time, often over many years. This challenge is offset by the database having many different uses that go beyond mapping and into the realms of effective conservation planning.

7.3.4.2 Significance of absence point records

While absence cannot be proved, as the database grows, absences increase in probability of being true. This is particularly the case when dragonfly species point records are associated with an exact description of the habitat and the date of recording. As site visits continue to be made and adult odonates, as well as other life stages, are recorded from an increasing number of sites, a clearer picture of a species preferred habitat, its geographic distribution, season population events, extreme weather events and climate change begin to emerge. However, this

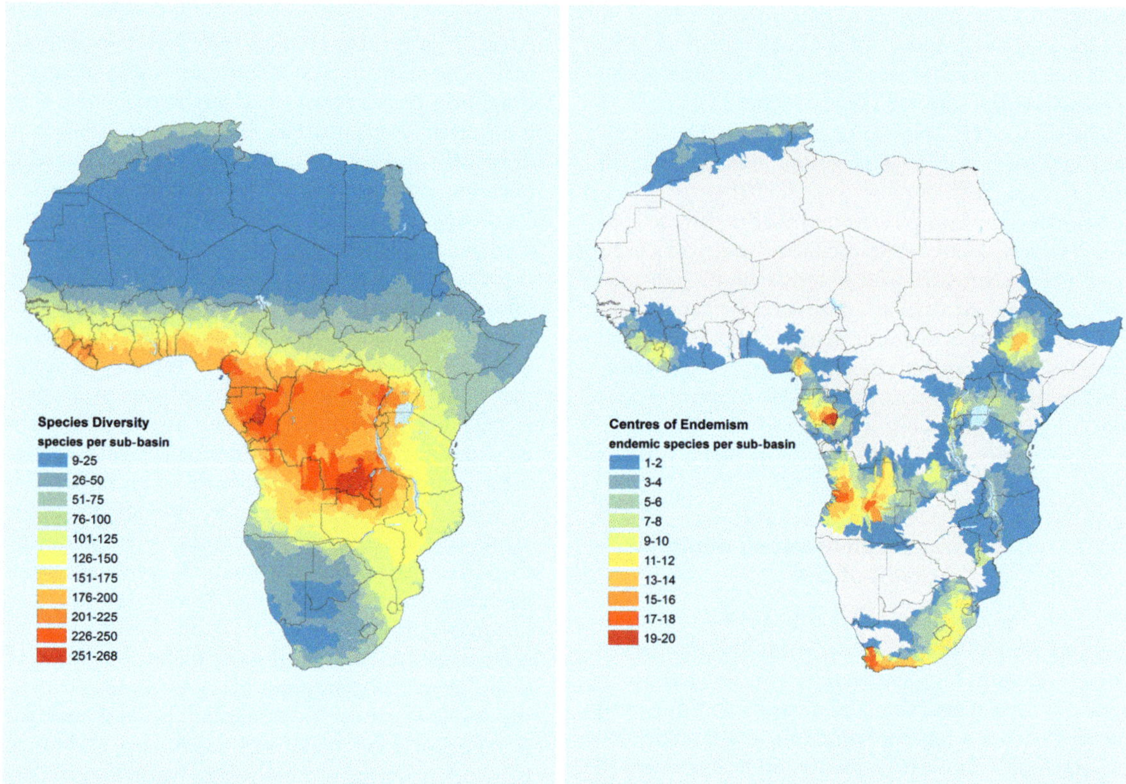

Fig. 7.14. Odonata diversity across the African continent as of November 2023, based on about 790 described and morphotyped species (left), and centres of endemism on the continent, based on 204 endemic or narrow-range restricted species (right). A sub-basin is a smaller working unit within a larger river basin or catchment area. For details of the mapping procedure, see Kipping *et al.* (2009) and Clausnitzer *et al.* (2012). (Maps compiled by Jens Kipping).

means that at each point location, sufficient time must be given to ensure that all species living in that location are recorded (Finch *et al.*, 2006).

7.3.4.3 Value of long-term mapping

Following several years of odonate data gathering across South Africa, it was possible to develop conservative deterministic models using point presence/absence data for each of the regions' 160 described species. These models were cross-validated with a coefficient of similarity used as an index of model performance, and a sensitivity analysis determining the influence of extreme values and errors in the data on predictive ability. Models identified disjunct distribution patterns and accurately predicted restricted ranges of habitat-specialist species. However, models overstated the distribution of habitat-generalist species and those with distinct outlier records. For accurate predictions of geographically broad species, a probabilistic approach is required. Nevertheless, basic distribution patterns generated in this way can be further applied to the study of species richness and diversity relating to conservation such as reserve design (Finch *et al.*, 2006) (Fig. 7.15).

7.3.4.4 Complementary analyses used in mapping

Prioritizing within existing protected areas is based on individual-scoring and complementarity-based algorithms, often guided by individual natural features rather than complementarity among protected areas. Yet, the efficiency of synthesizing individual-scoring and complementarity-based algorithms for protected areas in prioritization assessments can be valuable for conservation planning. Use of three algorithms for prioritizing protected areas included an

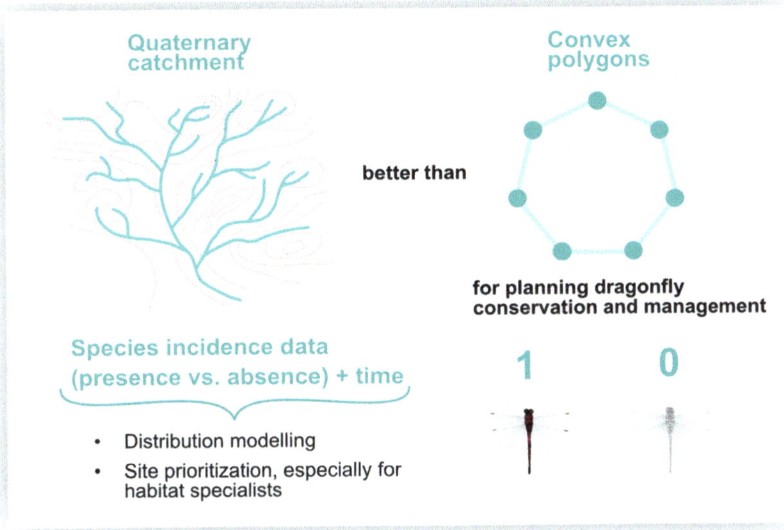

Fig. 7.15. Quaternary catchments are better than convex polygons for large-scale mapping of dragonflies for their conservation management. Also, incidence data, which involves noting species absences at well-surveyed sites, as well as presences, has great value for both modelling geographic distributions and for prioritizing sites, especially for habitat-specialist species over large areas. But this approach requires extensive and intensive surveys at many sites, which is resource intensive as it takes a long time to achieve high accuracy. (Illustration: Robyn Symons).

individual-scoring algorithm (ISA), a species-number complementarity algorithm (SNCA), and then a species-value complementarity algorithm (SVCA) synthesizing the other two algorithms. The average species number conservation efficiency in the SNCA was higher than in the ISA or the SVCA, while the average species conservation value efficiency in the SVCA was higher than in the SNCA and the ISA. In sum, this means that the SNCA could protect more species at a minimum cost when conservation planning is aimed only at protecting the number of species. However, when species conservation value is the focus of conservation planning, the SVCA is more cost-effective (Wang *et al.*, 2021).

7.4 Relative Value of the Different Dragonfly Life Stages in Assessment and Monitoring

7.4.1 Standardizing assessment and monitoring of adult dragonflies

7.4.1.1 Sampling methodology and abiotic environmental conditions

There is much debate over the best method for surveying odonates. The point perhaps is not whether one or other method is best, but rather which is best for the local conditions, feasibility, and behaviour and ecology of the various species in the local assemblages. Furthermore, sampling must also be tailored to the question requiring an answer and the conservation goal, given a finite amount of time and resources.

An important starting point is to sample at exactly the right time of year which can be critically important for poorly known, rare and elusive odonates with distinct flight periods yet in need of conservation, such as *Syncordulia* species in South Africa (Samways and Grant, 2007).

Nehalennia speciosa is a threatened species in central and eastern Europe. The spatial variation in abundance is driven by the vegetation type and physical characteristics of pools, although it is a difficult species to detect. The probability of detection further decreases under windy conditions (Orioli *et al.*, 2021). The point here is that great care must be taken when sampling odonate species at any one location to ensure that data gathering is appropriate and weather conditions are suitable, especially for crepuscular and nocturnal species in tropical and subtropical environments.

Understanding detection probability is important when accurately assessing adult population levels because non-detection of a target species at a site does not necessarily imply absence. In a comparative study of *Tramea onusta*, *Epitheca princeps*, *Pantala flavescens* and *Calopteryx maculata* in Iowa, USA, using standardized visual encounter assessments and a robust design occupancy model, detection probability along with site occupancy, site extinction and site colonization probabilities can be determined for each species. Although there were few significant effects of landscape variables among these species in terms of site occupancy, site extinction or site colonization, the landscape variable of a 600 m distance of individuals from breeding sites was significant. Importantly, however, temperature and cloud cover were variables affecting detection probability for all species. While it is intuitive that detection probabilities would be higher on days with higher temperatures and low levels of cloud cover, species–habitat relationships are significant because ignoring detection probability could lead to overestimation of the degree of site occupancy. Detection probabilities also vary by species, with *P. flavescens*, for example, having high detection scores as it patrols continuously. In sum, estimating detection probability provides a measure of the effectiveness of assessments, valuable when conducting targeted searches or when assessing assemblages (Harms *et al.* 2014).

7.4.1.2 Sampling efficiency

Using 20-metre line transects or 3-minute duration point counts at 62 survey sites, repeated monthly, in a localized area the UK, showed that ambient temperature was the most important weather variable affecting nine species in terms of their apparent abundance, total abundance and among the nine species' richness. Most of the spatial variation in the survey results was between water bodies, rather than between sites. This suggests that adult counts may indicate aspects of water body quality, although there were differences in these patterns when anisopterans are compared with zygopterans. The ease of sampling adults relatively infrequently but rapidly encourages citizen science participation while providing repeatable abundance and species richness data that contribute to assessment and monitoring (Pearce-Higgins and Chandler, 2020). However, when estimating abundance, there is always the risk of double-counting individuals when purely visual counts are undertaken (Darshetkar *et al.*, 2023).

In the Brazilian Cerrado, a savanna system, the most efficient way to sample odonates is to measure their richness and composition based on the time of sampling and the sampling effort over time and space. Most efficient is when sampling is performed between 12:00 h and 14:00 h, in transects measuring 100 m, with at least two sampling events at the same location (Batista *et al.*, 2021).

Repeat surveys are required to assess the full spectrum of odonate species at a defined location at any one time. At ponds, lakes and wetlands in North America, weekly 1-hour surveys of mature male odonates were repeated at least 15 times, then the weekly frequency was reduced to 75% (every 1.5 weeks), 50% (fortnightly), 33% (every 3 weeks), and 25% (monthly) and the 1-hour survey was reduced to 50, 40, 30, 20 and 10 minutes. Reducing the original effort by half (i.e. to 30 min, fortnightly) retained an average of about 80% of the species. The least effort (10 min, monthly) retained only about 49% of species, with the greatest information loss occurring between 20 and 10 min. Across-site analysis showed that data subsets correlated well with the original data set despite an up to 50% species loss. Very strong correlations remained with a 10–15% species loss. Interestingly, improvement in the efficiency of sampling of adult male dragonflies at these standing waters also suggests that non-recording of a few species across the area does not greatly affect richness and compositional comparisons among sites. The efficient sampling approach involves fortnightly surveys lasting 20–40 min at each site, providing a representative and cost-effective protocol (Bried *et al.*, 2012b).

In a study of a 19-species odonate assemblage in India, belt transect counts (slow walking along the riverbank) were compared with point counts (stopping at points along the riverbank and surveying the surrounding area) using binoculars. Belt transects had the highest coverage of species richness (80–90% of species) and were the most time-efficient technique. However, point counts were best for detecting species per unit area, but being intensive, were not time efficient (Darshetkar *et al.*, 2023) (Fig. 7.16).

Effective monitoring of changes in relative abundance of the rare and threatened *Petalura gigantea* in Australia involved using a predetermined transect route and width (about 12 m each side of and in front of the observer) repeated at regular intervals across the flight season. However, the success of the approach is contingent upon a clear understanding of the species' habitat preferences and

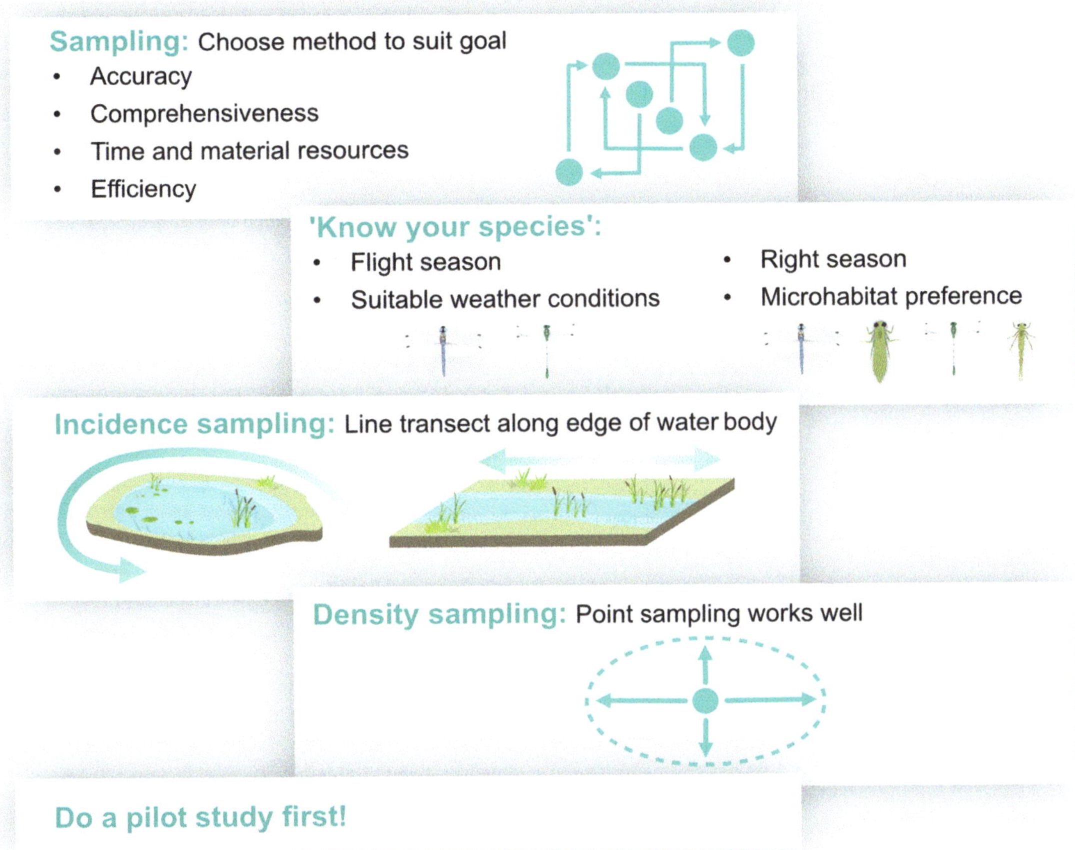

Fig. 7.16. Effective sampling of dragonflies means selecting the method to suit the ecological or conservation goal. However, there is a trade-off between accuracy and comprehensiveness, and availability of time and material resources, to select the most efficient method. It is important to 'know your species': the flight season and suitable weather for adults (A), and the right season and microhabitat preference for larvae and adults (L + A). Line transects along the running or still water body edge, which accounts for all possible microhabitats, is the favoured approach for incidence (presence) sampling of adults. These transects are tailored to local conditions such as microhabitat types and the accessibility of the site. While transect sampling can also be used for sampling the density of individuals, point sampling may be best. Importantly, do a pilot study first before embarking on a full assessment or a monitoring scheme. (Illustration: Robyn Symons).

behaviour, along with the careful choice of survey timing relative to the species' traits and weather conditions to satisfy minimum requirements for detectability, repeatability and replicability. This approach has potential for monitoring other patchily distributed odonates toward their effective conservation management (Baird and Burgin, 2023).

7.4.1.3 Significance of using multiple sampling methods

For a total species inventory, as opposed to comparative quantitative studies at different sites, it is important to use a variety of collecting methods when sampling adults. This is borne out by a study in the Parque Nacional da Serra do Cipó, Minas Gerais state, Brazil, at the ecotone between the Atlantic Forest and the Cerrado savanna. Three collecting methods, standard hand-held aerial nets, Malaise traps across rivers and light-at-night sheet traps, sampled very different odonate assemblages. Aerial nets captured 11 species, Malaise traps 14 species and light traps 7 species. However, the sampled sets were highly complementary in species composition. Gomphids (7 out of 26 Odonata species), including a new species of *Cyanogomphus*, were almost entirely captured by Malaise traps and light traps (De Almeida *et al.*, 2013). The message here is clear: for a full inventory of adult dragonflies in any one area, especially in warm regions, complementary collecting methods must be used.

7.4.1.4 Opportunistically versus systematically gathered adult dragonfly data

Citizen science records can contribute greatly to dragonfly conservation but often have the disadvantage that they are gathered opportunistically rather than systematically. Site-occupancy models help reduce biases in opportunistic databases, such as recording effort over time and incomplete sampling of odonate assemblages. Site-occupancy models estimate the probability of sites being occupied while also considering imperfect species detection. The models require presence/absence data from many sites. They are then applied to opportunistic species data gathered over time. This modelling approach was applied to a long-term data set of Dutch odonates and based on three levels of recording effort: one-species records, short daily species lists and comprehensive daily species lists. All three databases were compared to a fourth one, a standardized monitoring scheme which acted as a yardstick for comparison. Occupancy trends based on comprehensive daily species lists per site, in combination with species occupancy models, gave the best results and similar to those from the standardized monitoring scheme. The findings contrast with those based on one-species records and short daily species lists, which were too imprecise to have value. For conservation monitoring schemes using opportunistic data, good-quality daily lists have great value in that they accurately illustrate trends in odonate species distributions at a large spatial scale (van Strien *et al.*, 2010, 2013; van Strien and van Grunsven, 2023).

The turnover of *Enallagma* species across the USA indicates that opportunistically gathered presence/absence data accumulated over many years of sampling can produce results that are consistent with many standardized observational studies using the same type of analysis. This is despite only adult records being present in the opportunistic data sets compared to larval or resident occurrences in the systematic data sets. Although absences may often reflect detection failures, the data are still a basis for valuable presence-only information. The large volume and geographic extent of opportunistic presence-only records may partly compensate for any inherent biases, and potentially enable the use of large-scale inventory data and extensive mapping to help understand general patterns in spatial assemblage structure at broad spatial scales relative to species' geographic ranges (Bried and Siepielski, 2018) (Fig. 7.17).

7.4.1.5 Zygoptera/Anisoptera ratios

The ecophysiological hypothesis (EH) refers to the interaction of odonate thermoregulation, body size and level of sunlight availability in tropical forest streams, and the associated thermoregulatory restrictions. It predicts that large species are disfavoured and that streams will only support nested assemblages based on the body size of species. In the Brazilian tropical Cerrado, increasing channel width leads to an increase in abundance and species richness of anisopterans but a decrease in abundance of zygopterans. While maximum thorax size is positively correlated to greater channel width, there is not that relationship for the minimum thorax size, with no lower size constraint on small-sized species. There are also nested assemblages related to body size, indicating the EH as an ecological assembly rule based on abiotic factors. Forest cover determines which species occupy a certain site, with the EH indicating the highest threats to small-bodied zygopterans in response to riparian forest loss in tropical streams (De Marco Júnior *et al.*, 2015).

Adult odonates were sampled in Brazilian Amazonian streams to evaluate whether the relative abundance and species richness of zygopterans versus anisopterans (the Z/A ratio) can be used to measure environmental impacts on these streams following an assessment of stream ecological conditions using the Habitat Integrity Index (HII) (see below). Habitat modification results in a decrease in zygopteran relative to anisopteran diversity compared to the historic state. A decline of only 0.1 in the HII score results in an increase of about 13 individuals and 11 species of anisopterans, with the opposite effect for zygopterans. Overall, sites where more than 54% of species were zygopterans could be classified as preserved sites, while those dominated by anisopterans (more than 59% of total species) could be considered as degraded. The implication for environmental impact assessments is that using these ratios reduces sampling effort and overcomes the effect of any assessor inexperience on assessment outcomes by not

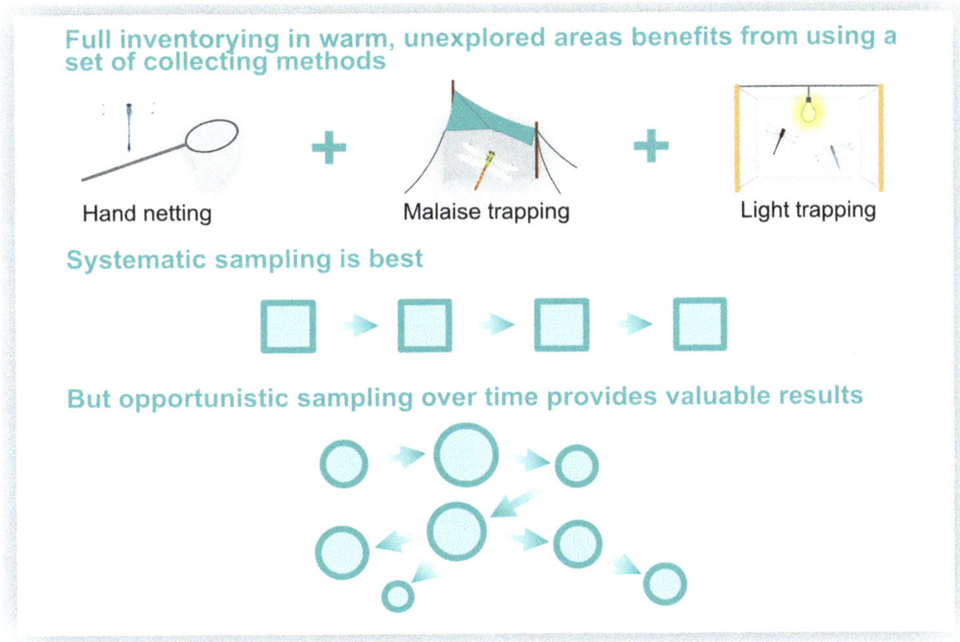

Fig. 7.17. A full inventory of dragonfly adults in warm unexplored areas benefits greatly from using a set of different collecting methods. While systematic sampling is the most preferred, opportunistic sampling, especially over long periods of time, can produce valuable results. (Illustration: Robyn Symons).

relying on specialist knowledge, also given that members of the two suborders are easily distinguished in the field (Oliveira-Junior and Juen, 2019) (Fig. 7.18).

In an evaluation of loss of natural vegetation cover on the diversity and morphology (length of thorax and abdomen) of Amazonian stream odonates, there was a positive relationship between high natural vegetation cover and greater species richness of odonates overall and proportion of zygopteran species richness. However, forest loss benefitted the anisopterans, especially those with longer abdomens. Six zygopteran species were associated with high natural forest cover, and six species (three anisopterans and three zygopterans) were associated with reduced vegetation cover, emphasizing the importance of preserving both stream quality and forest cover within 500 m of the streams (Calvão *et al.*, 2023).

The Z/A ratio of adults can be an effective tool in measuring ecological status in the Atlantic Forest streams in Brazil, with the HII being used to represent the state of the habitat. The Z/A ratio indicates that when the proportions for Anisoptera constitute 67% species richness and 52% abundance, the streams are transformed. In contrast, when the proportions for Zygoptera constitute 54% species richness and 67% abundance, the streams are in a natural or near-natural state. The advantage of this Z/A ratio approach, at least for Brazilian wet forests, is its ease of identifying the specimens and simplicity of interpretation. This means that it is highly tractable for assessing and monitoring streams by non-specialists (Ribeiro *et al.*, 2021).

Among the 97 odonate species recorded along streams in Pará, Brazil, 9 were indicators of ecological condition, 4 species of which were associated with more degraded streams (3 anisopterans, 1 zygopteran) and 5 with streams in historic condition (all species were zygopterans). Anisopterans indicate more degraded environments as they have more efficient heat regulation and are highly mobile, enabling them to tolerate a wide range of environmental conditions. In contrast, zygopterans are better indicators of more intact environments with high levels of environmental heterogeneity. This is because of their smaller body size, small home ranges and more restricted ecophysiology.

Fig. 7.18. There is a shift in the predominance of Zygoptera species in intact Brazilian tropical forest (top left and right) to that of Anisoptera species when the forest canopy is opened up through human activity (bottom left and right). (Photos: Leandro Juen).

For small Amazonian streams, odonate assemblage composition is strongly associated with the structure of riparian vegetation, with agricultural activities determining changes in assemblage composition, as measured by the HII. Changes in assemblage composition away from the historic state indicate riparian habitat loss and disturbance. Moderately degraded areas are characterized by habitat-generalist species, and although disturbed areas may have high species richness, few species are zygopterans. To maintain historic assemblages requires intact dense riparian vegetation in the region (Oliveira-Junior *et al.*, 2015).

Among streams in the eastern Amazon in Brazil, while using the HII to grade the ecological state of the streams, odonate suborders and families were used to determine stream ecological condition. The ability of the families to represent the diversity of the two suborders is influenced by the abundance of individuals and species richness of the family. While stream ecological integrity affects diversity, the cumulative measure of abundance has some merit as an assessment tool. The similarity of the responses at species and family levels supports the use of odonate families for the detection of ecological thresholds in these streams, suggesting that the family-level approach is effective in detecting stream ecological transformation of habitats and ecological integrity (Mendoza-Penagos *et al.*, 2021).

In Neotropical savanna headwater streams, sites over a human disturbance gradient indicated that species richness of zygopteran larvae was highest in shaded sites with canopy cover taller than 5 m, while bare ground without riparian vegetation was important for anisopteran species richness and abundance. Overall, odonate larvae have the same distribution, species richness and abundance patterns as adults, while noting that disturbances related to the removal or reduction of riparian vegetation favours anisopterans over zygopteran larval assemblages. This emphasizes the need to conserve riparian canopy cover to maintain the historical diversity of zygopterans both in the water as larvae and along the banks as adults (Silva *et al.*, 2021).

In the Cerrado savanna region of Brazil, zygopterans discriminate with regard to the ecological integrity of streams better than do anisopterans, based in part on their precise ecological requirements of both the forest canopy and the understorey, and that zygopterans generally have poorer dispersal ability. Certain species are especially sensitive to either low or high levels of vegetation cover. Within Zygoptera, small-sized species in this region do not always occur in shaded locations, as they have traits related to needs for resources associated with territoriality and oviposition, attracting them to open areas (Dutra and De Marco, 2015).

This microhabitat preference is also seen among South African zygopterans, with some of the smallest species, as in the genus *Agriocnemis*, predominant in areas with no tree canopy. Being so small, they can easily find both refuge and resources low among grass, sedge or reed stems in marshy areas (Samways, 2008). This is not so much a continental difference as an ecosystem one, as shown by the predominance of shade-preferring zygopterans in the tropical forest region of Ghana, with some species present in open habitats (Acquah-Lamptey *et al.*, 2013).

It is important not to assume that shade is the only driver of changes in proportions of the two suborders. In seasonal tropical forest in Brazil, anisopteran larvae tolerate a wide range of environmental variables associated with degraded habitats. When local physicochemical variables like water temperature, pH, turbidity, electrical conductivity, dissolved oxygen, total dissolved solids and oxidation reduction potential are considered, they relate to changes in odonate species richness, abundance and composition. This is especially so for pH, electrical conductivity, dissolved oxygen, total dissolved solids and oxidation reduction potential. While anisopterans showed a significant relationship with these variables, zygopterans were only related to pH (Mendes *et al.*, 2018), though bearing in mind that zygopterans are greatly affected when tree cover is lost in the region.

The increased abundance of anisopteran species in disturbed tropical Brazilian forest suggests the exclusion of species requiring better quality microhabitats or the persistence of more tolerant zygopteran species. Anisopterans in this region benefit more from degraded environments that are more exposed to sunlight with semi-lentic characteristics, with an average of 15 species more than the number found in altered habitats, and 24 more than in historic areas (Juen *et al.*, 2014).

A meta-analysis of adult-based bioassessments on a global scale and with reference to tropical forest degradation, revealed that the Z/A ratio is in general a poor indicator of tropical forest state from a global perspective. These findings likely stem from ecological diversity within these two groups. However, ratios of Coenagrionidae/other Zygoptera and Libellulidae/other Anisoptera increased greatly along a disturbance gradient. This suggests that these two ratios have the potential to be effective indicators of well-preserved, altered or heavily degraded forest habitats at large spatial scales. While these results align with studies that emphasize the usefulness of adult odonates as versatile indicators for assessment and monitoring, the value of these two ratios involves classifying specimens only to morphological operational taxonomic units (MOTUs) so long as there is accurate identification to family level (classified as Coenagrionidae, Libellulidae, or 'other'), which could be performed by non-specialists with minimal training for assessing both freshwater and terrestrial habitat status (Šigutová *et al.*, 2019).

Assessment indices based on the Z/A ratio do not always show any significant pattern along a forest disturbance gradient. The reason for this is that some comparisons are based on HII quantifying the structural characteristics of stream banks (e.g. Monteiro-Júnior *et al.* (2014) in Brazil), while others were done along a more robust forest disturbance scale (e.g. Oppel (2005) in Papua New Guinea; Monteiro-Júnior *et al.* (2015) in Brazil; Seidu *et al.* (2017) in Ghana), or used a sampling methodology biased toward sampling indicator species (Seidu *et al.*, 2018).

Šigutová *et al.* (2019) however, used a unified approach with clear habitat rating, and the sorting of species on a disturbance scale according to exact criteria across various types of tropical forests and aquatic habitats (streams, rivers, swamps, dams, etc.) on a global scale. This suggests that the Z/A ratio may work only locally and/or on a more robust disturbance scale, suggesting that it cannot be recommended as a reliable measure for indicating the level of all types of tropical forest degradation worldwide.

In an intense assessment of Brazilian savanna headwater stream sites, 30 odonate larvae genera were sampled, with seven (*Argia*, *Brechmorhoga*, *Cacoides*, *Gomphoides*, *Phyllocycla*, *Progomphus* and *Mnesarete/Hetaerina*) identified as effective and robust indicators of environmental transformation. While most genera of zygopteran larvae are associated with less-disturbed sites and most anisopteran genera larvae were present in disturbed sites, not all genera corresponded with this tendency. This led to the recommendation that using odonate larvae at the genus level, rather than at the suborder level, gave the most accurate impact assessment of the stream sites (Silva *et al.*, 2023).

Streams in north-eastern Brazil were sampled during the dry season and disturbance scored using the HII. Odonate species composition was most effective for evaluating stream ecological integrity, being accurate and relatively inexpensive. Taxonomic diversity and distinctness also provided relatively reliable results, contributing additional information on the evolutionary relationships among the taxa. Fine assessment of assemblages is necessary to accurately detect stressors and habitat transformation, when seen against the considerable variation in the environmental requirements of the various species. Overall, the best metric for determining environmental gradients of disturbance in this geographical area is species composition, followed by species diversity and taxonomic distinctness (Miguel *et al.*, 2017) (Fig. 7.19).

7.4.1.6 The Dragonfly Biotic Index

The Dragonfly Biotic Index (DBI) is an aquatic- and terrestrial-oriented index of the overall quality of odonate habitats. It is based on three subindices per odonate species: extent of geographic distribution, IUCN Red List status (global and/or national) and habitat sensitivity. Each of these subindices has a score ranging from 0 to 3. Zero for geographic distribution is a widespread species, whereas 3 relates to a narrow-range endemic. Zero for Red List status refers to a species of Least Concern, whereas 3 relates to a species that is Vulnerable, Endangered or Critically Endangered. Zero for habitat sensitivity refers to a species that not only occurs in natural habitats but also a wide range of disturbed, perturbed and/or artificial habitats, whereas 3 refers to a species that is only ever found in natural habitats. Overall, it means that each species has an established score, ranging from 0 to 9, the sum of its three subindex scores. A species that has a score of 0 is a geographically widespread, not currently threatened, habitat generalist. A species with a score of 9 is a narrow-range endemic, that is distinctly threatened, and only occurs in natural habitats. Most species occur somewhere along this spectrum of 0 to 9, with few at the extremes. Examples in South Africa of the species scoring 0 are *Ceriagrion glabrum*, *Ischnura senegalensis*, *Crocothemis erythraea*, *Sympetrum fonscolombii* and *Trithemis arteriosa*. Examples in the same region that score 9 are *Chlorolestes apricans*, *Spesbona angusta*, *Metacnemis valida*, *Proischnura polychromatica* and *Orthetrum rubens*. A practical

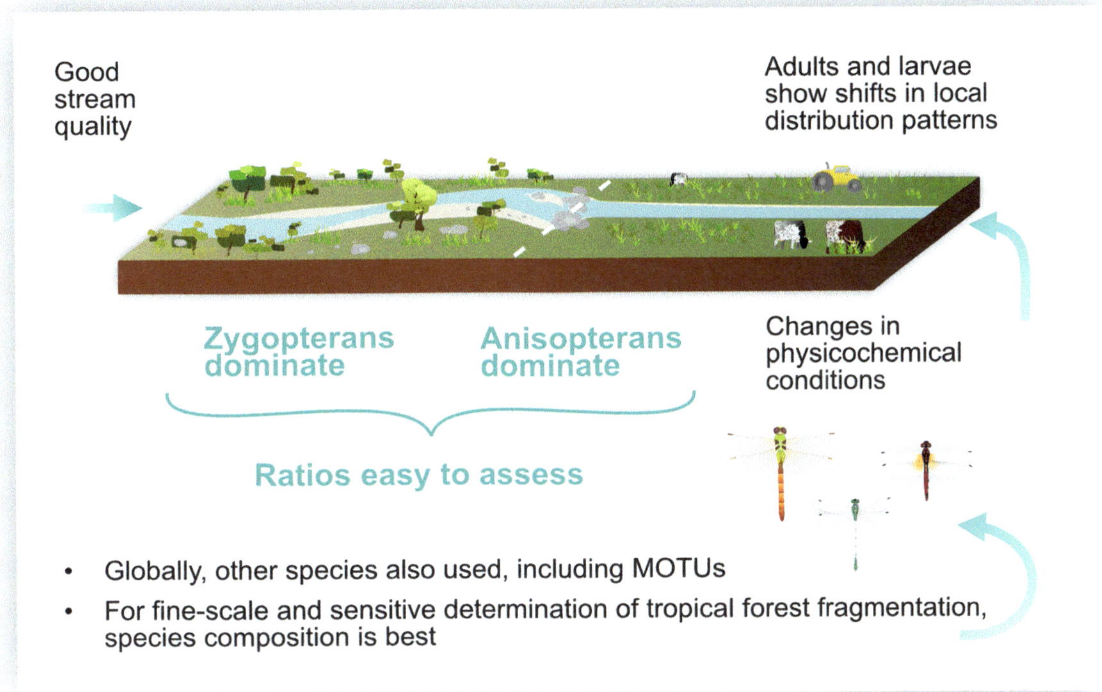

Fig. 7.19. The ratio of Zygoptera/Anisoptera species richness is an easy first approach to characterize tropical forest transformation and preservation (e.g. instigating 500-metre wide riparian buffer zones), especially in Brazil. Morphological operational taxonomic units (MOTUs) (morphospecies) have also been used, as have certain taxonomic subsets of species. However, from a global perspective, species composition is the preferred means of sensitive comparative assessment of landscape transformation, albeit a more resource-hungry approach. (Illustration: Robyn Symons).

manual has been developed for species recognition, and use and application of the index (Samways and Simaika, 2016) (Fig. 7.20).

The DBI was first developed specifically for conservation activities in South Africa (Simaika and Samways, 2009a). It can be used to assess all types of freshwater water body, running and still, fully natural to highly disturbed or artificial, and perennial or ephemeral. It can be used to compare sites or measure their deterioration versus recovery/restoration, as well as determining the conservation value of a set of sites. It can not only be used to assess small areas such as ponds or river reaches, but also to compare, for example, the comparative degree of naturalness of whole rivers. It can also be applied to all aspects of hydromorphology along rivers, such as side streams, flood margins, marshy pools and oxbow lakes, as well as the main channel.

The DBI can be used to assess or monitor the effect of certain stressors, whether physical like dams, loss of riparian cover or invasion by aliens such as trees or fish, or physicochemical variables like water velocity, turbidity and pollution. However, it is vital to measure the suspected stressor involved. Sometimes the stressor is clear, like the effect of a dam or invasive alien trees. At other times, the stressor is not so clear, especially when multiple diffuse stressors are involved, and their effect must be teased apart through sampling of suspected stressors.

For the DBI, dragonflies are sampled along a 100-metre-long site subdivided into 10 m contiguous or non-contiguous sections so that all microhabitats both in-water, on the margin and along the banks are included. Sampling is repeated until the species accumulation curve reaches an asymptote (in reality a near-asymptote), which may mean that sampling is also repeated in both spring and autumn, as well as summer, in some regions.

The spatial and/or temporal comparison of sites is based on the total of all the species scores, the site DBI score. A site with a much higher score than another implies it has many more rare and sensitive

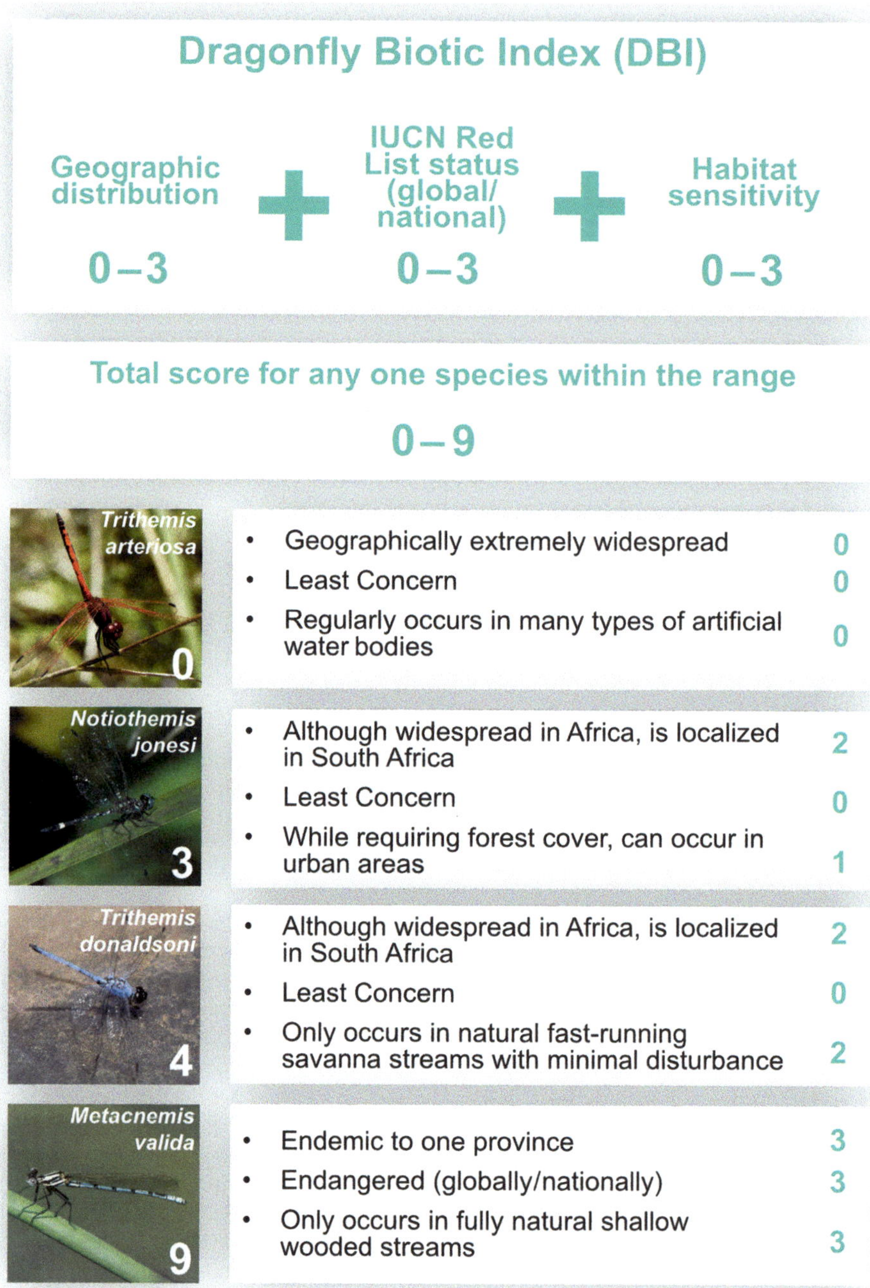

Fig. 7.20. The Dragonfly Biotic Index (DBI) was originally developed for easily assessing and monitoring the state of South African water bodies, both running and still. It uses dragonfly species composition to assess the state of an aquatic/terrestrial focal system based on the scores of each species. Each species' score is based on the total of 'subscores' from three subindices: extent of geographic distribution, IUCN Red List status (global/national) and habitat sensitivity. Each species subindex has a score from 0–3, making the DBI total score for any one species within the range 0–9. For example, *Trithemis arteriosa* scores 0, *Notiothemis jonesi* 3, *Trithemis donaldsoni* 4, *Metacnemis valida* 9. (Photos: Michael Samways; Illustration: Robyn Symons).

species. Standardization can be achieved by using the average score per taxon, i.e. the average of all the species scores at a site or even a set of sites divided by the number of species.

The DBI has been remarkably effective in showing recovery and rates of recovery once invasive alien trees are removed and the natural riparian vegetation is re-established. It is especially effective when there are some rare, endemic and red-listed species (Simaika and Samways, 2012).

The DBI is largely based on the presence of adult males. However, in essence, as it is about the local presence of all species, it means that information on female adults, eggs, larvae and exuviae can be added in, so long as the species-level identifications are correct. This allows for specific supplementary sampling, using different methodologies, of certain species to be included in the DBI. Even in the case of adult males, and when they have unusual behaviour, the results of that sampling can be included. For example, *Gynacantha* species that fly in the evening and often above the canopy are only rarely caught on the wing. However, they can be found relatively easily in the daytime in the dark and dense recesses of the forest vegetation adjacent to water bodies.

Following identification of a stressor or stressors, conservation remediation can take place. However, a key precondition for site recovery is that there must be a species pool in the immediate vicinity so that individuals are able to recolonize the restored site (Simaika and Samways, 2009a). Sometimes, there are surprises with extremely rare species appearing. At some restored sites in South Africa, some rare species appeared that had not been recorded for several decades, including *Spesbona angusta* (DBI score = 9), *Proischnura polychromatica* (score = 9) and *Pseudagrion newtoni* (score = 8) (Samways *et al.*, 2005) (Fig. 7.21).

The Dragonfly Recovery Score (DRS) refers to the percentage change in total site DBI when restoration has taken place, for example: a total score of 20 before restoration and a total score of 40 after restoration = a restoration score of 200%, meaning that the conservation value has doubled. The highest relative recovery has been on Table Mountain, Cape Town. After removal of alien pine trees choking the small streams on the mountain, while species richness doubled, the DRS was 600%, a six-fold increase in recovery due mainly to the reappearance of several localized endemic species. However, one species, *Orthetrum rubens* (score = 9) did not return, as it was not present in the local species pool, with its nearest population stronghold about 100 km away and with inhospitable habitat in between (Fig. 7.22). Elsewhere, in the savanna region with no narrow-range endemic species, and where invasive alien trees had greatly impoverished dragonfly assemblages, species richness also doubled. Importantly, however, following alien tree removal, the DRS was only 114% (i.e. a 1.14-fold improvement) as most species are widespread habitat generalists each having a low DBI score (see Fig. 6.21).

7.4.1.7 Other dragonfly-based indices

The Coefficient of Conservatism (CoC) assesses wetland condition relative to the adult odonate species present. It was developed using 135 odonate species across a wide range of wetlands in Rhode Island, USA, subject to various levels of human disturbance. It was first tested on a subset of wetlands of various qualities, following georeferenced point records that were subject to quantification and indicator analysis. The final CoC index relates very well to levels of human impact and wetland condition. It is also independent of sampling effort, wetland size and geomorphic class. Furthermore, it appears to have general applicability for other geographical areas (Kutcher and Bried, 2014).

The CoC was modified and applied to local conditions at ecotones between streams and open water in Florida, USA in the form of the Odonata Index of Wetland Integrity (OIWI). The index weighed species occurrences at a site within a habitat type by its observation frequency across samples to estimate CoC contributions of each species. The resulting CoC scores at each site–habitat combination were then averaged to obtain four OIWI values per habitat type. OIWI values for the larger wetland were calculated in the same way but referred to species occurrences across habitat types within a site. Derived OIWI values indicated that scoring of ecological integrity would not change when only species composition at ecotones is considered, while also indicating that stream species are highly sensitive to human disturbance. Low OIWI scores from open water habitats emphasize the inconsistent detection rates of species with moderate CoC values. Species tolerant of disturbed sites were driving the lower OIWI values despite high species richness in open water habitats. The many species at ecotones were consistently detected because of the similarity in environmental factors across the

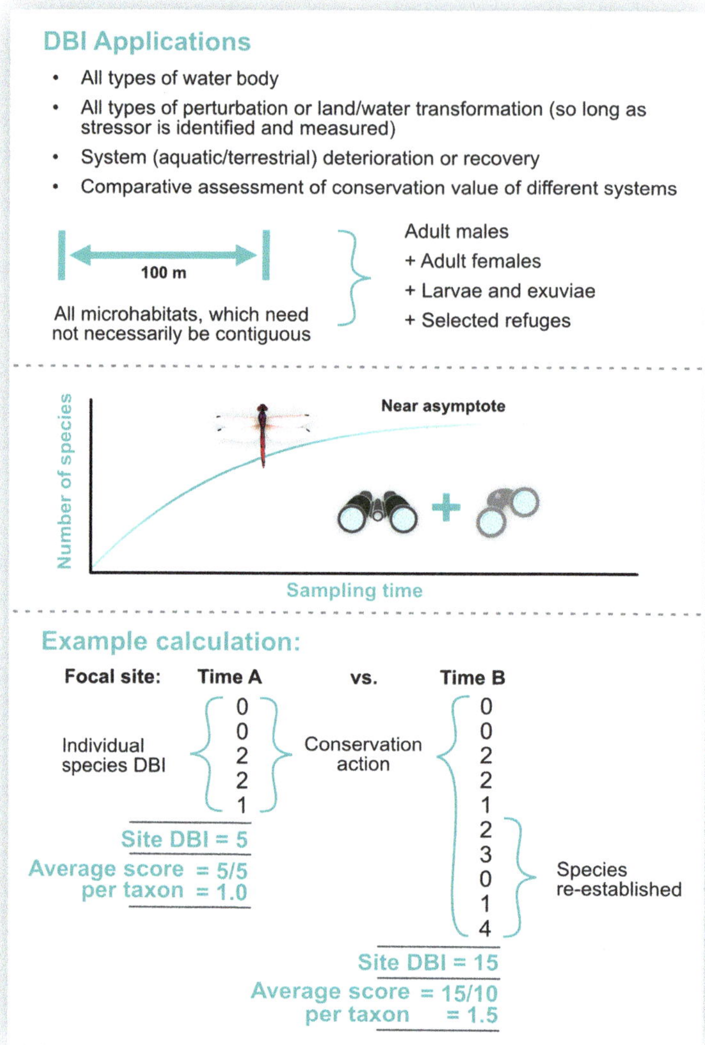

Fig. 7.21. The Dragonfly Biotic Index (DBI) has several uses. But importantly, a near-asymptote of all the species must be reached through adequate sampling. One sampled site is 100 m along the margin of the water body, which includes all microhabitats locally present. However, as it is not always possible to sample all microhabitats (because of difficulties in accessing the water margin), these microhabitats need not necessarily be contiguous so long as the sampled length of water margin adds up to 100 m. Use of close-focus binoculars is highly recommended, as is occasional capture of individuals for examination with a hand lens. While adult males are the primary target, adult females and young stages, as well as sampling of daytime refuges for crepuscular species or those that only forage away from water, can be used to supplement the males to ensure a complete species list, so long as they can be identified to species. A disturbed/transformed site at Time A has five species all with low DBI values, as they are disturbance/transformation-tolerant species. The site DBI is the sum of these values. The average score per taxon is the sum of the DBI values divided by the number of species sampled, which here is 1. Following conservation action (i.e. restoration), and assuming that the species are locally present to re-establish, the number of species increases to 10, giving a site DBI of 15. These recently added species are those that are generally more sensitive species with overall higher DBI values, increasing the average score per taxon to 1.5. This example is a measure of site recovery, but the DBI can also be used to measure site deterioration against a benchmark historic state, or it can be used to compare sites for assessing their comparative site values for prioritization or complementarity purposes. (Illustration: Robyn Symons).

Fig. 7.22. *Orthetrum rubens* (top left) is an extremely rare and threatened endemic species found in high-elevation marshes (top right) in the Cape Floristic Region, South Africa. Its type locality is Table Mountain, Cape Town. However, it was extirpated from Table Mountain, principally by the planting of alien pine trees. While this locality was restored through removal of the alien trees, and all other known species to the mountain returned, *O. rubens* never did. This was seemingly because of not only the large distance between Table Mountain (bottom, in the distance) and the nearest potential source populations (top right), but also because the intervening area is unsuitable habitat of hot and dry lowland and which is now also highly urbanized. (Photos: (top left and top right) Klaas-Douwe Dijkstra, (bottom) Michael Samways).

ecotones. Overall, the combined effect of a high complexity of aquatic vegetation, heightened odonate shade sensitivities, and high levels of species richness and abundance, together led to the similarity in OIWI scores between ecotones and the entire wetland system, emphasizing the significance of ecotones for conservation of dragonflies in wetlands (Cunningham-Minnick *et al.*, 2019).

In lowland central Europe, the Dragonfly Association Index (DAI) assesses both the water and morphological aspects of rivers following the general principles of the European Union's Water

Framework Directive (Copetti and Erba, 2024) which rates river ecological status as high, good, moderate, poor or bad against the historic baseline condition. The DAI is a developmental extension of an earlier Odonata Habitat Index (OHI) (Chovanec and Waringer, 2001; Chovanec *et al.*, 2004) but with greater consideration given to ease of use. The DAI incorporates physicochemical conditions, and importantly, aspects like flow velocity, hydroperiod, river size, extent of open water, submerged macrophyte cover, and margin and riparian cover and condition. Aerial photographs of ancient channels and old dragonfly records are used to help determine the historical state. Adult dragonfly needs in terms of species traits and assemblage composition are then associated with certain environmental conditions to categorize the state of the river. The DAI can also be used to measure the degree and extent of river and bank recovery following remedial action (Chovanec *et al.*, 2015).

The OHI was used to evaluate the habitat quality of a Danube river floodplain in Lower Austria in terms of 29 resident odonate species. Site-specific OHIs ranged from 1.72 to 3.67. The mean OHI for the whole floodplain was 2.79, indicating a relatively high level of habitat diversity. Comparing these figures with reference conditions from old species inventories dating back to the nineteenth century and before water regulation of the Danube, the status of the floodplain was ranked as class II (good ecological status) in a five-tiered classification scheme (Schultz *et al.*, 2003).

The Odonate River Index was developed to evaluate the ecological integrity of whole river corridors in northern Italy. Besides physicochemical conditions being recorded, all associated hydromorphological microsystems are included. Additionally, exuviae as well as adults are included in the assessments, leading to an emphasis on the importance of secondary channels and floodplain ponds for increasing local dragonfly diversity (Golfieri *et al.*, 2016).

The Odonata Community Index – Corsica has been used on that Mediterranean island to assess the ecological status of rivers there. It uses mostly adults, supplemented with records of larvae and exuviae, alongside physicochemical and hydromorphological data to classify rivers according to the EU's Water Framework Directive. Furthermore, this index also provides a rapid earlier warning of river deterioration in Corsica (Berquier *et al.*, 2016).

7.4.2 Comparative value of different assessment indices

A major consideration is whether any one assessment index using dragonflies, and often developed in one geographical area, is widely applicable across regions. A set of indices relating to ponds were compared in France, South Africa and Switzerland based on two important criteria. The first criterion relates to whether the IUCN Red List Categories are included or not. The second criterion relates to the scoring method that assesses the conservation value of freshwater assemblages. Scoring can be expressed at different levels, either at the assemblage level ('per assemblage'), or at the species level ('per species'), and mostly relating to a statistical mean of the species composing the assemblage. For example, the Species Quality Score (SQS) is expressed per assemblage and based on the sum of the threat levels of all species belonging to an assemblage. A modification of the SQS, the Species Rarity Index (SRI) is instead expressed per species and is the sum of the threat levels of all species belonging to the assemblage (i.e. the SQS) averaged by the number of species, the average score per taxon.

The results of the comparison were clear. The type of method used had a great effect on the assessment of the conservation value of the focal pond. Whether the score was expressed as per species or per assemblage, and the weight given to Red List categories, changed the value of a pond. This means that while different assessment/scoring methods might be applicable to different geographical areas and types of standing water body, they require standardization. This can come about by moving beyond the commonly used species-richness assessment and calculating conservation value as a mean per species. This is useful as it provides additional information. Overall, using assessment/scoring methods expressed as a mean per species and coupling this with the Red List gives the best performance (Rosset *et al.*, 2013).

The Dragonfly Biotic Index (DBI) is robust as it is based on per species trait scores as well as Red List status scores. However, this does not exclude other measures which might explain why, for example, a DBI site score has a certain value. Among the natural environmental variables to consider, for example, are local climate, especially hydroperiod, whether rivers have flooded margins or not, or for standing water bodies, their fluctuation in water levels. At small spatial scales, there are a whole host of different variables (aquatic, marginal, ecotonal, riparian, etc.) that play a role, often with different microhabitats contributing greatly to dragonfly residency (Hawking and New, 1999). Furthermore, there are then the human-induced stressors, which may be multiple, that must be identified and addressed for dragonfly and freshwater recovery (Fig. 7.23).

7.4.2.1 Modifications of the Dragonfly Biotic Index for use in other areas

The South African Dragonfly Biotic Index (DBI) was developed into a continental scale index, the African Dragonfly Biotic Index (ADBI) by adapting the subindices, especially the 'habitat sensitivity' subindex (Vorster *et al.*, 2020). The challenge at the continental scale compared to the national scale, was that there were not the data on range of suitable microhabitats of all the species. The habitat sensitivity subindex of the South African DBI is a quantitative and relative measure based on the extent to which each odonate species occupies natural versus human-modified or artificial freshwater systems. At the continental scale, although data for the two subindices, species' geographical distribution and Red List threat statuses, are available, this was not so for the habitat sensitivity subindex. This shortfall was addressed by using an alternative subindex, the 'species vulnerability subindex'. This subindex is based on expert knowledge of the vulnerabilities of each of the species to various types of landscape disturbance or transformation. This then enabled each species' vulnerability subindex score to be combined with the geographical distribution and Red List threat status subindex scores to develop ADBI scores for over 600 dragonfly species across continental Africa. The various ADBI scores then provided a tractable framework for determining running and still freshwater quality relative to human disturbance. The ADBI identifies threats to, and sensitivities of, African freshwater

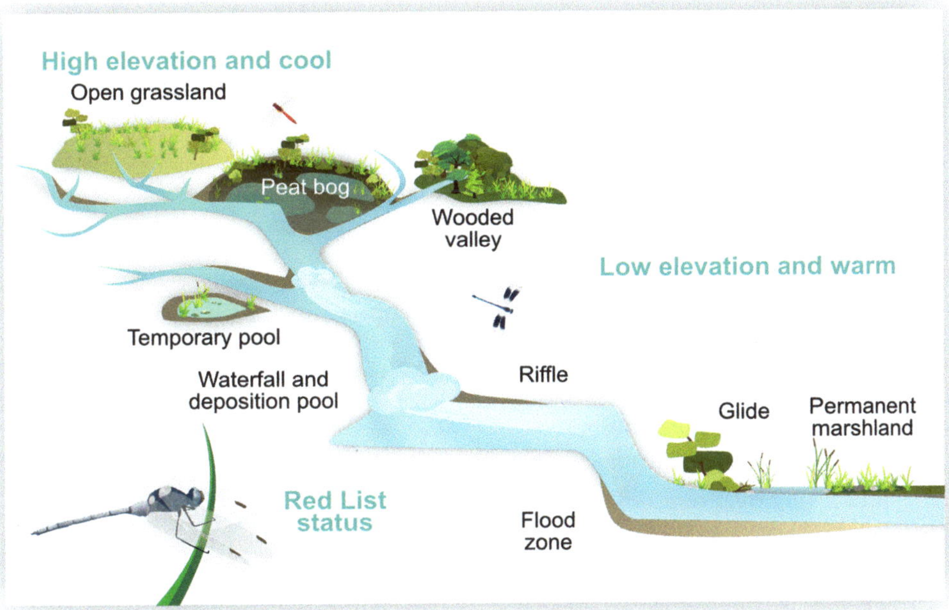

Fig. 7.23. It is desirable to have standardized methodology for assessment and monitoring of the conservation status of freshwater systems. While sampling time and space can be standardized at specific sites (e.g. 100 m transect/30 minutes sampling), it is not easy to standardize across wide geographical areas where factors like geology, topography, climate and hydroperiod all play a role in determining the species composition of the various dragonfly assemblages. Even within one catchment there can be a whole host of different microfeatures and microhabitats that can cause shifts in local assemblages, some of which are shown here. All these features need to be included in the assessment or monitoring scheme as the spatial scale is increased. However, doing so puts increasing demand on human resources. Comparative analysis of indices suggests that inclusion of Red List status as well as trait measurements increases the robustness of a compound index of the status or state of a system. (Illustration: Robyn Symons).

systems through assessment and monitoring of deteriorating or improving freshwater quality changes in the future.

The ADBI can then be used as a basis for developing other local DBIs. For example, the ADBI was tailored as the basis for the Rwanda Dragonfly Biotic Index (RDBI) to identify conservation priority sites and monitor their restoration. The RDBI is based on three subindices: distribution-based score (DBS), threat-based score (TBS) and sensitivity-based score (SBS). Species-level DBSs range from species that are widespread across all ecological zones (low score) to those that are restricted to only one ecological zone (high score). TBSs range from species that are red listed as Least Concern (low score) to those that are Critically Endangered (high score). Species SBSs range from those thriving in a highly disturbed habitat (low score) to those occurring only in a relatively intact habitat (high score). The index identifies odonate hotspot habitats and sites for restoration. Hotspots are those areas that are rich in species, have unique species and have high RDBI site scores. Candidate sites for restoration are the ones with the highest RDBI site scores in each ecological zone (Uyizeye *et al.*, 2021).

The DBI, with some national modification, has also been used to assess various types of water body associated with differing degrees of habitat protection, for example in Czechia (Harabiš and Dolný, 2015), Slovakia (Balázs *et al.*, 2022), Tunisia (Khelifa, 2019), Mediterranean-type ecosystems in Morocco, Algeria and Tunisia (Khelifa et al., 2021a), Nigeria (Adu *et al.*, 2019) and Indonesia (Ilhamdi *et al.*, 2020; Rachmawati *et al.*, 2023).

7.4.3 Integrating river morphology and dragonfly assessment

7.4.3.1 Measuring river morphology

The Habitat Integrity Index (HII) (Nessimian *et al.*, 2008) is used for assessing streams in Brazil relative to the degree of biotope transformation and the impact on biodiversity quality (see Fig. 7.18). It can also indicate the capacity of the local environment to support and maintain a set of species, its diversity and its functional organization relative to that of the regional natural habitat in the same area. The value of the HII lies in it being a physical habitat assessment method, with habitat structure being important in shaping local aquatic communities and odonate assemblages. The HII shows a significant relationship with tropical Brazilian stream insect assemblages, explaining 29% of variation in assemblage composition and structure. This indicates that the HII reflects local environmental quality and wider landscape conditions, making it a good predictor for aquatic and amphibiotic organisms (Brasil *et al.*, 2020a).

In equatorial Brazil, loss of natural primary forest and conversion to other land uses over many years changes the HII scores, and with them the composition of zygopteran versus anisopteran assemblages. The two suborders have different responses to habitat change, due largely to the different behavioural responses of the two groups. In general, anisopteran species in this area are attracted to sun-basking sites, while zygopteran species largely seek more sheltered microhabitats.

The HII is also valuable when integrated with odonate assessments to determine the effects of urbanization, for example, in Manaus in the Brazilian Amazon region. Such assessments help improve town planning around rivers and streams, especially when habitat-sensitive, easy-to-use adult zygopterans are used (Monteiro-Júnior *et al.*, 2014).

As in many biodiversity-rich parts of the world, it is difficult to sample the total species diversity across a whole region. However, the species richness of a taxon can be estimated to evaluate the completeness of a survey or sampling efficiency, giving a Species Richness Completeness Index. This is a strategic way forward through the adoption of a structured sampling approach (e.g. recording odonate species present in each of 20 segments of 5 m per unit time, such as one hour per stream, repeated across streams) along with use of biodiversity quality indices, such as the HII, for defined taxonomic indicator groups such as odonates. This structured approach enables confidence to be placed in local, regional, national and global assessments and comparisons. Among 68 Brazilian odonates, while the Species Richness Completeness Index under-recorded the number of species, the difference was only two or three species, with 80% of the species present being recorded. This shows the differential biodiversity quality response of the two suborders to changed and prevailing environmental conditions. A further point is that the setting of a biodiversity quality baseline index such as this allows resurveying and statistical comparison of assemblages at a future date. Interestingly, this study found that species identification is not necessary for recording biodiversity quality, as it only requires individual species (morphotypes) without specific identifications (Juen and Feest, 2019). Although this approach enables the assessment of biodiversity quality, from a conservation perspective, we still need to know what rare and/or threatened species are present but not recorded, as well as being alert to the possibility of certain genotypes or even cryptic species being present. However, this would require more in-depth study.

River morphology assessments are usually applied at large spatial scales, such as the scale of a river reach (often a few kilometres with homogenous morphological characteristics and margin conditions), and relative to the condition of the overall river corridor, both its channel and floodplain/margin. Both historical and current channel modifications are recorded on maps and remote sensing analysis is used to assess river processes such as sediment transport and bank erosion (Belletti *et al.*, 2015). When the Odonate River Index (ORI) is compared with a measure of river morphology, dragonflies appear as good indicators of river corridor condition and transformation. Like the Dragonfly Biotic Index (DBI), the ORI identifies which components of the river–floodplain system, such as main channel, secondary channels or floodplain wetlands, are most affected by human activity and are candidates for restoration. The important point here is to integrate assessment of the lateral dimension of the river corridor and not neglect the complexity of channel morphology and margins, which otherwise would bias the overall assessment of river status. It is also essential to select indices that are sensitive to various human-induced stressors at the appropriate spatial scale. Integrating reach-scale indices, such as a river morphology index and the ORI, is important while also considering other significant components of riverine systems, especially riparian vegetation condition, to improve management (Golfieri *et al.*, 2018).

There is great value in involving citizen scientists to provide data on, for example, small rivers. The reason is that the field-gathered data collectively adds up to a comprehensive assessment of river quality that can also be used alongside remote sensing data. The Modular River Survey (MoRPh) is designed to enable citizen scientists in the UK to monitor physical habitat mosaics and human pressures within short river reaches (up to 40 m) known as modules. MoRPh can detect notable differences in hydraulic, sediment, physical and vegetation habitat characteristics across a single river type. Furthermore, MoRPh underpins a multiscale Modular River Survey (MultiMoRPh) by providing local information, and when collected across 10 contiguous modules, and at scales of 400 m of river length, produces a MultiMoRPh river subreach survey. In turn, these subreach assessments contribute to a HydroMoRPh assessment of reaches extending over tens of kilometres of river length, based on remote data sources (Shuker *et al.*, 2017).

7.4.3.2 Multimetric indices

Disturbance levels at streams at multiple spatial scales can be determined by using three indices together. The first index is the Integrated Disturbance Index, which is based on eleven assessments of the presence and closeness of human disturbances at a focal stream site, such as human infrastructure, as well as parks and lawns, row crop agriculture, pastures, logging and mining. The second index is the Catchment Disturbance Index, calculated by adding the percentage land use, weighted by the potential for degradation that each has on an aquatic ecosystem. The third index is the Local Disturbance Index, calculated as the distance between the focal site and the origin of the disturbance according to the first two indices. The values expressed by these three indices together increase with the increasing levels of disturbance (Ligeiro *et al.*, 2013).

The combination of these three indices when applied to the species composition of odonate assemblages, as characterized by the various sets of morphological behavioural and physiological traits, can be used to determine thresholds of significant change to assemblage composition relative to environmental gradients of transformation. These changes are quantified by using Threshold Indicator Taxa Analysis which identifies abrupt changes in the frequency of occurrence and abundance of each of the species over environmental gradients by combining change-point analysis (nCPA) and indicator species analysis (IndVal), giving measures of species sensitivity. The determination of species' ecological thresholds enables detection of environmental change points before they become irreversible, allowing the instigation of mitigation measures. This methodology can be applied to odonate assemblages and effectively applied to Neotropical larval assemblages (Silva *et al.*, 2023).

Use of multimetric indices showed that pasture was the most extensive disturbance in the streams of the Brazilian Cerrado, affecting 41% of the stream length, followed by 40% for low levels of bed stability. The conservation value of these findings emphasizes that it is important to use and standardize multimetric indices for assessing which disturbances pose the most risk to aquatic assemblages against disturbance scores. The point is that these composite indices facilitate the selection of the most appropriate mitigation actions across a whole river basin (Martins, *et al.*, 2021).

Finally, Vadas *et al.* (2022) review the use and applicability of multimetric indices globally. They have concerns that are centred on the reference conditions used, aspects of sampling related to the degree of effort, methodology and season, trophic guild definition, aspects of metric comprehensiveness and lack of validation procedures. They also emphasize that these metrics would also benefit from greater understanding of the underlying ecological mechanisms related to the metrics proposed. Overall, they suggest more focus on study design, ecological drivers and metric development.

7.4.4 Estimating population sizes of adult dragonflies

7.4.4.1 *Capture–mark–recapture methods*

Effective site conservation of rare and threatened odonate populations requires information on abundance, effective population size, sex ratios and dispersal abilities relative to the condition of a species' habitat. The capture–mark–recapture (CMR) (also known as mark–release–recapture (MRR)) process is used to do this. It involves capturing a living individual, marking it with code and then recapturing it later. When this is done on many individuals in a population and the results are analysed along with recording appropriate habitat variables, it is possible to establish the chances of persistence of a population. The method is suitable for rare and threatened odonates, as they can often be easily captured and marked without doing them harm. Population size estimates can also be used alongside genetic studies to provide insight into population viability and to guide effective conservation management to avoid local extinction.

Coenagrion mercuriale and *Ischnura pumilio* are of conservation concern in the UK. An extensive CMR study of the two species provided survival and recapture probabilities and population size estimates across population strongholds in the UK. Males were mostly more abundant than females, while daily survival probability of both males and females was remarkably high. Interestingly, estimated population sizes were much greater than previously thought, suggesting that where good-quality habitat is maintained, their populations can be very large. The importance of these findings is that extensive tracts of optimal habitat is required to maintain such large populations and to avoid genetic inbreeding (Allen and Thompson, 2014).

For *C. mercuriale* in central Italy, the maximum daily number of estimated individuals was on average 507 individuals for males and 220 for females, and the estimated cumulative population size was about 4000 individuals, with a mean density of 1.2 individuals/m^2 at breeding sites, similar to findings from the northern parts of the species' range. Monitoring plans, with at least seven marking occasions, can provide consistent estimations of population size, important for evaluating population viability and for assessing the conservation status of the species (La Porta and Goretti, 2019).

Not all odonate populations are so easily assessed by CMR. A CMR and sighting study of a *Leucorrhinia dubia* population in the UK, found that survival between capture visits was very low, and that *L. dubia* shows a distinct preference for open moss habitat. It is recommended that encouraging repeated observations from recorders who visit sites on a regular basis would improve the estimates of population trends and habitat associations (Davies *et al.*, 2018). Using *L. dubia* data collected by volunteers over several years, Geary and von Hardenberg (2020) applied species distribution modelling (SDM) with a land cover data set, especially to determine the proportion of native coniferous forest within 1 km to predict all current populations. In Scotland in particular, models suggest that habitats characterized by coniferous forest and areas with high potential evapotranspiration represent the most suitable habitat for the species. They recommend that future monitoring of the species should be expanded to include areas in northern Scotland that are currently poorly surveyed but with high suitability. Furthermore, results suggest that *L. dubia* management should focus on maintaining sphagnum-rich pool complexes and the maintenance and restoration of native forests where these complexes occur.

CMR studies of fast-moving anisopterans are challenging as the low recapture rates can bias estimates of demographic parameters, compromising significant conservation decisions. However, high-speed videos (HSV) can be used to detect the experimental marks (using a permanent marker pen) on the hindwings of large dragonfly species such as *Anax junius* and *Rhionaeschna multicolor*. When this is done, and compared to conventional visual observations, there is an increase in resighting rates of 64% for *A. junius* and 48% for *R. multicolor*. These results improve estimates of both survival rates and the effects of demographic covariates on individual survival. Furthermore, including HSV results improved credible intervals for the resighting rate and survival probability by 190% for *A. junius* and 130% for *R. multicolor* (Khelifa *et al.*, 2021b) (Fig. 7.24).

7.4.4.2 Ecological niche modelling

Ecological niche models focus on the relationship between a certain species and the specific environment where it currently occurs. This approach is used to identify unexplored areas where the species might be present. *Cyanallagma bonariense* is endemic to the Pampas biome, South America, where there are sufficient records to undertake ecological niche modelling on the species. Results suggest that it may have a broader distribution throughout the Pampas, although subject to range contractions rather than range shifts in the future, as it is a habitat specialist of these grasslands. Of concern is that current protected areas in the future will be outside its predicted future geographic range. As perhaps an indicator species for many other specialist freshwater insects, protected areas in the region should be extended (Pires *et al.*, 2018).

Ecological niche modelling was also used to predict odonate geographic range shifts in Sweden in the coming years, with most species expected to expand their range northward at the average latitudinal rate of between 1.83–3.25 km/year by 2061–2080. Although most warm-dwelling species may increase their range, cold-dwelling species will likely show range contractions. Furthermore, while the present-day niche overlap patterns among species will stay largely the same, there are likely to be changes in co-occurrence patterns, with many pairs of species displaying increased co-occurrence, while other pairs will no longer co-occur because of range contractions (Pélissié *et al.*, 2022).

The ecological niche models of protected habitats for Brazilian Amazonian odonates in three classes of protected areas (strictly protected area, sustainable use area and indigenous territory) together support about 91.4% of the assessed species within the whole network. However, only 11.1% were in the strictly protected area network. Overall, the findings emphasize the importance of a network of all three classes of protected area for safeguarding most Amazonian odonate species. For under-represented or gap species (mostly species with small home ranges), it is necessary to instigate effective management and protection on private properties where land is set aside to ensure the

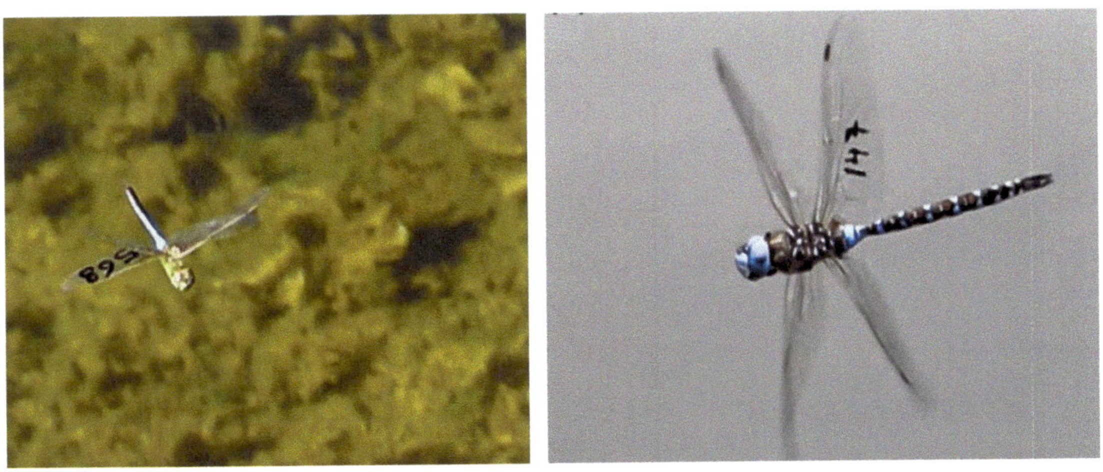

Fig. 7.24. Still images taken from high-speed videos displaying the mark code on the hindwings of the two dragonflies: *Anax junius* (left) and *Rhionaeshna multicolor* (right). These marks are usually not visible to the naked human eye when these dragonflies are in flight. (Photos: Rassim Khelifa).

protection of all species (Brasil *et al.*, 2021). The low figure of 11.1% here for strictly protected area species compares with 36% in a core Cerrado area of Brazil (Nóbrega and De Marco, 2011). The difference in figures is perhaps due to Amazonian species having smaller distributional range sizes compared with the Cerrado species, lowering their chances of being included within conservation units.

For the African mainland, priority areas for dragonfly conservation largely coincide with global biodiversity hotspots, with the Zambian swamps and woodlands, and the rainforests of the lower Guinea and Congo basins being important dragonfly diversity hotspots. Among the globally threatened species, 72% were recorded at least once in a protected area. However, although the reserve network covers 10.7% of the continental area, the proportional representation of species geographic distributions in protected areas is only 1.2%, making the reserve network inadequate for continental-scale conservation of dragonflies. As in Brazil, there is merit in campaigning for inclusion of priority areas that are not formally protected, especially those adjacent to the existing protected areas (Simaika *et al.*, 2013).

Ecological niche modelling has also been used to study the distinct biogeographic populations of the Neotropical giant helicopter damselfly *Megaloprepus caerulatus* relative to its specialized niche in mature lowland tropical forests. The potential distributions estimated by modelling are strongly fragmented with no geographic overlap of potential distributions among populations. However, there is geographic correspondence between populations with a close phylogenetic relationship. There is niche divergence between Corcovado (Costa Rica) population and the other three populations, Los Tuxtlas (Mexico), Barro Colorado (Panama) and La Selva (Panama), as well as between Barro Colorado (Panama) and La Selva (Costa Rica), which is evidence of strong ecological divergence in the Corcovado and Barro Colorado populations. While the modelling was based on broad ecological conditions, microhabitat specialization is also significant. Only certain tree species can support sufficiently large tree holes that contain more than 1 l of water necessary for maintaining an effective population size. Furthermore, the species, being a large-sized lowland humid forest specialist is highly susceptible to forest loss, making each of the subpopulations of conservation interest. The Los Tuxtlas (Mexico) population is of most concern as it is the one most threatened by deforestation (Encarnación-Luévano *et al.*, 2022) (Fig. 7.25).

7.4.5 The 'Exuviae Dilemma'

The exuvia is the cast-off exoskeleton of the final stage of the larva after the individual has emerged from the water to become an adult. Technically, it is not a life stage in its own right. The presence of an exuvia at a spot location suggests residence. But really its presence only means that a particular individual lived as a larva at that spot to emerge as an adult, which may not necessarily be optimal habitat for multiple generations.

The exuvia of some species are very cryptic and rarely seen. Others disappear very quickly either because of rain and/or wind (Reels, 2011) or are seemingly scavenged. In the greater Cape Floristic Region, South Africa, the rare and threatened *Syncordulia* spp. are extremely difficult to locate as larvae, and for unknown reasons, the exuviae are very rarely seen. Yet the adults are easily located, but only when they emerge as adults after which they largely leave the water and move into the rugged mountain terrain. This means that it is essential to know when adult emergence occurs, which varies greatly among the four species (Samways and Grant, 2007).

On the other hand, for some species, it is the adult that is more difficult to find, with the recording of the exuviae the best way to assess the presence and physical location of individuals. For example, intensive sampling of the exuviae of the rare *Oxygastra curtisii* is required to obtain estimates of effective population size in northern Europe (Ott *et al.*, 2007) (Fig. 7.26). This methodology has now been standardized along riverbanks in south-western France to effectively monitor population levels of *Macromia splendens*, *Gomphus graslinii* and *Oxygastra curtisii* by collecting exuviae along 100 m of riverbank over two visits to both maximize species detection and minimize the monitoring cost (Arguel *et al.*, 2022).

In the case of the rare and threatened *Ophiogomphus howei* in the USA, occasional records of adults stimulated a search for exuviae to establish residency at a potential new site. The new breeding population was discovered in southern Michigan many tens of kilometres distant from previously known breeding sites (Craves *et al.*, 2020).

However, exuviae elsewhere can be very scarce. At the southern tip of Africa, where sites were

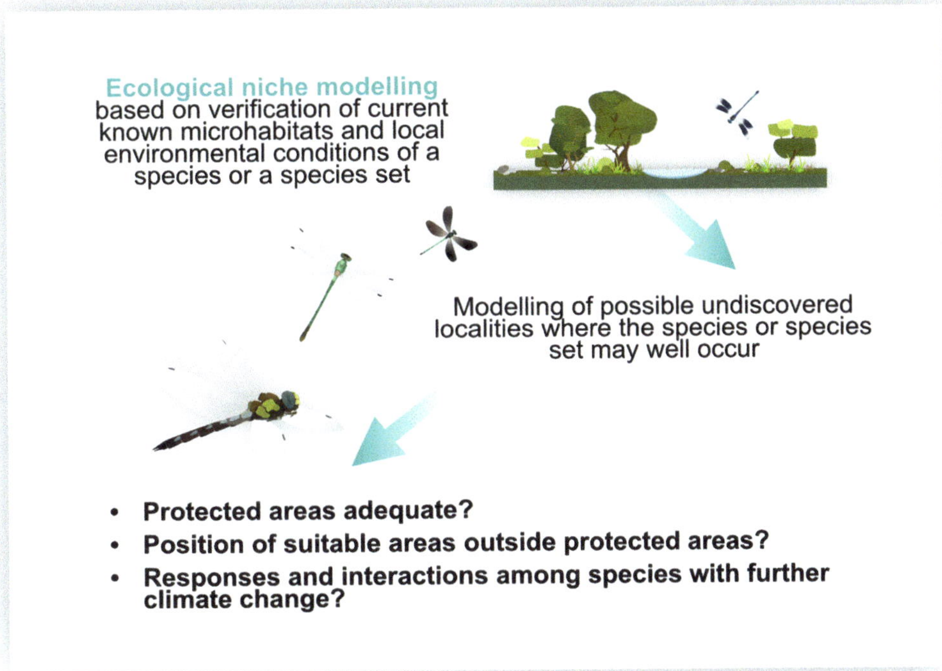

Fig. 7.25. Ecological niche modelling is proving to be an effective approach for identifying possible undiscovered localities for a species or species set. It helps identify the effectiveness of protected areas, the position of suitable areas outside protected areas, and possible responses and species interactions with further climate change. (Illustration: Robyn Symons).

Fig. 7.26. *Oxygastra curtisii* population levels can only realistically be determined by counting exuviae. Here is the oviposition site along a riverbank (left), the exuvia (centre) and the underwater microhabitat of the larva (right). (Photos: Jürgen Ott).

sampled intensely, the sampled ratios among the three developmental stages, among all odonate species, were 1268 adults, 1795 larvae, but only 118 exuviae, despite intensive searches, which equates to less than 10% of each of the two living stages (Kietzka *et al.*, 2021b).

Numbers like those of Kietzka *et al.* (2021b) fly in the face of Raebel *et al.*'s (2010) 'The dragonfly delusion: why it is essential to sample exuviae to avoid biased surveys', based on easily identifiable species and accessible English ponds with narrow margins in farmland which have experienced a shifting baseline. They also state that 'Ovipositing females were recorded at all ponds where exuviae were totally absent hence adult surveys over-estimate pond quality and low-quality ponds are functioning as ecological traps'. However, the absence of exuviae does not prove the presence of ecological traps. Finally, they claim 'Odonate biodiversity monitoring would benefit from applying the best survey method (exuviae) to avoid wasting valuable financial resources while providing unbiased data, necessary to achieve conservation objectives'.

Bried *et al.* (2012a) responded with a critique of Raebel *et al.* (2010), pointing out that their study drew this conclusion from limited surveys and detection-naive analysis while failing to acknowledge that exuvial richness is typically biased low. Quantifying exuvial bias can be done by using the two related metrics of species detectability from concurrent adult and exuvial surveys and estimated exuvial species richness based on imperfect detectability and the regional pool (the cumulative total across study sites) of exuvial species observed. Then using concurrent adult and exuvial data from French lakes, Bried *et al.* (2012a) showed detectability to be generally lower in 1-hour exuvial searches than in 20-minute adult searches, with exuvial surveys leading to strong negative bias in richness estimation. This finding emphasizes that the alleged delusion of adult surveys was exaggerated. This means that when controlling for species detection probability, it is essential to determine odonate species richness at a site and, by extension, compare adult and exuvial species richness. Exuviae sampling avoids positive bias, but not bias in general. The conclusion is that there must be a relatively intensive search effort for statistical accounting of false species absences, or acceptance of negatively biased species richness.

It is important to consider whole biology when relating ratios of adult odonates to their exuviae relative to hydroperiod and the presence of free water. 'Whole biology' means not only the mobility traits of a species but also the presence of dragonfly prey and its fish predators. Seasonal water bodies can be free of fish predators that cannot survive the dry period. However, as dragonflies are relatively opportunistic feeders, switching prey type is often not a problem for them.

In regions with frequent dry periods, adult individuals of some odonate and other species depart from drying habitats and colonize suboptimal habitats (Deacon *et al.*, 2019). Other individuals of other species remain and wait for the rains to come and breed again. Among amphibiotic taxa in general, there is remarkable tolerance among many species, including dragonflies, to fluctuating water levels. They can also occupy fluctuating water bodies in high abundance (Jooste *et al.*, 2020). Furthermore, several aquatic insect groups colonize fluctuating ponds with low predator pressure that are occupied by many prey species (e.g. larval Diptera), seemingly driven by a long history of adaptation to naturally extreme seasonal weather conditions and fluctuating water levels (Bird *et al.*, 2019).

There can be high functional redundancy among these taxa in drought-prone water bodies, making the assemblage overall resistant to hydrological drought (Boersma *et al.*, 2014). Dragonflies in drought-prone areas have rapid development, enabling them to outlive seasonal aquatic habitats while remaining at their natal ponds in readiness for the inevitable coming of the rains (Suhling *et al.*, 2015; Bird *et al.*, 2019). This means that there must be careful interpretation of the presence of only adults and not their younger stages, as suboptimal habitats may be functioning as waystations or stepping-stone habitats. In short, adults can ebb and flow across the landscape in search of locally more optimal habitat.

Deciding which odonate life stage/manifestation to use depends on issues such as the assessment goal, time constraints for sampling, denseness of vegetation preventing access to the water's edge, accessibility to special emergence sites, frequency of heavy rainstorms and scavenging of exuviae by other animals. There are also safety concerns, with wading in many parts of Africa simply too dangerous whether from water-borne pathogens or large and aggressive vertebrates. Also important, is that the level of taxonomic knowledge of exuviae is often poor in many parts of the world, especially for many tropical species where only the adult is known. Where the last-instar larva is known, it is mostly to genus. Furthermore, in many species-rich areas of the world, many of the genera are made up of many species, sometimes with very different habitat requirements, threat levels and sizes of geographic range, as in the case of *Pseudagrion*, *Trithemis* and *Orthetrum* in Africa.

Although exuviae have the most readily identifiable larval morphology and indicate life-cycle completion and confirm current residency, they can underestimate species richness and especially diversity. Among USA wetlands, the occurrence of exuviae as representative of adult abundance, detection frequency across surveys, teneral frequency and frequency of breeding behaviour while controlling for imperfect detectability was investigated. Exuviae occupancy probabilities suggested some reliable indicators of species residency. These probabilities were finding adults on ≥4 surveys, finding tenerals on ≥2 surveys and counting more than 20 adults on ≥1 surveys. The probability of exuviae occurrence when these conditions were met were about 9–18 times greater than when no adults were detected. The conclusion is that species residency may be accurately inferred by undertaking adult surveys. Furthermore, recording adult presence is more expedient for conservation purposes (Bried et al., 2015).

Use of the different dragonfly life stages to assess environmental conditions often gives contrasting results. From remediated ponds in Italy, all adults, breeding adults, larvae and exuviae were sampled from 22 species. Some species were detected only when using a certain method, while diversity measures using the four groups differed considerably. Alpha diversity was greatest for all adults and least for exuviae. Breeding adults and larvae had intermediate values. Beta diversity showed an inverse trend – the lowest value for all adults and higher values for all the other groups. Congruence among the assemblages of the four groups was generally low. In sum, results from the four different groups/life stages are not interchangeable but rather complementary (Giugliano *et al.*, 2012).

Little is known about the co-occurrence of larvae, exuviae and adults, or their respective response to local biotic and abiotic factors relative to spatial scale. These comparative aspects were tested in the Cape Floristic Region biodiversity hotspot, South Africa. The extent to which adult male dragonflies are congruent with their larvae was investigated at three independent sampling unit scales (small 10 m × 3 m, medium 90 m × 3 m and large 450 m × 3 m) along four small rivers. A subset of 40 randomly selected small-scale sites were also used to test for the effects of sampling design on outcomes of spatial scale analyses. Then, the extent to which adult males and larvae share similar responses to environmental variables was also established. At medium and large spatial scales, larvae and adults were highly congruent for abundance, species richness and DBI scores. Despite these congruencies, at the small spatial scale, only 15% of observations matched in the sense of having both adults and conspecific larvae, although this figure increased to 46% at the medium scale, and 60% at the large scale, neither of which were significantly different from the number of mismatches. Dragonfly species composition differed between larval and adult assemblages at the small and medium scales but did not differ at the large scale. Water temperature was the only environmental variable that resulted in similar responses in both larval and adult stages, and at all spatial scales. These results indicate that for assessments at medium or large spatial scales, such as determining the state of a river, either adults or larvae can be used. However, for comprehensive odonate surveys, both larvae and adults should be sampled, with exuviae providing supplementary information, albeit that in this environment exuviae abundance returns are less than 10% of those for larvae and adults (Kietzka *et al.*, 2021b) (Fig. 7.27).

At quarries with pools and plant communities at different stages of ecological succession in Czechia, both adult odonates and exuviae were studied over an 18-year period. The probability of species detection was highest when sampling adults. Although exuvial surveys do largely miss rare species, they are useful for indicating residency status. This means that to ascertain a complete assessment of odonate diversity in an area, long-term monitoring is required where both adults and exuviae/juvenile adults are sampled. Adults can be assessed semi-quantitatively through observation, and exuviae/juveniles qualitatively through ad hoc collection. Combining these two methods provides sufficient information without being too demanding on time or resources, making it an effective approach for long-term monitoring schemes (Dolný *et al.*, 2021).

7.4.6 Assessing and monitoring dragonfly larvae

Immature stages of insects are not well-known at the species level in many parts of the world, especially in biodiversity-rich and poorly explored areas. Even in temperate areas, there can be some surprises. An extensive study of previously surveyed adult and larval stages of dragonflies across the relatively well-known wetlands in Alberta, Canada revealed that sampling of larvae failed to collect or identify almost 60% of the known wetland adult species (25 anisopteran and 12 zygopteran species). Furthermore, as early instars are abundant, 82% of the processed Zygoptera collection and 62% of the processed Anisoptera collection was left at suborder (Bried and Hinchliffe, 2019).

Gómez-Tolosa *et al.* (2021) reviewed the use of odonates for evaluating and monitoring freshwater bodies and environmental change resulting from deforestation, urbanization, land-use intensity,

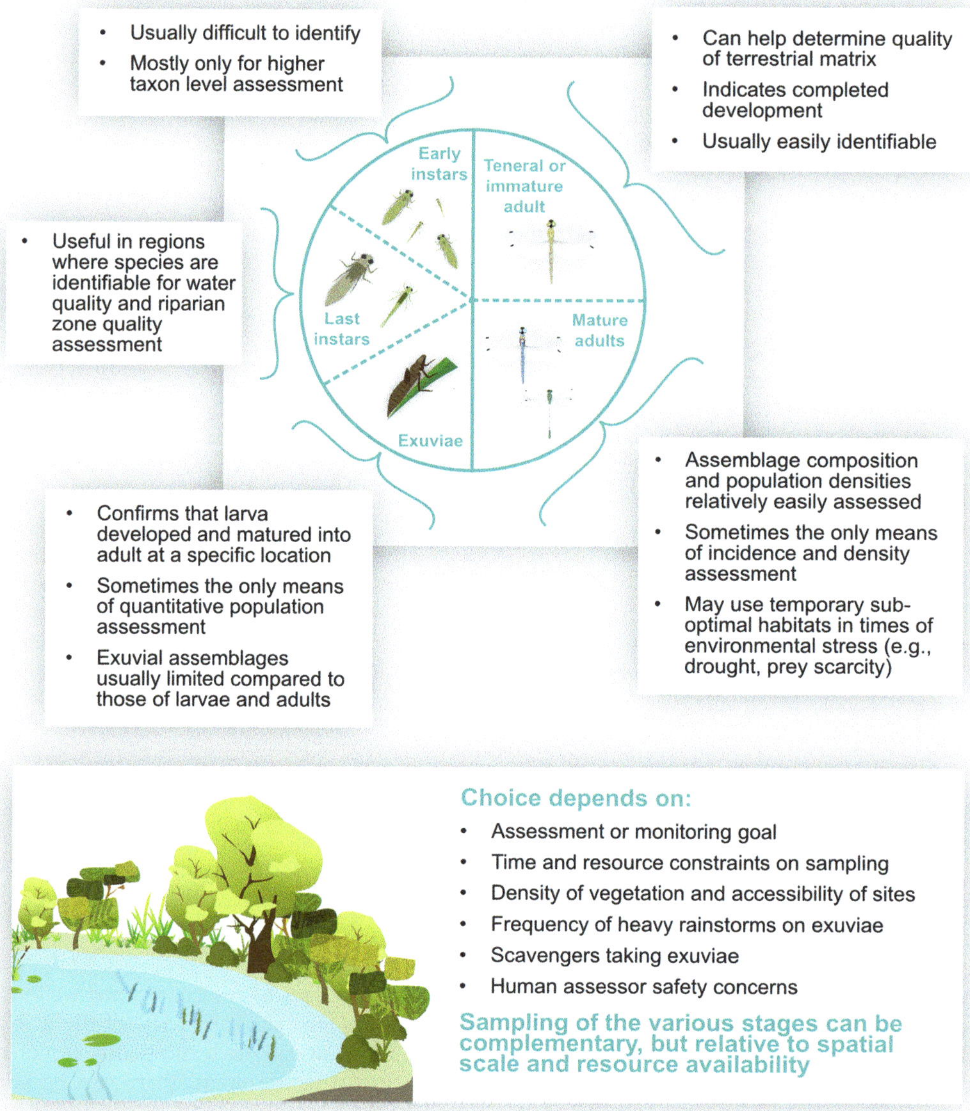

Fig. 7.27. The different life stages and their degree of maturity have various assessment or monitoring advantages or shortfalls according to species, degree of knowledge and the focal part of the world. Importantly, the choice of subject depends on various factors associated with the intended assessment or monitoring goal, as well as the feasibility of obtaining meaningful results. (Illustration: Robyn Symons).

climate change and management practices in the Neotropical region. While ecosystem health was the most frequently used type of environmental assessment, ecological integrity was the most used research objective. There was a positive correlation between biotic and habitat indices with sampling methods. Surveying geographical areas and transects are frequently used to select habitat indices, while biotic indices mostly use net/transect sampling. Adult life stage sampling is preferred over that of larvae and exuviae, while also being less costly and easy to employ. However, the inclusion of larval stages can provide a more accurate evaluation. This suggests that it is best to use the three life stages (larvae, exuviae, adults) to obtain a more accurate assessment, while bearing in mind practicalities.

Due to the great diversity of habitats and complexity in the Neotropical region, there is no single effective habitat assessment tool, while it is complicated to set up a unique protocol. This means that a way forward for environmental quality assessments would be to establish an index (Odonata Neotropical index) that encompasses the value of the various dragonflies and damselflies to assess environmental integrity and degradation.

Among variously disturbed streams in south-western Brazil, odonate larvae were sampled along 1 m segments of channel and adults along 100 m segments of channel margin. Environmental variables partly explained the congruence between the two life stages. Larvae genera and adult species or genera were 54% congruent. The abundance of adult genera and species were 94% congruent. In terms of the cost–benefit of taxonomic level and sampling effort, while bearing in mind that each genus was composed of only a few species, the result suggests that adult genera can be used for rapid assessments in this region (Valente-Neto *et al.*, 2015).

An assessment of larvae genera in a river and a network of streams in a river basin in Mato Grosso do Sul state, Brazil, produced 739 individuals belonging to 25 genera and 3 families. An asymptote was not reached, with genera accumulation curves on 29 samples estimating that '29.8' genera should be present, compared to 37 genera recorded to date from the state (Boneto *et al.*, 2017).

In Tunisia, odonates were used to assess the effectiveness of certain biodiversity indices relative to the different focal 'stages' (adult, ovipositing adult, larva, exuvia), the influence of water type (running versus standing) and the closeness to a diversity of wetlands. Three indices were used: Relative Taxonomic Distinctness (RTD) which is a measure of the taxonomic distance within an assemblage and characterized by whether an assemblage consists of closely related species (less diverse) or distantly related species (more diverse) based on presence/absence data; the Conservation Priority Index (CPI), derived from the Tunisian Stream Odonatological Index (TSOI) (Korbaa *et al.*, 2018) and combining them in the same equation for both taxonomic distance and its conservation priority; and the Dragonfly Biotic Index (DBI), notwithstanding that the DBI can include exuviae, larvae and mating behaviour (see Section 7.4.1.6). Unlike exuviae and larvae, use of adults gave inaccurate estimates of species richness in the case of the RTD and CPI but was reliable for the DBI. Significantly, recording the mating state (oviposition) of the adult improved the accuracy of RTD and CPI by about 40% and 60%, respectively. The estimation bias was higher in running than in standing waters and increases with the freshwater landscape complexity. In sum, applying a multiple life stage approach in biodiversity indices reveals site connectivity at the landscape level (Khelifa, 2019).

Along the whole length of an Australian river, where anisopterans dominated over zygopterans, larvae and adults gave congruent results. The river has distinct zones with their own assemblages. At smaller scales down to microhabitat, there is also strong congruency between the two groups. The conclusion is that systematic sampling of either larvae or adults is sufficient to assess the conservation values of individual sites along a river through comparative measures of relative species richness and diversity (Hawking and New, 1999).

A georeferenced database, along with a set of contemporary records of odonates from across Oklahoma, USA, was used to determine whether ecological niches modelled from records of adults (i.e. lacking evidence of breeding) differed from niches modelled for records indicating breeding in the form of tandem pairs, ovipositing females, larvae, exuviae or tenerals at surveyed sites. While adults were overall geographically congruent and showed similar assemblage diversity responses, environmental variables (niche dimensions) focused on verified breeding populations showed more ecological specialization among the various species. Adult data are easy to obtain, but they can give a misleading picture of the specific ecological niches of the various species. This means that it is important that evidence of breeding is provided, especially from tenerals or exuviae, to fine-tune ecological and conservation assessments (Patten *et al.*, 2015).

In the rainforest–savanna ecotone of Mato Grosso, Brazil, as similar ecological mechanisms at the site level affect both the larvae and the adults in the same way, either adults or larvae can provide an accurate diagnosis of habitat integrity for conservation management. To minimize costs and reduce the time necessary for the collection of data, either life stage can be used for monitoring, as they reflect the local impacts on terrestrial or aquatic habitats with the same degree of accuracy (Mendes *et al.*, 2017).

Which stage to choose depends on the ecological or conservation question at hand. This is emphasized with respect to the playa wetlands of Texas, USA, where it is essential to consider not just wetland margins but also the wider context of more distal land use, which has a major effect on odonate diversity in these systems, as clearly shown by adult surveys (Reece and McIntyre, 2009).

The invasive alien crayfish *Procambarus clarkii* has invaded certain Italian rivers where it preys on odonate larvae. Siesa *et al.* (2014) provide valuable insight into the impact of this invasive alien species on organisms like dragonflies with complex life cycles. They suggest that larvae and exuviae have more value for assessing the impact of the alien species at early stages of the invasions, while adults better describe the long-term consequences of invasion at the landscape scale. Overall, they conclude that considering multiple life-history stages of odonates improves our understanding of the impact of biological invasions in freshwaters (Fig. 7.28).

7.4.7 Dragonfly abundance metrics

7.4.7.1 *Issues of dragonfly population fluctuation*

It is well known that odonate populations at any one site can fluctuate considerably over time (Moore, 1991). Yet knowledge of abundance levels is central to the assessment of the ecological condition of a population and the evolutionary potential among its individuals. In the context of conservation, we need to know whether fluctuations are

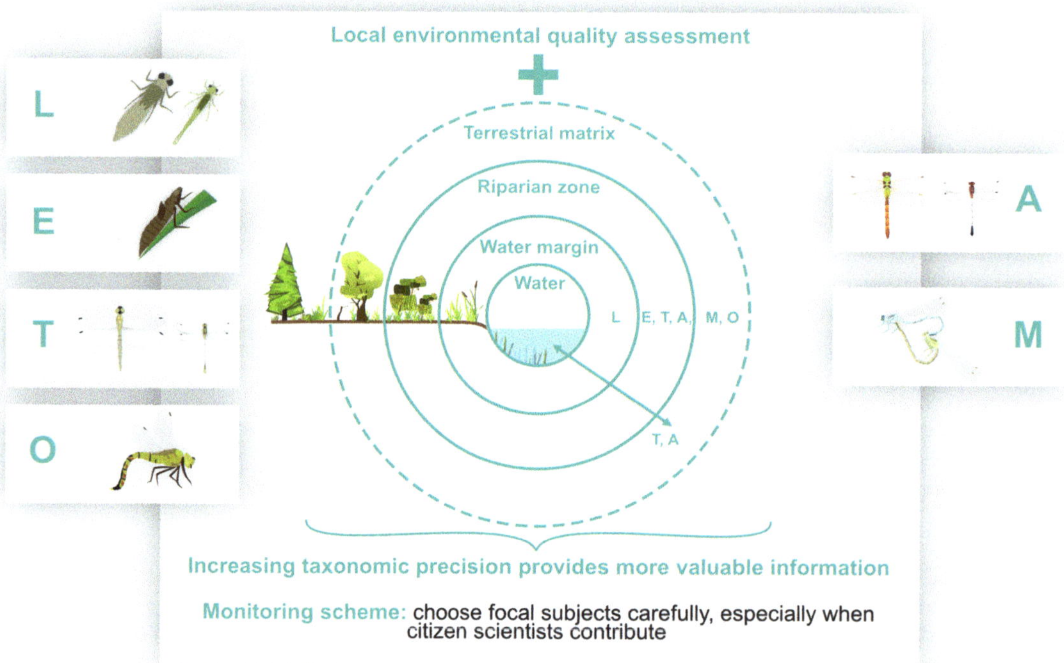

Fig. 7.28. A biotic index using dragonflies goes hand in hand with local environmental assessment involving measures of, for example, water physicochemical conditions, riparian vegetation condition and the transformation of the matrix around a water body. Depending on the level of local taxonomic knowledge, conservation task, spatial scale, and time and resources available, one or more life 'stages' can be chosen: larva (L), exuvia (E), teneral and immature adult (T), mature adult (A), mating pairs (M) and ovipositing female (O). There is, in general terms, a spectrum of available life 'stages' according to the spatial scale and aims of the assessment. All life 'stages' (L, E, T, A, M, O) have especially good complementary value when assessing the local water and margin environmental state or status, with an emphasis on L, E, M and O in support of use of A. Widening the assessment to the terrestrial matrix places more emphasis on T and A. Increasing taxonomic precision (i.e. species rather than genus or family) provides more valuable information, especially in geographic areas with species-rich genera. Careful choice of focal 'stages' becomes critical when embarking on a monitoring scheme, with its needs for much time and resources, often with highly valuable input from citizen scientists. (Illustration: Robyn Symons).

showing a continuous decline or not. Following conservation mitigation, we also need to know that the target population is increasing in a consistent way so that there is a strong effective population. An issue with many odonate abundance estimates is that they are limited in space and time, and so do not signal whether there is a persistent decline of conservation concern, a steady population level or an improvement in population level. This means that large fluctuations require long time series and large sample sizes to detect, assess and predict changes over time (Bried *et al.*, 2020) (Fig. 7.29).

7.4.7.2 Abundance/biomass comparisons

The abundance/biomass comparison (ABC) method (Warwick, 1986) hypothetically indicates that the distribution of the numbers of individuals versus biomass among species should differ when a community is perturbed following a human-induced change in state or condition and plotted as graphed curves. The curves rank species in order of importance on the horizontal *x*-axis and illustrate the cumulative percentage dominance of each species in the form of two curves, one for abundance and the other for biomass, on the vertical *y*-axis. When a community or assemblage is in its natural state the biomass curve lies above the abundance curve, as many large-sized species are present. However, under severe stress or disturbance to an ecosystem there is loss of the larger species, resulting in the biomass curve lying under the abundance curve. Furthermore, under severe stress or disturbance, communities or assemblages become increasingly dominated by fewer smaller species with some large species being no longer present, resulting in 'numerical diversity' being lower than 'biomass diversity'. Under moderate stress or disturbance, the curves coincide and may cross over each other.

When this hypothetical approach is applied to Brazilian odonates subject to the loss of riparian vegetation, while bearing in mind the phylogenetic constraints of Zygoptera versus Anisoptera, the hypothesis was partly supported. For Zygoptera, the biomass curve for both natural and transformed areas was above the abundance curve, although the difference in transformed areas was very small. For Anisoptera in transformed areas,

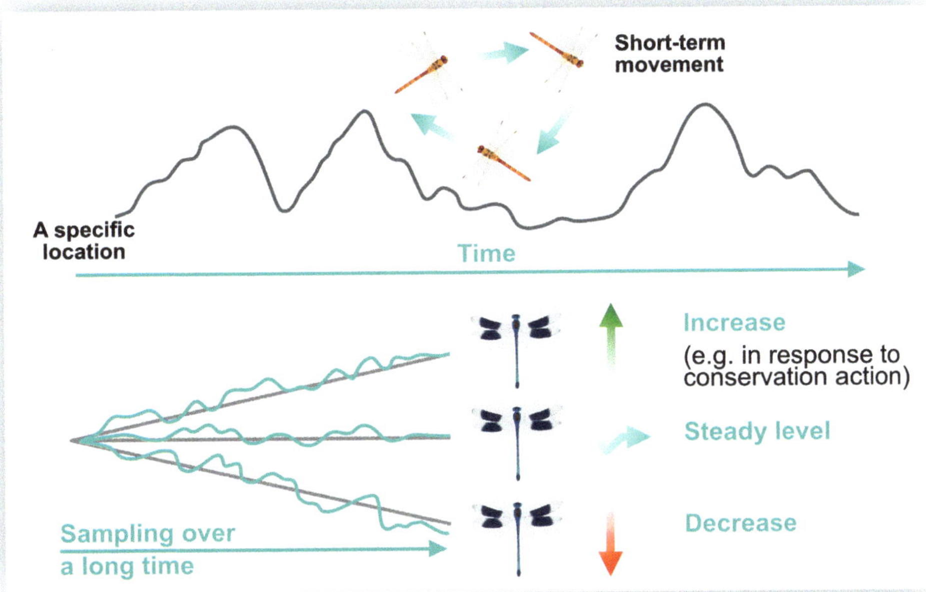

Fig. 7.29. As with many insect species, dragonfly populations can fluctuate greatly at a specific site. This fluctuation is often associated with movement by adults around the local landscape in response to changing ecological and environmental conditions as they seek optimal microhabitats for the various life functions and survival. Absolute population decline, as opposed to a steady level, or an increase in population level (perhaps in response to conservation action) can usually only be determined through long-term sampling. (Illustration: Robyn Symons).

there was dominance by fast growing and larger species, manifested by the biomass curve lying above the abundance curve. These results overall relate to the thermoregulation demands of the two suborders. However, these phylogenetic/ecophysiological constraints mean that in terms of the ABC hypothesis, the two suborders must be considered separately, and the hypothesis is not supported by Odonata as one group in that zygopterans are smaller than anisopterans. Rather, here the fast-growing, sun-loving anisopterans became more abundant, while the zygopterans less abundant, with loss of the natural shady habitats (Carvalho *et al.*, 2013).

7.4.8 Value of opportunistically gathered point data

Opportunistically gathered point data, while valuable, do not necessarily distinguish breeding habitat from that being used at other times in the life cycle. This suggests that it is important to estimate the possibility that a site supports a breeding population when only opportunistic data are available.

Using a georeferenced data set of odonates across Oklahoma, USA breeding occupancy was determined from records of tenerals/immature adults against those of all adults, adult females only and the incidence of breeding behaviour, such as ovipositing, mate guarding and tandem pairs. However, no single threshold fitted all odonates. Rather, thresholds of varying precision were identified for the two suborders and for families and certain genera in both suborders. Counts of females greatly reduced the required threshold, while breeding behaviour data reduced the threshold in some cases. Overall, it is possible to identify breeding occurrences in opportunistic adult Odonata records, so long as in addition to recording the presence of adults, the sex and breeding behaviour are also noted. Doing this opens up opportunities, for example, for citizen science to make a greater contribution to fine-scale ecological knowledge of odonates to benefit their conservation (Patten *et al.*, 2019).

7.5 Genetic Aspects of Dragonfly Assessment and Monitoring

7.5.1 Combining morphology and genomics

7.5.1.1 Barcoding and metabarcoding

A DNA barcode is a short section of DNA from a specific gene or genes used in species identification. Since 2003, an ~650 base pair sequence from the mitochondrial cytochrome c oxidase subunit I gene (COI) has been a valuable barcode for genetically identifying animal species, with results lodged with the Barcode of Life Data System (BOLD).

The value of barcoding lies in its focus on alpha taxonomy (identification and delimitation of species) and the discovery of cryptic species not easily revealed by morphology. It is important for inventorying both living and museum-preserved biodiversity, ecological studies of species interactions and community structure, and in conservation biology. Protocols for the barcoding analysis of bulk insect samples are improving rapidly (Iwaszkiewicz-Eggebrecht *et al.*, 2023).

Technological advances allow for faster acquisition of sequences at less cost, with multiplex sequencing with Illumina MiSeq being effective and inexpensive. However, single DNA barcode markers and the distance-based approach for identifying species are not sufficient for diagnosing phylogenetic relationships, with sequences from multiple genes and genomes being preferred for examining evolutionary divergence within and between species (Honeycutt, 2021).

Metabarcoding is the large-scale taxonomic identification of complex environmental samples through analysis of the various barcodes. While metabarcoding has value, nuclear copies of mitochondrial genes and other artefacts lead to taxonomic inflation, compromising the reliability of biodiversity inventories. However, information noise can be reduced, and non-authentic mitochondrial sequences filtered, using certain metabarcoding procedures. This involves whole organism community DNA (wocDNA) metabarcoding to build a local reference database with COI barcode sequences, leading to reliable haplotype data after the reduction of nuclear mitochondrial copies and other artefacts. This approach can lead to the discovery of new species records, new distribution records and validated phylogeographic relationships compared to traditional sequencing approaches (Arjona *et al.*, 2023).

An assessment may aim to determine which dragonfly species are present in a focal area, a process known as inventorying. While this is primarily based on morphology for determining the identity of the species, it also includes recording their DNA barcodes (Wilson *et al.*, 2017), even for ancient specimens in museums using next-generation sequencing (Prosser *et al.*, 2016).

Barcoding freshwater macroinvertebrates can increase the species inventory by 475% over morphological inventories (Sweeney *et al.* 2011), by uncovering evolutionarily distinct cryptic species which may be threatened, especially if they are localized endemics (Delić *et al.*, 2017). Barcoding can be improved when delineation procedures, such as identifying the four genetic markers mitochondrial COI, 16S rRNA, nuclear 28S and 18S rRNA, are employed (Jörger and Schrödl, 2013).

Molecular diagnosis is also useful for associating larvae with adults (Packer *et al.*, 2009), and important for associating odonate exuviae with their adults (Landmann *et al.*, 2021). The great advantage of this approach for dragonfly conservation is that it is not only a non-invasive method but also assists in identifying species with very similar larval morphology (Sittenthaler *et al.*, 2023).

Currently, however, interpretation of DNA barcoding data still lacks some statistical rigor, especially with respect to the DNA barcode gap (in congeneric COI sequences, the difference between mean intra- and interspecific sequence variability). Problems stem from three angles: 1) improper allocation of the specimen sampling effort necessary to capture adequate levels of within-species genetic variation; 2) the failure to firmly categorize intraspecific and interspecific genetic distances; and 3) inconsistent, inappropriate use or absence of statistical inference associated with DNA barcoding gap analyses. However, progress is being made to address these issues to make DNA barcoding a useful tool to confidently assess unknowns versus knowns related to the barcode gap (Phillips *et al.*, 2022).

Last and importantly, it is essential to have high-quality raw sequence data (16S rDNA and COI), as was shown for the Neotropical damselfly *Megaloprepus caerulatus*. If such data are not used, there can be many unresolved characters in the sequence alignments as well as internal gaps that lead to different outcomes in terms of species delimitations and population genetic relationships (Feindt and Hadrys, 2022). Lorenzo-Carballa *et al.* (2022) make a similar point. The species status of *Ischnura aurora* has long been debated. A morphological and mitochondrial plus nuclear sequence data analysis of specimens resulted in a split in the Asian and Australo–Pacific forms of *I. aurora* into two well-differentiated closely related species, while also suggesting the existence of a third and even fourth taxonomic unit. This means that there is a need to revise all described *I. aurora* subspecies material. Furthermore, as there are several questionable DNA sequences currently available in public repositories, it is important to link available DNA sequence data with voucher specimens while carefully examining DNA sequence data prior to inclusion in taxonomic studies.

Quantity as well as quality of genetic material can also make a difference in interpreting phylogenies and biogeographic barrier effects. In short, sound inferences require adequate sampling in terms of the number of individuals per population and the number of geographic populations sampled. A total of 186 samples from 56 localities of the Taiwanese endemic damselfly *Psolodesmus mandarinus* indicates a significant difference in the genetic structure of eastern and western populations separated by the country's central mountain range. Yet when only a subset of individuals is randomly selected for study, the geographic populations are shown as monophyletic. Furthermore, historical demographic expansion is only detectable when enough samples are used in the analyses (Wang *et al.*, 2022).

7.5.1.2 Genetic complementation with other determination methods

When genetic divergences are used alongside other data on morphology, traits, ecology and geography, together they are known as a 'taxonomic circle'. For dragonfly conservation, this genome variation across their geographic ranges indicates that some molecular operational taxonomic units (mOTUs) require more conservation focus than others (see Chapter 6, Section 6.6).

Barcoding approaches encompass the debate about what defines a species for conservation purposes (Garnett and Christidis, 2017). This means that we must also consider mOTUs to help resolve taxonomic issues, such as odonates in poorly explored areas like high-elevation tropical lakes in the Andes (Chiriboga-Ortega *et al.*, 2021).

It is essential to have good-quality short DNA reference libraries to use the method for undertaking biodiversity gap analysis for conservation prioritization across large areas. Even across Europe, and in line with the European Union Water Framework Directive, when data from 28,000 aquatic species were analyzed from data from the Barcode of Life Data System (BOLD) and NCBI GenBank, in general there are still considerable unknowns among many species and geographic areas (Weigand *et al.*, 2019). While both species and geographic coverage of odonates is reasonably good, groups like diatoms, at the base of dragonfly food chains, still require much more work.

A study of 103 species of European odonates, found that over 88% of the species can be identified using their DNA barcodes lodged in BOLD. Among the data set, however, unambiguous identification is hampered in 12 species due to weak mitochondrial differentiation and partial haplotype sharing. Cases where there was a lack of an unacceptably large barcode gap was only found in two groups of five species. But when geographic distributions are considered, the figure for unambiguous identification rose to over 95%. Furthermore, where there is considerable intraspecific divergence, as in *Cordulia aenea*, there is the possibility of raising currently labelled subspecies to the rank of species (Geigher *et al*., 2021).

In north-eastern Brazil, surveys of odonates led to the development of a DNA barcode database based on cytochrome c oxidase subunit I (COI) in 27 genera and 45 species, encompassing 70% of the odonate species in Paraíba state. The results, along with those from morphological determination, demonstrate that using the COI at a regional scale can help identify and delimit those species evaluated. Eight species (17%) showed a low percentage of differentiation (<2%) compared to other species already in the GenBank or BOLD. Overall, these data provide new insights into Neotropical odonate diversity (Koroiva *et al*., 2022). However, there can be a discrepancy between molecular and morphological results due to the presence of cryptic species, rapid and/or recent radiation events, and errors in identifying the deposited specimens, although with the positive effect of the discovery of new species (Koroiva *et al*., 2022).

7.5.1.3 Genetic assessment of macroinvertebrates and dragonflies

In Japan, sampling of stream macroinvertebrates produced 44 species recognized by DNA metabarcoding, with 35 species collapsing into 11 groups matching the morphologically identified taxa, with the recovery of a high rate (92%) of relative abundance detection of the morphologically identified samples. However, the recovery of samples by presence/absence counts was low, with a high rate of false-negative detection, especially for rare species. Nevertheless, metabarcoding provides a practical and cost-effective approach for rapid biological monitoring of freshwater macroinvertebrate assemblages. However, further improvements in the detection of scarce samples should be considered for increasing the detectability of most, if not all, of the species present at specific sites (Serrana *et al*., 2019).

Assessment of the effect of urbanization on the ecological status of German rivers using metabarcoding of macroinvertebrates (including odonates) showed distinct seasonal effects on the presence/absence of both molecular operational taxonomic unit (mOTU) and morphological operational taxonomic unit (MOTU) composition. At the heterogeneous urbanized sites there were strong site- and stressor-specific impacts, consistent between seasons. These results suggest that presence/absence metabarcoding data reliably measure macroinvertebrate assemblage composition and reflect environmental, natural or human pressure in both space and time (Zizka *et al*., 2020).

In the lower Danube river, barcoding showed that both *Gomphus flavipes* and *Erythromma viridulum* larvae prefer areas with reduced hydrological connectivity and dynamics, where there are high rates of sedimentation dominated by sandy substrates. These results support studies from larval morphological determination of their habitats (Menabit *et al*., 2022).

7.5.1.4 Genetic monitoring of dragonflies

Long-term genetic monitoring can help evaluate and provide insight into the impact of human activities on ecosystems at spatial scales ranging from local to global. Odonates are a highly tractable taxon for doing this. Valuable insight was gained from genetic monitoring of a population of *Orthetrum coerulescens* in an artificial irrigation ditch, rich in odonate species, in southern France over a 24-year period. During that time there was ditch and bank clearing, expected to greatly impact the larval and adult populations. Two mitochondrial markers (COI and ND1) and a panel of nuclear microsatellites revealed a major decline in genetic diversity during that time. However, in the latter years, the population stabilized but did not reach the degree of genetic variation present in the early years. This decline is not attributed to ditch clearing or any other direct human impact. Rather it is most likely that the population decay was due to extreme summer droughts causing reduced water levels and a change in the water variables, greatly affecting temperature-sensitive habitat specialists like *O. coerulescens*. The message here is that population genetic dynamics and metrics are not always congruent with traditional monitoring data such as abundance. This is an important consideration when developing management plans for transformed landscapes (Herzog and Hadrys, 2017).

7.5.1.5 *Genetics and capture–mark–recapture (CMR)*

Genetic studies are also providing valuable information that can be used directly for optimal conservation management. The rare *Coenagrion hylas* is a habitat specialist present as small, isolated populations. A combination of CMR and population genetics in a Tyrolean valley evaluated the state of the populations, and interestingly tested whether exuviae are usable as a DNA source. DNA was extracted from midleg tibiae and exuviae and genotyped with species-specific microsatellite markers, leading to the CMR and genetic approaches providing mutually supportive results. The species has a high tendency to disperse locally up to distances of 30 km, while also living longer than other zygopterans, with an average longevity of 12 days and a maximum lifespan of over 40 days. Low inbreeding coefficients and low ranges of genetic differentiation across sites suggested panmixia with minimal inbreeding. Exuviae deliver a sufficient amount of DNA, important for future monitoring. Results also suggested the improvement of management benefitting the species, such as the creation of stepping-stone habitats to maintain the current population status (Landmann *et al.*, 2021).

7.5.1.6 *Ensuring genetic diversity in a species*

The prioritization of populations for conservation is important but often difficult. One way of selecting the populations requiring conservation action can be by using genetic distinctiveness, especially as financial costs associated with next-generation sequencing have come down greatly. The thinking behind conserving genetically distinct populations is that they will contribute more to total within-species genetic variation and contain irreplaceable components of the species' evolutionary legacy, while contributing the significant additive genetic variation necessary for future adaptation. It has been suggested that there is no need identify the extent of adaptive variation in species of concern before prioritizing populations as using genome-wide genetic variation is a suitable proxy for identifying local adaptation (Fernandez-Fournier *et al.*, 2021) (Fig. 7.30).

7.5.2 Applying environmental DNA for freshwater conservation

7.5.2.1 *Principles of eDNA analysis*

Environmental DNA (eDNA) analysis follows the collection of bulk environmental samples such as water, sediment, or air to assess the genetic remnants that organisms have left behind in their environment. eDNA analysis can be used to obtain information on species presence in a local area. Its value for conservation is when informative genetic material is obtained from contemporary environmental samples where organisms have shed DNA, especially when they are not detectable or are missed using traditional methods. However, to date, the methodology of river macrobial eDNA for macroinvertebrates varies considerably among studies. Various methods are used for capturing eDNA from water or preservative ethanol, as well as from different sources of eDNA (i.e. sediment, biofilm), followed by the use of different primer pairs. These findings suggest that there is a need for standardization of some key steps of the eDNA metabarcoding process to increase the comparability of results and the robustness of methods for further incorporation into large-scale monitoring programs (Vourka *et al.*, 2023).

Today, eDNA analysis has become a powerful tool for ecological and conservation research and management, especially as the field samples of DNA can be investigated using high-throughput, new generation gene sequencing technology that enables metabarcoding with massive numbers of genetic sequences in an efficient, non-invasive and standardized way (Buchner *et al.*, 2021). The method can also be used alongside traditional morphological methods, but there must be careful selection of methods (Bush *et al.*, 2019). But when this is done, for example, for benthic macroinvertebrates, it has the potential for monitoring stream quality (Elbrecht *et al.*, 2017).

Pawlowski *et al.* (2020) advocate that the term 'eDNA' be used in its generic sense, as originally defined, encompassing the DNA of all organisms present in environmental samples, including microbial, meiofaunal and macrobial taxa. This means first specifying the environmental origin of the DNA sample: water eDNA, sediment eDNA or soil

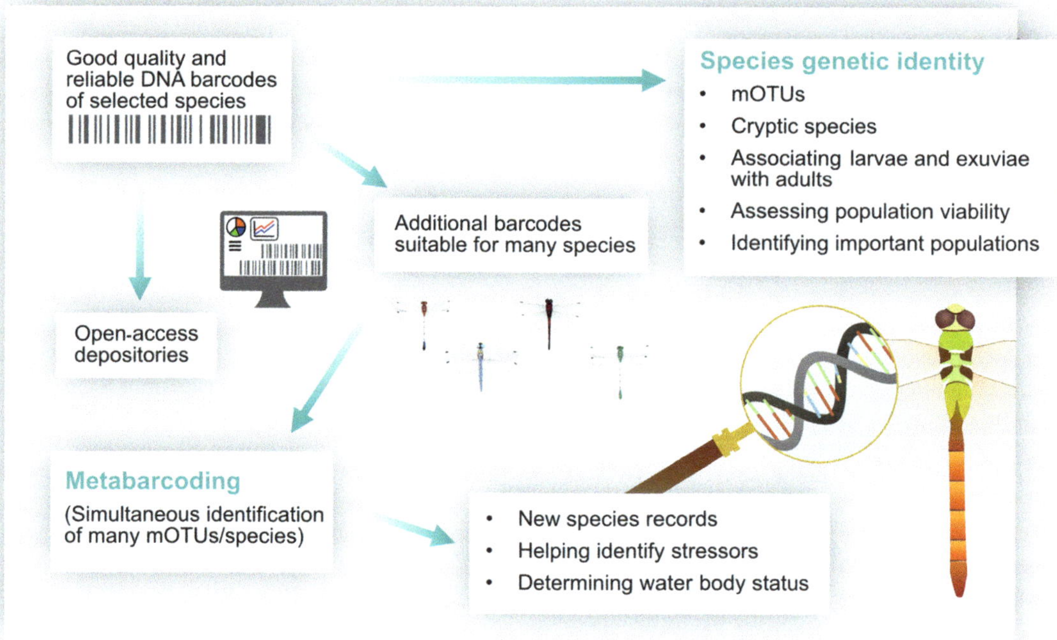

Fig. 7.30. Genetic analyses using specific genetic markers (barcodes) are playing an increasingly important role in dragonfly conservation, so long as there is good qualitive and quantitative data. These barcode data are lodged in DNA barcode open-access depositories where they support, along with voucher specimens, species identification (including molecular operational taxonomic units (mOTUs) and cryptic species) and the association of larval or exuvial identity with that of the adult. Examination of complex samples of multiple barcodes from many mOTUs/species (metabarcoding) is improved by using additional genetic markers. Together all the barcode advances are complementing morphological, ecological and geographic findings. This progress is also helping to identify stressor impacts, assess population viability and genetically identify distinct populations of conservation significance. (Illustration: Robyn Symons).

eDNA. This wide approach does not presume the specific state of the DNA sampled, i.e. intracellular versus extracellular. The taxonomic group of interest is then targeted through polymerase chain reaction amplification, such as invertebrate eDNA or odonate eDNA.

Furthermore, eDNA methodology can also be used to sample the historic state from biological material deposited many years ago and compared with modern day samples to determine assemblage or community changes over time. However, the presence of DNA remnants does not necessarily mean that a species is currently present and alive at a specific sampling site, as the DNA material could have been transported by water movement or have come from living or dead organisms upstream.

Although the absence of certain DNA material does not necessarily mean that a species is absent from the sampled site, eDNA protocols are proving to be most valuable in biodiversity assessments and for ongoing monitoring schemes toward conservation, both in running and standing waters (Barnes *et al.*, 2021; Monchamp *et al.*, 2023). However, there is still the need for more consistent experimental and modelling methods. These must account for spatiotemporal variation, and uncertainty in eDNA collection and analysis, as well as incorporating prior knowledge of sources of variability via Bayesian modelling approaches to account for uncertainties such as imperfect detection, generation of robust diversity estimates and efforts to increase the comparability of eDNA data sets for effective environmental monitoring across space and time (Mathieu *et al.*, 2020).

7.5.2.2 eDNA analyses of aquatic insects and macroinvertebrates

Combining eDNA extracted from stream water samples with a certain stream hydromorphology allows upscaling of aquatic insect diversity at high spatial resolution for determining the distribution patterns of taxa. Across a river basin in Switzerland and using a grain size resolution of 1-kilometre long stream sections, the matching of direct observations with known local species occurrences was 57–100% accurate. These findings suggest that eDNA can be used for high-resolution aquatic insect assessments in rivers with minimal prior knowledge of the system. This approach allows the identification of biodiversity hotspots that could be otherwise overlooked, enabling the implementation of focused conservation strategies (Carraro *et al.*, 2020).

In the Eastern Cape, South Africa, most DNA-based methods gave the same water quality assessment category as the standard morphology-based macroinvertebrate assessment. Bulk homogenized metabarcoded DNA samples recovered more freshwater invertebrate diversity than all other methods, suggesting the standardized rapid macroinvertebrate methods overlook certain taxa, although 90% of the reads were non-target taxa. Although eDNA recovered the fewest numbers of target macroinvertebrate taxa, it did detect some target taxa that all other methods failed to do. In sum, unsorted bulk metabarcoded samples can be used for water quality assessment and monitoring as they provide higher diversity estimates for macroinvertebrates than either rapid morphological or eDNA samples. Overall, incorporating DNA-based approaches into existing morphological rapid assessment protocols would provide additional taxonomic resolution for better assessment and monitoring (Pereira-da-Conceicoa *et al.*, 2021).

In an urban green space in Johannesburg, South Africa, and following five decades of restoration, the sampling of water and sediment for eDNA barcoding fragments of about 710 bp cytochrome c oxidase subunit I (COI) in both running and standing waters produced 26 phyla, 52 classes, 134 orders, 289 families, 380 genera and 522 known species. The results highlight the critical role that restored urban green areas play in protecting aquatic biodiversity (Webster *et al.*, 2020).

7.5.2.3 Barcode referencing and eDNA sampling

The development of national-level genetic databases is now underway. Fischer *et al.* (2018) outline the procedure for integrating genetic and morphological approaches for assessing Austrian odonate fauna. Before setting up standardized protocols, it is essential to first evaluate the potential of the methods. A starting point is to have morphologically correctly identified voucher specimens before successful application of eDNA metabarcoding. For rural Vienna, this also means including certain species of conservation concern, such as *Cordulegaster heros* and *Leucorrhinia pectoralis*. Species-specific DNA barcodes are generated from collected individuals which become reference voucher specimens. Against this reference database, eDNA extracted from water samples can then be analysed to detect single species and to metabarcode the whole odonate assemblage.

eDNA analysis holds much promise for pond biodiversity monitoring, but there are challenges specific to the pond environment, as in other freshwater habitats. These challenges must be addressed to obtain accurate, standardized tools that can be routinely and reproducibly implemented. A consensus is required on the amount of water and the number of samples taken from individual ponds to establish representativeness. This means having representative samples from water to understand horizontal, vertical and temporal patterns among species. Progress is being made on broad standardization of eDNA workflows that guide effective management and the conservation of pond biodiversity, without stifling innovation or development. The European DNAqua-net, a consortium of researchers and managers, has been developed to this end (Harper *et al.*, 2019) (Fig. 7.31).

Environmental DNA has been tested regarding its applicability to threatened *Somatochlora hineana* in North America where traditional monitoring costs are high. An essential step for the development of an eDNA monitoring scheme is to evaluate the dynamics of eDNA concentration under controlled conditions. Quantitative polymerase chain reaction (qPCR) was used to determine seasonal shifts in persistence and net-accumulation of eDNA from captive *S. hineana* larvae in experimental mesocosms at temperatures corresponding with their overwintering (5°C) and active (16°C) seasons. Environmental DNA persisted longer at 5°C but accumulated more readily at 16°C. Differences in the accumulation and persistence of eDNA reflect both seasonal differences in larval *S. hineana* behaviour, as well as differences in the longevity of eDNA at different temperatures, highlighting the importance of considering how seasonal changes in temperature influence not only eDNA shedding

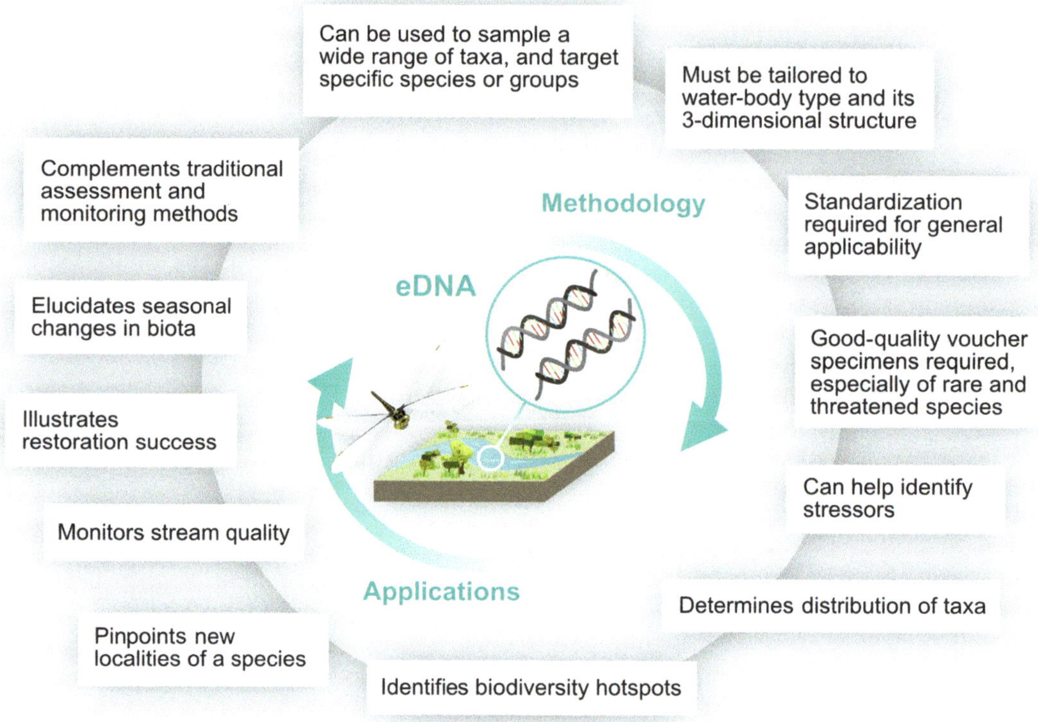

Fig. 7.31. Environmental DNA (eDNA) sampling is fast developing as a method with great value for ecology and conservation. The method is based on sampling a range of taxa across all trophic levels, while also applicable to target taxa. It requires honing for type of water body and its three-dimensional structure, as well as the ecological or conservation goal. For dragonflies at least, the method requires good-quality voucher specimens, especially of rare and threatened species. eDNA has various applications and can greatly complement traditional methods. (Illustration: Robyn Symons).

rates but also the speed of eDNA degradation of the species (Schmidt *et al.*, 2021).

7.6 Monitoring Freshwater Bodies Using Traditional Morphological Methods

7.6.1 Monitoring changes in macroinvertebrates and dragonflies

7.6.1.1 Monitoring macroinvertebrate changes

Monitoring using macroinvertebrates is widely used to assess and track changes in environmental conditions that reflect the ecological state or health of a freshwater system. Responses by the various species depend very much on the traits of species. In turn, the traits of a species determine its susceptibility to a stressor or multiple stressors.

In German rivers, 324 macroinvertebrate taxa, with a whole host of traits and vulnerabilities to stressors, were assessed in terms of their response to 21 environmental gradients. The most vulnerable species were those with aquatic larval stages, isolated cemented eggs, a life cycle of less than one year, aerial and aquatic active dispersal, or plastron respiration. In contrast, the most tolerant species were those with adult aquatic life stages, polyvoltinism, ovoviviparity, an egg deposited in vegetation, scavenging behaviour on dead animals or living macroinvertebrates, substrate preference for macrophytes, a preference for silt or mud, and a body size of more than 2–4 cm (Berger *et al.*, 2018).

Kadoya *et al.* (2011) assessed a range of Japanese ponds subject to various levels of stressors: eutrophication (represented by total phosphorus and nitrogen, suspended solids, chlorophyll *a* and cyanobacteria density), habitat condition (surrounding land-use intensity and level of bank protection) and the impact of invasive alien species (two species of fish, a crayfish and a bullfrog). Using aquatic macrophytes, larval and adult odonates, and benthic macroinvertebrates as indicators of pond quality, eutrophication emerged as the dominant stressor. Another finding was that total species richness, threatened species richness and functional diversity composed a practical working framework for pond assessment and for guiding management.

7.6.1.2 Monitoring dragonfly assemblages

As monitoring involves sustained time and resource investment, the focal group must be carefully chosen. This is especially the case when different aquatic insect species vary in their life history phenology (Cayrou and Céréghino, 2005). When red-listing dragonfly species, which is usually based on the adults, choosing the right time of year to sample is critical (Samways and Grant, 2007). However, dragonfly emergence patterns are also changing across many parts of the world in response to global warming (Hassall *et al.*, 2007; Dingemanse and Kalkman, 2008; Doi, 2008), which means that these emergence patterns must be watched carefully to detect any phenological shifts. Furthermore, individuals in recessed microhabitats or those that have difficult-to-sample life stages can be missed during assessments, only being discovered when sites are reassessed over time, sampled at different times of the year or various sampling methods are used. This is the case for many species of dragonfly larvae, with the taxonomic challenge greatest for the early instars.

7.6.1.3 Dragonfly traits associated with climate change

Many dragonfly species have shifted their geographic ranges in recent years, and some greatly (Chapter 3, Section 3.2), with others expected to have range shifts in the future (Chapter 3, Section 3.4). However, there are some technical issues associated with dragonfly monitoring that warrant some extra focus.

Certain UK odonates have expanded their geographic range northward in recent years, which could potentially affect the use of water quality assessments involving macroinvertebrates at the family level, including odonates. However, even though the odonate families used in monitoring are shifting their ranges poleward and species richness is increasing at most UK latitudes, these shifts have a negligible effect on water quality indicators. However, variation in odonate species richness, especially in species-poor areas, does have a significant effect on water quality assessments. This means that caution is warranted when using higher taxa as indicators, such as dragonfly families, when assessing freshwater quality (Hassall *et al.*, 2010).

Trait differences can explain variation in recent range shifts in accordance with life-history theory and invasion models, but the predictive capacity of these relationships has limitations. In the case of odonates with endophytic egg laying, habitat tended to range shift less, possibly driven by associated life-history characteristics (smaller clutches) and/or ecological specialization (reliance on appropriate host plants). But as so few odonate traits had significant effects, conclusions about the relative importance of different classes of traits (i.e. traits related to dispersal ability versus those related to establishment probability) or relative to the effect of traits on different types of range shifts (i.e. elevation versus latitude) are not possible. Overall, there are some significant relationships between certain traits and recent leading-edge range shifts, suggesting that these traits influence the ability of a species to colonize newly available habitat. However, the relatively poor explanatory power of the findings suggests that a trait-based approach has limited use for practical conservation, such as informing managers in identifying species with the greatest vulnerability to climate change or for designing reserves or corridors for species with different probabilities of range movement (Angert *et al.*, 2011).

These conclusions are borne out by more recent information on European odonates, where the extinction risk among 115 species showed no relationship with the various traits of body volume as measured by body size, colour lightness, wing area and geographic range size. There is no universal trait or evolutionary history that defines the extinction risk that they face. However, among odonates, there are idiosyncratic responses to stressors like pollution, land-use change and climate change. This means that both the relative importance of major stressors and threats, alongside mechanisms linking intrinsic traits to environmental factors, need to be assessed to understand odonate extinction risk and its mitigation. In short, species-specific information is required to mitigate the effects of ongoing stressors and threats (Franke *et al.*, 2022).

7.6.2 Dragonfly bioindicators

Food webs involving dragonflies and their prey while subject to pollutants and pesticides, have received detailed attention in the laboratory and to some extent in the field (Chapter 5, Sections 5.1 and 5.3). As mentioned in the introductory Section 7.4.1.6, to label a species or group of species as a bioindicator technically needs to be tested using a set of postulates. This test procedure is rarely done, with many species that are labelled as 'bioindicators' being really the most sensitive or insensitive 'responders' to a stressor such as vegetation transformation, pollution or pesticide loading.

7.6.2.1 Bioindication of habitat quality using dragonfly species

Besides the many studies showing shifts in dragonfly assemblage composition with human-induced changes in vegetation structure, there are some that highlight certain indicator species. In the Brazilian Amazon forest, among 23 genera and 35 species of odonates recorded from streams with various degrees of habitat disturbance, only *Psaironeura tenuissima* was an indicator of preserved sites, while *Argia* sp.1 and *Hetaerina* (*Mnesarete*) *smaragdina* were indicative of altered sites (Pereira-Moura *et al.*, 2021). Among southern coastal Brazilian ponds along an urbanization gradient, three species, *Erythrodiplax* sp.1, *Erythemis credula* and *Telebasis corallina*, were associated with non-urbanized areas, possibly in response to specific macrophytes and water chemistry (Sganzerla *et al.*, 2021).

Determining ecological thresholds in terms of the Zygoptera/Anisoptera ratio has important management implications for Brazilian streams. Canopy cover and stream width are predictors of taxonomic richness and abundance and the Z/A ratio in the Brazilian Cerrado–Caatinga ecotone. The turning points were determined by the Habitat Integrity Index (HII) (see Section 7.4.1.5), where below 0.38 there was an increase in habitat-generalist odonates and a decline in sensitive taxa. However, above 0.79 there was an increase in sensitive taxa and a decrease in generalists. Four indicator species were selected, two of which are associated with a negative response (*Perithemis tenera* and *Acanthagrion aepiolum*) and two with positive responses (*Epipleoneura metallica* and *Zenithoptera lanei*) for habitat integrity (Veras *et al.*, 2022).

Shifts in odonate assemblage composition are often associated with landscape transformation and habitat loss. Sometimes, certain species can characterize levels of transformation in areas of conservation significance. For example, in the Owabi Wildlife Sanctuary, Ghana, of the 29 odonate species recorded, *Pantala flavescens* characterizes high levels of human disturbance, *Orthetrum abbotti* and *Africallagma vaginale* moderate disturbance, and *Gynacantha bullata*, *Tramea limbata*, *Olpogastra lugubris*, *Neodythemis klingi*, *Brachythemis leucosticta*, *Chlorocypha curta*, *C. luminosa*, *C. radix* and *Agriocnemis* sp. characterize low disturbance (Manu *et al.*, 2023). While *P. flavescens* benefits from the more open environments induced by human activity, *B. leucosticta* is an interesting species that tolerates high native megaherbivore impact, but is in a way, a habitat specialist that is not so tolerant of human impact.

7.6.2.2 Bioindication of river type

In Austria, near-natural streams with canyon-like geological features that have limited riparian habitat types for dragonflies are characterized by two river-type-specific core indicator species (*Calopteryx virgo* and *Onychogomphus forcipatus*) and five river-type-specific species (*C. splendens*, *Platycnemis pennipes*, *Ischnura elegans*, *Gomphus vulgatissimus* and *Orthetrum brunneum*), with this last group indicating some influence from the immediate surroundings probably due to increased water temperatures (Chovanec, 2022).

7.6.2.3 Bioindication of pollutants

The effects of pollution adversely affect a dragonfly's metabolic health before becoming clear at the population or assemblage level. Significant metabolic changes are known as biomarkers which can be an early warning system ahead of changes to groups of whole organisms. Shifts in local stream conditions, such as canopy cover, water temperature, pH, dissolved oxygen and habitat integrity increase the exposure of dragonflies and other aquatic organisms to these stressors. Responses to these stressors include the production of greater concentrations of energy in the form of adenosine triphosphate to enhance the antioxidant defence system to avoid oxidative stress. Among the biomarkers for this metabolic processing is glutathione S-transferase (GST) which also contributes to the detoxification of several organic pollutants, such as pesticides in aquatic insects, with detoxification performance related to the adaptive fitness of the organism to certain environmental conditions. When natural habitats are transformed into oil palm plantations, there are changes, often co-related, in vegetation structure, canopy cover, dissolved oxygen, water temperature and pH. The calopterygid *Hetaerina* (*Mnesarete*) *aenea* expressed higher GST activity in pesticide-treated oil palm areas than in natural forest areas. These results suggest that this species might act as a sentinel for biomarker evaluations, while providing an early warning signal via its biochemical responses that could help to prevent changes at the population and community levels (Mendes *et al.*, 2020).

Wastewater effluent affects the behaviour and physiological traits of *Coenagrion hastulatum* larvae. Behavioural shifts include reduced activity and foraging, but a faster escape response. Effluent exposure causes a significant change in 14 individual compounds (4 amino acids, 3 carnitines, 3 lysolipids, 1 peptide, 2 sugar acids, 1 sugar), which together indicate an increase in protein metabolism and oxidative stress. When behavioural and physiological assessments are used together, it provides a means of early detection of the effects of wastewater effluent (Späth *et al.*, 2022).

In a watershed in Spain, human activity is characterized by changes in electrical conductivity, salinity and total dissolved solids in the water. Probability mapping using spatial uncertainty (kriging with external drifting, developed in the geosciences) shows that presence of *Ischnura graellsii*, although an endemic species of the Iberian Peninsula and North Africa, is a strong bioindicator of polluted sites (Campos *et al.*, 2016).

Mercury is a highly bioaccumulative metal in dragonfly tissues, although varying in levels according to aquatic environmental conditions (Nelson *et al.*, 2024). Across the USA, sampling dragonfly larvae with dip nets and then testing the samples for total mercury is an effective way of indicating mercury bioavailability in aquatic food webs. This approach can be used to develop an integrated impairment index of mercury contamination risk to aquatic ecosystems. Furthermore, by engaging citizen scientists, it can also be incorporated into a national landscape monitoring scheme for assessing degrees of risk to aquatic food webs and to human health (Eagles-Smith *et al.*, 2020).

Also in the USA, among eight odonate larvae genera, higher trace element concentrations accumulated in larvae from upstream areas close to contaminant sources. Body form and habitat type were more influential in trace element accumulation than phylogeny for several elements (nickel, barium, strontium, vanadium, beryllium, cadmium and chromium), with higher concentrations accumulating in sprawler and climber-sprawler genera, irrespective of family. In contrast, arsenic and selenium accumulated at variably higher levels in burrowers, although accumulation in sprawlers differed between sites. The greater variation between genera than within genera suggests genus as an acceptable unit of comparison among dragonfly larvae. In sum, taxonomic differences in trace element accumulation can be substantial, often exceeding variation between sites. These findings emphasize the element- and taxon-specific nature of trace element accumulation (Fletcher *et al.*, 2017).

Fletcher *et al.* (2020) then measured bioaccumulation of zinc, lead, chromium, cadmium and aluminium in the larvae of *Anax*, *Erythemis*, *Libellula*, *Pachydiplax*, *Tramea* and *Plathemis*, as well as the crayfish *Cambarus latimanus*, as an indicator of constructed wetland and stream quality. *Erythemis*, *Libellula* and *Pachydiplax* are especially good indicators of water condition, as they are associated with bottom sediment and were abundant.

The quality of sewerage treatment ponds in Canada was determined in terms of larval odonate species diversity and abundance over time. The presence of *Anax junius*, *Enallagma civile* and *Ischnura verticalis* alone was associated with poorer water quality. Better water quality was associated with higher diversity, including the appearance of *Lestes disjunctus*, *Leucorrhinia* spp. and *Erythemis simplicicollis*. An advantage of using these bioindicators is that they reflect a time-period rather than one point in time while they are also inexpensive to monitor (Catling, 2005).

In Turkish stream habitats, certain odonate larvae are effective indicators of heavy metal contamination. *Aeshna juncea* is an indicator species for levels of manganese, lead and nickel, while *Platycnemis pennipes* is an indicator species for cadmium, boron and iron, as well as total water hardness (Girgin *et al.*, 2010).

A suite of stressors can suggest different indicator species for different stressors. In Croatian lowland rivers and among the 19 odonate larvae species, water pollution (as measured by levels of chemical oxygen demand and total organic carbon), water temperature and oxygen concentration were the most influential drivers of assemblage composition. *Ischnura elegans* and *Platycnemis pennipes* are the most tolerant species of a wide range of ecological conditions. Overall, degraded rivers can support a relatively low number of species with broad ecological tolerances, while rare and specialist species are generally not able to reproduce there. A combination of *Gomphus vulgatissimus*, *Onychogomphus forcipatus*, *P. pennipes*, *I. elegans*, *Somatochlora meridionalis* and *Calopteryx virgo*, arguably a mix of habitat specialists and generalists, characterize the better-preserved sites. In turn, the heavily polluted sites were also overgrown by dense growth of aquatic macrophytes and algae associated with high nutrient and low oxygen levels, limiting certain odonate species. In contrast, less polluted sites with higher oxygen levels and lower water temperatures supported more species. Moreover, two species with rather differing ecological demands, *Aeshna cyanea* and *Calopteryx splendens*, were both present in polluted water with its higher temperatures and lower oxygen concentrations and medium levels of detritus. However, while *C. splendens* favours unshaded flowing water habitats, *A. cyanea* prefers partly shaded standing waters (Vilenica *et al.*, 2020).

Ostensible examples of dragonfly bioindication include the pollution status as per government standards in Indonesia (Hidayat *et al.*, 2019).

7.6.2.4 *Bioindication of pesticides*

While there has been much progress toward understanding the vulnerability of odonates to pesticide exposure (Chapter 5, Section 5.3), most studies have been limited to populations at a single latitude with one local thermal regime. There remains the question of how local thermal adaptation vis-à-vis pesticide sensitivity might vary across the entire geographic range of a species.

When *Ischnura elegans* larvae populations from low latitudes (southern France), mid-latitudes (Belgium) or high latitudes (southern Sweden) are compared in terms of their response to the widely used pesticide chlorpyrifos, the low latitude populations were more vulnerable to the pesticide at higher temperatures and pesticide concentrations, with evidence for the effect of local thermal adaptation on the vulnerability of larvae being weak. The results emphasize the need for spatially explicit bioassessment and conservation action, with management practices that mitigate pesticide run-off into aquatic ecosystems being especially important in agricultural areas at low latitudes (Dinh Van *et al.*, 2014).

Some adverse chemical impacts can be subtle. Even sublethal doses of the insecticide monosultap triazophos in the laboratory result in a fluctuating asymmetry of the first tibia length of *Ceriagrion* sp. larvae. This means that fluctuating asymmetry measurements can be used as an indicator reflecting the level of insecticide stress (Chang *et al.*, 2007) (Fig. 7.32).

7.7 Surrogates Associated with Dragonfly and Freshwater Conservation

7.7.1 Surrogacy at multiple spatial scales

The relationship between aquatic insects and environmental variables needs to be understood at various spatial scales. In the case of running water, spatial extent ranges from channel, through inclusion of the riparian zone, to regional waterscapes. Assessments are usually done through measurement of species richness and composition of assemblages and/or abundance or density of individuals. In many tropical regions that are rich in biodiversity and where forest loss is a major concern, multiscale assessments require much more attention (Brasil *et al.*, 2020b).

There is emerging evidence that sampling done on a hierarchical spatial scale (scale-linked, involving gradual addition of carefully chosen sites of interest) is more efficient than sampling at a single scale repeated on an ad hoc basis. For stream macroinvertebrates in the USA, solutions using a scale-linked approach were 5% less costly and required 4.7% fewer planning units than a single-scale approach. Scale-linked solutions were penalized 16% less than single-scale solutions. This suggests that scale-linked approaches can decrease cross-scale disparities and they better reflect hierarchical processes without sacrificing planning efficiency. The upshot is that scale-linked conservation planning may help ease implementation efforts while enhancing the long-term resilience and conservation effectiveness of landscapes surrounding protected areas (Brumm *et al.*, 2022).

7.7.2 Choosing effective surrogates

Knowledge progress can be achieved when there is greater information on how certain species or sets of species stand in for other groups, i.e. are surrogates.

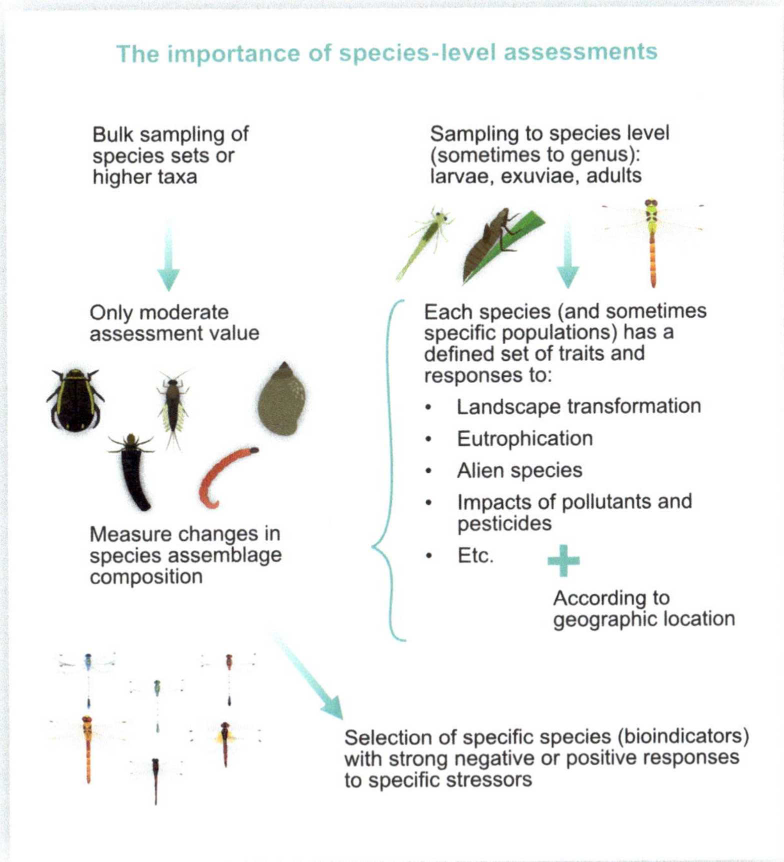

Fig. 7.32. What has emerged from focusing on the impact of stressors, such as landscape transformation, eutrophication, alien species, and pollutants and pesticides, is that using macroinvertebrates or dragonflies as bulk groups has far less value for assessment and monitoring compared to the unpacking of groups into their component species, and in some cases, genera. The reason for this is that traits of specific species (and even sometimes individual populations of a species) and their individual responses to stressors and/or environmental change, whether local or more general like climate change, vary greatly from one species to the next. In turn, the stressors may also vary greatly in type and impact from one location to another. This means that it is important to measure changes in assemblage species composition for accurate assessment and monitoring, whether for larvae, exuviae and/or adults. Certain species may show a particularly strong response, either negative or positive, to a stressor, and are considered bioindicators. (Illustration: Robyn Symons).

However, surrogate species or species sets that help select high-value conservation areas, both protected or novel, generally require good taxonomic resolution and good distributional data. In short, there must be sound verification of the extent of matching between the surrogate group and the target group. Surrogate insect species and insect species sets can be particularly useful in parts of the world where the local insect fauna is complex and/or poorly known.

7.7.2.1 Dragonfly species turnover relative to surrogacy

In regions with long evolutionary histories without major impoverishing climatic events, differences in compositional biodiversity can be acute. The Cape Floristic Region, South Africa has had no glacial attrition of biodiversity for over 200 million years, with no major geological uplifting during that time, resulting in a relatively stable topographic landscape.

The outcome has been much evolutionary divergence so that today there is high species turnover from one local river to the next. Each river has a different species signature with widely differing sets of dragonfly species and much localized endemism (Kietzka *et al.*, 2018).

In that same region, the pattern of turnover between rivers is also the same for other aquatic insect groups, which are largely congruent from one river to the next. These various groups have been honed by a long evolutionary history to the same moderating climatic events and are equally sensitive to significant novel threats, such as from invasive alien trees, while recovering in synchrony when these alien trees are removed (Samways *et al.*, 2011).

With spatial scale being strongly associated with the conservation goal, it is essential to carefully select the surrogate that truly represents the assemblage or community according to the conservation goal. Sometimes, use of both taxonomic and functional complementary groups (including plants) can be used to represent biodiversity overall. This is usually highly spatial-scale dependent. At the large spatial scale, broad geographic patterns show strong congruency among freshwater fish, molluscs, aquatic plants and dragonflies across African catchments (Holland *et al.*, 2011).

However, at a regional scale, especially in postglacial temperate river basins with different environmental variables, there can be poor congruence among freshwater macroinvertebrates at species, genus and family levels (Heino, 2014), with any one taxonomic group not well representing another (Heino, 2010). This emphasizes that while there may be similar levels of species, or higher-taxon turnover, this does not necessarily mean that one group can be comfortably used to represent another for the purposes of conservation management (Fig. 7.33).

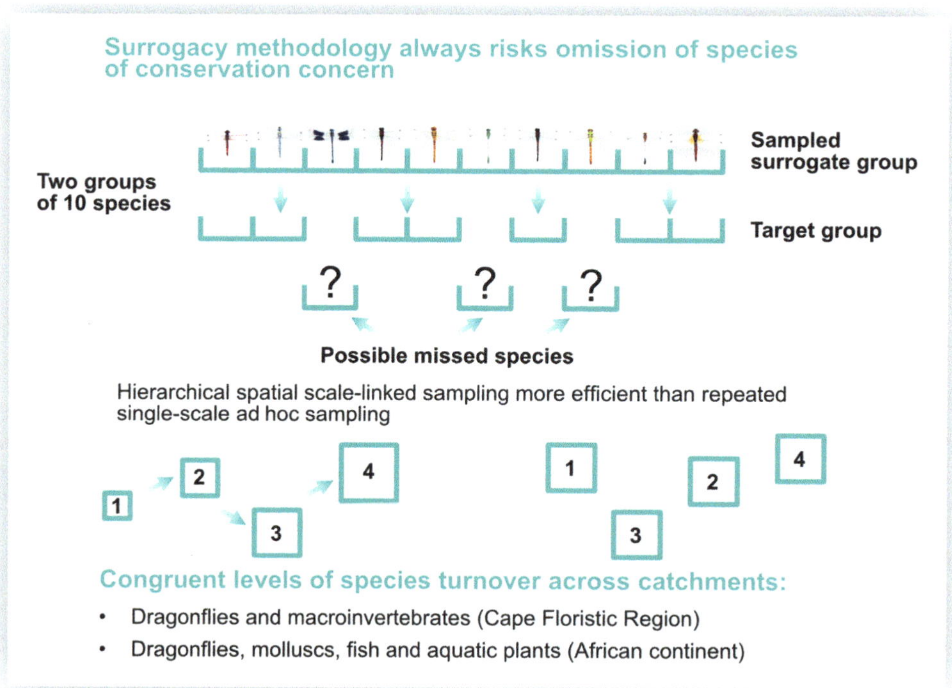

Fig. 7.33. Choosing surrogates for assessment and monitoring, while seemingly a good approach for the sake of sampling ease, always carries the risk that certain species of conservation interest are left out of the assessment among the target group. Sampling procedure also matters, with hierarchical spatial-scale sampling more efficient than ad hoc single-scale sampling. While there can be similar trends and levels of species turnover among different taxa across river catchments (whether for dragonflies and macroinvertebrates in the Cape Floristic Region, South Africa, or a range of taxa across Africa), it does not necessarily mean there is sufficient congruence for one species (or higher taxonomic group) to represent another for reliable conservation management at the local scale. (Illustration: Robyn Symons).

7.7.2.2 *Taxon sets representing wetland quality*

Nevertheless, sets of rare and threatened insect species can be surrogates for a wide range of other species, especially when there is a long history of shared evolutionary opportunities. For example, the threatened blue butterfly *Maculinea alcon* in Belgium can represent a wide range of other rare taxa in groups such as vascular plants, a grasshopper, dragonflies, other butterflies and birds. When all these taxa are used together, they constitute a good indicator set of typical wetland species diversity and habitat quality, while also displaying equal vulnerability to various stressors (Maes and Van Dyck, 2005).

7.7.3 Representativeness among dragonflies and other macroinvertebrates

7.7.3.1 *Dragonflies and certain other taxa*

A comparison of Trichoptera and Odonata larvae and adults, both morphologically and using barcodes, along Canadian rivers indicated that caddisflies do not fully reflect site-scale taxon richness or assemblage variation among odonates. Furthermore, there was much variation in the size of caddisfly and odonate assemblages across the catchments. However, when rarefied data were used, the magnitude of assemblage variation was reduced, resulting in stronger correlations in richness and assemblage variation. These increased correlations likely stemmed from differences in abundance among the two groups, perhaps owing to some incomplete taxonomy, especially among caddisflies. The conclusion is that conservation planning in riverine ecosystems based on surrogates should proceed cautiously until there is better understanding of how the results reflect biodiversity patterns in undersampled taxa and habitats. Future studies of surrogacy should consider both richness and assemblage variation as each provides valuable information for conservation-related decisions. Furthermore, taxonomic resolution, including using DNA-based taxonomy, and the relative abundance of comparison groups can potentially impact the strength, direction and statistical significance of patterns (Curry *et al.*, 2012).

Narrower, complementary species sets have been used to assess habitat quality. Damselflies in the family Coenagrionidae, alongside caddisflies in the family Limnephilidae, can represent the invertebrate conservation value of ponds in the UK (Briers and Biggs, 2003), odonates and water beetles can represent artificial pond quality in South Africa (Deacon *et al.*, 2018), and odonates, caddisflies and water beetles can represent the impact of environmental disturbance to canals in Poland (Buczyńska and Buczyńska, 2019). Across a subtropical region of South Africa, ponds under three types of land use (protected area, sugarcane-dominated, plantation forestry-dominated) are complementary in their odonate diversity. Furthermore, odonate assemblages among the various pond types are good surrogates for macrophyte and water beetle diversity (Briggs *et al.*, 2019a).

In Colombian wetlands, local distribution patterns and the abundance of larvae of the geographically widespread libellulid *Miathyria marcella* were correlated with macroinvertebrate composition relative to the macrophyte communities. The most abundant invertebrate groups were Neotaenioglossa (26%), Odonata (15%), Calanoida (10%) and Diptera (8%). There was a high positive correlation between the libellulid species and Odonata, Coleoptera, Basommatophora and Hemiptera, relative to the type of hydrological influences such as the sea, major rivers and local streams (Moreno Pallares *et al.*, 2023).

7.7.3.2 *Dragonflies and Ephemeroptera/Plecoptera/Trichoptera*

The diversity and abundance of insect groups such as dragonflies and Ephemeroptera/Plecoptera/Trichoptera (EPT) feature strongly in natural versus perturbed or transformed river systems (Gomes *et al.*, 2022), and are often associated with the proportional increase in abundance of chironomid flies relative to the other groups in perturbed systems (Carew *et al.*, 2011).

The interrelationship between EPT and dragonflies for freshwater assessment is potentially interesting as both groups show great value as amphibiotic responders to change, both in the water and on the adjacent land. The umbrella index was developed as a quantitative method to select the best surrogate species or taxon set for representing certain other species or taxon sets. This index was used to evaluate rivers experiencing a range of disturbances in the Cape Floristic Region, South Africa with its exceptional level of rare and endemic species. The umbrella index identified eight EPT species and seven Odonata species as potential surrogates. The odonates in descending order of umbrella index scores are *Pseudagrion furcigerum*, *Syncordulia gracilis*, *Zosteraeschna minuscula*, *Allocnemis leucosticta*, *Pinheyschna subpupillata*, *Pseudagrion draconis* and *Chlorolestes conspicuus*, all endemics to southern Africa and some endemic to the local region. Both groups co-occurred with high percentages of their own overall groups, i.e. EPT surrogates for overall EPT, and Odonata surrogates for overall Odonata. There was also a high level of surrogacy for the other group, i.e. EPT surrogates for overall Odonata, and Odonata surrogates for overall EPT. The index was highly flexible, performing well on species of conservation significance, even across spatial scales larger than one river, and across sites with various disturbance levels. In short, both EPT and Odonata are potential surrogates for quality assessment and conservation planning, and the umbrella index can be used in a biodiversity hotspot rich in endemic species, as well as for freshwater aquatic insects regularly used in freshwater assessments. Odonate adults can be easily recorded as named species, making them the preferred surrogate. Furthermore, EPT surrogates would conserve 87% of all the EPT species, and 92% of all the odonate species, while the Odonata surrogates perform better, conserving 96% of all the Odonata species, and 93% of all EPT species, along with only 50% of the sampling effort compared to EPT (Kietzka *et al.*, 2019).

In Mato Grosso do Sul, Brazil adult dragonflies are not a favoured group on their own for assessing and monitoring river condition. Although when combined with EPT, the two groups are complementary for different environmental variables, and together they may be used for monitoring river condition (Valente-Neto *et al.* 2018). Yet in the eastern Amazonian state of Pará, Brazil, genera of odonate adults are highly congruent with EPT, with each also well-representing benthic macroinvertebrates (Martins *et al.*, 2022).

Tippler *et al.* (2018) compared the sensitivity of macroinvertebrate assemblages (percentage EPT, family richness, abundance of individuals, family diversity, and order and family separation (two scores)) with those of odonate larvae (percentage of odonate taxa in the original sample, species richness, abundance of individuals and species diversity) associated with aquatic and riparian impacts on streams in Sydney, Australia, where water quality progressively declines with increased levels of urban development. The macroinvertebrate assemblage was impoverished relative to the level of urban development, especially the percentage of EPT taxa. However, macroinvertebrate abundance did not vary across the urban development gradient. In contrast, the odonate assemblage did not vary across the development gradient. These results contrast with those of another Australian urban study where odonate larvae responded strongly to pollution (Watson *et al.*, 1982). They also contradict a whole host of other studies from many other parts of the world (Chapter 5, Section 5.6). These differing results are puzzling and may reflect a shifting baseline where the remaining species are those adapted to current conditions, with the sensitive species lost many years ago (e.g. Küry and Christ, 2010; Adu *et al.*, 2022).

In a tropical dry karst region of Brazil, the richness and composition of certain species of terrestrial (frugivorous butterflies and dung beetles) and aquatic insects, sorted to order level (Plecoptera, Trichoptera, Ephemeroptera, Odonata and Coleoptera), were used to assess the quality of riparian zones. There was no correlation between the richness or composition of the groups and riparian zone integrity, possibly due to high variability in ecological requirements among different taxa. The sampling was too coarse to comment on any correlation or not between EPT and dragonflies (Roque *et al.*, 2017) (Fig. 7.34).

7.7.4 Value of dragonfly subsets for conservation

The Habitat Integrity Index (HII) was used in association with the assemblage composition of adult odonates at different taxonomic levels in south-eastern Mexico. The following groups all responded well in changes to the HII score: the two suborders, Zygoptera and Anisoptera, Coenagrionidae species and a set of 20 habitat-specialist species in the genera *Argia*, *Cora*, *Hetaerina* and *Heteragrion* (ACHH) all responded well to changes in the HII score. All taxonomic levels were included in the analyses as predictors of habitat integrity. However, the abundance, species richness and diversity of the ACHH group was the best predictor of the HII compared to the other taxonomic groups. In contrast, there was a negative and significant relationship between HII and Anisoptera abundance, richness and diversity (Gómez-Tolosa *et al.*, 2022b).

Of interest is the genus *Argia*, which is highly species diverse in the region, with the species richness and diversity of the group being strongly positively correlated with the ecological quality status

Taxonomic groups	Country	Representativeness
Odonata Trichoptera	Canada	Different responses to river biotopes
Coenagrionidae (Zygoptera) Limnephilidae (Trichoptera)	UK	Together represent well artificial pond value
Odonata Water beetles	South Africa	Together represent well artificial pondscape value
Odonata Water beetles Aquatic macrophytes	South Africa	Dragonflies represent well water beetle and macrophyte diversity across pond types
Odonata Water beetles Trichoptera	Poland	Together represent well canal disturbance
Odonata EPT	Brazil	Different results from different regions: one with high and another one with low congruency in representing river condition
Odonata EPT	South Africa	Very high congruency relative to river condition

Fig. 7.34. One aim of conservation management is to assess the quality of sites that are of high conservation value, or ones that may need conservation attention. To achieve this end, it is useful to know which groups of species are best for guiding this process. There are two perspectives for undertaking this task, each of which is at either end of a spectrum: 1) congruency (where the various species in one group respond in more or less the same way to environmental conditions as the species in another group) and 2) complementarity (where the various species in one group respond differently to environmental conditions compared to the species in another group). Dragonflies and various other groups of insects and plants have been used together as representatives of the state or quality of particular aquatic ecosystems with varying degrees of congruency and complementarity. Different local sites may also be congruent (i.e. they support the same or similar suites of species) or they may be complementary (i.e. they support dissimilar species sets). Such local sites may be parts of a stream or river, or they may be different ponds. Some examples are given here. EPT = Ephemeroptera/Plecoptera/Trichoptera. (Illustration: Robyn Symons).

of the riparian vegetation. This diverse group is also a good surrogate for a wide range of local zygopterans (Gómez-Tolosa *et al.*, 2022a).

In the Kafa Biosphere Reserve, Ethiopian Highlands, the national endemic species *Pseudagrion guichardi*, *Notogomphus cottarellii*, *N. ruppeli*, *Orthetrum kristenseni* and *Atoconeura aethiopica*, which require forested and clear rocky streams or rivers, can act as indicator species for the intact forest core zone of the reserve (Clausnitzer, 2017).

The damming of small rivers in the southern Atlantic Forest in southern Brazil leads to a loss of tree cover and a shift from running to still water. In turn, the change in environmental conditions leads to a huge shift in dragonfly diversity. The best performing indicators of this shift, as expressed in ratios, are the Coenagrionidae/other Zygoptera richness ratio, the Zygoptera/Anisoptera abundance ratio and the Libellulidae/other Anisoptera richness ratio (Araujo *et al.*, 2022).

Spring brooks in Namibia have a diverse yet unique odonate assemblage. There is a species set where all the species together characterize this special ecosystem in this arid landscape. Species include *Pseudagrion kersteni*, *Zosteraeschna minuscula*, *Anax speratus*, *Crocothemis sanguinolenta*, *Orthetrum julia* and *Trithemis stictica*. While *Z. minuscula* is confined to southern Africa, the rest are widespread across parts of Africa, indicating that a set of common and widespread species has value for characterizing an unusual system (Suhling *et al.*, 2006).

7.7.5 Plants as surrogates for dragonflies

Results can be highly variable when plants are used as a surrogate for odonates or aquatic insects. This variation seems to stem from differences in geographical location, spatial scale of measurement, methods of assessment, and focal taxonomic and functional groups.

Among several impoundments in northern Mississippi, USA, vascular plants and adult odonates were assessed for their umbrella potential for wetland conservation. While presenting some technical and statistical challenges, the umbrella approach provided evidence for the close spatial and functional relationship between wetland macrophytes and adult odonates. These findings suggest that such easily measurable assemblages can be efficiently used to set priorities for wetland conservation planning (Bried *et al.*, 2007).

In ponds and lakes across a wide range of landscapes in the UK, macrophyte richness, as measured by macrophyte diversity, was an effective surrogate for odonate, aquatic beetle and mollusc species richness in ponds after accounting for variation in physical variables, water chemistry and surrounding land use. However, only mollusc species richness could be predicted by macrophyte diversity in lakes, with no significant predicted effect on odonates or beetles. Overall, macrophyte diversity is a suitable surrogate of macroinvertebrate biodiversity across diverse landscapes, particularly in ponds and to a lesser extent in lakes, providing valuable insight for assessment and conservation action (Law *et al.*, 2019).

Ponds in South Africa, with a wide variety of plants and physical features, can greatly enhance the heterogeneity of the pondscape in terms of its aquatic insect diversity, including that of odonates. Well-vegetated ponds, both in the open water and along the margins, promote odonate diversity, as long as the marginal vegetation is not too dense and according to season. Odonate species richness increases up to 70% in sedge cover in spring, and up to 40% in bulrush cover in spring and summer, but above these density levels, odonate species richness decreases (Briggs *et al.* 2019b).

7.7.6 Vertebrate surrogates

Swiss ponds of high value for amphibian conservation were used to test whether amphibians are surrogates for odonates, aquatic beetles, aquatic gastropods and aquatic plants. Species richness, conservation value, and community composition of amphibians were weakly congruent with these measures for the other taxonomic groups, while paired comparisons among the five groups showed that for each metric, amphibians had the lowest level of congruence. The conclusion is that sites designated for amphibian conservation do not necessarily provide protection for other components of freshwater biodiversity as a whole. This means that to conserve freshwater species in general, it is necessary to include other taxonomic groups besides amphibians in site prioritization and designation (Ilg and Oertli, 2016).

An extensive survey across the USA indicated that fish and macroinvertebrate assemblages are only weakly correlated with each other, as a result of differential responses to various environmental variables. This finding supports the use of several assemblages as well as the sampling of multiple environmental factors for ecological assessments at large spatial scales (Herlihy *et al.*, 2020).

In a catchment-scale versus site-scale study of fish species and benthic macroinvertebrate families in Brazilian Cerrado streams, it was revealed that macroinvertebrate families and fish species assemblage richness both correlated well with the level of geophysical landscape and site habitat variability when analysed alone. However, macroinvertebrates were more sensitive to human land use and cover. It appears that the more restricted mobility and physiology of aquatic macroinvertebrates makes them more responsive to various stressors compared to the more mobile and physiologically adaptable fish (Macedo *et al.*, 2014).

Costing assessment and monitoring can help decisions on which method to use. Using threshold indicator taxa analysis in Brazil to identify six indicator taxa, four Ephemeroptera/Plecoptera/Trichoptera (EPT) genera and two fish species, the four EPT genera had low negative thresholds to loss of indigenous vegetation along streams, whereas the two fish species had positive thresholds. Thresholds were lower for the 100 m buffers than the 1000 m buffers for EPT. The most sensitive taxon (*Macronema*, Trichoptera) had a threshold of 0% indigenous vegetation loss in the 100 m buffers and 40% in the 1000 m buffers. While the total cost for EPT assessment was US$3616, for fish it was US$1901. Importantly, although fish incurred less cost than EPT, they did not respond negatively to vegetation loss. Furthermore, their positive threshold started at 48% vegetation loss, a level that was highly disruptive to EPT, making fish poor candidates for assessing changes in vegetation cover in Brazil. While EPT assessment costs were 52% more than those of fish, this taxon set was far more sensitive to vegetation loss, especially in the 100 m buffers (Valente-Neto *et al.*, 2021) (Fig. 7.35).

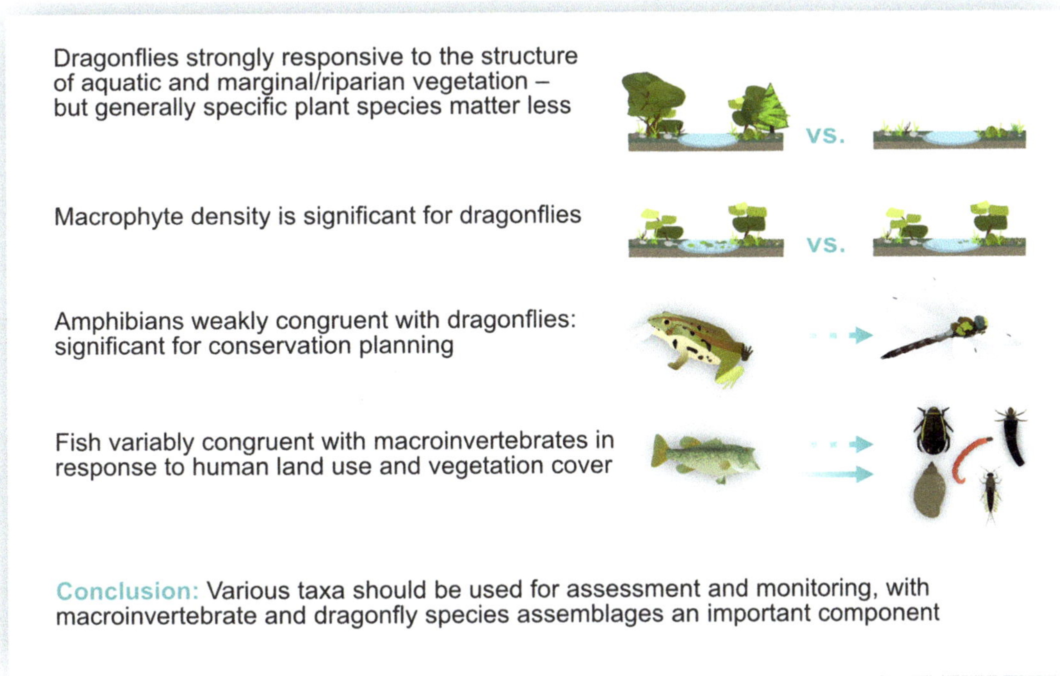

Fig. 7.35. While dragonfly larvae and adults are strongly responsive to vegetation structure and density, plant species identity matters less for the group overall. The exceptions are certain habitat-specialist species. At local scales, amphibians are poorly congruent with dragonflies. Fish and macroinvertebrates are variably congruent with dragonflies. Overall, it is important to sample a range of taxa for effective environmental assessment, notwithstanding that dragonflies are an important group both in terms of responsiveness and practical application. (Illustration: Robyn Symons).

7.8 Dragonfly Species-level Assessment and Monitoring

7.8.1 Aspects of mapping species distribution

Mapping species distributions is an essential part of conservation. The IUCN Red List of Threatened Species uses two spatial metrics to describe species distributions, both of which feed into extinction risk assessments. These are the extent of occurrence (EOO) and area of occupancy (AOO), which respectively represent the spatially large-scale and local-scale distributions of a species. They have a fundamentally different basis, with the EOO describing spread of the degree of risk and the AOO representing population size and local distribution (Fig. 7.36).

A supplementary mapping approach for IUCN Red List assessments is to use area of habitat (AOH). It is the area of potentially suitable ecological conditions for a species to reside within its geographic range. Doing this informs conservation action by targeting areas for field surveys, assessing the proportions of a species' habitat and protected areas, and monitoring habitat loss and landscape fragmentation. This means having good field knowledge of a species' habitat preferences so as to determine the conservation status of a species (Brooks *et al.*, 2019).

An inherent problem here is that maps based on current knowledge can be inaccurate due to errors of omission (the known presence of a species outside of the area being mapped) and commission (the known absence inside the area being mapped).

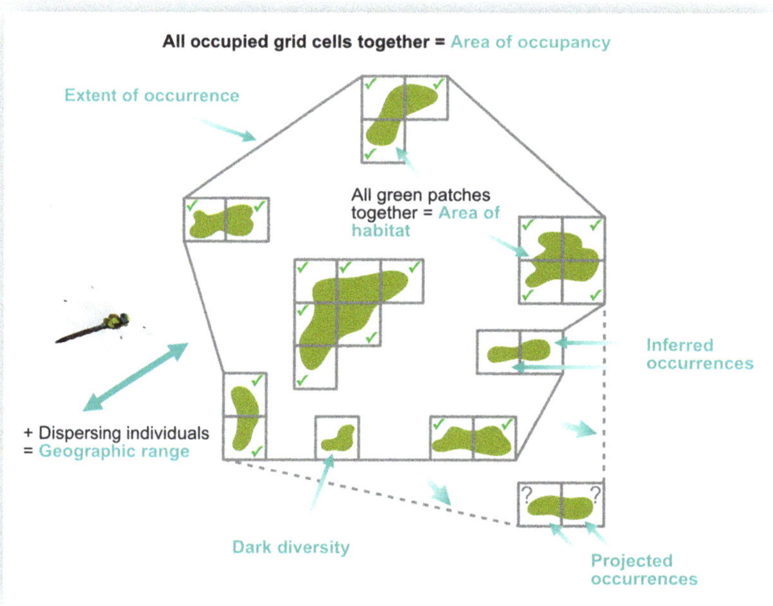

Fig. 7.36. The extent of occurrence (EOO) is the convex polygon that encompasses all known resident populations of a species. It differs from geographic range which refers to the total area where individuals have dispersed to and have been recorded. These individuals may be vagrants, but certain dispersal events may eventually lead to the establishment of a population at a new locality. Within the EOO, resident populations occur in suitable habitat (here in green) which altogether add up to a species' area of occupancy (AOO) (all ticked grid cells). However, optimal habitat (area of habitat: AOH) may be identified within the EOO but there are AOH locations where resident individuals have yet to be recorded but are very likely to occur there. These are inferred occurrences. In turn, suitable habitat may be identified outside the current known EOO. These are projected occurrences, and when found to support a population of a species, would extend the EOO. All these mapping procedures help determine the conservation status of a species. Furthermore, a dark diversity assessment, which is the probability of being absent from a habitat type with suitable ecological conditions, may also help hone conservation status assessment. NB This a conceptual illustration where the grid cell size is not in proportion to the EOO. (Illustration: Robyn Symons).

Although this is not a problem for many large and highly visible organisms and fairly discrete ecosystems like wetlands, it becomes increasingly significant for smaller, visually more cryptic and seasonally apparent organisms like insects. However, this gap is closing, with the information gained from expert knowledge and modelling procedures being complementary (Maddock and Samways, 2000).

Suspected absences of species in their seemingly optimal habitat can also help to hone species distributional assessments. This is being explored in plants based on presence/absence or abundance only, and then calculating for each species the dark diversity probability (the probability of being absent from a habitat type with suitable ecological conditions). Functional-based methods then allow recognition of larger sets of exclusive species, adding dominant species linked to the structure and functions of habitat subtypes, i.e. to the functional centroid. Dark diversity probability is equally distributed between diagnostic and non-diagnostic species identified by functional-based methods. Species with higher dark diversity probability among those associated with the functional centroid can be considered as early warning indicators of changes in habitat conditions. This approach enables species ranking based on measurable variables (functional association, dark diversity probability) and can be standardized for the identification of typical plant species for habitat types, as determined by their plant species (Dalle Fratte *et al.*, 2022).

7.8.2 Species distribution modelling

One approach that correlates environmental variables with known point occurrences of individuals of a species is species distribution modelling (SDM). For small animals like insects, there are means for overcoming data limitations using modern data sources and new modelling approaches. SDMs can be combined with dispersal models and increasingly available phylogenetic and functional data to provide insight into evolutionary shifts in geographic ranges and range-limiting traits over past and contemporary timescales (Mammola *et al.*, 2021). Using *Anax guttatus* as a model organism in India, De *et al.* (2020) suggest that SDM modelling should not rely on any single algorithm (e.g. generalized linear, maximum entropy, random forest and support vector machines in their case) but rather test performance of all available models for focal species and ecological variables, and then choose the best one for building an SDM.

In the core area of the Cerrado region of Brazil, odonate distribution data were used to predict the compositional resemblance based on niche-based species distribution models to address various questions including whether the current conservation network proposals protect odonate assemblages. Methods involved generating odonate species distribution predictions and maps derived from estimated species richness, beta diversity and gap analysis for all species, and comparing these results with current conservation units, land-use patterns and the position of conservation units. Odonate species records were of little value for setting conservation priorities, requiring the additional use of niche-based models. The outcome was that current conservation units overlapped with the areas of high predicted species richness and beta diversity, and were not preserving rare and restricted range species, while there was a direct linear correlation between species range size and the proportion of its range protected in the current network of conservation units. In short, current conservation units and proposed future units did not overlap with areas of high conservation value for odonates, and wide-ranging species were more protected than range-restricted or threatened species (Nóbrega and De Marco, 2011).

Species distribution modelling of odonate species has been used to predict the potential geographic distribution of species based on scattered records, to quantify hotspots for biodiversity and identify protected area gaps, to assess species' environmental requirements and limitations, to quantify the dispersal abilities of species with different life histories, to study niche conservatism among sympatric species, to model the effect of predicted climate change on species distributions, and to assess the efficacy of different modelling approaches (reviewed by Collins and McIntyre (2015)). There are some limitations attached to the use of SDMs for these aims, such as limited taxonomic coverage and limited spatial resolution at fine scales. This means that although SDM is a useful tool for understanding species–environment relationships and has increased our knowledge of variables that limit the distribution of odonates, there are still many more opportunities available for using this approach. SDMs will improve as more input data on the various species become available, and importantly when SDMs are associated with improved ground-referenced data and with follow-up projects that verify (or not) SDM results. To these ends there must be effective interaction between modellers and managers on the ground (Guisan *et al.*, 2013).

The integration of species-specific information on dispersal distances among odonate species improves SDMs in their estimation of potential geographic range changes. This improvement has implications for conservation management, as well as greatly increasing the accuracy of predicted range shifts. This is required for the development of targeted and efficient adaptation strategies for the conservation of threatened species in Europe, such as *Coenagrion mercuriale*, *C. ornatum*, *Leucorrhinia albifrons*, *L. caudalis*, *L. pectoralis* and *Ophiogomphus cecilia* (Jaeschke *et al.*, 2013).

SDMs involving extent of occurrence (EOO) and area of occupancy (AOO) have also been used to refine the IUCN Red List status of the three localized Mexican damselfly species in the genus *Paraphlebia*, with *P. quinta* and *P. zoe* being moved to the Endangered category, and *P. hyalina* to the Data Deficient category based on the reduced geographic range and historical loss of habitat. For *P. quinta*, future climatic projections suggest an initial reduction followed by an expansion in the years 2050 and 2080 in areas with suitable habitat, especially where there is shade, high humidity and perching sites. In the case of *P. zoe*, it is predicted to decrease in its geographic range (Cuevas-Yáñez *et al.*, 2015).

Other approaches to SDM are being applied, such as kriging (e.g. for *Ischnura graellsii* (Campos *et al.*, 2016)) and inverse distance weighting (IDW) (Palacio *et al.*, 2021). IDW is a deterministic, accessible, fast, easy-to-use method for producing species range maps, based on known point presences and absences to construct a surface of probability values for the occurrence of a species in an area. IDW assumes that a species is more likely to be found closer to presence points but farther from absences and the local influence of points (weighted average) diminishes with distance. This is a way of reflecting the naturally discontinuous distribution of populations of a species at high resolutions. IDW maps often coincide with SDM maps and also when environmental and ecological variables used for correlative modelling do not provide more explanatory power than the dispersion patterns of the point data.

Area of habitat (AOH) maps derived from IDW can be used for identifying areas for the protection of species, designing habitat corridors and identifying critical areas for conservation actions. While useful for local-scale applications, they are not intended for every application. The generated AOH maps refer to the known data, useful for systematic conservation planning, but not for exploration of areas likely to contain a species. Also, these range maps are not suitable for macroecological studies, such as determining endemism, or where coarser spatial resolutions are recommended and where expert maps may be more appropriate (Palacio *et al.*, 2021).

Boys *et al.* (2021) combined the predictions of two algorithms, maximum entropy and random forest, to estimate the current and future geographic distributions of two endemic dragonflies (*Gomphurus ozarkensis* and *Somatochlora ozarkensis*) which are nationally listed as Species of Greatest Conservation Need in the mountainous area of south-central USA. Currently suitable areas predicted by both algorithms mostly overlapped for each species, though different environmental variables were significant for predicting their distributions as well as for guiding further field surveys. Future climate change scenarios support maintaining currently suitable areas as still likely to be strongholds for the species, notwithstanding that there may be important local variables that are essential for the survival of the two species. Importantly, combining the outputs of both SDMs at stream-level, as well as perhaps other models, is most effective for fine-tuning projected distributions of these geographically limited and rare species.

SDMs were used to determine the effects of recent bushfires, land-cover change and climatic variables on the distribution of *Synthemis eustalacta*, a species that inhabits Australian forests that had undergone severe burning. While fire-specific variables were of negligible importance for the models, the percentage of tree and non-vegetative cover were most important. Significantly, as dynamic models are trained on occurrence data that match the environmental predictor variables over time, and so estimate variation in habitat suitability as time passes, they outperformed the static ones for determining the dragonfly's responses to fire. Model predictions indicated temporal variation in the area and spatial arrangement of suitable habitat, but suggested no patterns related to habitat expansion, contraction or shifting. The results highlight the efficacy of dynamic modelling to capture spatiotemporal variables such as vegetation cover for this dragonfly in response to fire, as well as providing a novel approach to mapping species distributions with relatively few locality records (Goodman *et al.*, 2023).

Climate change will redefine taxonomic, functional and phylogenetic diversity over space and time, leading to broad-scale species redistributions and possibly changing the structure and composition of assemblages and their host ecosystems. Among 107 European odonate species and using stacked SDMs, Cancellario *et al.* (2022) forecast widespread latitudinal and elevation rearrangements in odonate assemblage composition associated with a broad turnover of traits and evolutionary lineages. However, only body size and flight period partially correlate with observed range shifts. They also forecast variations in habitat availability for the species, with complex changes in their alpha and beta diversity at the continental scale and with respect to taxonomic, functional and phylogenetic diversity. However, the phylogenetic signal is low, with a poor relationship to the considered functional traits. Nevertheless, the assemblages will not reshuffle randomly, with habitat suitability increasing in the northern latitudes and upper elevations, along with contraction of suitable areas in the Mediterranean area for most species. Overall, it seems that odonate assemblages will experience future species sorting, feeding into changes in ecosystem dynamics and functioning. The upshot in terms of assessment and monitoring is that taxonomic diversity is a possible proxy for predicting changes in phylogenetic and functional diversity. These findings inform conservation strategies in that changes in biodiversity are related to non-linear species interactions in and among ecosystems, often involving multiple temporal and spatial scales. This complexity is not easily predictable using correlative methods, while increasing taxonomic, functional and phylogenetic diversity may bring in new evolutionary lineages to certain ecosystems and improve ecosystem resilience through novel functioning (Fig. 7.37).

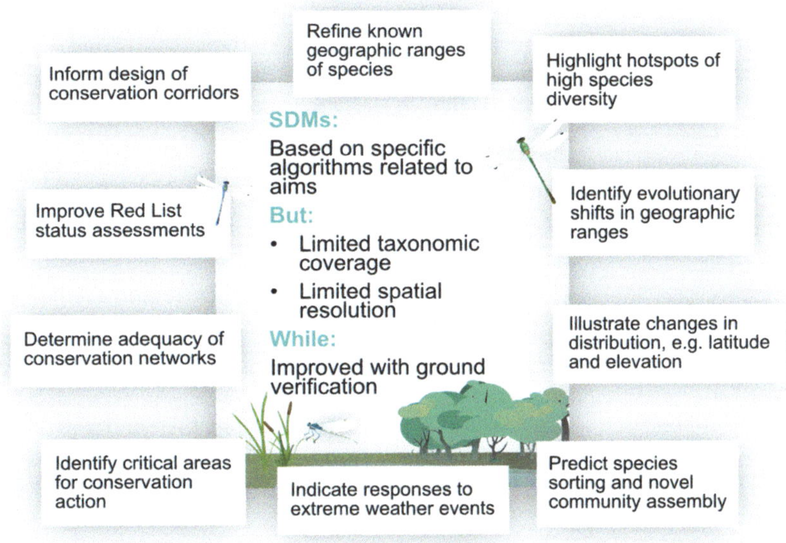

Fig. 7.37. Species distribution models (SDMs) are valuable for exploring species distributions in space and time, as well as addressing many enquiries related to biogeography, climate change and extreme weather events, and conservation. They can employ various algorithms that relate to various facets, including, for example, aspects of species' habitat types and niches, dispersion patterns and species traits. To date, they still have some limitations related to taxonomic coverage and spatial resolution of assessment, though they can be improved following ground verification. SDMs produce various results of great ecological and conservation significance. (Illustration: Robyn Symons).

7.9 Global Assessment and Monitoring

7.9.1 Global assessments that identify important regions and hotspots

7.9.1.1 *Ecoregions*

The world is composed of biogeographic ecoregions determined by ecological and evolutionary patterns. These ecoregions enable global and regional conservation planning, including threatened ecosystems in need of conservation action. Globally, there are over 800 terrestrial ecoregions (based on mammal assessments) (Olson *et al.*, 2001) and 426 freshwater ecoregions (based on fish assessments) (Abell *et al.*, 2008). Each ecoregion blends into the next with transition zones (Fiorentino *et al.*, 2018). While insects were not involved in the estimated demarcation of these ecoregions, they are conceptually useful for insect conservation action, notwithstanding that about 13% of the terrestrial ecoregions are without formally protected areas (Jenkins and Joppa, 2009).

7.9.1.2 *Biodiversity hotspots*

The world has several biodiversity hotspots, each with exceptional concentrations of endemic species and which are rich in threatened biodiversity (Myers *et al.*, 2000). Each hotspot has different mechanisms responsible for their origin while suffering from their own suite of stressors and requiring their own special conservation action (Orme *et al.*, 2005). However, all are exposed to the common stressor of climate change and extreme weather events (Malcolm *et al.*, 2006).

Currently, there are 36 global biodiversity hotspots, each attracting considerable conservation attention. One issue for freshwater and dragonfly conservation is that each hotspot was initially designated using endemic vertebrates and vascular plants. Furthermore, each has its own set of species diversity, level of endemism and number of threatened species. At the very large spatial scale of global freshwater assessment, amphibians have high levels of congruence with several other freshwater taxa, indicating that these are areas that may be of high diversity value for aquatic insects as well (Tisseuil *et al.*, 2013). However, congruency at smaller spatial scales, such as that of ponds and river reaches, can be poor.

The currently designated hotspots are not the only areas of focus for conservation action. It is becoming increasingly clear that there are many more refugia with endemic species and high heterogeneity, with the possibility of more hotspots being recognized (Noss *et al.*, 2015). Furthermore, conservation priority areas do not necessarily equate to areas of high phylogenetic value. This suggests conservation planning should perhaps prioritize actions rather than species and places to reduce species decline through considerations of complementarity among species and ecosystems, costs involved, feasibility of positive action and socio-economic context (Brown *et al.* 2015).

7.9.1.3 *IPBES assessments*

The Intergovernmental Science-Policy Platform on Biodiversity and Ecosystem Services (IPBES) is an independent intergovernmental body comprising over 130 member governments, providing policy makers with objective scientific assessments on the state of knowledge regarding the planet's biodiversity, ecosystems and the contributions they make to people, as well as options and actions to protect and sustainably use these vital natural assets. The Convention on Biological Diversity (CBD) and five other biodiversity-related conventions have adopted IPBES as their science–policy interface. These assessments are clarifying where the most vulnerable species live, the extent to which humanity changes the planet and what drives extinctions, along with estimates of current extinction rates (IPBES, 2019).

7.9.1.4 *IUCN Red List for Ecosystems*

The IUCN Red List for Ecosystems (RLE) (IUCN, 2016a) is a globally recognized standard for the conservation status of ecosystems along the lines of the IUCN Red List of Threatened Species (www.iucnredlist.org). It categorizes communities as Least Concern, Vulnerable, Endangered or Critically Endangered. The protocol used in the RLE introduces the concept of ecological collapse as an analogue of species extinctions. The RLE is based on five criteria: rates of decline in ecosystem distribution, restricted distributions with continuing declines or threats, rates of abiotic environmental degradation, rates of disruption of biotic processes, and quantitative estimates of the risk of ecosystem collapse. The potential for collapse enables an integrated assessment of multiple processes and provides a conceptual anchor for the other four criteria (Keith *et al.*, 2013). This means that it is necessary to demonstrate any downgrading of the risk of ecosystem collapse within the IUCN framework when communities recover to a lower threat level. This important task is challenging not least because the RLE is relatively new, and ecosystem recovery and most recovery programmes have not yet assessed the risk of ecosystem collapse (IUCN, 2016a).

Another challenge is that ecosystems are not always clearly delineated. This last point is being addressed with more specific definitions of the various global ecosystems using better methodology for tracking ecosystem recovery, along with newly mapped global ecosystem typology and an increasing list of ecosystem status assessments.

Despite all these developments, the same obstacles remain as with other programmes aimed at ecosystem restoration – the feasibility of returning an ecosystem to its historic state. This means that there must be an explicit understanding of the chance of ecosystem recovery within the boundaries of realistic and achievable goals that are supported by effective policy (Tierney, 2022).

7.9.1.5 Key Biodiversity Areas

Key Biodiversity Areas (KBAs) are locations contributing greatly to the maintenance of global biodiversity (Eken *et al.*, 2004). Guidelines for KBA identification have been developed (IUCN, 2016b). They are identified nationally using globally standardized criteria and thresholds. Initially developed from the species database of the IUCN Red List, they now also include species richness and compositional turnover suited to lesser-known groups such as insects (Ferrier *et al.*, 2004).

About 15,000 KBAs have been identified globally, aiming to conserve all the species, including insects. They are recognized through a process of prioritization and gap analysis using a set of conservation targets that are implemented nationally or regionally by local conservation agencies through the instrument of a country's National Biodiversity Strategy and Action Plan (Langhammer *et al.*, 2007).

KBAs include criteria on threatened, rare and range-restricted species as well as on ecosystems, meaning that the IUCN-designated threatened ecosystems will be a subset of the delineated KBAs. While both approaches can be used for country comparisons, KBAs are more comprehensive in their comparison of biodiversity status among different regions of the world.

KBAs have been identified mainly based on the presence of trigger species: those species that trigger either vulnerability and/or the irreplaceability criterion. However, many KBAs have not been validated. To test this premise, individuals of freshwater taxa, including larval and adult odonates, were sampled on the Iberian Peninsula and in Moroccan river basins in the Mediterranean Biodiversity Hotspot. These assessments were supplemented with environmental DNA (eDNA) analyses in the Moroccan KBAs. Results showed that there was a mismatch between the purported and actual presence of trigger species. None of the trigger species was found in 43% and 50% of all KBAs surveyed in the two regions respectively, owing largely to flawed or lack of distribution data for the trigger species. This unsatisfactory situation needs addressing with a more conservative approach to KBA selection using local expert knowledge and digital data to determine habitat quality, species distribution and potential threats, consideration of the subcatchment when delineating KBAs boundaries, and additional consideration of terrestrial special areas to delineate conservation areas, as well as periodic field validation (Nogueira *et al.*, 2021).

7.9.1.6 Alliance for Zero Extinction

The Alliance for Zero Extinction (AZE) is a consortium of about 70 non-governmental organizations around the world focusing on prevention of human-induced extinctions. AZE sites are subsets of Key Biodiversity Areas (KBAs) and are based on three strict criteria: 1) each location must support at least one Endangered or Critically Endangered species; 2) the location is irreplaceable in terms of the threatened species that is/are present at the location, or have their population there and nowhere else; and 3) the location has a definable ecological boundary. Insects are included in this initiative, with, for example, five dragonfly species being involved in AZE site designation in Africa (Holland *et al.*, 2011).

7.9.1.7 Essential Biodiversity Variables

Assessing and monitoring biodiversity change at the global level has challenges. Among these are insufficient and uneven coverage across the world, with bias toward developed countries, and typically assessments and monitoring that do not cover large geographic areas over long periods of time. The instigation of Essential Biodiversity Variables (EBVs) aims to address some of these points using *in situ* and *ex situ* data. *In situ* data covers: 1) extensive schemes which maximize geographic coverage, often with input from citizen scientists, and mostly by focusing on certain species sets and their changing abundance; 2) intensive schemes which include many taxa and details at the local scale; and 3) ecological studies, which measure changes in abundance, and which may include changes in ecosystem composition and function. *Ex situ* data comes from the remote sensing of, usually, vegetation cover. Together these two types of data cover four EBVs (with classes within these EBVs): 1) genetic composition; 2) species and abundance (species populations); 3) ecosystem extent and fragmentation (ecosystem structure); and 4) primary productivity (ecosystem function) (Proença *et al.*, 2017), with the aim to better monitor the status and trends of biodiversity to improve biodiversity conservation globally (Schmeller *et al.*, 2015). EBVs are not fixed and will evolve over time as technology improves, such as the use of satellite remote sensing and methods of data sharing coordination (Pettorelli *et al.*, 2016).

Hoban *et al.* (2022) explored the challenges of aggregation, standardization, the costs of operationalizing the genetic composition of EBVs, future directions and opportunities for maximizing the uptake of EBVs globally in research and policy. After reviewing the field of genetics and evaluating how each aspect contributes to global and regional genetic biodiversity monitoring with respect to theory, sampling logistics, metadata, archiving, data aggregation, modelling and technological advances, they propose four genetic EBVs: 1) genetic diversity; 2) genetic differentiation; 3) inbreeding; and 4) effective population size.

Regarding the EBVs associated with ecosystems, developments in remote sensing technology are enabling them to be monitored from space relative to the spatial scale of community composition, plant functional types, vegetation structure, canopy diversity, certain animal groups, fragmentation, human disturbances and as input for biodiversity modelling, as well as earth observation-based variables. These are all key components for biodiversity assessment and monitoring (Reddy *et al.*, 2021) (Fig. 7.38).

7.9.1.8 The Aichi Biodiversity Targets and beyond

The Convention on Biological Diversity (CBD) and over 190 parties (mostly nation states) to the convention agreed to: 1) conserve biodiversity; 2) use its components sustainably; and 3) engage in fair and equitable sharing of benefits arising from genetic resources. In short, the parties agreed to safeguard the planet's biodiversity. Later, the Convention of the Parties then agreed to a Strategic Plan for Biodiversity 2011–2020, which included the Aichi Biodiversity Targets which have five goals (A-E), each with several quantifiable targets: A) address underlying causes of biodiversity loss by mainstreaming biodiversity across government and society (made up of 4 targets); B) reduce the direct pressure on biodiversity and promote sustainable use (6 targets); C) improve the status of biodiversity by safeguarding ecosystems, species and genetic diversity (3 targets); D) enhance the benefits to all from biodiversity and ecosystem services (3 targets); and E) enhance implementation through participatory planning, knowledge management and capacity building (4 targets). The principal instruments for engaging the targets at the national level are over 170 National Biodiversity Strategies and Action Plans. With the passing of 2020, a new strategy is being developed under the name of the Kunming–Montreal Global Biodiversity Framework, as set out by the Conference of the Parties to the Convention on Biological Diversity (UNEP, 2022).

7.9.2 Global species-focused protocols

7.9.2.1 IUCN Red List of Threatened Species

The IUCN Red List of Threatened Species (www.iucnredlist.org) is more than simply a list of threatened species, but of all assessed species with respect to their global conservation status. It is a living database, which is continually updated, both in terms of new species being added, and including re-evaluations of species which had been previously evaluated. Most insect species are not assessed at all, and many of those that have been assessed are categorized as Least Concern. The Red List is arguably the most authoritative information source on the extinction risk faced by species globally. It also makes suggestions to reduce extinction risk. However, among the world's estimated described species to date (more than 2,130,000 species), only about 2% of described invertebrate species have been assessed.

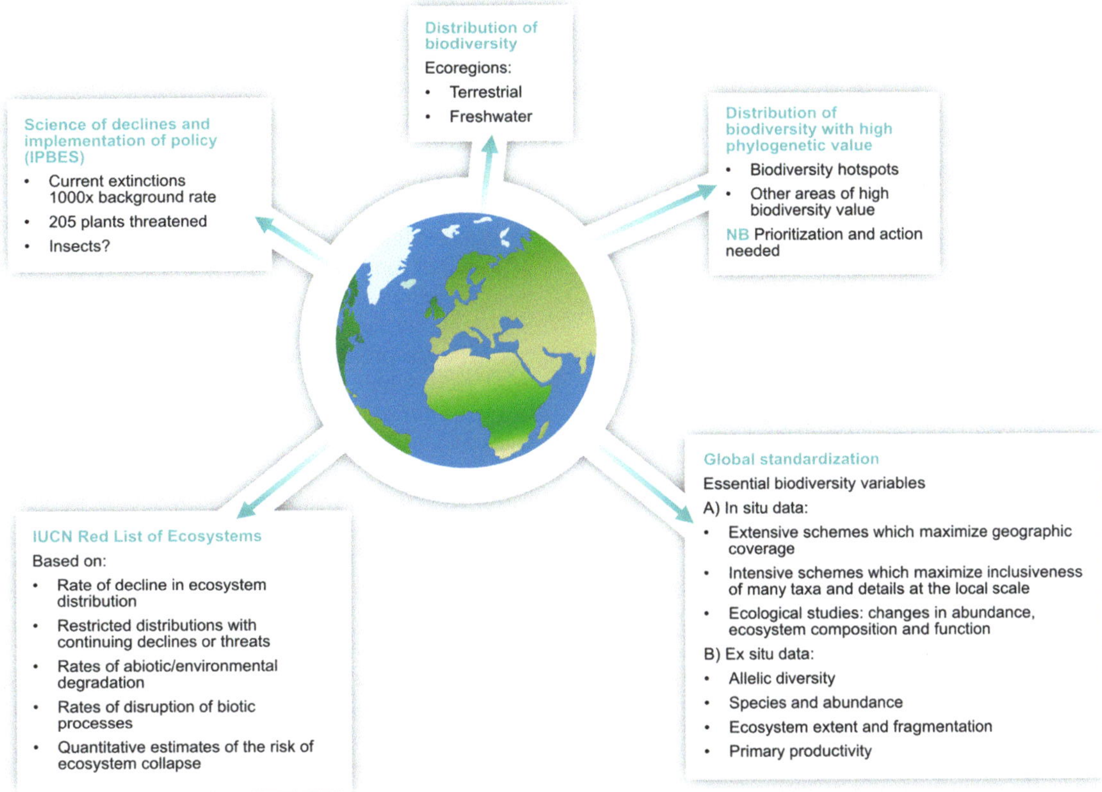

Fig. 7.38. There are several types of global assessments that are being updated and refined, all of which have a bearing on dragonflies and their habitats. (Illustration: Robyn Symons).

The mapping of species distributions is an essential base to red listing. The extent of occurrence (EOO) of a species is used in the red-listing process. It is defined as the area contained within the shortest continuous imaginary boundary that can be drawn as a polygon to encompass all the known, inferred or projected sites of the present occurrence of a species. Red listing, while starting out as a single assessment, usually of a group of species in one taxon or in a geographically well-defined area such as an island, is technically meant to be followed up with further reassessments done ostensibly every 10 years (Rondinini *et al.*, 2014). The aim of the follow-up assessment is to re-measure the current conservation status, either in terms of decline in the face of habitat deterioration, or improvement resulting from conservation action, as well as including the discovery of new sites for the species.

7.9.2.2 Fundamental aspects of red listing

The Red List has several categories and is based on species with resolved taxonomy, meaning that species without a verified scientific name cannot be evaluated. A species is assigned a threat status by going through a short series of decision branches. Red listing is essentially about the total population of a species globally, rather than how 'population' is used in other areas of ecology or conservation biology, where it refers to a group of potentially interbreeding individuals in a local area. Distinct ecological populations, often with different morphologies and genomes, in IUCN terminology are referred to as subpopulations.

Most species do not occupy all the geographic area of their extent of occurrence (EOO), as their optimal habitat rarely occurs throughout the entire EOO polygon (Maes *et al.*, 2015). Placing a grid over the EOO often shows a species occupying certain squares but not others, sometimes resulting in an overestimation of actual geographic range for narrow range and specialist species (Jetz *et al.*, 2008). This is highly significant for many threatened insect species with discontinuous distributions. This means that there are areas specifically occupied by a species which make up its area of occupancy (AOO). However, interpretation of the AOO depends on the species being assessed and the size of the areas being assessed, which at least theoretically could range from infinitely small to infinitely large. In practical terms, however, a species AOO is usually measured in grain size squares of 2 km x 2 km. This grain size is a practical indicator of the level of risk from threats to a species (Keith *et al.*, 2018). However, while occupied 2 km x 2 km grid cells are strongly correlated with the total number of records of species presence, it is likely that only a small percentage of species can be adequately assessed using this method, which means that it is an inadequate method when applied to poorly recorded species. In such cases of undersampled areas, species distribution models may provide more accurate assessments (Marsh *et al.*, 2023).

7.9.2.3 The Red List categories

The first decision branch in the red-listing process is simply about whether a species has been evaluated or not. Where it has been evaluated, then at the next branch a decision is made on whether there are sufficient data for it to be assigned to a particular threat category. If there are insufficient data to assign a category, then it is categorized as Data Deficient (DD). This process does not exclude a re-evaluation in the future when more data become available. Such a re-evaluation is important, as a species might actually be threatened. Each DD species ideally should be assigned a 'justification tag' (i.e. tagged as to why it is DD because of, for example, too few records, only old records are known, uncertain provenance or uncertain threats). The intention here is to ensure that DD species are not forgotten and consequently slip into extinction (Bland *et al.*, 2017).

When there are sufficient data, a species can then be assigned to one of five threat categories: Vulnerable (VU), Endangered (EN), Critically Endangered (CR), Extinct in the Wild (EW) or Extinct (EX). At the other end of the category spectrum and where the species is widespread and common, and assessed as not being threatened at the time of evaluation, it is assigned the category Least Concern (LC).

The category Extinct in the Wild (EW) refers to species that are extinct in their natural habitat but are being maintained in captivity or under carefully managed conditions to boost their numbers with the intention of reintroducing them back into their original provenance once the stressors have been reduced. Most of these species are vertebrates or plants. No dragonfly species fall into this category, in part because they are so difficult to maintain in captivity for any length of time. Most dragonfly reintroductions are from strong source populations into areas where they formerly occurred following the restoration of the receiving sites (Chapter 6, Section 6.6).

Species that are extinct altogether are categorized as Extinct (EX). However, this category is very conservative, so as not to raise any false alarm that a species is gone forever. Most often a species is only assigned to the category EX when, despite extensive and intensive searches, it has not been rediscovered alive after 50 years of focused searching. Only one dragonfly species, *Sympetrum dilatatum*, endemic to the small Atlantic island of St Helena, is currently listed as EX.

When a species has not been re-recorded in its natural habitat for many years yet is suspected of being extinct, but there is still some possibility that it is still extant somewhere, then the species is categorized as Critically Endangered (CR) (Possibly Extinct). The category CR (Possibly Extinct), rather than simply CR, is usually used for insects, even for those that have not been seen for more than 50 years, as they are small and may still be alive at a remote location that has been poorly explored. Furthermore, keeping options open that the species might still be alive encourages more searches for the species (and others at the same time) while also possibly discouraging further land and water transformation. When a species is not EX, EW or LC, then it may be assigned to one of the threatened species categories (CR, EN or VU), or to the category Near Threatened (NT).

An important consideration is that insect conservation is about being pre-emptive, i.e. employing the precautionary principle. In practical terms, it is not always wise to wait until every last shred of scientific evidence is gathered to assign the correct category. Besides, a category revision can always take place following further field evidence. Furthermore, a species may not be threatened at the time of the first assessment, yet there could be a looming threat which if not addressed soon, and if it continues unabated, will threaten the species in the near future. This is when the category NT is applied.

There are three threatened species categories pertaining to species that are still surviving in their natural habitat but are under threat from human activity: Vulnerable (VU), Endangered (EN) and Critically Endangered (CR). Assigning a species to the appropriate category is done using a set of criteria, known as A to E, and are detailed in https://www.iucnredlist.org/assessment/process. These criteria refer to: A = reduction in overall population size based on four types of measurement; B = geographic range and extent of occupation within that range; C = an estimation of global population size relative to decline or fluctuation; D = the overall size of a population in the sense that very small populations are at greatest risk; and E = probability of extinction in the wild.

During the criterion assignment process, the types of threats facing a species are identified as far as is possible. Doing this helps conservation action planning, which first targets the most threatening process/es. This emphasizes that the Red List is not just a species catalogue, but a framework for suggested action. When conservation actions are brought into play and conditions for the current and future survival of threatened insect species are improved, the species may be assigned to a new more optimistic category in response to the improvement, an action known as down-listing.

Although Red List reassessments on each species should be updated every 10 years, this is a challenging task for insects simply because there are so many species relative to the number of assessors. Nevertheless, this does not mean discontinuing the red-listing process for insect species, whether in terms of completely new assessments where feasible, or reassessments where possible, so as to continue to determine the overall status of insect species relative to the threats facing them as time goes by.

A further challenge is that it is difficult to calculate the degree of insect population decline in populations that fluctuate strongly. While populations with high temporal variability do present challenges, they can still be assigned to IUCN Red List categories based on certain criteria. However, even with sophisticated statistical methods, there may still be substantial uncertainties. In response, the Red List Guidelines emphasize acknowledging all uncertainties while encouraging the use of more than one Red List category where plausible. For example, where a species is listed as EN, the assessment may state VU and CR as plausible categories. But the uncertainty should not be used as a reason to avoid assessing a species or listing it as DD (unless plausible categories include extremes like both CR and LC) (Akçakaya *et al.*, 2021) (Fig. 7.39).

7.9.2.4 The Data Deficient debate

Data Deficient (DD) species on the Red List present special challenges in the case of highly speciose groups, such as insects, and especially in megadiverse or high-endemism regions. In response to this dearth of information, an 8-point strategy has been devised to close existing data gaps by: 1) reviving explorative field research on species distribution, abundance and ecology; 2) linking taxonomic research more closely with conservation; 3) improving global biodiversity databases by making the submission of spatially explicit data mandatory for scientific publications; 4) developing a global spatial database on threats to biodiversity to facilitate Red List assessments; 5) automating pre-assessments by integrating distribution and spatial threat data; 6) building capacity in taxonomy, ecology and monitoring in countries with high species richness or endemism; 7) creating species monitoring schemes for lesser-known taxa; and 8) developing sufficient funding mechanisms to reduce reliance on voluntary efforts (Hochkirch *et al.*, 2021). An odonate example of progress on a lesser-known group is the Red List reassessment of the Antillean endemic zygopteran genus *Hypolestes* (Torres-Cambas *et al.*, 2016).

While the IUCN Red List of Threatened Species is a valuable tool for conservation efforts, species classified as DD often mislead practitioners as the categorization does not provide information on extinction risk. Using machine learning-derived probabilities of being threatened by extinction for a set of DD species indicates that DD species as a group may in fact be overall much more threatened than data-sufficient species. An upshot from this analysis suggests, among other things, that the conservation relevance of biodiversity hotspots in South America, for example, may be boosted by up to 20% if DD species are acknowledged. Furthermore, the predicted probabilities for DD species are highly variable across both taxa and regions, implying

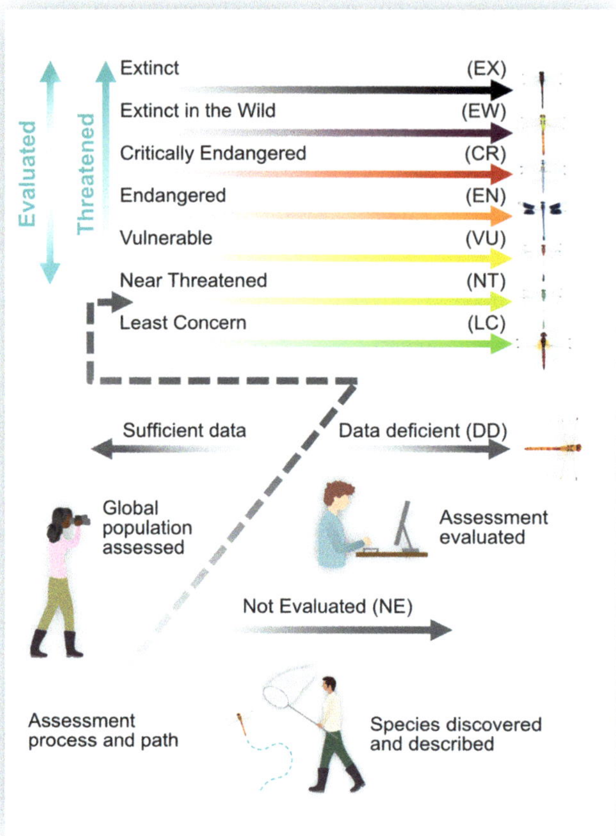

Fig. 7.39. There is an assessment process path for evaluating the Red List conservation status of a species. Once a species has been adequately scientifically described, it is then a candidate for red listing. The global IUCN-designated population is then assessed, notwithstanding that there may also be a national assessment as well. The field assessment is then vetted by an IUCN-appointed evaluator. The decision branch is whether there is sufficient data to make a sound decision on the conservation status of the species. If not, then the species is assigned to Data Deficient. If there is sufficient data, it is then assigned to either Least Concern or to one of the threat categories, according to prescribed criteria and increasing levels of threat of imminent extinction. The evaluation signals the degree and urgency of conservation action that is required. Only one dragonfly species is categorized as Extinct, while Extinct in the Wild does not yet apply to the group. (Illustration: Robyn Symons).

current Red List-derived indices and priorities may be biased (Borgelt *et al*., 2022).

7.9.2.5 Other Red List discussion points

With regard to extent, the Red List recognizes that most conservation action takes place at the local scale. This means that there is value in gaining knowledge and galvanizing conservation action for species that are threatened locally through the process of regional or national red listing. Guidelines for regional and national red listing are given on https://www.iucnredlist.org/assessment/process. Genetically and morphologically distinct, and often isolated populations of a certain species, termed subspecies by the IUCN, can be important at regional and national scales. These subspecies can also be Red List-evaluated in the same way as other nominal subspecies, on the assumption that scientific names have been designated.

By way of a corollary, there is a significant aspect of red listing insect species relative to the conservation of other species. As all species are listed with equal consideration on the Red List, in effect it means that any one insect species is given an equal degree of attention to any other species, no matter how big or small they are. This is highly beneficial for improving the profile of insects, especially in the eyes of the media and civil society. Insect species that may have lived in relative obscurity in the past, now benefit from considerable media exposure, often leading to a great increase in local conservation activity. However, this elevated exposure must be seen against the background that the IUCN does not prescribe any particular conservation action, but rather supplies the best scientific evidence available on the best way forward for the conservation of certain species. Furthermore, red listing often feeds into local community guardianship of insect species in addition to improving awareness of the plight of certain insect species, as well as of insects in general.

While most Red List assessments are conducted by professional conservationists formerly linked to the IUCN, there is a growing opportunity to include assessment activities into academic curricula. This would provide sorely needed data from megadiverse areas of the world, where there are many students able to take part (Böhm *et al.*, 2022).

Furthermore, the IUCN Red List of Threatened Species must be frequently updated, become more taxonomically and geographically representative, and be consistent within and among taxonomic groups, although currently hampered in achieving these ends by limited resources. While there are models and automated calculations that support Red List assessments, for instance, using citizen science or remote sensing data to predict extinction risk, there is still a major research implementation gap in the application of these methods. This gap could be bridged by providing assessors with easy access to the most relevant tools, hands-on training and better communication. Additionally, extra effort is needed to develop relevant methods to prioritize assessments or to better predict extinction risk (Cazalis *et al.*, 2021).

The current IUCN Red List Strategic Plan includes revising the IUCN Red List classification schemes for coding Threats, Habitats and Conservation Measures as a high priority for the current quadrennium. The IUCN SSC Red List Technical Working Group is examining the Threats Classification Scheme, where the larger Red List assessment community will provide feedback on how best to refine the current scheme. Possible improvements include adding new threat codes, redefining existing codes or better characterizing the existing codes in the classification scheme. Feedback from the Red List community and work with the Conservation Measures Partnership will together revise the scheme. The revisions will use standard classifications for describing the threats and conservation actions as proposed by Salafsky *et al.* (2008) to be useful for practitioners in three ways. First, the classifications will help practitioners identify the direct threats occurring at a particular site or impacting on a certain species and the actions needed to counter the threats. Second, the classifications will enable managers and policy makers to determine the frequency of threats or actions across projects at various organizational or spatial scales for setting priorities and allocating resources. Classifications will also enable researchers to combine threats and action summaries with other information for more detailed analyses of conservation situations and solutions. Third, and most importantly, the classifications will facilitate cross-project learning by permitting practitioners to precisely describe the chains linking targets, threats, contributing factors and actions. These chains are then shared on common databases of conservation practice. This will enable practitioners to share and compare experiences more easily, leading to the development of a more systematic science of biodiversity conservation.

7.9.2.6 The Red List Index

Emanating from the IUCN Red List of Threatened Species is the Red List Index (RLI). The RLI assesses change in the aggregate extinction risk of species and is a key indicator for tracking progress toward some of the Aichi targets and many of the proposed post-2020 Global Biodiversity Framework Targets.

The RLI was developed to meet the challenges posed for various taxonomic groups that are relatively poorly known (Butchart *et al.*, 2007). It uses a subset of all the global species known in a set drawn from potential candidates within the whole taxon and considered to represent the conservation status of the group as a whole. The RLI uses a representative sample of 1500 species from selected species-rich taxonomic groups, can be used to compare groups and can also be used to assess how the group as a whole is changing over time. As regards dragonflies globally, the RLI estimates that 10% of dragonflies are threatened with extinction, which is a relatively low percentage compared to that of other animal groups that have been similarly assessed (Clausnitzer *et al.*, 2009).

The RLI can be used as a tool for determining habitat change based on species threat assessment data. The RLIs of many species from several major taxa (including odonates) and the habitat types of each species were calculated for the whole of Finland to give a habitat-based RLI. The RLI changed over time (2000–2010) and among species grouped by their habitats, and were found to be negative for all habitat types except for forests and rural biotopes, which were stable. The trend for dragonflies was slightly positive over that time. The habitat-based RLI showed a different but complementary perspective to the ecosystem-based IUCN Red List for Ecosystems (RLE) (see Section 7.9.1.4). While some habitats did not improve, their constituent species showed positive trends due to other factors like climate change. The habitat-based RLI bridges the gap between the taxon-based RLI and the RLE (Juslén *et al.*, 2016) (Fig. 7.40).

7.9.2.7 IUCN Green Status of Species

The IUCN Red List of Threatened Species focuses on extinction risk, and suggests ways of reducing it – a beginning to a species conservation programme. Following conservation action, evidence is then needed to show that there has indeed been species population recovery. However, quantification of recovery requires a set of criteria that clearly illustrate the level of species recovery and the degree of long-term persistence. This task was tackled by the IUCN through the development of a Green Status of Species ('Green Status'), to complement the Red List (Akçakaya *et al.* 2018). The Green Status is a practical tool that focuses on species with high conservation dependence yet high potential for conservation improvement through effective conservation action. Species recovery can be ascribed in three ways: 1) viability of the global population of a species through habitat security, genetic adaptive capacity and resilience in the face of environmental change; 2) function of the global population in respect of its natural ecological interactions and functional contribution within an ecosystem; and 3) representation in terms of it being present in communities and ecosystems throughout its geographic range.

The Green Status framework quantifies recovery by defining a set of four conservation metrics that demonstrate: 1) current conservation effectiveness (conservation legacy); 2) dependence of a species' survival on conservation action (conservation dependence); 3) quantification of gains achieved from conservation action in the medium term (conservation gain); and 4) quantification of a species' recovery potential to achieve maximum recovery in the long term (recovery potential). Together these metrics illustrate the success of conservation action without dependence only on change in threat status, such as moving from Critically Endangered (CR) to Endangered (EN), or Endangered (EN) to Vulnerable (VU). The Green Status, like the Red List, does not set priorities, but rather has a positive outlook through informing on the level of conservation success, and through the development of categories that complement those of the Red List to provide a more complete assessment of a species' conservation status and its prospects for its continued survival.

The Green Status framework was tested on 181 species (including five dragonfly species: *Stylurus (Gomphus) flavipes*, *Leucorrhinia pectoralis*, *Sympetrum fonscolombii* (not globally threatened, only locally), *Syncordulia gracilis* and *S. venator*) representing diverse taxa, life histories, biomes and IUCN Red List categories (extinction risk). This led to seven species recovery categories: 1) fully recovered; 2) slightly depleted; 3) moderately depleted; 4) largely depleted; 5) critically depleted; 6) extinct in the wild; and 7) indeterminate. Of the tested species, 59% were assigned the categories largely or critically depleted. There was a negative relationship between extinction risk and species recovery score, although variation was considerable. Some species in lower risk categories were assessed as further from recovery than those at higher risk, highlighting that species recovery is conceptually different from extinction risk. These results emphasize the value of the Green Status approach to better understand species conservation status. Although extinction risk does not predict conservation legacy, conservation dependence or conservation gain, it does predict recovery potential. Very few of the tested species were categorized as zero across all four of the conservation metrics, indicating that conservation action can improve or at least maintain species status for most of the species. These findings led to an updated assessment framework, the IUCN Green Status of Species Standard, with the option of using a dynamic baseline to assess future conservation effectiveness in the short

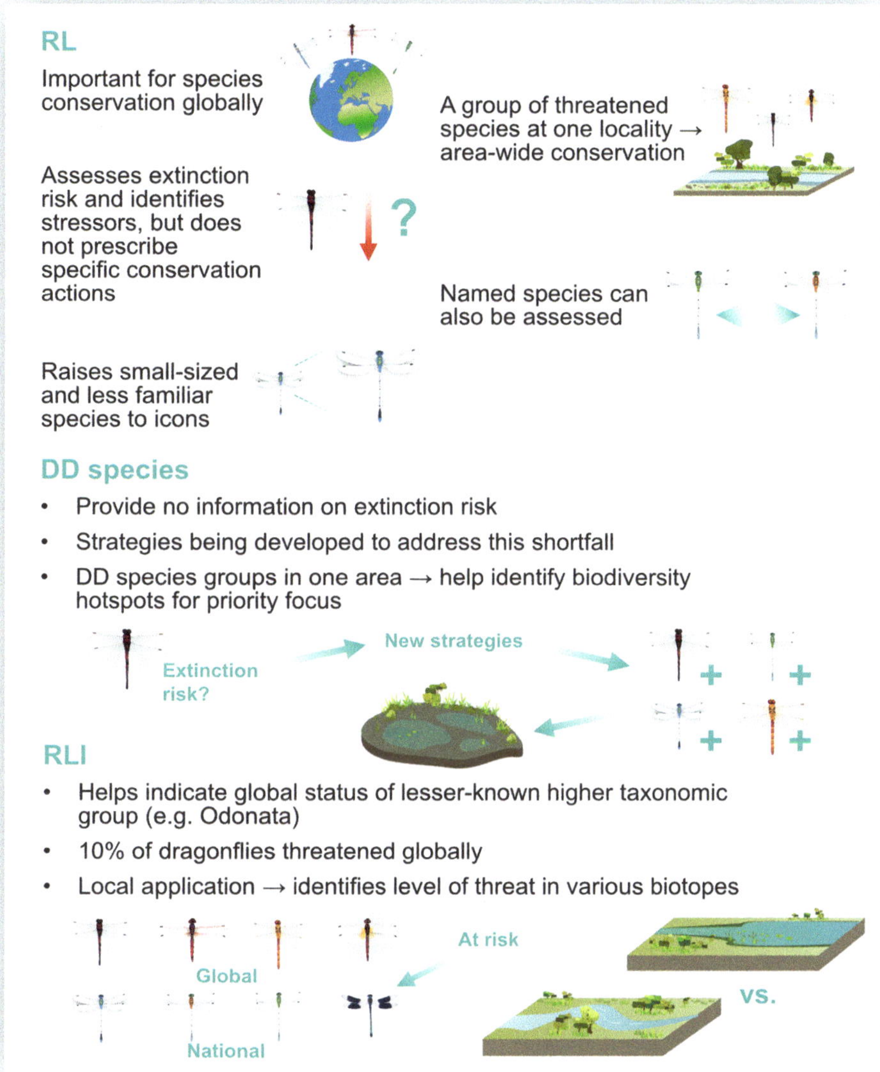

Fig. 7.40. The IUCN Red List of Threatened Species (RL) is an important instrument globally for assessing the extinction risk that specific species face. However, it is not prescriptive in terms of the precise conservation action that should be undertaken, as it simply identifies the type and level of threat to each species. Conservation of the species is then addressed by local and national authorities and interest groups. An important facet of red listing is that it raises a small-sized and less familiar species, such as a dragonfly, to iconic status. This helps galvanize conservation action on behalf of the threatened species. A high threat status of several species together at one location can promote area-wide conservation. Named 'subspecies' (morphological operational taxonomic units (MOTUs)) may also receive special conservation attention. Data Deficient (DD) species have no epithet of the extinction risk that they are experiencing. This challenge is being addressed with new strategies. Groups of DD species in the same location can, with the help of machine learning, be used to identify biodiversity hotpots and so feed into area-focused conservation action. The Red List Index (RLI) is a globally selected subset of species from a lesser-known taxonomic group employed to assess the overall threat to the group globally. In the case of dragonflies, 10% are globally threatened. When the RLI is then applied nationally, as in the case of dragonflies in Finland, certain biotopes can be identified where the various species are less or more threatened. (Illustration: Robyn Symons).

term to avoid misleading results that were uncovered in a few cases, and redefines 'short term' as 10 years to better align with conservation planning (Grace *et al.*, 2021, 2022) (Fig. 7.41).

7.9.3 Global assessments relating to abundance

7.9.3.1 Global assessment and monitoring of dragonfly abundance

Besides dragonflies being recorded and assessed for the global IUCN Red List of Threatened Species, there are other global initiatives where they are the focus of attention. One issue globally is that there is sparse data on their abundance. Efforts to overcome this challenge will involve citizen scientists as well as professional scientists. Although assessing odonates is commonplace in some parts of the world, current data do provide a global perspective on odonate abundance patterns and trends. In response, a global volunteer network to count odonate individuals is proposed to improve organizational structure, advertising and recruiting, as well as data collection, submission and synthesis (Bried *et al.*, 2020). An application that provides information on abundance involving simple 15-minute counts, which is much easier than formal transects and can be done anywhere, is being developed, first in Europe (R. van Grunsven, pers. comm.).

7.9.3.2 BioTIME

BioTIME is a global database set up to monitor changes in the abundance of species over time. There are to date about 9 million species abundance records

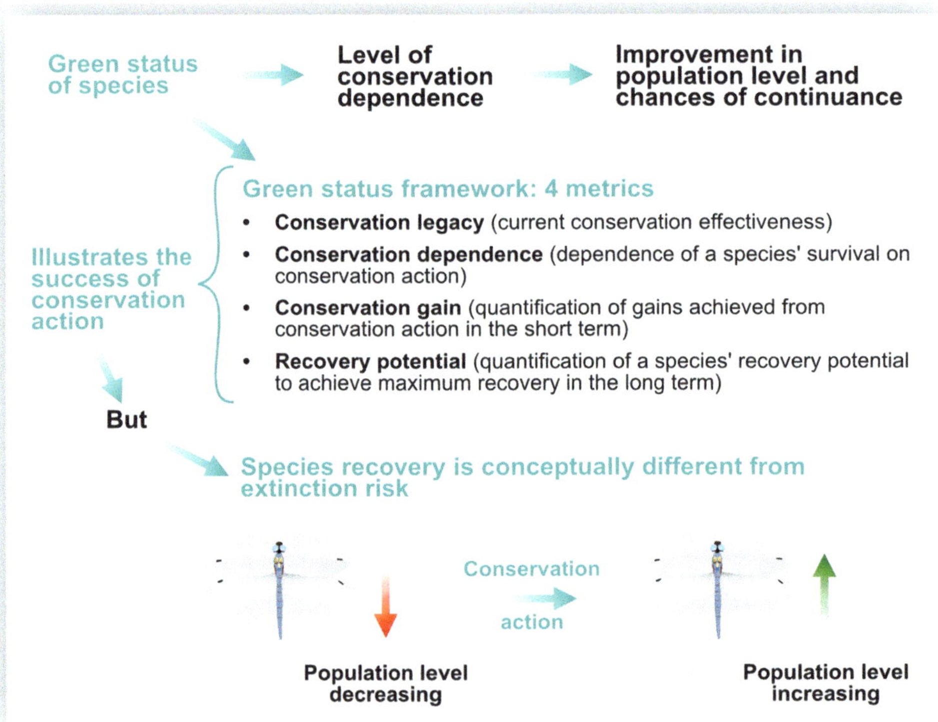

Fig. 7.41. The Green Status of Species ('Green Status') is a practical tool focusing on species with high conservation dependence yet having high potential for conservation improvement through effective conservation action. The Green Status framework quantifies recovery by defining a set of four conservation metrics relating to conservation legacy, conservation dependence, conservation gain and recovery potential. Together these metrics illustrate the success of conservation action without necessarily a change in Red List threat status. Tests on species with varying levels of threat illustrated that some species in lower risk categories were assessed as further from recovery than those at higher risk, highlighting that species recovery is conceptually different from extinction risk. (Illustration: Robyn Symons).

on more than 44,400 species (Dornelas *et al.*, 2018). These data are especially valuable when they are collectively gathered on many species. Doing this provides insight into species that are responding to various drivers or stressors, both negative (e.g. climate change, extreme weather, habitat loss, landscape fragmentation, pollution) and positive (e.g. transition from conventional agriculture to agroecology or sustainable forestry, restoration of ecosystems, improving functional connectivity).

7.9.4 Regional and national dragonfly assessment

7.9.4.1 What regional and national assessments have to offer

The highly dynamic natural environment and its constant flux means that conditions on the ground vary greatly from one area to another, and from one time to the next. Owing to the smallness of insects, we need to consider many spatial scales from microhabitat to landscape, as well as regional events, such as catchment change, and global occurrences, especially in response to climate oscillations, ongoing change and extreme weather events. Furthermore, spatial attributes are intimately related to temporal changes. A microhabitat, for example, is subject to short-term changes such as daily temperature change and short-term weather events. Mid-term changes involve events such as shortening of the fire interval, increasing flood intensity, and periodic and even ongoing high-intensity grazing activity. Long-term changes are those emanating from climate change such as increased temperature and storm intensity, prolonged drought and for coastal populations, sea-level rise.

Regional and national assessments have great value in providing detailed focus on local issues relating to both time and space, from microhabitat to larger geographic areas. Often, this is because citizen scientists are involved and provide many motivated eyes and feet on the ground, contributing an immense amount of detailed knowledge. The collective information gathered can provide considerable insight into the natural dynamics and changes taking place that would otherwise not be available to the analytical scientific fraternity.

While citizen science is playing an ever-increasing role in contributing to important databases (Deacon *et al.*, 2023), there can be levels of great variability (measurement error) or bias (e.g. spatiotemporal clustering) of the gathered data. Analytic approaches and machine learning can overcome these challenges (Bird *et al.*, 2014) leading to standardized outputs. However, it is essential to match the needs of science and public involvement with the right type of citizen science project and the right method of public participation. When this is done, the outcome is sound science, generation of high-quality data and the provision of conservation solutions (McKinley *et al.*, 2017), as well as the ability to compare data sets over time (Boieiro *et al.*, 2023).

Climate-induced range shifts are often recorded over long time-periods. However, the absence of standardized collecting procedures in some historical data sets introduces bias and skews data which can result in misleading conclusions. When different methods are compared to account for uneven recording effort using a single British Odonata data set and tested against changes in temperature over time to select the most accurate method, there is variation in results from the different methods. Geographic range statistics also exhibit different biases in varying recording effort. This means that a combination of existing methods is required to account for temporal variation in recording effort, which focuses on random resampling of the more intensely recorded time. A range statistic based on a gamma frequency distribution, which avoids the inherent bias of existing statistics, is suggested as a descriptor for range margins. When the most robust methods that control for uneven recording effort are combined with the most robust range statistics describing the range shifts, odonates were shown to track the isotherms between 1960–2005, indicating that this method should be used to predict future odonate range shifts (Hassall and Thompson, 2010).

There is also a more fundamental issue: that of taxonomic verification. Chandran and Sherif (2022) point this out in the case of records for Kerela, India, where citizen scientists post their observations on open-access biodiversity portals, such as iNaturalist and the India Biodiversity Portal, believing that their contributions will further assist the research and conservation of species and their habitats. However, certain records were not taxonomically verified, yet were earlier synthesized as an assessment of odonate species in Kerala and published in open access publications without due consideration to records already verified by expert opinion. They emphasize that if specimens are not available for taxonomic verification, detailed photographs showing taxonomic features are necessary to establish the presence of the species in the region.

With a sound citizen science base an adaptive management scheme for wetland restoration can be used to make scientific predictions, as was done in a Japanese floodplain. Data was provided by citizen scientists on odonate distribution patterns at a large site along with ecological information such as behaviour and habitat preferences. Based on these data, the species recovery trajectory at the wetland restoration site was predicted, and then tested by monitoring the species recovery to test the prediction. Colonization by odonate species at the restored site could be predicted using species prevalence on the regional scale based on the nestedness rule, i.e. by those species in the local species pool according to their traits and based on the premise that common species are widespread while rare species tend to occur only at the most suitable sites (Kadoya and Washitani, 2007).

There is an increasingly loud call for the adoption of a Global Biodiversity Framework that transforms conservation and sustainable use efforts worldwide. A major reason for this is the biodiversity paradox: that biodiversity is largely concentrated in the tropics, while human, institutional and financial resources are mostly located at higher latitudes, both north and south. Addressing the biodiversity paradox requires a global expansion and mobilization of human, institutional and financial resources, especially for aquatic insects (Valente-Neto *et al.*, 2022).

The IUCN Species Survival Commission (SSC) recognizes that most conservation action occurs at the national or local level. Various strategies are being applied to address the issue by partners of Reverse the Red: a global movement that involves strategic cooperation and science-based action to improve the chances of survival of wild species and natural ecosystems. The SSC contributes to Reverse the Red through two primary instruments: National Species Specialist Groups and Centres for Species Survival. Through building on existing expert networks and catalysing efforts with established local institutions, the aim is to greatly expand the capacity to implement conservation action at the national level and reverse the negative trends indicated by the Red List of Threatened Species and the Red List for Ecosystems (RLE) (Rodriguez *et al.*, 2022).

Red List Indices (RLIs) (see Section 7.9.2.6) are now being used to measure biodiversity trends at the national level. They have been tested in three countries (Brazil, Finland, South Africa) and encompass several taxa, including dragonflies. National trends of different taxa can be compared with each other and across countries. While dragonfly trends overall are largely stable in South Africa, they are improving in Finland, but decreasing in Brazil. Dragonflies are in much better shape, for example, than cycads in South Africa, indicating great differences among taxa within countries (Raimondo *et al.*, 2022).

7.9.4.2 Examples of regional dragonfly species assessments

There are regional Red List assessments and analyses of dragonflies in Europe (Kalkman *et al.*, 2010), the Mediterranean basin (Riservato *et al.*, 2009), continental Africa (Dijkstra *et al.*, 2011; Clausnitzer *et al.*, 2012), and North Africa (Samraoui *et al.*, 2010), as well as publications that highlight threats facing certain rare and localized species (e.g. Reels *et al.* (2012) for South Asia and Myanmar).

The highest overall odonate species richness and standing water species diversity occur in western and central Europe, while the highest strictly running water species diversity is in south-western France and parts of the Iberian Peninsula. The highest number of endemic species occurs in south-western France and the Iberian and Balkan Peninsulas. However, there is a mismatch between the patterns of threatened species listed in the European Union Habitats Directive and those listed on the European Red List. Threatened red-listed species occur mainly in the Mediterranean, while species legally protected by the EU are concentrated in western and central Europe. This mismatch has an historical origin in that species listed in the Habitats Directive were mostly selected in the 1970s and 1980s when water quality was poor in western and central Europe. However, since the 1990s, water and habitat quality in those parts of Europe has improved, while during the same period, pressure on Mediterranean aquatic habitats has increased greatly (Kalkman *et al.*, 2018).

7.9.4.3 Examples of national and local dragonfly species assessments

National red lists of dragonflies are often included in a country's National Biodiversity Strategy and Action Plan, and periodically updated as more information becomes available on species local distribution and as threat assessments are reassessed or fine-tuned. An earlier UK national dragonfly assessment (Daguet *et al.*, 2008) was later followed by the *State of Dragonflies in Britain and Ireland* (Taylor *et al.*, 2021). Similarly, an earlier South African national dragonfly assessment (Samways, 2006) was followed by an updated one (Samways and Simaika, 2016).

Determining conservation priorities depends on having sufficient spatial data. This was tested for Mexican odonates. The country's odonate inventory is considered near complete, with 205 anisopteran and 151 zygopteran species, but there had not been any formal exercise to test this. Sample-based and coverage-based rarefaction curves were used to verify distribution data. The curves indicated 100% completeness of species lists for all odonates and the two suborders separately. However, species' distribution data was patchy, with areas either well (e.g. central Mexico) or poorly (e.g. coast of Guerrero and Oaxaca) sampled. These results are valuable for researchers to focus odonate sampling in support of conservation assessment of threatened species, to refine knowledge on geographic range shifts with rising global temperatures, and to increase public interest and awareness in protected and touristic areas (Cuevas-Yañez *et al.*, 2017).

Some countries are so large that only parts of them have been assessed for the threat status of dragonflies and their habitats. The north-eastern USA (Virginia to Maine) is an odonate diversity hotspot while also a region of historical and growing threats to freshwater ecosystems, making it urgent that the region's 228 resident odonate species are protected. Using a prioritization framework, modified from NatureServe's approach for assessing conservation status, a list was drawn up of species of combined vulnerability and regional management responsibility. A single regional vulnerability metric (R-rank) was assigned reflecting each species' extinction risk in the region based on the following components: three rarity factors (geographic range extent, area of occupancy (AOO) and habitat specificity); one threat factor (vulnerability of occupied habitats); and one trend factor (relative change in range size), along with the degree of endemicity (percentage of the species' USA and Canadian range that falls within the region) as a proxy for regional responsibility. Overall, 18% of the region's odonate fauna is threatened (R1 and R2). In terms of ecosystems that are all exceptionally rich in species, peatlands, low-gradient streams and seeps, high-gradient headwaters and larger rivers are conservation priorities (White *et al.*, 2015).

The wildlife action plans for the USA aim to prevent wildlife from going into decline and becoming threatened to the extent where recovery becomes unlikely. The implementation of action plans depends on establishing Species of Greatest Conservation Need (SGCN), those species with small or declining populations or other characteristics that make them vulnerable. As of 2010, over half the states neglected to assign any odonate SGCN. Western and southern states listed proportionately fewer odonate SGCN species than states of the Great Lakes, the Mid-Atlantic and New England regions, seemingly reflecting geographic patterns of legal authority, available information and involvement by odonatologists. As dragonflies are a highly tractable group for inclusion on the SGCN list, the involvement of dragonfly specialists would be highly effective at improving the list. Additionally, there should be: 1) use of existing conservation lists; 2) inferences from field guides and other reviewed literature; 3) determination of patterns of endemism; and 4) application of species distribution modelling (Bried and Mazzacano, 2010).

Well-illustrated and accurate field guides are essential to take the step from 'catching to watching', and to support dragonfly conservation initiatives (Turnhout and Halffman, 2023). Similarly, dragonfly recognition and data capturing mobile applications, like Dragonfly Hunter CZ for species in Czechia, also have great conservation value for the citizen science community (Ozana *et al.*, 2019) (Fig. 7.42).

7.10 Currently Available Large Databases and Freshwater Conservation Initiatives

The value of large-scale collaborative management databases lies in the sharing, analysis and development of conservation strategies (Costello *et al.*, 2014). These databases are available to many users, and store information from inventories, often through crowdsourced data from citizen scientists (Lin *et al.*, 2015). Important among these databases are: the Index to Organism Names (www.organismnames.com), listing species and subspecies names, and authorities; Map of Life (www.mol.org), illustrates species geographical range maps and occurrence data, as well as trends in biodiversity knowledge, distribution and conservation;

Fig. 7.42. While global assessments and monitoring schemes provide information on planetary environmental changes taking place, conservation action largely takes place at regional, and especially at national and more local levels. These conservation initiatives, besides involving governmental and non-governmental agencies, draw upon valuable data gathered by citizen scientists, especially when there is standardized data collection and verification of species identities. Well-illustrated and accurate field guides, both hard copy and electronic, are essential for assuring that this happens. The 'biodiversity paradox' (that the species-richest parts of the world are the areas where human, institutional and financial resources are less prevalent) is being addressed by IUCN's Reverse the Red initiative. National assessments of species often form part of a country's National Biodiversity Strategy and Action Plan and can be updated with more exact information on species' statuses and population trends, and a major conservation platform. (Illustration: Robyn Symons).

Catalogue of Life (www.catalogueoflife.org), an authoritative and comprehensive database of known species of animals, plants, fungi and micro-organisms; Global Biodiversity Information Facility (www.gbif.org), makes available scientific data on biodiversity via the internet using web services; Encyclopedia of Life (https://eol.org), an online collaborative effort that is documenting all living species known to science; Tree of Life (www.tolweb.org/tree/), data provided by biologists and naturalists globally, providing information on biodiversity, features of organisms and their evolutionary history; re3data (https://www.re3data.org/), a database of images of organisms; iNaturalist (www.inaturalist.org), a photographic database of all living organisms, and for crowdsourcing; Wildscreen (www.wildscreen.org), a comprehensive multimedia guide to the world's threatened species; GenBank (www.ncbi.nlm.nih.gov/genbank/), an annotated collection of all publicly available nucleotide sequences and their protein translations; Barcode of Life Data System (www.boldsystems.org), a database of genctic sequences focusing on storing and analysing DNA sequences; and the Biodiversity Heritage Library (www.biodiversitylibrary.org), makes biodiversity literature openly available to the world.

As regards freshwater, there is the: Alliance for Freshwater Life (https://allianceforfreshwaterlife.org), which aims to engage with conservation professionals, civil society, government and policy makers, and to place freshwater species and their associated ecosystems on the global agenda as legitimate targets for conservation action; Freshwater Information Platform (www.freshwaterplatform.eu), which provides up-to-date information on freshwater science as well as an array of research resources and tools for the assessment and management of freshwater systems, with a focus on Europe, but with a global perspective; and the Global Lake Ecological Observatory Network (https://gleon.org/), a global community of practice, unified in the goal to improve biodiversity observations in freshwaters, while contributing to building the global capacity for monitoring freshwater biodiversity and conducting assessments.

A global living checklist of dragonfly species is the World Odonata List (https://www.pugetsound.edu/puget-sound-museum-natural-history/biodiversity-resources/insects/dragonflies).

References

Abell, R., Thieme, M.L., Revenga, C., Bryer, M., Kottelat, M. *et al.* (2008) Freshwater ecoregions of the world: a new map of biogeographic units for freshwater biodiversity conservation. *BioScience* 58, 403–414. DOI: 10.1641/B580507.

Acquah-Lamptey, D., Kyerematen, R. and Owusu, E.O. (2013) Using odonates as markers of the environmental health of water and its land related ecotone. *International Journal of Biodiversity and Conservation* 5, 761–769.

Adu, B.W., Amusan, B.O. and T.O. Oke, T.O. (2019) Assessment of the water quality and Odonata assemblages in three waterbodies in Ilara-Mokin, south-western Nigeria. *International Journal of Odonatology* 22, 101–114. DOI: 10.1080/13887890.2019.1593889.

Adu, B., Dada, D. and Tunwase, V. (2022) An ecological study of freshwater ecosystem and its colligation to Odonates assemblages in Ipogun, Southwest Nigeria. *Bulletin of the National Research Centre* 46, e86. DOI: 10.1186/s42269-022-00774-4.

Akçakaya, H.R., Bennet, E.L., Brooks, T.M., Grace, M.K., Heath, A. *et al.* (2018) Quantifying species recovery and conservation success to develop an IUCN Green List of species. *Conservation Biology* 32, 1128–1138. DOI: 10.1111/cobi.13112.

Akçakaya, H.R., Hochkirch, A., Bried, J.T., van Grunsven, R.H.A, Simaika, J.P. *et al.* (2021) Calculating population reductions of invertebrate species for IUCN Red List assessments. *Journal of Insect Conservation* 25, 377–382. DOI:10.1007/s10841-021-00303-0.

Allen, K.A. and Thompson, D.J. (2014) Population size and survival estimates for the rare damselflies, *Coenagrion mercuriale* and *Ischnura pumilio*. *Insect Conservation and Diversity* 7, 241–251. DOI: 10.1111/icad.12047.

Angert, A.L., Crozier, L.G., Rissler, L.J., Gilman, S.E., Tewksbury, J.J. and Chunco, A.J. (2011) Do species' traits predict recent shifts at expanding range edges? *Ecology Letters* 14, 677–689. DOI: 10.1111/j.1461-0248.2011.01620.x.

Araujo, B.R., Pinto, Â.P. and Padial, A.A. (2022) Influence of landscape homogenization due to river damming on dragonfly (Odonata) community structuring in a subtropical forest in the southern Atlantic Forest. *Ecohydrology* 15, e2419. DOI: 10.1002/eco.2419.

Arguel, L., Denis, A.S., Danflous, S., Gouix, N., Santoul, F. *et al.* (2022) Detection and monitoring of riverine dragonfly of community interest (Insecta: Odonata): proposal for a standardised protocol based on exuviae collection. *Diversity* 14, e728. DOI: 10.3390/d14090728.

Arjona, Y., Arribas, P., Salces-Castellano, A., López, H., Emerson, B.C. and Andújar, C. (2023) Metabarcoding for biodiversity inventory blind spots: a test case using the beetle fauna of an insular cloud forest. *Molecular Ecology* 32, 6130–6146. DOI: 10.1111/mec.16716.

Azevedo-Santos, V.M., Frederico, R.G., Fagundes, C.K., Pompeu, P.S., Pelicice, F.M. *et al.* (2019) Protected areas: a focus on Brazilian freshwater biodiversity. *Diversity and Distributions* 25, 442–448. DOI: 10.1111/ddi.12871.

Bacca, J.C., Cararo, E.R., Lima-Rezende, C.A., Martins, R.T., Macedo-Reis, L.E. *et al.* (2023) Land-use effects on aquatic macroinvertebrate diversity in subtropical highland grasslands streams. *Limnetica* 42, 215–231. DOI: 10.23818/limn.42.16.

Baird, I.R.C. and Burgin, S. (2023) Monitoring within-patch, flying season population dynamics of an endangered mire-dwelling dragonfly, *Petalura gigantea* (Odonata: Petaluridae). *Australian Zoologist* 43, 368–389. DOI: 10.7882/AZ.2023.038.

Balázs, A., Šipoš, J., Matúšová, Z., Hamerlík, L. Novikmec, M. and Svitok, M. (2022) Comparison of conservation values among man-made aquatic habitats using Odonata communities in Slovakia. *Biologia* 77, 2549–2561. DOI:10.1007/s11756-022-01129-0.

Ball-Damerow, J.E., M'Gonigle, L.K. and Resh, V.H. (2014) Changes in occurrence, richness, and biological traits of dragonflies and damselflies (Odonata) in California and Nevada over the past century. *Biodiversity and Conservation* 23, 2107–2126. DOI: 10.1007/s10531-014-0707-5.

Barnes, M.A., Klymus, K. and Yamanaka, H. (2021) Environmental DNA innovations for conservation. *Frontiers in Ecology and Evolution* 9, e785077. DOI: 10.3389/fevo.2021.785077.

Barry, D.A., Liardon, J.-L., Paccaud, P., Klaus, P., Gujja Shaik, N.S. *et al.* (2019) A low-cost, autonomous mobile platform for limnological investigations, supported by high resolution mesoscale airborne imagery. *PLoS ONE* 14, e0210562. DOI: 10.1371/journal. pone.0210562.

Batista, J.D., Ferreira, V.R.S., Cabette, H.S.R, de Castro, L.A., De Marco, P. and Juen, L. (2021) Sampling efficiency of a protocol to measure Odonata diversity in tropical streams. *PLoS ONE* 16, e0248216. DOI: 10.1371/journal. pone.0248216.

Bazzi, G., Galimberti, A., Foglini, C., Bani, L., Bazzi, L. *et al.* (2023) Odonate diversity of a highly urbanised region: an annotated checklist of the damselflies and dragonflies (Insecta, Odonata) of Lario and Brianza (Lombardy, N Italy). *Biodiversity Data Journal* 11, e111358. DOI: 10.3897/BDJ.11.e111358.

Belletti, B., Renaldi, M., Buijse, A.D., Gurnell, A.M. and Mosselman, E. (2015) A review of assessment methods for river hydromorphology. *Environmental Earth Sciences* 73, 2079–2100. DOI: 10.1007/s12665-014-3558-1.

Berger, E., Haase, P., Schäffer, R.B. and Sundermann, A. (2018) Towards stressor-specific macroinvertebrate indices: which traits and taxonomic groups are associated with vulnerable and tolerant taxa? *Science of the Total Environment* 619–620, 144–154. DOI: 10.1016/j.scitotenv.2017.11.022.

Berquier, C., Orsini, A., Ferrat, L. and Andrei-Ruiz, M.-C. (2016) "*Odonata* Community Index-Corsica" (OCIC): a new biological Index based on adult odonate populations for assessment of the ecological status of watercourses in Corsica. *Ecological Indicators* 66, 163–172. DOI: 10.1016/j.ecolind.2016.01.022.

Bird, M.S., Mlambo, M.C., Wasserman, R.J., Dalu, T., Holland, A.J. *et al.* (2019) Deeper knowledge of shallow waters: reviewing the invertebrate fauna of southern African temporary wetlands. *Hydrobiologia* 827, 89–121. DOI: 10.1007/s10750-018-3772-z.

Bird, T.J., Bates, A.E., Lefcheck, J.S., Hill, N.A., Thomson, R.J. *et al.* (2014) Statistical solutions for error and bias in global citizen science datasets. *Biological Conservation* 173, 144–154. DOI: 10.1016/j.biocon.2013.07.037.

Biswas, J.K., Mondal, B., Priyadarshini, P., Abhilash, P.C., Biswas, S. and Bhatnagar, A. (2022) Formulation of Water Sustainability Index for India as a performance gauge for realizing the United Nations Sustainable Development Goal 6. *Ambio* 51, 1569–1587. DOI: 10.1007/s13280-021-01680-1.

Bland, L.M., Bielby, J., Kearney, S., Orme, C.D.L., Watson, J.E.M. and Collen, B. (2017) Toward reassessing data-deficient species. *Conservation Biology* 31, 531–539. DOI: 10.1111/cobi.12850.

Boersma, K.S., Bogan, M.T., Henrichs, B.A. and Lytle, D.A. (2014) Invertebrate assemblages of pools in arid-land streams have high functional redundancy and are resistant to severe drying. *Freshwater Biology* 59, 491–501. DOI: 10.1111/fwb.12280.

Böhm, M. and Collen, B. (2015) Toward equality of biodiversity knowledge through technology transfer. *Conservation Biology* 29, 1290–1302. DOI: 10.1111/cobi.12529.

Böhm, M., Waldien, D.L., Setliff, G.P., Abenis, K.O., Aguirre, L.F. *et al.* (2022) Catalyzing Red List assessments of underrepresented taxa through partner networks and student engagement. *Diversity* 14, e723. DOI: 10.3390/d14090723.

Boieiro, M., Antunes, S., Figueiredo, H., Soares, A., Lopes, A. *et al.* (2023) Standardised inventories of lepidopterans and odonates from Serra da Estrela Natural Park (Portugal): setting the scene for mountain biodiversity monitoring. *Biodiversity Data Journal* 11, e99558. DOI: 10.3897/BDJ.11.e99558.

Boneto, D.D., Batista-Silva, V.F., Soares, J.A.C., Kashiwaqui, E.A.L. and Oliveira, I.A.D.V. (2017) Immature Odonata-Anisoptera in the Iguatemi river basin, upper Paraná River, Mato Grosso do Sul State, Brazil. *Acta Scientiarum* 39, 211–217. DOI: 10.4025/actascibiolsci.v39i2.30769.

Borgelt, J., Dorber, M., Høiberg, M.A. and Verones, F. (2022) More than half of data deficient species predicted to be threatened by extinction. *Communications Biology* 5, e679. DOI: 10.1038/s42003-022-03638-9.

Boys, W.A., Siepielski, A.M., Smith, B.D., Patten, M.A. and Bried, J.T. (2021) Predicting the distributions of regional endemic dragonflies using a combined model approach. *Insect Conservation and Diversity* 14, 52–66. DOI: 10.1111/icad.12444.

Brasil, L.S., Oliveira-Júnior, J.M., Calvão, L.B., Carvalho, F.G., Monteiro-Júnior, C.S. *et al.* (2018) Spatial, biogeographic and environmental predictors of diversity in Amazonian Zygoptera. *Insect Conservation and Diversity* 11, 174–184. DOI: 10.1111/icad.12262.

Brasil, L.S., Luiz de Lima, E., Spigoloni, Z.A., Ribeiro-Brasil, D.R G. and Juen, L. (2020a) The habitat integrity index and aquatic insect communities in tropical streams: a meta-analysis. *Ecological Indicators* 116, e106495. DOI: 10.1016/j.ecolind.2020.106495.

Brasil, L.S., Luiza-Andrade, A., Batista, L. Calvão, L.B., Dias-Silva, K. *et al.* (2020b) Aquatic insects and their environmental predictors: a scientometric study focused on environmental monitoring in lotic environmental. *Environmental Monitoring and Assessment* 192, e194. DOI: 10.1007/s10661-020-8147-z.

Brasil, L.S., de Andrade, A.F.A., Ribeiro, B.R., Spigoloni, Z.A., Juen, L. and De Marco Jr, P. (2021) A niche-based gap analysis for the conservation of odonate species in the Brazilian Amazon. *Aquatic Conservation: Marine and Freshwater Ecosystems* 31, 1150–1157. DOI:10.1002/aqc.3599.

Brazner, J.C., Danz, N.P., Trebitz, A.S., Niemi, G.J., Regal, R.R. *et al.* (2007) Responsiveness of Great Lakes wetland indicators to human disturbances at multiple spatial scales: a multi-assemblage assessment. *Journal of Great Lakes Research* 33 (Special Issue 3), 42–66.

Bried, J.T. and Hinchliffe, R.P. (2019) Improving taxonomic resolution in large-scale freshwater biodiversity monitoring: an example using wetlands and Odonata. *Insect Conservation and Diversity* 12, 9–17. DOI: 10.1111/icad.12323.

Bried, J.T and Mazzacano, C.A. (2010) National review of state wildlife action plans for Odonata species of greatest conservation need. *Insect Conservation and Diversity* 3, 61–71. DOI: 10.1111/j.1752-4598.2010.00081.x.

Bried, J.T. and Siepielski, A.M. (2018) Opportunistic data reveal widespread species turnover in *Enallagma*

damselflies at biogeographical scales. *Ecography* 41, 958–970. DOI: 10.1111/ecog.03419.
Bried, J.T., Herman, B.D. and Ervin, G.N. (2007) Umbrella potential of plants and dragonflies for wetland conservation: a quantitative case study using the umbrella index. *Journal of Applied Ecology* 44, 833–842. DOI: 10.1111/j.1365-2664.2007.01299.x.
Bried, J.T., D'Amico, F. and Samways, M.J. (2012a) A critique of the dragonfly delusion hypothesis: why sampling exuviae does not avoid bias. *Insect Conservation and Diversity* 5, 398–402. DOI: 10.1111/j.1752-4598.2011.00171.x.
Bried, J.T., Hager, B.J., Hunt, P.D., Fox, J.N. Jensen, H.J. and Vowels, K.M. (2012b) Bias of reduced-effort community surveys for adult Odonata of lentic waters. *Insect Conservation and Diversity* 5, 213–222. DOI: 10.1111/j.1752-4598.2011.00156.x.
Bried, J.T., Hassall, C., Simaika, J.P., Corser, J.D. and Ware, J. (2015) Directions in dragonfly applied ecology and conservation science. *Freshwater Science* 34, 1020–1022. DOI: 10.1086/682393.
Bried, J., Ries, L., Smith, B., Patten, M., Abbott, J. *et al.* (2020) Towards global volunteer monitoring of odonate abundance. *BioScience* 70, 914–923. DOI: 10.1093/biosci/biaa092.
Briers, R.A. and Biggs, J. (2003) Indicator taxa for the conservation of pond invertebrate diversity. *Aquatic Conservation: Marine and Freshwater Ecosystems* 13, 323–330. DOI: 10.1002/aqc.576.
Briggs, A., Pryke, J.S., Samways, M.J. and Conlong, D.E. (2019a) Macrophytes promote insect conservation in artificial ponds. *Aquatic Conservation: Marine and Freshwater Ecosystems* 29, 1190–1201. DOI: 10.1002/aqc.3157.
Briggs, A.J., Pryke, J.S., Samways, M.J. and Conlong, D.E. (2019b) Complementarity among dragonflies across a pondscape in a rural landscape mosaic. *Insect Conservation and Diversity* 12, 241–250. DOI: 10.1111/icad.12339.
Brooks, T.M., Pimm, S.L., Akçakaya, R., Buchanan, G.M., Butchart, S.H.M. *et al.* (2019) Measuring terrestrial Area of Habitat (AOH) and its utility for the IUCN Red List. *Trends in Ecology and Evolution* 34, 977–986. DOI: 10.1016/j.tree.2019.06.009.
Brown, C.J., Bode, M., Venter, O., Barnes, M.D., McGowan, J. *et al.* (2015) Effective conservation requires clear objectives and prioritizing actions, not places or species. *Proceedings of the National Academy of Sciences of the USA* 112, e4342. DOI: 10.1073/pnas.1509189112.
Brumm, K.J., Hanks, R.D., Baldwin, R.F. and Peoples, B.K. (2022) A scale-linked conservation planning framework for freshwater ecosystems. *Landscape Ecology* 37, 2589–2605. DOI: 10.1007/s10980-022-01505-w.
Buchner, D., Haase, P. and Leese, F. (2021) Wet grinding of invertebrate bulk samples: a scalable and cost-efficient protocol for metabarcoding and metagenomics. *Metabarcoding and Metagenomics* 5, 73–81. DOI: 10.3897/mbmg.5.67533.
Buczyńska, E. and Buczyńska, P. (2019) Survival under anthropogenic impact: the response of dragonflies (Odonata), beetles (Coleoptera) and caddisflies (Trichoptera) to environmental disturbances in a two-way industrial canal system (central Poland). *PeerJ* 6, e6215. DOI: 10.7717/peerj.6215.
Bush, A., Compson, Z.G., Monk, W.A., Porter, T.M., Steeves, R. *et al.* (2019) Studying ecosystems with DNA metabarcoding: lessons from biomonitoring of aquatic macroinvertebrates. *Frontiers in Ecology and Evolution* 7, e434. DOI: 10.3389/fevo.2019.00434.
Butchart, S.H.M., Akçakaya, H.R., Chanson, J., Baillie, J.E.M., Collen, B. *et al.* (2007) Improvements to the Red List Index. *PLoS ONE* 2, e140. DOI: 10.1371/journal.pone.0000140.
Calvão, L.B., Brito, J.S., Ferreira, D., Cunha, E.J., Oliveira-Junior, J.M.B. and Juen, L. (2023) Effects of the loss of forest cover on odonate communities in eastern Amazonia. *Journal of Insect Conservation* 27, 205–218. DOI: 10.1007/s10841-022-00444-w.
Campos, F., Velasco, T., Sanz, G., Casanueva, P., Albuqueque, M.T.D. and Antunes, I.M.H.R. (2016) *Ischnura graellsii* (Insecta: Odonata) a water pollution biovulnerability indicator: probability mapping using spatial uncertainty. *River Research and Applications* 32, 483–489. DOI: 10.1002/rra.2839.
Cancellario, T., Miranda, R., Baquero, E., Fontaneto, D., Martínez, A. and Mammola, S. (2022) Climate change will redefine taxonomic, functional, and phylogenetic diversity of Odonata in space and time. *npj Biodiversity* 1, e1. DOI: 10.1038/s44185-022-00001-3.
Cano-Cobos, Y., Bota-Sierra, C.A. and Medoza-Penagos, C. (2023) Ten new records of Odonata for Colombia (Coenagrionidae, Aeshnidae). *Biota Colombiana* 24, e1112. DOI: 10.21068/c001.
Carew, M.E., Miller, A.D. and Hoffmann, A.A. (2011) Phylogenetic signals and ecotoxicological responses: potential implications for aquatic biomonitoring. *Ecotoxicology* 20, 595–606. DOI: 10.1007/s10646-011-0615-3.
Carraro, L., Mächler, E., Wüthrich, R. and Altermatt, F. (2020) Environmental DNA allows upscaling spatial patterns of biodiversity in freshwater ecosystems. *Nature Communications* 11, e3585. DOI: 10.1038/s41467-020-17337-8.
Carvalho, F.G., Oliveira Junior, J.M.B., Farias, A.P.J. and Juen, L. (2013) Use of the ABC curve as a method to detect the effect of anthropogenic change on the Odonata (Insecta) assembly. *Interciencia* 38, 516–522.
Catling, P.M. (2005) A potential for the use of dragonfly (Odonata) diversity as a bioindicator of the efficiency of sewage lagoons. *The Canadian Field-Naturalist* 119, 233–236.

Cayrou, J. and Céréghino, R. (2005) Life cycle phenology of some aquatic insects: implications for pond conservation. *Aquatic Conservation: Marine and Freshwater Ecosystems* 15, 559–571.

Cazalis, V., Di Marco, M., Butchart, S.H.M., Akçakaya, H.R., González-Suárez, M. *et al.* (2021) Bridging the research-implementation gap in IUCN Red List assessments. *Trends in Ecology and Evolution* 37, 359–370. DOI: 10.1016/j.tree.2021.12.002.

Cezário, R.R., Firme, P.P., Pestana, G.C., Vilela, D.S., Juen, L. *et al.* (2021) Sampling methods for dragonflies and damselflies. In: Santos, J.C. and Fernandes, G.W. (eds) *Measuring Arthropod Biodiversity*. Springer Nature, Switzerland, pp. 223–240. DOI: 10.1007/978-3-030-53226-0_9.

Chadd, R. and Extence, C. (2004) The conservation of freshwater macroinvertebrate populations: a community-based classification scheme. *Aquatic Conservation: Marine and Freshwater Ecosystems* 14, 597–624. DOI: 10.1002/aqc.630.

Chahl, J., Chitsaz, N., McIvor, B., Ogunwa, T., Kok, J.-M. *et al.* (2021) Biomimetic drones inspired by dragonflies will require a systems based approach and insights from biology. *Drones* 5, e24. DOI: 10.3390/drones5020024.

Chandran, A.V. and Sherif, K.M. (2022) Comments on 'The dragonflies and damselflies (Odonata) of Kerala: status and distribution'. *Journal of Threatened Taxa* 14, 21282–21284. DOI: 10.11609/jott.7989.14.6.21282-21284.

Chang, X., Zhai, B., Wang, M. and Wang, B. (2007) Relationship between exposure to an insecticide and fluctuating asymmetry in a damselfly (Odonata, Coenagriidae). *Hydrobiologia* 586, 213–220. DOI: 10.1007/s10750-007-0620-y.

Cheri, C.R. and Finn, D.S. (2023) Odonata as indicators? Dragonflies and damselflies respond to riparian conditions along Ozark spring streams. *Hydrobiology* 2, 260–276. DOI: 10.3390/ hydrobiology2010017.

Chiriboga-Ortega, R., Van der Heyden, C., Burgos, M.E.S., Ortega, S., Oña, T. *et al.* (2021) Molecular diversity of dragonflies in three high-elevation Andean tropical lakes through DNA barcoding. *Biotropica* 53, 354–358. DOI: 10.1111/btp.12927.

Chirwa, E.R. and Chilima, L. (2017) Use of benthic macroinvertebrate indices as bioindicators of ecosystem health for the detection of degraded landscapes in Malawi. *International Journal of Agriculture, Forestry and Fisheries* 5, 128–134. ISSN: 2381-4373.

Chovanec, A. (2022) The assessment of the dragonfly fauna (Insecta: Odonata) as a tool for the detailed typological characterisation of running waters. *Acta ZooBot Austria* 158, 129–147.

Chovanec, A. and Waringer, J. (2001) Ecological integrity of river-floodplain systems: assessment by dragonfly surveys (Insecta: Odonata). *Regulated Rivers: Research and Management* 17, 493–507. DOI: 10.1002/rrr.664.

Chovanec, A., Waringer, J., Raab, R. and Laister, G. (2004) Lateral connectivity of a fragmented large river system: assessment on a macroscale by dragonfly surveys (Insecta: Odonata). *Aquatic Conservation: Marine and Freshwater Ecosystems* 14, 163–178. DOI: 10.1002/aqc.596.

Chovanec, A., Schindler, M., Waringer, J. and Wimmer, R. (2015) The dragonfly association index (Insecta: Odonata): a tool for type-specific lowland rivers. *River Research and Applications* 31, 627–638. DOI: 10.1002/rra.2760.

Clausnitzer, V. (2017). Dragonflies and damselflies (Odonata) at the Kafa Biosphere Reserve. In: Committee on NABU's Biodiversity Assessment at the Kafa Biosphere Reserve (eds) *The Nature and Biodiversity Conservation Union (NABU)*, Berlin, Addis Ababa, pp. 160–189.

Clausnitzer, V., Kalkman, V.J., Ram, M., Collen, B., Baillie, J.E.M. *et al.* (2009) Odonata enter the biodiversity crisis debate: the first global assessment of an insect group. *Biological Conservation* 142, 1864–1869. DOI: 10.1016/j.biocon.2009.03.028.

Clausnitzer, V., Dijkstra, K.-D.B., Koch, R., Boudot, J.-P., Darwall, W.R.T. *et al.* (2012) Focus on African freshwaters: hotspots of dragonfly diversity and conservation concern. *Frontiers in Ecology and Environment* 10, 129–134. DOI: 10.1890/110247.

Collins, S.D. and McIntyre, N.E. (2015) Modeling the distribution of odonates: a review. *Freshwater Science* 34, 1144–1158. DOI: 10.1086/682688.

Copetti, D. and Erba, S. (2024) A bibliometric review on the Water Framework Directive twenty years after its birth. *Ambio* 53, 95–108. DOI: 10.1007/s13280-023-01918-0.

Costello, M.J., Appeltans, W., Bailly, N., Berendsohn, W.G., de Jong, Y. *et al.* (2014) Strategies for the sustainability of online open-access biodiversity databases. *Biological Conservation* 173, 155–165. DOI: 10.1016/j.biocon.2013.07.042.

Craves,J.A., O'Brien, D.S., and Marvin, D.A. (2020) New population of the rare dragonfly *Ophiogomphus howei* (Odonata: Gomphidae) in Southern Michigan, United States. *Journal of Insect Science* 20, 1–9. DOI: 10.1093/jisesa/ieaa125.

Cuevas-Yáñez, K., Rivas, M., Muñoz, J. and Córdoba-Aguilar, A. (2015) Conservation status assessment of *Paraphlebia* damselflies in Mexico. *Insect Conservation and Diversity* 8, 517–524. DOI: 10.1111/icad.12132.

Cuevas-Yañez, K., Espinosa-Rivera, J.C., Martínez-Falcón, A.P. and Córdoba-Aguilar, A. (2017) Are all Mexican odonate species documented? An assessment of species richness. *Systematics and Biodiversity* 15, 253–258. DOI: 10.1080/14772000.2016.1249529.

Cunningham-Minnick, M.J., Meyer, T.B. and Crist, T.O. (2019) Shifts in dragonfly community structure across aquatic ecotones. *International Journal of Odonatology* 22, 121–133. DOI: 10.1080/13887890.2019.1615006.

Curry, C.J., Zhou, X. and Baird, D.J. (2012) Congruence of biodiversity measures among larval dragonflies and caddisflies from three Canadian rivers. *Freshwater Biology* 57, 628–639. DOI: 10.1111/j.1365-2427.2011.02724.x.

Daguet, C.A., French, G.C. and Taylor, P. (2008) The Odonata Red Data List for Great Britain. *Species Status* 11, 1–34. Joint Nature Conservation Committee, Peterborough, UK.

Dallas, H.F. (2021) Rapid bioassessment protocols using aquatic macroinvertebrates in Africa: considerations for regional adaptation of existing biotic indices. *Frontiers in Water* 3, e628227. DOI: 10.3389/frwa.2021.628227.

Dallas, H. and Mosepele, B. (2020) Spatial variability of aquatic macroinvertebrate assemblages in the Okavango Delta, Botswana: considerations for developing a rapid bioassessment tool. *African Journal of Aquatic Science* 45, 350–363. DOI: 10.2989/16085914.2019.1704215.

Dalle Fratte, M., Caccianiga, M., Ricotta, C. and Cerabolini, B.E.L. (2022) Identifying typical and early warning species by the combination of functional-based diagnostic species and dark diversity. *Biodiversity and Conservation* 31, 1735–1753. DOI: 10.1007/s10531-022-02427-4.

Darshetkar, A., Patwardhan, A. and P. Koparde (2023) A comparison of four sampling techniques for assessing species richness of adult odonates at riverbanks. *Journal of Threatened Taxa* 15, 22471–22478. DOI: 10.11609/jott.7259.15.1.22471-22478.

Davies, R., von Hardenberg, A. and Geary, M. (2018) Recapture rates and habitat associations of *Leucorrhinia dubia* (White-faced Darter; Vander Linden) on Fenn's and Whixall Moss, Shropshire, UK. *Journal of the British Dragonfly Society* 34, 89–101.

De Almeida, M.V.O., Pinto, Â.P., do L. Carvalho, A. and Takiya, D.M. (2013) When rare is just a matter of sampling: unexpected dominance of clubtail dragonflies (Odonata, Gomphidae) through different collecting methods at Parque Nacional da Serra do Cipó, Minais Gerais, Brazil. *Revista Brasileira de Entomologia* 57, e4. DOI: 10.1590/S0085-56262013005000042.

De Almeida, M.V.O., Pinto, Â.P. and Carvalho, A.L. (2021) Digitizing primary data on biodiversity to protect natural history collections against catastrophic events: the type material of dragonflies (Insecta: Odonata) from Museu Nacional of Brazil. *Biodiversity Information Science and Standards* 5, e75284. DOI: 10.3897/biss.5.75284.

De Marco Júnior, P., Batista, J.D. and Cabette, H.S.R. (2015) Community assembly of adult odonates in tropical streams: an ecophysiological hypothesis. *PLoS ONE* 10, e0123023. DOI: 10.1371/journal.pone.0123023.

de Vries, J., Kraak, M.H.S., Skeffington, R.A., Wade, A.J. and Verdonschot, P.F.M. (2021) A Bayesian network to simulate macroinvertebrate responses to multiple stressors in lowland streams. *Water Research* 194, e116952. DOI: 10.1016/j.watres.2021.116952.

De, K., Ali, S.Z., Dasgupta, N., Uniyal, V.P., Johnson, J.A. and Hussain, S.A. (2020) Evaluating performance of four species distribution models using Blue-tailed Green Darner *Anax guttatus* (Insecta: Odonata) as model organism from the Gangetic riparian zone. *Journal of Threatened Taxa* 12, 16962–16970. DOI: 10.11609/jott.6106.12.14.16962-16970.

Deacon, C. and Samways, M.J. (2017) Conservation planning for the extraordinary and Endangered Spesbona damselfly. *Journal of Insect Conservation* 21, 121–128. DOI: 10.1007/s10841-017-9960-8.

Deacon, C., Samways, M.J. and Pryke, J.S. (2018) Artificial reservoirs complement natural ponds to improve pondscape resilience in conservation corridors in a biodiversity hotspot. *PLoS ONE* 13, e0204148. DOI: 10.1371/journal. pone.0204148.

Deacon, C., Samways, M.J. and Pryke, J.S. (2019) Aquatic insect decline in abundance and occupy low-quality artificial habitats to survive hydrological droughts. *Freshwater Biology* 64, 1643–1654. DOI: 10.1111/fwb.13360.

Deacon, C., Govender, S. and Samways, M.J. (2023) Overcoming biases and identifying opportunities for citizen science to contribute more to global macroinvertebrate conservation. *Biodiversity and Conservation* 32, 1789–1806. DOI: 10.1007/s10531-023-02595-x.

Deitsch, J. and O'Grady, P. (2022) *A Catalog of Odonata Material in the Cornell University Insect Collection*. Cornell University, USA.

Del Palacio, A., Lozano, F., Ramos, L.S., Navarro, M.d.l.M. and Muzón, J. (2022) Odonata from Iberá Wetland System (Corrientes, Argentina): are regional biogeographic schemes useful to assess Odonata biodiversity and its conservation? *Diversity* 14, e842. DOI: 10.3390/d14100842.

Delić, T., Trontelj, P., Rendoš, M. and Fišer, C. (2017) The importance of naming cryptic species and the conservation of endemic subterranean amphipods. *Scientific Reports* 7, e3391. DOI: 10.1038/s41598-017-02938-z.

Dias-Silva, K., Vieira, T.B., de Matos, T.P., Juen, L., Simião-Ferreira, J. *et al.* (2021) Measuring stream habitat conditions: can remote sensing substitute for field data? *Science of the Total Environment* 788, e147617. DOI: 10.1016/j.scitotenv.2021.147617.

Dijkstra, K.-D.B., Boudot, J.-P., Clausnitzer, V., Kipping, J., Kisakye, J.J. *et al.* (2011) Dragonflies and damselflies of Africa (Odonata): history, diversity, distribution, and conservation. In: Darwall, W., Smith, K., Allen, D., Holland, R., Harrison, I. and Brooks, E. (eds) *The Diversity of Life in African Freshwaters: Underwater, Under Threat*. IUCN, Gland, Switzerland and Cambridge, pp. 126–173.

Dingemanse, N.J. and Kalkman, V.J. (2008) Changing temperature regimes have advanced the phenology of Odonata in the Netherlands. *Ecological Entomology* 33, 394–402. DOI: 10.1111/j.1365-2311.2007.00982.x.

Dinh Van, K., Janssens, L., Debecker, S. and Robby Stoks, R. (2014) Temperature- and latitude-specific individual growth rates shape the vulnerability of damselfly larvae to a widespread pesticide. *Journal of Applied Ecology* 51, 919–928. DOI: 10.1111/1365-2664.12269.

Doi, H. (2008) Delayed phenological timing of dragonflies emerging in Japan over five decades. *Biology Letters* 4, 388–391. DOI: 10.1098/rsbl.2008.0234.

Dolný, A., Pyszo, P. and Šigutová, H. (2021) Community changes in odonate monitoring: why are long-term studies so relevant? *Insect Conservation and Diversity* 14, 597–608. DOI: 10.1111/icad.12491.

Dornelas, M., Antão, L.H., Moyes, F., Bates, A.E., Magurran, A.E. *et al.* (2018) BioTIME: a database of biodiversity time series for the Anthropocene. *Global Ecology and Biogeography*, 27, 760–786. DOI: 10.1111/geb.12729.

Duffy, J.P., Anderson, K., Fawcett, D., Curtis, R.J. and Maclean, I.M.D. (2021) Drones provide spatial and volumetric data to deliver new insights into microclimate modelling. *Landscape Ecology* 36, 685–702. DOI: 10.1007/s10980-020-01180-9.

Dusabe, M.C., Neubauer, T.A., Muvundja, F.A., Hyangya, B.L. and Albrecht, C. (2022) Family-level bioindication does not detect the impacts of dams on macroinvertebrate communities in a low-diversity tropical river. *Frontiers in Environmental Science* 10, e902246. DOI: 10.3389/fenvs.2022.902246.

Dutra, S. and De Marco, P. (2015) Bionomic differences in odonates and their influence on the efficiency of indicator species. *Ecological Indicators* 49, 132–142. DOI: 10.1016/j.ecolind.2014.09.016.

Eagles-Smith, C.A., Willacker, J.J., Nelson, S.J., Flanagan Pritz, C.M., Krabbenhoft, D.P. *et al.* (2020) A national-scale assessment of mercury bioaccumulation in United States National Parks using dragonfly larvae as biosentinels through a citizen-science framework. *Environmental Science and Technology* 54, 8779–8790. DOI: 10.1021/acs.est.0c01255.

Eken, G., Bennun, L., Brooks, T.M., Darwall, W., Fishpool, L.D.C. *et al.* (2004) Key biodiversity areas as site conservation targets. *BioScience* 54, 1110–1118. DOI: 10.1641/0006-3568(2004)054[1110:KBAASC]2.0.CO;2.

Elbrecht, V., Vamos, E.E., Meissner, K. Aroviita, J. and Leese, F. (2017) Assessing strengths and weaknesses of DNA metabarcoding-based macroinvertebrate identification for routine stream monitoring. *Methods in Ecology and Evolution* 8, 1265–1275. DOI: 10.1111/2041-210X.12789.

Encarnación-Luévano, A., Escoto-Moreno, J.A. and Villalobos-Jiménez, G. (2022) Evaluating potential distribution and niche divergence among populations of the world's largest living damselfly, *Megaloprepus caerulatus* (Drury, 1782). *Diversity* 14, e84. DOI: 10.3390/d14020084.

Feindt, W. and Hadrys, H. (2022) The quality of sequence data affects biodiversity and conservation perspectives in the Neotropical damselfly *Megaloprepus caerulatus*. *Diversity* 14, e1056. DOI: 10.3390/d14121056.

Fennessy, M.S., Jacobs, A.D., and Kentula, M.E. (2007) An evaluation of rapid methods for assessing the ecological condition of wetlands. *Wetlands* 27, 543–560.

Fernandez-Fournier, P., Lewthwaite, J.M.M. and Mooers, A.Ø. (2021) Do we need to identify adaptive genetic variation when prioritizing populations for conservation? *Conservation Genetics* 22, 205–216. DOI: 10.1007/s10592-020-01327-w.

Ferrier, S., Powell, G.V.N., Richardson, K.S., Manion, G., Overton, J.M. *et al.* (2004) Mapping more of terrestrial biodiversity for global conservation assessment. *BioScience* 54, 1101–1109. DOI: 10.1641/0006-3568(2004)054[1101:MMOTBF]2.0.CO;2.

Finch, J.M., Samways, M.J., Hill, T.R., Piper, S.E. and Taylor, S. (2006) Application of predictive distribution modelling to invertebrates: Odonata in South Africa. *Biodiversity and Conservation* 15, 4239–4251. DOI: 10.1007/s10531-005-3577-z.

Fiorentino, D., Lecours, V. and Brey, T. (2018) On the art of classification in spatial ecology: fuzziness as an alternative for mapping uncertainty. *Frontiers in Ecology and Evolution* 6, e231. DOI: 10.3389/fevo.2018.00231.

Fischer, I., Sittenthaler, M., Traugott, M., Thalinger, B., Zangl, L. *et al.* (2018) Monitoring and DNA barcoding of dragonflies (Insecta: Odonata) in Vienna, Austria. *Acta ZooBot Austria* 155, 39–42.

Fletcher, D.E., Lindell, A.H., Stillings, G.K., Blas, S.A. and McArthur, J.V. (2017) Trace element accumulation in lotic dragonfly nymphs: genus matters. *PLoS ONE* 12, e0172016. DOI: 10.1371/ journal.pone.0172016.

Fletcher, D.E., Lindell, A.H., Stankus, P.T., Fletcher, N.D., Lindell, B.E. and McArthur, J.V. (2020) Metal accumulation in dragonfly nymphs and crayfish as indicators of constructed wetland effectiveness. *Environmental Pollution* 256, e113387. DOI: 10.1016/j.envpol.2019.113387.

Franke, S., Pinkert, S., Brandl, R. and Thorn, S. (2022) Modeling the extinction risk of European butterflies and odonates. *Ecology and Evolution* 12, e9465. DOI: 10.1002/ece3.9465.

Frazier, A.E. (2023) Scope and its role in advancing a science of scaling in landscape ecology. *Landscape Ecology* 38, 637–643. DOI: 10.1007/s10980-022-01403-1.

Ganguly, I., Patnaik, L. and Nayak, S. (2018) Macroinvertebrates and its impact in assessing water quality of riverine system: a case study of Mahanadi river, Cuttack, India. *Journal of Applied and Natural Science* 10, 958–963. DOI: 10.31018/jans.v10i3.1817.

García-Girón, J., Bini, L.M. and Heino, J. (2023) Shortfalls in our understanding of the causes and consequences of functional and phylogenetic variation of freshwater communities across continents. *Biological Conservation* 282, e110082. DOI: 10.1016/j.biocon.2023.110082.

Garnett, S.T. and Christidis, L. (2017) Taxonomy anarchy hampers conservation. *Nature* 546, 25–27. DOI: 10.1038/546025a.

Geary, M. and von Hardenberg, A. (2020) White-faced darter distribution is associated with coniferous forests in Great Britain. *Insect Conservation and Diversity* 14, 15–25. DOI: 10.1111/icad.12438.

Geigher, M., Koblmüller, S., Assandri, G., Chovanec, A., Ekrem, T. *et al.* (2021) Coverage and quality of DNA barcode references for Central and Northern European Odonata. *PeerJ* 9, e11192. DOI: 10.7717/peerj.11192.

Girgin, S., Kazanci, N. and Dügel, M. (2010) Relationship between aquatic insects and heavy metals in an urban stream using multivariate techniques. *International Journal of Environmental Science and Technology* 7, 653–664. DOI: 10.1007/BF03326175l.

Giugliano, L., Hardersen, S. and Santini, G. (2012) Odonata communities in retrodunal ponds: a comparison of sampling methods. *International Journal of Odonatology* 15, 13–23. DOI: 10.1080/13887890.2012.660403.

Godreau, V., Bornette, G., Frochot, B., Amoros, C., Castella, E. *et al.* (1999) Biodiversity in the floodplain of Saône: a global approach. *Biodiversity and Conservation* 8, 839–864.

Golfieri, B., Hardersen, S., Maiolini, B. and Surian, N. (2016) Odonates as indicators of the ecological integrity of the river corridor: development and application of the Odonate River Index in northern Italy. *Ecological Indicators* 61, 234–247. DOI: 10.1016/j.ecolind.2015.09.022.

Golfieri, B., Surian, N. and Hardersen, S. (2018) Towards a more comprehensive assessment of river corridor conditions: a comparison between the Morphological Quality Index and three biotic indices. *Ecological Indicators* 84, 525–534. DOI: 10.1016/j.ecolind.2017.09.011.

Gomes, P.G.S., Lima, E.L., Silva, S.R., Juen, L. and Brasil, L.S. (2022) Does land use and land cover affect adult communities of Ephemeroptera, Plecoptera and Trichoptera (EPT)? A systematic review with meta-analysis. *Environmental Monitoring and Assessment* 194, e697. DOI: 10.1007/s10661-022-10352-w.

Gómez-Tolosa, M., Rivera-Velázquez, G., Rioja-Paradela, T.M. Mendoza-Cuenca, L.F., Tejeda-Cruz, C. and López, S. (2021) The use of Odonata species for environmental assessment: a meta-analysis for the Neotropical region. *Environmental Science and Pollution Research* 28, 1381–1396. DOI: 10.1007/s11356-020-11137-9.

Gómez-Tolosa, M., González-Soriano, E., Mendoza-Cuenca, L.F., Pérez-Munguía, R.M., Rioja-Paradela, T.M. *et al.* (2022a) The use of highly diverse clades as a surrogate for habitat integrity analysis: *Argia* damselflies as a practical tool for rapid assessments. *Environmental Science and Pollution Research* 29, 24334–24347. DOI: 10.1007/s11356-022-18815-w.

Gómez-Tolosa, M., Mendoza-Cuenca, L. Rivera-Velázquez, G., Rioja-Paradela, T.M., Tejeda-Cruz, C. *et al.* (2022b) Using the ecological relationships of Odonata with a habitat integrity index to test the biodiversity ecosystem function framework. *Journal of Insect Conservation* 26, 191–203. DOI: 10.1007/s10841-022-00377-4.

Goodman, A.M., Kass, J.M. and Ware, J. (2023) Dynamic distribution modelling of the swamp tigertail dragonfly *Synthemis eustalacta* (Odonata: Anisoptera: Synthemistidae) over a 20-year bushfire regime. *Ecological Entomology* 48, 209–225. DOI: 10.1111/een.13216.

Grace, M.K., Akçakaya, H.R., Bennett, E.L., Brooks, T.M., Heath, A. *et al.* (2021) Testing a global standard for quantifying species recovery and assessing conservation impact. *Conservation Biology* 35, 1833–1849. DOI: 10.1111/cobi.13756.

Grace, M.K., Akçakaya, H.R., Bennett, E.L., Boyle, M.J.W., Hilton-Taylor, C. *et al.* (2022) The impact of spatial delineation on the assessment of species recovery outcomes. *Diversity* 14, e742. DOI: 10.3390/d14090742.

Grether, G.F., Beninde, J., Beraut, E., Chumchim, N., Escalona, M. *et al.* (2023) Reference genome for the American rubyspot damselfly, *Hetaerina americana*. *Journal of Heredity* 114, 385–394. DOI: 10.1093/jhered/esad031.

Grigoropoulou, A., Ab Hamid, S., Acosta, R., Akindele, E.O., Al-Shami, S.A. *et al.* (2023) The global EPTO database: worldwide occurrences of aquatic insects. *Global Ecology and Biogeography* 32, 642–655. DOI: 10.1111/geb.13648.

Guisan, A., Tingley, R., Baumgartner, J.B., Naujokaitis-Lewis, H., Sutcliffe, P.R. *et al.* (2013) Predicting species distributions for conservation decisions. *Ecology Letters* 16, 1424–1435. DOI:10.1111/ele.12189.

Habel, J.C., Teucher, M., Ulrich, W., Bauer, M. and Rödder, D. (2016) Drones for butterfly conservation: larval habitat assessment with an unmanned aerial vehicle. *Landscape Ecology* 31, 2385–2395. DOI: 10.1007/s10980-016-0409-3.

Harabiš, F. and Dolný, A. (2015) Odonates need natural disturbances: how human-induced dynamics affect the diversity of dragonfly assemblages. *Freshwater Science* 34, 1050–1057. DOI: 10.1086/682234.

Harms, T.M., Kinkead, K.E. and Dinsmore, S.J. (2014) Evaluating the effects of landscape configuration on site occupancy and movement dynamics of odonates in Iowa. *Journal of Insect Conservation* 18, 307–315. DOI: 10.1007/s10841-014-9638-4.

Harper, L.R., Buxton, A.S., Rees, H.C., Bruce, K., Brys, R. *et al.* (2019) Prospects and challenges of environmental

DNA (eDNA) monitoring in freshwater ponds. *Hydrobiologia* 826, 25–41. DOI: 10.1007/s10750-018-3750-5.

Hassall, C. and Thompson, D.J. (2010) Accounting for recorder effort in the detection of range shifts from historical data. *Methods in Ecology and Evolution* 1, 343–350. DOI: 10.1111/j.2041-210X.2010.00039.x.

Hassall, C., Thompson, D.J., French, G.C. and Harvey, I.F. (2007) Historical changes in the phenology of British Odonata are related to climate. *Global Change Biology* 13, 933–941. DOI: 10.1111/j.1365-2486.2007.01318.x.

Hassall, C., Thompson, D.J. and Harvey, I.F. (2010) The impact of climate-induced distributional changes on the validity of biological water quality metrics. *Environmental Monitoring and Assessment* 160, 451–456. DOI: 10.1007/s10661-008-0709-4.

Hawking, J.H. and New, T.R. (1999) The distribution patterns of dragonflies (Insecta: Odonata) along the Kiewa River, Australia, and their relevance in conservation assessment. *Hydrobiologia* 392, 249–260.

Heino, J. (2010) Are indicator groups and cross-taxon congruence useful for predicting biodiversity in aquatic systems. *Ecological Indicators* 10, 112–117. DOI: 101016/j.ecolind. 2009.04.013.

Heino, J. (2014) Taxonomic surrogacy, numerical resolution and responses of stream macroinvertebrate communities to ecological gradients: are the inferences transferable among regions? *Ecological Indicators* 36, 186–194. DOI: 10.1016/j.ecolind.2013.07.022.

Herlihy, A.T., Sifneos, J.C., Hughes, R.M., Peck, D.V. and Mitchell. R.M. (2020) The relation of lotic fish and benthic macroinvertebrate condition indices to environmental factors across the conterminous USA. *Ecological Indicators* 112, e105958. DOI: 10.1016/j.ecolind.2019.105958.

Herzog. R. and Hadrys, H. (2017) Long-term genetic monitoring of a riverine dragonfly, *Orthetrum coerulescens* (Odonata: Libellulidae): direct anthropogenic impact versus climate change effects. *PLoS ONE* 12, e0178014. DOI: 10.1371/journal.pone.0178014.

Hidayat, S., Husnia, F., Rohmah, E. and Mukhlishoh, S. (2019) Study on diversity of dragonfly (Odonata) as a bioindicator of water quality in Mount Muria area, Central Java. *Journal of Natural Sciences and Mathematics Research* 5, 53–61.

Hoban, S., Archer, F.I, Bertola, L.D., Bragg, J.G., Breed, M.F. *et al.* (2022) Global genetic diversity status and trends: towards a suite of Essential Biodiversity Variables (EBVs) for genetic composition. *Biological Reviews* 97, 1511–1538. DOI: 10.1111/brv.12852.

Hochkirch, A., Samways, M.J., Gerlach, J., Böhm, M., Williams, P. *et al.* (2021) A strategy for the next decade to address data deficiency in neglected biodiversity. *Conservation Biology* 35, 502–509. DOI: 10.1111/cobi.13589.

Holland, R.A., Garcia, N., Brooks, E.G.E., Juffe, D., Howard, G. *et al.* (2011) Synthesis for all taxa. In: Darwall, W.R.T., Smith, K.G., Allen, D.J., Holland, R.A., Harrison, I.J. and Brooks, E.G.E. (eds) *The Diversity of Life in African Freshwaters: Under Water, Under Threat. An Analysis of the Status and Distribution of Freshwater Species throughout Mainland Africa.* IUCN, Cambridge and Gland, Switzerland, pp. 228–269.

Honeycutt, R.L. (2021) DNA barcodes: controversies, mechanisms, and future applications. *Frontiers in Ecology Evolution* 9, e718865. DOI: 10.3389/fevo.2021.718865.

Houle, D. (2010) Numbering the hairs on our heads: the shared challenge and promise of phenomics. *PNAS* 107, 1793–1799. DOI: 10.1073/pnas.0906195106.

Høye, T.T., Ärje, J., Bjerge, K., Hansen, O.L.P., Iosifidis, A. *et al.* (2021) Deep learning and computer vision will transform entomology. *PNAS* 118, e2002545117. DOI: 10.1073/pnas.2002545117.

Ilg, C. and Oertli, B. (2016) Effectiveness of amphibians as biodiversity surrogates in pond conservation. *Conservation Biology* 31, 437–445. DOI: 10.1111/cobi.12802.

Ilhamdi, M.L., Idrus, A.A. and Santoso, D. (2020) Biotic index of dragonflies in the Suranadi Natural Park Area West Lombok. *Jurnal Pijar MIPA* 15, 424–428. DOI: 10.29303/jpm.v15i4.2201.

IPBES (2019) Summary for policymakers of the global assessment report on biodiversity and ecosystem services of the Intergovernmental Science-Policy Platform on Biodiversity and Ecosystem Services. Díaz, S., Settele, J., Brondízio, E.S., Ngo, H.T., Guèze, M. *et al.* (eds) *IPBES Secretariat*, Bonn, Germany. 56 pp. DOI: 10.5281/zenodo.3553579.

IUCN (2016a) *An Introduction to the IUCN Red List for Ecosystems: The Categories and Criteria for Assessing Risks to Ecosystems*. Gland, Switzerland: Version 2016–1.

IUCN (2016b) *A Global Standard for the Identification of Key Biodiversity Areas, Version 1.0.* First edition. Gland, Switzerland.

Iwaszkiewicz-Eggebrecht, E., Łukasik, P., Buczek, M., Deng, J., Hartop, E.A. *et al.* (2023) FAVIS: fast and versatile protocol for nondestructive metabarcoding of bulk insect samples. *PLoS ONE* 18, e0286272. DOI: 10.1371/journal.pone.0286272.

Jackson, M.C., Weyl, O.L.F., Altermatt, F., Durance, I., Friberg, N. *et al.* (2016) Recommendations for the next generation of global freshwater biological monitoring tools. *Advances in Ecological Research* 55, 615–636. DOI; 10.1016/bs.aecr.2016.08.008.

Jaeschke, A., Bittner, T., Reineking, B. and Beierkuhnlein, C. (2013) Can they keep up with climate change? Integrating specific dispersal abilities of protected Odonata in species distribution modelling. *Insect*

Conservation and Diversity 6, 93–103. DOI: 10.1111/j.1752-4598.2012.00194.x.

Jenkins, C.N. and Joppa, L. (2009) Expansion of the global terrestrial protected area system. *Biological Conservation* 142, 2166–2174. DOI: 10.1016/j.biocon.2009.04.016.

Jetz, W., Sekercioglu, C.H. and Watson, J.E.M. (2008) Ecological correlates and conservation implications of overestimating species geographic ranges. *Conservation Biology* 22, 110–119. DOI: 10.1111/j.1523-1739.2007.00847.x.

Johnson, N.F. (2017) Insect biodiversity informatics. In: Foottit, R.G. and Adler, P.H. (eds) *Insect Biodiversity: Science and Society*, Vol. 1, 2nd edn. Wiley-Blackwell, Oxford, pp. 593–602.

Jooste, M.L., Samways, M.J. and Deacon, C. (2020) Fluctuating pond water levels and aquatic insect persistence in a drought-prone Mediterranean-type climate. *Hydrobiologia* 847, 1315–1326. DOI: 10.1007/s10750-020-04186-1.

Jörger, K.M. and Schrödl, M. (2013) How to describe a cryptic species? Practical challenges of molecular taxonomy. *Frontiers in Zoology* 10, e59. DOI: 10.1186/1742-9994-10-59.

Juen, L. and Feest, A. (2019) Evaluating the biodiversity quality response of tropical odonata to tree clearance. *International Journal of Tropical Insect Science* 39, 45–52. DOI: 10.1007/s42690-019-00011-y.

Juen, L., de Oliveira-Júnior, J.M.B., Shimano, Y., Mendes, T.P. and Cabette, H.S.R. (2014) Composition and richness of Odonata (Insecta) in streams with different levels of conservation in a Cerrado-Amazonian forest ecotone. *Acta Amazonica* 44, 223–233. DOI: 10.1590/S0044-59672014000200008.

Juslén, A., Pykälä, J., Kuusela, S., Kaila, L., Kullberg, J. *et al.* (2016) Application of the Red List Index as an indicator of habitat change. *Biodiversity and Conservation* 25, 569–585. DOI 10.1007/s10531-016-1075-0.

Kadoya, T. and Washitani, I. (2007) An adaptive management scheme for wetland restoration incorporating participatory monitoring into scientific predictions using dragonflies as an indicator taxon. *Global Environmental Research* 11, 179–185.

Kadoya, T., Akasaka, M., Aoki, T. and Takamura, N. (2011) A proposal of framework to obtain an integrated biodiversity indicator for agricultural ponds incorporating the simultaneous effects of multiple pressures. *Ecological Indicators* 11, 1396–1402. DOI: 10.1016/j.ecolind.2011.03.001.

Kalkman, V.J., Boudot, J.-P., Bernard, R., Conze, K.-J., De Knijf, G. *et al.* (2010) *European Red List of Dragonflies. Publications Office of the European Union, Luxembourg.* DOI: 10.2779/84650.

Kalkman, V.J., Boudot, J.-P., Bernard, R., De Knijf, G., Suhling, F. and Termaat, T. (2018) Diversity and conservation of European dragonflies and damselflies (Odonata). *Hydrobiologia* 811, 269–282. DOI: 10.1007/s10750-017-3495-6.

Keith, D.A., Rodríguez, J.P., Rodríguez-Clark, K.M., Nicholson, E., Aapala, K. *et al.* (2013) Scientific foundations for an IUCN Red List of Ecosystems. *PLoS ONE* 8, e62111. DOI: 10.1371/journal.pone.0062111.

Keith, D.A., Akçakaya, H.R. and Murray, N.J. (2018) Scaling range sizes to threats for robust predictions of risks to biodiversity. *Conservation Biology* 32, 322–332. DOI: 10.1111/cobi.12988.

Khelifa, R. (2019) Sensitivity of biodiversity indices to life history stage, habitat type and landscape in Odonata community. *Biological Conservation* 237, 63–69. DOI: 10.1016/j.biocon.2019.06.010.

Khelifa, R., Deacon, C., Mahdjoub, H., Suhling, F., Simaika, J.P. and Samways, M.J. (2021a) Dragonfly conservation in the increasingly stressed African Mediterranean-type ecosystems. *Frontiers in Environmental Science* 9, e660163. DOI: 10.3389/fenvs. 2021.660163.

Khelifa, R., Mahdjoub, H., Baaloudj, A., Cannings, R.A. and Samways, M.J. (2021b) Effects of both climate change and human water demand on a highly threatened damselfly. *Scientific Reports* 11, e7725. DOI: 10.1038/s41598-021-86383-z.

Kietzka, G.J., Pryke, J.S. and Samways, M.J. (2018) Comparative effects of urban and agricultural land transformation on Odonata assemblages in a biodiversity hotspot. *Basic and Applied Ecology* 33, 89–98. DOI: 10.1016/j.baae.2018.08.008.

Kietzka, G.J., Pryke, J.S., Gaigher, R. and Samways, M.J. (2019) Applying the umbrella index across aquatic insect taxon sets for freshwater assessment. *Ecological Indicators* 107, e105655. DOI: 10.1016/j.ecolind.2019.105655.

Kietzka, G.J., Pryke, J.S., Gaigher, R. and Samways, M.J. (2021a) 32 years of essential management to retain value of an urban dragonfly awareness pond. *Urban Ecosystems* 24, 1295–1304. DOI: 10.1007/s11252-021-01115-5.

Kietzka, G.J., Pryke, J.S., Gaigher, R. and Samways, M.J. (2021b) Congruency between adult male dragonflies and their larvae in river systems is relative to spatial grain. *Ecological Indicators* 124, e107390. DOI: 10.1016/j.ecolind.2021.107390.

Kietzka, G.J., Deacon, C. and Patten, M.A. (2023) Odonata as surrogates of biodiversity. In: Córdoba-Aguilar, A., Beatty, C.D. and Bried, J.T (eds) *Dragonflies and Damselflies: Model Organisms for Ecological and Evolutionary Research* 2nd edn. Oxford University Press, Oxford, pp. 359–369. DOI:10.1093/oso/9780192898623.003.0025.

Kindsvater, H.K., Dulvy, N.K., Horswill, C., Juan-Jordá, M.-J., Mangel, M. and Matthiopoulos, J. (2018) Overcoming the data crisis in biodiversity conservation. *Trends in Ecology and Evolution* 33, 676–688. DOI: 10.1016/j.tree.2018.06.004.

Koperski, P. (2023) It is not only data: freshwater invertebrates misused in biological monitoring. *Animals* 13, e2570. DOI: 10.3390/ani13162570.

Korbaa, M., Ferreras-Romero, M., Ruiz-García, A. and Boumaiza, M. (2018) TSOI: a new index based on Odonata populations to assess the conservation relevance of watercourses in Tunisia. *Odonatologica* 47, 43–72. DOI: 10.5281/zenodo.1239945.

Koroiva, R., Gomes, V.G.N. and Vilela, D.S. (2022) DNA Barcoding and new records of odonates (Insecta: Odonata) from Paraíba State, Brazil. *Diversity* 14, e203. DOI: 10.3390/d14030203.

Küry, D. and Christ, J. (2010) Odonata fauna and conservation in the canton Basel-Stadt (NW Switzerland). *Mitteilungen der Naturforschenden Gesellschaften beider Basel* 12, 105–118.

Kutcher, T.E. and Bried, J.T. (2014) Adult Odonata conservatism as an indicator of freshwater wetland condition. *Ecological Indicators* 38, 31–39. DOI: 10.1016/j.ecolind.2013.10.028.

La Porta, G. and Goretti, E. (2019) Investigation on the declining Southern Damselfly (*Coenagrion mercuriale*, Odonata) in a Mediterranean population: survival rate and population size. *Journal of Insect Conservation* 23, 667–675. DOI: 10.1007/s10841-019-00160-y.

Landmann, M., Schlick-Steiner, B.C., Steiner, F.M. and Landmann, A. (2021) Connectivity within isolation: dispersal, population genetics, and conservation of the rarest European damselfly. *Insect Conservation and Diversity* 14, 800–813. DOI: 10.1111/icad.12516.

Langhammer, P.F., Bakarr, M.I., Bennun, L.A., Brooks, T.M., Clay, R.P. *et al.* (2007) *Identification and Gap Analysis of Key Biodiversity Areas: Targets for Comprehensive Protected Area Systems*. IUCN, Gland, Switzerland.

Law, A., Baker, A., Sayer, C., Foster, G., Gunn, I.D.M. *et al.* (2019) The effectiveness of aquatic plants as surrogates for wider biodiversity in standing fresh waters. *Freshwater Biology* 64, 1664–1675. DOI: 10.1111/fwb.13369.

Leal, C.G., Lennox, G.D., Ferraz, S.F.B., Ferreira, J., Gardner, T.A. *et al.* (2020) Integrated terrestrial-freshwater planning doubles conservation of tropical aquatic species. *Science* 370 (6512), 117–121. DOI: 10.1126/science.aba7580.

Leitão, P.J. and Santos, M.J. (2019) Improving models of species ecological niches: a remote sensing overview. *Frontiers in Ecology and Evolution* 7, e9. DOI: 10.3389/fevo.2019.00009.

Ligeiro, R., Hughes, R.M., Kaufmann, P.R., Macedo, D.R., Firmiano, K.R. *et al.* (2013) Defining quantitative stream disturbance gradients and the additive role of habitat variation to explain macroinvertebrate taxa richness. *Ecological Indicators* 25, 45–57. DOI: 10.1016/j.ecolind.2012.09.004.

Lin, Y.-P., Deng, D., Lin, W.-C., Lemmens, R., Crossman, N.D. *et al.* (2015) Uncertainty analysis of crowd-sourced and professionally collected field data used in species distribution models of Taiwanese moths. *Biological Conservation* 181, 102–110. DOI: 10.1016/j.biocon.2014.11.012.

Lindenmeyer, D.B., Woinarski, J., Legge, S., Maron, M., Garnett, S.T. *et al.* (2022) Eight things you should never do in a monitoring program: an Australian perspective. *Environmental Monitoring and Assessment* 194, e701. DOI: 10.1007/s10661-022-10348-6.

Löcken, H., Fischer, O.W., Selz, J. and Boppré, M. (2020) 'Drone-netting' for sampling live insects. *Journal of Insect Science* 20 (3), 1–3. DOI: 10.1093/jisesa/ieaa086.

Lorenzo-Carballa, M.O., Sanmartín-Villar, I. and Cordero-Rivera, A. (2022) Molecular and morphological analyses support different taxonomic units for Asian and Australo-Pacific forms of *Ischnura aurora* (Odonata, Coenagrionidae). *Diversity* 14, e606. DOI: 10.3390/d14080606.

Lozano, F., Palacio, A., Ramos, L.S., Granato, L., Drozd, A. and Muzón, J. (2022) Recovery of local dragonfly diversity following restoration of an artificial lake in an urban area near Buenos Aires. *Basic and Applied Ecology* 58, 88–97. DOI: 10.1016/j.baae.2021.11.006.

Ma, Y., Yu, Z., Jia, S., Wu, N., Yin, K. *et al.* (2023) Multiple anthropogenic stressors influence the taxonomic and functional homogenization of macroinvertebrate communities on the mainstream of an urban-agricultural river in China. *Journal of Environmental Management* 341, e118017. DOI: 10.1016/j.jenvman.2023.118017.

Macedo, D.R., Hughes, R.M., Ligeiro, R., Ferreira, W.R., Castro, M.A. *et al.* (2014) The relative influence of catchment and site variables on fish and macroinvertebrate richness in cerrado biome streams. *Landscape Ecology* 29, 1001–1016. DOI: 10.1007/s10980-014-0036-9.

Maddock, A. and Samways, M.J. (2000) Planning for biodiversity conservation based on the knowledge of biologists. *Biodiversity and Conservation* 9, 1153–1169. DOI: 10.1023/A:1008992602151.

Maes, D. and Van Dyck, H. (2005) Habitat quality and biodiversity indicator performances of a threatened butterfly versus a multispecies group for wet heathlands in Belgium. *Biological Conservation* 123, 177–187. DOI: 10.1016/j.biocon.2004.11.005.

Maes, D., Isaac, N.J.B., Harrower, C.A., Collen, B., van Strien, A.J. and Roy, D.B. (2015) The use of opportunistic data for IUCN Red List assessments. *Biological Journal of the Linnean Society* 115, 690–706. DOI: 10.1111/bij.12530.

Malcolm, J.R., Liu, C., Neilson, R.P., Hansen, L. and Hannah, L. (2006) Global warming and extinctions of endemic species from global hotspots. *Conservation Biology* 20, 538–548. DOI: 10.1111/j.1523-1739.2006.00364.x.

Mammola, S., Pétillon, J., Hacala, A., Monsimet, J., Marti, S.-L. *et al.* (2021) Challenges and opportunities of species distribution modelling of terrestrial

arthropod predators. *Diversity and Distributions* 27, 2596–2614. DOI: 10.1111/ddi.13434.
Manu, M.K., Ashiagbor, G., Seidu, I., Groen, T., Gyimah, T. and Toxopeus, B. (2023) Odonata as bioindicator for monitoring anthropogenic disturbance of Owabi wetland sanctuary, Ghana. *Aquatic Insects* 44, 151–169. DOI: 10.1080/01650424.2022.2108844.
Marsh, C.J., Syfert, M.M., Aletrari, E., Gavish, Y., Kunin, W.E. and Brummitt, N. (2023) The effect of sampling effort and methodology on range size estimates of poorly-recorded species for IUCN Red List assessments. *Biodiversity and Conservation* 32, 1105–1123. DOI: 10.1007/s10531-023-02543-9.
Marshall, S.A. (2018) Digital photography and the democratization of biodiversity information. In: Foottit, R.G. and Adler, P.H. (eds) *Insect Biodiversity: Science and Society*, Volume 2, 2nd edition. Wiley-Blackwell, Oxford, pp. 839–850.
Martins, I., Macedo, D.R., Hughes, R.M. and Callisto, M. (2021) Major risks to aquatic biotic condition in a Neotropical Savanna River basin. *River Research and Applications* 37, 858–868. DOI: 10.1002/rra.3801.
Martins, R.T., Brito, J., Dias-Silva, K., Leal, C.G., Leitão, R.P. *et al.* (2022) Congruence and responsiveness in the taxonomic compositions of Amazonian aquatic macroinvertebrate and fish assemblages. *Hydrobiologia* 849, 2281–2298. DOI: 10.1007/s10750-022-04867-z.
Masese, F.O., Wanderi, E.W., Nyakeya, K., Achieng, A.O., Fouchy, K. and McClain, M.E. (2023), Bioassessment of multiple stressors in Afrotropical rivers: evaluating the performance of a macroinvertebrate based index of biotic integrity, diversity, and regional biotic indices. *Frontiers in Environmental Science* 11, e1015623. DOI: 10.3389/fenvs.2023.1015623.
Mathieu, C., Hermans, S.M., Lear, G., Buckley, T.R., Lee, K.C. and Buckley, H.L. (2020) A systematic review of sources of variability and uncertainty in eDNA data for environmental monitoring. *Frontiers in Ecology and Evolution* 8, e135. DOI: 10.3389/fevo.2020.00135.
Matthews, T.J., Cottee-Jones, H.E. and Whittaker, R.J. (2014) Habitat fragmentation and the species–area relationship: a focus on total species richness obscures the impact of habitat loss on habitat specialists. *Diversity and Distributions* 20, 1136–1146. DOI: 10.1111/ddi.12227.
McKinley, D.C., Miller-Rushing, A.J., Ballard, H.L., Bonney, R., Brown, H. *et al.* (2017) Citizen science can improve conservation science, natural resource management, and environmental protection. *Biological Conservation* 208, 15–28. DOI: 10.1016/j.biocon.2016.05.015.
Menabit, S., Iancu, L., Pavel, A.B., Popa, A., Lupascu, N. and Purcarea, C. (2022) Molecular identification and distribution of insect larvae in the Lower Danube River. *Oceanological and Hydrological Studies* 51, 74–89. DOI: 10.26881/oahs-2022.1.07.
Mendes, T.P., Oliveira-Junior, J.M.B., Cabette, H.S.R., Batista, J.D. and Juen, L. (2017) Congruence and the biomonitoring of aquatic ecosystems: are odonate larvae or adults the most effective for the evaluation of impacts. *Neotropical Entomology* 46, 631–641. DOI: 10.1007/s13744-017-0503-5.
Mendes, T.P., Luiza-Andrade, A., Cabette, H.S.R. and Juen, L. (2018) How does environmental variation affect the distribution of dragonfly larvae (Odonata) in the Amazon-Cerrado transition zone in Central Brazil? *Neotropical Entomology* 47, 37–45. DOI: 10.1007/s13744-017-0506-2.
Mendes, T.P., Amado, L.L., Ribeiro, R.A.B. and Juen, L. (2020) Morphological diversity of Odonata larvae (Insecta) and abiotic variables in oil palm plantation areas in the Eastern Amazon. *Hydrobiologia* 847, 161–175. DOI: 10.1007/s10750-019-04079-y.
Mendoza-Penagos, C.C., Calvão, L.B. and Juen, L. (2021) A new biomonitoring method using taxonomic families as substitutes for the suborders of the Odonata (Insecta) in Amazonian streams. *Ecological Indicators* 124, e107388. DOI: 10.1016/j.ecolind.2021.107388.
Menetrey, N., Oertli, B. and Lachavine, J.-B. (2011) The CIEPT: a macroinvertebrate-based multimetric index for assessing the ecological integrity of Swiss lowland ponds. *Ecological Indicators* 11, 590–600. DOI: 10.1016/j.ecolind.2010.08.005.
Miguel, T.B., Oliveira-Junior, J.M.B., Ligeiro, R. and Juen, L. (2017) Odonata (Insecta) as a tool for the biomonitoring of environmental quality. *Ecological Indicators* 81, 555–566. DOI: 10.1016/j.ecolind.2017.06.010.
Monchamp, M.-E., Taranu, Z.E., Garner, R.E., Rehill, T., Morissette, O. *et al.* (2023) Prioritizing taxa for genetic reference database development to advance inland water conservation. *Biological Conservation* 280, e109963. DOI: 10.1016/j.biocon.2023.109963.
Monteiro-Júnior, C.S., Juen, L. and Hamada, N. (2014) Effects of urbanization on stream habitats and associated adult dragonfly and damselfly communities in central Brazilian Amazonia. *Landscape and Urban Planning* 127, 28–40. DOI: 10.1016/j.landurbplan.2014.03.006.
Monteiro-Júnior, C.S., Juen, L. and Hamada, N. (2015) Analysis of urban impacts on aquatic habitats in the central Amazon basin: adult odonates as bioindicators of environmental quality. *Ecological Indicators* 48, 303–311. DOI: 10.1016/j.ecolind.2014.08.021.
Moore, N.W. (1991) The development of dragonfly communities and the consequences of territorial behaviour: a 27 year study on small ponds at Woodwalton Fen, Cambridgeshire, United Kingdom. *Odonatologica* 20, 203–231.
Moreno Pallares, M.I., Bonilla Gómez, M.A., Guillot Monroy, G.H. and Torregroza-Espinosa, A.C. (2023) Macroinvertebrate composition as a determinant of larval abundance in the dragonfly, *Miathyria marcella* in tropical wetlands. *Global Journal of Environmental Science and Management* 9, 1–12. DOI: 10.22034/gjesm.2023.01.

Mu, Y., Li, X., Liang, C., Li, P., Guo, Y. *et al.* (2021) Rapid landscape assessment for conservation effectiveness of wetland national nature reserves across the Chinese mainland. *Global Ecology and Conservation* 31, e01842. DOI: 10.1016/j.gecco.2021.e01842.

Myers, N., Mittermeier, R.A., Mittermeier, C.G., da Fonseca, G.A.B. and Kent, J. (2000) Biodiversity hotspots for conservation priorities. *Nature* 403, 853–858. DOI: 10.1038/35002501.

Ndatimana, G., Nantege, D. and Arimoro, F.O (2023) A review of the application of the macroinvertebrate-based multimetric indices (MMIs) for water quality monitoring in lakes. *Environmental Science and Pollution Research* 30, 73098–73115. DOI: 10.1007/s11356-023-27559-0.

Nelson, S.J., Willacker, J., Eagles-Smith, C., Pritz, C.F., Chene, C.Y. *et al.* (2024) Habitat and dissolved organic carbon modulate variation in the biogeochemical drivers of mercury bioaccumulation in dragonfly larvae at the national scale. *Science of the Total Environment* 912, e169396. DOI: 10.1016/j.scitotenv.2023.169396.

Nessimian, J.L., Venticinque, E.M., Zuanon, J., De Marco Jr, P., Gordo, M. *et al.* (2008) Land use, habitat integrity, and aquatic insect assemblages in Central Amazonian streams. *Hydrobiologia* 614, 117–131. DOI: 10.1007/s10750-008-9441-x.

Nicolet, P., Biggs, J., Fox, G., Hodson, M.J., Reynolds, C. *et al.* (2004) The wetland plant and macroinvertebrate assemblages of temporary ponds in England and Wales. *Biological Conservation* 120, 261–278. DOI: 10.1016/j.biocon.2004.03.010.

Niwa, H. and Hirata, T. (2022) A new method for surveying the world's smallest class of dragonfly in wetlands using unoccupied aerial vehicles. *Drones* 6, e427. DOI: 10.3390/ drones6120427.

Nóbrega, C.C. and De Marco, P. (2011) Unprotecting the rare species: a niche-based gap analysis for odonates in a core Cerrado area. *Diversity and Distributions* 17, 491–505. DOI: 10.1111/j.1472-4642.2011.00749.x.

Nogueira, J.G., Sousa, R., Benaissa, H., De Knijf, G., Ferreira, S. *et al.* (2021) Alarming decline of freshwater trigger species in western Mediterranean key biodiversity areas. *Conservation Biology* 35, 1367–1379. DOI: 10.1111/cobi.13810.

Noss, R.F., Platt, W.J., Sorrie, B.A., Weakley, A.S., Means, D.B. *et al.* (2015) How global biodiversity hotspots may go unrecognized: lessons from the North American Coastal Plain. *Diversity and Distributions* 21, 236–244. DOI: 10.1111/ddi.12278.

Oliveira-Junior, J.M.B. and Juen, L. (2019) Structuring of dragonfly communities (Insecta: Odonata) in Eastern Amazon: effects of environmental and spatial factors in preserved and altered streams. *Insects* 10, e322. DOI: 10.3390/insects10100322.

Oliveira-Junior, J.M.B., Shimano, Y., Gardner, T.A., Hughes, R.M., De Marco Júnior, P. and Juen, L. (2015) Neotropical dragonflies (Insecta: Odonata) as indicators of ecological condition of small streams in the eastern Amazon. *Austral Ecology* 40, 733–744. DOI: 10.1111/aec.12242.

Olson, D.M., Dinerstein, E., Wikramanayake, E.D., Burgess, N.D., Powell, G.V.N. *et al.* (2001) Terrestrial ecoregions of the world: a new map of life on earth. *BioScience* 51, 933–938.DOI:10.1641/0006-3568(2001)051[0933:TEOTWA]2.0.CO;2.

Oppel, S. (2005) Habitat associations of an Odonata community in a lower montane rainforest in Papua New Guinea. *International Journal of Odonatology* 8, 243–257. DOI: 10.1080/13887890.2005.9748256.

Orioli, V., Gentili, R., Bani, L. and Aguzzi, S. (2021) Microhabitat selection and population density of *Nehalennia speciosa* Charpentier, 1840 (Odonata: Coenagrionidae) in a peripheral microrefugium. *Wetlands* 41, e86. DOI: 10.1007/s13157-021-01483-w.

Orme, C.D.L., Davies, R.G., Burgess, M., Eigenbrod, F., Pickup, N. *et al.* (2005) Global hotspots of species richness are not congruent with endemism or threat. *Nature* 436, 1016–1019. DOI: 10.1038/nature03850.

Ott, J., Schorr, M., Trockur, B. and Lingenfelder, U. (2007) Species protection programme for the Orange-spotted Emerald (*Oxygastra curtisii*, Insecta: Odonata) in Germany: the example of the River Our population. *Invertebrate Ecology and Conservation Monographs*, Vol. 3. Pensoft, Sophia, Bulgaria.

Ozana, S., Burda, M., Hykel, M., Malina, M., PrÃ¡sek, M. and BÃ¡rta, D. (2019) Dragonfly Hunter CZ: mobile application for biological species recognition in citizen science. *PLoS ONE* 14, e0210370. DOI: 10.1371/journal.pone.0210370.

Packer, L., Gibbs, J., Sheffield, C. and Hanner, R. (2009) DNA barcoding and the mediocrity of morphology. *Molecular Ecology Resources* 9 (Suppl. 1), 42–50. DOI: 10.1111/j.1755-0998.2009.02631.x.

Painter, D. (1999) Macroinvertebrate distributions and the conservation value of aquatic Coleoptera, Mollusca and Odonata in the ditches of traditionally managed and grazing fen at Wicken Fen, UK. *Journal of Applied Ecology* 36, 33–48.

Palacio, R.D., Negret, P.J., Valásquez-Tibatá, J. and Jacobson, A.P. (2021) A data-driven geospatial workflow to map species distributions for conservation assessments. *Diversity and Distributions* 27, 2559–2570. DOI: 10.1111/ddi.13424.

Patten, M.A., Bried, J.T. and Smith-Patten, B.D. (2015) Survey data matter: predicted niche of adult vs breeding Odonata. *Freshwater Science* 34, 1114–1122. DOI: 10.1086/682676.

Patten, M.A., Hjalmarson, E.A., Smith-Patten, B.D. and Bried, J.T. (2019) Breeding thresholds in opportunistic Odonata records. *Ecological Indicators* 106, e105460. DOI: 10.1016/j.ecolind.2019.105460.

Pawlowski, J., Apothéloz-Perret-Gentil, L. and Altermatt, F. (2020) Environmental DNA: what's behind the term? Clarifying the terminology and recommendations for its future use in biomonitoring. *Molecular Ecology* 29, 4258–4264. DOI: 10.1111/mec.15643.

Pearce-Higgins, J.W. and Chandler, D. (2020) Do surveys of adult dragonflies and damselflies yield repeatable data? Variation in monthly counts of abundance and species richness. *Journal of Insect Conservation* 24, 877–889. DOI: 10.1007/s10841-020-00260-0.

Pélissié, M., Johansson, F. and Hyseni, C. (2022) Pushed northward by climate change: range shifts with a chance of co-occurrence reshuffling in the forecast for Northern European odonates. *Environmental Entomology* 51, 910–921. DOI: 10.1093/ee/nvac056.

Peñas, F.J., Álvarez-Cabria, M., Sáinz-Bariáin, M., Mata-Campo, M.P., Pérez-Haase, A. *et al.* (2023) An evaluation of freshwater monitoring programs in ILTER nodes and mountain national parks: identifying key variables to monitor global change effects. *Biodiversity and Conservation* 32, 65–94. DOI: 10.1007/s10531-022-02466-x.

Pereira-da-Conceicoa, L., Elbrecht, V., Hall, A., Briscoe, A., Barber-James, H. and Price, B. (2021) Metabarcoding unsorted kick-samples facilitates macroinvertebrate-based biomonitoring with increased taxonomic resolution, while outperforming environmental DNA. *Environmental DNA* 3, 353–371. DOI: 10.1002/edn3.116.

Pereira-Moura, L., Sena, W.S., Neiss, U.G. and Couceiro, S.R.M. (2021) Environmental integrity as a modeler of the composition of the Odonata community. *Environmental Monitoring and Assessment* 193, e160. DOI: 10.1007/s10661-021-08957-8.

Pettorelli, N., Wegmann, M., Skidmore, A., Mücher, S., Dawson, T.P. *et al.* (2016) Framing the concept of satellite remote sensing essential biodiversity variables: challenges and future directions. *Remote Sensing in Ecology and Conservation* 2, 122–131. DOI: 10.1002/rse2.15.

Phillips, J.D., Gillis, D.J and Hanner, R.H. (2022) Lack of statistical rigor in DNA barcoding likely invalidates the presence of a true species' barcode gap. *Frontiers in Ecology and Evolution* 10, e859099. DOI: 10.3389/fevo.2022.859099.

Pires, M.M., Périco, E., Renner, S. and Sahlén, G. (2018) Predicting the effects of future climate change on the distribution of an endemic damselfly (Odonata, Coenagrionidae) in subtropical South American grasslands. *Journal of Insect Conservation* 22, 303–319. DOI: 10.1007/s10841-018-0063-y.

Proença, V., Martin, L.J., Pereira, H.M., Fernandez, M., McRae, L. *et al.* (2017) Global biodiversity monitoring: from data sources to Essential Biodiversity Variables. *Biological Conservation*, 213, 256–263. DOI: 10.1016/j.biocon.2016.07.014.

Prosser, S.W.J., deWaard, J.R., Miller, S.E. and Hebert, P.D.N. (2016) DNA barcodes from century-old type specimens using next-generation sequencing. *Molecular Ecology Resources*, 16, 487–497. DOI: 10.1111/1755-0998.12474.

Rachmawati, A., Yustian, I., Pujiastuti, Y., Suparman S. and Arinafril (2023) Biotic and dragonfly diversity indices as ecological quality evaluation in Lahat District Rivers, South Sumatra, Indonesia. *Biodiversitas* 24, 6059–6068. DOI: 10.13057/biodiv/d241127.

Raebel, E.M., Merckx, T., Riordan, P., Macdonald, D.W. and Thompson, D.J. (2010) The dragonfly delusion: why it is essential to sample exuviae to avoid biased surveys. *Journal of Insect Conservation* 14, 523–533. DOI: 10.1007/s10841-010-9281-7.

Raimondo, D., Young, B.E., Brooks, T.M., Cardoso, P., van der Colff, D. *et al.* (2022) Using Red List Indices to monitor extinction risk at national scales. *Conservation Science and Practice* 5, e12854. DOI: 10.1111/csp2.12854.

Reddy, C.S., Kurian, A., Srivastava, G., Singhal, J. Varghese, A.O. *et al.* (2021) Remote sensing enabled essential biodiversity variables for biodiversity assessment and monitoring: technological advancement and potentials. *Biodiversity and Conservation* 30, 1–14. DOI: 10.1007/s10531-020-02073-8.

Reece, B.A. and McIntyre, N.E. (2009) Community assemblage patterns of odonates inhabiting a wetland complex influenced by anthropogenic disturbance. *Insect Conservation and Diversity* 2, 73–80. DOI: 10.1111/j.1752-4598.2008.00044.x.

Reels, G.T. (2011) Emergence patterns and adult flight season of Anisoptera at a managed wetland site in Hong Kong, southern China. *International Journal of Odonatology* 14, 33–48. DOI: 10.1080/13887890.2011.570155.

Reels, G., Dow, R., Hämäläinen, M. and Manh Cuong, D. (2012) The status and distribution of dragonflies and damselflies (Odonata) in Indo-Burma. In: Allen, D.J., Smith, K.G. and Darwall, W.R.T (eds) *The Status and Distribution of Freshwater Biodiversity in Indo-Burma*. IUCN, Cambridge and Gland, Switzerland, pp. 90–101.

Resende, B.O., Ferreira, V.R.S., Juen, L. and Cabette, H.S.R. (2020) Emergence trap for the collection of exuviae and adults of Odonata. *Oecologia Australis* 24, 742–747. DOI: 10.4257/oeco.2020.2403.18.

Ribeiro, C., Juen, L. and Rodrigues, M.E. (2021). The Zygoptera/Anisoptera ratio as a tool to assess anthropogenic changes in Atlantic Forest streams. *Biodiversity and Conservation* 30, 1315–1329. DOI: 10.1007/s10531-021-02143-5.

Riservato, E., Boudot, J.-P., Ferreira, S., Jović, M., Kalkman, V.J. *et al.* (2009). *The Status and Distribution of Dragonflies of the Mediterranean Basin*. IUCN, Gland, Switzerland and Malaga, Spain.

Riva, F. and Nielsen, S.E. (2021) A functional perspective on the analysis of land use and land cover data in

ecology. *Ambio* 50, 1089–1100. DOI: 10.1007/s13280-020-01434-5.
Rivera-Pérez, J.M., Shimano, Y., Luiza-Andrade, A., Pinto, N.S., Dias, L.G. *et al*. (2023) Effect of mining on the EPT (Ephemeroptera, Plecoptera and Trichoptera) assemblage of Amazonian streams based on their environmental specificity. *Hydrobiologia* 850, 645–664. DOI: 10.1007/s10750-022-05111-4.
Robinson, J.L., Fordyce, J.A. and Parker, C.R. (2016) Conservation of aquatic insect species across a protected area network: null model reveals shortfalls of biogeographical knowledge. *Journal of Insect Conservation* 20, 565–581. DOI: 10.1007/s10841-016-9889-3.
Rocchini, D., Boyd, D.S., Féret, J.-B., Foody, G.M., He, K.S. *et al*. (2016) Satellite remote sensing to monitor species diversity: potential and pitfalls. *Remote Sensing in Ecology and Conservation* 2, 25–36. DOI: 10.1002/rse2.9.
Rodríguez, J.P., Sucre, B., Mileham, K., Sánchez-Mercado, A., De Andrade, N. *et al*. (2022) Addressing the Biodiversity Paradox: mismatch between the co-occurrence of biological diversity and the human, financial and institutional resources to address its decline. *Diversity* 14, e708. DOI: 10.3390/d14090708.
Rondinini, C., Di Marco, M., Visconti, P., Butchart, S.H.M. and Boitani, L. (2014) Update or outdate: long-term viability of the IUCN Red List. *Conservation Letters* 7, 126–130. DOI: 10.1111/conl.12040.
Roque, F.O., Corrêa, E.C., Valente-Neto, F., Stefan, G., Schulz, G. *et al*. (2017) Idiosyncratic responses of aquatic and terrestrial insects to different levels of environmental integrity in riparian zones in a karst tropical dry forest region. *Austral Entomology* 56, 459–465. DOI: 10.1111/aen.12259.
Rosset, V., Simaika, J.P., Arthaud, F., Bornette, G., Vallod, D. *et al*. (2013) Comparative assessment of scoring methods to evaluate the conservation value of pond and small lake biodiversity. *Aquatic Conservation: Marine and Freshwater Ecosystems* 23, 23–36. DOI: 10.1002/aqc.2287.
Rydhmer, K., Jansson, S., Still, L., Beck, B.D., Chatzaki, V. *et al*. (2024) Photonic sensors reflect variation in insect abundance and diversity across habitats. *Ecological Indicators* 158, e11483. DOI: 10.1016/j.ecolind.2023.111483.
Salafsky, N., Salzer, D., Stattersfield, A.J., Hilton-Taylor, C. Neugarten, R. *et al*. (2008) A standard lexicon for biodiversity conservation: unified classifications of threats and actions. *Conservation Biology* 22, 897–911. DOI: 10.1111/j.1523-1739.2008.00937.x.
Sály, P., Dolezsai, A., Lukács, B., Várbíró, G. and Erős, T. (2020) Characterizing surrogacy performance in the systematic conservation planning of riverine networks. *Aquatic Conservation: Marine and Freshwater Ecosystems* 30, 246–259. DOI: 10.1002/aqc.3261.
Samraoui, B., Boudot, J.P., Ferreira, S., Riservato, E., Jović, M. *et al*. (2010) The status and distribution of dragonflies. In: García, N., Cuttelod, A. and Abdul Malak, D. (eds) *The Status and Distribution of Freshwater Biodiversity in Northern Africa*. IUCN, Gland, Switzerland, Cambridge and Malaga, Spain, pp. 51–70.
Samways, M.J. (2006) National Red List of South African dragonflies (Odonata). *Odonatologica* 35, 341–368.
Samways, M.J. (2008) *Dragonflies and Damselflies of South Africa*. Pensoft, Sophia, Bulgaria.
Samways, M.J. and Grant, P.B.C. (2007) Honing Red List assessments of lesser-known taxa in biodiversity hot-spots. *Biodiversity and Conservation* 16, 2575–2586. DOI: 10.1007/s10531-006-9023-z.
Samways, M.J. and Simaika, J.P. (2016) *Manual of Freshwater Assessment: Dragonfly Biotic Index*. South African National Biodiversity Institute, Pretoria, South Africa.
Samways, M.J., Taylor, S. and Tarboton, W. (2005) Extinction reprieve following alien removal. *Conservation Biology* 19, 1329–1330. DOI: 10.1111/j.1523-1739.2005.00176.x.
Samways, M.J., McGeoch, M.A. and New, T.R. (2010) *Insect Conservation: Approaches and Methods*. Oxford University Press, Oxford.
Samways, M.J., Sharratt, N.J. and Simaika, J.P. (2011) Effect of alien riparian vegetation and its removal on a highly endemic river macroinvertebrate community. *Biological Invasions* 13, 1305–1324. DOI: 10.1007/s10530-010-9891-8.
Santos Ferreira, V.R., Oliveira de Resende, B., Costa Bastos, R., Silva da Brito, J., de Carvalho, F.G. *et al*. (2023) Amazonian Odonata Trait Bank. *Ecology and Evolution* 13, e10149. DOI: 10.1002/ece3.10149.
Scheibler, E.E., Montemayor, S.I. and Melo, M.C. (2022) Macroinvertebrate communities in high mountain desert wetlands: building biological indexes to address the vulnerability of species and communities. *Insect Conservation and Diversity* 15, 714–724. DOI: 10.1111/icad.12597.
Schmeller, D.S., Julliard, R., Bellingham, P.J., Böhm, M., Brummitt, N. *et al*. (2015) Towards a global terrestrial species monitoring program. *Journal for Nature Conservation* 25, 51–57. DOI: 10.1016/j.jnc.2015.03.003.
Schmidt, K.J., Soluk, D.A., Mays Maestas, S.E. and Britten, H.B. (2021) Persistence and accumulation of environmental DNA from an endangered dragonfly. *Scientific Reports* 11, e18987. DOI: 10.1038/s41598-021-98099-1.
Schultz, H., Waringer, J.A. and Chovanec, A. (2003) Assessment of the ecological status of Danubian flood-plains at Tulln (Lower Austria) based on the Odonata Habitat Index (OHI). *Odonatologica* 32, 355–370.
Seidel, D, Stiers, M., Ehbrecht, M, Werning, M. and Annighöfer, P. (2021) On the structural complexity of

central European agroforestry systems: a quantitative assessment using terrestrial laser scanning in single-scan mode. *Agroforestry Systems* 95, 669–685. DOI: 10.1007/s10457-021-00620-y.

Seidu, I., Danquah, E., Nsor, C.A., Kwarteng, D.A. and Lancaster, L.T. (2017) Odonata community structure and patterns of land use in the Atewa Range Forest Reserve, Eastern Region (Ghana). *International Journal of Odonatology* 20, 173–189. DOI: 10.1080/13887890.2017.1369179.

Seidu, I., Nsor, C.A., Danquah, E. and Lancaster, L.T. (2018) Odonata assemblages along an anthropogenic disturbance gradient in Ghana's Eastern Region. *Odonatologica* 47, 73–100. DOI: 10.5281/zenodo1239947.

Serrana, J.M., Miyake, Y., Gamboa, M. and Watanabe, K. (2019) Comparison of DNA metabarcoding and morphological identification for stream macroinvertebrate biodiversity assessment and monitoring. *Ecological Indicators* 101, 963–972. DOI: 10.1016/j.ecolind.2019.02.008.

Sganzerla, C., Dalzochio, M.S., Prass, G.S. and Périco, E. (2021) Effects of urbanization on the fauna of Odonata on the coast of southern Brazil. *Biota Neotropica* 21, e20201122. DOI: 10.1590/16760611-BN-2020-1122.

Shirk, A.J., Jones, G.M., Yang, Z., Davis, R.J., Ganey, J.L. *et al.* (2023) Automated habitat monitoring systems linked to adaptive management: a new paradigm for species conservation in an era of rapid environmental change. *Landscape Ecology* 38, 7–22. DOI: 10.1007/s10980-022-01457-1.

Shuker, L.J., Gurnell, A.M., Wharton, G., Gurnell, D.J., England, J. *et al.* (2017) MoRPh: a citizen science tool for monitoring and appraising physical habitat changes in rivers. *Water and Environment Journal* 31, 418–424. DOI: 10.1111/wej.12259.

Siesa, M.E., Padoa-Schioppa, E., Ott, J., De Bernardi, F. and Ficetola, G.F. (2014) Assessing the consequences of biological invasions on species with complex life cycles: impact of the alien crayfish *Procambarus clarkii* on Odonata. *Ecological Indicators* 46, 70–77. DOI: 10.1016/j.ecolind.2014.05.036.

Šigutová, H., Šipoš, J. and Dolný, A. (2019) A novel approach involving the use of Odonata as indicators of tropical forest degradation: when family matters. *Ecological Indicators* 104, 229–236. DOI: 10.1016/j.ecolind.2019.05.001.

Šigutová, H., Dolný, A., Samways, M.J., Hardersen, S., Oliveira-Junior, J.M. *et al.* (2023) Odonata as indicators of pollution, habitat quality, and landscape disturbance. In: Córdoba-Aguilar, A., Beatty, C.D. and Bried, J.T (eds) *Dragonflies and Damselflies: Model Organisms for Ecological and Evolutionary Research* 2nd edn. Oxford University Press, Oxford, pp. 371–384. DOI:10.1093/oso/9780192898623.003.0026.

Silva, D.R.O., Herlihy, A.T., Hughes, R.M. and Callisto, M. (2017) An improved macroinvertebrate multimetric index for the assessment of wadeable streams in the neotropical savanna. *Ecological Indicators* 81, 514–525. DOI: 10.1016/j.ecolind.2017.06.017.

Silva, L.F.R., Castro, D.M.P., Juen, L., Callisto, M., Hughes, R.M. and Hermes, M.G. (2021) A matter of suborder: are Zygoptera and Anisoptera larvae influenced by riparian vegetation in Neotropical Savanna streams? *Hydrobiologia* 848, 4433–4443. DOI: 10.1007/s10750-021-04642-6.

Silva, L.F.R., Castro, D.M.P., Juen, L., Callisto, M., Hughes, R.M. and Hermes, M.G. (2023) Ecological thresholds of Odonata larvae to anthropogenic disturbances in Neotropical savanna headwater streams. *Hydrobiologia* 851, 313–326. DOI: 10.1007/s10750-022-05097-z.

Simaika, J.P. and Samways, M.J. (2009a) An easy-to-use index of ecological integrity for prioritizing freshwater sites and for assessing habitat quality. *Biodiversity and Conservation* 18, 1171–1185. DOI: 10.1007/s10531-008-9484-3.

Simaika, J.P. and Samways, M.J. (2009b) Reserve selection using Red Listed taxa in three global biodiversity hotspots: dragonflies in South Africa. *Biological Conservation* 142, 638–651. DOI: 10.1016/j.biocon.2008.11.012.

Simaika, J.P. and Samways, M.J. (2010) Large-scale estimators of threatened freshwater catchment species relative to practical conservation management. *Biological Conservation* 143, 311–320. DOI: 10.1016/j.biocon.2009.10.012.

Simaika, J.P. and Samways, M.J. (2011) Comparative assessment of indices of freshwater habitat conditions using different invertebrate taxon sets. *Ecological Indicators* 11, 370–378. DOI: 10.1016/j.ecolind.2010.06.005.

Simaika, J.P. and Samways, M.J. (2012) Using dragonflies to monitor and prioritize lotic systems: a South African perspective. *Organisms, Diversity and Evolution* 12, 251–259. DOI: 10.1007/s13127-012-0104-4.

Simaika, J.P., Samways, M.J., Kipping, J., Suhling, F., Dijkstra, K.-D.B. *et al.* (2013) Continental-scale conservation prioritization of African dragonflies. *Biological Conservation* 157, 245–254. DOI: 10.1016/j.biocon.2012.08.039.

Simons, A.L., Mazor, R., Stein, E.D., and Nuzhdin, S. (2019) Using alpha, beta, and zeta diversity in describing the health of stream-based benthic macroinvertebrate communities. *Ecological Applications* 29, e01896. DOI: 10.1002/eap.1896.

Sittenthaler, M., Fischer, I., Chovanec, A., Koblmüller, S., Macek, O. *et al.* (2023) DNA barcoding of exuviae for species identification of Central European damselflies and dragonflies (Insecta: Odonata). *Journal of Insect Conservation* 27, 435–450. DOI: 10.1007/s10841-023-00467-x.

Smith, J., Samways, M.J. and Taylor, S. (2007) Assessing riparian quality using two complementary sets of bioindicators. *Biodiversity and Conservation* 16, 2695–2713. DOI: 10.1007/s10531-006-9081-2.

Solimini, A.G., Bazzanti, M., Ruggiero, A. and Carchini, G. (2008) Developing a multimetric index of ecological integrity based on macroinvertebrates of mountain ponds in central Italy. *Hydrobiologia*, 597, 109–123. DOI: 10.1007/s10750-007-9226-7.

Späth, J., Fick, J., McCallum, E., Cerveny, D., Nording, M.L. and Brodin, T. (2022) Wastewater effluent affects behaviour and metabolomic endpoints in damselfly larvae. *Scientific Reports* 12, e6830. DOI: 10.1038/s41598-022-10805-9.

Stephenson, P.J., Ntiamoa-Baidu, Y. and Simaika, J.P. (2020) The use of traditional and modern tools for monitoring wetlands biodiversity in Africa: challenges and opportunities. *Frontiers in Environmental Science* 8, e61. DOI: 10.3389/fenvs.2020.00061.

Suhling, F., Sahlén, G., Martens, A., Marais, E. and Schütte, C. (2006) Dragonfly assemblages in arid tropical environments: a case study from western Namibia. *Biodiversity and Conservation* 15, 311–332. DOI: 10.1007/s10531-005-2007-6.

Suhling, F., Suhling, I. and Richter, O. (2015) Temperature response of growth of dragonflies: an overview. *International Journal of Odonatology* 18, 15–30. DOI: 10.1080/13887890.2015.1009392.

Sundar, S., Heino, J., Roque, F.O., Simaika, J.P., Melo, A.S. *et al.* (2019) Conservation of freshwater macroinvertebrate biodiversity in tropical regions. *Aquatic Conservation: Marine and Freshwater Ecosystems* 30, 1238–1250. DOI: 10.1002/aqc.3326.

Sweeney, B.W., Battle, J.M., Jackson, J.K. and Dapkey, T. (2011) Can DNA barcodes of stream macroinvertebrates improve descriptions of community structure and water quality? *Journal of the North American Benthological Society* 30, 195–216. DOI: 10.1899/10-016.1.

Taylor, P., Smallshire, D. J., Parr, A.J., Brooks, S.J., Cham, S.A. *et al.* (2021) *State of Dragonflies in Britain and Ireland*. British Dragonfly Society, Old Weston, Huntingdon, UK.

Theron, K.J., Pryke, J.S. and Samways, M.J. (2022a) Identifying managerial legacies within conservation corridors using remote sensing and grasshopper bioindicators. *Ecological Applications* 32, e02496. DOI: 10.1002/eap.2496.

Theron, K.J., Pryke, J.S. and Samways, M.J. (2022b) Maintaining functional connectivity in grassland corridors between plantation forests promotes high-quality habitat and conserves range-restricted grasshoppers. *Landscape Ecology* 37, 2081–2097. DOI: 10.1007/s10980-022-01471-3.

Theron, K.J., Pryke, J.S., Latte, N. and Samways, M.J. (2022c) Mapping an alien invasive shrub within conservation corridors using super-resolution satellite imagery. *Journal of Environmental Management* 321, e116023. DOI: 10.1016/j.envman.2022.116023.

Thomson, S.A., Pyle, R.L., Ahyong, S.T., Alonso-Zarazaga, M., Ammirati, J. *et al.* (2018) Taxonomy based on science is necessary for global conservation. *PLoS Biology* 16, e2005075. DOI: 10.1371/journal.pbio.2005075.

Tierney, D.A. (2022) Linking restoration to the IUCN red list for ecosystems: a case study of how we might track the Earth's ecosystems. *Austral Ecology* 47, 852–866. DOI: 10.1111/aec.13168.

Tippler, C., Wright, I.A., Davies, P.J. and Evans, C.R. (2018) Are Odonata nymph adversely affected by impaired water quality in urban streams. *Austral Ecology* 43, 890–902. DOI: 10.1111/aec.12630.

Tisseuil, C., Cornu, J.-F., Beauchard, O., Brosse, S., Darwall, W. *et al.* (2013) Global diversity patterns and cross-taxa convergence in freshwater systems. *Journal of Animal Ecology* 82, 365–376. DOI: 10.1111/1365-2656.12018.

Tolman, E.R., Bruchim, O.R., Driever, E.S., Jordan, D., Kohli, M.K. *et al.* (2023) Changes in effective population size of Odonata in response to climate change revealed through genomics. *International Journal of Odonatology*, 26, 205–211. DOI:10.48156/1388.2023.1917241.

Torres-Cambas, Y., Cabana-Ortero, M., Lorenzo-Carballa, M.O. and Cordero-Rivera, A. (2016) Conservation status and protection of three Antillean endemic damselflies. *Journal of Insect Conservation* 20, 277–284. DOI: 10.1007/s10841-016-9862-1.

Turnhout, S. and Halffman, W. (2023) Readjusting observational grids in dragonfly field guides. *Social Studies of Science* e03063127231183011. DOI: 10.1177/03063127231183011.

UNEP (2022) United Nations Environment Programme/Convention on Biological Diversity/15.4. Kunming–Montreal Global Biodiversity Framework. 15 pp.

Uyizeye, E., Clausnitzer, V., Kipping, J., Dijkstra, K.-D.B., Willey, L. and Kaplin, B.A. (2021) Developing an odonate-based index for prioritizing conservation sites and monitoring restoration of freshwater ecosystems in Rwanda. *Ecological Indicators* 125, e107586. DOI: 10.1016/j.ecolind.2021.107586.

Vadas, R.L., Hughes, R.M., Bae, Y.J., Back, M.J., Gonzáles, O.C.B. *et al.* (2022) Assemblage-based biomonitoring of freshwater ecosystem health via multimetric indices: a critical review and suggestions for improving their applicability. *Water Biology and Security* 1, e100054. DOI: 10.1016/j.watbs.2022.100054.

Valente-Neto, F., Roque, F.O., Rodrigues, M.E., Juen, L. and Swan, C.M. (2015) Toward a practical use of Neotropical odonates as bioindicators: testing congruence across taxonomic resolution and life stages. *Ecological Indicators* 61, 952–959. DOI: 10.1016/j.ecolind.2015.10.052.

Valente-Neto, F., Rodrigues, M.E. and de Oliveira Roque, F. (2018) Selecting indicators based on biodiversity surrogacy and environmental response in a riverine network: bringing operationality to biomonitoring.

Ecological Indicators 94, 198–206. DOI: 10.1016/j.ecolind.2018.06.066.
Valente-Neto, F., Martínez, B.T., Hughes, R.M., Ferreira, A., Severo-Neto, F. *et al.* (2021) Incorporating costs, thresholds and spatial extents for selecting stream bioindicators in an ecotone between two Brazilian biodiversity hotspots. *Ecological Indicators* 127, e107761. DOI: 10.1016/j.ecolind.2021.107761.
Valente-Neto, F., Piovezan-Borges, A.C., Urbieta, G.L., Samways, M.J. and Roque, F.O. (2022) Research networks should improve connectivity for halting freshwater insect extinctions. *Ecological Entomology* 47, 63–75. DOI: 10.1111/een.13091.
van Klink, R., August, T., Bas, Y., Bodesheim, P., Bonn, A. *et al.* (2022) Emerging technologies revolutionise insect ecology and monitoring. *Trends in Ecology and Evolution* 37, 872–885. DOI: 10.1016/j.tree.2022.06.001.
van Strien, A.J. and van Grunsven, R.H.A. (2023) In the past 100 years dragonflies declined and recovered by habitat restoration and climate change. *Biological Conservation* 277, e109865. DOI: 10.1016/j.biocon.2022.109865.
van Strien, A.J., Termaat, T., Groenendijk, D., Mensing, V. and Kéry, M. (2010) Site-occupancy models may offer new opportunities for dragonfly monitoring based on daily species lists. *Basic and Applied Ecology* 11, 495–503. DOI: 10.1016/j.baae.2010.05.003.
van Strien, A.J., van Swaay, C.A.M. and Termaat, T. (2013) Opportunistic citizen science data of animal species produce reliable estimates of distribution trends if analysed with occupancy models. *Journal of Applied Ecology* 50, 1450–1458. DOI: 10.1111/1365-2664.12158.
Vega-Sánchez, Y.M., Mendoza-Cuenca, L.F. and González-Rodríguez, A. (2019) Complex evolutionary history of the American Rubyspot damselfly, *Hetaerina americana* (Odonata): evidence of cryptic speciation. *Molecular Phylogenetics and Evolution* 139, e106536. DOI: 10.1016/j.ympev.2019.106536.
Vega-Sánchez, Y.M., Mendoza-Cuenca, L.F. and González-Rodríguez, A. (2020) *Hetaerina calverti* (Odonata: Zygoptera: Calopterygidae) sp. nov., a new cryptic species of the American Rubyspot complex. *Zootaxa* 4766, 485–497. DOI: 10.11646/zootaxa.4766.3.7.
Veras, D.S., Pinto, N.S., Calvão, L., Lustosa, G.S., Azevêdo, C.A.S. and Juen, L. (2022) Environmental thresholds of dragonflies and damselflies from a Cerrado-Caatinga ecotone. *Environmental Monitoring and Assessment* 194, e614. DOI: 10.1007/s10661-022-10310-6.
Vilenica, M. and Mihaljević, Z. (2022) Odonata assemblages in anthropogenically impacted habitats in the Drava River: a long-term study. *Water* 14, e3119. DOI: 10.3390/ w14193119.
Vilenica, M., Pozojević, I., Vučković, N. and Mihaljević, Z. (2020) How suitable are man-made water bodies as habitats for Odonata? *Knowledge and Management of Aquatic Ecosystems* 421, e3. DOI: 10.1051/kmae/2020008.
Vorster, C., Samways, M.J., Simaika, J.P., Kipping, J., Clausnitzer, V. *et al.* (2020) Development of a new continental-scale index for freshwater assessment based on dragonfly assemblages. *Ecological Indicators* 109, e105819. DOI: 10.1016/j.ecolind.2019.105819.
Vourka, A., Karaouzas, I. and Parmakelis, A. (2023) River benthic macroinvertebrates and environmental DNA metabarcoding: a scoping review of eDNA sampling, extraction, amplification and sequencing methods. *Biodiversity and Conservation* 32, 4221–4238. DOI: 10.1007/s10531-023-02710-y.
Wang, L.-J., Chou, Y.-W. and Huang, J.-P. (2022) Testing the effect of sampling effort on inferring phylogeographic history in *Psolodesmus mandarinus* (Calopterygidae, Odonata). *Diversity* 14, e809. DOI: 10.3390/d14100809.
Wang, Y., Yang, F., Hua, C., Hu, J., Wang, J. *et al.* (2021) Comparing the efficiencies of individual-scoring and complementarity-based algorithms in prioritizing existing protected areas. *Ecological Indicators* 125, e107539. DOI: 10.1016/j.ecolind.2021.107539.
Warwick, R. M. (1986). A new method for detecting pollution effects on marine macrobenthic communities. *Marine Biology* 92, 557–562.
Watson, J.A.L., Arthington, A.H. and Conrick, D.L. (1982) Effect of sewage effluent on dragonflies (Odonata) of Bulimba Creek, Brisbane. *Australian Journal of Marine and Freshwater Research* 33, 517–528.
Webster, H.J., Emami-Khoyi, A., van Dyk, J.C., Teske, P.R. and Jansen van Vuuren, B. (2020) Environmental DNA metabarcoding as a means of estimating species diversity in an urban aquatic ecosystem. *Animals* 10, e2064. DOI: 10.3390/ani10112064.
Webster, M.T., Beaurepaire, A., Neumann, P. and Stolle, E. (2023) Population genomics for insect conservation. *Annual Review of Animal Biosciences* 11, 115–140. DOI: 10.1146/annurev-animal-122221075025.
Weigand, H., Beermann, A.J., Čiampor, F., Costa, F.O., Csabai, Z. *et al.* (2019) DNA barcode libraries for the monitoring of aquatic biota in Europe: gap-analysis and recommendations for future work. *Science of the Total Environment* 678, 499–524. DOI: 10.1016/j.scitotenv.2019.04.247.
Wheeler, Q.D. and Miller, K.B. (2017) The science of insect taxonomy: prospects and needs. In: Foottit, R.G. and Adler, P.H. (eds) *Insect Biodiversity: Science and Society*, Vol. 1, 2nd edn. Wiley-Blackwell, Oxford, pp. 499–526.
White, E.L., Hunt, P.D., Schlesinger, M.D., Corser, J.D. and deMaynadier, P.G. (2015) Prioritizing Odonata for conservation action in the northeastern USA. *Freshwater Science* 34, 1079–1093. DOI: 10.1086/682287.
Williams, P., Whitfield, M., Biggs, J., Bray, S., Fox, G. *et al.* (2004) Comparative biodiversity of rivers,

streams, ditches and ponds in an agricultural landscape in Southern England. *Biological Conservation* 115, 329–341. DOI: 10.1016/S0006-3207(3)00153-8.

Wilson, J.-J., Sing, K-W., Floyd, R.M. and Hebert, P.D.N. (2017) DNA barcodes and insect biodiversity. In: Foottit, R.G. and Adler, P.H. (eds) *Insect Biodiversity: Science and Society*, Vol. 1, 2nd edn. Wiley-Blackwell, Oxford, pp. 575–592.

Yoshioka, A., Shimizu, A, Oguma, H, Kumada, N., Fukasawa, K. *et al.* (2020) Development of a camera trap for perching dragonflies: a new tool for freshwater environmental assessment. *PeerJ* 8, e9681. DOI: 10.7717/peerj.9681.

Yoshioka, A., Mitamura, T., Matsuki, N., Shimizu, A., Ouchi, H. *et al.* (2023) Camera-trapping estimates of the relative population density of *Sympetrum* dragonflies: application to multihabitat users in agricultural landscapes. *PeerJ* 11, e14881. DOI: 10.7717/peerj.14881.

Zizka, V.M.A., Geiger, M.F. and Leese, F. (2020) DNA metabarcoding of stream invertebrates reveals spatiotemporal variation but consistent status class assessments in a natural and urban river. *Ecological Indicators* 115, e106383. DOI: 10.1016/j.ecolind.2020.106383.

8 Future-proofing Freshwaters and Their Dragonfly Sentinels

'Conservation of dragonflies: sentinels for freshwater conservation' as interpreted by artificial intelligence and crafted by Marlon Samways.

DOI: 10.1079/9781789248395.0008

Abstract

The fate of dragonflies, freshwaters and people are entwined, with dragonflies being key for assessing the condition and changes in freshwaters. Having healthy water and a heterogeneous riparian zone improves human well-being, stimulating more care for nature and conservation action.

People need freshwater biodiversity and their ecosystem services, with nature-based solutions now helping sustain service delivery.

Major stressors on freshwater biodiversity are being addressed by some key conservation actions, especially using future-proofing approaches, improving environmental water flows and the development of a nature futures framework.

Different types of freshwaters require certain specific conservation actions. But more conciliation is needed between freshwater conservation and threatened species conservation through improved information sharing among policy makers, management practitioners and scientists. New funding models are being proposed to achieve this end equitably globally.

Being amphibiotic, dragonflies and other sentinel taxa are paving the way for measuring freshwater/riparian improvement, and a better world for all.

8.1 Value of Freshwater Biodiversity

8.1.1 Dragonflies–freshwaters–people

The last few chapters have shown that dragonflies, freshwaters and people share an entwined fate, whether relating to responses to stressors or to conservation action. In turn, assessment and monitoring are crucial for determining the state of and change in freshwater ecosystems and their adjoining land surfaces. Dragonflies, being amphibiotic, can contribute greatly to these assessments and monitoring schemes.

We shall now briefly review some of the recent important thrusts relating to the conservation of freshwater biodiversity, then move on to focus down to some local educational and outreach initiatives associated with dragonflies. Finally, we end with an important call for action from the seven Chairs of the International Union for the Conservation of Nature (IUCN), who emphasize that we must urgently address the critically important *joint* issue of biodiversity loss and climate change.

8.1.2 Freshwater biodiversity and human well-being

Despite much evidence that being in harmony with nature has multiple benefits for both humans and nature, these benefits are not fully considered by health and conservation policy makers. Evidence points to physical connection with nature improving human cognition, social skills, and physical and mental health, as well as the psychological connection to nature, leading to pro-environmental behaviours and values, which will contribute to achieving the United Nations Sustainable Development Goals (Barragan-Jason *et al.*, 2023). Furthermore, the perception of biodiversity loss as a distant threat is related to weaker engagement in pro-environmental behaviours (Bosone and Bertoldo, 2022).

A way forward is to improve the human–nature connection, where easy access to living nature greatly improves the well-being of city dwellers, whom otherwise might be out of touch with wild nature (Colding *et al.*, 2020). An interrelated factor is that human well-being is improved by elevating aesthetic benefits through landscape design, planning and management. Riparian aesthetic quality is determined by the quality of both vegetation and the water body. Higher vegetation quality depends on higher plant abundance, more vegetation patches and more vegetation patch types, while higher water-body quality depends on clear and clean water and a larger water surface. Vertical vegetation configuration, diversity of patch type and patch shape, and shrub species diversity contribute most to the attributes of naturalness, complexity and plant species diversity. This suggests that riparian management can contribute to the positive feedback cycle of a more aesthetic environment, leading to a greater care of nature (Zhang *et al.*, 2022) (Fig. 8.1).

8.1.3 Freshwater ecosystem services

People need freshwater biodiversity via fundamental ecosystem services that fall into three main categories: material (food, health and genetic resources, and material goods), non-material (culture, education and science, and recreation), and regulating (catchment integrity, climate regulation, water purification and nutrient cycling). When freshwater biodiversity is protected, conserved and restored in an integrated manner, as well as being appreciated by people, it will continue to contribute to human well-being and a sustainable future (Lynch *et al.* 2023a).

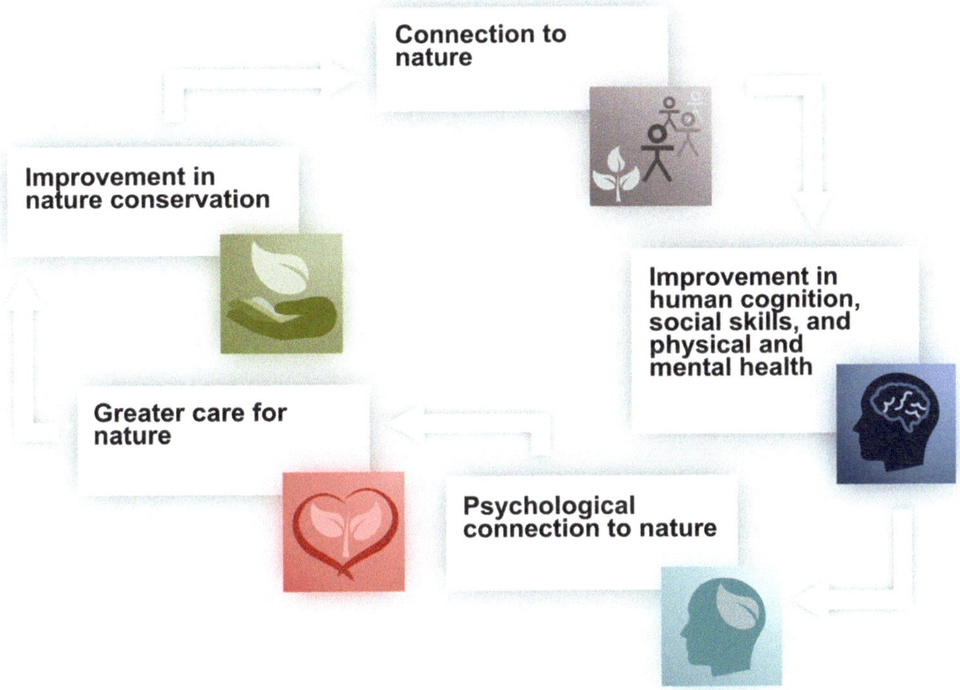

Fig. 8.1. The positive feedback cycle between nature conservation and human well-being. (Illustration: Robyn Symons).

Nature-based solutions (NBS) refer to naturally occurring, restored and artificial ecosystems that deliver ecosystem services to address key societal needs, while also including management actions that protect or enhance the ecosystems and their service provision. NBS are both beneficial for society and the environment through restoring or protecting habitats while promoting the sustainable delivery of key ecosystem services. NBS play a major role in restoring degraded freshwater and floodplain ecosystems, enhancing in-stream water quality and improving hydrological connectivity among freshwater ecosystems. However, system-level monitoring is required to ensure that freshwater NBS produce their promised benefits for ecosystems and species (van Rees *et al.*, 2023) (Fig. 8.2).

8.2 Current Freshwater Biodiversity Conservation Challenges and Solutions

8.2.1 Global risks

The World Economic Forum (The Global Risks Report, 2023) identifies the ten most severe risks facing the planet in the next 10 years, based on a global scale in order of importance as: 1) failure to mitigate climate change; 2) failure of climate-change adaptation; 3) natural disasters and extreme weather events; 4) biodiversity loss and ecosystem collapse; 5) large-scale involuntary human migration; 6) natural resource crises; 7) erosion of social cohesion and societal polarization; 8) widespread cybercrime and cyber insecurity; 9) geoeconomic confrontation; and 10) large-scale environmental damage incidents. These stressors are interconnected, meaning that all must be tackled together. Water is a central theme among some of these risks, which must be addressed all around the globe as it is critical to the future of biodiversity and human well-being in an ever-changing world.

Some of the major impacts on freshwater biodiversity are great changes in soil erosion–deposition dynamics, changes in hydrological regimes via water impoundment and diversion, more extensive and intensive land-use conversion, increased chemical and nutrient pollution, and an increase in the number and impact of invasive alien species (Flitcroft *et al.*, 2019).

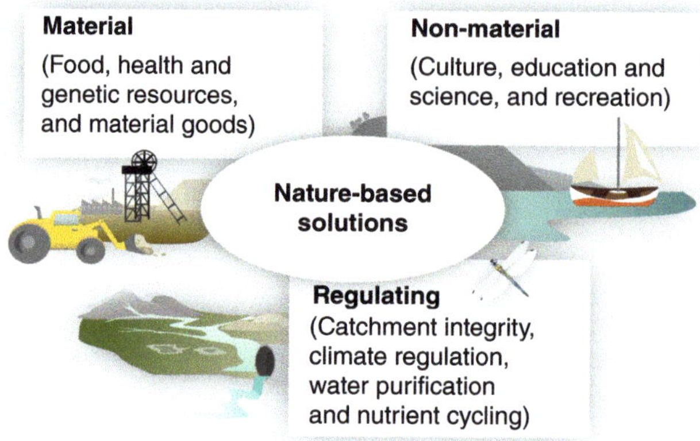

Fig. 8.2. The three major categories of freshwater biodiversity service provision revolve around nature-based solutions that lead to human well-being and a sustainable future. (Illustration: Robyn Symons).

The combined impacts on aquatic systems tend to homogenize biodiversity, with an important overall goal now being to maintain natural spatial environmental heterogeneity, especially where there has been biotic homogenization (Agra *et al.* 2023).

8.2.2 Mitigating the risks to freshwaters

With the decline of freshwater species due to the disturbance and transformation of freshwater ecosystems, there is now focus on the protection of freshwater quality and ecological integrity, stimulating conservation and restoration efforts (van Rees *et al.* 2021). To this end, Tickner *et al.* (2020) devised an Emergency Recovery Plan (ERP) with six key actions to help reverse freshwater biodiversity loss: 1) accelerate the implementation of environmental flows; 2) improve water quality to sustain aquatic life; 3) protect and restore critical habitats; 4) manage the exploitation of freshwater species and riverine aggregates (e.g. sand, gravel and crushed rock); 5) prevent and control non-native species invasions in freshwater habitats; and 6) safeguard and restore freshwater connectivity. The ERP's recommendations are aligned with the strategic plan of the Convention on Biological Diversity/Kunming–Montreal Global Biodiversity Framework (UNEP, 2022) to restore ecosystems and recover biodiversity over the next three decades.

These six actions can be put in place using future-proofing approaches. These approaches anticipate future risks and adversely synergistic effects while minimizing likely stressors so that freshwater biodiversity is more resilient in the face of possible future stressors, including climate change. The aim of future-proofing freshwater ecosystems is to account for the uncertainty of future conditions that have no historical baseline, yet the level of uncertainty associated with future conditions is unprecedented. Future-proofing for the ERP for freshwater biodiversity requires anticipating future changes and developing and implementing conservation actions to address the future changes. Based on what has been learned from various local case studies, meaningful future-proofing approaches that are likely to be effective and successful can be implemented to support the ERP with science-informed choices, even where there are potentially uncertain and new future conditions. Importantly, this current time is an inflection point for global freshwater biodiversity loss, which means that learning from the defeats and successes of the past can support improved actions toward a sustainable future (Lynch *et al.*, 2023b) (Fig. 8.3).

Furthermore, there is ever-increasing concern over the extent to which human-induced changes to water flow are greatly impacting freshwater biodiversity. This concern is being translated into a focus on restoring environmental flows (e-flows) of water aimed at mitigating the threat of both altered hydrological regimes in river systems and the current functional disconnection between water bodies.

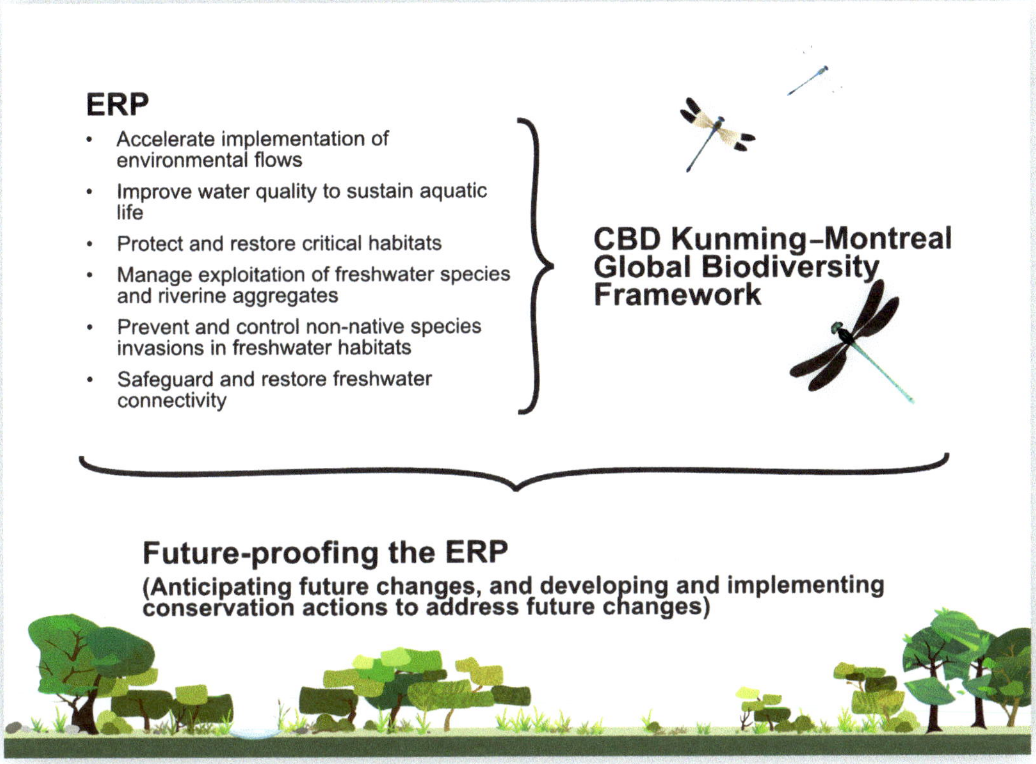

Fig. 8.3. The Emergency Recovery Plan (ERP) for freshwater biodiversity aims to reverse biodiversity loss based on six key actions, which align with the Convention on Biological Diversity's (CBD) Kunming–Montreal Global Biodiversity Framework. The ERP also requires a future-proofing strategy. (Illustration: Robyn Symons).

Improved e-flows can support river conservation and help to restore the biodiversity and resilience of hydrologically altered and water-stressed rivers. It is an adaptive management approach involving several factors: 1) legislation and governance; 2) financial and human resourcing; 3) stakeholder engagement and coproduction of knowledge; 4) collaborative monitoring of ecological and social-economic outcomes; 5) capacity training and research; 6) exploration of trade-offs among water users; 7) changing of water infrastructure to facilitate e-flows and connectivity; and 8) adaptation to climate change. The intention is to equip practitioners with the latest tools and insights to enable adaptive environmental water management globally. This involves restoring rivers and developing adaptiveness to climate change and extreme weather events under the funding commitments of the Kunming–Montreal Global Biodiversity Framework (Arthington *et al.*, 2023).

The Nature Futures Framework (NFF) aims to advance the contribution of scenarios and models for environmental assessments. It places the diverse relationships between people and nature at its core, identifying three value perspectives: 1) Nature for Nature – intrinsic value: nature has value in and of itself, while maintaining its ability to function autonomously, and the preservation of nature's diversity and functions is of primary importance; 2) Nature for Society – instrumental value: nature is primarily valued, and sustainably managed for the benefit of humans; and 3) Nature as Culture – relational value: humans are perceived as an integral part of nature, where societies, cultures, traditions and faiths are intricately intertwined with nature, while including important relational values such as those reflecting cultural identities and ways of life. The NFF is used to explore what desirable freshwater futures would look like from each of the above perspectives. Furthermore, it considers scenario strategies and current models to determine how freshwater modelling can be linked to the NFF in terms of its aims and outcomes. It also identifies the aspects of the NFF that are not yet captured in current freshwater models and suggests possible ways to include them. Overall, the NFF provides future directions for a more holistic freshwater model and scenario development, while also demonstrating how society can benefit from freshwater models that are integrated with the value perspectives of the NFF (Kramer *et al.* 2023) (Fig. 8.4).

8.2.3 Assessment and monitoring of specific freshwater systems

Regarding rivers, past challenges and technological barriers have reduced our capacity to assess and monitor river health globally. Inhibiting factors include relatively poor conservation awareness for freshwater systems, a focus on the extractive rather than ecosystem value of rivers, monitoring methods restricted to local or national conditions and challenges in applying remote sensing monitoring methods to rivers. However, currently available large databases combined with technological advancements can now support the development of an indicator framework that represents the biophysical health of rivers at a global scale. Additionally, a widespread commitment to a framework requires the development of a cohesive monitoring approach, advancing research findings and data syntheses, and improvement in the accuracy of these indicators over time. Through integration of bottom-up and top-down approaches, an effective global framework for river conservation and restoration will develop for use at a wide range of spatial scales, from local to international (Kuehne *et al.*, 2023).

In turn, small water bodies, including ponds, small lakes, feeder streams, ditches and springs, being the most numerous freshwater environments globally, are important not only for freshwater biodiversity but also for ecosystem service delivery, especially carbon cycling. They are among the best remaining examples of intact freshwater ecosystems and the most likely to remain unpolluted, while often being a refuge for species which have disappeared from

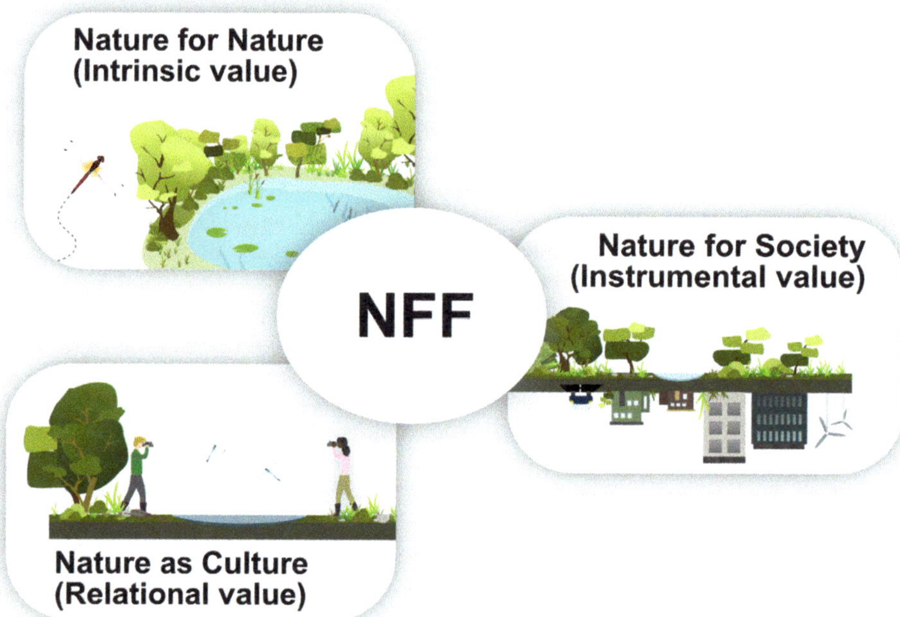

Fig. 8.4. The Nature Futures Framework (NFF) is based on combining three aspects of nature with complementary value systems. (Illustration: Robyn Symons).

larger and more transformed water bodies. The caveat is that they are among the least investigated of freshwater types. Research priorities for these small water bodies require policy actions to better integrate small waters into the management of catchments and landscapes. A first step is to install more monitoring programmes that are effective toward developing measures to protect the biodiversity and the ecosystem services that they provide (Biggs *et al.*, 2017).

In the case of ponds, there are practical steps for the effective incorporation or enhancement of ponds within five policy areas: 1) through open water sustainable urban drainage systems in urban planning; 2) increased incentives in agri-environmental schemes; 3) curriculum inclusion in education; 4) emphasis on ecological scale in mitigation measures following human infrastructure development; and 5) the inclusion of pondscapes in biodiversity conservation policy (Hill *et al.*, 2018).

However, there can be lack of convergence when the management of ponds is compared to the conservation of a threatened species, such as *Coenagrion mercuriale*, a damselfly which is highly dependent on water management actions and protected by the European Habitats Directive in Germany. Nature conservation interests are only a minor component of water management decision-making. Furthermore, knowledge exchange is insufficient between policy makers, management practice and science. Conciliation between water management and nature conservation would enhance the conservation status of the key protected species of ponds under the Habitats Directive (Idilbi *et al.*, 2023) (Fig. 8.5).

8.3 Complementary Aspects to Freshwater Biodiversity Conservation

8.3.1 Recognizing the importance of the terrestrial matrix

Conservation of freshwater must be viewed relative to its terrestrial matrix. Forested watersheds provide

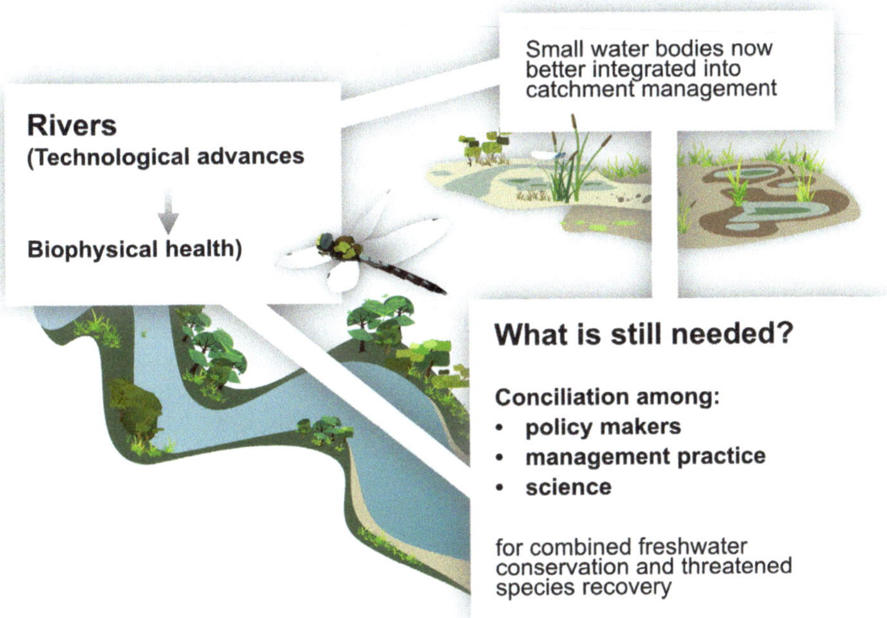

Fig. 8.5. Technological advances are leading to a much better understanding of the biophysical health of rivers. Small water bodies, both running and standing, are now being better integrated into the management of catchments. However, there still needs to be more conciliation between the three main activity areas of conservation, especially with regard to conservation of freshwaters versus that of threatened species. (Illustration: Robyn Symons).

75% of the world's freshwater, delivering water to over half the global human population. This crucial service comes about because healthy forests and trees are integral to the global water cycle as they regulate water quality and quantity, protect soils and provide protection from extreme weather events. This means that it is important to improve the global information base on the vital protective functions of forests for soil and water, while embracing emerging techniques and methodologies that provide guidance and recommendations on managing forests for their water ecosystem services. The point is that intact native forests and well-managed planted forests can be a relatively inexpensive approach to water management while providing many other benefits, meaning that water-focused forests can offer nature-based solutions that ensure global water resilience (FAO, IUFRO and USDA, 2021).

8.3.2 Value of information sharing

The sharing of information and networking has enormous power for better informed actions (Scharf, 2021). There have been increasingly loud calls for greater internationalization of conservation science and practice through greater collaboration and knowledge transfer among people, institutions and countries. The aim is to address, among other things, the biodiversity paradox, where overall the least financial resources for nature conservation, yet the greatest biodiversity, is in the global south (Cardoso *et al.*, 2022). As regards Neotropical insect conservation, Duffus *et al.* (2023) suggest some improvements, such as evaluating the extinction risk of more Neotropical insects to better focus on at-risk species with species-specific policies and to conserve their habitats, incorporating measurable and achievable insect conservation targets into international policies and conventions, and integrating more community engagement and enhanced public awareness to improve insect conservation policies.

For aquatic insects globally, there is a need for increased research collaboration for status and solution studies as many megadiverse countries are not present in the research networks (Valente-Neto *et al.* 2022). While in the case of non-floodplain wetlands, better collaboration, including improved cohesion in classification and terminology, and unified approaches to modelling and simulation will enhance conservation policy (Chen *et al.*, 2022).

8.3.3 Funding perspectives

As pointed out by Rodríguez *et al.* (2023) 'We know how to do conservation – we just need to do more of it!' Besides effective conservation being evidence-based conservation, there must also be significant expansion in the distribution of the funding and the resource base.

The Conference of the Parties (COP27 in 2022) to the United Nations Framework Convention on Climate Change (UNFCCC) ratified a decision to establish a loss and damage fund to compensate low- and middle-income countries suffering negative impacts from climate change. The fund aims to address the Global Adaptation Gap (GAG), which addresses the rising cost of adaptation needed to cope with climate change impacts arising from delayed action to curb greenhouse gas emissions. Given that the implementation of the loss and damage fund still needs negotiation and commitment from signatories to the UNFCCC, certain important knowledge gaps on how loss and damage can be measured, quantified, valued, understood, communicated and adapted to, have been identified. These relate to understanding the complex interactions between people, politics, nature and climate in an interdisciplinary context (Balzter *et al.*, 2023).

At the species level there are developing strategies to put correct monetary value on species within wider economies. This approach is viewed by economists as key to reversing many of the pressures that cause species decline and extinction, while helping reduce the risk of future species extinctions. There is also a fundamentally important change taking place within environmental economics, with economies now being viewed as being embedded in the environment. This fundamental change in emphasis means that environmental economics as well as economics in general are now beginning to view species extinction in a different light. Species losses are now viewed less as one-off events with limited repercussions, but rather as a warning that current economic activities are unsustainable and pushing the environment beyond its ability to support future generations. This means that the conservation community should embrace the positive contribution that economics can make toward averting species extinction through transitioning to nature-positive economies (Fraser *et al.*, 2023) (Fig. 8.6).

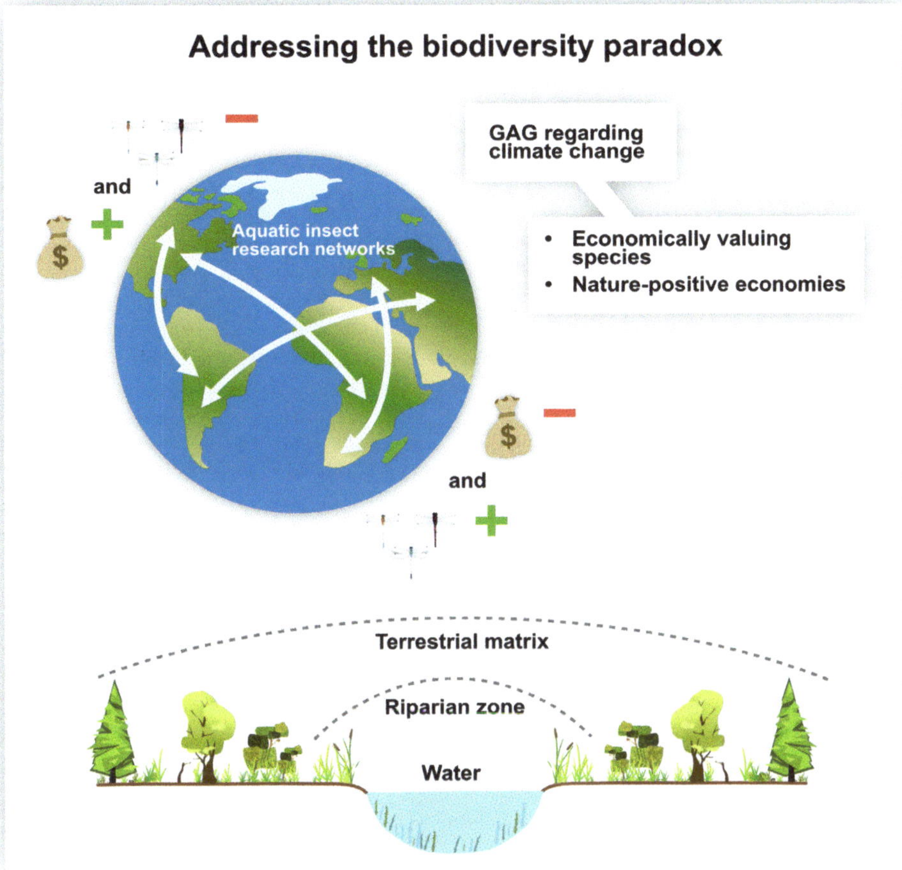

Fig. 8.6. The biodiversity paradox refers to finances being less, yet biodiversity more, in the global south: an equitability issue that requires far more attention. As regards climate change, this paradox is being addressed by the Global Adaptation Gap (GAG), and with new economic valuing models for species and the transition to nature-positive economies. Even among scientific networks associated with aquatic insects, there needs to be much more research cooperation and sharing of information to help save global freshwater diversity. This diversity refers not only to the realm of the water but also that of the riparian zone and the terrestrial matrix. (Illustration: Robyn Symons).

8.4 Conservation Value of Dragonflies

Throughout this book, we have seen that the conservation of dragonflies and their habitats is intimately related to the wider perspective of the conservation of freshwater systems. This relationship largely stems from the main ecosystem service that dragonflies provide as indicators of the state or status of freshwater systems and their neighbouring terrestrial matrices, i.e. they have instrumental value in this regard. They are of course also conservation subjects, i.e. they have intrinsic value.

There are, however, two important corollaries. First, it is odonate species diversity or assemblage composition that is usually the most valuable metric for measuring human impact or recovery following conservation action. Second, dragonflies are not necessarily the sole taxon to use for assessment and monitoring. Plants, both aquatic and terrestrial, are important as a co-variable, as can be other invertebrate taxa and fish. This means that there is always the opportunity to employ a wider spectrum of biodiversity for assessment and monitoring.

While conservation of aquatic systems and freshwater has attracted much big science, which is feeding into policy and management, we still need accurate data gathered locally. Citizen science is contributing greatly to this progress, with dragonfly data now being gathered and collated in many countries across the world. This thrust is being stimulated by education and outreach. We now end with some examples of this.

Conservation action for insects is often improved with public education and outreach. Lampert *et al.* (2023a) developed a framework called Action Competence for Insect Conservation as a tool for improving education, outreach and citizen action for better insect diversity conservation. The framework is applicable in various contexts and settings in both formal (e.g. schools, universities) and informal (e.g. outreach) education, to design and improve educational approaches, develop social interventions for insect conservation and develop instruments to assess such interventions. The framework sits on the educational concept of action competence, which is beyond traditional theory-based education, and rather encourages peoples' action-oriented knowledge, confidence in their actions and willingness to take positive action. The action focus helps overcome divides between knowledge and action implementation, with the framework being applicable to a wide range of local environments while addressing various conservation needs, including that of dragonflies. Furthermore, to improve educational measures, a research tool (the Self-perceived Action Competence for Insect Conservation scale) has been developed to monitor people's competence to sustain insect biodiversity. The scale provides insights into people's self-perceived knowledge, confidence and willingness to take specific actions for sustaining insect biodiversity through assessing, improving or designing educational measures that fit learners' needs (Lampert *et al.*, 2023b).

There are opportunities for employing practical interdisciplinary approaches that combine ecological research and environmental ethics to enhance aquatic insect conservation. In southern Chile, for example, a Field Environmental Philosophy (FEP) approach uses a four-step cycle: 1) transdisciplinary research; 2) poetic communication through the composition of metaphors; 3) design of field activities with an ethical and ecological orientation; and 4) *in situ* conservation, to contribute to biocultural conservation. This FEP-integrated approach has fostered freshwater insect awareness, conservation and value among many people while raising awareness for the underappreciated aquatic insects that they live with. This methodology also promotes educational practices that encourage direct encounters with aquatic insects, their habitats and their lifestyles, making them appreciated and valued as real but small co-inhabitants with whom we share the planet, stimulating greater aquatic insect conservation awareness (Contador *et al.*, 2023).

Whether conservation of freshwater ecosystems, threatened species or insects in general, there is much evidence pointing to the importance of involving education initiatives, especially of younger citizens who are influential in seeing the risks facing the future of the world. Not surprisingly, with dragonflies being so iconic, associated with freshwater, easy to identify and indicators of ecosystem health, they can be easily employed as environmental sentinels and indicators of freshwater health, providing an effective tool for environmental impact assessments and freshwater monitoring. This has led to the development of various education programmes to this effect around the world (Clausnitzer *et al.*, 2017).

Social games are common in all societies and are a way to spend leisure time, reduce stress, promote creativity, learn new skills, train the mind and create social relationships. Integrating games into environmental education promotes communication of complex scientific knowledge to a large audience of different ages and social backgrounds. In turn, games offer many opportunities to understand some of the essential and important concepts of ecology, evolution and conservation. Against this background a game, EcoDragons, was developed. It educates players about the basic principles of ecology and conservation from the perspective of dragonflies. The game offers 133 cards in 3 categories and a landscape with 5 aquatic habitat types. It includes species cards (50 species: 25 zygopterans, 25 anisopterans), action cards (ecological, evolutionary and conservation processes) and anthropogenic disturbance cards (climate change, invasive species and habitat fragmentation). The aim of the game is to colonize empty habitats with species, increase species diversity and solve randomly-arising environmental disturbance with various action cards that involve ecological and evolutionary processes, as well as different conservation measures. The winner of the game is the one who occupies the most habitats, establishes and maintains the largest number of species, and finds solutions to the most human disturbances. While the current EcoDragons is based on European species, the game can be adapted, with a change in species names, for dragonfly assemblages in other regions. Besides the game's educational value, it also fosters public engagement in biodiversity and environmental conservation as well as community science (Khelifa and Mahdjoub, 2021) (Fig. 8.7).

8.5 The Urgency for Change for the Better

Climate change and biodiversity loss are two of the most important risks and challenges facing humans. The two are inseparable, interdependent and mutually reinforcing. In response, the seven Chairs of the IUCN's expert Commissions issued a statement (https://www.iucn.org/crossroads-blog/202311/we-must-unite-our-efforts-fight-climate-change-and-biodiversity-loss; accessed 29 November 2023) emphasizing that the window of opportunity to address these two joint challenges is rapidly closing, noting that the current approaches fall short of what scientific evidence indicates is needed to address them. The statement ends by emphasizing that: 'Solving the linked challenges of climate change and biodiversity loss demands bold and transformative efforts. Responsibility falls on us all.' This is tantamount to all of us doing something positive, such as surveying dragonflies as sentinels for freshwater conservation, even if it is only a small brick in a large wall (Fig. 8.8).

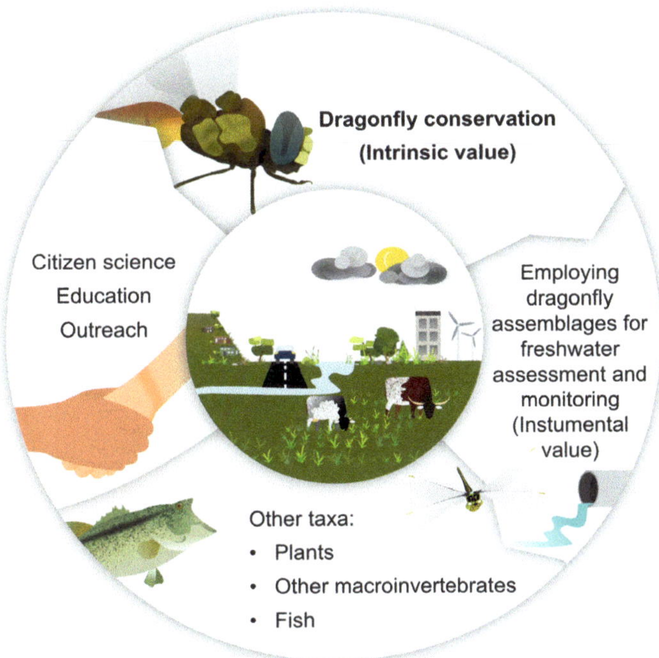

Fig. 8.7. Dragonfly conservation is essentially an intrinsic value activity, yet highly compatible with their instrumental value for measuring the health status of freshwaters and terrestrial margins. Citizen science, education and outreach are all improving the connection between intrinsic and instrumental values, especially when dragonflies are valued alongside other biodiversity in the same local area. (Illustration: Robyn Symons).

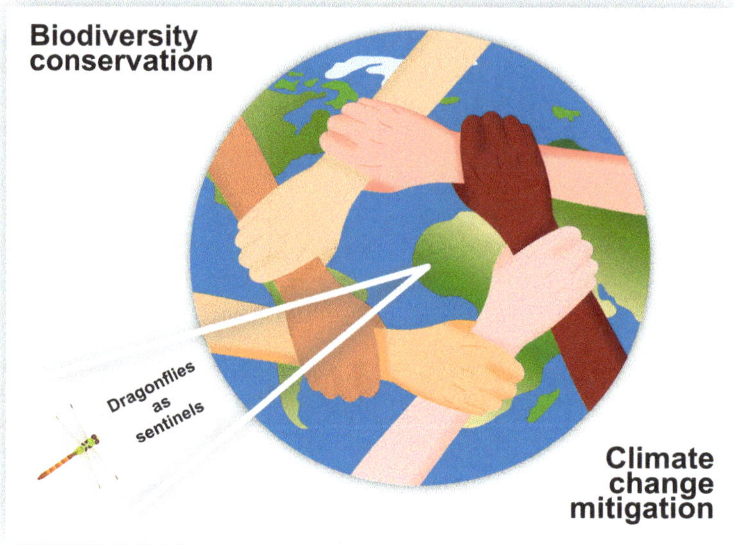

Fig. 8.8. Conserving global biodiversity and addressing climate change go hand in hand. We all have a responsibility to make some contribution toward the protection of Earth's ecosystems. Using dragonflies as freshwater sentinels is one step in that direction. (Illustration: Robyn Symons).

References

Agra, J., Cornelissen, T., Viana-Junior, A.B. and Callisto, M. (2023) A global synthesis and meta-analysis of the environmental heterogeneity effects on the freshwater biodiversity *Oikos* 2023, e10186. DOI: 10.1111/oik.10186.

Arthington, A.H., Tickner, D., McClain, M.E., Acreman, M.C., Anderson, E.P. *et al.* (2023) Accelerating environmental flows implementation to bend the curve of global freshwater biodiversity loss. *Environmental Reviews* (in press). DOI: 10.1139/er-2022-0126.

Balzter, H., Macul, M., Delaney, B., Tansey, K., Espirito-Santo, F. *et al.* (2023) Loss and damage from climate change: knowledge gaps and interdisciplinary approaches. *Sustainability* 15, e11864. DOI: 10.3390/su151511864.

Barragan-Jason, G., Loreau, M., de Mazancourt, C., Singer, M.C. and Parmesan, C. (2023) Psychological and physical connections with nature improve both human well-being and nature conservation: a systematic review of meta-analyses. *Biological Conservation* 277, e109842. DOI: 10.1016/j.biocon.2022.109842.

Biggs, J., von Fumetti, S. and Kelly-Quinn, M. (2017) The importance of small waterbodies for biodiversity and ecosystem services: implications for policy makers. *Hydrobiologia* 793, 3–39. DOI: 10.1007/s10750-016-3007-0.

Bosone, L. and Bertoldo, R. (2022) The greater the contact, the closer the threat: the influence of contact with nature on the social perception of biodiversity loss and the effectiveness of conservation behaviours. *Sustainability* 14, e16490. DOI: 10.3390/su142416490.

Cardoso, P., Fukushima, C.S. and Mammola, S. (2022) Quantifying the internationalization and representativeness in research. *Trends in Ecology and Evolution* 37, 725–728. DOI: 10.1016/j.tree.2022.06.011.

Chen, W., Thorslund, J., Nover, D.M., Rains, M.C., Li, X. *et al.* (2022) A typological framework of non-floodplain wetlands for global collaborative research and sustainable use. *Environmental Research Letters* 17, e113002. DOI:10.1088/1748-9326/ac9850.

Clausnitzer, V., Simaika, J.P., Samways, M.J. and Daniel, B.A. (2017) Dragonflies as flagships for sustainable use of water resources in environmental education. *Applied Environmental Education and Communication* 16, 196–209. DOI: 10.1080/1533015X.2017.1333050.

Colding, J., Giusti, M., Haga, A., Wallhagen, M. and Barthel, S. (2020) Enabling relationships with nature in cities. *Sustainability* 12, e4394. DOI: 10.3390/su12114394.

Contador, T., Rendoll-Cárcamo, J., Gañan, M., Ojeda, J., Kennedy, J., Convey, P. and Rozzi, R. (2023) Underwater with a hand lens: ecological sciences and environmental ethics to value freshwater biodiversity. In: Rozzi, R., Tauro, A., Avriel-Avni, N., Wright, T. and May Jr, R.H. (eds) *Field Environmental Philosophy: Education for Biocultural Conservation*

Springer, Switzerland, pp. 53–69. DOI: 10.1007/978-3-031-23368-5.

Duffus, N.E, Echeverri, A., Dempewolf, L., Noriega, J.A., Furumo, P.R. and Morimoto, J. (2023) The present and future of insect biodiversity conservation in the Neotropics: policy gaps and recommendations. *Neotropical Entomology* 52, 407–421. DOI: 10.1007/s13744-023-01031-7.

FAO, IUFRO and USDA (2021) *A Guide to Forest-water Management*. FAO Forestry Paper No. 185. Rome. DOI: 10.4060/cb647en.

Flitcroft, R., Cooperman, M.S., Harrison, I.J., Juffe-Bignoli, D. and Boon, P.J. (2019) Theory and practice to conserve freshwater biodiversity in the Anthropocene. *Aquatic Conservation: Marine and Freshwater Ecosystems* 29, 1013–1021. DOI: 10.1002.aqc.3187.

Fraser, I.M., Roberts, D.L. and Brock, M. (2023). The economics of species extinction: an economist's viewpoint. *Cambridge Prisms: Extinction* 1, e20. DOI:10.1017/ext.2023.18.

Hill, M.J., Hassall, C., Oertli, B., Fahrig, L., Robson, B.J. *et al.* (2018) New policy directions for global pond conservation. *Conservation Letters* 11, e12447. DOI: 10.1111/conl.12447.

Idilbi, I., Ssymank, A. and Martens, A. (2023) Current water management of small lotic waterbodies in the context of nature conservation in Germany. *Environmental Management* 73, 162–176. DOI: 10.1007/s00267-023-01904-y.

Khelifa, R. and Mahdjoub, H. (2021) EcoDragons: a game for environmental education and public outreach. *Insects* 12, e776. DOI: 10.3390/ insects12090776.

Kramer, L., Teurlincx, S., Rashleigh, B., Janssen, A.B.G., Janse, J.H. *et al.* (2023) New paths for modelling freshwater nature futures. *Sustainability Science* (in press). DOI: 10.1007/s11625-023-01341-0.

Kuehne, L.M., Dickens, C., Tickner, D., Messager, M.L., Olden, J.D. *et al.* (2023) The future of global river health monitoring. *PLOS Water* 2, e0000101. DOI: 10.1371/journal.pwat.0000101.

Lampert, P., Goulson, D., Olsson, D., Piccolo, J. and Gericke, N. (2023a) Sustaining insect biodiversity through Action Competence: an educational framework for transformational change. *Biological Conservation* 283, e110094. DOI: 10.1016/j.biocon.2023.110094.

Lampert, P., Olsson, D. and Gericke, N. (2023b) A research instrument to monitor people's competence to sustain insect biodiversity: the Self-Perceived Action Competence for Insect Conservation scale (SPACIC), *International Journal of Science Education, Part B* (in press). DOI: 10.1080/21548455.2023.2281931.

Lynch, A.J., Cooke, S.J., Arthington, A.H., Baigun, C. Bossenbroek, L. *et al.* (2023a) People need freshwater biodiversity. *WIRES Water* 10, e1633. DOI: 10.1002/wat2.1633.

Lynch, A.J., Hyman, A.A., Cooke, S.J., Capon, S.J., Franklin, P.A. *et al.* (2023b) Future-proofing the emergency recovery plan for freshwater biodiversity. *Environmental Reviews* (in press). DOI: 10.1139/er-2022-0116.

Rodríguez, J.P., Sucre, B., Mileham, K., Sánchez-Mercado, A., De Andrade, N. and Jabado, R.W. (2023) We know how to do conservation. We just need to do more of it! *Diversity* 15, e443. DOI: 10.3390/d15030443.

Scharf, C. (2021) *The Ascent of Information: How Data rules the World*. Riverhead Books, New York.

The Global Risks Report (2023) *The Global Risks Report 2023*, 18th Edition Insight Report. World Economic Forum, Geneva, Switzerland.

Tickner, D., Opperman, J.J., Abell, R., Acreman, M., Arthington, A.H. *et al.* (2020) Bending the curve of global freshwater biodiversity loss: an emergency recovery plan. *BioScience* 70, 330–342. DOI: 10.1093/biosci/biaa002.

UNEP (2022) United Nations Environment Programme/Convention on Biological Diversity/15.4. Kunming–Montreal Global Biodiversity Framework. 15 pp.

Valente-Neto, F., Piovezan-Borges, A.C., Urbieta, G.L., Samways, M.J. and Roque, F.O. (2022) Research networks should improve connectivity for halting freshwater insect extinctions. *Ecological Entomology* 47, 63–75. DOI: 10.1111/een.13091.

van Rees, C.B., Waylen, K.A., Schmidt-Kloiber, A., Thackeray, S.J., Kalinkat, G. *et al.* (2021) Recommendations for the new global biodiversity framework from the European experience. *Conservation Letters* 14, e12771. DOI: 10.1111/conl.12771.

van Rees, C.B., Jumani, S., Abera, L., Rack, L., McKay, S.K. and Wenger, S.J. (2023) The potential for nature-based solutions to combat the freshwater biodiversity crisis. *PLOS Water* 2, e0000126. DOI:10.1371/journal.pwat.0000126.

Zhang, N., Zheng, X. and Wang, X. (2022) Assessment of aesthetic quality of urban landscapes by integrating objective and subjective factors: a case study for riparian landscapes. *Frontiers in Ecology and Evolution* 9, e735905. DOI: 10.3389/fevo.2021.735905.

Glossary

Abdominal appendages Projections at the end of the abdomen, assisting males in grabbing hold of females during mating, and important for both sexes in obtaining information on the immediate environment.

Abiotic environmental variables Aspects of the environment pertaining to non-living factors, such as temperature, humidity and air characteristics, as well as stressors such as barrages, pollutants and pesticides.

Abstraction The drawing of water by human agency, but when water resources are so depleted that there is ecosystem change, it is known as overabstraction.

Abutments Supporting structures of bridges.

Accessory (secondary) genitalia The second set of genitalia of male odonates at the base of the abdomen to which sperm is moved from the first (primary) genitalia just before mating.

Activity traits Traits associated with movement and dispersal, for example, foraging for prey among dragonfly larvae, or the flying adult locating a new habitat.

Adaptive capacity A combination of a species' genetic characteristics, behavioural and physiological disposition, and traits like life-history type and dispersal/colonization abilities for short-term survival and long-term evolutionary potential.

Adaptive evolution (capacity) The selection of adaptive traits for survival in prevailing environmental conditions.

Adaptive learning A change of behaviour that has survival benefits after learning from previous experiences.

Adaptive peak Where natural and/or sexual selection has reached an optimum in prevailing environmental conditions.

Adversely synergistic effects The interactive effects of two or more stressors, often with one stressor augmenting the effects of another stressor.

Aeshnids Dragonflies in the family Aeshnidae.

Aestivation Prolonged torpor or dormancy during a long hot and/or dry period.

African Dragonfly Biotic Index (ADBI) An adaptation of the Dragonfly Biotic Index used for continental-scale assessment of the status or state of a locality using dragonfly species presence, but with the three subindices, especially the 'habitat sensitivity' subindex, modified for large-scale use.

Agonistic encounters/behaviour Behavioural activities which cause, threaten to cause or seek to reduce physical damage, involving threat behaviour and/or aggression on the one hand and submission and/or retreat on the other.

Agroecology The study and application of ecological and adaptive processes, and the development of strategies for biodiversity conservation and the improvement of ecosystem services in plant and animal production systems.

Algal blooms The rapid growth of aquatic algae or cyanobacteria in response to high nutrient input, visible as coloured scum on the surface, which can change local aquatic food webs.

Alien species Non-native species.

Allele Any of the different forms of a gene at a given locus on the chromosome.

Allelic diversity A general term used to encompass the total heterozygosity and adaptation potential of a species or evolutionarily significant unit, often used in the context of overall biodiversity and complementing the structural, compositional and functional (bio)diversity of a specified area, or also used as a general expression to consider all genes and their alleles.

Alliance for Zero Extinction (AZE) A consortium of about 70 non-governmental organizations around the world focusing on the prevention of human-induced extinctions, with AZE sites being subsets of Key Biodiversity Areas.

Allopatric species Species that do not occur in the same physical space.

Alpha diversity Species richness or diversity at the local scale, as measured within a defined plot or an ecological unit such as a pond or part of a stream or river.

Amphibiotic Species that have an aquatic larval form and a terrestrial adult form.

Anisoptera An Odonata suborder ('true' or 'familiar' dragonflies).

Anisopteran Pertaining to taxa in the suborder Anisoptera.

Anisozygoptera An Odonata suborder with few species, confined to Asia, that show some characteristics between the suborders Anisoptera and Zygoptera.

Anisozygopteran A member of the Odonata suborder Anisozygoptera.

Antennal sensilla Very fine extensions on the surface of the antennae used to detect chemical cues, motion, humidity and/or temperature.

Anthropogenic Being caused by human activity.

Apparency The visibility of organisms, as opposed to those present when assessed by various technical methods.

Aquatic vegetation Plants living in water, totally submergent or reaching the water surface.

Area of habitat (AOH) A supplementary approach for IUCN Red List assessments that identifies an area of potentially suitable ecological conditions for a species, enabling it to reside within its geographic range and so informs conservation action.

Area of occupancy (AOO) A localized area occupied by a resident population of a taxon and scattered across its extent of occurrence, excluding cases of vagrancy, and reflecting the fact that a taxon will not usually occur throughout the whole of its extent of occurrence, which may, for example, contain unsuitable habitats.

Artificial ponds Small water bodies constructed through human activity.

Assemblage A group of species, usually focal or sampled species, often in a specific geographic area, and belonging to one taxon (e.g. dragonflies) (cf. Community).

Assemblage composition The identified set of species in a specified group (e.g. dragonflies) recorded as present at a specified locality or spatial scale. Often used interchangeably with 'species composition'.

Assemblage structure The identified set of species in a specified group (e.g. dragonflies) and their relative abundances recorded at a specified locality or spatial scale.

Assembly rules At a newly-created site (such as a pond), the hypothesis that various factors including historical patterns of speciation, migration of species, dispersal, abiotic environmental factors and biotic interactions, none of which are mutually exclusive, all contribute to how species arrive and establish at a site. Often used in the context of whether the arrival and establishment at a site by one species has some effect on a later arriving species. To date, there is no evidence of this among odonate species, although predation by larger individuals can affect the survival of younger, later-developing ones of the same or different species.

Assessment The measurement of the state or status of a focal subject or system at one time at a defined spatial scale (grain) and/or at specified location(s). However, the larger the spatial scale or the greater the inclusion of more subjects, systems and/or states, the more challenging it is to undertake a complete and accurate assessment.

Asymmetric competition A situation where individuals of one species exploit disproportionately greater amounts of resources than another, competing species.

Augmentative biological control Where a potential biological control agent is reared and released in large numbers, aiming to artificially boost its numbers in the wild to control a pest.

Backcrossing Mating of a hybrid dragonfly (offspring of genetically unlike parents) with one of its parents or one genetically similar to the parent, and useful for determining certain characteristics in a related group of dragonflies.

Barcoding A subset of DNA sequencing technology aimed at comparing the same part of an organism's genome among species. A specific barcode is assigned to a specific, identified sample using classic taxonomy.

Baseflow The regular long-term flowing conditions of a river, excluding the increase in discharge and flow velocity after extreme rainfall events.

Baseline condition An initial measurement of a condition that is taken at an early point in time and compared against subsequent measurements to identify changes, for example, to measure the improvement in the odonate assemblage over the course of a restoration project.

Bayesian network A compact, flexible and interpretable representation of a joint probability distribution learned from data, and used to represent causal relations between variables.

Behavioural syndrome A correlated suite of behavioural traits, often measured across multiple contexts, usually manifested as a distinctly recognizable pattern of behaviour.

Benthic macroinvertebrate An aquatic invertebrate, easily visible to the naked eye, which spends at least part of its life cycle on the bottom vegetation or substratum.

Benthic substrate granularity Referring to the coarseness of the substratum in terms of the deposited geological material.

Benthic zone The bottom zone of a water body.

Bergmann's rule Referring to endotherms (animals that physiologically generate heat to warm their bodies) that are postulated to have larger body size in colder environments, whether toward the poles or with increasing elevation.

Beringia The land and sea area encompassing eastern Russia, Alaska and northern Canada, that has periodically in the past been connected by a land/ice bridge and then separated by sea transgressions.

Berm A flat strip of land, raised bank or terrace bordering a river or channel.

Beta diversity The rate of species turnover over space, such as between two sites. There are two aspects: nestedness (the difference in the number of species at each site and when the number of species at the second site is a subset of those at the first site, useful for determining the effect of landscape fragmentation or identifying refuges) and turnover (when species richness or diversity are similar, but the species identities (species composition) are different, useful for determining environmental or biogeographic gradients). The term is also used for changes over time (see Temporal beta diversity).

Bioaccumulation The gradual accumulation of substances from all environmental sources, such as pesticides or other chemicals, in the bodies of organisms, and absorbed faster than can be broken down or excreted.

Biochemical oxygen demand The amount of oxygen used in the biodegradation of organic material per unit of time.

Bioconcentration The accumulation of waterborne substances, such as pesticides or other chemicals, in the bodies of organisms, and absorbed faster than can be broken down or excreted.

Biodiversity gap Species richness is highest in the least explored areas of the world, particularly the tropics, yet often lowest in the most explored regions of the world.

Biodiversity hotspots Thirty-six areas of the world, each with exceptional concentrations of endemic species, which are rich in biodiversity that is threatened.

Biodiversity paradox Biodiversity is largely concentrated in the tropics, while human, institutional and financial resources are mostly located at higher latitudes both north and south.

Bioindicator A species or species set used to indicate the state or status of diversity or the environment, and especially to reveal the level of stress in the system or alternatively the recovery from stress following conservation action.

Biological control The control of a pest species using other species, predatory and/or parasitic.

Biological control agent A species used in an attempt to control a pest species.

Biological traits Inherent characteristics associated with the functioning of a species, related to life history and mobility, and may be physiological, behavioural or morphological.

Biomass The total quantity or weight of a defined set of organisms in a given area or volume.

Bioregion A region defined by the characteristics of the natural environment and its biota rather than by political divisions, and although usually portrayed as having a distinct boundary, in reality has a fuzzy boundary with neighbouring bioregions.

Biosphere reserve An important international initiative of the *Man and the Biosphere Programme* (United Nations Educational, Scientific and Cultural Organization). Each biosphere reserve consists of three zones: 1) A central core zone where there is full biodiversity preservation; 2) A buffer zone around the core with only agroecological farming as well as natural areas; and 3) An outer, transition zone where there is more extensive agroecological farming. This zoning approach constrains stressors such as avoiding conventional agricultural conversion and limiting agroecological activities and infrastructure development to specific areas.

Biotic environmental variables Environmental factors that are biological, usually meaning living organisms, such as vegetation type and structure, competitors, parasites, predators, etc., that have an influence on the subject dragonfly species, dragonfly assemblages or communities, and that determine ecosystems.

Biotic homogenization An impoverished subset of the naturally occurring organisms (e.g. dragonflies) due to human activity, while also implying the replacement of habitat specialists with habitat generalists across a disturbed or transformed landscape.

BioTIME A global database set up to monitor changes in the abundance of species over time.

Biotope A small, defined, physical area where insects live, and usually defined by the structure of the vegetation and physical features (cf. Habitat).

Bivoltine Two broods in one season.

Bog A freshwater wetland of soft, spongy ground with much decayed plant matter (peat), mostly in cool, northern climates, largely naturally formed from poorly drained lake basins after glacial retreat.

Boreal forests Wet coniferous forests of mostly pines, spruces and larches.

Boreo-alpine Referring to species distributed over the more northern parts of Europe, while also occurring in the higher mountains.

Boundary layer The lowest layer of the atmosphere that directly feels the effect of the earth's surface and so moderates local atmospheric behaviour according to surface topography, vegetation type and urban development.

Bromeliad A plant in the family Bromeliaceae of the tropical and subtropical Americas, and having rosettes of stiff, spiny leaves, which in many species can hold sufficient water (tank bromeliads) to be a habitat for certain odonate species.

Buffer zone A protection zone that is the area on the bank and adjacent to a water body that is vegetated and aimed at reducing any adverse effects of human activity farther away from the water body.

Burst swimming A sudden forward movement generated through anisopteran larvae rapidly expelling water from the anus. Usually to escape predators, such as fish and large aeshnid larvae.

Cabruca An agroforestry system where crops are grown under the canopy of native trees, e.g. cacao in Brazil.

Calopterygid A damselfly in the Calopterygidae/Calopterygidoidea.

Cannibalism The consumption of individuals of the same species.

Cape Floristic Region Located at the southern tip of South Africa, it is the smallest of the world's six floral kingdoms, supporting extremely high diversity and endemism, with over 9000 vascular plant species, 69% of which are endemic. The diversity is largely associated with the fynbos biome, a Mediterranean-type, fire-prone shrubland. It also supports endemic dragonflies that amount to about one-fifth of the country's total species numbers.

Capture–mark–recapture (CMR) (also known as mark–release–recapture (MRR)) The capture of a living individual, marking it with code, and then recapturing it later. When done on many individuals in a population and the results are analysed along with recorded and appropriate habitat variables, it is possible to establish the chances of persistence of a population.

Catchment (of a river) An area of land where all flowing surface water converges to a single point, whether a lake, swamp or sea.

Caudal lamellae The leaf-like projections at the tip of the abdomen of damselfly larvae used largely for respiration, but occasionally for swimming.

Centinelan extinction (silent extinction) The extinction of a species before it is scientifically described.

Cerrado A South American savanna-type biome of various ecosystems, including dry forests, grasslands, wetlands, shrublands, gallery forests and some wet forests.

Channelization The restriction or redirection of a natural watercourse through the building of an artificial water channel.

Character displacement Occurs when individuals most dissimilar from the average resource-use or reproductive phenotypes of another species are more successful at obtaining resources and having successful reproduction than other individuals in the population.

Chironomids Non-biting midges in the family Chironomidae.

Chromatocytes Cells containing a coloured pigment, and together they largely give an animal like a dragonfly its body colour.

Citizen scientist A person in civil society who collects and sometimes analyses data relating to the natural world, usually while collaborating with professional scientists on a collective project.

Cladocera A diverse group of small crustaceans common in aquatic habitats, and important as food items for odonate larvae.

Clementsian pattern of species distribution Whole assemblages showing turnover along particular environmental gradients.

Climate connectivity failure Characteristic of protected areas, that may not be adequately spaced across larger landscapes and regions to be able to give refuge to species that may need to move in response to climate change.

Climatic debt The lag between rising temperatures and the responsive changes in population levels, species distribution, assemblage composition, community structure and/or ecosystem function.

Clinal variation Typically, the gradation in one or more characteristics within a species, especially between different populations, over a large geographic area.

Cloud forest (montane rainforest) The vegetation of tropical mountainous regions where rainfall is often heavy, and condensation is persistent through the cooling of moisture-laden air currents deflected upward by the mountains.

Coenagrionid A damselfly in the family Coenagrionidae.

Cohort A group of individuals in the same species, recognized by common characteristics, especially at the same time.

Cohort splitting The separation of groups of individuals of the same cohort, such as isolation mediated by elevation.

Colour morphs Alternative body colours among adults of the same sex.

Commensal An organism that has a close association with another without any adverse effect on either.

Community An interacting or potentially interacting group of organisms from a whole host of taxa living together in one local physical area (cf. Assemblage).

Complementarity (in ecology) Where two or more areas each bring extra biodiversity into the overall area.

Conductivity (in water) A measure of the propensity of water to pass an electrical current, which is increased through elevation in levels of dissolved salts and other inorganic chemicals.

Congeners Different species in the same genus.

Congruence (cross-taxon) When diversity patterns of different taxa typically co-vary in space.

Connectivity (functional) The extent to which two or more areas are joined while allowing the organisms to move, disperse and flow across the landscape/waterscape as they would naturally.

Connectivity (physical) The extent to which two or more areas are joined, for example, by a conservation corridor.

Conspecifics Individuals in the same species.

Core resident species The typical collective of species present from one year to the next as ongoing breeding residents, excluding those that breed temporarily or are vagrants.

Critically Endangered (CR) A category on the IUCN Red List of Threatened Species containing those species that on the best available evidence indicates that they meet any of the criteria A to E for Critically Endangered, and are therefore considered to be facing a high risk of extinction in the wild (see https://www.iucnredlist.org).

Critically Endangered (possibly Extinct) As the IUCN Red List of Threatened Species does not consider a species extinct until extensive, targeted surveys have been conducted, species that are possibly extinct are still listed as Critically Endangered, but where there is a strong suspicion that they are no longer extant, they are given the epithet 'possibly Extinct'.

Crown group (taxa) The smallest clade that includes all living members of a specified group.

Crypsis Being camouflaged against the background.

Cryptic (biological) species Different species that are morphologically similar, and sometimes even seemingly identical, but their genomes confirm that they are different species subject to certain criteria.

Damselflies The Odonata suborder Zygoptera.

Dark diversity Species that are missing from a sample of species from a certain physical area with seemingly suitable habitat, yet are present in the immediate surroundings or elsewhere with the same ecological conditions (sampled/observed diversity + dark diversity = actual diversity).

Data Deficient (DD) A category on the IUCN Red List of Threatened Species containing those species where there is inadequate information to make a direct, or indirect, assessment of their risk of extinction based on their distribution and/or population status.

Decomposer An organism that breaks down organic material.

Demographic (population) dynamics Changes in the size and composition of populations over time.

Deposition pool A localized area where the water velocity along a stream or river drops to near zero (usually from a small natural impoundment such as rocks, or below a waterfall), and where organic matter, such as leaves and twigs, accumulates.

Detritivores Organisms that feed on dead organic material, usually meaning that of plants.

Developmental plasticity Permanent morphological, behavioural and/or physiological changes in the development of an organism during its lifespan that are carryover effects of external environmental drivers on the interactions between the genotype and the environment.

Developmental polymorphism The different forms and functions of the different life stages of an insect (e.g. a dragonfly larva adapted to living in the water and feeding on underwater prey, and an aerial flying adult which feeds on prey in the aerial realm).

Discharge rate (of a stream or river) The amount of water delivered by flowing water, and the product of its velocity (length of travel per unit of time) multiplied by water depth and then by water width.

Discoidal area (of the hindwing) An expanded area of the trailing edge of the hindwing, important for maintaining flight stability and for gliding.

Disjunctive (distribution) Split up into separate areas.

Disparity (in palaeontology) The range or significance of morphology in a given sample of fossil specimens.

Dispersal Movement (by an individual or individuals) to a different area.

Dispersal ability Adult anisopterans have high mobility powered by strong flight muscles, although they may or may not disperse over long distances. However, the dispersal ability of some species is increased when they make use of prevailing winds to carry them farther than is possible using only powered flight. There is some tentative evidence that some damselflies, although having weak powered flight, are able to disperse far when they move out of the boundary level into the higher air stream to reach distant suitable habitat. Furthermore, larvae of some species may drift downstream using the current to disperse farther than they are able to walk.

Disruptive wing colouration Coloured areas on the wings that break up its outline and so reduce visibility to would-be predators.

DNA barcoding A method of species identification using a short and defined section of DNA from a specific gene or set of genes.

Donor site In restoration and species reintroduction programmes, the good-quality site from which individuals are drawn and then introduced elsewhere.

Dragonfly Confusingly, it can mean a 'true' dragonfly (Anisoptera) but when used in the plural (dragonflies) it can mean all of the order Odonata.

Dragonfly Biotic Index (DBI) A practical measure and tool based on total dragonfly species counts (adults mostly but supplemented with larvae and exuviae) present at a defined running or still water locality. The index measures the state of the local biotope which includes both the water and the riparian zone based on the sum of three subindices per odonate species: extent of geographic distribution, IUCN Red List status (global and/or national) and sensitivity to change in habitat quality.

Drone netting The catching of flying insects, like dragonfly adults, using a drone with a net attached.

Dytiscid diving beetles Predatory aquatic beetles in the family Dytiscidae, often large in size.

Ecdysis During development, the casting off of the old cuticle.

Ecological integrity An ecosystem has integrity when it is deemed characteristic of a natural, undisturbed area, including the composition and abundance of native species and biological communities, rates of change and supporting processes.

Ecological landscaping Landscaping of a focal area using ecological principles, with special emphasis on establishing habitats for indigenous species and supporting local ecosystem processes.

Ecological memory (when referring to impacts, legacy effects/events) How past stressors influence current and future responses by organisms.

Ecological niche modelling Use of machine-learning algorithms to discover possible new locations of a species over wide geographic space, based on the traits and known habitat preferences of the species.

Ecological succession The process where the structure of a biological community evolves over time, often with special focus on the changing plant community.

Ecological surrogate A process or element (species, ecosystem or abiotic factors) that represents another aspect of an ecosystem or area, known as the 'target'.

Ecological threshold A point or zone at which rapid change occurs from one ecological condition, state or status to another.

Ecological trait Referring to the biotope specialization and habitat preference of a population or a species.

Ecological trap Poor-quality habitat (often the result of human activity), which may be temporary suitable habitat, but cannot sustain an effective population. Also, where preferred habitat by a species confers lower fitness than another available habitat.

Ecophysiology The study of the response of an organism's physiology to environmental conditions.

Ecoregions Biogeographic areas of the world determined by particular ecological and evolutionary patterns, and used in global and regional conservation planning, including threatened ecosystems in need of conservation action. Globally, there are over 800 terrestrial ecoregions, and 426 freshwater ecoregions.

Ecosystem function Structures and interactions among individuals and species in an ecosystem, and usually measured against the natural or historic state. Sometimes also used to describe the capacity of natural processes and components to provide goods and services for human benefit.

Ecotone The transition area between two biological communities or ecosystems, where species usually display turnover.

Ectothermic Body temperature is largely determined by the temperature of the immediate external environment.

Edge effects Changes in the population levels of a species at the boundary of two or more habitats, or of community diversity at the boundary of two or more ecosystems.

Effective population The size of a population that is contributing to the next generation, usually referring to the number of breeding individuals required to maintain a viable population in the long term.

El Niño–Southern Oscillation (ENSO) A periodic variation in winds and sea surface temperatures over the tropical eastern Pacific Ocean, affecting the climate of much of the tropics and subtropics.

Electromagnetic radiation Waves of the electromagnetic (EM) field, which propagate through space and carry momentum and electromagnetic radiant energy.

Embalsados Dense floating mats of the plant *Pontederia crassipes* (water hyacinth) in South America.

Emergence The appearance of the adult from the last developmental stage of the larva following the larva leaving the water, and leaving behind the exuvia.

Emergence traps Physical tent-like traps that are placed at the water surface and used to sample insects emerging from the water as adults.

Emergent event An event requiring action to prevent or mitigate deterioration to an ecosystem.

Emergent macrophyte A plant that is easily visible and extends its growth from beneath the water surface to above.

Endangered (EN) A category on the IUCN Red List of Threatened Species containing those species that on the best available evidence indicate that they meet any of the criteria A to E for Endangered, and are therefore considered to be facing a high risk of extinction in the wild (see https://www.iucnredlist.org/).

Endophytic oviposition The laying of eggs inside a substrate, usually living or dead plant tissue.

Endothermic Generating body warmth by physical and physiological activity.

Enemy-free space The physical space where predators, parasitoids and/or pathogens are at their lowest or are not present.

Environmental bottleneck A time when environmental conditions are at their harshest for the survival of a population or a community of organisms, and for which the majority of organisms may be poorly adapted.

Environmental filter The set of abiotic factors that prevent the establishment or persistence of species at a particular location.

Environmental flow (e-flow) Managed changes to a river's flow pattern to maintain or improve river health through mimicking the natural flow pattern.

Environmental gradient (climate gradient) The change in abiotic factors over space, and ultimately over time as well.

Environmental heterogeneity The presence of unrelated or differing parts within a particular environment, often referred to as 'high' when there is much variety, and 'low' when there is little variety.

Environmental legacy Environmental change resulting from earlier human disturbance such as changes in land use and cover, fire regime, water diversion, introduction of chemicals or pollutants, and/or other disruptions to natural systems.

Environmental surrogate A natural feature of the landscape that stands in for a set of species, such as a series of small wetland patches standing in for a set of rare dragonflies.

Ephemeral (temporary/seasonal) pools Pools that are not present all year round as they periodically dry out. Dry phases may last from a few months to several years.

Epiphyte A plant that grows on another, larger plant such as a tree.

Epiphytic oviposition The laying of eggs on the outside of a substrate, usually plant tissue.

EPT An abbreviation for the collective group: Ephemeroptera (mayflies)/Plecoptera (stoneflies)/ Trichoptera (caddisflies), together popularly known as riverflies.

Equilibrium dynamics A system in a steady state is one where forward reaction and backward reaction occur at the same rate. However, this is rarely the case in natural ecosystems, but for a while at least an ecosystem may be close to such a system and is best referred to as in 'dynamic equilibrium'.

Essential Biodiversity Variables (EBVs) A globally standardized set of variables that aim to better monitor the status and trends of biodiversity to improve biodiversity conservation globally, which fall into four main groups: 1) genetic composition; 2) species and abundance (species populations); 3) ecosystem extent and fragmentation (ecosystem structure); and 4) primary productivity (ecosystem function).

Eutrophication The process where a water body becomes increasingly enriched with minerals and nutrients, especially nitrogen and phosphorus, while stimulating plant growth, especially algae, often with associated drops in oxygen levels in the water.

Evapotranspiration The sum of evaporation from both the land and water surface and the transpiration from plants.

Evenness The relative abundance of different species at a specified location.

Evidence-based conservation The application of evidence arising from conservation policy making and management actions accrued from the rational assessment of scientific information from research publications, manuals, expert and practitioner feedback, independent expert assessment and indigenous knowledge to determine the current effectiveness of management interventions, identify emergent threats and challenges, and assess economic viability.

Evolutionary potential (in conservation) Conservation action that ensures that population size and genetic diversity is sufficient for natural selection to still take place and to lead to evolutionary change, even though there may be some landscape transformation.

Evolutionary relatedness (phylogenetic signal) Recency to a common ancestor: if species A and B diverged before species A and C, A and C can be considered as more closely related than A and B, which have had a longer time to evolve in their own directions.

Exophytic oviposition The dropping of eggs onto the water surface, or the eggs are washed from the tip of the abdomen during flight, without the eggs being placed directly into or onto a solid substrate.

Extant Alive today.

Extent of occurrence (EOO) Used especially in the IUCN red-listing process: the area contained within the shortest continuous imaginary boundary that can be drawn as a polygon to encompass all the known, inferred or projected sites of the present occurrence of a species.

Extinct (EX) A category on the IUCN Red List of Threatened Species containing those species when there is no reasonable doubt that the last individual has died. This classification follows when exhaustive surveys in known and/or expected habitat at the appropriate times of the day or season throughout their historic range have failed to record any individual in the last 50 years.

Extinction debt The possible local extinction of species due to past and current ongoing disturbance events.

Extirpate To eradicate completely, such as a local population of a species.

Extirpated Eradicated or destroyed completely in a defined area.

Exuvia (plural: exuviae) The cast-off husk of the last stage of the larva once the adult has emerged from it.

Feasibility analysis An assessment of the practicality of a proposed plan, project or scheme.

Feeder stream Small stream, often at a high elevation, that feeds into a larger stream, river or still water body.

Fens (fenland) Low and marshy, frequently flooded areas of land, with alkaline, neutral or only slightly acid peaty soil.

Final instar The last stage of larval development that eventually emerges from the water to become a teneral adult, leaving behind the exuvia.

Fishponds Ponds, usually artificial or intensely managed for the production of fish biomass.

Flagship species A species with a high profile, often representing an important ecosystem.

Fliers A casual term referring to species of anisopterans that spend most of their time in flight.

Floating macrophytes Visible plants that have growth that floats on the water surface, such as water lilies.

Focal organism (population, species, assemblage, community, ecosystem, landscape, waterscape) The group that is the focus of a particular study.

Food web All the food chains in an ecosystem, and the paths along which energy and nutrients flow.

Founder effect When a gravid female or a few individuals of a species become established in a new area, such as an island, the very low numbers of individuals and associated reduced level of genetic diversity leads to a particular evolutionary trajectory in response to the novel conditions.

Fragmentation model A hypothetical situation where the landscape/waterscape is viewed as parcels of land and/or water and the occurrence of organisms are similarly in separate parcels (cf. Variegation model).

Functional alpha diversity The functional diversity of the various species at the local spatial scale (i.e. the role that species play, often interactively as they are in close proximity to each other).

Functional connectivity The extent to which landscapes facilitate or impede the movement of organisms between areas of habitat (cf. Structural connectivity).

Functional morphology The relationship between the form of an organism and its behaviour and dynamics.

Functional traits Measurable characteristics of a species that influence its performance, as well as its interactions with other species.

Fundamental niche (of a species) All the environmental conditions in which a species can survive, reproduce and leave viable offspring.

Future-proofing Accounting for the uncertainty of future conditions that have no historical baseline, even though the level of uncertainty associated with future conditions is unprecedented, and requires the anticipation of future changes and the development and implementation of conservation actions to address the impacts of those future changes.

Gallery forest Where trees and other vegetation line rivers and other waterways in a savanna-type landscape.

Gamma diversity A measure of the overall biological diversity, or that of a specified species group or assemblage, within a large area or geographic region.

Gene flow In dragonflies, the transfer of genetic material among breeding individuals that come into contact through physical movement, and when high, leads to fairly even distribution of allele frequencies throughout the local population.

Genetic bottleneck The decrease in genetic diversity when population size decreases greatly, often due to one or a combination of a sudden catastrophic event, the effects of a long term stressor and/or long-term geographic isolation.

Genetic diversity All the genetic characteristics in the genetic constitution (genome) of a species, varying among individuals and among different populations. High genetic diversity increases the ability of species and populations to resist adverse stressors, while, in contrast, low genetic diversity usually predisposes populations to be more susceptible to the stressors.

Genetic drift A change in frequency of alleles through random effects in small populations with limited breeding opportunities.

Genetic load The presence of unfavourable material among the genes in a population, often decreasing the chances of long-term population viability.

Genetic marker A DNA sequence with a known physical location on a chromosome that represents observed variation among individuals or species, and in dragonfly biology the genetic markers ND1 and CO1 are especially important for species identification.

Genetic plasticity The collective of genomes among individuals of a population, expressed as phenotypic diversity in terms of their physiology, behaviour and morphology. These enable individuals to survive certain environmental changes.

Genetic rescue A mitigation strategy aimed at restoring genetic diversity and reducing extinction risks in small, isolated and inbred populations.

Genetic resilience Having sufficient genetic variation to tolerate or overcome certain defined or potential stressors or impacts.

Genetic viability Being able to avoid problems of inbreeding by having a sufficient number of interbreeding individuals in a population.

Genomics The interdisciplinary field of biology focusing on the structure, function, evolution, mapping and editing of genomes.

Genotype An organism's complete set of genetic material in terms of its alleles or genetic variants in a particular gene or genetic location.

Geographic range (biogeographic range) The total overall area across which a species occurs, although not usually at every geographical location within that area.

Geomorphology The study of the physical features of the surface of the earth and their relation to its geological structures.

Geotectonic depression A sunken landform arising from tectonic activity.

Glide Where a shallow river runs fairly fast, and there are distinct water movements on the water surface, which remains more or less flat.

Gomphid A member of the anisopteran family Gomphidae.

Grain (in area assessment) The size of units used in spatial scale measurements for describing the geographic distribution of a species, an assemblage, a community, an ecosystem or a landscape/waterscape. Grain is used to describe the relative size or extent of something and has dimension(s) such as 30 m (the scale or resolution of Landsat imagery) or 2 km (the practical scale unit used in species area of occupancy studies).

Grain (in conservation) The size of the individual units at which a species distribution or extent of population, community, ecosystem, landscape or phenomena is studied, and having dimensions determined according to a defined task at hand.

Green roof (in architecture) The roof of a building that is partially or completely covered with vegetation in a growing medium, planted over a waterproofing membrane, and that may include a vegetated pond for dragonflies.

Green Status of Species (IUCN) A practical tool that focuses on species with high conservation dependence yet high potential for conservation improvement through effective conservation action, and used to complement the IUCN Red List of Threatened Species.

Greening (regreening) The planting of vegetation to re-establish a green covering of the land and/or partial recovering of the water surface, and not necessarily according to ecological principles, but rather for aesthetic and/or engineering ones.

Greenscape The result of a landscaping practice that improves the health and appearance of urban areas while protecting and preserving natural capital and interactions, so harmonizing nature and human residency.

Gregarine (gut parasite) Large single-celled parasitic organism (Phylum Apicomplexa) that is parasitic inside a dragonfly body.

Guild A group of organisms each of which feeds in a similar way, e.g. pollinator guild, leaf-chewing guild, predator guild.

Habitat A specified area necessary for a species to carry out all its life functions (sometimes used in the context of a defined area) (cf. Biotope).

Habitat fragmentation The breaking up by human activity of the area where a species carries out all its life functions (cf. Landscape fragmentation).

Habitat generalist A species that can live in many different types of local environment.

Habitat integrity (intactness) A measure or expression of the relative intactness of a species' total habitat conditions against the benchmark of the natural historic state, with high habitat intactness being a state close to the historic state and low habitat intactness being a greatly altered state relative to the historic state.

Habitat Integrity Index (HII) A measure for assessing streams, particularly in Brazil, relative to the degree of biotope transformation and the impact on biodiversity quality, while indicating the capacity of the local environment to support and maintain a set of species, its diversity and its functional organization relative to that of the regional natural habitat in the same area.

Habitat loss The loss of the whole area where a species carries out all its life functions, and usually used in terms of human activity.

Habitat occupancy Referring to whether a suitable habitat of species is actually occupied by that species or not.

Habitat patches Patches of suitable habitat that support a particular species.

Habitat specialist A species that can live in one or very few slightly different type(s) of local environment.

Habitat suitability modelling The use of machine-learning algorithms to predict where further suitable habitat of a species might occur over geographic space, using the environmental data of known habitat preferences and distribution records of the species.

Habitats Directive (European Union) Aims to protect over a thousand species from various taxonomic groups along with 230 characteristic habitat types to ensure their continuance in the long term.

Haemolymph A fluid equivalent of blood that circulates in the body of dragonflies and many other invertebrates.

Haplotype A set of DNA variants close together along a single chromosome and usually inherited together, with recombination between these variants being rare.

Hawking A behaviour among the 'flier' group of dragonflies that involves the flying back and forth across the landscape, usually along stream or pond edges, and mostly in search of flying insect food items.

Headwater streams Water flows emanating from seeps and springs feeding into streams at higher elevations of a river close to or forming part of its source.

Heat island effect The increased ambient temperature as a result of much energy transfer in cities.

Heliophilic Favours areas with much inflow of sunlight.

Herbivore An animal that feeds on living plant material.

Heterogeneity (in ecology) In a defined area, the variation in species composition or communities, which may range from low to high, and is usually driven by factors like geochemistry, topography, rainfall pattern, human impact, etc.

Heterospecific pairing The mating of pairs of individuals from different species.

High latitudes Those latitudes closest to the poles and farthest from the equator.

Historic state (condition) The original or natural state before human modification or transformation, which may not always be determinable (see 'Shifting baseline').

Holarctic The zoogeographical region comprising the Nearctic and Palaearctic regions combined.

Homogenization (of a landscape) When an area is disturbed or changed by human activity, and variation in the physical structure of the landscape is reduced. As a result, the biota is only a subset of the historic state, both in the number and type of species present and their relative abundances.

Homozygosity An increasing proportion of identical genes present on both homologous chromosomes.

Hybrid zones Geographic areas where genetically distinct populations meet, mate and leave offspring of mixed ancestry.

Hybridization The bonding together of two complementary single-stranded DNA and/or RNA molecules to form a double-stranded molecule, dependent on appropriate base-pairing across the two single-stranded molecules.

Hydrodynamic The motion of water and the forces acting on immersed solid bodies.

Hydrogeomorphic patches Rivers that share similar hydrologic, geomorphic and climatic conditions.

Hydrologic drought Localized drying out of the land adjacent to a constructed hard surface channel.

Hydrologic region Areas of land and water that share a common climate, and geological and topographical structure.

Hydrological connectivity The magnitude, duration and timing of water-mediated transfer of materials.

Hydromorphology The physical character and water content of water bodies, either natural or in relation to how human activity influences the hydrology and geomorphology of water bodies and the ecosystems that they support.

Hydroperiod The length of time during the year that a swale, pool or pond has standing water or is waterlogged.

Immature adult The developmental phase between the recently emerged teneral adult and the fully mature, sexually active adult, usually characterized by not yet having mature adult body colouring, especially among males.

Impoundment A barrier built to hold back and retain water.

Inbreeding depression The loss of fitness and vigour in individuals from the loss of variety of alleles (i.e. increased homozygosity).

Incipient evolution A group of organisms, usually geographically isolated, that is about to become a separate species from other, related individuals.

Indicator surrogate A surrogate that provides information about ecological systems. For example, measuring selected components of dragonfly assemblage diversity to represent the relative naturalness of an ecosystem, such as rare habitat specialists of natural bogs.

Indigenous species Those species native to a specified area.

Information Stimuli that have meaning in some context for its receiver, which may, for example, be a molecule, an individual or a system.

Instrumental value Represents the value of species or ecosystems with some use for people, and often measured in terms of service or in monetary terms (cf. Intrinsic value).

Insular effect The number of species occurring in an undisturbed insular environment, such as an oceanic island, determined by relative rates of immigration and local extinction.

Interaction network A description of the interactions between species within a community, such as competition, mutualism and predation, especially relative to the structure and stability of the network.

Interference-free space Voluminous space where there is no behavioural interference from other individuals of the same or different species.

Interfluves Areas between major rivers.

Intergovernmental Science-Policy Platform on Biodiversity and Ecosystem Services (IPBES) An independent intergovernmental body of over 130 member governments, providing policy makers with objective scientific assessments on the state of knowledge regarding the planet's biodiversity and ecosystems, and the contributions they make to people, as well as options and actions to protect and sustainably use these vital natural assets.

Interspecific Occurring between two or more species.

Interspecific competition Competition among different species for resources, and sometimes for territory.

Intervention ecology (restoration) A management activity that aims to alter the restoration trajectory according to the initial target, which is usually the historic condition.

Intraspecific Occurring within one species.

Intraspecific competition Competition among individuals within one species for resources, territorial space, mates, emergence space or oviposition sites.

Intrinsic value A perspective that considers that nature has value in and of its own right, independent of human uses, and posits a view from an ecocentric or biocentric standpoint (cf. Instrumental value).

Introgressive hybridization (introgression) The transfer of genetic material from one gene pool into another through repeated backcrossing of an interspecific hybrid with one of its parents, leading to a variably complex mixture of genes in the offspring.

Invasive alien species Species that have been introduced accidentally or deliberately by humans to outside their indigenous range. These species then establish and spread into and across natural ecosystems in the reception area, often with adverse impacts on certain local indigenous populations.

Irradiation (solar radiation) The energy emitted by the sun through electromagnetic waves.

Irreplaceable (in conservation) Referring to a remnant population, species, community, ecosystem, or landscape at a specific location that if lost would be lost forever.

Ischnurid Referring to a damselfly in the genus *Ischnura*.

Isotherm A weather map line joining places which have the same temperature.

Isotope analysis The identification of an isotopic signature, and the abundance of certain stable isotopes of chemical elements within organic and inorganic compounds.

IUCN International Union for Conservation of Nature.

IUCN Red List of Threatened Species (www.iucnredlist.org) The world's most comprehensive inventory of the global conservation status of biological species.

Kairomone A semiochemical emitted by an organism. In interspecific interactions, it benefits the receiver over the emitter.

Karst A landscape underlain by limestone which has been eroded by dissolution from water activity.

Key Biodiversity Areas (KBAs) Locations contributing greatly to the maintenance of global biodiversity, with about 15,000 identified globally.

Key species Either a species of special focus, or a keystone species, one that contributes greatly to defining an ecosystem while also being essential for ecosystem functioning.

Kunming–Montreal Global Biodiversity Framework Supports the achievement of the Sustainable Development Goals and builds on the Convention on Biological Diversity's previous Strategic Plans. It sets out an ambitious pathway to reach the global vision of a world living in harmony with nature by 2050. Among the Framework's key elements are 4 goals for 2050 and 23 targets for 2030.

Labium The lower lip of the insect's mouthparts, positioned toward the back of the head which helps hold food in place when the insect feeds. In dragonfly larvae, it is extendable and bears two moveable hooks, as well as other dentations and setae.

Lagg zone The transition zone along the outer margin of a raised bog, represented by a swamp or fen plant community, and which contributes to sustaining a high water table within the peat area of the bog.

Lake productivity The trophic state of a lake, i.e. a lake's level of ability to support plant and animal life.

Landscape (scale) The dynamic surface of the land and freshwater over a scale of tens of metres to several kilometres, which may encompass more than one ecosystem.

Landscape configuration The spatial arrangement, distribution and organization of natural or agricultural landscape elements.

Landscape element The smallest area on the landscape with a typical vegetation type and land use (e.g. remnant patch, wheat field, rice paddy, paddock, plantation block, farm buildings, urban garden, green node).

Landscape fragmentation The conversion of a historically contiguous landscape into one composed of various landscape elements, and sometimes leaving behind remnant patches of natural habitat.

Landscape heterogeneity The variety of habitats, ecosystems and landscape elements across the landscape, often used to mean 'quality heterogeneity', i.e. not including heterogeneity that comes about from high landscape disturbance and degradation.

Larva (sometimes referred to as a nymph or naiad) The young growing stage of a dragonfly, developmentally between egg and adult.

Latitude Diversity Gradient rule A hypothesis that there is an increase in species richness toward the equator.

Least Concern (LC) A category on the IUCN Red List of Threatened Species containing those species that have been evaluated against Red List criteria and do not qualify for Critically Endangered, Endangered, Vulnerable or Near Threatened.

Legacy conditions (effects/events) Conditions from the past. Natural ones which may include glaciations, droughts, fires or floods, and more recent human-induced perturbations such as persistent insecticide input or the impact of invasive alien species.

Lek An aggregation of males that engage in competitive displays and courtship rituals to entice visiting females which are surveying prospective partners with which to mate.

Lentic Pertaining to standing water.

Lestids Individuals or species of damselfly in the family Lestidae.

Libellulid An anisopteran in the family Libellulidae.

Libelluloid An anisopteran in the super-family Libelluloidea.

Light sheet An artificially illuminated white sheet suspended between two upright supports, used for attracting insects that fly at dusk or at night.

Lineages (genetic) A series of mutations or changes in the genotype connecting an ancestor's genetic code to that of their descendants.

Littoral zone (for freshwater) The part of a lake, pond or river that is close to the shore.

Longwall mining A type of underground coal mining where a long wall of coal is mined in a single slice.

Lotic Pertaining to running water.

Low latitudes Those latitudes closest to the equator and farthest from the poles.

Macroecology The study of relationships between organisms and their environment at large spatial scales to characterize and explain statistical patterns of abundance, distribution and diversity.

Macroinvertebrate (aquatic) An aquatic invertebrate easily visible to the naked eye, which spends at least part of its life cycle in water.

Macroinvertebrate rapid assessment protocols The use of macroinvertebrate higher taxa (e.g. family) to assess water quality.

Macrophyte An aquatic plant (submerged, floating or emergent from the water surface) large enough to be seen by the naked eye.

Malaise trap A tent-like structure used for trapping flying insects, such as certain adult dragonflies.

Managed (management) restoration While there is inevitably a passive restoration component to restoration (i.e. allowing a site to regenerate naturally while largely drawing on the natural capital in the immediate vicinity), it may be necessary to help the recovery trajectory on its way by introducing certain managed actions. Actions include ensuring that water levels are at the right levels or the planting of certain vegetation at appropriate densities. Furthermore, there may still be certain management actions required to steer the restoration trajectory in the right direction according to the restoration targets (cf. Passive restoration).

Management surrogate A species or group of species that represents another aspect of a system so as to inform effective conservation management, such as using dragonfly diversity to determine the success or not of a restoration project.

Marginal vegetation Vegetation at the interface between land and water.

Matrix The physical area surrounding an area of focus, often referring to a conservation area surrounded by transformed areas.

Meganeurid A member of the extinct family of giant dragonflies, the Meganeuridae, commonly known as griffin flies.

Mesocosm An artificial water tank for rearing dragonflies and studying them closely.

Mesohabitat A slightly larger area than a microhabitat covering a broader set of environmental conditions.

Meta-analysis The analysis of data from many independent studies of the same subject to determine overall trends.

Metabarcoding Simultaneous comparison of genome sequences among many species in a mixed sample.

Metacommunity A set of local communities connected by species dispersal.

Metapopulation A group of partially isolated populations where there is some movement of certain individuals between some of the populations, often resulting in vitally important outbreeding.

Metapopulation dynamics The patterns and timing of movements among individuals making up a metapopulation.

Microhabitat A very small area where an insect is living when observed, and nested within its larger habitat. Microhabitats are often important for shelter or survival for a critical life stage of an insect (e.g. the interstices between boulders on the stream bed in the case of certain dragonfly larvae).

Microsporidia Spore-forming fungal unicellular parasite.

Migration Large-scale movement of individual members of a species (migrants) to a different geographical area, often on a regular basis and according to season.

Mire A stretch of swampy or boggy terrain.

Mitogenome Mitochondrial genome.

Mobility The level of capability of movement of a whole organism.

Molecular operational taxonomic unit (mOTU) A taxonomic unit determined by its genome, and representing an ostensible species without a formal scientific name.

Monitoring Ongoing reassessment over time to detect any changes that may have occurred in the state or status of a focal subject or system at a defined spatial scale (grain) or specified location(s). However, the larger the spatial scale or the greater the inclusion of more subjects, systems and/or states, and the longer the time over which monitoring is undertaken, the more challenging it is to sustain accurate reassessments on a regular basis.

Monophyletic A group of organisms descended from a common evolutionary ancestor or ancestral group, especially one not shared with any other group.

Morphological operational taxonomic unit (MOTU) A taxonomic unit determined by its morphology, and representing an ostensible species without a formal scientific name (sometimes referred to as a morphospecies).

Morphotyping The sorting of morphological operational taxonomic units (MOTUs).

Motion camouflage Where the pursuer is hidden from the pursued individual's sight, even though the pursuer is in motion.

Multiple stressors Many stressors acting together and affecting biota and ecosystems.

mya Millions of years ago.

Narrow-range endemism Confined to a specific, relatively small, defined geographic area.

National Biodiversity Strategy and Action Plan (NBSAP) A nation state's strategy to conserve and sustainably use biodiversity according to the international Convention on Biological Diversity.

National Red List An IUCN Red List of Threatened Species in a defined country.

Nationally recorded (relating to taxa) Taxa, such as species and subspecies, observed in a particular country.

Nationally threatened species A species that is under threat in a particular country but may not be threatened elsewhere across its geographic range.

Natural biological control A major component of agroecology, the process which involves harnessing the predatory ability of potential biocontrol agents to control pests, often by manipulating the local landscape.

Natural capital The remaining species and communities in a refuge in a disturbed area, and often source populations from which individuals recolonize an area that is being restored.

Natural selection The process where individuals that are more or best adapted to their local environment are more likely to survive and pass on their genes that underpinned their success, causing species to change and diverge over time – the mechanism of evolution.

Nature-based solutions (NBS) Naturally occurring, restored and artificial ecosystems that deliver ecosystem services to address key societal needs, while also including management actions that protect or enhance ecosystems and their service provision.

Near Threatened (NT) A category on the IUCN Red List of Threatened Species containing those species that have been evaluated against the Red List criteria and do not qualify for Critically Endangered, Endangered or Vulnerable now, but are close to qualifying for, or are likely to qualify for, a threatened category in the near future.

Neonative A species that has expanded its residential geographic range beyond trivial or seasonal movements, with climate change a major driver. The term does not include those species that now have reoccupied an area from which they were earlier extirpated through human agency. It is applied to a latitudinal shift of at least 100 km, and an elevation shift of several hundreds of metres.

Nestedness rule Those species in the local species pool according to their traits and based on the premise that common species are widespread, while rare species tend to occur only at the most suitable sites.

Niche The role an organism plays in a community.

Niche space The specific region of ecospace (the multidimensional and spatiotemporal space of conditions and resources) occupied by a specific species, usually differing from that of another species in the same physical area through ecological adaptation and competitive exclusion.

Niche specialization and niche marginality Measures of niche spacing along an environmental gradient. In the tropics, niche spaces are often narrowly defined along the local environmental gradients, although often more widely defined toward the temperate zone.

Niche-based natural selection Against a background of genetic variation among individuals in a species, a certain niche space may provide an opportunity for a particular genetic variant. This leads to it being the dominant phenotype in that environment. This process may be advantageous in certain human modified and/or transformed landscapes.

Nitrification The conversion of ammonia or reduced nitrogen compounds into the easily absorbable forms of nitrogen (nitrates and nitrites), and which, through fertilizer use in particular, artificially adds nitrogen as a pollutant to natural ecosystems.

Normalized Difference Vegetation Index (NDVI) A remote sensing index that measures vegetation structure and change on the ground through detection of the spectral values of plant chlorophyll. The index is best used when it is challenging to measure biotope structure in the field.

Novel landscape A human disturbed landscape that nevertheless has some value for certain species.

Novel stressor An environmental pressure on a population or community, and while it can be natural, it is usually a stressor considered induced by direct or indirect human activity, leading to conditions that the population or community has not previously experienced.

Nutrient loading The quantity of nutrients entering an ecosystem in a given period of time.

Obelisk position The raising of the adult anisopteran body, especially the abdomen, into a near-vertical or vertical position to keep cool on a hot day through minimizing the body surface area exposed to direct sunlight.

Oceanic islands Islands with a sea matrix.

Odonata (odonates) The insect order that contains the suborders Zygoptera, Anisoptera and Anisozygoptera.

Olfaction The action of smelling.

Oligohaline Water bodies where salt concentrations are typically between 0.5–5 practical salinity units, i.e. that are low in salt content.

Oligotrophic Describing lakes and ponds with a low accumulation of dissolved nutrient salts, supporting little algal growth and other organisms, and having a high oxygen content from low organic matter accumulation.

Ommatidia Eye units that make up the compound eye of the dragonfly.

Omnigenic concept of complex traits The concept where heritability is driven by peripheral genes that trans-regulate core genes.

Opportunistic species Species that can quickly exploit new resources as they arise, such as rapidly colonizing a new environment.

Opsin genes Genes that are responsible for light receptors.

Optimal habitat An area where resources are well-provisioned for adequately sustaining all life stages of a species, including for breeding adults, leading to maintenance of an effective population.

Ornament expression The phenotypic expression of aspects of morphology related to body morphology and wing colouration.

Orthologous groups Referring to genes in different species that stem from vertical descent from a single gene in the last common ancestor.

Outbreeding The mating and reproduction between individuals of a population that are less related than the average in that population, often contributing greatly to the maintenance of a genetically viable population.

Outflow The stream or river discharge of water from a water body, such as a lake.

Oviposition The laying of eggs.

Oviposition site The exact location where eggs are laid.

Oxbow (lake) The cutting off of a river meander, straightening the river, but leaving behind a curved lake, which then undergoes succession as if now a standing water system.

Oxygen demand (of tissues) The amount of oxygen transported to tissues.

Pampa biome A grassland region of South America, principally in Argentina.

Parasite An organism that lives inside or on the outside of a host organism to its benefit while often harming the host, especially when parasite loads are high.

Parasitoid A parasitic species that lives inside its host, usually killing it, such as a hymenopteran inside a host insect.

Passive restoration Allowing a restoration site to regenerate naturally while largely drawing on the natural capital in the immediate vicinity, and after management of the most significant adverse stressors (e.g. allowing forest to regenerate after logging, removal of invasive alien species).

Peatland Land consisting largely of peat (mostly partially decayed vegetation) or peat bogs.

Peninsula effect Referring to the prediction that the number of species should decline from the base of a peninsula to its tip.

Perchers A casual term for dragonflies which spend a lot of time on perches, darting out to catch prey, defend territories and pursue potential mates.

Peri-urban area An urban settlement on the outskirts of a major town or a city, often a transition zone between rural and urban ecosystems.

Permeability (in ecology) In relative terms, the ecotone between two ecosystems where individuals of some species readily cross but others do not.

Pesticide drift The movement of pesticide particles or droplets beyond the target area, especially when there are air currents, and when foggers, mistblowers or ultra-low volume applicators are used.

Petalurid Dragonfly in the family Petaluridae.

Petiolate Of wings, stalked and paddle-shaped.

pH (potential of hydrogen) A measure of the concentration of hydrogen ions, for example, in water, and a measure of its acidity (low pH), neutrality (medium pH) or alkalinity (high pH).

Phenology The study of periodic events in organismal life cycles and how they are influenced by seasonal and interannual variations in climate and habitat, regularly used in dragonfly studies to mean the time of emergence of the adult.

Phenotype An individual's observable or measurable traits.

Phenotypic plasticity The extent to which organisms express different phenotypes according to prevailing and potential environmental conditions (i.e. the environmentally sensitive production of alternative phenotypes by a given genotype), affecting an organism's response to, and its ability to survive, changes in local environmental conditions, including global climate change. Variation among the traits of an organism emanate from different allele expressions among individuals in a population. Phenotypic plasticity enables survival in the face of environmental fluctuation through plastic populations retaining novel adaptive traits.

Phenotypic resilience Referring to the suite of observable traits that enable an individual to overcome a certain short-term stressor.

Philopatry The tendency of individuals of a species to remain at, or return to, a particular habitat or area.

Phylogenetic composition The composition of the various taxa (i.e. their specified identity) in a defined area at a certain time.

Phylogenetic constraint (inertia, conservatism) Where the ability to make a specific phenotype has not yet evolved or has been lost, and so the species has no mechanistic way of producing a phenotype under current environmental conditions.

Phylogenetic correction Statistically controlling for evolutionary relatedness.

Phylogenetic diversity A measure of biodiversity that incorporates phylogenetic differences among species.

Phylogeny The evolutionary history and relationships among or within groups of organisms according to observed heritable traits, such as DNA sequences, protein amino acid sequences and/or morphology.

Physicochemical The combined action of both physical and chemical processes.

Phytotelmata Cavities in terrestrial plants that collect water, and are formed from modified leaves, leaf axils, flowers, stem holes or depressions, open fruits or fallen leaves.

Plastic behaviour Behavioural plasticity is the change in an individual's behaviour following exposure to stimuli, such as changing environmental conditions, and can change more rapidly than is the case for most morphological traits and many physiological traits.

Plastron Bumps and setae on the body surface of an insect used to trap air around the body for breathing when the adult descends under the water (e.g. in diving beetles). Present in some zygopterans to lay eggs below the water surface.

Polarized light pollution Intense horizontal polarization of reflected visible and ultraviolet light from artificial surfaces (often stronger than that from the water surface).

Poleward In the direction of one of the global poles.

Pollutant loading The amount of stress placed upon an ecosystem by pollution released into it, physical or chemical, and usually meant in terms of human agency.

Polychromatism Alternative colour forms in the adults of some odonate species.

Polyphenism A trait for which there are multiple discrete phenotypes arising from a single genotype in response to differing environmental conditions, and therefore a type of phenotypic plasticity.

Pondscape A group of various types of ponds that differentially contribute to increased biodiversity over that of a single pond.

Population All the individuals of one species or morphological/molecular operational taxonomic unit living and potentially interacting in a prescribed area (in IUCN terminology, and in relation to red listing, 'a population' encompasses all the individuals globally).

Population differentiation A measure that describes the partitioning of genetic variation within and among populations.

Potholes Water-filled holes scoured out of the rock substratum over long periods of time along rapid rivers.

Primary forest Natural, undisturbed forest.

Prolarva The tiny short-lived developmental stage that emerges from the egg.

Protected area A formally designated area set aside for nature and biodiversity conservation.

Protists A general name for a broad group of mostly unicellular eukaryotic organisms that are not plants, animals or fungi.

Protoneurid A damselfly in the family Protoneuridae.

Provenance In biology, the original native geographic area of a species.

Pruinescence Waxy covering on the body surface, strikingly so on the abdomen of some adult odonate species, and a strong reflector of ultraviolet light.

Pterostigma A thickened and coloured area in the outer, leading edge wing area of dragonflies, which assists gliding flight, and in some species, is an important identification aid in living individuals.

Putative species Named species that are presumed to be actual species.

Rain pool A temporary pool that appears following heavy rains, usually small in size.

Rank abundance curve Pictorial representation of the relative abundance of a set of focal species.

Rank order (of species) A commonness–rarity spectrum, represented as a graph, ranging from being common through to being rare in a specified area.

Rapoport's rule A proposition that the size of the geographic range of a species of the same lineage is smaller toward the equator.

Rarefaction curve A plot of the number of species against the number of samples, with the resultant curve being created by randomly resampling the pool of many samples several times and then plotting the average number of species that are estimated to occur in each sample.

Reach (of a river) Usually a stretch of a river, a few kilometres long, with relatively homogenous morphological characteristics and margin conditions.

Realized adaptive capacity The ability of a species to occupy a range of habitats which for some species becomes greater under human-induced landscape change but for many others it is reduced.

Realized niche (of a species) The special conditions when and where a species is able to survive various constraints, such as competition, predation, disease and certain environmental conditions, according to the traits and phenotypic plasticity it possesses.

Recession (of water) The retreat of water exposing dry areas.

Recipient site During a reintroduction programme, the site receiving individuals from a donor site.

Reclamation The process of repairing severely degraded land or water sites (e.g. former mine sites, drained marshland) to a more ecologically resilient ecosystem, either for sustainable human use or for nature conservation.

Reconciliation ecology The new reality of maintaining and even improving novel ecosystems for high-grade biodiversity and ecological resilience.

Rectal gills In anisopteran and anisozygopteran larvae, the breathing apparatus are located inside the rectum, over which water is drawn in and out.

Red List for Ecosystems (RLE) An IUCN initiative that introduces the concept of ecological collapse as an analogue of species extinctions. It is based on five criteria that relate to the potential for an ecosystem to collapse.

Red List Index (RLI) A globally selected subset of species from a lesser-known taxonomic group employed to assess the overall threat to the group globally, which in the case for dragonflies is an estimated 10% are globally threatened.

Red-listed species Technically, being red listed means that the conservation status of a species has been assessed and included on the IUCN Red List of Threatened Species, although the expression is often used to mean only the threatened species and not those of Least Concern.

Refuge A place of retreat and safety in the face of some major change in local environmental conditions.

Refugium (plural Refugia) A geographic region that has remained relatively unaltered by the past climatic events which affected the surrounding regions and is a centre for relict fauna and flora.

Regionally Extinct (RE) A category on the IUCN Red List of Threatened Species, referring to a taxon that is now extinct from a particular region but still occurs in its natural range outside that region.

Rehabilitation Management actions that aim to reinstate some level of ecosystem functioning at degraded sites, and not necessarily specifically for biodiversity conservation typical of a native reference ecosystem.

Reintroduction (in species conservation) The artificial movement of individuals from a site with a strong population to one where the population has been lost due to human impact. It assumes that the recipient site has been restored prior to the reintroduction, and that essential ongoing management continues after reintroduction.

Remediation liming The adding of lime (an alkaline substance consisting of calcium oxide and obtained by heating limestone) to counteract post-mining acidification.

Remnant pool (in a river) When a river dries down during the dry season or a drought, a pool that remains in a deep depression of the riverbed.

Rescuing the extinction of experience Experiential programmes aimed at enabling mostly young people to gain first-hand contact with nature.

Resilience (ecological) Of a population, assemblage, ecosystem or landscape: the ability to withstand stressors or impacts, natural and/or human (e.g. drought, floods, climate change, and harvesting), and return to, or near to, former levels after any or several of these stressors or impacts have been lifted or mitigated.

Resilience foresight (for dragonflies) A component of conservation strategic management based in part on what we know about dragonfly biogeographic range changes, coupled with a knowledge of their traits across their ranges.

Resource availability hypothesis The theory that a species' physical size is determined by its energetic requirements, so that regions with greater resource availability, as in the tropics, should support larger-sized species than regions with lower productivity.

Resource bottleneck A point in time where natural and essential resources are limited and put pressure on a population, assemblage, community or ecosystem, and may jeopardize long-term continuance.

Restoration (ecological) Putting restoration ecology into practice to bring back the functional and compositional aspects of the historic (original, natural) ecosystem.

Restoration ecology The scientific activity of enabling the recovery of a disturbed or greatly altered ecosystem.

Restoration trajectory The gradual change in community and ecosystem over time once a site has received restoration action. The trajectory may be altered using specific intervention management (e.g. removal of some trees, partial dredging of the benthic zone, adding of organic matter such as logs).

Retroposition A cellular molecular process in which mRNA is transcribed into DNA and reintegrated into the genome.

Retrospective analysis A solution-based approach where the attainment of a goal ('the solution') is first defined and visualized. Only then are the strategically necessary data gathered, while also applying known and workable conservation solutions in association with essential biological knowledge of the focal organisms.

Reverse the Red An IUCN-based global movement that involves strategic cooperation and science-based action to improve the chances of survival of wild species and natural ecosystems. Working through National Species Specialist Groups and Centres for Species Survival the movement builds on existing expert networks and catalyses efforts with established local institutions, so as to greatly expand the capacity to implement conservation action at the national level and reverse the negative trends indicated by the Red List of Threatened Species and the Red List for Ecosystems.

Revetment A constructed wall to hold back or redirect water.

Rewilding A type of management restoration of landscapes and waterscapes to their natural uncultivated state, with the aim of creating the right ecological conditions prior to the reintroduction of wild animals that had earlier been locally extirpated.

Riffle Where a shallow river runs fast over a stony bottom and the water surface is uneven, often producing some splashing on the surface of the water.

Riparian zone The interface between a river or stream and the land, characterized by plant communities (riparian vegetation) that are hydrophilic (water loving) (sometimes also used in connection with standing water bodies).

Riprap Loose stones used to form a foundation for a water body.

River continuum The concept that a river is a continuum of contiguous ecosystems that can be recognized and classified.

River discharge The volume of water flowing through a river channel, measured at any given point in cubic metres per second.

River flow velocity The speed at which water flows along a river at defined points.

River intermittency When parts of rivers have lower discharge rates or dry out during parts of the year.

Riverine aggregates Geological resources, such as sand, gravel and crushed rock, extracted from rivers.

Runnel A small drainage stream.

Sand bar A long and often narrow sandbank that is exposed when the water level of certain rivers drops.

Sea regressions A geological process occurring when areas of submerged seafloor are exposed above the sea level.

Secondary forest Forested areas that have regenerated naturally, or have been restored following disruptive or destructive natural weather events and/or human activity.

Secondary sexual trait Physical characteristic developing at maturity which distinguishes the sexes but is not directly involved in reproduction (e.g. a dragonfly's coloured wing markings).

Sedge A grass-like plant with triangular stems and small flowers, typically growing on damp or wet ground.

Sediment load The amount of sediment (solid matter) that is transported through a stream cross-section over one year.

Seepage (Seep) A localized area where water comes out of the ground.

Semiochemical A pheromone or other chemical that conveys a signal from one individual to another and that modifies the behaviour of the recipient individual.

Semivoltine A species that takes more than one year to complete its life cycle.

Sentinel Someone or something that is keeping watch over a potentially changing situation, i.e. in a surveillance role.

Setae The fine, mostly hair-like projections on the body of a dragonfly.

Sexual dimorphism Where the sexes of the same species have different morphological features, particularly characteristics not directly involved in reproduction. They often relate to body size and body colouring.

Sexual selection The evolution of certain conspicuous physical traits, such as distinct wing colouration and increased body size in males, that increase the chances of mating success.

Shifting baseline Referring to the measurement of a current state (e.g. of an ecosystem) against a previous state, which may itself be a changed state from an earlier one, such as before influential human activity.

Sinkhole ponds The blocking of sinkholes (collapsed ground) with surface materials and the depression then fills with water.

Site occupancy model A model used to account for the imperfect detection of individuals in surveys and to determine the probability of the true presence or absence of a species at a site. Achieved by quantifying the detection probability of a species at a site based on a validated data set.

Sky islands Terrestrial islands across the land surface that differ greatly in their ecosystems from the surrounding landscape matrix, often manifested as mountains with forested streams in warm climates, with localized endemic dragonflies.

Source population Usually a large population in an optimum habitat from which occasional individuals show wandering dispersal (perhaps from overcrowding) to colonize newly restored or other nearby and healthy habitats.

Spate (of a river) Flowing high and fast after high levels of rainfall.

Spatial scale The measurement and determination of two-dimensional spatial extent, often portrayed as a map.

Spatiotemporal dynamics (in ecology) The changing structures and interactions taking place among organisms both across land and water and over time.

Spatiotemporal scale The measurement and determination of three-dimensional spatial extent and time, often portrayed as a three-dimensional graph with space on the two horizontal dimensions and time on the vertical dimension.

Species (general use in conservation) The whole population of similar groups of individuals that can all potentially mate and leave viable offspring.

Species–area curve The pictorial representation of the increase in species richness at increasing spatial scales.

Species–area effect (curve, hypothesis, relationship) The relationship between the size of the area of the land/water system and the number of species present within that area, with more species present with the increasing size of the area.

Species assessments (for the IUCN Red List of Threatened Species) The determination of the locations where a species is present and the geographic range of a species, to ascertain the level of threat to a species.

Species composition The identified species in a specified group recorded as present at a specified locality or spatial scale. Often used interchangeably with 'assemblage composition'.

Species distribution modelling The use of machine-learning algorithms to predict the distribution of a species over geographic space using environmental data of known occurrences of the species.

Species divergence (diversification) The evolutionary process whereby an extensive population of species diverges into two or more descendant species, which become more dissimilar over time.

Species diversity Species composition combined with the relative abundance (evenness) in a specified group recorded as present at a specified locality or spatial scale. Sometimes, the expression is used in the meaning of 'species composition', i.e. the identified species present at a specified locality or spatial scale with no reference to their relative abundance.

Species nestedness A measure of structure in an ecosystem, usually pertaining to when the number and identity of species present is only a subset of those with more species, as often occurs when a natural ecosystem is disturbed or transformed by human activity.

Species of Greatest Conservation Need (SGCN) (in USA) Those species with small or declining populations or other characteristics that make them vulnerable.

Species packing The number of species occupying a defined area.

Species recovery The increase in the abundance, area of occupancy and sometimes also genetic diversity of a species following conservation action.

Species replacement The verified substitution of one species by another.

Species richness A measure of the number of species present in a defined area without them necessarily being identified, which may include the presence of species non-native to the area.

Species set A group of specified species of focal interest.

Sphagnum A large genus of mosses that usually grow in wet landscapes and become compacted with other plant matter to form peat.

Spring (of water) Pertaining to water pressure causing a natural flow of groundwater onto the earth's surface, especially when rainwater enters and recharges the aquifer and water pressure increases.

Standing water A body of still water, that may be recharged by an inflow of water and maintained at the same level by water outflow

Stepping-stone habitats Patches of land or vegetation that improve the functional connectivity of the landscape, which are used by dragonflies for dispersing across the landscape, but not necessarily for maintaining an effective population.

Stormwater pond (pool) A retention pond to accommodate excess water run-off, mostly from roads, so as to avoid soil erosion.

Stressor (in conservation) An anthropogenic influence, and usually used in a negative sense, on a range of ecological hierarchies from populations to ecosystem processes.

Structural connectivity The physical relationship between landscape elements, usually used in the context of facilitating or impeding movement (cf. Functional connectivity).

Submerged macrophyte A plant, visible to the naked eye, that lives underwater.

Suboptimal habitat An area where resources are poorly provisioned for sustaining all the life stages of a species, and may be a temporary refuge in times of climatic adversity, or a stepping-stone habitat used as a stopping-off point during wider dispersal of the adult.

Subpopulations Separate populations of the same species between which there is no or very little gene flow that have slightly different genomes and may also differ in morphology. Following genetic analyses, subpopulations are often redefined as subspecies.

Subsidy flux Caused by organisms or materials that cross over habitat patch boundaries, subsidizing the resident populations.

Succession trajectory The direction (type) of change in the structure of a biological community over time, usually with a focus on the changing plant community. Often used in the context of site restoration.

Successional stage The structure of the vegetation at one point in the succession trajectory.

Sunflecks Brief occurrences where sunshine penetrates through the forest canopy to reach the understorey, usually referring to earth rotation relative to the sun during the day.

Surrogate Something that represents something else, i.e. is a proxy (see also Environmental surrogate, Indicator surrogate, Management surrogate, and Target).

Survivorship curves A visual representation of the proportion of individuals surviving versus dying from one stage to the next, with one of the main aims of conservation being to ensure that survivorship is as high as possible across all life stages.

Swale A low area, usually referring to a marshy depression between hills or ridges.

Swamp An area of low-lying uncultivated ground where water accumulates. Swamps support much vegetation, especially in wet and warm climates.

Symbiont The relationship or interaction between two dissimilar types of organism, usually referring to a situation where both organisms benefit from the relationship, such as the gut bacterial symbionts inside dragonflies.

Sympatric species Species living in the same local space.

Synlestid A damselfly in the family Synlestidae.

Systematic conservation planning Provision of a clear, comprehensive guide to the process of creating and maintaining a conservation area network for regions, that will best represent the biodiversity of those regions in the most cost-effective way.

Tandem position (pair) When the male clasps and flies with the female, and/or in some damselfly species, when both descend underwater together.

Tank bromeliad Epiphytic plants in the Bromeliaceae that form a hollow tube that collects water and serves as a habitat for some odonate species.

Target (in ecological assessment and monitoring) The process or element (species, ecosystem or abiotic factors) being represented by an ecological surrogate.

Taxonomic circle The use of all lines of evidence that combine behavioural, ecological, biogeographical, morphological and molecular characters in an integrated way for characterizing a species.

Telemetry Collection in the field of data at remote points and their automatic transmission to receiving equipment for monitoring geographic position and movement.

Temporal beta diversity Relative changes in species composition in different assemblages over time.

Temporary pond (pool) (ephemeral pool or vernal pool) A standing water body that is usually shallow and dries out at the driest time of the year, but fills when freshwater becomes available, especially after heavy rainfall (i.e. they have a distinct hydroperiod).

Teneral (adult) Developmental state immediately following the emergence of the adult from the last instar larva, when the adult body is soft and immature in colour.

Terrestrial matrix The land area surrounding a freshwater body, usually referring to more than just the bank or riparian zone.

Territorial behaviour Protection by adult male dragonflies of an area of land and/or water from incursions by other males of its species, and sometimes also males of other species. When the territory is optimal habitat, this leads to successful breeding, improving the chances of an effective population.

Thelytokous parthenogenesis Females produce only female offspring, preventing formation of lethal recessive mutations.

Thermal adaptation The effect of environmental drivers related to temperature on trait evolution.

Thermal (niche) conservatism Where individuals regularly seek near-to-optimum temperatures among macrohabitats at different temperatures.

Thermal plasticity The effect of environmental drivers related to temperature on trait expression.

Thermophilic Having a preference for warm areas.

Thermoregulation Mechanism by which certain animals (e.g. mammals) maintain body temperature with tightly controlled self-regulation, independent of external temperatures (i.e. they are endothermic). Odonates, although ectothermic, can in the case of many large anisopterans generate body heat by wing whirring to warm up their thoracic muscles ready for powered flight.

Threatened (endangered, imperiled) species Those species at risk of a decline in numbers of individuals (often through loss of suitable habitat) and heading toward local, regional or total extinction.

Tibial spines The spines on the lower part of the leg (tibia) used to form a basket in flight to improve prey capture. In some species they may also be used to enhance courtship behaviour.

Time stress In terms of traits, the progress of the season that enables active feeding.

Tracheae The network of tubes permeating the body of dragonflies and used for breathing and transporting gasses to and from major organs.

Trait co-variation As each gene can have several variants (alleles), which code for different variants of a particular trait, operationalization of the various different alleles for different trait expressions at the same time results in several trait variations also being expressed together, i.e. they co-vary.

Traits The characteristics of an organism, determined genetically but made operational by physiology, behaviour and ecology relative to prevailing environmental conditions, though honed by phylogenetic constraints.

Transgression (of water) The long-term flooding of dry land by water.

Translocation (of individuals of a species) The movement of individuals from one location to another.

Tufa A type of porous rock composed of calcium carbonate and formed by precipitation from water, often around mineral springs.

Turbidity (of water) The degree to which water is cloudy, opaque or has suspended matter.

Ultraviolet light The portion of the electromagnetic spectrum extending from the violet (short-wavelength) to the end of the light range (before the X-ray region), and visible to dragonflies.

United Nations Sustainable Development Goals A collection of seventeen interlinked objectives designed to serve as a shared blueprint for peace and prosperity for people and the planet, now and into the future.

Univoltine Producing one brood in a season.

Variegation model A hypothetical situation where the landscape/waterscape is viewed as areas of land and/or water that blend into each other (i.e. have ecotones), and where the occurrence of organisms is similarly spread across the overall area according to their habitat preferences and other traits (cf. Fragmentation model).

Voltinism The number of generations within one year. However, some dragonfly species or populations take more than one year to complete their life cycle and are described as being semivoltine.

Vulnerable (VU) A category on the IUCN Red List of Threatened Species containing those species that on the best available evidence indicates that they meet any of the criteria A to E for Vulnerable, and are therefore considered to be facing a high risk of extinction in the wild (see https://www.iucnredlist.org/).

Wallow A place where large mammals drink and bathe, making a naturally highly disturbed pool that mostly dries out during the dry season.

Water filter gardens Ecologically landscaped lakes instigated to purify wastewater.

Water recharge The process of increasing the water volume in a specified area.

Waterscape (bluescape) The aquatic equivalent of a 'landscape' but referring specifically to the various aquatic ecosystems. The dynamic surface of freshwater over a scale of tens of metres to several kilometres, which may encompass more than one ecosystem. It covers other, more specific terms, like 'pondscape' and 'river reach'.

Watershed An area or ridge of land that separates waters flowing into different rivers or basins.

Wheel position When the male dragonfly grasps the female at the back of the head, and the female curls her abdomen under her body to pick up sperm from the male's secondary genitalia at the base of his abdomen.

Wing iridescence When the of wings dragonflies appear to change in brightness and colour relative to the angle of view or of illumination.

Wing spot allometry Differences in the proportional size of coloured wing spots in different individuals.

Wing whirring The shivering of muscles by some large dragonflies, while remaining on a solid surface, as a means of warming up the thoracic muscles for optimal flight performance.

Zeta diversity A measured degree of overlap in the type of taxa present between a set of observed communities.

Zygoptera (zygopterans) An Odonata suborder (damselflies).

Index

Note

All index entries refer to dragonflies *unless* otherwise specified. 'Species' or 'assemblages' or 'diversity' etc. refer to dragonfly species/assemblages/diversity.

The index is in Word-by-Word order, such that 'farm ponds' precedes 'farming'.

Page numbers in **bold** refer to figures and their captions.